FACTORS FOR UNIT CONVERSIONS

W9-AQI-844

Quantity	Equivalent Values
Mass	$1 \text{ kg} = 1000 \text{ g} = 0.001 \text{ metric ton} = 2.20462 \text{ lb}_m = 35.27392 \text{ oz}$ $1 \text{ lb}_m = 16 \text{ oz} = 5 \times 10^{-4} \text{ ton} = 453.593 \text{ g} = 0.453593 \text{ kg}$
Length	$1 \text{ m} = 100 \text{ cm} = 1000 \text{ mm} = 10^6 \text{ microns } (\mu\text{m}) = 10^{10} \text{ angstroms (Å)}$ $\quad = 39.37 \text{ in.} = 3.2808 \text{ ft} = 1.0936 \text{ yd} = 0.0006214 \text{ mile}$ $1 \text{ ft} = 12 \text{ in.} = 1/3 \text{ yd} = 0.3048 \text{ m} = 30.48 \text{ cm}$
Volume	$1 \text{ m}^3 = 1000 \text{ L} = 10^6 \text{ cm}^3 = 10^6 \text{ mL}$ $\quad = 35.3145 \text{ ft}^3 = 220.83 \text{ imperial gallons} = 264.17 \text{ gal}$ $\quad = 1056.68 \text{ qt}$ $1 \text{ ft}^3 = 1728 \text{ in.}^3 = 7.4805 \text{ gal} = 0.028317 \text{ m}^3 = 28.317 \text{ L}$ $\quad = 28{,}317 \text{ cm}^3$
Force	$1 \text{ N} = 1 \text{ kg·m/s}^2 = 10^5 \text{ dynes} = 10^5 \text{ g·cm/s}^2 = 0.22481 \text{ lb}_f$ $1 \text{ lb}_f = 32.174 \text{ lb}_m\text{·ft/s}^2 = 4.4482 \text{ N} = 4.4482 \times 10^5 \text{ dynes}$
Pressure	$1 \text{ atm} = 1.01325 \times 10^5 \text{ N/m}^2 \text{ (Pa)} = 101.325 \text{ kPa} = 1.01325 \text{ bar}$ $\quad = 1.01325 \times 10^6 \text{ dynes/cm}^2$ $\quad = 760 \text{ mm Hg at } 0°\text{C (torr)} = 10.333 \text{ m H}_2\text{O at } 4°\text{C}$ $\quad = 14.696 \text{ lb}_f/\text{in.}^2 \text{ (psi)} = 33.9 \text{ ft H}_2\text{O at } 4°\text{C}$ $\quad = 29.921 \text{ in. Hg at } 0°\text{C}$
Energy	$1 \text{ J} = 1 \text{ N·m} = 10^7 \text{ ergs} = 10^7 \text{ dyne·cm}$ $\quad = 2.778 \times 10^{-7} \text{ kW·h} = 0.23901 \text{ cal}$ $\quad = 0.7376 \text{ ft-lb}_f = 9.486 \times 10^{-4} \text{ Btu}$
Power	$1 \text{ W} = 1 \text{ J/s} = 0.23901 \text{ cal/s} = 0.7376 \text{ ft·lb}_f/\text{s} = 9.486 \times 10^{-4} \text{ Btu/s}$ $\quad = 1.341 \times 10^{-3} \text{ hp}$

Example: The factor to convert grams to lb_m is $\left(\dfrac{2.20462 \text{ lb}_m}{1000 \text{ g}} \right)$.

ELEMENTARY PRINCIPLES OF CHEMICAL PROCESSES

ELEMENTARY PRINCIPLES OF CHEMICAL PROCESSES

Third Edition

Richard M. Felder

Department of Chemical Engineering
North Carolina State University
Raleigh, North Carolina

Ronald W. Rousseau

School of Chemical Engineering
Georgia Institute of Technology
Atlanta, Georgia

John Wiley & Sons, Inc.

New York / Chichester
Brisbane / Toronto
Singapore

ACQUISITIONS EDITOR	Wayne Anderson
MARKETING MANAGER	Katherine Hepburn
FREELANCE PRODUCTION MANAGER	Jeanine Furino
SENIOR PRODUCTION EDITOR	Robin Factor
SENIOR DESIGNER	Dawn Stanley
ILLUSTRATION COORDINATOR	Sigmund Malinowski
COVER PHOTO	Rosenfeld Images Ltd./Photo Researchers, Inc.

This book was set in Times Roman by Publication Services and printed and bound by Quebecor Printing. The cover was printed by Phoenix Color.

This book is printed on acid free paper. ∞

Copyright © 2000 John Wiley & Sons, Inc. All rights reserved.

No part of this publication may be reproduced, stored in a retrieval system or transmitted in any form or by any means, electronic, mechanical, photocopying, recording, scanning or otherwise, except as permitted under Sections 107 or 108 of the 1976 United States Copyright Act, without either the prior written permission of the Publisher, or authorization through payment of the appropriate per-copy fee to the Copyright Clearance Center, 222 Rosewood Drive, Danvers, MA 01923, (508) 750-8400, fax (508) 750-4470. Requests to the Publisher for permission should be addressed to the Permissions Department, John Wiley & Sons, Inc., 605 Third Avenue, New York, NY 10158-0012, (212) 850-6011, fax (212) 850-6008, E-mail: PERMREQ@WILEY.COM. To order books please call 1(800)225-5945.

Library of Congress Cataloging-in-Publication Data:

Felder, Richard M., 1939–   Elementary principles of chemical processes / Richard M. Felder,
   Ronald W. Rousseau. — 3rd ed.
      p.   cm.
   Includes index.
   ISBN 0-471-53478-1 (alk. paper)
   1. Chemical processes.   I. Rousseau, Ronald W., 1943–   .
   II. Title.
   TP155.7.F44 1999
   660′ .28–dc21

99-14252
CIP

Printed in the United States of America

10 9 8 7 6

We dedicate this book to our first and most important teachers, our parents: Shirley Felder, and the late Robert Felder, Dorothy Rousseau, and Ivy John Rousseau.

About the Authors

Richard M. Felder is Hoechst Celanese Professor of Chemical Engineering at North Carolina State University. He has authored or coauthored well over 100 papers on chemical process engineering and engineering education, and presented seminars, workshops, and short courses on both topics to industrial and research institutions and universities in the United States and abroad. Since 1990, he has codirected the National Effective Teaching Institute under the auspices of the American Society for Engineering Education. He has won a number of awards for his contributions to engineering education, including the 1989 Chemical Manufacturers Association National Catalyst Award, the 1988, 1989, and 1996 Wickenden Awards for the outstanding paper in the *Journal of Engineering Education* (ASEE), the 1986 and 1993 Corcoran Awards for the outstanding paper in *Chemical Engineering Education* (ASEE Chemical Engineering Division), and the 1997 Carlson Award for innovations in engineering education (ASEE). He was designated the AIChE Institute Lecturer in 1991, and was named a Fellow of the ASEE in 1996.

Ronald W. Rousseau is Professor and Chair of the School of Chemical Engineering at the Georgia Institute of Technology. He has published more than 140 journal articles and technical reports, and has made more than 170 presentations on his research and related activities. His research and teaching focus is separation process technology. His research has emphasized investigations of crystal nucleation and growth and the roles these phenomena play in determining crystal quality. His contributions were recognized through the 1996 Clarence G. Gerhold Award of the Separations Division of the American Institute of Chemical Engineers. Dr. Rousseau is a consulting editor for the *AIChE Journal,* a member of the Publication Board of *Chemical Engineering Education,* and an associate editor of *Journal of Crystal Growth,* and he has served as a member of the International Advisory Board of *Separations Technology.* He has provided consulting services for over 45 organizations and given over 130 industrial short courses. Dr. Rousseau has served as Chair of the Council for Chemical Research, and he is a Fellow and served as Director of the American Institute of Chemical Engineers. He is a graduate of Louisiana State University and was elected to the LSU Engineering Hall of Distinction.

Preface to the First Edition

An introductory stoichiometry course traditionally plays several important roles in the chemical engineering curriculum. On the most obvious level, it prepares the student to formulate and solve material and energy balances on chemical process systems and lays the foundation for subsequent courses in thermodynamics, unit operations, kinetics, and process dynamics. More fundamentally, it introduces the engineering approach to solving process-related problems: breaking a process down into its components, establishing the relations between known and unknown process variables, assembling the information needed to solve for the unknowns using a combination of experimentation, empiricism, and the application of natural laws, and, finally, putting the pieces together to obtain the desired problem solution.

We have tried in this book to fulfill each of these functions. Moreover, recognizing that the stoichiometry course is often the students' first real encounter with what they think may be their chosen profession, we have attempted to provide in the text a realistic, informative, and positive introduction to the practice of chemical engineering.

We begin the book with a qualitative discussion of the kinds of problems engineers must face in connection with several apparently dissimilar processes and, after a brief introduction to fundamental techniques of engineering calculations, systematically develop the structure of elementary process analysis: what process variables are and how they are expressed, measured, and calculated; laws of nature that govern the performance of processes; and physical properties of process materials that must be determined in order to design a new process or analyze an existing one.

The chemical process constitutes the structural and motivational framework for the presentation of all of the text material. When we bring in concepts from physical chemistry—for example, vapor pressure, solubility, and compressibility—we introduce them as quantities whose values are required to determine process variables or to perform material and/or energy balance calculations on a process. When we discuss computational techniques such as curve-fitting, root-finding methods, and numerical integration, we present them on the same need-to-know basis in the context of process analysis.

An important feature of the book is a set of industrial process case studies, which demonstrate the role of single-unit calculations in the analysis of multiple-unit processes. We have designed the case studies to be worked on as term projects by individuals or (preferably) small teams of students, beginning after the students have completed the introductory chapter on material balances (Chapter 4). In each study, the students are asked to produce a flowchart of a moderately complex process from a given description and to perform material and energy balance calculations on the process; in addition, they are called on to answer questions that require them to think about how the overall process is structured and why it might be structured that way.

Knowing the problems associated with the case study, the students tend to be on the lookout for information in the formal course material that will help them obtain the required solutions. The case study thus provides both a motivation for learning the text material and a feeling for the contextual significance of this material; moreover, it introduces the common engineering discipline of starting with a large, multifaceted problem and systematically building up the blend of information and technique needed to solve it.

In writing this book we have tried to avoid pedanticism without sacrificing thoroughness or rigor. The writing style is largely informal; all solution techniques are illustrated by examples; and short "Test Yourself" questions of the type used effectively in the self-paced approach to education help focus the students' attention and reinforce the main points in each section. Most of the problems concern real processes and contain realistic data; they are designed to provide practice in all of the methods discussed in the chapters they follow and, in addition, illustrate the range of activities encompassed by chemical engineering, both in the traditional areas of chemical processing and in such fields as environmental sciences and technology and biomedicine.

The SI System of units is used widely but not exclusively throughout the text, and extensive SI data tables, including steam tables, are contained in appendices. Computer programming is not covered explicitly, but applications of computers in process analysis are discussed, and problems for which computer solutions are appropriate are given.

We acknowledge with gratitude the many contributions of colleagues and friends who have helped us in the preparation of the book. Proceeding in no particular order, our thanks go to Professors David Marsland, of North Carolina State University, and Richard Seagrave, John Stevens, and George Burnet, of Iowa State University, who read the manuscript and offered many helpful suggestions for its improvement; Russ O'Dell, who worked out solutions and corrected several dozen mistakes in the manuscript after the authors had convinced themselves there were none left; Jim Ferrell, our beloved leader, who gave us much moral and financial support throughout the seemingly unending years in which the book was being written; Bobbie and Tess, for more than we could summarize here; Kenneth Felder, who performed with great diligence the horrible job of proofreading data tables; Ron Jr., David, Brett, Elena, and Gary for just being there; and Thurman Poston, the engaging engineering editor, who provided us with many broad perspectives, deep insights, and free lunches. We also thank students who served as guinea pigs during the formative stages of the book's development, and we sincerely apologize for all those problems they involuntarily and painfully debugged for us; in particular, we raise our glasses to the Lost Generation of Fall 1973, who had the miserable fortune to get the first draft as a course text.

We are especially indebted to Dr. James Fair of the Monsanto Company, Mr. Norman Kaplan of the U.S. Environmental Protection Agency, and Dr. Ray E. Harrison of Westvaco, Inc., who reviewed draft copies of the case studies and suggested changes to bring the process descriptions closer in line with industrial practice. In some cases we chose to retain deviations from practice for pedagogic reasons; any such deviations should be attributed to us and not our reviewers.

Last, and most of all, we thank Magnificent Mary Wade, who uncomplainingly and with great good humor typed revision after revision, until the authors, unable to stand any more, declared the book done.

Richard M. Felder

Ronald W. Rousseau

Preface to the Third Edition

Because the laws of conservation of mass and energy have not been repealed in the 13 years since the second edition appeared, users of that edition will find a comfortable familiarity in the basic content and structure of this edition. Two noteworthy changes in chemical engineering education have occurred since 1986, however. First, prior to 1986, most new graduates went to work in the chemical and petroleum industries, whereas today a large and growing number are finding jobs in such fields as environmental engineering, biotechnology, and microelectronics. Second, in 1986 computer programming in languages such as Fortran and Pascal still played an important role in the curriculum, whereas now most computation in chemical engineering courses involves spreadsheets, symbolic and numerical mathematics packages, and flowsheet simulators.

This edition of the text reflects these patterns of change in chemical engineering education. The chapter-end problems include some that involve biochemistry, biomedicine, chemical vapor deposition of silicon films on semiconductor chips, assessment and control of pollutant emission levels in plant discharges, and plant and laboratory safety. Structured computer programming problems have largely been replaced by problems that call for the use of spreadsheets and/or equation-solving software.

Additional changes have been made throughout the text. We now provide study aids in the form of instructional objectives and chapter-end summaries for all chapters, as well as a nomenclature list and glossary of common process-related terms used in the text. We place greater emphasis on degree-of-freedom analysis in formulating solutions to process analysis problems. We have eliminated the confusing conversion factor g_c from formulas for weight and potential and kinetic energy, treating force unit conversions (e.g., from kg·m/s^2 to newtons) like all other unit conversions. We have increased our use of extents of reaction in the analysis of reactive process systems rather than relying almost exclusively on atomic species balances. We offer explicit guidance in the use of spreadsheets (particularly the goalseek tool) to solve nonlinear equations such as the SRK equation of state and energy balances on adiabatic reactors. (We have retained the appendices on iterative numerical methods like regula-falsi and Newton's rule, however, for instructors who wish to include numerical analysis in their topical coverage.) We have added a section in Chapter 6 on the commercially important unit operation of adsorption. Many chapter-end problems now call on the students to explain process operations and interpret results of calculations instead of asking them simply to perform the calculations.

A change that we have deliberately chosen *not* to make is to emphasize the use of symbolic mathematics programs such as Maple® and Mathematica® in the solution of material and energy balance problems. We appreciate the power of such programs and understand that they can be used to solve virtually every problem in this text once the correct set of equations has been derived. We believe, however, that deriving those equations systematically—drawing and fully labeling a flowchart, carrying out a degree-of-freedom analysis, and examining the structure of the process to determine an efficient problem-solving strategy—should be the focus of the first quantitative course in chemical process analysis. Once this pre-solution analysis has been performed, whether the system equations are solved algebraically or with software is simply a matter of convenience. In any case, there is no need for us to provide explicit instruction on using equation-solving software: instructors who wish to do so need only add the phrase

"Use _____ (*insert name of software*)" as a prefix to any problem statement that calls for a calculation.

Today, most students either have their own computers or have easy access to them, which makes it possible to offer additional resources to instructors and students in the form of instructional software. This edition includes a CD containing *Interactive Chemical Process Principles*, a software package developed by Intellipro, Inc. The package contains instructional tutorials developed by Richard Felder, a powerful but student-friendly algebraic and differential equation-solver, and an *Encyclopedia of Chemical Process Equipment* developed by Dr. Susan Montgomery of the University of Michigan. (See Page xix for a more detailed description.) In addition, updates to the text and additional resources to support the teaching and learning process will be provided periodically at www.wiley.com/college/engin/epcp.

We are indebted to numerous faculty colleagues and students who suggested many of the changes in this edition and who helped with its preparation. With apologies to those whom we might inadvertently omit, we thank David Allen, Tim Anderson, D. F. Bagster, Mort Barlaz, Stanley Barnett, Ken Beatty, Bob Bird, Brice Carnahan, Paul Chan, Paulette Clancy, Bill Conger, Cam Crowe, Max Deibert, Noel de Nevers, Larry Dodd, Chuck Eckert, Larry Erickson, P. T. Eubank, Ken Felder, Arthur Fontijn, Paul Frymier, Gary Gilleskie, Carole Heath, Debbie Kaufman, Paul Kohl, Joe Lemanski, Octave Levenspiel, Pete Ludovice, Susan Montgomery and her students, Peter Rasmussen, Mary Rezac, Skip Rochefort, Ken Sampson, Amyn Teja, and Hank Van Ness for their suggestions; Gary Huvard, Norm Kaplan, Brian Keyes, and Jack Winnick for their assistance in the development of the case studies; Philippe Marchal and his Intellipro colleagues for their efforts in developing the supplementary software; and Matt Burke, Swapnil Chhabra, Jun Gao, Linda Holm, Concepción Jimenez-Gonzalez, Amit Khandelwal, Stephanie Manfredi, Janette Mendez-Santiago, Amy Michel, Dong Niu, Amitabh Sehgal, Jim Semler, Kai Wang, Esther Wilcox, Tao Wu, and Jian Zhou for invaluable assistance with the solution manual. We thank Rebecca and Sandra, who encouraged and persevered with us throughout the preparation of this edition. Finally, we thank our students who took the trouble to point out errors in previous editions and early drafts of this edition. We know they did it out of a sense of professional responsibility and not just to collect the quarters.

Richard M. Felder

Ronald W. Rousseau

To the Instructor

The organization of this text has been planned to provide enough flexibility to accommodate classes with diverse backgrounds within the scope of a one-semester or two-quarter course. We anticipate that semester-long courses in which most students have traditional first-year engineering backgrounds will cover most of the first nine chapters, possibly augmented with one case study. A one-quarter course should cover Chapters 1 through 6. Students who have been exposed to dimensional analysis and elementary data correlation can skip or skim Chapter 2, and students whose freshman chemistry courses provided a detailed coverage of process variable definitions and the systematic use of units to describe and analyze chemical processes may omit Chapter 3. The time gained as a result of these omissions may be used to cover additional sections in Chapters 4 through 9, to add Chapter 10 on computer-aided balances or Chapter 11 on transient balances, or to cover appended material on numerical analysis.

A criticism sometimes leveled at the stoichiometry course is its emphasis on routine solution methods, an emphasis that gives the student little idea of the wide range of problems likely to be encountered by practicing engineers and the creativity needed to solve some of them. Unfortunately, much of the drill-like aspect of the course is necessary. We have found that the only way to teach students to use the engineering approach to process analysis is to have them continually practice it, and many of the chapter-end problems are structured to provide this practice. Within each topic, the problems gradually build in complexity, so that basic problem-solving techniques can be practiced in the context of increasingly realistic problems. Representative assignment schedules are given in the instructor's manual, and there is enough duplication of problem types for the schedules to be varied considerably from one course offering to another.

On the other hand, we have tried in this edition to provide a variety of more open-ended problems that focus on conceptual understanding and creative thinking, both imbedded within chapter-end problems and as separate "Creativity Exercises." We urge instructors to assign these open-ended problems on a regular basis and perhaps to include similar problems on tests after ample practice has been provided in assignments. The problems can be introduced in a variety of ways: as focal points for in-class brainstorming sessions, as parts of regular or extra-credit homework assignments, or as individual or group projects with rewards (e.g., bonus points on subsequent tests) for the solutions exhibiting the greatest fluency (quantity of solutions), flexibility (variety of solutions), and originality (novelty or cleverness of solutions). Far more than the algorithmic drills, these exercises convey a sense of the challenging and intellectually stimulating possibilities in a chemical engineering career. Conveying this sense may be the most valuable task that can be accomplished in the introductory chemical engineering course.

We have discussed in the Preface the motivational aspects of the case studies and the way they complement the formal text material. An additional benefit occurs if the assignments are made to groups, an approach we regularly use in our classes. We invariably see the groups starting out in a state of semianarchy and then developing cohesion as the weeks go by. By the end of the term the students have learned how to divide the labor appropriately and to learn from one another, because they know they are liable to be tested on any part of the project, not just the part for which they were principally responsible. This is the part of the course the students usually say they enjoyed most. We have also found that periodic conferences between the groups and the instructor to discuss the case studies provide added educational benefits to all parties concerned.

RMF
RWR

Interactive Chemical Process Principles

The CD that accompanies this edition of the text contains a variety of resources for students and instructors collected under the title *Interactive Chemical Process Principles* (ICPP). Some of the components of ICPP are instructional aids for the stoichiometry course, and others are computational and reference tools that should prove useful throughout the chemical engineering curriculum. One or more of the ICPP tools can be effectively applied to almost every example and problem in the book.

In this section, we provide an overview of ICPP and some thoughts on how it might be used effectively as an adjunct to the text. This outline and the Preface are the only places in this book where the courseware will be mentioned. If we had chosen to refer explicitly to the courseware whenever it could be used, references would have been included in almost every example and chapter-end problem in the book.

We encourage you to read through this outline and then explore the tools for yourself. If you are a student, you will soon be able to recognize when you can use the tools for problem solving; if you are an instructor, you will see when suggestions for using the tools might be helpful in your lecture notes or assignments.

Instructional Tutorials

ICPP contains six interactive tutorials, designed for students to work through after they have progressed to certain points in the text. In each tutorial, process descriptions are given, questions are posed, and students enter answers and receive affirmation or corrective feedback and then go on to additional questions. Process simulations are also included, so that students may predict how process systems will respond to changes in certain system variables and then explore the effects of those changes experimentally. The active involvement and immediate feedback implicit in these exercises can significantly reinforce learning.

Once students can successfully work through a tutorial from beginning to end, they may be confident that they have mastered a significant portion of the material covered in that tutorial. If they have repeated trouble with a part of the tutorial, they will be able to identify gaps in their understanding of the course material and get help with them.

The tutorials and the points in the course when they may be completed are as follows:

1. Basic process calculations and process system variables (end of Chapter 3).
2. Material balances on nonreactive single-unit processes (end of Section 4.3).
3. Material balances on reactive multiple-unit processes (end of Chapter 4).
4. Material balances on multiphase systems (end of Chapter 6).
5. Material and energy balances on nonreactive processes (end of Chapter 8).
6. Material and energy balances on reactive processes. (End of Chapter 9).

E-Z Solve

E-Z Solve is a powerful and user-friendly equation-solving program designed and constructed by Intellipro Inc., the company that produced ICPP. It can be used to obtain numerical solutions of sets of linear and nonlinear algebraic equations of the types that occur in almost every chapter-end problem in Chapters 4 through 10 of the text, and it can also solve ordinary differential equations of the types that occur in Chapter 11. Examples of applications of E-Z Solve to representative stoichiometry problems are provided on the CD. E-Z Solve is convenient to use whenever a problem calls for solving three or more simultaneous linear algebraic equations or any number of nonlinear algebraic equations and ordinary differential equations.

Visual Encyclopedia of Chemical Engineering Equipment

Most of the examples and problems in the text refer to items of equipment commonly found in chemical processes, such as reactors, heat exchangers, distillation columns, absorption towers, crystallizers, filters, and centrifuges. In some cases, brief explanations of these equipment items are given; in others the terms are simply used. *The Visual Encyclopedia of Chemical Engineering Equipment*, created by Dr. Susan Montgomery of the University of Michigan, is a reference that can be used whenever a specific type of equipment is mentioned. The encyclopedia provides photographs, cutaway diagrams, movies and animations, and explanations of how the different equipment items work. You may consult it to clarify process descriptions in this text and to improve your understanding of most unit operations you will encounter as a student and as a graduate in chemical engineering.

Physical Property Database

The physical property database of ICPP contains easily accessed values of molecular weights, specific gravities, phase transition points, critical constants, enthalpies, and latent heats for many species. The values retrieved from the database may easily be incorporated into process calculations performed using E-Z Solve.

Index of Learning Styles

Students learn in a variety of ways. For example, some students are concrete and practical. These students appreciate many illustrations, examples, and applications of course material and are uncomfortable with abstract mathematical presentations. Other students are much more comfortable with abstraction and are easily bored by repetitive calculations and examples. Some learn visually, getting much more from pictures and diagrams than they do from words and formulas, while others benefit more from verbal explanations than from visual representations. The list of such differences is endless.

A student's *learning style* is a set of preferences for certain ways of taking in and processing information. The **Index of Learning Styles** (ILS) is an instrument that enables students to assess several of their learning style preferences. We suggest that the first thing they do when they begin to explore ICPP is to complete the ILS, determine their learning style profile, and then read the accompanying material that suggests study strategies that might work effectively for them in the stoichiometry course and in subsequent courses.

Nomenclature

The variables to be listed will be expressed in SI units for illustrative purposes, but they could equally well be expressed in any dimensionally consistent units.

a, b, c, d	Either arbitrary constants or coefficients of a polynomial expression for heat capacity, such as those listed in Appendix B.2.
C_p [kJ/(mol·K)], C_v [kJ/(mol·K)]	Heat capacities at constant pressure and constant volume, respectively.
E_k(kJ), $\dot{E}_k$(kJ/s)	Kinetic energy, rate of kinetic energy transport by a flowing stream.
E_p(kJ), $\dot{E}_p$(kJ/s)	Potential energy, rate of potential energy transport by a flowing stream.
g(m/s^2)	Gravitational acceleration constant, equal to 9.8066 m/s^2 or 32.174 ft/s^2 at sea level.
H (kJ), $\dot{H}$(kJ/s), $\hat{H}$(kJ/mol)	Enthalpy of a system (H), rate of transport of enthalpy by a process stream ($\dot{H}$), specific enthalpy ($\hat{H}$), all determined relative to a specified reference state.
m, M (kg), $\dot{m}$(kg/s)	Mass (m or M) or mass flow rate ($\dot{m}$) of a process stream or stream component.
n(mol), $\dot{n}$(mol/s)	Number of moles (n) or molar flow rate ($\dot{n}$) of a process stream or stream component.
p_A(N/m^2)	Partial pressure of species A in a mixture of gaseous species, $= y_A P$.
$p_A^*(T)$(N/m^2)	Vapor pressure of species A at temperature T.
P(N/m^2)	Total pressure of a system. Unless specifically told otherwise, assume that P is absolute pressure and not gauge pressure.
P_c(K)	Critical pressure. Values of this property are listed in Table B.1.
Q (kJ), $\dot{Q}$(kJ/s)	Total heat transferred to or from a system (Q), rate of heat transfer to or from a system ($\dot{Q}$). Q is defined to be positive if heat is transferred to the system.
R [kJ/(mol·K)]	Gas constant, given in different units on the inside back cover of the text.

SCMH, SCLH, SCFH	Abbreviations for standard cubic meters per hour [m³(STP)/h], standard liters per hour [L(STP)/h], and standard cubic feet per hour [ft³(STP)/h], respectively: the volumetric flow rate of a gas stream if the stream were brought from its actual temperature and pressure to standard temperature and pressure (0°C and 1 atm).
SG	Specific gravity, or ratio of the density of a species to the density of a reference species. The abbreviation is always used for liquids and solids in this text and usually refers to species for which specific gravities are listed in Table B.1.
t(s)	Time
T(K)	Temperature
T_{mp}, T_{bp}, T_c(K)	Melting point temperature, boiling point temperature, and critical temperature, respectively. Values of these properties are listed in Table B.1.
U (kJ), $\dot{U}$(kJ/s), $\hat{U}$(kJ/mol)	Internal energy of a system (U), rate of transport of internal energy by a process stream ($\dot{U}$), specific internal energy ($\hat{U}$), all relative to a specified reference state.
V (m³), $\dot{V}, \dot{v}$ (m³/s), $\hat{V}$(m³/mol)	Volume (V) of a fluid or process unit, volumetric flow rate ($\dot{V}$ or $\dot{v}$) of a process stream, specific volume ($\hat{V}$) of a process material.
W (kJ), $\dot{W}_s$(kJ/s)	Work transferred to or from a system (W), rate of transfer of shaft work to or from a continuous process system ($\dot{W}_s$). W is defined to be positive (in this text) if work is transferred from a system to its surroundings.
x, y, z	Mass fraction or mole fraction of a species in a mixture. (Subscripts are usually used to identify the species.) In liquid-vapor systems, x usually denotes fraction in the liquid and y denotes fraction in the vapor. z may also denote the compressibility factor of a gas.

Greek letters

Δ	In batch (closed) systems, ΔX denotes the difference $X_{\text{final}} - X_{\text{initial}}$, where X is any system property. In continuous (open) systems, $\Delta \dot{X}$ denotes the difference $\dot{X}_{\text{output}} - \dot{X}_{\text{input}}$.
$\Delta \hat{H}_c, \Delta \hat{H}_f$(kJ/mol)	Heats of combustion and formation, respectively. Values of these properties at 25°C and 1 atmosphere are listed in Table B.1.
$\Delta \hat{H}_m, \Delta \hat{H}_v$(kJ/mol)	Heats of melting (fusion) and vaporization, respectively. Values of these properties at the

	normal melting and boiling points are listed in Table B.1.
ν_A	Stoichiometric coefficient of species A in a chemical reaction, defined to be positive for products, negative for reactants. For $N_2 + 3H_2 \rightarrow 2NH_3$, $\nu_{N_2} = -1$, $\nu_{H_2} = -3$, $\nu_{NH_3} = 2$.
ξ(mol)	Extent of reaction. If n_{A0}(mol) of reactive species A is initially present in a reactor and n_A(mol) is present some time later, then the extent of reaction at that time is $\xi = (n_{A0} - n_A)/\nu_A$, where ν_A is the stoichiometric coefficient of A in the reaction (see preceding definition). The value of ξ is the same regardless of which reactant or product is chosen as species A.
$\dot{\xi}$(mol/s)	Extent of reaction for a continuous process at steady state. If $\dot{n}_{A0}$(mol/s) of reactive species A enters the reactor and $\dot{n}_A$(mol/s) exits, then the extent of reaction is $\dot{\xi} = (\dot{n}_{A0} - \dot{n}_A)/\nu_A$, where ν_A is the stoichiometric coefficient of A in the reaction. The value of $\dot{\xi}$ is the same regardless of which reactant or product is chosen as species A.
ρ(kg/m^3)	Density.

Other Symbols

$\dot{}$ (e.g., $\dot{m}$)	Flow rate, such as mass flow rate.
$\hat{}$ (e.g., $\hat{U}$)	Specific property, such as specific internal energy.
( )	Parentheses are used to express functional dependence, as in $p^*(T)$ to denote a vapor pressure that depends on temperature, and also to enclose units of variables, as in m(g) to denote a mass expressed in grams. The intended use can usually be easily seen in context.

Glossary of Chemical Process Terms

Absorption   A process in which a gas mixture contacts a liquid solvent and a component (or several components) of the gas dissolves in the liquid. In an *absorption column* or *absorption tower* (or simply *absorber*), the solvent enters the top of a column, flows down, and emerges at the bottom, and the gas enters at the bottom, flows up (contacting the liquid), and leaves at the top.

Adiabatic   A term applied to a process in which no heat is transferred between the process system and its surroundings.

Adsorption   A process in which a gas or liquid mixture contacts a solid (the *adsorbent*) and a mixture component (the *adsorbate*) adheres to the surface of the solid.

Barometer   A device that measures atmospheric pressure.

Boiler   A process unit in which tubes pass through a combustion furnace. *Boiler feedwater* is fed into the tubes, and heat transferred from the hot combustion products through the tube walls converts the feedwater to steam.

Boiling point (at a given pressure)   For a pure species, the temperature at which the liquid and vapor can coexist in equilibrium at the given pressure. When applied to the heating of a mixture of liquids exposed to a gas at the given pressure, the temperature at which the mixture begins to boil.

Bottoms product   The product that leaves the bottom of a distillation column. The bottoms product is relatively rich in the less volatile components of the feed to the column.

Bubble point (of a mixture of liquids at a given pressure)   The temperature at which the first vapor bubble appears when the mixture is heated.

Calibration (of a process variable measurement instrument)   A procedure in which an instrument is used to measure several independently known process variable values, and a *calibration curve* of known variable values versus the corresponding instrument readings is plotted. Once the instrument has been calibrated, readings obtained with it can be converted to equivalent process variable values directly from the calibration curve.

Catalyst   A substance that significantly increases the rate of a chemical reaction, although it is neither a reactant nor a product.

Compressibility factor $z = PV/nRT$ for a gas. If $z = 1$, then $PV = nRT$ (the ideal gas equation of state) and the gas is said to behave ideally.

Compressor   A device that raises the pressure of a gas.

Condensation   A process in which an entering gas is cooled and/or compressed, causing one or more of the gas components to liquefy. Uncondensed gases and liquid *condensate* leave the condenser as separate streams.

Critical pressure, P_c   The highest pressure at which distinct vapor and liquid phases can coexist for a species.

Critical temperature, T_c   The highest temperature at which distinct vapor and liquid phases can coexist for a species. The critical temperature and pressure, collectively referred to as the *critical constants,* are listed for various species in Table B.1.

Crystallization   A process in which a liquid solution is cooled or solvent is evaporated to an extent that solid crystals of solute form. The crystals in the *slurry* (suspension of solids in a liquid) leaving the crystallizer may subsequently be separated from the liquid in a filter or centrifuge.

Decanter   A device in which two liquid phases or liquid and solid phases separate by gravity.

Degrees of freedom   When applied to a general process, the difference between the number of unknown process variables and the number of equations relating those variables; the number of unknown variables for which values must be specified before the remaining values can be calculated. When applied to a system at equilibrium, the number of intensive system variables for which values must be specified before the remaining values can be calculated. The degrees of freedom in the second sense is determined using the Gibbs Phase Rule.

Dew point (of a gas mixture)   The temperature at which the first liquid droplet appears when the mixture is cooled at constant pressure.

Distillation   A process in which a mixture of two or more species is fed to a vertical column that contains either a series of vertically spaced horizontal plates or solid packing through which fluid can flow. Liquid mixtures of the feed components flow down the column and vapor mixtures flow up. Interphase contact, partial condensation of the vapor, and partial vaporization of the liquid all take place throughout the column. The vapor flowing up the column becomes progressively richer in the more volatile components of the feed, and the liquid flowing down becomes richer in the less volatile components. The vapor leaving the top of the column is condensed: part of the condensate is taken off as the *overhead product* and the rest is recycled to the reactor as *reflux,* becoming the liquid stream that flows down the column. The liquid leaving the bottom of the column is partially vaporized: the vapor is recycled to the reactor as *boilup,* becoming the vapor stream that flows up the column, and the residual liquid is taken off as the *bottoms product.*

Drying   A process in which a wet solid is heated or contacted with a hot gas stream, causing some or all of the liquid wetting the solid to evaporate. The vapor and the gas it evaporates into emerge as one outlet stream, and the solid and remaining residual liquid emerge as a second outlet stream.

Enthalpy (kJ)   Property of a system defined as $H = U + PV$, where $U = $ internal energy, $P = $ absolute pressure, and $V = $ volume of the system.

Evaporation (vaporization)   A process in which a pure liquid, liquid mixture, or solvent in a solution is vaporized.

Extraction (liquid extraction)   A process in which a liquid mixture of two species (the *solute* and the *feed carrier*) is contacted in a mixer with a third liquid (the *solvent*) that is immiscible or nearly immiscible with the feed carrier. When the liquids are contacted, solute transfers from the feed carrier to the solvent. The combined mixture is then allowed to settle into two phases that are then separated by gravity in a decanter.

Filtration   A process in which a *slurry* of solid particles suspended in a liquid passes through a porous medium. Most of the liquid passes through the medium (e.g., a filter) to form the *filtrate,* and the solids and some entrained liquid are retained on the filter to form the *filter cake.* Filtration may also be used to separate solids or liquids from gases.

Flash vaporization   A process in which a liquid feed at a high pressure is suddenly exposed to a lower pressure, causing some vaporization to occur. The vapor product is rich in the more volatile components of the feed and the residual liquid is rich in the less volatile components.

Flue gas   See stack gas.

Heat   Energy transferred between a system and its surroundings as a consequence of a temperature difference. Heat always flows from a higher temperature to a lower one.

Heat exchanger   A process unit through which two fluid streams at different temperatures flow on opposite sides of a metal barrier. Heat is transferred from the stream at the higher temperature through the barrier to the other stream.

Internal energy (U)   The total energy possessed by the individual molecules in a system (as opposed to the kinetic and potential energies of the system as a whole). U is a strong function of temperature, phase, and molecular structure and a weak function of pressure (it is independent of pressure for ideal gases). Its absolute value cannot be determined, so it is always expressed relative to a reference state at which it is defined to be zero.

Membrane   A thin solid or liquid film through which one or more species in a process stream can permeate.

Overhead product   The product that leaves the top of a distillation column. The overhead product is relatively rich in the most volatile components of the feed to the column.

Pump   A device used to propel a liquid or slurry from one location to another, usually through a pipe or tube.

Scrubber   An absorption column designed to remove an undesirable component from a gas stream.

Settler   See decanter.

Shaft work   All work transferred between a continuous system and its surroundings other than that done by or on the process fluid at the system entrance and exit.

Stack gas   The gaseous products exiting from a combustion furnace.

Stripping   A process in which a liquid containing a dissolved gas flows down a column and a gas (stripping gas) flows up the column at conditions such that the dissolved gas comes out of solution and is carried off with the stripping gas.

Vapor pressure   The pressure at which pure liquid A can coexist with its vapor at a given temperature. In this text, vapor pressures can be determined from tabulated data (e.g., Tables B.3 and B.5–B.7 for water), the Antoine equation (Table B.4), or the Cox chart (Figure 6.1-4).

Volume percent (% v/v)   For liquid mixtures, the percentage of the total volume occupied by a particular component; for ideal gases, the same as mole percent. For nonideal gases, the volume percent has no meaningful physical significance.

Work   Energy transferred between a system, and its surroundings as a consequence of motion against a restraining force, electricity or radiation, or any other driving force except a temperature difference.

Contents

Chapter 8   **Balances on Nonreactive Processes   357**

Chapter 9   **Balances on Reactive Processes   440**

Chapter 10   **Computer-Aided Balance Calculations   504**

Chapter 11   **Balances on Transient Processes   543**

Part One

Engineering Problem Analysis

Chapter 1

What Some Chemical Engineers Do for a Living

Last May, chemical engineering seniors at a large university took their last final examination, attended their graduation ceremonies, flipped their tassels and threw their mortarboards in the air, enjoyed their farewell parties, said goodbye to one another and promised faithfully to stay in touch, and headed off in an impressive variety of geographical and career directions.

Since you bought this book, you are probably thinking about following in the footsteps of those graduates—spending the next few years learning to be a chemical engineer and possibly the next 40 applying what you learn in a career. Even so, it is a fairly safe bet that, like most people in your position, you have only a limited idea of what chemical engineering is or what chemical engineers do. A logical way for us to begin this book might therefore be with a definition of chemical engineering.

Unfortunately, no universally accepted definition of chemical engineering exists, and almost every type of skilled work you can think of is done somewhere by people educated as chemical engineers. We will therefore abandon the idea of formulating a simple definition and instead take a closer look at what those recent graduates did, either immediately after graduation or following a well-earned vacation. Consider these examples and see if any of them sound like the sort of career you can see yourself pursuing and enjoying.

- About 45% of the class went to work for large chemical, petrochemical, pulp and paper, plastics and other materials, or textile manufacturing firms.
- Another 35% went to work for government agencies and design and consulting firms (many specializing in environmental regulation and pollution control) and for companies in fields such as microelectronics and biotechnology that have not traditionally been associated with chemical engineering.
- About 10% of the class went directly into graduate school in chemical engineering. The masters degree candidates will get advanced education in traditional chemical engineering areas (thermodynamics, chemical reactor analysis and design, fluid dynamics, mass and heat transfer, and chemical process design and control), and in about two years most of them will graduate and get jobs doing process or control systems design or product development. The doctoral degree candidates will get advanced education and work on major research projects, and in four to five years most will graduate and either go into industrial research and development or join a university faculty.
- The remaining 10% of the class went into graduate school in an area other than chemical engineering, such as medicine, law, and business.

- Several graduates went to work for companies manufacturing specialty chemicals—pharmaceuticals, paints and dyes, and cosmetics, among many other products. All of these companies used to hire only chemists to design and run their production processes, but in the past few decades they discovered that if they wanted to remain competitive they would have to pay attention to such things as mixing efficiency, heat transfer, automatic temperature and liquid level control, statistical quality control, and control of pollutant emissions. They also discovered that those are areas in which chemical engineers are educated and chemists are not, at which point these industries became an increasingly important job market for chemical engineers.

- Some went to work for companies that manufacture integrated semiconductor circuits. A critical step in the production of (for example) computer chips involves coating small silicon wafers with extremely thin and uniform layers of silicon-containing semiconducting materials. The technique used for this process is *chemical vapor deposition*, in which the coating material is formed in a gas-phase reaction and then deposited on the surface of the wafer. The graduates working in this area may be called on to identify reactions that can be used to produce the desired films, determine the best conditions at which to run the reactions, design the reactors, and continue to improve their operation.

- Some took elective courses in biochemistry and microbiology and got jobs with small but rapidly growing biotechnology firms. One graduate works on the design of pharmaceutical production processes that involve *immobilized enzymes,* biological chemicals that can make specific reactions go orders of magnitude faster than they would in the absence of the enzymes. Several others work on processes that involve *genetic engineering,* in which recombinant DNA is synthesized and used to produce valuable proteins and other medicinal and agricultural chemicals that would be hard to obtain by any other means.

- Some joined companies that manufacture *polymers* (plastics). One is working on the development of membranes for desalination of seawater (fresh water passes through, salt is kept out) and for gas separations (hydrogen passes through and hydrocarbons are kept out, or vice versa); another is developing membranes to be used in hollow-tube artificial kidneys (blood flows from the patient's body through thin-walled tubes; metabolic wastes in the blood pass through the tube walls but proteins and other important body chemicals remain in the blood, and the purified blood is returned to the body).

- Four of the graduates went to medical school. (Chemical engineering graduates who take several electives in the biological sciences have a strong record of success in gaining medical school admission.) One went to law school. Three enrolled in Master of Business Administration programs and after graduation will probably move into management tracks in chemical-related industries.

- One graduate joined the Peace Corps for a two-year stint in East Africa helping local communities develop sanitary waste disposal systems and also teaching science and English in a rural school. When she returns, she will complete a Ph.D. program, join a chemical engineering faculty, write a definitive book on environmental applications of chemical engineering principles, quickly rise through the ranks to become a full professor, resign after ten years to run for the United States Senate, win two terms, and eventually become head of a large and highly successful private foundation dedicated to improving education in economically deprived communities. She will attribute her career successes to the problem-solving skills she acquired in her undergraduate training in chemical engineering.

- At various points in their careers, some of the graduates will work in chemical or biochemical or biomedical or material science laboratories doing research and development or quality engineering, at computer terminals designing processes and products and control systems, at field locations managing the construction and startup of manufacturing plants, on production floors supervising and troubleshooting and improving operations, on the road doing technical sales and service, in executive offices performing administrative functions, in government agencies responsible for environmental and occupational health and safety, in hospitals and clinics practicing medicine or biomedical engineering, in law offices specializing in chemical process-related patent work, and in classrooms teaching the next generation of chemical engineering students.

Even the chemical engineering graduates who go into a traditional chemical manufacturing process end up performing a wide variety of different tasks. Consider the following example, and see if any of the problems described seem to present the sort of challenge you can see yourself taking on and enjoying.

EXAMPLE 1

A chemist in your company's research and development division has discovered that if he mixes two reactants in a certain proportion at an elevated temperature, he obtains a product significantly more valuable than both reactants. The company contemplates manufacturing the product using a process based on this reaction. At this point the matter becomes an engineering problem or, more precisely, hundreds of engineering problems.

1. What type of reactor should be used? A long pipe? A large tank? Several smaller tanks? An extremely large test tube? How large? Made of what? Does it have to be heated? If so, how much and how? With an electrical heater inside or outside the reactor? By passing a hot fluid through a heating coil in the reactor? By heating the reactants before they get into the reactor? Does the reaction supply its own heat, so that heating is needed only for startup? If so, can the reactor "run away" and possibly explode? Should control measures be introduced to prevent this? What kind?

2. Where should the reactants be obtained? Buy them, or make them? In what proportions should they be fed to the reactor?

3. Should the reactor effluent, which contains the product and unconsumed reactants, be sold as is, or should the product be separated from the reactants and the latter be sent back to the reactor? If separation is desirable, how can it be accomplished? Heat the mixture and draw off and condense the vapor, which will be richer in the more volatile substances than the original mixture? Add another substance that extracts the product and is immiscible with the reactants, and then separate the two phases mechanically? If all of the process materials are gases at the reaction temperature, can the mixture be cooled to a temperature at which the product condenses but the reactants do not, or vice versa, or if they are liquids can the mixture be cooled to a temperature at which the product crystallizes? If one of these alternatives is chosen, what kind of equipment is needed? What size? What materials? What are the heating or cooling requirements? Are controls needed to keep the operation of the process within rigid limits? What kind of controls? Should they be manual or automatic?

4. How should the reactant and product streams be moved to and from the reactor and any heating, cooling, and separation equipment involved in the process? By gravity from a raised feed tank? With pumps, or blowers, or compressors, or conveyor belts? What kinds? How big? In pipes made of what?

5. Is enough known about the reaction system to be able to answer all of these questions, or should additional laboratory studies be carried out? What studies? Can the laboratory data be used directly to design the industrial plant, or should a smaller pilot plant be constructed first to test the design? How much smaller?

6. What can possibly go wrong with the process, and what can be done if and when it does?

7. Are waste products produced by the process? In what quantities? Are they potentially harmful if released untreated into the environment? If so, in what way? What should be done to reduce pollution hazards? Chemically treat the wastes? Dump liquid and solid wastes into containers, seal and cart them out to sea? Disperse gases in the atmosphere with a high stack? Precipitate solids electrostatically from gas exhausts?

8. How much of the process should be automated, and how should the automation be done?

9. How much will all of this cost? For how much can the product be sold, and to whom? How much money will the process net each year? Is it enough to make it worthwhile? If so, where should the plant be built?

10. Once the plant has been built, what procedure should be followed for startup?

11. Six months later when startup has been achieved, why is the product not coming out the way it did in the laboratory? Is it an equipment malfunction or a change in conditions somewhere between the laboratory and industrial process? How can we find out? What can be done to correct the problem? Is it necessary to shut down the operation for modifications?

12. Is it significant or just a coincidental series of bad breaks that there have been three explosions and four fires within six months in the reactor unit? In either case, how do we stop them from recurring?

13. All sorts of other things are going wrong with the process operation. Why weren't they on the list of things that could possibly go wrong? What can be done about them?

14. When the process finally starts working perfectly and the next day an order comes down to change the product specifications, how can it be done without redesigning the entire process? Why didn't they think of this before they built the plant?

The different careers and tasks described in this chapter are clearly too diverse to fall into a single category. They involve disciplines including physics, chemistry, biology, environmental science, medicine, applied mathematics, statistics, computer science, economics, management and information science, research, design, construction, sales and service, production supervision, and business administration. The single feature they have in common is that chemical engineers can be found doing them. Some of the specific knowledge needed to carry out the tasks will be presented later in the chemical engineering curriculum, a small amount of it is contained in this text, and most of it must be learned after graduation. There are, however, fundamental techniques that have been developed for setting up and attacking technical problems that are independent of the particular problem under consideration. What some of these techniques are and how and when to use them are the subjects of this book.

Chapter 2

Introduction to Engineering Calculations

Chapter 1 suggests the range of problems encompassed by chemical engineering, both in traditional areas of chemical processing and in relatively new fields such as environmental science and engineering, bioengineering, and semiconductor manufacturing. Differences between the systems mentioned in the chapter—chemical manufacturing processes, genetic engineering laboratories, pollution control facilities, and the others—are obvious. In this book, we examine the similarities.

One similarity is that all of the systems described involve **processes** designed to transform raw materials into desired products. Many of the problems that arise in connection with the design of a new process or the analysis of an existing one are of a certain type: given amounts and properties of the raw materials, calculate amounts and properties of the products, or vice versa.

The object of this text is to present a systematic approach to the solution of problems of this type. This chapter presents basic techniques for expressing the values of system variables and for setting up and solving equations that relate these variables. In Chapter 3 we discuss the variables of specific concern in process analysis—temperatures, pressures, chemical compositions, and amounts or flow rates of process streams—describing how they are defined, calculated, and, in some cases, measured. Parts Two and Three of the book deal with the laws of conservation of mass and energy, which relate the inputs and outputs of manufacturing systems, power plants, and the human body. The laws of nature constitute the underlying structure of all of process design and analysis; in the same way, the techniques that we present in this chapter underlie all applications of the laws.

2.0 INSTRUCTIONAL OBJECTIVES

After completing this chapter, you should be able to do the following:

- Convert a quantity expressed in one set of units into its equivalent in any other dimensionally consistent units using conversion factor tables. [For example, convert a heat flux of 235 $kJ/(m^2 \cdot s)$ into its equivalent in $Btu/(ft^2 \cdot h)$.]
- Identify the units commonly used to express both mass and weight in SI, CGS, and American Engineering units. Calculate weights from given masses in either natural units (e.g., $kg \cdot m/s^2$ or $lb_m \cdot ft/s^2$) or defined units (N, lb_f).
- Identify the number of significant figures in a given value expressed in either decimal or scientific notation and state the precision with which the value is known based on its significant figures. Determine the correct number of significant figures in the result of a series of arithmetic operations (adding, subtracting, multiplying, and dividing).

- Validate a quantitative problem solution by applying back-substitution, order-of-magnitude estimation, and the test of reasonableness.
- Given a set of measured values, calculate the sample mean, range, sample variance, and sample standard deviation. Explain in your own words what each of the calculated quantities means and why it is important.
- Explain the concept of dimensional homogeneity of equations. Given the units of some terms in an equation, use this concept to assign units to other terms.
- Given tabulated data for two variables (x and y), use linear interpolation between two data points to estimate the value of one variable for a given value of the other. Sketch a plot of y versus x and use it to illustrate how and when linear interpolation can lead to significant errors in estimated values.
- Given two points on a straight-line plot of y versus x, derive the expression for $y(x)$. Given tabulated data for x and y, fit a straight line by visual inspection.
- Given a two-parameter expression relating two variables [such as $y = a \sin(2x) + b$ or $P = 1/(aQ^3 + b)$] and two adjustable parameters (a and b), state what you would plot versus what to generate a straight line. Given data for x and y, generate the plot and estimate the parameters a and b.
- Given a power-law or exponential expression involving two variables (such as $y = ax^b$ or $k = ae^{b/T}$), state what you would plot versus what on rectangular, semilog, or logarithmic axes that would generate a straight line. Given a linear plot involving two variables on any of the three types of axes and two points on the line, determine the expression relating the two variables and the values of the two parameters.

2.1 UNITS AND DIMENSIONS

A measured or counted quantity has a numerical **value** (2.47) and a **unit** (whatever there are 2.47 of). It is useful in most engineering calculations—and essential in many—to write both the value and the unit of each quantity appearing in an equation:

$$2 \text{ meters}, \quad \tfrac{1}{3} \text{ second}, \quad 4.29 \text{ kilograms}, \quad 5 \text{ gold rings}$$

A **dimension** is a property that can be measured, such as length, time, mass, or temperature, or calculated by multiplying or dividing other dimensions, such as length/time (velocity), length^3 (volume), or mass/length^3 (density). Measurable units (as opposed to countable units) are specific values of dimensions that have been defined by convention, custom, or law, such as grams for mass, seconds for time, and centimeters or feet for length.

Units can be treated like algebraic variables when quantities are added, subtracted, multiplied, or divided. *The numerical values of two quantities may be added or subtracted only if the units are the same.*

$$3 \text{ cm} - 1 \text{ cm} = 2 \text{ cm} \qquad (3x - x = 2x)$$

but

$$3 \text{ cm} - 1 \text{ mm (or 1 s)} = ? \qquad (3x - y = ?)$$

On the other hand, *numerical values and their corresponding units may always be combined by multiplication or division.*

$$3 \text{ N} \times 4 \text{ m} = 12 \text{ N} \cdot \text{m}$$

$$\frac{5.0 \text{ km}}{2.0 \text{ h}} = 2.5 \text{ km/h}$$

$$7.0 \frac{\text{km}}{\text{h}} \times 4 \text{ h} = 28 \text{ km}$$

$$3 \text{ m} \times 4 \text{ m} = 12 \text{ m}^2$$

$$6 \text{ cm} \times 5 \ \frac{\text{cm}}{\text{s}} = 30 \text{ cm}^2/\text{s}$$

$$\frac{6 \text{ g}}{2 \text{ g}} = 3 \qquad (3 \text{ is a } \textit{dimensionless} \text{ quantity})$$

$$\left(5.0 \ \frac{\text{kg}}{\text{s}}\right) \bigg/ \left(0.20 \ \frac{\text{kg}}{\text{m}^3}\right) = 25 \text{ m}^3/\text{s} \qquad (\text{Convince yourself})$$

2.2 CONVERSION OF UNITS

A measured quantity can be expressed in terms of any units having the appropriate dimension. A particular velocity, for instance, may be expressed in ft/s, miles/h, cm/yr, or any other ratio of a length unit to a time unit. The numerical value of the velocity naturally depends on the units chosen.

The equivalence between two expressions of the same quantity may be defined in terms of a ratio:

$$\frac{1 \text{ cm}}{10 \text{ mm}} \qquad (1 \text{ centimeter per 10 millimeters}) \qquad \textbf{(2.2-1)}$$

$$\frac{10 \text{ mm}}{1 \text{ cm}} \qquad (10 \text{ millimeters per centimeter}) \qquad \textbf{(2.2-2)}$$

$$\left[\frac{10 \text{ mm}}{1 \text{ cm}}\right]^2 = \frac{100 \text{ mm}^2}{1 \text{ cm}^2} \qquad \textbf{(2.2-3)}$$

Ratios of the form of Equations 2.2-1, 2.2-2, and 2.2-3 are known as **conversion factors.**

To convert a quantity expressed in terms of one unit to its equivalent in terms of another unit, multiply the given quantity by the conversion factor (new unit/old unit). For example, to convert 36 mg to its equivalent in grams, write

$$(36 \text{ mg}) \times \left(\frac{1 \text{ g}}{1000 \text{ mg}}\right) = 0.036 \text{ g} \qquad \textbf{(2.2-4)}$$

(Note how the old units cancel, leaving the desired unit.) An alternative way to write this equation is to use a vertical line instead of the multiplication symbol:

$$\frac{36 \text{ mg} \ \bigg| \ 1 \text{ g}}{\bigg| \ 1000 \text{ mg}} = 0.036 \text{ g}$$

Carrying along units in calculations of this type is the best way of avoiding the common mistake of multiplying when you mean to divide and vice versa. In the given example, the result is known to be correct because milligrams cancel leaving only grams on the left side, whereas

$$\frac{36 \text{ mg} \ \bigg| \ 1000 \text{ mg}}{\bigg| \ 1 \text{ g}} = 36{,}000 \text{ mg}^2/\text{g}$$

is clearly wrong. (More precisely, it is not what you intended to calculate.)

If you are given a quantity having a compound unit [e.g., miles/h, cal/(g·°C)], and you wish to convert it to its equivalent in terms of another set of units, set up a **dimensional equation**: write the given quantity and its units on the left, write the units of conversion factors that cancel the old units and replace them with the desired ones, fill in the values of the conversion factors, and carry out the indicated arithmetic to find the desired value. (See Example 2.2-1.)

TEST
YOURSELF

1. What is a conversion factor?
2. What is the conversion factor for s/min (s = second)?
3. What is the conversion factor for min^2/s^2 (See Equation 2.2-3.)
4. What is the conversion factor for m^3/cm^3?

EXAMPLE 2.2-1   *Conversion of Units*

Convert an acceleration of 1 cm/s^2 to its equivalent in km/yr^2.

SOLUTION

$$\frac{1 \text{ cm}}{\text{s}^2} \left| \frac{3600^2 \text{ s}^2}{1^2 \text{ h}^2} \right| \frac{24^2 \text{ h}^2}{1^2 \text{ day}^2} \left| \frac{365^2 \text{ day}^2}{1^2 \text{ yr}^2} \right| \frac{1 \text{ m}}{10^2 \text{ cm}} \left| \frac{1 \text{ km}}{10^3 \text{ m}} \right.$$

$$= \frac{(3600 \times 24 \times 365)^2}{10^2 \times 10^3} \frac{\text{km}}{\text{yr}^2} = \boxed{9.95 \times 10^9 \text{ km/yr}^2}$$

A principle illustrated in this example is that raising a quantity (in particular, a conversion factor) to a power raises its units to the same power. The conversion factor for h^2/day^2 is therefore the square of the factor for h/day:

$$\left(\frac{24 \text{ h}}{1 \text{ day}} \right)^2 = 24^2 \frac{\text{h}^2}{\text{day}^2}$$

2.3 SYSTEMS OF UNITS

A system of units has the following components:

1. **Base units** for mass, length, time, temperature, electrical current, and light intensity.
2. **Multiple units,** which are defined as multiples or fractions of base units such as minutes, hours, and milliseconds, all of which are defined in terms of the base unit of a second. Multiple units are defined for convenience rather than necessity: it is simply more convenient to refer to 3 yr than to 94,608,000 s.
3. **Derived units,** obtained in one of two ways:
   (a) By multiplying and dividing base or multiple units (cm^2, ft/min, kg·m/s^2, etc.). Derived units of this type are referred to as **compound units**.
   (b) As defined equivalents of compound units (e.g., 1 erg $\equiv$ (1g·cm/s^2), 1 lb$_f$ $\equiv$ 32.174 lb$_m$·ft/s^2).

The "Système Internationale d'Unités," or **SI** for short, has gained widespread acceptance in the scientific and engineering community. Two of the base SI units—the ampere for electrical current and the candela for luminous intensity—will not concern us in this book. A third, the kelvin for temperature, will be discussed later. The others are the meter (m) for length, the kilogram (kg) for mass, and the second (s) for time.

Prefixes are used in SI to indicate powers of ten. The most common of these prefixes and their abbreviations are mega (M) for 10^6 (1 megawatt = 1 MW = 10^6 watts), kilo (k) for 10^3, centi (c) for 10^{-2}, milli (m) for 10^{-3}, micro (μ) for 10^{-6}, and nano (n) for 10^{-9}. The conversion factors between, say, centimeters and meters are therefore 10^{-2} m/cm and 10^2 cm/m. The principal SI units and prefixes are summarized in Table 2.3-1.

The **CGS system** is almost identical to SI, the principal difference being that grams (g) and centimeters (cm) are used instead of kilograms and meters as the base units of mass and length. The principal units of the CGS system are shown in Table 2.3-1.

The base units of the **American engineering system** are the foot (ft) for length, the pound-mass (lb$_m$) for mass, and the second (s) for time. This system has two principal difficulties. The first is the occurrence of conversion factors (such as 1 ft/12 in), which, unlike those in the metric systems, are not multiples of 10; the second, which has to do with the unit of force, is discussed in the next section.

Factors for converting from one system of units to another may be determined by taking ratios of quantities listed in the table on the inside front cover of this book. A larger table of conversion factors is given on pp. 1-4 through 1-20 of *Perry's Chemical Engineers' Handbook*.[1]

---

[1]R. H. Perry and D. W. Green, Eds., *Perry's Chemical Engineers' Handbook,* 7th Edition, McGraw-Hill, New York, 1997.

Table 2.3-1 SI and CGS Units

Base Units		
Quantity	Unit	Symbol
Length	meter (SI)	m
	centimeter (CGS)	cm
Mass	kilogram (SI)	kg
	gram (CGS)	g
Moles	gram-mole	mol or g-mole
Time	second	s
Temperature	kelvin	K
Electric current	ampere	A
Light intensity	candela	cd

Multiple Unit Preferences	
tera (T) = 10^{12}	centi (c) = 10^{-2}
giga (G) = 10^{9}	milli (m) = 10^{-3}
mega (M) = 10^{6}	micro (μ) = 10^{-6}
kilo (k) = 10^{3}	nano (n) = 10^{-9}

Derived Units			
Quantity	Unit	Symbol	Equivalent in Terms of Base Units
Volume	liter	L	0.001 m^3
			1000 cm^3
Force	newton (SI)	N	1 kg·m/s^2
	dyne (CGS)		1 g·cm/s^2
Pressure	pascal (SI)	Pa	1 N/m^2
Energy, work	joule (SI)	J	$1 \text{ N·m} = 1 \text{ kg·m}^2/\text{s}^2$
	erg (CGS)		$1 \text{ dyne·cm} = 1 \text{ g·cm}^2/\text{s}^2$
	gram-calorie	cal	$4.184 \text{ J} = 4.184 \text{ kg·m}^2/\text{s}^2$
Power	watt	W	$1 \text{ J/s} = 1 \text{ kg·m}^2/\text{s}^3$

TEST YOURSELF

1. What are the factors (numerical values and units) needed to convert
   (a) meters to millimeters?
   (b) nanoseconds to seconds?
   (c) square centimeters to square meters?
   (d) cubic feet to cubic meters (use the conversion factor table on the inside front cover)?
   (e) horsepower to British thermal units per second?
2. What is the derived SI unit for velocity? The velocity unit in the CGS system? In the American engineering system?

EXAMPLE 2.3-1   *Conversion Between Systems of Units*

Convert 23 $\text{lb}_m \cdot \text{ft/min}^2$ to its equivalent in kg·cm/s^2.

SOLUTION   As before, begin by writing the dimensional equation, fill in the units of conversion factors (new/old) and then the numerical values of these factors, and then do the arithmetic. The

result is

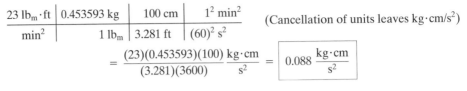

$$= \frac{(23)(0.453593)(100)}{(3.281)(3600)} \frac{kg \cdot cm}{s^2} = \boxed{0.088 \; \frac{kg \cdot cm}{s^2}}$$

2.4   FORCE AND WEIGHT

According to Newton's second law of motion, force is proportional to the product of mass and acceleration (length/time²). *Natural force units* are, therefore, kg·m/s² (SI), g·cm/s² (CGS), and $lb_m \cdot ft/s^2$ (American engineering). To avoid having to carry around these complex units in all calculations involving forces, *derived force units* have been defined in each system. In the metric systems, the derived force units (the **newton** in SI, the **dyne** in the CGS system) are defined to equal the natural units:

$$1 \text{ newton (N)} \equiv 1 \text{ kg·m/s}^2 \qquad \text{(2.4-1)}$$

$$1 \text{ dyne} \equiv 1 \text{ g·cm/s}^2 \qquad \text{(2.4-2)}$$

In the American engineering system, the derived force unit—called a **pound-force** (lb_f)—is defined as the product of a unit mass (1 lb_m) and the acceleration of gravity at sea level and 45° latitude, which is 32.174 ft/s²:

$$1 \text{ lb}_f \equiv 32.174 \text{ lb}_m \cdot ft/s^2 \qquad \text{(2.4-3)}$$

Equations 2.4-1 through 2.4-3 define conversion factors between natural and derived force units. For example, the force in newtons required to accelerate a mass of 4.00 kg at a rate of 9.00 m/s² is

$$F = \frac{4.00 \text{ kg} \; | \; 9.00 \text{ m} \; | \; 1 \text{ N}}{| \; s^2 \; | \; 1 \text{ kg·m/s}^2} = 36.0 \text{ N}$$

The force in lb_f required to accelerate a mass of 4.00 lb_m at a rate of 9.00 ft/s² is

$$F = \frac{4.00 \text{ lb}_m \; | \; 9.00 \text{ ft} \; | \; 1 \text{ lb}_f}{| \; s^2 \; | \; 32.174 \text{ lb}_m \cdot ft/s^2} = 1.12 \text{ lb}_f$$

Factors needed to convert from one force unit to another are summarized in the table on the inside front cover. The symbol g_c is sometimes used to denote the conversion factor from natural to derived force units: for example,

$$g_c = \frac{1 \text{ kg·m/s}^2}{1 \text{ N}} = \frac{32.174 \text{ lb}_m \cdot ft/s^2}{1 \text{ lb}_f}$$

We will not use this symbol in the text, but if you should encounter it elsewhere remember that it is simply a conversion factor (not to be confused with gravitational acceleration, which is usually denoted by g).

The **weight** of an object is the force exerted on the object by gravitational attraction. Suppose that an object of mass m is subjected to a gravitational force W (W is by definition the weight of the object) and that if this object were falling freely its acceleration would be g. The weight, mass, and free-fall acceleration of the object are related by Equation 2.4-4:

$$W = mg \qquad \text{(2.4-4)}$$

The gravitational acceleration (g) varies directly with the mass of the attracting body (the earth, in most problems you will confront) and inversely with the square of the distance between the centers of mass of the attracting body and the object being attracted. The value of g at sea level and 45° latitude is given below in each system of units:

$$
\begin{aligned}
g &= 9.8066 \text{ m/s}^2 \\
&= 980.66 \text{ cm/s}^2 \\
&= 32.174 \text{ ft/s}^2
\end{aligned}
$$

(2.4-5)

The acceleration of gravity does not vary much with position on the earth's surface and (within moderate limits) altitude, and the values in Equation 2.4-5 may accordingly be used for most conversions between mass and weight.

TEST YOURSELF

1. What is a force of 2 kg·m/s² equivalent to in newtons? What is a force of 2 $\text{lb}_m \cdot \text{ft/s}^2$ equivalent to in lb_f?
2. If the acceleration of gravity at a point is $g = 9.8$ m/s² and an object is resting on the ground at this point, is this object accelerating at a rate of 9.8 m/s²?
3. Suppose an object weighs 9.8 N at sea level. What is its mass? Would its mass be greater, less, or the same on the moon? How about its weight?
4. Suppose an object weighs 2 lb_f at sea level. What is its mass? Would its mass be greater, less, or the same at the center of the earth? How about its weight? (Careful!)

EXAMPLE 2.4-1 *Weight and Mass*

Water has a density of 62.4 lb_m/ft^3. How much does 2.000 ft³ of water weigh (1) at sea level and 45° latitude and (2) in Denver, Colorado, where the altitude is 5374 ft and the gravitational acceleration is 32.139 ft/s²?

SOLUTION

The mass of the water is

$$
M = \left(62.4 \, \frac{\text{lb}_m}{\text{ft}^3}\right) (2 \text{ ft}^3) = 124.8 \text{ lb}_m
$$

The weight of the water is

$$
W = (124.8 \text{ lb}_m)g \left(\frac{\text{ft}}{\text{s}^2}\right)\left(\frac{1 \text{ lb}_f}{32.174 \text{ lb}_m \cdot \text{ft/s}^2}\right)
$$

1. At sea level $g = 32.174$ ft/s², so that $W = 124.8 \text{ lb}_f$.
2. In Denver, $g = 32.139$ ft/s², and $W = 124.7 \text{ lb}_f$.

As this example illustrates, the error incurred by assuming that $g = 32.174$ ft/s² is normally quite small as long as you remain on the earth's surface. In a satellite or on another planet it would be a different story.

2.5 NUMERICAL CALCULATION AND ESTIMATION

2.5a Scientific Notation, Significant Figures, and Precision

Both very large and very small numbers are commonly encountered in process calculations. A convenient way to represent such numbers is to use **scientific notation**, in which a number is expressed as the product of another number (usually between 0.1 and 10) and a power of 10. Examples:

$$
123,000,000 = 1.23 \times 10^8 \text{ (or } 0.123 \times 10^9)
$$
$$
0.000028 = 2.8 \times 10^{-5} \text{ (or } 0.28 \times 10^{-4})
$$

The **significant figures** of a number are the digits from the first nonzero digit on the left to either (a) the last digit (zero or nonzero) on the right if there is a decimal point, or (b) the last nonzero digit of the number if there is no decimal point. For example,

2300 or 2.3×10^3 has two significant figures.
2300. or 2.300×10^3 has four significant figures.
2300.0 or 2.3000×10^3 has five significant figures.
23,040 or 2.304×10^4 has four significant figures.
0.035 or 3.5×10^{-2} has two significant figures.
0.03500 or 3.500×10^{-2} has four significant figures.

(***Note:*** The number of significant figures is easily shown and seen if scientific notation is used.)

The number of significant figures in the reported value of a measured or calculated quantity provides an indication of the precision with which the quantity is known: the more significant figures, the more precise is the value. Generally, if you report the value of a measured quantity with three significant figures, you indicate that the value of the third of these figures may be off by as much as a half-unit. Thus, if you report a mass as 8.3 g (two significant figures), you indicate that the mass lies somewhere between 8.25 and 8.35 g, whereas if you give the value as 8.300 g (four significant figures) you indicate that the mass lies between 8.2995 and 8.3005 g.

Note, however, that this rule applies only to measured quantities or numbers calculated from measured quantities. If a quantity is known precisely—like a pure integer (2) or a counted rather than measured quantity (16 oranges)—its value implicitly contains an infinite number of significant figures (5 cows really means 5.0000 . . . cows).

When two or more quantities are combined by multiplication and/or division, the number of significant figures in the result should equal the lowest number of significant figures of any of the multiplicands or divisors. If the initial result of a calculation violates this rule, you must round off the result to reduce the number of significant figures to its maximum allowed value, although if several calculations are to be performed in sequence it is advisable to keep extra significant figures of intermediate quantities and to round off only the final result. Examples:

$$\overset{(3)}{(3.57)}\overset{(4)}{(4.286)} = \overset{(7)}{15.30102} \Longrightarrow \overset{(3)}{15.3}$$

$$\overset{(2)}{(5.2 \times 10^{-4})}\overset{(4)}{(0.1635 \times 10^7)}/\overset{(3)}{(2.67)} = \overset{(9)}{318.426966} \Longrightarrow \overset{(2)}{3.2 \times 10^2} = \overset{(2)}{320}$$

(The raised quantities in parentheses denote the number of significant figures in the given numbers.) *Warning:* If you calculate, say, 3×4, and your calculator or computer gives you an answer like 11.99999, and you copy this answer and hand it in, your instructor may become violent!

The rule for addition and subtraction concerns the position of the last significant figure in the sum—that is, the location of this figure relative to the decimal point. The rule is: *When two or more numbers are added or subtracted, the positions of the last significant figures of each number relative to the decimal point should be compared. Of these positions, the one farthest to the left is the position of the last permissible significant figure of the sum or difference.*

Several examples of this rule follow, in which an arrow ($\downarrow$) denotes the last significant figure of each number.

$$\begin{array}{r} \downarrow \\ 1530 \;\downarrow \\ -\;2.56 \\ \hline 1527.44 \Longrightarrow 1530 \\ \uparrow \end{array}$$

$$\overset{\downarrow}{1.0000} + \overset{\downarrow}{0.036} + \overset{\downarrow}{0.22} = \overset{\downarrow}{1.2560} \Longrightarrow 1.26$$

$$2.75 \times 10^6 + 3.400 \times 10^4 = (2.75 + 0.03400) \times 10^6$$

$$= \overset{\downarrow}{2.784000} \times 10^6 \Longrightarrow 2.78 \times 10^6$$

Finally, a rule of thumb for rounding off numbers in which the digit to be dropped is a 5 is always to make the last digit of the rounded-off number even:

$$1.35 \implies 1.4$$
$$1.25 \implies 1.2$$

TEST YOURSELF

1. Express the following quantities in scientific notation and indicate how many significant figures each has.
   (a) 12,200   (b) 12,200.0   (c) 0.003040
2. Express the following quantities in standard decimal form and indicate how many significant figures each has.
   (a) 1.34×10^5 (b) 1.340×10^{-2} (c) 0.00420×10^6
3. How many significant figures would the solution of each of the following problems have? What are the solutions of (c) and (d)?
   (a) $(5.74)(38.27)/(0.001250)$ (c) $1.000 + 10.2$
   (b) $(1.76 \times 10^4)(0.12 \times 10^{-6})$ (d) $18.76 - 7$
4. Round off each of the following numbers to three significant figures.
   (a) 1465   (b) 13.35   (c) 1.765×10^{-7}
5. When the value of a number is given, the significant figures provide an indication of the uncertainty in the value; for example, a value of 2.7 indicates that the number lies between 2.65 and 2.75. Give ranges within which each of the following values lie.
   (a) 4.3   (d) 2500
   (b) 4.30   (e) 2.500×10^3
   (c) 2.778×10^{-3}

2.5b Validating Results

Every problem you will ever have to solve—in this and other courses and in your professional career—will involve two critical questions: (1) How do I get a solution? (2) When I get one, how do I know it's right? Most of this book is devoted to Question 1—that is, to methods of solving problems that arise in the design and analysis of chemical processes. However, Question 2 is equally important, and serious problems can arise when it is not asked. All successful engineers get into the habit of asking it whenever they solve a problem and they develop a wide variety of strategies for answering it.

Among approaches you can use to validate a quantitative problem solution are *back-substitution, order-of-magnitude estimation,* and *the test of reasonableness.*

- Back-substitution is straightforward: after you solve a set of equations, substitute your solution back into the equations and make sure it works.
- Order-of-magnitude estimation means coming up with a crude and easy-to-obtain approximation of the answer to a problem and making sure that the more exact solution comes reasonably close to it.
- Applying the test of reasonableness means verifying that the solution makes sense. If, for example, a calculated velocity of water flowing in a pipe is faster than the speed of light or the calculated temperature in a chemical reactor is higher than the interior temperature of the sun, you should suspect that a mistake has been made somewhere.

The procedure for checking an arithmetic calculation by order-of-magnitude estimation is as follows:

1. Substitute simple integers for all numerical quantities, using powers of 10 (scientific notation) for very small and very large numbers.

$$27.36 \rightarrow 20 \text{ or } 30 \text{ (whichever makes the subsequent arithmetic easier)}$$
$$63,472 \rightarrow 6 \times 10^4$$
$$0.002887 \rightarrow 3 \times 10^{-3}$$

2. Do the resulting arithmetic calculations by hand, continuing to round off intermediate answers.

$$\frac{(36,720)(0.0624)}{0.00478} \approx \frac{(4 \times 10^4)(5 \times 10^{-2})}{5 \times 10^{-4}} = 4 \times 10^{(4-2+4)} = 4 \times 10^6$$

The correct solution (obtained using a calculator) is 4.78×10^6. If you obtain this solution, since it is of the same magnitude as the estimate, you can be reasonably confident that you haven't made a gross error in the calculation.

3. If a number is added to a second, much smaller, number, drop the second number in the approximation.

$$\frac{1}{4.13 + 0.04762} \approx \frac{1}{4} = 0.25$$

The calculator solution is 0.239.

EXAMPLE 2.5-1 *Order-of-Magnitude Estimation*

The calculation of a process stream volumetric flow rate has led to the following formula:

$$\dot{V} = \left[\frac{254}{(0.879)(62.4)} + \frac{13}{(0.866)(62.4)} \right] \times \frac{1}{(31.3145)(60)}$$

Estimate $\dot{V}$ without using a calculator. (The exact solution is 0.00230.)

SOLUTION

$$\dot{V} \approx \left[\frac{250}{50} + \frac{10}{60} \right] \times \frac{1}{(4 \times 10^1)(6 \times 10^1)} \approx \frac{5}{25 \times 10^2} \approx 0.2 \times 10^{-2} = 0.002$$

The third way to check a numerical result—and perhaps the first thing you should do when you get one—is to see if the answer is reasonable. If, for example, you calculate that a cylinder contains 4.23×10^{32} kg of hydrogen when the mass of the sun is only 2×10^{30} kg, it should motivate you to redo the calculation. You should similarly be concerned if you calculate a reactor volume larger than the earth (10^{21} m³) or a room temperature hot enough to melt iron (1535°C). If you get in the habit of asking yourself, "Does this make sense?" every time you come up with a solution to a problem—in engineering and in the rest of your life—you will spare yourself considerable grief and embarrassment.

2.5c Estimation of Measured Values: Sample Mean

Suppose we carry out a chemical reaction of the form A → Products, starting with pure A in the reactor and keeping the reactor temperature constant at 45°C. After two minutes we draw a sample from the reactor and analyze it to determine X, the percentage of the A fed that has reacted.

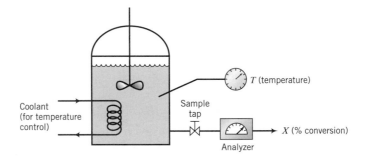

In theory X should have a unique value; however, in a real reactor X is a *random variable,* changing in an unpredictable manner from one run to another at the same experimental conditions. The values of X obtained after 10 successive runs might be as follows:

Run	1	2	3	4	5	6	7	8	9	10
$X(\%)$	67.1	73.1	69.6	67.4	71.0	68.2	69.4	68.2	68.7	70.2

Why don't we get the same value of X in each run? There are several reasons.

- It is impossible to replicate experimental conditions exactly in successive experiments. If the temperature in the reactor varies by as little as 0.1 degree from one run to another, it could be enough to change the measured value of X.
- Even if conditions were identical in two runs, we could not possibly draw our sample at exactly $t = 2.000\ldots$ minutes both times, and a difference of a second could make a measurable difference in X.
- Variations in sampling and chemical analysis procedures invariably introduce scatter in measured values.

We might ask two questions about the system at this point.

1. *What is the true value of X?*

In principle there may be such a thing as the "true value"—that is, the value we would measure if we could set the temperature exactly to $45.0000\ldots$ degrees, start the reaction, keep the temperature and all other experimental variables that affect X perfectly constant, and then sample and analyze with complete accuracy at exactly $t = 2.0000\ldots$ minutes. In practice there is no way to do any of those things, however. We could also define the true value of X as the value we would calculate by performing an infinite number of measurements and averaging the results, but there is no practical way to do that either. The best we can ever do is to *estimate* the true value of X from a finite number of measured values.

2. *How can we estimate of the true value of X?*

The most common estimate is the *sample mean* (or *arithmetic mean*). We collect N measured values of X $(X_1, X_2, \ldots, X_N)$ and then calculate

Sample Mean: $$\overline{X} = \frac{1}{N}(X_1 + X_2 + \cdots + X_N) = \frac{1}{N}\sum_{j=1}^{N} X_j \qquad \textbf{(2.5-1)}$$

For the given data, we would estimate

$$\overline{X} = \frac{1}{10}(67.1\% + 73.1\% + \cdots + 70.2\%) = 69.3\%$$

Graphically, the data and sample mean might appear as shown below. The measured values scatter about the sample mean, as they must.

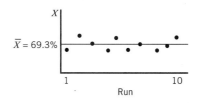

The more measurements of a random variable, the better the estimated value based on the sample mean. However, even with a huge number of measurements the sample mean is at best an approximation of the true value and could in fact be way off (e.g., if there is something wrong with the instruments or procedures used to measure X).

TEST YOURSELF

The weekly production rates of a pharmaceutical product over the past six weeks have been 37, 17, 39, 40, 40, and 40 batches per week.

1. Think of several possible explanations for the observed variation in the weekly production rate.
2. If you used the sample mean of the given data as a basis, what would you predict the next weekly production rate to be?
3. Come up with a better prediction, and explain your reasoning.

2.5d   Sample Variance of Scattered Data

Consider two sets of measurements of a random variable, X—for example, the percentage conversion in the same batch reactor measured using two different experimental techniques. Scatter plots of X versus run number are shown in Figure 2.5-1. The sample mean of each set is 70%, but the measured values scatter over a much narrower range for the first set (from 68% to 73%) than for the second set (from 52% to 95%). In each case you would estimate the true value of X for the given experimental conditions as the sample mean, 70%, but you would clearly have more confidence in the estimate for Set (a) than in that for Set (b).

Three quantities—the *range*, the *sample variance*, and the *sample standard deviation*—are used to express the extent to which values of a random variable scatter about their mean value. The *range* is simply the difference between the highest and lowest values of X in the data set:

Range:
$$R = X_{max} - X_{min} \qquad \textbf{(2.5-2)}$$

In the first plot of Figure 2.5-1 the range of X is 5% (73% − 68%) and in the second plot it is 43% (95% − 52%).

The range is the crudest measure of scatter: it involves only two of the measured values and gives no indication of whether or not most of the values cluster close to the mean or scatter widely around it. The *sample variance* is a much better measure. To define it we calculate the *deviation* of each measured value from the sample mean, $X_j - \overline{X}$ ($j = 1, 2, \ldots, N$), and then calculate

Sample Variance:
$$s_X^2 = \frac{1}{N-1}[(X_1 - \overline{X})^2 + (X_2 - \overline{X})^2 + \cdots + (X_N - \overline{X})^2] \qquad \textbf{(2.5-3)}$$

The degree of scatter may also be expressed in terms of the *sample standard deviation,* by definition the square root of the sample variance:

Sample Standard Deviation:
$$s_X = \sqrt{s_X^2} \qquad \textbf{(2.5-4)}$$

The more a measured value (X_j) deviates from the mean, either positively or negatively, the greater the value of $(X_j - \overline{X})^2$ and hence the greater the value of the sample variance and sample standard deviation. If these quantities are calculated for the data sets of Figure 2.5-1, for example, relatively small values are obtained for Set (a) ($s_X^2 = 0.30, s_X = 0.55$) and large values are obtained for Set (b) ($s_X^2 = 50, s_X = 7.1$).

For typical random variables, roughly two-thirds of all measured values fall within one standard deviation of the mean; about 95% fall within two standard deviations; and about

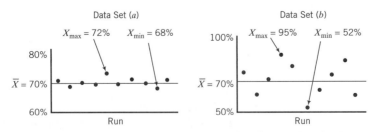

Figure 2.5-1   Scatter plots for two data sets with different levels of scatter.

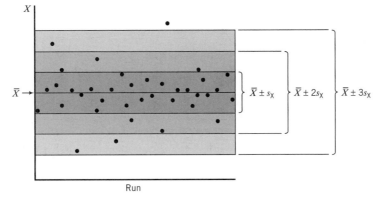

Figure 2.5-2 Data scatter about the mean.

99% fall within three standard deviations.[2] A graphical illustration of this statement is shown in Figure 2.5-2. Of the 37 measured values of X, 27 fall within one standard deviation of the mean, 33 within two standard deviations, and 36 within three standard deviations.

Values of measured variables are often reported with error limits, such as $X = 48.2 \pm 0.6$. This statement means that a single measured value of X is likely to fall between 47.6 and 48.8. The midpoint of the range ($X = 48.2$) is almost always the mean value of the data set used to generate this result; however, the significance of the given error limits (± 0.6) is not obvious unless more information is given. The interval between 47.6 and 48.8 may represent the range of the data set ($X_{max} - X_{min}$) or ± 0.6 might represent $\pm s_X, \pm 2s_X,$ or $\pm 3s_X$. (There are other possibilities, but they rarely occur.) If you report a variable value in this manner, make clear what your error limits mean.

TEST YOURSELF

The volumetric flow rate of a process fluid, $\dot{V}(cm^3/s)$, is measured five times, with the following results:

Measurement	1	2	3	4	5
$\dot{V}(cm^3/s)$	232	248	227	241	239

(a) Calculate the sample mean ($\overline{V}$), range, sample variance (s_V^2), and sample standard deviation (s_V).
(b) There is a high probability (above 90%) that a measured value of $\dot{V}$ will fall within two standard deviations of the mean. Report the value of $\dot{V}$ in the form $\dot{V} = a \pm b$, choosing the values of a and b to define this range.

EXAMPLE 2.5-2 *Statistical Quality Control*

Five hundred batches of a pigment are produced each week. In the plant's quality assurance (QA) program, each batch is subjected to a precise color analysis test. If a batch does not pass the test, it is rejected and sent back for reformulation.

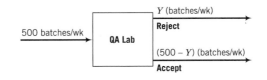

---

[2]The exact percentages depend on how the measured values are distributed about the mean—whether they follow a Gaussian distribution, for example—and how many points are in the data set used to calculate the mean and standard deviation.

Let Y be the number of bad batches produced per week, and suppose that QA test results for a 12-week base period are as follows:

Week	1	2	3	4	5	6	7	8	9	10	11	12
$\bar{Y}$	17	27	18	18	23	19	18	21	20	19	21	18

The company policy is to regard the process operation as normal as long as the number of bad batches produced in a week is no more than three standard deviations above the mean value for the base period (i.e., as long as $Y \leq \bar{Y} + 3s_Y$). If Y exceeds this value, the process is shut down for remedial maintenance (a long and costly procedure). Such large deviations from the mean might occur as part of the normal scatter of the process, but so infrequently that if it happens the existence of an abnormal problem in the process is considered the more likely explanation.

1. How many bad batches in a week would it take to shut down the process?
2. What would be the limiting value of Y if two standard deviations instead of three were used as the cutoff criterion? What would be the advantage and disadvantage of using this stricter criterion?

SOLUTION

1. From Equations 2.5-1, 2.5-3, and 2.5-4, the sample mean, sample variance, and sample standard deviation of Y during the base period are

$$\bar{Y} = \frac{1}{12} \sum_{j=1}^{12} (17 + 27 + \cdots + 18) = 19.9 \text{ batches/wk}$$

$$s_Y^2 = \frac{1}{11}[(17 - 19.9)^2 + (27 - 19.9)^2 + \cdots + (18 - 19.9)^2] = 7.9 \text{ (batches/wk)}^2$$

$$s_Y = \sqrt{7.9} = 2.8 \text{ batches/wk}$$

The maximum allowed value of Y is

$$\bar{Y} + 3s_Y = 19.9 + (3)(2.8) = \boxed{28.3}$$

If 29 or more bad batches are produced in a week, the process must be shut down for maintenance.

2. $\bar{Y} + 2s_Y = 19.9 + (2)(2.8) = \boxed{25.5}$. If this criterion were used, 26 bad batches in a week would be enough to shut down the process. The advantage is that if something *has* gone wrong with the process the problem will be corrected sooner and fewer bad batches will be made in the long run. The disadvantage is that more costly shutdowns may take place when nothing is wrong, the large number of bad batches simply reflecting normal scatter in the process.

2.6 DIMENSIONAL HOMOGENEITY AND DIMENSIONLESS QUANTITIES

We began our discussion of units and dimensions by saying that quantities can be added and subtracted only if their units are the same. If the units are the same, it follows that the dimensions of each term must be the same. For example, if two quantities can be expressed in terms of grams/second, both must have the dimension (mass/time). This suggests the following rule:

Every valid equation must be dimensionally homogeneous: that is, all additive terms on both sides of the equation must have the same dimensions.

Consider the equation

$$u(\text{m/s}) = u_0(\text{m/s}) + g(\text{m/s}^2)t(\text{s}) \tag{2.6-1}$$

This equation is dimensionally homogeneous, since each of the terms u, u_0, and gt has the same dimensions (length/time). On the other hand, the equation $u = u_0 + g$ is not dimensionally homogeneous (why not?) and therefore cannot possibly be valid.

Equation 2.6-1 is both dimensionally homogeneous *and* consistent in its units, in that each additive term has the units m/s. If values of u_0, g, and t with the indicated units are substituted into the equation, the addition may be carried out to determine the value of u. If an equation is dimensionally homogeneous but its additive terms have inconsistent units, the terms (and hence the equation) may be made consistent simply by applying the appropriate conversion factors.

For example, suppose that in the dimensionally homogeneous equation $u = u_0 + gt$ it is desired to express the time (t) in minutes and the other quantities in the units given above. The equation can be written as

$$u(\text{m/s}) = u_0(\text{m/s}) + g(\text{m/s}^2)t(\text{min})(60 \text{ s/min})$$
$$= u_0 + 60gt$$

Each additive term again has units of m/s (verify this), so the equation is consistent.

The converse of the given rule is not necessarily true—an equation may be dimensionally homogeneous and invalid. For example, if M is the mass of an object, then the equation $M = 2M$ is dimensionally homogeneous, but it is also obviously incorrect except for one specific value of M.

EXAMPLE 2.6-1   *Dimensional Homogeneity*

Consider the equation

$$D(\text{ft}) = 3t(\text{s}) + 4$$

1. If the equation is valid, what are the dimensions of the constants 3 and 4?
2. If the equation is consistent in its units, what are the units of 3 and 4?
3. Derive an equation for distance in meters in terms of time in minutes.

SOLUTION

1. For the equation to be valid, it must be dimensionally homogeneous, so that each term must have the dimension of length. The constant 3 must therefore have the dimension $\boxed{\text{length/time}}$, and 4 must have the dimension $\boxed{\text{length}}$.
2. For consistency, the constants must be $\boxed{3 \text{ ft/s}}$ and $\boxed{4 \text{ ft}}$.
3. Define new variables $D'(\text{m})$ and $t'(\text{min})$. The equivalence relations between the old and new variables are

$$D(\text{ft}) = \frac{D'(\text{m})}{} \left| \frac{3.2808 \text{ ft}}{1 \text{ m}} \right. = 3.28D'$$

$$t(s) = \frac{t'(\text{min})}{} \left| \frac{60 \text{ s}}{1 \text{ min}} \right. = 60t'$$

Substitute these expressions in the given equation

$$3.28D' = (3)(60t') + 4$$

and simplify by dividing through by 3.28

$$\boxed{D'(\text{m}) = 55t'(\text{min}) + 1.22}$$

Exercise: What are the units of 55 and 1.22?

Example 2.6-1 illustrates a general procedure for rewriting an equation in terms of new variables having the same dimensions but different units:

1. Define new variables (e.g., by affixing primes to the old variable names) that have the desired units.

2. Write expressions for each old variable in terms of the corresponding new variable.
3. Substitute these expressions in the original equation and simplify.

A **dimensionless quantity** can be a pure number $(2, 1.3, \frac{5}{2})$ or a multiplicative combination of variables with no net dimensions:

$$\frac{M(\text{g})}{M_o(\text{g})} \qquad \frac{D(\text{cm})u(\text{cm/s})\rho(\text{g/cm}^3)}{\mu[\text{g/(cm·s)}]}$$

A quantity such as M/M_o or $Du\rho/\mu$ is also called a **dimensionless group**.

Exponents (such as the 2 in X^2), transcendental functions (such as log, exp $\equiv e$, and sin), and arguments of transcendental functions (such as the X in sin X) must be dimensionless quantities. For example, 10^2 makes perfect sense, but $10^{2 \text{ ft}}$ is meaningless, as is log (20 s) or sin (3 dynes).

EXAMPLE 2.6-2   *Dimensional Homogeneity and Dimensionless Groups*

A quantity k depends on the temperature T in the following manner:

$$k\left(\frac{\text{mol}}{\text{cm}^3 \cdot \text{s}}\right) = 1.2 \times 10^5 \exp\left(-\frac{20,000}{1.987T}\right)$$

The units of the quantity 20,000 are cal/mol, and T is in K (kelvin). What are the units of 1.2×10^5 and 1.987?

SOLUTION

Since the equation must be consistent in its units and exp is dimensionless, 1.2×10^5 should have the same units as k, mol/(cm^3·s). Moreover, since the argument of exp must be dimensionless, we can write

$$\frac{20,000 \text{ cal}}{\text{mol}} \left| \frac{1}{T(\text{K})} \right| \frac{\text{mol} \cdot \text{K}}{1.987 \text{ cal}} \qquad \text{(All units cancel)}$$

The answers are thus

$$\boxed{1.2 \times 10^5 \text{ mol/(cm}^3 \cdot \text{s)} \quad \text{and} \quad 1.987 \text{ cal/(mol·K)}}$$

TEST YOURSELF

1. What is a dimensionally homogeneous equation? If an equation is dimensionally homogeneous, is it necessarily valid? If an equation is valid, is it necessarily dimensionally homogeneous?
2. If $y(\text{m/s}^2) = az(\text{m}^3)$, what are the units of a?
3. What is a dimensionless group? What multiplicative combination of $r(\text{m})$, $s(\text{m/s}^2)$, and $t(\text{s})$ would constitute a dimensionless group?
4. If $z(\text{lb}_\text{f}) = a \sin (Q)$, what are the units of a and Q?

2.7 PROCESS DATA REPRESENTATION AND ANALYSIS

The operation of any chemical process is ultimately based on the measurement of process variables—temperatures, pressures, flow rates, concentrations, and so on. It is sometimes possible to measure these variables directly, but, as a rule, indirect techniques must be used.

Suppose, for example, that you wish to measure the concentration, C, of a solute in a solution. To do so, you normally measure a quantity, X—such as a thermal or electrical conductivity, a light absorbance, or the volume of a titer—that varies in a known manner with C, and then calculate C from the measured value of X. The relationship between C and X is determined in a separate **calibration** experiment in which solutions of known concentration are prepared and X is measured for each solution.

Consider a calibration experiment in which a variable, y, is measured for several values of another variable, x:

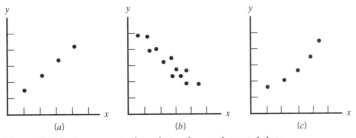

Figure 2.7-1 Representative plots of experimental data.

x	1.0	2.0	3.0	4.0
y	0.3	0.7	1.2	1.8

In the terms of the first paragraph, y might be a reactant concentration or some other process variable and x would be a readily measured quantity (such as conductivity) whose value correlates with the value of y. Our object is to use the calibration data to estimate the value of y for a value of x between tabulated points (**interpolation**) or outside the range of the table data (**extrapolation**).

A number of interpolation and extrapolation methods are commonly used, including two-point linear interpolation, graphical interpolation, and curve fitting. Which one is most appropriate depends on the nature of the relationship between x and y.

Figure 2.7-1 shows several illustrative (x, y) plots. If the plot of a given data set looks like that shown in plot (a) or (b) of this figure, a straight line would probably be fitted to the data and used as the basis for subsequent interpolation or extrapolation. On the other hand, if the plot is distinctly curved as in plot (c), a curve could be drawn in by inspection and used as the basis of interpolation, or straight-line segments could be fitted to successive pairs of points, or a nonlinear function $y(x)$ that fits the data could be sought.

The technique of drawing a line or curve through the data by inspection is self-explanatory. The other methods are reviewed in the sections that follow.

2.7a Two-Point Linear Interpolation

The equation of the line through (x_1, y_1) and (x_2, y_2) on a plot of y versus x is

$$y = y_1 + \frac{x - x_1}{x_2 - x_1}(y_2 - y_1) \qquad \textbf{(2.7-1)}$$

(Can you prove it?) You may use this equation to estimate y for an x between x_1 and x_2; you may also use it to estimate y for an x outside of this range (i.e., to extrapolate the data), but with a much greater risk of inaccuracy.

If the points in a table are relatively close together, linear interpolation should provide an accurate estimate of y for any x and vice versa; on the other hand, if the points are widely separated or if the data are to be extrapolated, one of the curve-fitting techniques to be outlined in the next section should be used.

TEST YOURSELF

1. Values of a variable (f) are measured at several times (t):

f	1	4	8
t	1	2	3

Show that if two-point linear interpolation is used (a) $f(t = 1.3) \approx 1.9$; (b) $t(f = 5) \approx 2.25$.

2. If a function $y(x)$ appears as shown in each of the diagrams shown below, would two-point linear interpolation yield estimates of y that are too high, too low, or correct? If the two-point linear interpolation formula (Equation 2.7-1) were used to estimate $y(x_3)$ from the tabulated values of (x_1, y_1) and (x_2, y_2) in plot (b), would the estimated value be too high or too low?

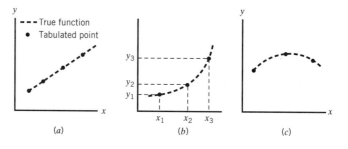

(a)   (b)   (c)

2.7b   Fitting a Straight Line

A convenient way to indicate how one variable depends on another is with an equation:

$$y = 3x + 4$$
$$y = 4.24(x - 3)^2 - 23$$
$$y = 1.3 \times 10^7 \sin(2x)/(x^{1/2} + 58.4)$$

If you have an analytical expression for $y(x)$ like those shown above, you can calculate y for any given x or (with a somewhat greater effort) determine x for any given y, or you can program a computer to perform these calculations.

Suppose the values of a dependent variable y have been measured for several values of an independent variable x, and a plot of y versus x on rectangular coordinate axes yields what appears to be a straight line. The equation you would use to represent the relationship between x and y is then

$$y = ax + b \qquad \text{(2.7-2)}$$

If the points show relatively little scatter, like those in Figure 2.71a, a line may then be drawn through them by inspection, and if (x_1, y_1) and (x_2, y_2) are two points—which may or may not be data points—on the line, then

Slope: $\qquad\qquad\qquad\qquad a = \dfrac{y_2 - y_1}{x_2 - x_1} \qquad\qquad$ **(2.7-3)**

Intercept: $\qquad\qquad\qquad b \begin{cases} = y_1 - ax_1 \\ = y_2 - ax_2 \end{cases} \qquad\qquad$ **(2.7-4)**

Once a has been calculated from Equation 2.7-3 and b has been determined from either of Equations 2.7-4, it is good practice to check the result by verifying that Equation 2.7-2 is satisfied at the point—(x_1, y_1) or (x_2, y_2)—*not* used in the calculation of b.

EXAMPLE 2.7-1   ***Fitting a Straight Line to Flowmeter Calibration Data***

Rotameter calibration data (flow rate versus rotameter reading) are as follows:

Flow Rate $\dot{V}$(L/min)	Rotameter Reading R
20.0	10
52.1	30
84.6	50
118.3	70
151.0	90

1. Draw a calibration curve and determine an equation for $\dot{V}(R)$.
2. Calculate the flow rate that corresponds to a rotameter reading of 36.

SOLUTION

1. The calibration curve appears as follows:

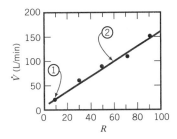

A line drawn through the data by visual inspection passes through the points ($R_1 = 10$, $\dot{V}_1 = 20$) and ($R_2 = 60, \dot{V}_2 = 101$). Therefore,

$$\dot{V} = aR + b \quad \text{(Since the data fall on a line)}$$

$$a = \frac{\dot{V}_2 - \dot{V}_1}{R_2 - R_1} = \frac{101 - 20}{60 - 10} = 1.62 \quad \text{(From Equation 2.7-3)}$$

$$b = \dot{V}_1 - aR_1 = 20 - (1.62)(10) = 3.8 \quad \text{(From Equation 2.7-4)}$$

The result is, therefore,

$$\boxed{\dot{V} = 1.62R + 3.8}$$

Check: At point ②,

$$aR_2 + b = (1.62)(60) + 3.8 = 101 = \dot{V}_2$$

2. At $R = 36, \dot{V} = (1.62)(36) + 3.8 = \boxed{62.1 \text{ L/min}}$.

2.7c   Fitting Nonlinear Data

During a recent week at a major university, 423 experimenters separately measured and plotted data and found that their data points did not lie on straight points; 416 of these people shrugged their shoulders, said "Close enough," and drew a line anyway; and the other seven went about finding an equation other than $y = ax + b$ to relate the variables.

Fitting a nonlinear equation (anything but $y = ax + b$) to data is usually much harder than fitting a line; however, with some nonlinear equations you can still use straight-line fitting if you plot the data in a suitable manner. Suppose, for example, that x and y are related by the equation $y^2 = ax^3 + b$. A plot of measured y versus x data would clearly be curved; however, a plot of y^2 versus x^3 would be a straight line with a slope a and intercept b. More generally, if any two quantities are related by an equation of the form

$$\text{(Quantity 1)} = a\,\text{(Quantity 2)} + b$$

then a plot of the first quantity (y^2 in the above example) versus the second (x^3) on rectangular coordinates yields a straight line with slope a and intercept b.

Here are several additional examples of plots that yield straight lines:

1. $y = ax^2 + b$. Plot y versus x^2.
2. $y^2 = \frac{a}{x} + b$. Plot y^2 versus $\frac{1}{x}$.
3. $\frac{1}{y} = a(x + 3) + b$. Plot $\frac{1}{y}$ versus $(x + 3)$.
4. $\sin y = a(x^2 - 4)$. Plot $\sin y$ versus $(x^2 - 4)$. The line through the data must be drawn through the origin. (Why?)

Even if the original equation is not in a proper form to generate a linear plot, you can sometimes rearrange it to get it into such a form:

5. $y = \dfrac{1}{C_1 x - C_2} \implies \dfrac{1}{y} = C_1 x - C_2$

Plot $\dfrac{1}{y}$ versus x. Slope $= C_1$, intercept $= -C_2$.

6. $y = 1 + x(mx^2 + n)^{1/2} \implies \dfrac{(y-1)^2}{x^2} = mx^2 + n$

Plot $\dfrac{(y-1)^2}{x^2}$ versus x^2. Slope $= m$, intercept $= n$.

Let us summarize the procedure. If you have (x, y) data that you wish to fit with an equation that can be written in the form $f(x, y) = ag(x, y) + b$,

1. Calculate $f(x, y)$ and $g(x, y)$ for each tabulated (x, y) point, and plot f versus g.
2. If the plotted points fall on a straight line, the equation fits the data. Choose two points on the line—(g_1, f_1) and (g_2, f_2)—and calculate a and b as outlined in the previous section.

$$a = \frac{f_2 - f_1}{g_2 - g_1} \qquad b = f_1 - ag_1 \qquad \text{or} \qquad b = f_2 - ag_2$$

EXAMPLE 2.7-2   *Linear Curve-Fitting of Nonlinear Data*

A mass flow rate $\dot{m}$(g/s) is measured as a function of temperature T(°C).

T	10	20	40	80
$\dot{m}$	14.76	20.14	27.73	38.47

There is reason to believe that $\dot{m}$ varies linearly with the square root of T:

$$\dot{m} = aT^{1/2} + b$$

Use a straight-line plot to verify this formula and determine a and b.

SOLUTION   If the formula is correct, a plot of $\dot{m}$ versus $T^{1/2}$ should be linear, with slope $= a$ and intercept $= b$. The data table is augmented by adding a $T^{1/2}$ row:

T	10	20	40	80
$T^{1/2}$	3.162	4.472	6.325	8.944
$\dot{m}$	14.76	20.14	27.73	38.47

and $\dot{m}$ is plotted versus $T^{1/2}$.

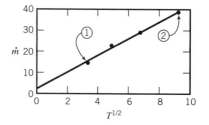

Since the plot is linear, the proposed formula is verified. A line drawn through the data points passes through the first and last points, so these points may be used to calculate the slope and intercept:

$$\dot{m} = aT^{1/2} + b$$

$$(T_1^{1/2} = 3.162, \ \dot{m}_1 = 14.76)$$
$$(T_2^{1/2} = 8.944, \ \dot{m}_2 = 38.47)$$

Slope: $a = \dfrac{\dot{m}_2 - \dot{m}_1}{T_2^{1/2} - T_1^{1/2}} = \dfrac{38.47 - 14.76}{8.944 - 3.162} = 4.10 \ \mathrm{g/(s \cdot {}^\circ C^{1/2})}$

Intercept: $b = \dot{m}_1 - aT_1^{1/2} = 14.76 - (4.10)(3.162) = 1.80 \ \mathrm{g/s}$

(verify the units), so that

$$\boxed{\dot{m} = 4.10T^{1/2} + 1.80}$$

Check: At point ②, $4.10T_2^{1/2} + 1.80 = (4.10)(8.944) + 1.80 = 38.47 = \dot{m}_2$.

Two nonlinear functions that often occur in process analysis are the **exponential function,** $y = ae^{bx}$ [or $y = a \exp(bx)$], where $e \approx 2.7182818$, and the **power law,** $y = ax^b$. Before we describe how the parameters of these functions may be determined by linear curve-fitting, let us review some algebra.

The natural logarithm (ln) is the inverse of the exponential function:

$$P = e^Q \iff \ln P = Q \tag{2.7-5}$$

It follows that

$$\ln[e^Q] = Q \quad \text{and} \quad e^{\ln P} = P \tag{2.7-6}$$

The natural logarithm of a number may be calculated from the common logarithm ($\log_{10}$ or just log) using the relation

$$\ln x = 2.302585 \log_{10} x \tag{2.7-7}$$

The familiar rules for taking logarithms of products and powers are applicable to natural logarithms: if $y = ax$ then $\ln y = \ln a + \ln x$, and if $y = x^b$ then $\ln y = b \ln x$. These properties suggest ways to fit exponential and power law functions to (x, y) data:

$$\left\{ \begin{array}{l} y = a \exp(bx) \implies \ln y = \ln a + bx \\ \text{Plot } \ln y \text{ versus } x. \text{ Slope} = b, \text{ intercept} = \ln a. \end{array} \right\} \tag{2.7-8}$$

$$\left\{ \begin{array}{l} y = ax^b \implies \ln y = \ln a + b \ln x \\ \text{Plot } \ln y \text{ versus } \ln x. \text{ Slope} = b, \text{ intercept} = \ln a. \end{array} \right\} \tag{2.7-9}$$

Once you have determined $\ln a$ as the intercept of either of these plots, you can calculate a from Equation 2.7-6 as $\exp(\ln a)$; for example, if $\ln a = 3.00$, then $a = \exp(3.00) = 20.1$.

TEST YOURSELF

1. The following plot is generated from experimental (x, y) data:

What equation would you use to relate x and y?

2. How would you plot (x, y) data to get a straight line, and how would you determine a and b for each of the following functions?

   (a) $y = a\sqrt{x} + b$

   Solution: Plot y versus $\sqrt{x}$; let $(\sqrt{x_1}, y_1)$ and $(\sqrt{x_2}, y_2)$ be two points on the line; calculate $a = (y_2 - y_1)/(\sqrt{x_2} - \sqrt{x_1})$, $b = y_1 - a\sqrt{x_1}$

    (b) $1/y = a(x - 3)^2 + b$ **(e)** $y = ae^{bx}$

    (c) $y = (ax^2 - b)^{1/3}$ **(f)** $y = ax^b$

    (d) $\sin(y) = x(ax + b)^{-2}$

2.7d Logarithmic Coordinates

Suppose you wish to fit an exponential function $y = a \exp(bx)$ to measured (x, y) data. If there are many data points, calculating the logarithm of each y value (needed to plot $\ln y$ versus x) could take more time than the fitting procedure itself. However, suppose that an additional scale were drawn parallel to the $\ln y$ axis, on which values of y were shown adjacent to the corresponding values of $\ln y$ on the first scale. (See Figure 2.7-2.) Now, instead of having to calculate $\ln y$ for each tabulated y to locate the data points on the graph, you can find the y values on the second scale and locate the points directly. If the same type of scale (called a **logarithmic scale**) were drawn parallel to the other axis, you could also plot $\ln x$ values without having to calculate them from tabulated x values. A plot with logarithmic scales on both axes is called a **log plot**, and a plot with one logarithmic and one rectangular (equal interval) axis is called a **semilog plot**. *Log paper* is graph paper with logarithmic scales on both axes, and *semilog paper* has one logarithmic axis and one rectangular axis. Most computer plotting packages allow selection of log or semilog scales as alternatives to rectangular scales.

When you plot values of a variable y on a logarithmic scale you are in effect plotting the logarithm of y on a rectangular scale. Suppose, for example, that y and x are related by the equation $y = a \exp(bx)$ ($\ln y = \ln a + bx$). To determine a and b you may plot y versus x on a semilog plot, choosing two points (x_1, y_1) and (x_2, y_2) on the resulting line, or you may plot $\ln y$ versus x on rectangular axes, drawing the line through the corresponding two points $(x_1, \ln y_1)$ and $(x_2, \ln y_2)$. In either case, b and a are obtained as

$$b = \frac{\ln y_2 - \ln y_1}{x_2 - x_1} = \frac{\ln(y_2/y_1)}{x_2 - x_1}$$

$$\ln a = \ln y_1 - bx_1$$

$$\text{or} \qquad\qquad \Longrightarrow [a = \exp(\ln a)]$$

$$\ln a = \ln y_2 - bx_2$$

In summary,

    1. If y versus x data appear linear on a semilog plot, then $\ln y$ versus x would be linear on a rectangular plot, and the data can therefore be correlated by an exponential function $y = a \exp(bx)$. (See Equation 2.7-8.)

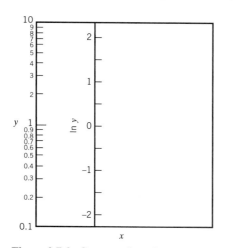

Figure 2.7-2 Construction of a logarithmic scale.

2. If y versus x data appear linear on a log plot, then $\ln y$ versus $\ln x$ would be linear on a rectangular plot, and the data can therefore be correlated by a power law $y = ax^b$. (See Equation 2.7-9.)
3. When you plot values of a variable z on a logarithmic axis and your plot yields a straight line through two points with coordinate values z_1 and z_2, replace $z_2 - z_1$ with $\ln(z_2/z_1)$ $(= \ln z_2 - \ln z_1)$ in the formula for the slope.
4. *Do not plot values of $\ln z$ on a logarithmic scale and expect anything useful to result.*

EXAMPLE 2.7-3   *Curve Fitting on Semilog and Log Plots*

A plot of F versus t yields a line that passes through the points $(t_1 = 15, F_1 = 0.298)$ and $(t_2 = 30, F_2 = 0.0527)$ on (1) a semilog plot and (2) a log plot. For each case, calculate the equation that relates F and t.

SOLUTION

1. *Semilog plot*

$$\ln F = bt + \ln a \qquad \text{(since the plot appears linear)}$$

$$\Downarrow$$

$$F = ae^{bt}$$

$$b = \frac{\ln(F_2/F_1)}{t_2 - t_1} = \frac{\ln(0.0527/0.298)}{(30 - 15)} = -0.1155$$

$$\ln a = \ln F_1 - bt_1 = \ln(0.298) + (0.1155)(15) = 0.5218$$

$$\Downarrow$$

$$a = \exp(0.5218) = 1.685$$

or

$$\boxed{F = 1.685\exp(-0.1155t)}$$

Check: $F(t_2) = 1.685\exp(-0.1155 \times 30) = 0.0527$.

2. *Log plot*

$$\ln F = b \ln t + \ln a \qquad \text{(since the plot appears linear)}$$

$$\Downarrow$$

$$F = at^b$$

$$b = \frac{\ln(F_2/F_1)}{\ln(t_2/t_1)} = \frac{\ln(0.0527/0.298)}{\ln(30/15)} = -2.50$$

$$\ln a = \ln F_1 - b \ln t_1 = \ln(0.298) + 2.5\ln(15) = 5.559$$

$$\Downarrow$$

$$a = \exp(5.559) = 260$$

or

$$\boxed{F = 260t^{-2.5}}$$

Check: $F(t_2) = 260(30)^{-2.5} = 0.0527$.

TEST YOURSELF

1. The following plots yield straight lines. What are the equations that relate the variables?
   (a) P versus t on rectangular coordinates.
   (b) P (logarithmic axis) versus t on a semilog plot.

(c) P versus t on a log plot.

(d) $y^2 - 3$ (logarithmic axis) versus $1/x^2$ on a semilog plot. (Express the answer as an exponential function.)

(e) $1/F$ versus $t^2 - 4$ on a log plot. (Express the answer as a power law.)

2. What would you plot against what on what kind of axes to get a straight line for the following relationships (a and b are constants)?

(a) $P = a \exp(bt)$ **(c)** $P^2 = \exp(at^3 + b)$

(b) $P = at^b$ **(d)** $1/P = a(t - 4)^{-b}$

2.7e Fitting a Line to Scattered Data

There is little problem fitting a line to data that look like this:

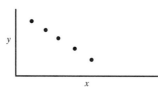

Life being the way it is, however, you are much more likely to come up with something more like this:

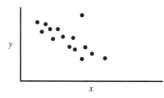

When data points are as scattered as these, you can draw any number of lines that appear to fit the points equally well (or equally poorly, depending on your point of view). The question is which line to use.

A number of statistical techniques exist for fitting a function to a set of scattered data. The application of the most common of these techniques—*linear regression* or the *method of least squares*—to the fitting of a straight line to a series of y versus x data points is outlined and illustrated in Appendix A.1, and the use of this technique is required for the solution of Problems 2.39 through 2.42 at the end of this chapter.

2.8 SUMMARY

This chapter introduces some fundamental problem-solving tools that you will need in the rest of this course, in subsequent engineering and science courses, and almost every time in your career when you perform mathematical calculations. The main points of the chapter are as follows.

- You can convert a quantity expressed in one set of units into its equivalent in other dimensionally consistent units using conversion factors, like those in the table on the inside front cover of the text.

- A *weight* is the force exerted on an object by gravitational attraction. The weight of an object of mass m may be calculated as $W = mg$, where g is the acceleration of gravity at the location of the object. At sea level on the earth, $g = 9.8066$ m/s^2 = 32.174 ft/s^2. To convert a weight (or any force) in natural units like kg·m/s^2 or lb$_m$·ft/s^2 to its equivalent in a derived force unit like N or lb$_f$, use the conversion factor table.

- The significant figures (s.f.) with which a number is reported specify the precision with which the number is known. For instance, $x = 3.0$ (2 s.f.) states that x is somewhere between 2.95

and 3.05, while $x = 3.000$ (4 s.f.) states that it is between 2.9995 and 3.0005. When you multiply and divide numbers, the number of significant figures of the result equals the lowest number of significant figures of any of the factors. In complex calculations, keep the maximum number of significant figures until the final result is obtained, then round off.

- If X is a measured process variable, the *sample mean* of a set of measured values, $\bar{X}$, is the average of the set (the sum of the values divided by the number of values). It is an estimate of the true mean, the value that would be obtained by averaging an infinite number of measurements. The *sample variance* of the set, s_X^2, is a measure of the spread of the measured values about the sample mean. It is calculated from Equation (2.5-3). The *sample standard deviation*, s_X, is the square root of the sample variance.

- If $\bar{X}$ and s_X are determined from a set of normal process runs and a subsequently measured value of X falls more than $2s_X$ away from $\bar{X}$, the chances are that something has changed in the process—there is less than a 10% chance that normal scatter can account for the deviation. If the deviation is greater than $3s_X$, there is less than a 1% chance that normal scatter is the cause. The exact percentages depend on how the measured values are distributed about the mean—whether they follow a Gaussian distribution, for example—and how many points are in the data set used to calculate the mean and standard deviation.

- Suppose you are given a set of values of a dependent variable, y, corresponding to values of an independent variable, x, and you wish to estimate y for a specified x. You can either assume a straight-line dependence for the two data points that bracket the specified x and use two-point linear interpolation (Equation 2.7-1) or fit a function to the data points and use it for the desired estimation.

- If (x,y) data appear to scatter about a straight line on a plot of y versus x, you may fit a line using Equations (2.7-3) and (2.7-4) or, for greater precision and an estimate of the goodness of the fit, use the method of least squares (Appendix A.1). If a plot of y versus x is nonlinear, you may try to fit various nonlinear functions by plotting functions of x and y in a manner that should yield a straight line. For example, to fit a function $y^2 = a/x + b$ to (x,y) data, plot y^2 versus $1/x$. If the fit is good, the plot should be a straight line with slope a and intercept b.

- Plotting y (log scale) versus x (linear scale) on a semilog plot is equivalent to plotting $\ln y$ versus x on rectangular axes. If the plot is linear in either case, x and y are related by an exponential function, $y = ae^{bx}$.

- Plotting y versus x on logarithmic axes is equivalent to plotting $\ln y$ versus $\ln x$ on rectangular axes. If the plot is linear in either case, x and y are related by a power law function, $y = ax^b$.

PROBLEMS

2.1. Using dimensional equations, convert
(a) 3 wk to milliseconds.   (c) 554 m⁴/(day·kg) to cm⁴/(min·g).
(b) 38.1 ft/s to miles/h.

2.2. Using the table of conversion factors on the inside front cover, convert
(a) 760 miles/h to m/s.   (c) 5.37×10^3 kJ/min to hp.
(b) 921 kg/m³ to lb$_m$/ft³.

2.3. Using a single dimensional equation, estimate the number of golf balls it would take to fill your classroom.

2.4. Using a single dimensional equation, estimate the number of steps it would take you, walking at your normal stride, to walk from the Earth to Alpha Centauri, a distance of 4.3 light-years. The speed of light is 1.86×10^5 miles/s.

2.5. A frustrated professor once claimed that if all the reports she had graded in her career were stacked on top of one another, they would reach from the Earth to the moon. Assume that an average report is the thickness of about 10 sheets of printer paper and use a single dimensional equation to estimate the number of reports the professor would have had to grade for her claim to be valid.

2.6. You are trying to decide which of two automobiles to buy. The first is American-made, costs $14,500, and has a rated gasoline mileage of 28 miles/gal. The second car is of European manufacture, costs

$21,700, and has a rated mileage of 19 km/L. If the cost of gasoline is $1.25/gal and if the cars actually deliver their rated mileage, estimate how many miles you would have to drive for the lower fuel consumption of the second car to compensate for the higher cost of this car?

2.7. A supersonic aircraft consumes 5320 imperial gallons of kerosene per hour of flight and flies an average of 14 hours per day. It takes roughly seven tons of crude oil to produce one ton of kerosene. The density of kerosene is 0.965 g/cm³. How many planes would it take to consume the entire annual world production of 4.02×10^9 metric tons of crude oil?

2.8. Calculate
   (a) the weight in lb_f of a 25.0-lb_m object.
   (b) the mass in kg of an object that weighs 25 newtons.
   (c) the weight in dynes of a 10-ton object (not metric tons).

2.9. A waste treatment pond is 50 m long and 15 m wide, and has an average depth of 2 m. The density of the waste is 85.3 lb_m/ft^3. Calculate the weight of the pond contents in lb_f, using a single dimensional equation for your calculation.

2.10. Five hundred lb_m of nitrogen is to be charged into a small metal cylinder at 25°C, at a pressure such that the gas density is 11.5 kg/m³. Without using a calculator, estimate the required cylinder volume. Show your work.

2.11. According to Archimedes' principle, the mass of a floating object equals the mass of the fluid displaced by the object. Use this principle to solve the following problems.
   (a) A wooden cylinder 30.0 cm high floats vertically in a tub of water (density = 1.00 g/cm³). The top of the cylinder is 14.1 cm above the surface of the liquid. What is the density of the wood?
   (b) The same cylinder floats vertically in a liquid of unknown density. The top of the cylinder is 20.7 cm above the surface of the liquid. What is the liquid density?

2.12. A right circular cone of base radius R, height H, and known density ρ_s floats base down in a liquid of unknown density ρ_f. A height h of the cone is above the liquid surface. Derive a formula for ρ_f in terms of ρ_s, R, and h/H, simplifying it algebraically to the greatest possible extent. [Recall Archimedes' principle, stated in the preceding problem, and note that the volume of a cone equals (base area)(height)/3.]

2.13. A horizontal cylindrical drum is 2.00 m in diameter and 4.00 m long. The drum is slowly filled with benzene (density = 0.879 g/cm³). Derive a formula for W, the weight in newtons of the benzene in the tank, as a function of h, the depth of the liquid in centimeters.

2.14. A **poundal** is the force required to accelerate a mass of 1 lb_m at a rate of 1 ft/s², and a **slug** is the mass of an object that will accelerate at a rate of 1 ft/s² when subjected to a force of 1 lb_f.
   (a) Calculate the mass in slugs and the weight in poundals of a 175 lb_m man (i) on earth and (ii) on the moon, where the acceleration of gravity is one-sixth of its value on earth.
   (b) A force of 355 poundals is exerted on a 25.0-slug object. At what rate (m/s²) does the object accelerate?

2.15. The **fern** is defined as the unit of force required to accelerate a unit of mass, called the **bung**, with the gravitational acceleration on the surface of the moon, which is one-sixth of the normal gravitational acceleration on earth.
   (a) What is the conversion factor that would be used to convert a force from the natural unit to the derived unit in this system? (Give both its numerical value and its units.)
   (b) What is the weight in ferns of a 3-bung object on the moon? What does the same object weigh in Lizard Lick, North Carolina?

2.16. Perform the following calculations. In each case, first estimate the solution without using a calculator, following the procedure outlined in Section 2.5b, and then do the calculation, paying attention to significant figures.
   (a) $(2.7)(8.632)$ **(c)** $2.365 + 125.2$
   (b) $(3.600 \times 10^{-4})/45$ **(d)** $(4.753 \times 10^4) - (9 \times 10^2)$

2.17. The following expression has occurred in a problem solution:

$$R = \frac{(0.6700)(264,980)(6)(5.386 \times 10^4)}{(3.14159)(0.479 \times 10^7)}$$

The factor 6 is a pure integer. Estimate the value of R without using a calculator, following the procedure outlined in Section 2.5b. Then calculate R, expressing your answer in both scientific and decimal notation and making sure it has the correct number of significant figures.

2.18. Two thermocouples (temperature measurement devices) are tested by inserting their probes in boiling water, recording the readings, removing and drying the probes, and then doing it again. The results of five measurements are as follows:

$T(°C)$—Thermocouple A	72.4	73.1	72.6	72.8	73.0
$T(°C)$—Thermocouple B	97.3	101.4	98.7	103.1	100.4

(a) For each set of temperature readings, calculate the sample mean, the range, and the sample standard deviation.

(b) Which thermocouple readings exhibit the higher degree of scatter? Which thermocouple is more accurate?

2.19. Product quality assurance (QA) is a particularly tricky business in the dye manufacturing industry. A slight variation in reaction conditions can lead to a measurable change in the color of the product, and since customers usually require extremely high color reproducibility from one shipment to another, even a small color change can lead to rejection of a product batch.

Suppose the various color frequency and intensity values that comprise a color analysis are combined into a single numerical value, C, for a particular yellow dye. During a test period in which the reactor conditions are carefully controlled and the reactor is thoroughly cleaned between successive batches (not the usual procedure), product analyses of 12 batches run on successive days yield the following color readings:

Batch	1	2	3	4	5	6	7	8	9	10	11	12
C	74.3	71.8	72.0	73.1	75.1	72.6	75.3	73.4	74.8	72.6	73.0	73.7

(a) The QA specification for routine production is that a batch that falls more than two standard deviations away from the test period mean must be rejected and sent for reworking. Determine the minimum and maximum acceptable values of C.

(b) A statistician working in quality assurance and a production engineer are having an argument. One of them, Frank, wants to raise the QA specification to three standard deviations and the other, Joanne, wants to lower it to one. Reworking is time-consuming, expensive, and very unpopular with the engineers who have to do it. Who is more likely to be the statistician and who the engineer? Explain.

(c) Suppose that in the first few weeks of operation relatively few unacceptable batches are produced, but then the number begins to climb steadily. Think of up to five possible causes, and state how you might go about determining whether or not each of them might in fact be responsible for the drop in quality.

*2.20. Your company manufactures plastic wrap for food storage. The tear resistance of the wrap, denoted by X, must be controlled so that the wrap can be torn off the roll without too much effort but it does not tear too easily when in use.

In a series of test runs, 15 rolls of wrap are made under carefully controlled conditions and the tear resistance of each roll is measured. The results are used as the basis of a *quality assurance specification* (see Problem 2.19). If X for a subsequently produced roll falls more than two standard deviations away from the test period average, the process is declared out of specification and production is suspended for routine maintenance.

---

*Computer problem.

The test series data are as follows:

Roll	1	2	3	4	5	6	7	8	9	10	11	12	13	14	15
X	134	131	129	133	135	131	134	130	131	136	129	130	133	130	133

(a) Write a spreadsheet to take as input the test series data and calculate the sample mean ($\bar{X}$) and sample standard deviation (s_X), preferably using built-in functions for the calculations.

(b) The following tear resistance values are obtained for rolls produced in 14 consecutive production runs subsequent to the test series: 128, 131, 133, 130, 133, 129, 133, 135, 137, 133, 137, 136, 137, 139. On the spreadsheet (preferably using the spreadsheet plotting capability), plot a control chart of X versus run number, showing horizontal lines for the values corresponding to $\bar{X}$, $\bar{X} - 2s_X$, and $\bar{X} + 2s_X$ from the test period, and show the points corresponding to the 14 production runs. (See Figure 2.5-2.) Which measurements led to suspension of production?

(c) Following the last of the production runs, the chief plant engineer returns from vacation, examines the plant logs, and says that routine maintenance was clearly not sufficient and a process shutdown and full system overhaul should have been ordered at one point during the two weeks he was away. When would it have been reasonable to take this step, and why?

2.21. A variable, Q, is reported to have a value of 2.360×10^{-4} kg·m²/h.

(a) Write a dimensional equation for Q', the equivalent variable value expressed in American engineering units, using seconds as the unit for time.

(b) Estimate Q' without using a calculator, following the procedure outlined in Section 2.5b. (Show your calculations.) Then determine Q' with a calculator, expressing your answer in both scientific and decimal notation and making sure it has the correct number of significant figures.

2.22. The **Prandtl number**, N_{Pr}, is a dimensionless group important in heat transfer calculations. It is defined as $C_p \mu / k$, where C_p is the heat capacity of a fluid, μ is the fluid viscosity, and k is the thermal conductivity. For a particular fluid, $C_p = 0.583$ J/(g·°C), $k = 0.286$ W/(m·°C), and $\mu = 1936$ lb$_m$/(ft·h). Estimate the value of N_{Pr} without using a calculator (remember, it is dimensionless), showing your calculations; then determine it with a calculator.

2.23. The **Reynolds number** is a dimensionless group defined for a fluid flowing in a pipe as

$$Re = Du\rho / \mu$$

where D is pipe diameter, u is fluid velocity, ρ is fluid density, and μ is fluid viscosity. When the value of the Reynolds number is less than about 2100, the flow is *laminar*—that is, the fluid flows in smooth streamlines. For Reynolds numbers above 2100, the flow is *turbulent*, characterized by a great deal of agitation.

Liquid methyl ethyl ketone (MEK) flows through a pipe with an inner diameter of 2.067 inches at an average velocity of 0.48 ft/s. At the fluid temperature of 20°C the density of liquid MEK is 0.805 g/cm³ and the viscosity is 0.43 centipoise [1 cP = 1.00×10^{-3} kg/(m·s)]. Without using a calculator, determine whether the flow is laminar or turbulent. Show your calculations.

2.24. The following empirical equation correlates the values of variables in a system in which solid particles are suspended in a flowing gas:

$$\frac{k_g d_p y}{D} = 2.00 + 0.600 \left(\frac{\mu}{\rho D} \right)^{1/3} \left(\frac{d_p u \rho}{\mu} \right)^{1/2}$$

Both $(\mu/\rho D)$ and $(d_p u \rho / \mu)$ are dimensionless groups; k_g is a coefficient that expresses the rate at which a particular species transfers from the gas to the solid particles; and the coefficients 2.00 and 0.600 are dimensionless constants obtained by fitting experimental data covering a wide range of values of the equation variables.

The value of k_g is needed to design a catalytic reactor. Since this coefficient is difficult to determine directly, values of the other variables are measured or estimated and k_g is calculated from the given correlation. The variable values are as follows:

$$d_p = 5.00 \text{ mm}$$

$$y = 0.100 \quad \text{(dimensionless)}$$

$$D = 0.100 \text{ cm}^2/\text{s}$$
$$\mu = 1.00 \times 10^{-5} \text{ N·s/m}^2$$
$$\rho = 1.00 \times 10^{-3} \text{ g/cm}^3$$
$$u = 10.0 \text{ m/s}$$

(a) What is the estimated value of k_g? (Give its value and units.)

(b) Why might the true value of k_g in the reactor be significantly different from the value estimated in part (a)? (Give several possible reasons.)

*(c) Create a spreadsheet in which up to five sets of values of the given variables (d_p through u) are entered in columns and the corresponding values of k_g are calculated. Test your program using the following variable sets: (i) the values given above; (ii) as above, only double the particle diameter d_p (making it 10.00 mm); (iii) as above, only double the diffusivity D; (iv) as above, only double the viscosity μ; (v) as above, only double the velocity u. Report all five calculated values of k_g.

2.25. A seed crystal of diameter D (mm) is placed in a solution of dissolved salt, and new crystals are observed to nucleate (form) at a constant rate r (crystals/min). Experiments with seed crystals of different sizes show that the rate of nucleation varies with the seed crystal diameter as

$$r(\text{crystals/min}) = 200D - 10D^2 \quad (D \text{ in mm})$$

(a) What are the units of the constants 200 and 10? (Assume the given equation is valid and therefore dimensionally homogeneous.)

(b) Calculate the crystal nucleation rate in crystals/s corresponding to a crystal diameter of 0.050 inch.

(c) Derive a formula for r(crystals/s) in terms of D(inches). (See Example 2.6-1.) Check the formula using the result of part (b).

2.26. The density of a fluid is given by the empirical equation

$$\rho = 70.5 \exp(8.27 \times 10^{-7} P)$$

where ρ is density (lb_m/ft^3) and P is pressure ($\text{lb}_f/\text{in.}^2$).

(a) What are the units of 70.5 and 8.27×10^{-7}?

(b) Calculate the density in g/cm^3 for a pressure of $9.00 \times 10^6 \text{ N/m}^2$.

(c) Derive a formula for $\rho(\text{g/cm}^3)$ as a function of $P(\text{N/m}^2)$. (See Example 2.6-1.) Check your result using the solution of part (b).

2.27. The volume of a microbial culture is observed to increase according to the formula

$$V(\text{cm}^3) = e^t$$

where t is time is seconds.

(a) Calculate the expression for $V(\text{in.}^3)$ in terms of $t(\text{h})$.

(b) Both the exponential function and its argument must be dimensionless. The given equation seems to violate both of these rules, and yet the equation is valid. Explain this paradox. [*Hint:* Observe the result of part (a).]

2.28. A concentration C (mol/L) varies with time (min) according to the equation

$$C = 3.00 \exp(-2.00t)$$

(a) What are the units of 3.00 and 2.00?

(b) Suppose the concentration is measured at $t = 0$ and $t = 1$ min. Use two-point linear interpolation or extrapolation to estimate $C(t = 0.6 \text{ min})$ and $t(C = 0.10 \text{ mol/L})$ from the measured values, and compare these results with the true values of these quantities.

(c) Sketch a curve of C versus t, and show graphically the points you determined in part (b).

*2.29. The vapor pressures of 1-chlorotetradecane at several temperatures are tabulated here.

$T(^\circ\text{C})$	98.5	131.8	148.2	166.2	199.8	215.5
$p*$ (mm Hg)	1	5	10	20	60	100

---

*Computer problem.

(a) Use two-point linear interpolation to estimate the value of $p*$ at $T = 185°C$.

(b) Write a computer subroutine to estimate the vapor pressure of 1-chlorotetradecane for any temperature between 98.5°C and 215.5°C using two-point linear interpolation. The subroutine must determine which two tabulated temperatures bracket the given temperature, and apply the interpolation to estimate $p* (T)$. Then write a main program to read and store the values of $p*$ and T given in the table and to generate a table of vapor pressures at temperatures $T = 100°C, 105°C, 110°C, \ldots, 215°C$, calling your subroutine to estimate $p*$ at each temperature. Check your program using the result of part (a).

2.30. Sketch the plots described below and calculate the equations for $y(x)$ from the given information. The plots are all straight lines. Note that the given coordinates refer to abscissa and ordinate values, not x and y values. [The solution of part (a) is given as an example.]

(a) A plot of $\ln y$ versus x on rectangular coordinates passes through (1.0, 0.693) and (2.0, 0.0) (i.e., at the first point $x = 1.0$ and $\ln y = 0.693$).

$$\text{Solution}: \ln y = bx + \ln a \implies y = ae^{bx}$$

$$b = (\ln y_2 - \ln y_1)/(x_2 - x_1) = (0 - 0.693)/(2.0 - 1.0) = -0.693$$

$$\ln a = \ln y_1 - bx_1 = 0.693 + 0.693 * 1.0 = 1.386 \implies a = e^{1.386} = 4.00$$

$$\Downarrow$$

$$\boxed{y = 4.00e^{-0.693x}}$$

(b) A semilog plot of y (logarithmic axis) versus x passes through (1, 2) and (2, 1).

(c) A log plot of y versus x passes through (1, 2) and (2, 1).

(d) A semilog plot of xy (logarithmic axis) versus y/x passes through (1.0, 40.2) and (2.0, 807.0).

(e) A log plot of y^2/x versus $(x - 2)$ passes through (1.0, 40.2) and (2.0, 807.0).

2.31. State what you would plot to get a straight line if experimental (x, y) data are to be correlated by the following relations, and what the slopes and intercepts would be in terms of the relation parameters. If you could equally well use two different kinds of plots (e.g., rectangular or semilog), state what you would plot in each case. [The solution to part (a) is given as an example.]

(a) $y^2 = ae^{-b/x}$.

Solution: Construct a semilog plot of y^2 versus $1/x$ or a plot of $\ln (y^2)$ versus $1/x$ on rectangular coordinates. Slope $= -b$, intercept $= \ln a$.

(b) $y^2 = mx^3 - n$

(c) $1/\ln(y - 3) = (1 + a\sqrt{x})/b$

(d) $(y + 1)^2 = [a(x - 3)^3]^{-1}$

(e) $y = \exp(a\sqrt{x} + b)$

(f) $xy = 10^{[a(x^2+y^2)+b]}$

(g) $y = [ax + b/x]^{-1}$

2.32. A **hygrometer**, which measures the amount of moisture in a gas stream, is to be calibrated using the apparatus shown here:

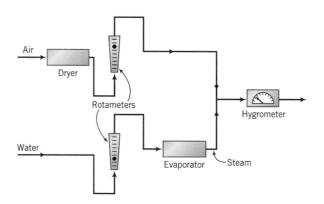

Steam and dry air are fed at known flow rates and mixed to form a gas stream with a known water content, and the hygrometer reading is recorded; the flow rate of either the water or the air is changed to produce a stream with a different water content and the new reading is recorded, and so on. The following data are taken:

Mass Fraction of Water, y	Hygrometer Reading, R
0.011	5
0.044	20
0.083	40
0.126	60
0.170	80

(a) Draw a calibration curve and determine an equation for $y(R)$.
(b) Suppose a sample of a stack gas is inserted in the sample chamber of the hygrometer and a reading of $R = 43$ is obtained. If the mass flow rate of the stack gas is 1200 kg/h, what is the mass flow rate of water vapor in the gas?

2.33. The temperature in a process unit is controlled by passing cooling water at a measured rate through a jacket that encloses the unit.

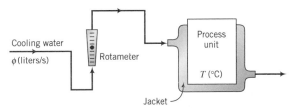

The exact relationship between the unit temperature $T(°C)$ and the cooling flow water flow rate ϕ (L/s) is extremely complex, and it is desired to derive a simple empirical formula to approximate this relationship over a limited range of flow rates and temperatures. Data are taken for T versus ϕ. Plots of T versus ϕ on rectangular and semilog coordinates are distinctly curved (ruling out $T = a\phi + b$ and $T = ae^{b\phi}$ as possible empirical functions), but a log plot appears as follows:

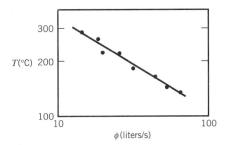

A line drawn through the data goes through the points ($\phi_1 = 25, T_1 = 210$) and ($\phi_2 = 40, T_2 = 120$).
(a) What is the empirical relationship between ϕ and T?
(b) Using your derived equation, estimate the cooling water flow rates needed to maintain the process unit temperature at 85°C, 175°C, and 290°C.
(c) In which of the three estimates in part (b) would you have the most confidence and in which would you have the least confidence? Explain your reasoning.

2.34. A chemical reaction A→B is carried out in a closed vessel. The following data are taken for the concentration of A, C_A(g/L), as a function of time, t(min), from the start of the reaction:

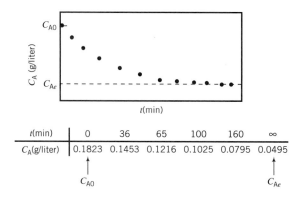

t(min)	0	36	65	100	160	∞
C_A(g/liter)	0.1823	0.1453	0.1216	0.1025	0.0795	0.0495

A proposed reaction mechanism predicts that C_A and t should be related by the expression

$$\ln \frac{C_A - C_{Ae}}{C_{A0} - C_{Ae}} = -kt$$

where k is the **reaction rate constant**.

(a) Do the data support this prediction? If so, determine the value of k. (Use a semilog plot in your solution of this problem.)

(b) If the tank volume is 30.5 gallons and there is no B in the tank at $t = 0$, how much B(g) does the tank contain after two hours?

2.35. The climactic moment in the film "The Eggplant That Ate New Jersey" comes when the brilliant young scientist announces his discovery of the equation for the volume of the eggplant:

$$V(\text{ft}^3) = 3.53 \times 10^{-2} \exp(2t^2)$$

where t is the time in hours from the moment the vampire injected the eggplant with a solution prepared from the blood of the beautiful dental hygienist.

(a) What are the units of 3.53×10^{-2} and 2?

(b) The scientist obtained the formula by measuring V versus t and determining the coefficients by linear regression. What would he have plotted versus what on what kind of coordinates? What would he have obtained as the slope and intercept of his plot?

(c) The European distributor of the film insists that the formula be given for the volume in m^3 as a function of t(s). Derive the formula.

2.36. The relationship between the pressure P and volume V of the air in a cylinder during the upstroke of a piston in an air compressor can be expressed as

$$PV^k = C$$

where k and C are constants. During a compression test, the following data are taken:

P(mm Hg)	760	1140	1520	2280	3040	3800
V(cm^3)	48.3	37.4	31.3	24.1	20.0	17.4

Determine the values of k and C that best fit the data. (Give both numerical values and units.)

2.37. In modeling the effect of an impurity on crystal growth, the following equation was derived:

$$\frac{G - G_L}{G_0 - G} = \frac{1}{K_L C^m}$$

where C is impurity concentration, G_L is a limiting growth rate, G_o is the growth rate of the crystal with no impurity present, and K_L and m are model parameters.

In a particular experiment, $G_0 = 3.00 \times 10^{-3}$ mm/min, and $G_L = 1.80 \times 10^{-3}$ mm/min. Growth rates are measured for several impurity concentrations C (parts per million, or ppm), with the following results:

C (ppm)	50.0	75.0	100.0	125.0	150.0
G(mm/min) $\times 10^3$	2.50	2.20	2.04	1.95	1.90

(For example, when $C = 50.0$ ppm, $G = 2.50 \times 10^{-3}$ mm/min).
(a) Determine K_L and m, giving both numerical values and units.
(b) A solution is fed to a crystallizer in which the impurity concentration is 475 ppm. Estimate the expected crystal growth rate in (mm/min). Then state why you would be extremely skeptical about this result.

2.38. A process instrument reading, Z(volts), is thought to be related to a process stream flow rate $\dot{V}$(L/s) and pressure P(kPa) by the following expression:

$$Z = a\dot{V}^b P^c$$

Process data have been obtained in two sets of runs—one with $\dot{V}$ held constant, the other with P held constant. The data are as follows:

Point	1	2	3	4	5	6	7
$\dot{V}$(L/s)	0.65	1.02	1.75	3.43	1.02	1.02	1.02
P(kPa)	11.2	11.2	11.2	11.2	9.1	7.6	5.4
Z(volts)	2.27	2.58	3.72	5.21	3.50	4.19	5.89

(a) Suppose you had only performed runs 2, 3, and 5. Calculate a, b, and c algebraically from the data for these three runs.
(b) Now use a graphical method and all the data to calculate a, b, and c. Comment on why you would have more confidence in this result than in that of part (a). (*Hint:* You will need at least two plots.)

2.39. Fit (a) a line and (b) a line through the origin to the following data using the method of least squares (Appendix A.1):

x	0.3	1.9	3.2
y	0.4	2.1	3.1

On a single plot, show both fitted lines and the data points.

2.40. A solution containing hazardous waste is charged into a storage tank and subjected to a chemical treatment that decomposes the waste to harmless products. The concentration of the decomposing waste, C, has been reported to vary with time according to the formula

$$C = 1/(a + bt)$$

When sufficient time has elapsed for the concentration to drop to 0.01 g/L, the contents of the tank are discharged into a river that passes by the plant.
The following data are taken for C and t:

t(h)	1.0	2.0	3.0	4.0	5.0
C(g/L)	1.43	1.02	0.73	0.53	0.38

(a) If the given formula is correct, what plot would yield a straight line that would enable you to determine the parameters a and b?
(b) Estimate a and b using the method of least squares (Appendix A.1). Check the goodness of fit by generating a plot of C versus t that shows both the measured and predicted values of C.
(c) Using the results of part (b), estimate the initial concentration of the waste in the tank and the time required for C to reach its discharge level.
(d) You should have very little confidence in the time estimated in part (c). Explain why.

(e) There are other potential problems with the whole waste disposal procedure. Suggest several of them.

2.41. The following (x, y) data are recorded:

x	0.5	1.4	84
y	2.20	4.30	6.15

(a) Plot the data on logarithmic axes.
(b) Determine the coefficients of a power law expression $y = ax^b$ using the method of least squares. (Remember what you are really plotting— there is no way to avoid taking logarithms of the data point coordinates in this case.)
(c) Draw your calculated line on the same plot as the data.

*2.42. A published study of a chemical reaction, A→P, indicates that if the reactor initially contains A at a concentration $C_{A0}(g/L)$ and the reaction temperature, T, is kept constant, then the concentration of P in the reactor increases with time according to the formula

$$C_P(g/L) = C_{A0}(1 - e^{-kt})$$

The *rate constant*, $k(s^{-1})$, is reportedly a function only of the reaction temperature.
   To test this finding, the reaction is run in four different laboratories. The reported experimental results are given below.

$t(s)$	Lab 1 $T = 275°C$ $C_{A0} = 4.83$	Lab 2 $T = 275°C$ $C_{A0} = 12.2$	Lab 3 $T = 275°C$ $C_{A0} = 5.14$	Lab 4 $T = 275°C$ $C_{A0} = 3.69$
	C_P (g/L)			
0	0.0	0.0	0.0	0.0
10	0.287	1.21	0.310	0.245
20	0.594	2.43	0.614	0.465
30	0.871	3.38	0.885	0.670
60	1.51	5.89	1.64	1.20
120	2.62	8.90	2.66	2.06
240	3.91	11.2	3.87	3.03
360	4.30	12.1	4.61	3.32
480	4.62	12.1	4.89	3.54
600	4.68	12.2	5.03	3.59

(a) What plot would yield a straight line if the given equation is correct?
(b) Enter the given data into a spreadsheet. For each data set (C_P versus t), generate the plot of part (a) and determine the corresponding value of k. (Your spreadsheet program probably has a built-in function to perform a linear regression on the data in two specified columns.)
(c) Use the results in part (b) to come up with a good estimate of the value of k at 275°C. Explain how you did it.
(d) If you did the calculation in part (b) correctly, one of the calculated values of k should be considerably out of line with the others. Think of as many possible explanations for this result as you can (up to 10).

2.43. Suppose you have n data points $(x_1, y_1), (x_2, y_2), \ldots, (x_n, y_n)$ and you wish to fit a line through the origin $(y = ax)$ to these data using the method of least squares. Derive Equation A.1-6 (Appendix A.1) for the slope of the line by writing the expression for the vertical distance d_i from the ith data point (x_i, y_i) to the line, then writing the expression for $\phi = \sum d_i^2$, and finding by differentiation the value of a that minimizes this function.

---

*Computer problem.

*2.44. Write a computer program to fit a straight line $y = ax + b$ to tabulated (x, y) data, assuming that no more than 100 data points will be taken in any one run. Your program should read in and store the data, evaluate the slope a and intercept b of the best line through the data using Equations A.1-3 through A.1-5 in Appendix A, then print out the measured values of x and y and calculated values of $y (= ax + b)$ for each tabulated value of x.

Test your program by fitting a line to the data in the following table:

x	1.0	1.5	2.0	2.5	3.0
y	2.35	5.53	8.92	12.15	15.38

2.45. The rate at which a substance passes through a semipermeable membrane is determined by the *diffusivity* $D(\text{cm}^2/\text{s})$ of the gas. D varies with the membrane temperature $T(K)$ according to the *Arrhenius equation*:

$$D = D_0 \exp\left(-E/RT\right)$$

where D_0 = the *preexponential factor*
E = the *activation energy* for diffusion
R = 1.987 cal/(mol·K)

Diffusitivities of SO_2 in a fluorosilicone rubber tube are measured at several temperatures, with the following results:

T (K)	$D\,(\text{cm}^2/\text{s}) \times 10^6$	
347.0	1.34	← (so that $D = 1.34 \times 10^{-6}\text{cm}^2/\text{s}$)
374.2	2.50	
396.2	4.55	
420.7	8.52	
447.7	14.07	
471.2	19.99	

(a) What are the units of D_0 and E?
(b) How should the data be plotted to obtain a straight line on rectangular coordinates?
(c) Plot the data in the manner indicated in part (b), and determine D_0 and E from the resulting line.
*(d) Write a computer program or spreadsheet to read in the (T, D) data and to calculate D_0 and E using the method of least squares (Appendix A.1). Then run the program and print out the results.

---

*Computer problem.

Chapter 3

Processes and Process Variables

A **process** is any operation or series of operations by which a particular objective is accomplished. In this textbook, we address those operations that cause a physical or chemical change in a substance or mixture of substances. The material that enters a process is referred to as the **input** or **feed**, and that which leaves is the **output** or **product**. It is common for processes to consist of multiple steps, each of which is carried out in a **process unit**, and each process unit has associated with it a set of input and output **process streams**.

As a chemical engineer, you might be called upon to *design* or *operate* a process. **Design** includes formulation of a process flowsheet (layout) as well as specification of individual process units (such as reactors, separation equipment, heat exchangers) and associated operating variables, and **operation** involves the day-to-day running of the process. The process and all the equipment must produce product at a designated rate and with specified characteristics. Moreover, you may be responsible for maintaining the economic competitiveness of the process by identifying cost-cutting measures that reduce the use of raw materials or energy. At times the process may function poorly and you will embark on a *troubleshooting* exercise to find the problem. Market conditions may dictate a production rate greater than that which can be accommodated by the existing equipment at the current operating conditions and *debottlenecking* will be in order; sometimes the reverse happens and the production rate needs to be reduced and *turndown* becomes a concern.

The links among all of the activities and functions described in the preceding paragraph are the process streams connecting process units and forming the process flowsheet. Performance of the functions requires knowledge of the amounts, compositions, and conditions of the process streams and materials within the process units. You must be able to measure or calculate such information for existing units or specify and calculate such information for units being designed.

In this chapter we present definitions, illustrative measurement techniques, and methods of calculating variables that characterize the operation of processes and individual process units. In later chapters, we discuss how you can use the measured values of some of these variables to calculate process-related quantities that cannot be measured directly but must be known before the process can be fully designed or evaluated.

3.0   INSTRUCTIONAL OBJECTIVES

After completing this chapter, you should be able to do the following:

- Explain in your own words and without the use of jargon (a) the difference between density and specific gravity; (b) the meaning of gram-mole, lb-mole, mol, and kmol; (c) at least two methods for measuring temperature and at least two for measuring fluid pressure; (d) the meaning of the terms absolute pressure and gauge pressure; (e) why atmospheric pressure is not necessarily 1 atm.
- Calculate the density in g/cm^3 or lb_m/ft^3 of a liquid or solid species from a knowledge of the specific gravity, and vice versa.
- Calculate two of the quantities mass (or mass flow rate), volume (or volumetric flow rate), and moles (or molar flow rate) from a knowledge of the third quantity for any species of known density and molecular weight.
- Given the composition of a mixture expressed in terms of mass fractions, calculate the composition in terms of mole fractions, and vice versa.
- Determine the average molecular weight of a mixture from the mass or molar composition of the mixture.
- Convert a pressure expressed as a head of a fluid to the equivalent pressure expressed as a force per unit area, and vice versa.
- Convert a manometer reading into a pressure difference for an open-end manometer, a sealed-end manometer, and a differential manometer.
- Convert among temperatures expressed in K, °C, °F, and °R.

3.1   MASS AND VOLUME

The **density** of a substance is the mass per unit volume of the substance (kg/m^3, g/cm^3, lb_m/ft^3, etc.) The **specific volume** of a substance is the volume occupied by a unit mass of the substance; it is the inverse of density. Densities of pure solids and liquids are essentially independent of pressure and vary relatively slightly with temperature. The temperature variation may be in either direction: the density of liquid water, for example, increases from 0.999868 g/cm^3 at 0°C to 1.00000 g/cm^3 at 3.98°C, and then decreases to 0.95838 g/cm^3 at 100°C. Densities of many pure compounds, solutions, and mixtures may be found in standard references (such as *Perry's Chemical Engineers' Handbook*,[1] pp. 2-7 through 2-47 and 2-91 through 2-120). Methods of estimating densities of gases and mixtures of liquids are given in Chapter 5 of this book.

The density of a substance can be used as a conversion factor to relate the mass and the volume of a quantity of the substance. For example, the density of carbon tetrachloride is 1.595 g/cm^3: the mass of 20.0 cm^3 of CCl_4 is therefore

$$\frac{20.0 \text{ cm}^3}{} \left| \frac{1.595 \text{ g}}{\text{cm}^3} \right. = 31.9 \text{ g}$$

and the volume of 6.20 lb_m of CCl_4 is

$$\frac{6.20 \text{ lb}_m}{} \left| \frac{454 \text{ g}}{1 \text{ lb}_m} \right| \frac{1 \text{ cm}^3}{1.595 \text{ g}} = 1760 \text{ cm}^3$$

The **specific gravity** of a substance is the ratio of the density ρ of the substance to the density ρ_{ref} of a reference substance at a specific condition:

$$SG = \rho / \rho_{ref} \tag{3.1-1}$$

---

[1] R. H. Perry and D. W. Green, Eds., *Perry's Chemical Engineers' Handbook*, 7th Edition, McGraw-Hill, New York, 1997.

The reference most commonly used for solids and liquids is water at 4.0°C, which has the following density:

$$
\begin{aligned}
\rho_{H_2O(l)}(4°C) &= 1.000 \text{ g/cm}^3 \\
&= 1000. \text{ kg/m}^3 \\
&= 62.43 \text{ lb}_m/\text{ft}^3
\end{aligned}
$$

(3.1-2)

Note that the density of a liquid or solid in g/cm^3 is numerically equal to the specific gravity of that substance. The notation

$$
SG = 0.6\frac{20°}{4°}
$$

signifies that the specific gravity of a substance at 20°C with reference to water at 4°C is 0.6.

If you are given the specific gravity of a substance, multiply it by the reference density in any units to get the density of the substance in the same units. For example, if the specific gravity of a liquid is 2.00, its density is 2.00×10^3 kg/m^3 or 2.00 g/cm^3 or 125 lb$_m$/ft^3. Specific gravities of selected liquids and solids are given in Table B.1.

Note: Special density units called degrees Baumé (°Bé), degrees API (°API), and degrees Twaddell (°Tw) are occasionally used, particularly in the petroleum industry. Definitions of and conversion factors for these units are given on p. 1-20 of *Perry's Chemical Engineers' Handbook.*

TEST YOURSELF

1. What are the units of specific gravity?
2. A liquid has a specific gravity of 0.50. What is its density in g/cm^3? What is its specific volume in cm^3/g? What is its density in lb$_m$/ft^3? What is the mass of 3.0 cm^3 of this liquid? What volume is occupied by 18 g?
3. If substance A and substance B each have a density of 1.34 g/cm^3, must 3 cm^3 of A have the same mass as 3 cm^3 of B?
4. If substance A and substance B each have a specific gravity of 1.34, must 3 cm^3 of A have the same mass as 3 cm^3 of B? Why not?
5. Freezing a sealed full bottle of water leads to a broken bottle and freezing a sealed full flexible-walled container of *n*-butyl alcohol leads to a container with concave walls. What can you conclude about the densities of the solid and liquid forms of these two substances?
6. Does the density of liquid mercury increase or decrease with increasing temperature? Justify your answer using a thermometer as an illustration.

EXAMPLE 3.1-1 *Mass, Volume, and Density*

Calculate the density of mercury in lb$_m$/ft^3 from a tabulated specific gravity, and calculate the volume in ft^3 occupied by 215 kg of mercury.

SOLUTION

Table B. 1 lists the specific gravity of mercury at 20°C as 13.546. Therefore,

$$
\rho_{Hg} = (13.546)\left(62.43 \ \frac{\text{lb}_m}{\text{ft}^3}\right) = \boxed{845.7 \ \frac{\text{lb}_m}{\text{ft}^3}}
$$

$$
V = \frac{215 \text{ kg}}{} \left| \frac{1 \text{ lb}_m}{0.454 \text{ kg}} \right| \frac{1 \text{ ft}^3}{845.7 \text{ lb}_m} = \boxed{0.560 \text{ ft}^3}
$$

As stated earlier, temperature and pressure do not have large influences on the densities of solids and liquids. Nevertheless, the fact that mercury in a thermometer rises or falls with changing temperature shows that the effect of temperature on liquid density is measurable. Coefficients of linear and cubic (volume) thermal expansion of selected liquids and solids are given as empirical polynomial functions of temperature on pp. 2-128 to 2-131 of *Perry's Chemical Engineers' Handbook*. For example, the *Handbook* gives the dependence of the volume of mercury on temperature as

$$V(T) = V_0\left(1 + 0.18182 \times 10^{-3}T + 0.0078 \times 10^{-6}T^2\right) \quad \text{(3.1-3)}$$

where $V(T)$ is the volume of a given mass of mercury at temperature $T(°C)$ and V_0 is the volume of the same mass of mercury at $0°C$.

EXAMPLE 3.1-2   *Effect of Temperature on Liquid Density*

In Example 3.1-1, 215 kg of mercury was found to occupy 0.560 ft^3 at $20°C$. (1) What volume would the mercury occupy at $100°C$? (2) Suppose the mercury is contained in a cylinder having a diameter of 0.25 in. What change in height would be observed as the mercury is heated from $20°C$ to $100°C$?

SOLUTION

1. From Equation 3.1-3

$$V(100°C) = V_0[1 + 0.18182 \times 10^{-3}(100) + 0.0078 \times 10^{-6}(100)^2]$$

and

$$V(20°C) = 0.560 \text{ ft}^3 = V_0[1 + 0.18182 \times 10^{-3}(20) + 0.0078 \times 10^{-6}(20)^2]$$

Solving for V_0 from the second equation and substituting it into the first yields

$$V(100°C) = \boxed{0.568 \text{ ft}^3}$$

2. The volume of the mercury equals $\pi D^2 H/4$, where D is the cylinder diameter and H is its height. Since D is constant,

$$H(100°C) - H(20°C) = \frac{V(100°C) - V(20°C)}{\pi D^2/4}$$

$$\Big\Downarrow D = (0.25/12) \text{ ft}$$

$$= \boxed{23.5 \text{ ft}}$$

3.2 FLOW RATE

3.2a Mass and Volumetric Flow Rate

Most processes involve the movement of material from one point to another—sometimes between process units, sometimes between a production facility and a transportation depot. The rate at which a material is transported through a process line is the **flow rate** of that material.

The flow rate of a process stream may be expressed as a **mass flow rate** (mass/time) or as a **volumetric flow rate** (volume/time). Suppose a fluid (gas or liquid) flows in the cylindrical pipe shown below, where the shaded area represents a section perpendicular to the direction

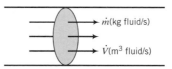

of flow. If the mass flow rate of the fluid is $\dot{m}(kg/s)$,[2] then every second m kilograms of the fluid pass through the cross section. If the volumetric flow rate of the fluid at the given cross section is $\dot{V}(m^3/s)$, then every second V cubic meters of the fluid pass through the cross section. However, the mass m and the volume V of a fluid—in this case, the fluid that passes through the cross section each second—are not independent quantities but are related through the fluid density, ρ:

$$\rho = m/V = \dot{m}/\dot{V} \qquad (3.2\text{-}1)$$

Thus, *the density of a fluid can be used to convert a known volumetric flow rate of a process stream to the mass flow rate of that stream or vice versa.*

The mass flow rates of process streams must be known for many process calculations, but it is frequently more convenient to measure volumetric flow rates. A common procedure is therefore to measure $\dot{V}$ and calculate $\dot{m}$ from $\dot{V}$ and the density of the stream fluid.

**TEST
YOURSELF**

1. The mass flow rate of *n*-hexane ($\rho = 0.659$ g/cm^3) in a pipe is 6.59 g/s. What is the volumetric flow rate of the hexane?
2. The volumetric flow rate of CCl$_4$ ($\rho = 1.595$ g/cm^3) in a pipe is 100.0 cm^3/min. What is the mass flow rate of the CCl$_4$?
3. Suppose a gas is flowing through a cone-shaped pipe.

How do the mass flow rates of the gas at the inlet and outlet compare? (Remember the law of conservation of mass.) If the density of the gas is constant, how do the volumetric flow rates at these two points compare? What if the density decreases from inlet to outlet?

3.2b Flow Rate Measurement

A **flowmeter** is a device mounted in a process line that provides a continuous reading of the flow rate in the line. Two commonly used flowmeters—the **rotameter** and the **orifice meter**—are shown schematically in Figure 3.2-1. *Perry's Chemical Engineers' Handbook*, pp. 5-7 through 5-17, describes many others.

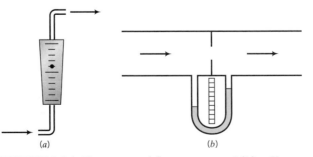

(a)   (b)

FIGURE 3.2-1 Flowmeters: (*a*) rotameter and (*b*) orifice meter.

---

[2]Variables whose symbols include a dot (·) are rates; for example, $\dot{m}$ is mass flow rate and $\dot{V}$ is volumetric flow rate.

The rotameter is a tapered vertical tube containing a float; the larger the flow rate, the higher the float rises in the tube. The orifice meter is an obstruction in the flow channel with a narrow opening through which the fluid passes. The fluid pressure drops (decreases) from the upstream side of the orifice to the downstream side; the pressure drop (which may be measured with a number of devices, including a differential manometer, which is discussed in the next section) varies with the flow rate—the greater the flow rate, the larger the pressure drop.

Problems at the end of this chapter illustrate the calibration and use of both types of flowmeters.

TEST YOURSELF

1. A steadily flowing steam of water is funneled into a graduated cylinder for exactly 30 s, during which time 50 mL is collected. What is the volumetric flow rate of the stream? The mass flow rate?
2. What is a rotameter? An orifice meter?
3. A rotameter calibration curve (flow rate versus float position) obtained using a liquid is mistakenly used to measure a gas flow rate. Would you expect the gas flow rate determined in this manner to be too high or too low?

CREATIVITY EXERCISE

Given below is the first of a series of exercises contained in this book that we call Creativity Exercises. These exercises differ considerably from the kinds of problems you are accustomed to seeing in homework assignments and quizzes. In the latter, you are generally given information and asked to find the single correct solution to a problem. In the creativity exercises, you are asked to think of many possible responses without spending too much time doing so. There is no such thing as the "correct" response or even a "good" or "bad" response. The idea is to go for quantity rather than quality, imagination (even humor) rather than "correctness." Try to suspend your critical judgment completely and come up with as many ideas as you can, regardless of how likely they are to be efficient, cost-effective, or even to work at all.

In a real industrial situation, this "brainstorming" approach to creative problem solving is frequently used as the first step in solving the most difficult problems facing the company. Doing these exercises will help sharpen the skills you will need for successful brainstorming and at the same time will give you a broader understanding of the concepts contained in the text.

Here, then, is the first exercise. Invent as many devices as you can that might function as flowmeters for gases and/or liquids. In each case, describe the device and state what would be measured. (*Example:* Put a propeller in a flowing stream and measure its rotational speed.)

3.3   CHEMICAL COMPOSITION

Most materials encountered in nature and in chemical process systems are mixtures of various species. The physical properties of a mixture depend strongly on the mixture composition. In this section we will review different ways to express mixture compositions, and later in the book we will outline methods of estimating physical properties of a mixture from the properties of the pure components.

3.3a   Moles and Molecular Weight

The **atomic weight** of an element is the mass of an atom on a scale that assigns ^{12}C (the isotope of carbon whose nucleus contains six protons and six neutrons) a mass of exactly 12. The atomic weights of all the elements in their naturally occurring isotopic proportions are listed in the table at the back of the book. The **molecular weight** of a compound is the sum of the atomic weights of the atoms that constitute a molecule of the compound: atomic oxygen (O), for example, has an atomic weight of approximately 16, and therefore molecular oxygen (O_2)

has a molecular weight of approximately 32. Molecular weights of a number of compounds are given in Table B.1.

A **gram-mole** (g-mole, or **mol** in SI units) of a species is the amount of that species whose mass in grams is numerically equal to its molecular weight. (If the species is an element, it is technically correct to refer to a gram-atom rather than a gram-mole. We will not observe this distinction, but will use moles for both elements and compounds.) Other types of moles (e.g., kg-moles or kmol, lb-moles, ton-moles) are similarly defined. Carbon monoxide (CO), for example, has a molecular weight of 28; 1 mol of CO therefore contains 28 g, 1 lb-mole contains 28 lb_m, 1 ton-mole contains 28 tons, and so on.

If the molecular weight of a substance is M, then there are M kg/kmol, M g/mol, and M lb_m/lb-mole of this substance. The molecular weight may thus be used as a conversion factor that relates the mass and the number of moles of a quantity of the substance. For example, 34 kg of ammonia (NH_3: $M = 17$) is equivalent to

$$\frac{34 \text{ kg } NH_3 \quad | \quad 1 \text{ kmol } NH_3}{| \quad 17 \text{ kg } NH_3} = 2.0 \text{ kmol } NH_3 \qquad \textbf{(3.3-1)}$$

and 4.0 lb-moles of ammonia is equivalent to

$$\frac{4.0 \text{ lb-moles } NH_3 \quad | \quad 17 \text{ lb}_m \text{ } NH_3}{| \quad 1 \text{ lb-mole } NH_3} = 68 \text{ lb}_m \text{ } NH_3 \qquad \textbf{(3.3-2)}$$

(It is often helpful in mass-mole conversions to include the chemical formula in the dimensional equation, as illustrated above.)

The same factors used to convert masses from one unit to another may be used to convert the equivalent molar units: there is 454 g/lb_m, for example, and therefore there is 454 mol/lb-mole, regardless of the substance involved. (Prove it—convert 1 lb-mole of a substance with molecular weight M to gram-moles.)

One gram-mole of any species contains approximately 6.02×10^{23} *(Avogadro's number) molecules of that species.*

EXAMPLE 3.3-1   *Conversion Between Mass and Moles*

How many of each of the following are contained in 100.0 g of CO_2 ($M = 44.01$)? (1) mol CO_2; (2) lb-moles CO_2; (3) mol C; (4) mol O; (5) mol O_2; (6) g O; (7) g O_2; (8) molecules of CO_2.

SOLUTION

1.
$$\frac{100.0 \text{ g } CO_2 \quad | \quad 1 \text{ mol } CO_2}{| \quad 44.01 \text{ g } CO_2} = \boxed{2.273 \text{ mol } CO_2}$$

2.
$$\frac{2.273 \text{ mol } CO_2 \quad | \quad 1 \text{ lb-mol}}{| \quad 453.6 \text{ mol}} = \boxed{5.011 \times 10^{-3} \text{ lb-mole } CO_2}$$

Each molecule of CO_2 contains one atom of C, one molecule of O_2, and two atoms of O. Therefore, each 6.02×10^{23} molecules of CO_2 (1 mol) contains 1 mol C, 1 mol O_2, and 2 mol O. Thus,

3.
$$\frac{2.273 \text{ mol } CO_2 \quad | \quad 1 \text{ mol C}}{| \quad 1 \text{ mol } CO_2} = \boxed{2.273 \text{ mol C}}$$

4.
$$\frac{2.273 \text{ mol } CO_2 \quad | \quad 2 \text{ mol O}}{| \quad 1 \text{ mol } CO_2} = \boxed{4.546 \text{ mol O}}$$

5.
$$\frac{2.273 \text{ mol } CO_2 \quad | \quad 1 \text{ mol } O_2}{| \quad 1 \text{ mol } CO_2} = \boxed{2.273 \text{ mol } O_2}$$

6.
$$\frac{4.546 \text{ mol O} \mid 16.0 \text{ g O}}{\mid 1 \text{ mol O}} = \boxed{72.7 \text{ g O}}$$

7.
$$\frac{2.273 \text{ mol O}_2 \mid 32.0 \text{ g O}_2}{\mid 1 \text{ mol O}_2} = \boxed{72.7 \text{ g O}_2}$$

8.
$$\frac{2.273 \text{ mol CO}_2 \mid 6.02 \times 10^{23} \text{ molecules}}{\mid 1 \text{ mol}} = \boxed{1.37 \times 10^{24} \text{ molecules}}$$

Note: Part 7 could also be done by observing from the molecular formula that each 44.0 g of CO_2 contains 32.0 g of O_2 or O, so that

$$\frac{100.0 \text{ g CO}_2 \mid 32.0 \text{ g O}_2}{\mid 44.0 \text{ g CO}_2} = 72.7 \text{ g O}_2$$

The molecular weight of a species can be used to relate the mass flow rate of a continuous stream of this species to the corresponding molar flow rate. For example, if carbon dioxide (CO_2: $M = 44.0$) flows through a pipeline at a rate of 100 kg/h, the molar flow rate of the CO_2 is

$$\frac{100 \text{ kg CO}_2 \mid 1 \text{ kmol CO}_2}{\text{h} \mid 44.0 \text{ kg CO}_2} = 2.27 \frac{\text{kmol CO}_2}{\text{h}} \qquad \textbf{(3.3-3)}$$

If the output stream from a chemical reactor contains CO_2 flowing at a rate of 850 lb-moles/min, the corresponding mass flow rate is

$$\frac{850 \text{ lb-moles CO}_2 \mid 44.0 \text{ lb}_m \text{ CO}_2}{\text{min} \mid \text{lb-mole CO}_2} = 37,400 \frac{\text{lb}_m\text{CO}_2}{\text{min}} \qquad \textbf{(3.3-4)}$$

TEST YOURSELF

1. What is a mol of a species of molecular weight M, in terms of (a) a number of molecules? (b) a mass?
2. What is a ton-mole of a species?
3. How many lb-moles and lb_m of (a) H_2 and (b) H are contained in 1 lb-mole of H_2O?
4. How many gram-moles of C_3H_8 are contained in 2 kmol of this substance?
5. One hundred kilograms of molecular hydrogen (H_2) is fed into a reactor each hour. What is the molar flow rate of this stream in gram-moles/hour?

3.3b Mass and Mole Fractions and Average Molecular Weight

Process streams occasionally contain one substance, but more often they consist of mixtures of liquids or gases, or solutions of one or more solutes in a liquid solvent.

The following terms may be used to define the composition of a mixture of substances, including a species A.

Mass fraction: $\quad x_A = \dfrac{\text{mass of A}}{\text{total mass}} \left(\dfrac{\text{kg A}}{\text{kg total}} \text{ or } \dfrac{\text{g A}}{\text{g total}} \text{ or } \dfrac{\text{lb}_m \text{ A}}{\text{lb}_m \text{ total}} \right) \qquad \textbf{(3.3-5)}$

Mole fraction: $\quad y_A = \dfrac{\text{moles of A}}{\text{total moles}} \left(\dfrac{\text{kmol A}}{\text{kmol}} \text{ or } \dfrac{\text{mol A}}{\text{mol}} \text{ or } \dfrac{\text{lb-moles A}}{\text{lb-mole}} \right) \qquad \textbf{(3.3-6)}$

The **percent by mass** of A is $100x_A$, and the **mole percent** of A is $100y_A$.

EXAMPLE 3.3-2 *Conversions Using Mass and Mole Fractions*

A solution contains 15% A by mass ($x_A = 0.15$) and 20 mole% B ($y_B = 0.20$).

1. Calculate the mass of A in 175 kg of the solution.

$$\frac{175 \text{ kg solution}}{} \left| \frac{0.15 \text{ kg A}}{\text{kg solution}} \right| = \boxed{26 \text{ kg A}}$$

2. Calculate the mass flow rate of A in a stream of solution flowing at a rate of 53 lb_m/h.

$$\frac{53 \text{ lb}_m}{h} \left| \frac{0.15 \text{ lb}_m \text{ A}}{\text{lb}_m} \right| = \boxed{8.0 \frac{\text{lb}_m \text{ A}}{h}}$$

(If a mass or molar unit—such as lb_m in 53 lb_m/h—is not followed by the name of a species, the unit should be understood to refer to the total mixture or solution rather than to a specific component.)

3. Calculate the molar flow rate of B in a stream flowing at a rate of 1000 mol/min.

$$\frac{1000 \text{ mol}}{\text{min}} \left| \frac{0.20 \text{ mol B}}{\text{mol}} \right| = \boxed{200 \frac{\text{mol B}}{\text{min}}}$$

4. Calculate the total solution flow rate that corresponds to a molar flow rate of 28 kmol B/s.

$$\frac{28 \text{ kmol B}}{s} \left| \frac{1 \text{ kmol solution}}{0.20 \text{ kmol B}} \right| = \boxed{140 \frac{\text{kmol solution}}{s}}$$

5. Calculate the mass of the solution that contains 300 lb_m of A.

$$\frac{300 \text{ lb}_m \text{ A}}{} \left| \frac{1 \text{ lb}_m \text{ solution}}{0.15 \text{ lb}_m \text{ A}} \right| = \boxed{2000 \text{ lb}_m \text{ solution}}$$

Note that *the numerical value of a mass or a mole fraction does not depend on the mass units in the numerator and denominator as long as these units are the same.* If the mass fraction of benzene (C_6H_6) in a mixture is 0.25, then $x_{C_6H_6}$ equals 0.25 kg C_6H_6/kg total, 0.25 g C_6H_6/g total, 0.25 lb_m C_6H_6/lb_m total, and so on.

A set of mass fractions may be converted to an equivalent set of mole fractions by (a) assuming as a **basis of calculation** a mass of the mixture (e.g., 100 kg or 100 lb_m); (b) using the known mass fractions to calculate the mass of each component in the basis quantity, and converting these masses to moles; and (c) taking the ratio of the moles of each component to the total number of moles. An analogous procedure is followed to convert mole fractions to mass fractions, differing only in that a total number of moles (e.g., 100 mol or 100 lb-moles) is taken as a basis of calculation.

EXAMPLE 3.3-3 *Conversion from a Composition by Mass to a Molar Composition*

A mixture of gases has the following composition by mass:

O_2	16%	($x_{O_2} = 0.16$ g O_2/g total)
CO	4.0%	
CO_2	17%	
N_2	63%	

What is the molar composition?

SOLUTION

Basis: 100 g of the mixture.

A convenient way to perform the calculations is to set them up in tabular form.

Component i	Mass Fraction x_i (g i/g)	Mass (g) $m_i = x_i m_{total}$	Molecular Weight M_i (g/mol)	Moles $n_i = m_i/M_i$	Mole Fraction $y_i = n_i/n_{total}$
O_2	0.16	16	32	0.500	0.150
CO	0.04	4	28	0.143	0.044
CO_2	0.17	17	44	0.386	0.120
N_2	0.63	63	28	2.250	0.690
Total	1.00	100		3.279	1.000

The mass of a species is the product of the mass fraction of that species and total mass (basis of 100 g). The number of moles of a species is the mass of that species divided by the molecular weight of the species. Finally, the mole fraction of a species is the number of moles of that species divided by the total number of moles (3.279 mol).

The **average molecular weight** (or mean molecular weight) of a mixture, $\overline{M}$ (kg/kmol, lb$_m$/lb-mole, etc.), is the ratio of the mass of a sample of the mixture (m_t) to the number of moles of all species (n_t) in the sample. If y_i is the mole fraction of the ith component of the mixture and M_i is the molecular weight of this component, then

$$\overline{M} = y_1 M_1 + y_2 M_2 + \cdots = \sum_{\substack{\text{all} \\ \text{components}}} y_i M_i \qquad \textbf{(3.3-7)}$$

(*Exercise:* Derive Equation 3.3-7 by taking a basis of 1 mol of the mixture and calculating m_t following the procedure of Example 3.3-3.) If x_i is the mass fraction of the ith component, then

$$\frac{1}{\overline{M}} = \frac{x_1}{M_1} + \frac{x_2}{M_2} + \cdots = \sum_{\substack{\text{all} \\ \text{components}}} \frac{x_i}{M_i} \qquad \textbf{(3.3-8)}$$

(Prove it.)

EXAMPLE 3.3-4

Calculation of an Average Molecular Weight

Calculate the average molecular weight of air (1) from its approximate molar composition of 79% N_2, 21% O_2 and (2) from its approximate composition by mass of 76.7% N_2, 23.3% O_2.

SOLUTION

1. From Equation 3.3-7, with $y_{N_2} = 0.79$, $y_{O_2} = 0.21$,

$$\overline{M} = y_{N_2} M_{N_2} + y_{O_2} M_{O_2}$$

$$= \frac{0.79\ \text{kmol}\ N_2}{\text{kmol}} \left| \frac{28\ \text{kg}\ N_2}{\text{kmol}} \right. + \frac{0.21\ \text{kmol}\ O_2}{\text{kmol}} \left| \frac{32\ \text{kg}\ O_2}{\text{kmol}} \right.$$

$$= \boxed{29\ \frac{\text{kg}}{\text{kmol}}} \left(= 29\ \frac{\text{lb}_m}{\text{lb-mole}} = 29\ \frac{\text{g}}{\text{mol}} \right)$$

2. From Equation 3.3-8,

$$\frac{1}{\overline{M}} = \frac{0.767 \text{ g } N_2/\text{g}}{28 \text{ g } N_2/\text{mol}} + \frac{0.233 \text{ g } O_2/\text{g}}{32 \text{ g } O_2/\text{mol}} = 0.035 \frac{\text{mol}}{\text{g}}$$

$$\Downarrow$$

$$\boxed{\overline{M} = 29 \text{ g/mol}}$$

Note: Air contains small amounts of carbon dioxide, argon, and other gases that have been neglected in this calculation, but whose presence does not significantly affect the calculated value of $\overline{M}$.

**TEST
YOURSELF**

1. The molecular weight of atomic hydrogen is approximately 1, and that of atomic bromine is 80. What are (a) the mass fraction and (b) the mole fraction of bromine in pure HBr?
2. If 100 lb_m/min of A (M_A = 2) and 300 lb_m/min of B (M_B = 3) flow through a pipe, what are the mass fractions and mole fractions of A and B, the mass flow rate of A, the molar flow rate of B, the total mass flow rate, and the total molar flow rate of the mixture?

3.3c Concentration

The **mass concentration** of a component of a mixture or solution is the mass of this component per unit volume of the mixture (g/cm^3, lb_m/ft^3, $kg/in.^3$,...). The **molar concentration** of a component is the number of moles of the component per unit volume of the mixture ($kmol/m^3$, lb-moles/ft^3,...). The **molarity** of a solution is the value of the molar concentration of the solute expressed in gram-moles solute/liter solution (e.g., a 2-molar solution of A contains 2 mol A/liter solution).

The concentration of a substance in a mixture or solution can be used as a conversion factor to relate the mass (or moles) of a component in a sample of the mixture to the sample volume, or to relate the mass (or molar) flow rate of a component of a continuous stream to the total volumetric flow rate of the stream. Consider, for example, a 0.02-molar solution of NaOH (i.e., a solution containing 0.02 mol NaOH/L): 5 L of this solution contains

$$\frac{5 \text{ L} \mid 0.02 \text{ mol NaOH}}{\text{L}} = 0.1 \text{ mol NaOH}$$

and if a stream of this solution flows at a rate of 2 L/min, the molar flow rate of NaOH is

$$\frac{2 \text{ L} \mid 0.02 \text{ mol NaOH}}{\text{min} \mid \text{L}} = 0.04 \frac{\text{mol NaOH}}{\text{min}}$$

**TEST
YOURSELF**

A solution with volume V(L) contains n(mol) of a solute A with a molecular weight of M_A(g A/mol). In terms of V, n, and M_A:

1. What is the molar concentration of A?
2. What is the mass concentration of A?

In terms of C_A(mol A/L) and c_A(g A/L):

3. What volume of the solution contains 20 mol of A?
4. What is the mass flow rate of A in a stream whose volumetric flow rate is 120 L/h?

EXAMPLE 3.3-5   *Conversion Between Mass, Molar, and Volumetric Flow Rates of a Solution*

A 0.50-molar aqueous solution of sulfuric acid flows into a process unit at a rate of 1.25 m^3/min. The specific gravity of the solution is 1.03. Calculate (1) the mass concentration of H_2SO_4 in kg/m^3, (2) the mass flow rate of H_2SO_4 in kg/s, and (3) the mass fraction of H_2SO_4.

SOLUTION   1.

$$c_{H_2SO_4}\left(\frac{kg\ H_2SO_4}{m^3}\right) = \frac{0.50\ mol\ H_2SO_4}{L}\left|\frac{98\ g}{mol}\right|\frac{1\ kg}{10^3\ g}\left|\frac{10^3\ L}{1\ m^3}\right.$$

$$= \boxed{49\ \frac{kg\ H_2SO_4}{m^3}}$$

2.

$$\dot{m}_{H_2SO_4}\left(\frac{kg\ H_2SO_4}{s}\right) = \frac{1.25\ m^3}{min}\left|\frac{49\ kg\ H_2SO_4}{m^3}\right|\frac{1\ min}{60\ s} = \boxed{1.0\ \frac{kg\ H_2SO_4}{s}}$$

3. The mass fraction of H_2SO_4 equals the ratio of the mass flow rate of H_2SO_4—which we know—to the total mass flow rate, which can be calculated from the total volumetric flow rate and the solution density.

$$\rho_{solution} = (1.03)\left(\frac{1000\ kg}{m^3}\right) = 1030\ \frac{kg}{m^3}$$

$$\Downarrow$$

$$\dot{m}_{solution}\left(\frac{kg}{s}\right) = \frac{1.25\ m^3\ solution}{min}\left|\frac{1030\ kg}{m^3\ solution}\right|\frac{1\ min}{60\ s} = 21.46\ \frac{kg}{s}$$

$$\Downarrow$$

$$x_{H_2SO_4} = \frac{\dot{m}_{H_2SO_4}}{\dot{m}_{solution}} = \frac{1.0\ kg\ H_2SO_4/s}{21.46\ kg\ solution/s} = \boxed{0.048\ \frac{kg\ H_2SO_4}{kg\ solution}}$$

CREATIVITY EXERCISE

Itemize as many ways as you can think of to measure the concentration of a solute in a solution. (*Example:* If the solute absorbs light of a specific wavelength, pass a beam of light of this wavelength through the solution and measure the fractional light absorption.)

3.3d Parts per Million and Parts per Billion

The units **parts per million (ppm)** and **parts per billion (ppb)**[3] are used to express the concentrations of *trace species* (species present in minute amounts) in mixtures of gases or liquids. The definitions may refer to mass ratios (usual for liquids) or mole ratios (usual for gases) and signify how many parts (grams, moles) of the species are present per million or billion parts (grams, moles) of the mixture. If y_i is the fraction of component i, then by definition

$$ppm_i = y_i \times 10^6 \tag{3.3-9}$$

$$ppb_i = y_i \times 10^9 \tag{3.3-10}$$

For example, suppose air in the vicinity of a power plant is said to contain 15 ppm SO_2 (15 parts per million sulfur dioxide). Assuming that a molar basis has been used (customary for gases), this statement means that every million moles of air contains 15 moles of SO_2, or equivalently, that the mole fraction of SO_2 in the air is 15×10^{-6}. Units such as ppm and ppb have become increasingly common in recent years as public concern about potentially hazardous trace species in the environment has grown.

---

[3]We are using the standard American definition of a billion as 10^9 or 1000 million, as opposed to the English definition of 10^{12}.

TEST YOURSELF

A liquid that is almost entirely water is reported to contain 125 ppb phenol (mass basis).

1. What is the mass fraction of phenol in the liquid?
2. How many milligrams of phenol are contained in one kilogram of the liquid?
3. What is the approximate concentration of phenol in the liquid in g/L?

3.4 PRESSURE

3.4a Fluid Pressure and Hydrostatic Head

A **pressure** is the ratio of a force to the area on which the force acts. Accordingly, pressure units are force units divided by area units (e.g., N/m^2, dynes/cm^2, and $lb_f/in.^2$ or psi). The SI pressure unit, N/m^2, is called a **pascal** (Pa).

Consider a fluid (gas or liquid) contained in a closed vessel or flowing through a pipe, and suppose that a hole of area A is made in the wall of the containing vessel, as in Figure 3.4-1. The **fluid pressure** may be defined as the ratio F/A, where F is the minimum force that would have to be exerted on a frictionless plug in the hole to keep the fluid from emerging.

We must introduce an additional definition of fluid pressure to explain the concept of atmospheric pressure and to discuss common methods for measuring pressures of fluids in tanks and pipes. Suppose a vertical column of fluid is h(m) high and has a uniform cross-sectional area $A(m^2)$. Further suppose that the fluid has a density of $\rho(kg/m^3)$, and that a pressure $P_0(N/m^2)$ is exerted on the upper surface of the column. (See Figure 3.4-2.) The pressure P of the fluid at the base of the column—called the **hydrostatic pressure** of the fluid—is, by definition, the force F exerted on the base divided by the base area A. F thus equals the force on the top surface plus the weight of the fluid in the column. It is not difficult to show that:

$$P = P_0 + \rho g h \qquad (3.4\text{-}1)$$

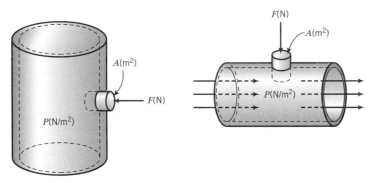

FIGURE 3.4-1 Fluid pressure in a tank and a pipe.

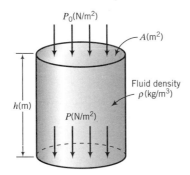

FIGURE 3.4-2 Pressure at the base of a fluid column.

(See if you can derive this equation.) Since A does not appear in this equation, the formula is applicable to a column of fluid as thin as a test tube or as wide as the ocean.

In addition to being expressible as a force per unit area, a pressure may be expressed as a **head** of a particular fluid—that is, as the height of a hypothetical column of this fluid that would exert the given pressure at its base if the pressure at the top were zero. You can thus speak of a pressure of 14.7 psi or equivalently of a pressure (or head) of 33.9 ft of water (33.9 ft H_2O) or 76 cm of mercury (76 cm Hg). *The equivalence between a pressure P (force/area) and the corresponding head P_h (height of a fluid) is given by Equation 3.4-1 with $P_0 = 0$:*

$$P\left(\frac{\text{force}}{\text{area}}\right) = \rho_{\text{fluid}}\, g\, P_h (\text{head of fluid}) \tag{3.4-2}$$

EXAMPLE 3.4-1 *Calculation of a Pressure as a Head of Fluid*

Express a pressure of 2.00×10^5 Pa in terms of mm Hg.

SOLUTION Solve Equation 3.4-2 for P_h(mm Hg), assuming that $g = 9.807$ m/s^2 and noting that the density of mercury is 13.6×1000 kg/m^3 = 13,600 kg/m^3.

$$P_h = \frac{P}{\rho_{\text{Hg}}g}$$

$$= \frac{2.00 \times 10^5 \text{ N}}{\text{m}^2} \left|\frac{\text{m}^3}{13{,}600 \text{ kg}}\right| \frac{\text{s}^2}{9.807 \text{ m}} \left|\frac{1 \text{ kg·m/s}^2}{\text{N}}\right| \frac{10^3 \text{ mm}}{\text{m}} = \boxed{1.50 \times 10^3 \text{ mm Hg}}$$

The relationship between the pressure at the base of a column of fluid of height h and the pressure at the top is particularly simple if these pressures are expressed as heads of the given fluid: if the column is mercury, for example, then

$$\boxed{P_h(\text{mm Hg}) = P_0(\text{mm Hg}) + h(\text{mm Hg})} \tag{3.4-3}$$

Any other length unit and chemical species may be substituted for mm Hg in this equation.

The conversion table on the inside front cover of this book lists values of a pressure expressed in several common force/area units and as heads of mercury and water. The use of this table for pressure unit conversion is illustrated by the conversion of 20.0 psi to cm Hg:

$$\frac{20.0 \text{ psi}}{} \left|\frac{76.0 \text{ cm Hg}}{14.696 \text{ psi}}\right. = 103 \text{ cm Hg}$$

EXAMPLE 3.4-2 *Pressure Below the Surface of a Fluid*

What is the pressure 30.0 m below the surface of a lake? Atmospheric pressure (the pressure at the surface) is 10.4 m H_2O, and the density of water is 1000.0 kg/m^3. Assume that g is 9.807 m/s^2.

SOLUTION First, the hard way, using Equation 3.4-1:

$$P_h = P_0 + \rho g h$$

$$\Downarrow$$

$$P_h = \frac{10.4 \text{ m } H_2O}{} \left|\frac{1.013 \times 10^5 \text{ N/m}^2}{10.33 \text{ m } H_2O}\right. + \frac{1000.0 \text{ kg/m}^3}{} \left|\frac{9.807 \text{ m}}{\text{s}^2}\right| \frac{30.0 \text{ m}}{} \left|\frac{1 \text{ N}}{1 \text{ kg·m/s}^2}\right.$$

$$= \boxed{3.96 \times 10^5 \text{ N/m}^2 \text{ (Pa)}}$$

or

$$P_h = 396 \text{ kPa}$$

Next, the easy way, using Equation 3.4-3:

$$P_h = 10.4 \text{ m H}_2\text{O} + 30.0 \text{ m H}_2\text{O} = \boxed{40.4 \text{ m H}_2\text{O}}$$

(Verify that the two calculated pressures are equivalent.)

Note: We will henceforth use an unsubscripted P to denote pressure expressed as either (force/area) or as the head of a fluid.

TEST YOURSELF

1. Define (a) the pressure of a fluid flowing in a pipe, (b) hydrostatic pressure, and (c) a head of a fluid corresponding to a given pressure.
2. Consider the tank in Figure 3.4-1. Does the pressure at the plug depend on the height of the opening in the tank? (*Hint:* Yes.) Why? Would you expect the difference between the pressure at the top and that at the bottom to be very large if the fluid were air? How about water? Mercury?
3. Suppose the pressure in the tank of Figure 3.4-1 is given as 1300 mm Hg. Does this tell you anything about the height of the tank? If you were given the area of the hole (say, 4 cm^2), how would you calculate the force needed to hold the plug in the hole?
4. Suppose the pressure at a point within a column of mercury in a tube is 74 mm Hg. What is the pressure 5 mm below this point? (If this takes you more than one second, you are probably doing it wrong.)

3.4b   Atmospheric Pressure, Absolute Pressure, and Gauge Pressure

The pressure of the atmosphere can be thought of as the pressure at the base of a column of fluid (air) located at the point of measurement (e.g., at sea level). Equation 3.4-1 can be used to calculate atmospheric pressure, assuming that the pressure at the top of the column (P_0) equals 0 and ρ and g are average values of the density of air and the acceleration of gravity between the top of the atmosphere and the measurement point.

A typical value of the atmospheric pressure at sea level, 760.0 mm Hg, has been designated as a standard pressure of 1 atmosphere. The conversion table on the inside front cover lists equivalent values of this pressure in various units.

The fluid pressures referred to so far are all **absolute pressures**, in that a pressure of zero corresponds to a perfect vacuum. Many pressure-measuring devices give the **gauge pressure** of a fluid, or the pressure relative to atmospheric pressure. A gauge pressure of zero indicates that the absolute pressure of the fluid is equal to atmospheric pressure. The relationship for converting between absolute and gauge pressure is

$$P_{\text{absolute}} = P_{\text{gauge}} + P_{\text{atmospheric}} \qquad \textbf{(3.4-4)}$$

The abbreviations psia and psig are commonly used to denote absolute and gauge pressure in $\text{lb}_f/\text{in.}^2$. Also, it is common to refer to negative gauge pressures (absolute pressures less than atmospheric) as positive amounts of vacuum: for example, a gauge pressure of -1 cm Hg (75.0 cm Hg absolute if atmospheric pressure is 76.0 cm Hg) may also be called 1 cm of vacuum.

TEST YOURSELF

1. Is atmospheric pressure always equal to 1 atm?
2. What is absolute pressure? Gauge pressure?

3. The gauge pressure of a gas is -20 mm Hg at a point where atmospheric pressure is 755 mm Hg. How else can the gas pressure be expressed in terms of mm Hg? (Give two values.)

4. A mercury column is open to the atmosphere on a day when atmospheric pressure is 29.9 in. Hg. What is the gauge pressure 4 in. below the surface? The absolute pressure? (Give the answers in in. Hg.)

3.4c Fluid Pressure Measurement

Perry's Chemical Engineers' Handbook (pp. 8-47 to 8-48) categorizes pressure-measurement devices as:

- elastic-element methods—Bourdon tubes, bellows, or diaphragms
- liquid-column methods—manometers
- electrical methods—strain gauges, piezoresistive transducers, and piezoelectric transducers

We limit our discussion here to Bourdon gauges and manometers but recognize the importance of other methods in modern process sensors.

The most common mechanical device used for pressure measurement is a **Bourdon gauge**, which is a hollow tube closed at one end and bent into a C configuration. The open end of the tube is exposed to the fluid whose pressure is to be measured. As the pressure increases, the tube tends to straighten, causing a pointer attached to the tube to rotate. The position of the pointer on a calibrated dial gives the *gauge* pressure of the fluid. A schematic diagram of a Bourdon gauge is shown in Figure 3.4-3.

Bourdon gauges are used to measure fluid pressures from nearly perfect vacuums to about 7000 atm. More accurate measurements of pressures below about 3 atm are provided by **manometers**.

A manometer is a U-shaped tube partially filled with a fluid of known density (the **manometer fluid**). When the ends of the tube are exposed to different pressures, the field level drops in the high-pressure arm and rises in the low-pressure arm. The difference between the pressures can be calculated from the measured difference between the liquid levels in each arm.

Manometers are used in several different ways, as shown in Figure 3.4-4. In each diagram, pressure P_1 is greater than pressure P_2.

Figure 3.4-4a shows an **open-end manometer**: one end is exposed to a fluid whose pressure is to be measured, and the other is open to the atmosphere. Figure 3.4-4b shows a **differential manometer**, which is used to measure the pressure difference between two points in a process line. Figure 3.4-4c shows a **sealed-end manometer**, which has a near-vacuum enclosed at one end. (Some of the enclosed fluid will vaporize into the empty space, thereby preventing the existence of a perfect vacuum.) If the open end of a sealed-end manometer is exposed to the atmosphere ($P_1 = P_{atm}$) the device functions as a **barometer**.

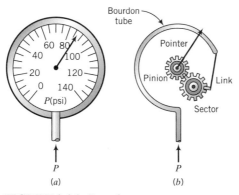

FIGURE 3.4-3 Bourdon gauge.

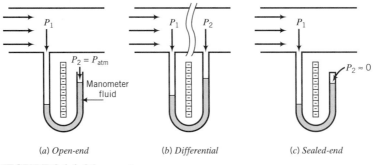

(a) Open-end   (b) Differential   (c) Sealed-end

FIGURE 3.4-4 Manometers.

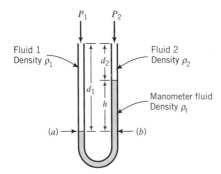

FIGURE 3.4-5 Manometer variables.

The formula that relates the pressure difference $P_1 - P_2$ to the difference in manometer fluid levels is based on the principle that the fluid pressure must be the same at any two points at the same height in a continuous fluid. In particular, *the pressure at the height of the lower surface of a manometer fluid is the same in both arms of the manometer.* (See Figure 3.4-5.) Writing and equating expressions for the pressures at points (*a*) and (*b*) in Figure 3.4-5 yields the general manometer equation

General Manometer Equation:
$$P_1 + \rho_1 g d_1 = P_2 + \rho_2 g d_2 + \rho_f g h \qquad \textbf{(3.4-5)}$$

In a differential manometer, fluids 1 and 2 are the same, and consequently $\rho_1 = \rho_2 = \rho$. The general manometer equation then reduces to

Differential Manometer Equation:
$$P_1 - P_2 = (\rho_f - \rho)gh \qquad \textbf{(3.4-6)}$$

If either fluid 1 or 2 is a gas at a moderate pressure (e.g., if one arm is open to the atmosphere), the density of this fluid is 100 to 1000 times lower than the density of the manometer fluid, so that the corresponding $\rho g d$ term in Equation 3.4-5 may be neglected. If *both* fluids are gases, then the equation becomes

$$P_1 - P_2 = \rho_f g h$$

and if both P_1 and P_2 are expressed as heads of the manometer fluid, then

Manometer Formula for Gases:
$$P_1 - P_2 = h \qquad \textbf{(3.4-7)}$$

If P_2 is atmospheric pressure, then the gauge pressure at point 1 is simply the difference in the levels of the manometer fluid.

EXAMPLE 3.4-3   *Pressure Measurement with Manometers*

1. A differential manometer is used to measure the drop in pressure between two points in a process line containing water. The specific gravity of the manometer fluid is 1.05. The measured levels in each arm are shown below. Calculate the pressure drop between points 1 and 2 in dynes/cm^2.

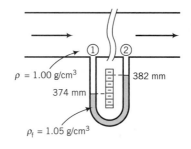

2. The pressure of gas being pulled though a line by a vacuum pump is measured with an open-end mercury manometer. A reading of −2 in. is obtained. What is the gas gauge pressure in inches of mercury? What is the absolute pressure if P_{atm} = 30 in. Hg?

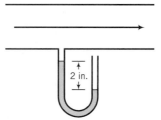

SOLUTION

1. $h = (382 − 374)$ mm = 8 mm. From Equation 3.4-6,

$$P_1 − P_2 = (\rho_f − \rho)gh$$

$$= \frac{(1.05 − 1.00)\ \text{g}}{\text{cm}^3} \left| \frac{980.7\ \text{cm}}{\text{s}^2} \right| \frac{1\ \text{dyne}}{1\ \text{g·cm/s}^2} \left| \frac{8\ \text{mm}}{} \right| \frac{1\ \text{cm}}{10\ \text{mm}}$$

$$= \boxed{40\ \frac{\text{dynes}}{\text{cm}^2}}$$

2. From Equation 3.4-7 and the definition of a gauge pressure,

$$P_1 − P_{atm} = P_{gauge} = \boxed{−2\ \text{in. Hg}}$$

$$\Downarrow$$

$$P_1 = P_{atm} + P_{gauge} = (30 − 2)\ \text{in. Hg} = \boxed{28\ \text{in. Hg}}$$

TEST YOURSELF

1. What is a Bourdon gauge? What range of pressures can it be used to measure? As it is normally calibrated, does it measure gauge or absolute pressure?
2. What is an open-end manometer? A differential manometer? A sealed-end manometer?

3. True or false?
   (a) An open-end manometer provides a direct reading of the gauge pressure of a gas.
   (b) A sealed-end manometer provides a direct reading of the absolute pressure of a gas, provided that the gas pressure in the sealed end may be neglected.
   (c) The reading of a differential manometer does not depend on the density of the fluid in the pipeline but only on that of the manometer fluid.
4. The pressure of a gas in a pipeline is measured with an open-end mercury manometer. The mercury level in the arm connected to the line is 14 mm *higher* than the level in the open arm. What is the gauge pressure of the gas in the line?

CREATIVITY EXERCISE

Think of several devices that might be used to measure fluid pressures, being as imaginative as you can. (*Example:* Allow a gas at the pressure to be measured to fill a calibrated balloon, and measure the final balloon diameter.)

3.5 TEMPERATURE

The temperature of a substance in a particular state of aggregation (solid, liquid, or gas) is a measure of the average kinetic energy possessed by the substance molecules. Since this energy cannot be measured directly, the temperature must be determined indirectly by measuring some physical property of the substance whose value depends on temperature in a known manner. Such properties and the temperature-measuring devices based on them include electrical resistance of a conductor (**resistance thermometer**), voltage at the junction of two dissimilar metals (**thermocouple**), spectra of emitted radiation (**pyrometer**), and volume of a fixed mass of fluid (**thermometer**).

Temperature scales can be defined in terms of any of these properties, or in terms of physical phenomena, such as freezing and boiling, that take place at fixed temperatures and pressures. You might refer, for example, to "the temperature at which the resistivity of a copper wire is 1.92×10^{-6} ohms/cm^3" or to "the temperature two-thirds of the way from the boiling point of water at 1 atm to the melting point of NaCl."

It is convenient to have, in addition to these physical scales, a simple numerical temperature scale—among other reasons, so that you do not have to use 25 words to give the value of a single temperature. A defined temperature scale is obtained by arbitrarily assigning numerical values to two reproducibly measurable temperatures; for example, assign a value of 0 to the freezing point of water and a value of 100 to the boiling point of water at 1 atm. The assigned values completely specify the scale, since in addition to locating the two points they specify that the length of a unit temperature interval (called a **degree**) is $\frac{1}{100}$ of the distance between the two reference points on the scale.

The two most common temperature scales are defined using the freezing point (T_f) and boiling point (T_b) of water at a pressure of 1 atm.

Celsius (or centigrade) scale: T_f is assigned a value of 0°C, and T_b is assigned a value of 100°C. *Absolute zero* (theoretically the lowest temperature attainable in nature) on this scale falls at −273.15°C.

Fahrenheit scale: T_f is assigned a value of 32°F, and T_b is assigned a value of 212°F. Absolute zero falls at −459.67°F.

The **Kelvin** and **Rankine** scales are defined such that absolute zero has a value of 0 and the size of a degree is the same as a Celsius degree (Kelvin scale) or a Fahrenheit degree (Rankine scale).

The following relationships may be used to convert a temperature expressed in one defined scale unit to its equivalent in another:

$$T(K) = T(°C) + 273.15 \tag{3.5-1}$$

$$T(°R) = T(°F) + 459.67 \tag{3.5-2}$$

$$T(°R) = 1.8T(K) \tag{3.5-3}$$

$$T(°F) = 1.8T(°C) + 32 \tag{3.5-4}$$

Equations like these always have the form of the equation of a line ($y = ax + b$). If ($°A$) and ($°B$) stand for any two temperature units, to derive the equation for $T(°B)$ in terms of $T(°A)$ you must know equivalent values on each scale of two temperatures—say, T_1 and T_2. Then

1. Write $T(°B) = aT(°A) + b$
2. Substitute $T_1(°B)$ and $T_1(°A)$ in the equation—you then have one equation in two unknowns (a and b). Substitute $T_2(°B)$ and $T_2(°A)$ to get the second equation in the two unknowns, and solve for a and b.

EXAMPLE 3.5-1   *Derivation of a Temperature Conversion Formula*

Derive Equation 3.5-4 for $T(°F)$ in terms of $T(°C)$. Use $T_1 = 0°C$ (32°F) and $T_2 = 100°C$ (212°F).

SOLUTION

$$T(°F) = aT(°C) + b$$

Substitute T_1: $32 = (a)(0) + b \Longrightarrow b = 32$

Substitute T_2: $212 = (a)(100) + 32 \Longrightarrow a = 1.8$

$$\Downarrow$$

$$T(°F) = 1.8T(°C) + 32$$

A degree is both a temperature and a temperature interval, a fact that sometimes leads to confusion. Consider the temperature interval from 0°C to 5°C. There are nine Fahrenheit and nine Rankine degrees in this interval, and only five Celsius degrees and five Kelvin. An interval of 1 Celsius degree or Kelvin therefore contains 1.8 Fahrenheit or Rankine degrees, leading to the conversion factors

$$\frac{1.8°F}{1°C}, \frac{1.8°R}{1\,K}, \frac{1°F}{1°R}, \frac{1°C}{1\,K} \tag{3.5-5}$$

$T(°C) \rightarrow$	0		1		2		3		4		5
$T(K) \rightarrow$	273		274		275		276		277		278

$T(°F) \rightarrow$	32	33	34	35	36	37	38	39	40	41
$T(°R) \rightarrow$	492	493	494	495	496	497	498	499	500	501

Note: These conversion factors refer to temperature intervals, not temperatures.[4] For example, to find the number of Celsius degrees between 32°F and 212°F you can say that

$$\Delta T(°C) = \frac{(212 - 32)°F}{} \left| \frac{1°C}{1.8°F} \right. = 100°C$$

but to find the Celsuis temperature corresponding to 32°F you must use Equation 3.5-4; you cannot say

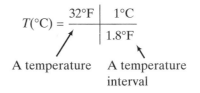

A temperature   A temperature
interval

TEST YOURSELF

1. Suppose you were given a glass tube with mercury in it but no scale markings, and you were given only a beaker of water, a freezer, and a bunsen burner with which to work. How would you calibrate the thermometer to provide readings in °C?
2. Which is warmer, a temperature of 1°C or 1°F?
3. Which reflects a greater change in temperature, an increase of 1°C or 1°F?

EXAMPLE 3.5-2   *Temperature Conversion*

Consider the interval from 20°F to 80°F.

1. Calculate the equivalent temperatures in °C and the interval between them.
2. Calculate directly the interval in °C between the temperatures.

SOLUTION

1. From Equation 3.5-4,

$$T(°C) = \frac{T(°F) - 32}{1.8}$$

so that

$$T_1(20°F) = \left(\frac{20 - 32}{1.8} \right)°C = -6.7°C$$

$$T_2(80°F) = \left(\frac{80 - 32}{1.8} \right)°C = 26.6°C$$

and

$$T_2 - T_1 = (26.6 - (-6.7))°C = 33.3°C$$

2. From Equation 3.5-5,

$$\Delta T(°C) = \frac{\Delta T(°F)}{} \left| \frac{1°C}{1.8°F} \right. = \frac{(80 - 20)°F}{} \left| \frac{1°C}{1.8°F} \right. = 33.3°C$$

---

[4]Some authors have proposed varying the position of the degree symbol to indicate whether a temperature or temperature interval is denoted; that is, 5°C refers to a temperature of five degrees Celsius, and 5C° means an interval of five Celsius degrees. This idea, while an excellent one in principle, has not caught on, so you will have to get used to making the distinction yourself from the context in which the unit appears.

EXAMPLE 3.5-3   *Temperature Conversion and Dimensional Homogeneity*

The heat capacity of ammonia, defined as the amount of heat required to raise the temperature of a unit mass of ammonia by precisely 1° at a constant pressure, is, over a limited temperature range, given by the expression

$$C_p\left(\frac{\text{Btu}}{\text{lb}_m\cdot{}^\circ\text{F}}\right) = 0.487 + 2.29 \times 10^{-4}T(^\circ\text{F})$$

Determine the expression for C_p in $\text{J}/(\text{g}\cdot{}^\circ\text{C})$ in terms of $T(^\circ\text{C})$.

SOLUTION

The $^\circ$F in the units of C_p refers to a temperature interval, while the unit of T is a temperature. The calculation is best done in two steps.

1. Substitute for $T(^\circ\text{F})$ and simplify the resulting equation:

$$C_p\left(\frac{\text{Btu}}{\text{lb}_m\cdot{}^\circ\text{F}}\right) = 0.487 + 2.29 \times 10^{-4}[1.8T(^\circ\text{C}) + 32]$$

$$= 0.494 + 4.12 \times 10^{-4}T(^\circ\text{C})$$

2. Convert to the desired temperature interval unit using Equation 3.5-5:

$$C_p\left(\frac{\text{J}}{\text{g}\cdot{}^\circ\text{C}}\right) = [0.494 + 4.12 \times 10^{-4}T(^\circ\text{C})]\frac{(\text{Btu})}{(\text{lb}_m\cdot{}^\circ\text{F})}\left|\frac{1.8^\circ\text{F}}{1.0^\circ\text{C}}\right|\frac{1\,\text{J}}{9.486 \times 10^{-4}\,\text{Btu}}\left|\frac{1\,\text{lb}_m}{454\,\text{g}}\right.$$

$$\Downarrow$$

$$\boxed{C_p\left(\frac{\text{J}}{\text{g}\cdot{}^\circ\text{C}}\right) = 2.06 + 1.72 \times 10^{-3}T(^\circ\text{C})}$$

CREATIVITY EXERCISES

1. Invent several temperature-measuring devices. For each, describe the device, and state what you would measure. (*Example:* Put a guinea pig on a treadmill in a room and measure the rate at which he runs to keep warm.) (Well, it *could* work.)
2. Think of as many ways as you can to use a solid block as a temperature-measuring device. (*Example:* Put it in a furnace equipped with a window and observe the color with which it glows.)

3.6 SUMMARY

In this chapter, we have described how amounts of material, flow rates, compositions, pressures, and temperatures are determined from direct measurements or calculated from measurements and physical properties. We also have described how to convert between different methods of expressing these variables. Here are some highlights.

- The *density* of a substance is the ratio of its mass to its volume. For example, the density of liquid acetone at 20°C is 0.791 g/cm^3, so that one cubic centimeter of liquid acetone at 20°C has a mass of 0.791 gram. Density can be thought of as a conversion factor between mass and volume or between mass flow rate and volumetric flow rate.
- The *specific gravity* of a substance is the ratio of the density of the substance to the density of a reference material (typically water at 4°C). Specific gravities of many liquids and solids are given in Table B.1, with the reference density being that of liquid water at 4°C (1.00 g/cm^3, 1.00 kg/L, 62.43 lb$_m$/ft^3). The density of a substance is the product of its specific gravity and the reference density in the desired units.
- The *atomic weight* of an element is the mass of an atom of that element on a scale that assigns ^{12}C a mass of exactly 12. Atomic weights of the elements in their naturally occurring

isotopic proportions are listed in the table at the back of the book. The *molecular weight* of a compound is the sum of the atomic weights of the atoms that constitute a molecule of that compound.

- A *gram-mole* or *mol* of a compound is the molecular weight of the compound in grams; for example, 1 mol H_2O has a mass of 18.01 grams. A *pound-mole* or *lb-mole* is the molecular weight in pounds mass; for example, 1 lb-mole H_2O has a mass of 18.01 lb_m. The molecular weight of water may therefore be expressed as 18.01 g/mol, 18.01 lb_m/lb-mole, and so on, and it may be used to convert masses to moles or mass flow rates to molar flow rates and vice versa.

- The *mass fraction* of a component in a mixture is the ratio of the mass of the component to the total mass of the mixture. If 100 grams of a mixture contains 30 grams of nitrogen, the mass fraction of nitrogen is 0.30 g N_2/g mixture. (The word "mixture" usually is omitted.) The mass fraction is also 0.30 kg N_2/kg and 0.30 lb_m N_2/lb_m, and the *percent by mass* or *weight percent* of nitrogen is 30%. The *mole fraction* of a component is defined similarly. If 10.0 kmol of a mixture contains 6.0 kmol of methanol, the mole fraction of methanol is 0.60 kmol CH_3OH/kmol (= 0.60 lb-mole CH_3OH/lb-mole), and the mole percent of methanol is 60%.

- The *average molecular weight* of a mixture is the ratio of the total mass to the total number of moles of all species.

- The *concentration* of a component in a mixture is the ratio of the mass or moles of the component to the total mixture volume. The *molarity* of a component of a solution is the concentration of the component expressed in mol/L.

- The *pressure* at a point in a fluid (gas or liquid) is the force per unit area that the fluid would exert on a plane surface passing through the point. Standard units of fluid pressure are N/m^2, (pascal, or Pa) in the SI system, $dyne/cm^2$ in the CGS system, and lb_f/ft^2 in the American engineering system. The unit $lb_f/in.^2$ (psi) is also common in the American engineering system.

- The pressure at the base of a vertical column of fluid of density ρ and height h is given by the expression

$$P = P_o + \rho g h \qquad \text{(3.4-1)}$$

where P_o is the pressure exerted on the top of the column and g is the acceleration of gravity. This result gives rise to two ways of expressing fluid pressure: as force per unit area (e.g., $P = 14.7 \, lb_f/in.^2$) or as an equivalent *pressure head*, $P_h = P/\rho g$ (e.g., $P_h = 760$ mm Hg), the height of a column of the specified fluid with zero pressure at the top that would exert the specified pressure at the bottom.

- The earth's atmosphere can be considered a column of fluid with zero pressure at the top. The fluid pressure at the base of this column is *atmospheric pressure* or *barometric pressure*, P_{atm}. Although atmospheric pressure varies with altitude and weather conditions, its value at sea level is always close to $1.01325 \times 10^5 \, N/m^2$ (= 14.696 $lb_f/in.^2$ = 760 mm Hg). This pressure value has been designated *1 atmosphere*. Other equivalents to 1 atm in different units are given on the inside front cover of this text.

- The *absolute pressure* of a fluid is the pressure relative to a perfect vacuum ($P = 0$). The *gauge pressure* is the pressure relative to atmospheric pressure: $P_{gauge} = P_{abs} - P_{atm}$. Common pressure gauges like a Bourdon gauge and an open-end manometer provide a direct reading of gauge pressure. If atmospheric pressure is not known from a weather report or a barometer reading, a value of $P_{atm} = 1$ atm is usually reasonable to assume when converting between absolute and gauge pressure.

- *Temperature scales* are obtained by assigning numerical values to two experimentally reproducible temperatures. For example, the Celsius scale is obtained by assigning a value of 0°C to the freezing point of pure water at 1 atm and a value of 100°C to the boiling point of pure water at 1 atm. A temperature of 40°C is consequently shorthand for "the temperature 40% of the way from the freezing point of water at 1 atm to the boiling point of water at 1 atm."

- The four most common temperature scales are Celsius (°C), Fahrenheit (°F), and the absolute temperature scales Kelvin (K) and Rankine (°R). Temperatures expressed in one of these scales may be converted to equivalent temperatures in another scale using Equations 3.5-1 through 3.5-4.

- Temperatures should not be confused with temperature intervals. For example, a temperature of 10°C is equivalent to a temperature of 50°F (from Equation 3.5-4), but a temperature interval of 10°C (e.g., the interval between $T = 10°C$ and $T = 20°C$) is equivalent to a temperature interval of 18°F (the interval between 50°F and 68°F). An interval of 1 Celsius degree or 1 Kelvin is equivalent to 1.8 Fahrenheit or Rankine degrees.

PROBLEMS

3.1. Perform the following estimations *without using a calculator.*
  (a) Estimate the mass of water (kg) in an Olympic-size swimming pool.
  (b) A drinking glass is being filled from a pitcher. Estimate the mass flow rate of the water (g/s).
  (c) Twelve heavyweight boxers coincidentally get on the same elevator in Great Britain. Posted on the elevator wall is a sign that gives the maximum safe combined weight of the passengers, W_{max}, in stones (1 stone = $14 \, lb_m \approx 6$ kg). If you were one of the boxers, estimate the lowest value of W_{max} for which you would feel comfortable remaining on the elevator.
  (d) An oil pipeline across Alaska is 4.5 ft in diameter and 800 miles long. How many barrels of oil are required to fill the pipeline?
  (e) Estimate the volume of your body (cm^3) in two different ways. (Show your work.)
  (f) A solid block is dropped into water and very slowly sinks to the bottom. Estimate its specific gravity.

3.2. Calculate densities in lb_m/ft^3 of the following substances:
  (a) a liquid with density of 995 kg/m^3. Use (i) conversion factors from the table on the inside front cover and (ii) Equation 3.1-2.
  (b) a solid with a specific gravity of 5.7.

3.3. The specific gravity of gasoline is approximately 0.70.
  (a) Determine the mass (kg) of 50.0 liters of gasoline.
  (b) The mass flow rate of gasoline exiting a refinery tank is 1150 kg/min. Estimate the volumetric flow rate in liters/s.
  (c) Estimate the average mass flow rate (lb_m/min) delivered by a gasoline pump.
  (d) Gasoline and kerosene (specific gravity = 0.82) are blended to obtain a mixture with a specific gravity of 0.78. Calculate the volumetric ratio (volume of gasoline/volume of kerosene) of the two compounds in the mixture, assuming $V_{blend} = V_{gasoline} + V_{kerosene}$.

3.4. Assume the price of gasoline in France is approximately 5 French francs per liter and the exchange rate is 5.22 francs per U.S. dollar. How much would you pay, in dollars, for 50.0 kg of gasoline in France, assuming gasoline has a specific gravity of 0.70. What would the same quantity of gasoline cost in the United States at a rate of $1.20 per gallon?

3.5. Liquid benzene and liquid *n*-hexane are blended to form a stream flowing at a rate of 700 lb_m/h. An on-line *densitometer* (an instrument used to determine density) indicates that the stream has a density of 0.850 g/mL. Using specific gravities from Table B.1, estimate the mass and volumetric feed rates of the two hydrocarbons to the mixing vessel (in American engineering units). State at least two assumptions required to obtain the estimate from the recommended data.

3.6. At 25°C, an aqueous solution containing 35.0 wt% H_2SO_4 has a specific gravity of 1.2563. A quantity of the 35% solution is needed that contains 195.5 kg of H_2SO_4.
  (a) Calculate the required volume (L) of the solution using the given specific gravity.
  (b) Estimate the percentage error that would have resulted if pure-component specific gravities of H_2SO_4 (SG = 1.8255) and water had been used for the calculation instead of the given specific gravity of the mixture.

3.7. A rectangular block of solid carbon (graphite) floats at the interface of two immiscible liquids. The bottom liquid is a relatively heavy lubricating oil, and the top liquid is water. Of the total block volume, 54.2% is immersed in the oil and the balance is in the water. In a separate experiment, an empty flask is weighed, 35.3 cm^3 of the lubricating oil is poured into the flask, and the flask is reweighed. If the scale reading was 124.8 g in the first weighing, what would it be in the second weighing? (*Suggestion:* Recall Archimedes' principle, and do a force balance on the block.)

3.8. A rectangular block floats in pure water with 0.5 in. above the surface and 1.5 in. below the surface. When placed in an aqueous solution, the block of material floats with 1 in. below the surface. Estimate the specific gravities of the block and the solution. (*Suggestion:* Call the horizontal cross-sectional area of the block A. A should cancel in your calculations.)

3.9. An object of density ρ_a, volume V_a, and weight W_a is thrown from a rowboat floating on the surface of a small pond and sinks to the bottom. The weight of the rowboat without the jettisoned object is W_b. Before the object was thrown out, the depth of the pond was h_{p1}, and the bottom of the boat was a distance h_{b1} above the pond bottom. After the object sinks, the values of these quantities are h_{p2} and h_{b2}. The area of the pond is A_p; that of the boat is A_b. A_b may be assumed constant, so that the volume of water displaced by the boat is $A_b(h_p - h_b)$.

(a) Derive an expression for the change in the pond depth $(h_{p2} - h_{p1})$. Does the liquid level of the pond rise or fall, or is it indeterminate?

(b) Derive an expression for the change in the height of the bottom of the boat above the bottom of the pond $(h_{b2} - h_{b1})$. Does the boat rise or fall relative to the pond bottom, or is it indeterminate?

3.10. Limestone (calcium carbonate) particles are stored in 50-L bags. The **void fraction** of the particulate matter is 0.30 (liter of void space per liter of total volume) and the specific gravity of solid calcium carbonate is 2.93.

(a) Estimate the **bulk density** of the bag contents (kg $CaCO_3$/liter of total volume).

(b) Estimate the weight (W) of the filled bags. State what you are neglecting in your estimate.

(c) The contents of three bags are fed to a **ball mill**, a device something like a rotating clothes dryer containing steel balls. The tumbling action of the balls crushes the limestone particles and turns them into a powder. (See p. 20-31 of *Perry's Chemical Engineers' Handbook*, 7th ed.) The limestone coming out of the mill is put back into 50-L bags. Would the limestone coming out of the mill (i) just fill three bags, (ii) fall short of filling three bags, or (iii) fill more than three bags? Briefly explain your answer.

3.11. A useful measure of an individual's physical condition is the fraction of his or her body that consists of fat. This problem describes a simple technique for estimating this fraction by weighing the individual twice, once in air and once submerged in water.

(a) A man has body mass m_b = 122.5 kg. If he stands on a scale calibrated to read in newtons, what would the reading be? If he then stands on a scale while he is totally submerged in water at 30°C (specific gravity = 0.996) and the scale reads 44.0 N, what is the volume of his body (liters)? (*Hint:* Recall from Archimedes' principle that the weight of a submerged object equals the weight in air minus the buoyant force on the object, which in turn equals the weight of water displaced by the object. Neglect the buoyant force of air.) What is his body density, ρ_b (kg/L)?

(b) Suppose the body is divided into fat and nonfat components, and that X_f (kilograms of fat/kilograms of total body mass) is the fraction of the total body mass that is fat:

$$x_f = \frac{m_f}{m_b}$$

Prove that

$$x_f = \frac{\dfrac{1}{\rho_b} - \dfrac{1}{\rho_{nf}}}{\dfrac{1}{\rho_f} - \dfrac{1}{\rho_{nf}}}$$

where ρ_b, ρ_f, and ρ_{nf} are the average densities of the whole body, the fat component, and the non-fat component, respectively. [*Suggestion:* Start by labeling the masses (m_f and m_b) and volumes (V_f and V_b) of the fat component of the body and the whole body, and then write expressions for the three densities in terms of these quantities. Then eliminate volumes algebraically and obtain an expression for m_f/m_b in terms of the densities.[5]]

(c) If the average specific gravity of body fat is 0.9 and that of nonfat tissue is 1.1, what fraction of the man's body in part (a) consists of fat?

(d) The body volume calculated in part (a) includes volumes occupied by gas in the digestive tract, sinuses, and lungs. The sum of the first two volumes is roughly 100 mL and the volume of the

---

[5]If you can't work out the proof, take the given formula as valid and proceed to the next part.

lungs is roughly 1.2 liters. The mass of the gas is negligible. Use this information to improve your estimate of x_f.

3.12. Aqueous solutions of the amino acid L-isoleucine (Ile) are prepared by putting 100.0 grams of pure water into each of six flasks and adding different precisely weighed quantities of Ile to each flask. The densities of the solutions at $50.0 \pm 0.05°C$ are then measured with a precision densitometer, with the following results:

r (g Ile/100 g H_2O)	0.0000	0.8821	1.7683	2.6412	3.4093	4.2064
ρ (g solution/cm^3)	0.98803	0.98984	0.99148	0.99297	0.99439	0.99580

(a) Plot a calibration curve showing the mass ratio, r, as a function of solution density, ρ, and fit a straight line to the data to obtain an equation of the form $r = a\rho + b$.

(b) The volumetric flow rate of an aqueous Ile solution at a temperature of 50°C is 150 L/h. The density of a sample of the stream is measured and found to be 0.9940 g/cm^3. Use the calibration equation to estimate the mass flow rate of Ile in the stream (kg Ile/h).

(c) It has just been discovered that the thermocouple used to measure the stream temperature was poorly calibrated and the temperature was actually 47°C. Would the Ile mass flow rate calculated in part (b) be too high or too low? State any assumption you make and briefly explain your reasoning.

3.13. Before a rotameter can be used to measure an unknown flow rate, a **calibration curve** of flow rate versus rotameter reading must be prepared. A calibration technique for liquids is illustrated below. A flow rate is set by adjusting the pump speed; the rotameter reading is recorded, and the effluent from the rotameter is collected in a graduated cylinder for a timed interval. The procedure is carried out twice for each of several pump settings.

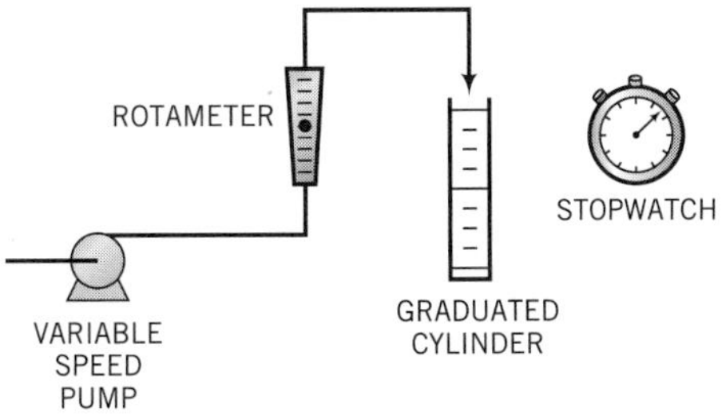

Rotameter Reading	Collection Time (min)	Volume Collected (cm^3)
2	1	297
2	1	301
4	1	454
4	1	448
6	0.5	300
6	0.5	298
8	0.5	371
8	0.5	377
10	0.5	440
10	0.5	453

(a) Assuming the liquid is water at 25°C, draw a calibration curve of mass flow rate, $\dot{m}$ (kg/min), versus rotameter reading, R, and use it to estimate the mass flow rate of a water stream for which the rotameter reading is 5.3.

(b) The **mean difference between duplicates**, $\overline{D_i}$, provides an estimate of the standard deviation of a single measurement, which was given the symbol s_x on p. 18 of Chapter 2:

$$s_x \approx \frac{\sqrt{\pi}}{2} \overline{D_i} = 0.8862 \overline{D_i}$$

Moreover, **confidence limits** on measured values can be estimated to a good approximation using the mean difference between duplicates. For example, if a single measurement of Y yields a value Y_{measured}, then there is a 95% probability that the true value of Y falls within the 95% confidence limits $(Y_{\text{measured}} - 1.74\overline{D_i})$ and $(Y_{\text{measured}} + 1.74\overline{D_i})$.[6] For a measured flow rate of 610 g/min, estimate the 95% confidence limits on the true flow rate.

---

[6]W. Volk, *Applied Statistics for Engineers*, McGraw-Hill, New York, pp. 113–115, 1958.

3.14. How many of the following are found in 15.0 kmol of benzene (C_6H_6)? (a) kg C_6H_6; (b) mol C_6H_6; (c) lb-mole C_6H_6; (d) mol (g-atom) C; (e) mol H; (f) g C; (g) g H; (h) molecules of C_6H_6.

3.15. Liquid toluene is flowing through a pipe at a rate of 175 m³/h.
   (a) What is the mass flow rate of this stream in kg/min?
   (b) What is the molar flow rate in mol/s?
   (c) In fact, the answer to part (a) is only an approximation that is almost certain to be slightly in error. What did you have to assume to obtain the answer?

3.16. A mixture of methanol and methyl acetate contains 15.0 wt% methanol.
   (a) Using a single dimensional equation, determine the g-moles of methanol in 200.0 kg of the mixture.
   (b) The flow rate of methyl acetate in the mixture is to be 100.0 lb-mole/h. What must the mixture flow rate be in lb_m/h?

3.17. The feed to an ammonia synthesis reactor contains 25 mole% nitrogen and the balance hydrogen. The flow rate of the stream is 3000 kg/h. Calculate the rate of flow of nitrogen into the reactor in kg/h. (*Suggestion:* First calculate the average molecular weight of the mixture.)

3.18. A suspension of calcium carbonate particles in water flows through a pipe. Your assignment is to determine both the flow rate and the composition of this slurry. You proceed to collect the stream in a graduated cylinder for 1.00 min; you then weigh the cylinder, evaporate the collected water, and reweigh the cylinder. The following results are obtained:

   Mass of empty cylinder: 65.0 g
   Mass of cylinder + collected slurry: 565 g
   Volume collected: 455 mL
   Mass of cylinder after evaporation: 215 g

   Calculate
   (a) the volumetric flow rate and mass flow rate of the suspension.
   (b) the density of the suspension.
   (c) the mass fraction of $CaCO_3$ in the suspension.

3.19. A mixture is 10.0 mole% ethyl alcohol, 75.0 mole% ethyl acetate ($C_4H_8O_2$), and 15.0 mole% acetic acid. Calculate the mass fractions of each compound. What is the average molecular weight of the mixture? What would be the mass (kg) of a sample containing 25.0 kmol of ethyl acetate?

3.20. Certain solid substances, known as **hydrated** compounds, have well-defined molecular ratios of water to some other species, which often is a salt. For example, calcium sulfate dihydrate (commonly known as *gypsum,* $CaSO_4 \cdot 2H_2O$), has 2 moles of water per mole of calcium sulfate; alternatively, it may be said that 1 mole of gypsum consists of 1 mole of calcium sulfate and 2 moles of water. The water in such substances is called **water of hydration**. (More information about hydrated salts is given in Chapter 6.)

   Solid gypsum is formed in a crystallizer and leaves that unit as a *slurry* (a suspension of solid particles in a liquid) of solid gypsum particles suspended in an aqueous $CaSO_4$ solution. The slurry flows from the crystallizer to a filter in which the particles are collected as a *filter cake*. The filter cake, which is 95.0 wt% solid gypsum and the remainder $CaSO_4$ solution, is fed to a dryer in which all water (including the water of hydration in the crystals) is driven off to yield anhydrous (water-free) $CaSO_4$ as product. A flowchart and relevant process data are given below.

   Solids content of slurry leaving crystallizer: 0.35 kg $CaSO_4 \cdot 2H_2O$/L slurry
   $CaSO_4$ content of slurry liquid: 0.209 g $CaSO_4$/100 g H_2O
   Specific gravities: $CaSO_4 \cdot 2H_2O$(s), 2.32; liquid solutions, 1.05

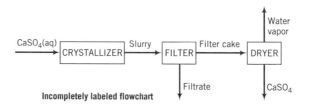

Incompletely labeled flowchart

(a) Briefly explain in your own words the functions of the three units (crystallizer, filter, and dryer).

(b) Take a basis of one liter of solution leaving the crystallizer and calculate the mass (kg) and volume (L) of solid gypsum, the mass of $CaSO_4$ in the gypsum, and the mass of $CaSO_4$ in the liquid solution.

(c) Calculate the percentage recovery of $CaSO_4$—that is, the percentage of the total $CaSO_4$ (precipitated plus dissolved) leaving the crystallizer recovered as solid anhydrous $CaSO_4$.

3.21. Things were going smoothly at the Breaux Bridge Drug Co. pilot plant during the midnight to 8 a.m. shift until Thérèse Lagniappe, the reactor operator, let the run instruction sheet get too close to the Coleman stove that was being used to heat water to prepare Lagniappe's bihourly cup of Community Coffee. What followed ended in a total loss of the run sheet, the coffee, and a substantial portion of the novel Lagniappe was writing.

Remembering the less than enthusiastic reaction she got the last time she telephoned her supervisor in the middle of the night, Lagniappe decided to rely on her memory of the required flow-rate settings. The two liquids being fed to a stirred-tank reactor were circulostoic acid (CSA: MW = 75, SG = 0.90) and flubitol (FB: MW = 90, SG = 0.75). The product from the system was a popular over-the-counter drug that simultaneously cures high blood pressure and clumsiness. The molar ratio of the two feed streams had to be between 1.05 and 1.10 mol CSA/mol FB to keep the contents of the reactor from forming a solid plug. At the time of the accident, the flow rate of CSA was 45.8 L/min. Lagniappe set the flow of flubitol to the value she thought had been in the run sheet: 55.2 L/min. Was she right? If not, how would she have been likely to learn of her mistake? (*Note:* The reactor was stainless steel, so she could not see the contents.)

3.22. A mixture of ethanol (ethyl alcohol) and water contains 60.0% water by mass.

(a) Assuming volume additivity of the components, estimate the specific gravity of the mixture at 20°C. What volume (in liters) of this mixture is required to provide 150 mol of ethanol?

(b) Repeat part (a) with the additional information that the specific gravity of the mixture at 20°C is 0.93518 (making it unnecessary to assume volume additivity). What percentage error results from the volume additivity assumption?

3.23. A mixture of methane and air is capable of being ignited only if the mole percent of methane is between 5% and 15%. A mixture containing 9.0 mole% methane in air flowing at a rate of 700. kg/h is to be diluted with pure air to reduce the methane concentration to the lower flammability limit. Calculate the required flow rate of air in mol/h and the percent *by mass* of oxygen in the product gas. (*Note:* Air may be taken to consist of 21 mole% O_2 and 79% N_2 and to have an average molecular weight of 29.0.)

3.24. A liquid mixture is prepared by combining N different liquids with densities $\rho_1, \rho_2, \ldots, \rho_N$. The volume of component i added to the mixture is V_i and the mass fraction of this component in the mixture is x_i. The components are completely miscible.

Determine which of the following two formulas should be used to estimate the density of the liquid mixture, $\bar{\rho}$, if the volume of the mixture equals the sum of the pure-component volumes.[7]

$$\bar{\rho} = \sum_{i=1}^{N} x_i \rho_i \quad \text{(A)} \qquad \frac{1}{\bar{\rho}} = \sum_{i=1}^{N} \frac{x_i}{\rho_i} \quad \text{(B)}$$

Determine whether (A) or (B) is the correct formula (show your proof), and then use the correct formula to estimate the density (g/cm^3) of a liquid mixture containing 60.0 wt% acetone, 25.0 wt% acetic acid, and 15.0 wt% carbon tetrachloride.

3.25. A gaseous mixture of CO, CO_2, CH_4, and N_2 is analyzed with a gas chromatograph (see Problem 3.26). The output appears on a strip-chart recorder, as shown here.

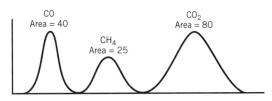

---

[7]This is an approximation for most liquids, unlike the exact statement that the mass of the mixture is the sum of the component masses.

For each of the three species, the area under the peak is approximately proportional to the number of moles of the indicated substance in the sample. From other information, it is known that the molar ratio of methane (CH_4) to nitrogen is 0.200.

(a) What are the mole fractions of the four species in the gas?

(b) What is the average molecular weight of the gas?

*3.26. A **gas chromatograph** (GC) is a device used to separate the components of a sample of a gas or liquid mixture and to provide a measure of the amount of each component in the sample. The output from a chromatographic analysis typically takes the form of a series of peaks on a strip-chart recorder. (See the preceding problem.) Each peak corresponds to a specific component, and the area under the peak is proportional to the amount of that component in the sample [$n_i(\text{mol}) = k_i A_i$, where A_i is the area of the peak corresponding to the ith species]. The proportionality constants (k_i) are determined in separate calibration experiments in which known amounts of the components are injected into the GC sample port and the corresponding peak areas are measured.

(a) Prepare a spreadsheet to calculate the composition of a mixture from a set of peak areas obtained from a chromatograph. The spreadsheet should appear as follows:

Sample	Species	MW	k	Peak Area	Mole Fraction	Mass Fraction
1	CH4	16.04	0.150	3.6	—	—
	C2H6	30.07	0.287	2.8	—	—
	C3H8	—	0.467	2.4	—	0.353
	C4H10	—	0.583	1.7	—	—
2	CH4	16.04	0.150	7.8	—	—
	C2H6	—	—	2.4	—	—
⋮	⋮	⋮	⋮	⋮	⋮	⋮

You may use additional columns to store intermediate quantities in the calculation of the mass and mole fractions. In the actual spreadsheet, the dashes (—) would be replaced by numbers.

Test your program on data for five mixtures of methane, ethane, propane, and n-butane. The k values for these species are those given in the above table, and the measured peaks are given below. For example, the area of the methane peak for the first mixture is 3.6, the area of the ethane peak for the same mixture is 2.8, and so on.

Sample	A_1	A_2	A_3	A_4
1	3.6	2.8	2.4	1.7
2	7.8	2.4	5.6	0.4
3	3.4	4.5	2.6	0.8
4	4.8	2.5	1.3	0.2
5	6.4	7.9	4.8	2.3

(b) Write a computer program (not a spreadsheet) to perform the same task—that is, calculate mole and mass fractions from measured chromatographic peak areas. The program should perform the following steps:

i. read in N, the number of species to be analyzed;

ii. read in $M_1, M_2, M_3, \ldots, M_N$, the molecular weights of the species;

iii. read in $k_1, k_2, k_3, \ldots, k_N$, the calibration constants for the species;

iv. read in N_d, the number of chromatographic analyses performed;

v. for the first analysis, read in the measured peak areas $A_1, A_2, A_3, \ldots, A_N$;

---

*Computer problem.

     vi. calculate and print out the sample number, the mole fractions of each species in the sample, and the mass fractions of each species in the sample;

    vii. Repeat steps v and vi for each of the remaining analyses. Test your program on the five-sample data set given in part (a).

3.27. *Biomass combustion*—burning of forests, grasslands, agricultural wastes, and other biological matter—is recognized as a serious threat to the environment.[8] The table below shows the distribution of carbon-containing compounds released to the atmosphere worldwide from all combustion sources as well as the portion coming from biomass burning.

Compound	Metric Tons C, All Sources	Metric Tons C, % from Biomass
CO_2	8700	40
CO	1100	26
CH_4	380	10

The numbers in the middle column reflect annual quantities of carbon released to the atmosphere in the indicated compound; for example, 8700 metric tons of carbon (8.7×10^6 kg C) was released in carbon dioxide.

(a) Determine the combined annual release (in metric tons) of all three species resulting from biomass combustion and the average molecular weight of the combined gases.

(b) Find a reference on atmospheric pollution and list the environmental hazards associated with CO and CO_2 release. What other elements might be released in environmentally hazardous forms if biomass is burned?

3.28. A 5.00-wt% aqueous sulfuric acid solution ($\rho = 1.03$ g/mL) flows through a 45-m long pipe with a 6.0 cm diameter at a rate of 87 L/min.

(a) What is the molarity of sulfuric acid in the solution?

(b) How long (in seconds) would it take to fill a 55-gallon drum, and how much sulfuric acid (lb_m) would the drum contain? (You should arrive at your answers with two dimensional equations.)

(c) The mean velocity of a fluid in a pipe equals the volumetric flow rate divided by the cross-sectional area normal to the direction of flow. Use this information to estimate how long (in seconds) it takes the solution to flow from the pipe inlet to the outlet.

3.29. A gas stream contains 18.0 mole% hexane and the remainder nitrogen. The stream flows to a condenser, where its temperature is reduced and some of the hexane is liquefied. The hexane mole fraction in the gas stream leaving the condenser is 0.0500. Liquid hexane condensate is recovered at a rate of 1.50 L/min.

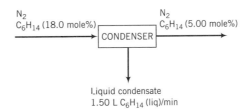

(a) What is the flow rate of the gas stream leaving the condenser in mol/min? (*Hint:* First calculate the molar flow rate of the condensate and note that the rates at which C_6H_{14} and N_2 enter the unit must equal the total rates at which they leave in the two exit streams.)

(b) What percentage of the hexane entering the condenser is recovered as a liquid?

3.30. The little-known rare earth element nauseum (atomic weight = 172) has the interesting property of being completely insoluble in everything but 12-year-old bourbon. This curious fact was discovered in the laboratory of Professor Ludwig von Schlimazel, the eminent German chemist whose

[8]*Chemical & Engineering News*, **68**, 4(March 26, 1990).

invention of the bathtub ring won him the Nobel Prize. Having unsuccessfully tried to dissolve nauseum in 7642 different solvents over a 10-year period, Schlimazel finally came to the 30 mL of Old Aardvark Bottled-in-Bond that was the only remaining liquid in his laboratory. Always willing to suffer personal loss in the name of science, Schlimazel calculated the amount of nauseum needed to make up a 0.03 molar solution, put the Aardvark bottle on the desk of his faithful technician Edgar P. Settera, weighed out the calculated amount of nauseum and put it next to the bottle, and then wrote the message that has become part of history:

"Ed Settera. Add nauseum!"

How many grams of nauseum did he weigh out? (Neglect the change in liquid volume resulting from the nauseum addition.)

3.31. The reaction A $\rightarrow$ B is carried out in a laboratory reactor. According to a published article the concentration of A should vary with time as follows:

$$C_A = C_{AO} \exp(-kt)$$

where C_{AO} is the initial concentration of A in the reactor and k is a constant.
(a) If C_A and C_{AO} are in lb-moles/ft^3 and t is in minutes, what are the units of k?
(b) The following data are taken for $C_A(t)$:

t(min)	C_A(lb-mole/ft^3)
0.5	1.02
1.0	0.84
1.5	0.69
2.0	0.56
3.0	0.38
5.0	0.17
10.0	0.02

Verify the proposed rate law graphically (first determine what plot should yield a straight line), and calculate C_{AO} and k.
(c) Convert the formula with the calculated constants included to an expression for the molarity of A in the reaction mixture in terms of t(seconds). Calculate the molarity at $t = 200$ s.

3.32. Perform the following pressure conversions, assuming when necessary that atmospheric pressure is 1 atm. Unless otherwise stated, the given pressures are absolute.
(a) 2600 mm Hg to psi
(b) 275 ft H_2O to kPa
(c) 3.00 atm to N/cm^2
(d) 280 cm Hg to dyne/m^2
(e) 20 cm Hg of vacuum to atm (absolute)
(f) 25.0 psig to mm Hg (gauge)
(g) 25.0 psig to mm Hg (absolute)
(h) 325 mm Hg to mm Hg gauge
(i) 35.0 psi to cm of carbon tetrachloride

3.33. A storage tank containing oil (SG = 0.92) is 10.0 meters high and 16.0 meters in diameter. The tank is closed, but the amount of oil it contains can be determined from the gauge pressure at the bottom.
(a) A pressure gauge connected to the bottom of the tank was calibrated with the top of the tank open to the atmosphere. The calibration curve is a plot of height of oil, h(m), versus P_{gauge}(kPa). Sketch the expected shape of this plot. What height of oil would lead to a gauge reading of 68 kPa? What would be the mass (kg) of oil in the tank corresponding to this height?
(b) An operator observes that the pressure gauge reading is 68 kPa and notes the corresponding liquid height from the calibration curve. What he did not know was that the absolute pressure above the liquid surface in the tank was 115 kPa when he read the gauge. What is the actual height of the oil? (Assume atmospheric pressure is 101 kPa.)

3.34. A rectangular block of height L and horizontal cross-sectional area A floats at the interface between two immiscible liquids, as shown below.

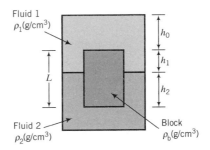

(a) Derive a formula for the block density, ρ_b, in terms of the fluid densities ρ_1 and ρ_2, the heights h_0, h_1, and h_2, and the cross-sectional area A. (It is not necessary that all of these variables appear in the final result.)

(b) Force balances on the block can be calculated in two ways: (i) in terms of the weight of the block and the hydrostatic forces on the upper and lower block surfaces; and (ii) in terms of the weight of the block and the buoyant force on the block as expressed by Archimedes' principle. Prove that these two approaches are equivalent.

3.35. The viewing window in a diving suit has an area of roughly 65 cm^2. If an attempt were made to maintain the pressure on the inside of the suit at 1 atm, what force (N and lb_f) would the window have to withstand if the diver descended to a depth of 150 m. Take the specific gravity of the water to be 1.05.

3.36. The great Boston molasses flood occurred on January 15, 1919. In it, 2.3 million gallons of crude molasses flowed from a 30-foot high storage tank that ruptured, killing 21 people and injuring 150. The estimated specific gravity of crude molasses is 1.4. What were the mass of molasses in the tank in lb_m and the pressure at the bottom of the tank in $lb_f/in.^2$? Give at least two possible causes of the tragedy.

3.37. The chemical reactor shown below has a cover (called a *head*) that is held in place by a series of bolts. The head is made of stainless steel (SG = 8.0), is 3 in. thick, has a diameter of 24 in., and covers and seals an opening 20 in. in diameter. During *turnaround*, when the reactor is taken out of service for cleaning and repair, the head was removed by an operator who thought the reactor had been depressurized using a standard venting procedure. However, the pressure gauge had been damaged in an earlier process upset (the reactor pressure had exceeded the upper limit of the gauge), and instead of being depressurized completely, the vessel was under a gauge pressure of 30 psi.

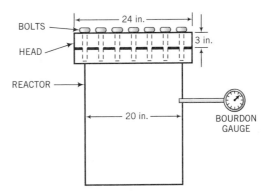

(a) What force (lb_f) were the bolts exerting on the head before they were removed? (*Hint:* Don't forget that a pressure is exerted on the top of the head by the atmosphere.) What happened when the last bolt was removed by the operator? Justify your prediction by estimating the initial acceleration of the head upon removal of the last bolt.

(b) Propose an alteration in the turnaround procedure to prevent recurrence of an incident of this kind.

3.38. In the movie *The Drowning Pool*, private detective Lew Harper (played by Paul Newman) is trapped by the bad guy in a room containing a swimming pool. The room may be considered rectangular, 5 meters wide by 15 meters long, with an open skylight window 10 meters above the floor. There is a single entry to the room, reached by a stairway: a locked 2-m high by 1-m wide door, whose bottom is 1 meter above the floor. Harper knows that his enemy will return in eight hours and decides he can escape by filling the room with water and floating up to the skylight. He plugs the drain with his clothes, turns on the water valves, and prepares to put his plan into action.

(a) Prove that if the door is completely under water and h is the distance from the top of the door to the surface of the water, then the net force exerted on the door satisfies the inequality

$$F > \rho_{H_2O} g h A_{door}$$

(Don't forget that a pressure is also exerted on the door by the outside air.)

(b) Assume that water enters the room at about five times the rate at which it enters an average bathtub and that the door can withstand a maximum force of 4500 newtons (about 1000 lb$_f$). Estimate (i) whether the door will break before the room fills and (ii) whether Harper has time to escape if the door holds. State any assumptions you make.

3.39. A housing development is served by a water tower with the water level maintained between 20 and 30 meters above the ground, depending on demand and water availability. Responding to a resident's complaint about the low flow rate of water at his kitchen sink, a representative of the developer came and measured the water pressure at the tap above the kitchen sink and at the junction between the water main (a pipe connected to the bottom of the water tower) and the feed pipe to the house. The junction is 5 m below the level of the kitchen tap. All water valves in the house were turned off.

(a) If the water level in the tower was 25 m above tap level, what should be the gauge pressures (kPa) at the tap and junction?

(b) Suppose the pressure measurement at the tap was lower than your estimate in part (a), but the measurement at the junction was as predicted. State a possible explanation.

(c) If pressure measurements corresponded to the predictions in part (a), what else could be responsible for the low water flow to the sink?

3.40. Two mercury manometers, one open-end and the other sealed-end, are attached to an air duct. The reading on the open-end manometer is 25 mm and that on the sealed-end manometer is 800 mm. Determine the absolute pressure in the duct, the gauge pressure in the duct, and the atmospheric pressure, all in mm Hg.

3.41. Three different liquids are used in the manometer shown here.

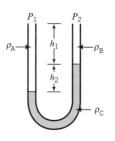

(a) Derive an expression for $P_1 - P_2$ in terms of ρ_A, ρ_B, ρ_C, h_1, and h_2.

(b) Suppose fluid A is methanol, B is water, and C is a manometer fluid with a specific gravity of 1.37; pressure $P_2 = 121.0$ kPa; $h_1 = 30.0$ cm; and $h_2 = 24.0$ cm. Calculate P_1 (kPa).

3.42. The level of toluene (a flammable hydrocarbon) in a storage tank may fluctuate between 10 and 400 cm from the top of the tank. Since it is impossible to see inside the tank, an open-end manometer with water or mercury as the manometer fluid is to be used to determine the toluene level. One leg of the manometer is attached to the tank 500 cm from the top. A nitrogen blanket at atmospheric pressure is maintained over the tank contents.

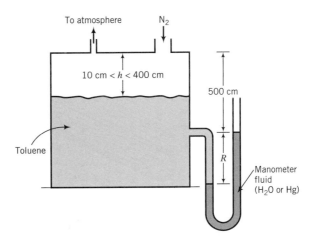

(a) When the toluene level in the tank is 150 cm below the top ($h = 150$ cm), the manometer fluid level in the open arm is at the height of the point where the manometer connects to the tank. What manometer reading, R (cm), would be observed if the manometer fluid is (i) mercury, (ii) water? Which manometer fluid would you use, and why?

(b) Briefly describe how the system would work if the manometer were simply filled with toluene. Give several advantages of using the fluid you chose in part (a) over using toluene.

(c) What is the purpose of the nitrogen blanket?

3.43. A fluid of unknown density is used in two manometers—one sealed-end, the other across an orifice in a water pipeline. The readings shown here are obtained on a day when barometric pressure is 756 mm Hg.

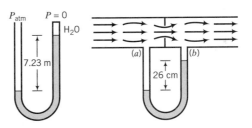

What is the pressure drop (mm Hg) from point (a) to point (b)?

3.44. An open-end mercury manometer is connected to a low-pressure pipeline that supplies a gas to a laboratory. Because paint was spilled on the arm connected to the line during a laboratory renovation, it is impossible to see the level of the manometer fluid in this arm. During a period when the gas supply is connected to the line but there is no gas flow, a Bourdon gauge connected to the line downstream from the manometer gives a reading of 7.5 psig. The level of mercury in the open arm is 900 mm above the lowest part of the manometer.

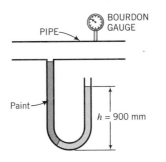

    (a) When the gas is not flowing, the pressure is the same everywhere in the pipe. How high above the bottom of the manometer would the mercury be in the arm connected to the pipe?

    (b) When gas is flowing, the mercury level in the visible arm drops by 25 mm. What is the gas pressure (psig) at this moment?

3.45. An **inclined manometer** is a useful device for measuring small pressure differences.

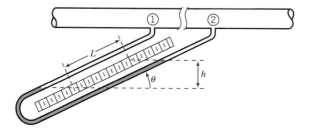

The formula given in Section 3.4 for the pressure difference in terms of the liquid-level difference h remains valid, but while h would be small and difficult to read for a small pressure drop if the manometer were vertical, L can be made quite large for the same pressure drop by making the angle of the inclination, θ, small.

    (a) Derive a formula for h in terms of L and θ.

    (b) Suppose the manometer fluid is water, the process fluid is a gas, the inclination of the manometer is $\theta = 15°$, and a reading $L = 8.7$ cm is obtained. What is the pressure difference between points ① and ②?

3.46. An open-end mercury manometer is to be used to measure the pressure in an apparatus containing a vapor that reacts with mercury. A 10 cm layer of silicon oil (SG = 0.92) is placed on top of the mercury in the arm attached to the apparatus. Atmospheric pressure is 765 mm Hg.

    (a) If the level of mercury in the open end is 365 mm below the mercury level in the other arm, what is the pressure (mm Hg) in the apparatus?

    (b) When the instrumentation specialist was deciding on a liquid to put in the manometer, she listed several properties the fluid should have and eventually selected silicon oil. What might the listed properties have been?

3.47. An orifice meter (see Figure 3.2-1) is to be calibrated for the measurement of the flow rate of a stream of liquid acetone. The differential manometer fluid has a specific gravity of 1.10.

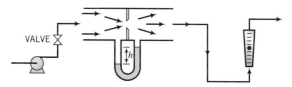

The calibration is accomplished by connecting the orifice meter in series with a rotameter that has previously been calibrated for acetone, adjusting a valve to set the flow rate, and recording the flow rate (determined from the rotameter reading and the rotameter calibration curve) and the differential manometer reading, h. The procedure is repeated for several valve settings to generate an orifice meter calibration curve of flow rate versus h. The following data are taken.

Manometer Reading h(mm)	Flow Rate $\dot{V}$ (mL/s)
0	0
5	62
10	87
15	107
20	123
25	138
30	151

(a) For each of the given readings, calculate the pressure drop across the orifice, ΔP(mm Hg).

(b) The flow rate through an orifice should be related to the pressure drop across the orifice by the formula

$$\dot{V} = K(\Delta P)^n$$

Verify graphically that the given orifice calibration data are correlated by this relationship, and determine the values of K and n that best fit the data.

(c) Suppose the orifice meter is mounted in a process line containing acetone and a reading $h = 23$ mm is obtained. Determine the volumetric, mass, and molar flow rates of acetone in the line.

3.48. Convert the temperatures in parts (a) and (b) and temperature intervals in parts (c) and (d):

(a) $T = 85°F$ to $°R, °C, K$

(b) $T = -10°C$ to $K, °F, °R$

(c) $\Delta T = 85°C$ to $K, °F, °R$

(d) $\Delta T = 150°R$ to $°F, °C, K$

3.49. A temperature scale that never quite caught on was formulated by the Austrian chemist Johann Sebastian Farblunget. The reference points on this scale were $0°FB$, the temperature below which Farblunget's postnasal drip began to bother him, and $1000°FB$, the boiling point of beer. Conversions between $°C$ and $°FB$ can be accomplished with the expression

$$T(°C) = 0.0940T(°FB) + 4.00$$

Louis Louis, Farblunget's French nephew, attempted to follow in his uncle's footsteps by formulating his own temperature scale. He defined the degree Louie using as reference conditions the optimum serving temperature of marinated snails ($100°L$ corresponding to $15°C$) and the temperature at which the elastic in his briefs began to relax ($1000°L$ corresponding to $43°C$).

(a) At what temperature in $°F$ does beer boil?

(b) What is the temperature interval of 10.0 Farblunget degrees equivalent to in $°C, K, °F,$ and $°R$?

(c) Derive equations for $T(°C)$ in terms of $T(°L)$ (see Example 3.5-1) and $T(°L)$ in terms of $T(°FB)$.

(d) What is the boiling point of ethane at 1 atm (Table B.1) in $°F, K, °R, °FB,$ and $°L$?

(e) What is a temperature interval of 50.0 Louie degrees equivalent to in Celsius degrees, Kelvin degrees, Fahrenheit degrees, Rankine degrees, and Farblunget degrees?

3.50. A **thermocouple** is a temperature-measurement device that consists of two dissimilar metal wires joined at one end. An oversimplified diagram follows.

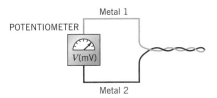

A voltage generated at the metal junction is read on a potentiometer or millivoltmeter. When certain metals are used, the voltage varies linearly with the temperature at the junction of the two metals:

$$V(mV) = aT(°C) + b$$

An iron–constantan thermocouple (constantan is an alloy of copper and nickel) is calibrated by inserting its junction in boiling water and measuring a voltage $V = 5.27$ mV, and then inserting the junction in silver chloride at its melting point and measuring $V = 24.88$ mV.

(a) Derive the linear equation for V (mV) in terms of T (°C). Then convert it to an equation for T in terms of V.

(b) If the thermocouple is mounted in a chemical reactor and the voltage is observed to go from 10.0 mV to 13.6 mV in 20 s, what is the average value of the rate of change of temperature, dT/dt, during the measurement period?

3.51. A thermostat control with dial markings from 0 to 100 is used to regulate the temperature of an oil bath. A calibration plot on logarithmic coordinates of the temperature, $T(°F)$, versus the dial setting, R, is a straight line that passes through the points ($R_1 = 20.0, T_1 = 110.0°F$) and ($R_2 = 40.0, T_2 = 250.0°F$).

(a) Derive an equation for $T\,(°F)$ in terms of R.

(b) Estimate the thermostat setting needed to obtain a temperature of 320°F.

(c) Suppose you set the thermostat to the value of R calculated in part (b) and the reading of a thermocouple mounted in the bath equilibrates at 295°F instead of 320°F. Suggest several possible explanations.

3.52. As will be discussed in detail in Chapter 5, the **ideal gas equation of state** relates absolute pressure, $P\,(\text{atm})$; gas volume, V (liters); number of moles of gas, $n(\text{mol})$; and absolute temperature, $T\,(\text{K})$:

$$PV = 0.08206nT$$

(a) Convert the equation to one relating $P(\text{psig})$, $V(\text{ft}^3)$, $n(\text{lb-mole})$, and $T(°F)$.

(b) A 30.0 mole% CO and 70.0 mole% N_2 gas mixture is stored in a cylinder with a volume of 3.5 ft^3 at a temperature of 85°F. The reading on a Bourdon gauge attached to the cylinder is 500 psi. Calculate the total amount of gas (lb-mole) and the mass of CO (lb_m) in the tank.

(c) Approximately to what temperature (°F) would the cylinder have to be heated to increase the gas pressure to 3000 psig, the rated safety limit of the cylinder? (The estimate would only be approximate because the ideal gas equation of state would not be accurate at pressures this high.)

3.53. Streams of methane and air (79 mole% N_2, the balance O_2) are combined at the inlet of a combustion furnace preheater. The pressures of each stream are measured with open end mercury manometers, the temperatures are measured with resistance thermometers, and the volumetric flow rates are measured with orifice meters.

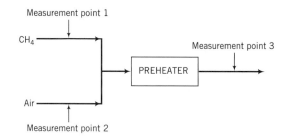

Data:

Flowmeter 1: $V_1 = 947\ \text{m}^3/\text{h}$
Flowmeter 2: $V_2 = 195\ \text{m}^3/\text{min}$
Manometer 1: $h_1 = 232\ \text{mm}$
Manometer 2: $h_2 = 156\ \text{mm}$
Manometer 3: $h_3 = 74\ \text{mm}$
Resistance thermometer 1: $r_1 = 26.159\ \text{ohms}$
Resistance thermometer 2: $r_2 = 26.157\ \text{ohms}$
Resistance thermometer 3: $r_3 = 44.789\ \text{ohms}$
Atmospheric pressure: A sealed-end mercury manometer reads $h = 29.76$ in.

The resistance thermometers were calibrated by measuring their resistances at the freezing and boiling points of water, with the following results:

$$T = 0°C: \quad r = 23.624\ \text{ohms}$$
$$T = 100°C: \quad r = 33.028\ \text{ohms}$$

A straight-line relationship between T and r may be assumed.

The relationship between the total molar flow rate of a gas and its volumetric flow rate is, to a good approximation, given by a form of the ideal gas equation of state:

$$\dot{n}\left(\frac{\text{kmol}}{s}\right) = \frac{12.186P(\text{atm})\dot{V}(\text{m}^3/\text{s})}{T(\text{K})}$$

where P is the *absolute* pressure of the gas.

(a) Derive the resistance thermometer calibration formula for $T(°C)$ in terms of r(ohm).

(b) Convert the given gas law expressions to an expression for $\dot{n}$(kmol/min) in terms of P(mm Hg), $T(°C)$, and $\dot{V}$(m³/min).

(c) Calculate the temperatures and pressures at points 1, 2, and 3.

(d) Calculate the molar flow rate of the combined gas stream.

(e) Calculate the reading of flowmeter 3 in m³/min.

(f) Calculate the total mass flow rate and the mass fraction of the methane at point 3.

*3.54. You are performing an experiment in which the concentration, C_A, of a reactive species is measured as a function of time, t, at several temperatures, T. At a fixed temperature, C_A varies with t according to the relation

$$1/C_A = 1/C_{A0} + kt \tag{1}$$

where C_A(mol/liter) is the concentration of A at time t(min), C_{A0}(mol/liter) is the inital concentration of A, and k[L/(mol·min)] is the reaction **rate constant**. The rate constant in turn depends on temperature, according to the formula

$$k = k_0 \exp[-E/(8.314T)] \tag{2}$$

where k_0 is a contant, T(K) is the reactor temperature, and E (J/mol) is the reaction *activation energy*.

Write a computer program that will carry out the following tasks:

(a) Read in M_A, the molecular weight of A, and N_T, the number of temperatures at which measurements were made.

(b) For the first temperature, read in the value of T in °C, the number of data points, N; and the concentrations and times (t_1, C_{A1}), (t_2, C_{A2}), ... , (t_n, C_{An}), where the times are in minutes and the concentrations are in grams of A/liter.

(c) Convert the temperature to kelvin and the concentrations to mol A/L.

(d) Use the method of least squares (Appendix A.1) in conjunction with Equation 1 to find the value of k that best fits the data. (*Hint:* First cast the equation in the form $y = kx + b$.) Store the values of k and T in arrays.

(e) Print out in a neat format the values of T(K), the converted concentrations (mol/L) and times, and k.

(f) Repeat steps (b) through (d) for the other temperatures.

[*For extra credit:* Use the method of least squares again in conjunction with Equation 2 to determine the value of E that best fits the calculated (T, k) values. Again, start by casting Equation 2 in the form $y = ax + b$.] It will be convenient to perform the least-squares slope calculation in a subroutine, since it must be done repeatedly. Test your program on the following data:

$$M_A = 65.0 \text{ g/mol}$$

t(min)	$T = 94°C$ C_A(g/L)	$T = 110°C$ C_A(g/L)	$T = 127°C$ C_A(g/L)	$T = 142°C$ C_A(g/L)
10	8.1	3.5	1.5	0.72
20	4.3	1.8	0.76	0.36
30	3.0	1.2	0.50	0.24
40	2.2	0.92	0.38	0.18
50	1.8	0.73	0.30	0.15
60	1.5	0.61	0.25	0.12

---

*Computer problem.

Part Two

Material Balances

Chapter 4

Fundamentals of
Material Balances

Certain restrictions imposed by nature must be taken into account when designing a new process or analyzing an existing one. You cannot, for example, specify an input to a reactor of 1000 g of lead and an output of 2000 g of lead or gold or anything else. Similarly, if you know that 1500 lb_m of sulfur is contained in the coal burned each day in a power plant boiler, you do not have to analyze the ash and stack gases to know that on the average 1500 lb_m of sulfur per day leaves the furnace in one form or another.

The basis for both of these observations is the *law of conservation of mass,* which states that mass can neither be created nor destroyed. (We will not be concerned in this book with the almost infinitesimal conversions between mass and energy associated with chemical reactions.) Statements based on the law of conservation of mass such as "total mass of input = total mass of output" or "(lb_m sulfur/day)$_{in}$ = (lb_m sulfur/day)$_{out}$" are examples of **mass balances** or **material balances**. The design of a new process or analysis of an existing one is not complete until it is established that the inputs and outputs of the entire process and of each individual unit satisfy balance equations.

Part Two of this book, which begins with this chapter, outlines procedures for writing material balances on individual process units and multiple-unit processes. In this chapter we present methods for organizing known information about process variables, setting up material balance equations, and solving these equations for unknown variables. In Chapters 5 and 6 we introduce various physical properties and laws that govern the behavior of process materials and indicate how these properties and laws are taken into account (as they must be) in formulating material balances.

4.0 INSTRUCTIONAL OBJECTIVES

After completing this chapter, you should be able to do the following:

- Briefly and clearly explain in your own words the meaning of the following terms: (a) *batch, semibatch, continuous, transient,* and *steady-state* processes; (b) *recycle* (and its purposes); (c) *purge* (and its purpose); (d) *degrees of freedom*; (e) *fractional conversion* of a limiting reactant; (f) *percentage excess* of a reactant; (g) *yield* and *selectivity*; (h) *dry-basis composition* of a mixture containing water; and (i) *theoretical air* and *percent excess air* in a combustion reaction.
- Given a process description, (a) draw and fully label a flowchart; (b) choose a convenient basis of calculation; (c) for a multiple-unit process, identify the subsystems for which balances might be written; (d) perform the degree-of-freedom analysis for the overall system

and each possible subsystem; (e) write in order the equations you would use to calculate specified process variables; and (f) perform the calculations. You should be able to do these computations for single-unit and multiple-unit processes and for processes involving recycle, bypass, or purge streams. If the system involves reactions, you should be able to use molecular species balances, atomic species balances, or extents of reaction for both the degree-of-freedom analysis and the process calculations.

• Given a combustion reactor and information about the fuel composition, calculate the feed rate of air from a given percent excess or vice versa. Given additional information about the conversion of the fuel and the absence or presence of CO in the product gas, calculate the flow rate and composition of the product gas.

4.1 PROCESS CLASSIFICATION

Chemical processes may be classified as **batch**, **continuous**, or **semibatch** and as either **steady-state** or **transient**. Before writing material balances for a process system, you must know into which of these categories the process falls.

1. *Batch process.* The feed is charged (fed) into a vessel at the beginning of the process and the vessel contents are removed sometime later. No mass crosses the system boundaries between the time the feed is charged and the time the product is removed. *Example:* Rapidly add reactants to a tank and remove the products and unconsumed reactants sometime later when the system has come to equilibrium.

2. *Continuous process.* The inputs and outputs flow continuously throughout the duration of the process. *Example:* Pump a mixture of liquids into a distillation column at a constant rate and steadily withdraw product streams from the top and bottom of the column.

3. *Semibatch process.* Any process that is neither batch nor continuous. *Examples:* Allow the contents of a pressurized gas container to escape to the atmosphere; slowly blend several liquids in a tank from which nothing is being withdrawn.

If the values of all the variables in a process (i.e., all temperatures, pressures, volumes, flow rates) do not change with time, except possibly for minor fluctuations about constant mean values, the process is said to be operating at **steady state**. If any of the process variables change with time, **transient** or **unsteady-state** operation is said to exist. By their nature, batch and semibatch processes are unsteady-state operations (why?), whereas continuous processes may be either steady-state or transient.

Batch processing is commonly used when relatively small quantities of a product are to be produced on any single occasion, while continuous processing is better suited to large production rates. Continuous processes are usually run as close to steady state as possible; unsteady-state (transient) conditions exist during the start-up of a process and following changes—intentional or otherwise—in process operation conditions.

TEST YOURSELF

Classify the following processes as batch, continuous, or semibatch, and transient or steady-state.

1. A balloon is filled with air at a steady rate of 2 g/min.
2. A bottle of milk is taken from the refrigerator and left on the kitchen table.
3. Water is boiled in an open flask.
4. Carbon monoxide and steam are fed into a tubular reactor at a steady rate and react to form carbon dioxide and hydrogen. Products and unused reactants are withdrawn at the other end. The reactor contains air when the process is started up. The temperature of the reactor is constant, and the composition and flow rate of the entering reactant stream are also independent of time. Classify the process (a) initially and (b) after a long period of time has elapsed.

4.2   BALANCES

4.2a   The General Balance Equation

Suppose methane is a component of both the input and output streams of a continuous process unit, and that in an effort to determine whether the unit is performing as designed, the mass flow rates of methane in both streams are measured and found to be different ($\dot{m}_{in} \neq \dot{m}_{out}$).[1]

There are several possible explanations for the observed difference between the measured flow rates:

1. Methane is being consumed as a reactant or generated as a product within the unit.
2. Methane is accumulating in the unit—possibly adsorbing on the walls.
3. Methane is leaking from the unit.
4. The measurements are wrong.

If the measurements are correct and there are no leaks, the other possibilities—generation or consumption in a reaction and accumulation within the process unit—are all that can account for a difference between the input and output flow rates.

A **balance** on a conserved quantity (total mass, mass of a particular species, energy, momentum) in a system (a single process unit, a collection of units, or an entire process) may be written in the following general way:

input	+	*generation*	−	*output*	−	*consumption*	=	*accumulation*
(enters		(produced		(leaves		(consumed		(buildup
through		within		through		within		within
system		system)		system		system)		system)
boundaries)				boundaries)				

$$(4.2\text{-}1)$$

The meaning of each term of the equation is illustrated in the following example.

EXAMPLE 4.2-1   *The General Balance Equation*

Each year 50,000 people move into a city, 75,000 people move out, 22,000 are born, and 19,000 die. Write a balance on the population of the city.

SOLUTION   Let P denote people:

$$\text{input} + \text{generation} - \text{output} - \text{consumption} = \text{accumulation}$$

$$50{,}000\,\frac{P}{\text{yr}} + 22{,}000\,\frac{P}{\text{yr}} - 75{,}000\,\frac{P}{\text{yr}} - 19{,}000\,\frac{P}{\text{yr}} = A\left(\frac{P}{\text{yr}}\right)$$

$$\Downarrow$$

$$A = -22{,}000\,\frac{P}{\text{yr}}$$

Each year the city's population decreases by 22,000 people.

---

[1]We will generally use the symbol m to denote a mass, $\dot{m}$ a mass flow rate, n a number of moles, and $\dot{n}$ a molar flow rate.

Two types of balances may be written:

1. *Differential balances,* or balances that indicate what is happening in a system at an instant in time. Each term of the balance equation is a **rate** (rate of input, rate of generation, etc.) and has units of the balanced quantity unit divided by a time unit (people/yr, g SO_2/s, barrels/day). This is the type of balance usually applied to a continuous process. (See Example 4.2-1.)
2. *Integral balances,* or balances that describe what happens between two instants of time. Each term of the equation is an **amount** of the balanced quantity and has the corresponding unit (people, g SO_2, barrels). This type of balance is usually applied to a batch process, with the two instants of time being the moment after the input takes place and the moment before the product is withdrawn.

We are concerned in this text primarily with differential balances applied to continuous steady-state systems and integral balances applied to batch systems between their initial and final states. In Chapter 11, we consider general balances on unsteady-state systems and show how integral and differential balances are related—in fact, how each can be derived from the other.

The following rules may be used to simplify the material balance equation:

- *If the balanced quantity is total mass, set generation = 0 and consumption = 0.* Except in nuclear reactions, mass can neither be created nor destroyed.
- *If the balanced substance is a nonreactive species (neither a reactant nor a product), set generation = 0 and consumption = 0.*
- *If a system is at steady state, set accumulation = 0, regardless of what is being balanced.* By definition, in a steady-state system nothing can change with time, including the amount of the balanced quantity.

4.2b Balances on Continuous Steady-State Processes

For continuous processes at steady-state, the accumulation term in the general balance equation, Equation 4.2-1, equals zero, and the equation simplifies to

$$\boxed{\text{input} + \text{generation} = \text{output} + \text{consumption}} \tag{4.2-2}$$

If the balance is on a nonreactive species or on total mass, the generation and consumption terms equal zero and the equation reduces to *input = output*.

EXAMPLE 4.2-2 *Material Balances on a Continuous Distillation Process*

One thousand kilograms per hour of a mixture of benzene (B) and toluene (T) containing 50% benzene by mass is separated by distillation into two fractions. The mass flow rate of benzene in the top stream is 450 kg B/h and that of toluene in the bottom stream is 475 kg T/h. The operation is at steady state. Write balances on benzene and toluene to calculate the unknown component flow rates in the output streams.

SOLUTION The process can be depicted schematically as follows:

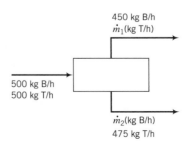

450 kg B/h
$\dot{m}_1$(kg T/h)

500 kg B/h
500 kg T/h

$\dot{m}_2$(kg B/h)
475 kg T/h

Since the process is at steady state there can be no buildup of anything in the system, so the accumulation term equals zero in all material balances. In addition, since no chemical reactions occur, there can be no nonzero generation or consumption terms. For all balances, Equation 4.2-2 therefore takes the simple form *input = output.*

Benzene Balance $\qquad$ 500 kg B/h = 450 kg B/h + $\dot{m}_2$

$$\Downarrow$$

$$\boxed{\dot{m}_2 = 50 \text{ kg B/h}}$$

Toluene Balance $\qquad$ 500 kg T/h = $\dot{m}_1$ + 475 kg T/h

$$\Downarrow$$

$$\boxed{\dot{m}_1 = 25 \text{ kg T/h}}$$

Check the calculation:

Total Mass Balance $\qquad$ 1000 kg/h = 450 + $\dot{m}_1$ + $\dot{m}_2$ + 475   (all kg/h)

$$\Downarrow \dot{m}_1 = 25 \text{ kg/h}, \ \dot{m}_2 = 50 \text{ kg/h}$$

1000 kg/h = 1000 kg/h ✔

4.2c Integral Balances on Batch Processes

Ammonia is produced from nitrogen and hydrogen in a batch reactor. At time $t = 0$ there are n_0 mol of NH_3 in the reactor, and at a later time t_f the reaction terminates and the contents of the reactor, which include n_f mol of ammonia, are withdrawn. Between t_0 and t_f no ammonia enters or leaves through the reactor boundaries, so the general balance equation (Equation 4.2-1) is simply *generation = accumulation.* Moreover, the quantity of ammonia that builds up (accumulates) in the reactor between t_0 and t_f is simply $n_f - n_0$, the final amount minus the initial amount.

The same reasoning may be applied to any substance participating in a batch process to obtain

$$\text{accumulation} = \text{final output} - \text{initial input} \quad \text{(by definition)}$$
$$= \text{generation} - \text{consumption} \quad \text{(from Equation 4.2-1)}$$

Equating these two expressions for the accumulation yields

$$\boxed{\text{initial input} + \text{generation} = \text{final output} + \text{consumption}} \qquad \textbf{(4.2-3)}$$

This equation is identical to Equation 4.2-2 for continuous steady-state processes, except that in this case the input and output terms denote the initial and final amounts of the balanced substance rather than flow rates of the balanced substance in continuous feed and product streams. The words "initial" and "final" may be left out for brevity, as long as you don't lose sight of what "input" and "output" mean in the context of batch processes.

EXAMPLE 4.2-3 **Balances on a Batch Mixing Process**

Two methanol–water mixtures are contained in separate flasks. The first mixture contains 40.0 wt% methanol, and the second contains 70.0 wt% methanol. If 200 g of the first mixture is combined with 150 g of the second, what are the mass and composition of the product?

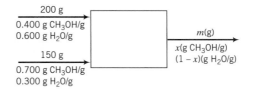

SOLUTION

Observe that the input and output "streams" shown on the chart denote the initial and final states for this batch process. Since no reactions are involved, the generation and consumption terms of Equation 4.2-3 may be omitted so that all balances have the simple form "*input = output*."

Total Mass Balance $200 \text{ g} + 150 \text{ g} = m$

$$\Downarrow$$

$$\boxed{m = 350 \text{ g}}$$

Methanol Balance

$$\frac{200 \text{ g}}{} \left| \frac{0.400 \text{ g CH}_3\text{OH}}{\text{g}} + \frac{150 \text{ g}}{} \right| \frac{0.700 \text{ g CH}_3\text{OH}}{\text{g}} = \frac{m(\text{g})}{} \left| \frac{x(\text{g CH}_3\text{OH})}{(\text{g})} \right.$$

$$\Downarrow m = 350 \text{ g}$$

$$\boxed{x = 0.529 \text{ g CH}_3\text{OH/g}}$$

We now know everything about the product, including the mass fraction of water (what is it?). A water balance serves only to check the solution.

Water Balance   (Verify that each additive term has the unit g H_2O.)

$$\text{input} = \text{output}$$

$$(200)(0.600) + (150)(0.300) = (350)(1 - 0.529) \quad (\textit{Verify!})$$

$$\Downarrow$$

$$165 \text{ g } H_2O = 165 \text{ g } H_2O \; \checkmark$$

4.2d Integral Balances on Semibatch and Continuous Processes

Integral balances can also be written for semibatch and continuous processes. The procedure is to write a differential balance on the system and then to integrate it between two instants of time. (A general discussion of the procedure in presented in Chapter 11.) In most cases the required calculations are more complex than those we have seen so far; however, some problems of this type are relatively straightforward, such as the one in the next example.

EXAMPLE 4.2-4   *Integral Balance on a Semibatch Process*

Air is bubbled through a drum of liquid hexane at a rate of 0.100 kmol/min. The gas stream leaving the drum contains 10.0 mole% hexane vapor. Air may be considered insoluble in liquid hexane. Use an integral balance to estimate the time required to vaporize 10.0 m^3 of the liquid.

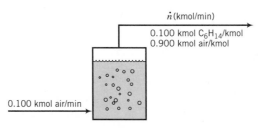

SOLUTION

We begin with a differential balance on air. Since we assume that air neither dissolves in the liquid (*accumulation* = 0) nor reacts with hexane in the process unit (*generation* = *consumption* = 0), the balance reduces to *input* = *output*:

$$0.100 \, \frac{\text{kmol air}}{\text{min}} = \frac{0.900 \text{ kmol air}}{\text{kmol}} \Bigg| \frac{\dot{n}(\text{kmol})}{(\text{min})} \implies \dot{n} = 0.111 \text{ kmol/min}$$

We next write an integral hexane balance, proceeding from time $t = 0$ to $t = t_f(\text{min})$, the time we seek to calculate. The balance has the form *accumulation* = −*output* (*verify*). The accumulation term, which is the total change in the moles of liquid hexane in the system during time t_f, must be negative since hexane is being lost from the system. Since the total number of moles of hexane evaporated occupied a liquid volume of 10.0 cubic meters and (from Table B.1) the specific gravity of liquid hexane is 0.659, the accumulation term equals

$$\Delta n = \frac{-10.0 \text{ m}^3}{} \Bigg| \frac{0.659 \text{ kg}}{\text{L}} \Bigg| \frac{10^3 \text{ L}}{\text{m}^3} \Bigg| \frac{1 \text{ kmol}}{86.2 \text{ kg}} = -76.45 \text{ kmol C}_6\text{H}_{14}$$

The output term in the balance is the rate at which hexane is leaving the system [$0.100\dot{n}$ (kmol C_6H_{14}/min)] times the total process time, $t_f(\text{min})$. The balance (*accumulation* = −*output*) is therefore

$$-76.45 \text{ kmol C}_6\text{H}_{14} = -0.100\dot{n}t_f$$

$$\Bigg\Downarrow \dot{n} = 0.111 \text{ kmol/min}$$

$$\boxed{t_f = 6880 \text{ min}}$$

TEST YOURSELF

Balances are to be written for each of the quantities listed below for a continuous process. For each case, state the conditions under which the balance equation takes the simple form "input = output." (The solutions to the first two parts are given as illustrations.)

1. Total mass. (Steady state)
2. Mass of species A. (Steady state, A is nonreactive)
3. Total moles.
4. Moles of species A.
5. Volume. (The answer provides an indication of why volumes should be converted to masses or moles before balances are written.)

CREATIVITY EXERCISE

Sulfur dioxide is contained in the feed and effluent streams of a chemical reactor, but it is neither a reactant nor a product. The volumetric flow rates of both streams (L/min) are measured with rotameters, and the concentrations of SO_2 in both streams (mol/L) are determined with a gas chromatograph. The molar flow rate of SO_2 in the reactor effluent (determined as the product of volumetric flow rate and concentration) is 20% lower than the molar flow rate of SO_2 in the feed. Think of as many possible explanations for the discrepancy as you can.

4.3 MATERIAL BALANCE CALCULATIONS

All material balance problems are variations on a single theme: given values of some input and output stream variables, derive and solve equations for others. Solving the equations is usually a matter of simple algebra, but deriving them from a description of a process and a collection of process data may present considerable difficulties. It may not be obvious from the problem statement just what is known and what is required, for example, and it is not uncommon to find students (particularly on tests) scratching their heads and staring vacantly for an hour at a problem that should take ten minutes to solve completely.

In this section we outline a procedure for reducing a description of a process to a set of equations that can be solved for unknown process variables. The approach to be outlined is not the only way to attack material balance problems, but it always works and it keeps the time spent on vacant staring and head scratching to a minimum.

4.3a   Flowcharts

In this book and in years to come, you will be confronted with prose of the following type.

> *The catalytic dehydrogenation of propane is carried out in a continuous packed-bed reactor. One thousand kilograms per hour of pure propane is preheated to a temperature of 670°C before it passes into the reactor. The reactor effluent gas, which includes propane, propylene, methane, and hydrogen, is cooled from 800°C to 110°C and fed to an absorption tower, where the propane and propylene are dissolved in oil. The oil then goes to a stripping tower in which it is heated, releasing the dissolved gases; these gases are recompressed and sent to a distillation column in which the propane and propylene are separated. The propane stream is recycled back to join the feed to the reactor preheater. The product stream from the distillation column contains 98% propylene, and the recycle stream is 97% propane. The stripped oil is recycled to the absorption tower.*

When you are given process information like this and asked to determine something about the process, it is essential to organize the information in a way that is convenient for subsequent calculations. The best way to do this is to draw a **flowchart** of the process, using boxes or other symbols to represent process units (reactors, mixers, separation units, etc.) and lines with arrows to represent inputs and outputs.[2]

For example, suppose a gas containing N_2 and O_2 is combined with propane in a batch combustion chamber in which some (but not all) of the O_2 and C_3H_8 react to form CO_2 and H_2O, and the product is then cooled, condensing the water. The flowchart of this two-unit process might appear as shown in Figure 4.3-1.

Used properly, the flowchart of a process can help get material balance calculations started and keep them moving. To do so, the chart must be fully *labeled* when it is first drawn, with values of known process variables and symbols for unknown variables being written for each input and output stream. Thereafter, the chart functions as a scoreboard for the problem solution: as each unknown variable is determined its value is filled in, so that the chart provides a continuous record of where the solution stands and what must still be done.

Several suggestions follow for labeling a flowchart to get the greatest possible benefit from it in material balance calculations.

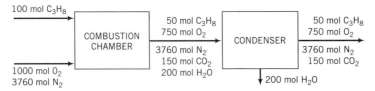

Figure 4.3-1   Flowchart of a combustion-condensation process.

---

[2]In professionally drawn flowcharts, special symbols are used to denote different types of process units such as distillation columns and heat exchangers. We will not generally use these symbols in this text, since our main purpose is to show you how to perform material and energy balance calculations. Simple boxes are perfectly adequate to represent process units on the flowcharts you draw for these calculations.

1. *Write the values and units of all known stream variables at the locations of the streams on the chart.* For example, a stream containing 21 mole% O_2 and 79% N_2 at 320°C and 1.4 atm flowing at a rate of 400 mol/h might be labeled

400 mol/h
$\longrightarrow$

0.21 mol O_2/mol
0.79 mol N_2/mol
T = 320°C, P = 1.4 atm

When you have done this for every stream on the chart, you have a summary of the known information about the process, each item being conveniently associated with the part of the process to which it relates.

The stream variables of primary interest in material balance problems are those that indicate how much of each component is present in the stream (for a batch process) or the flow rate of each component (for a continuous process). This information can be given in two ways: as the total amount or flow rate of the stream and the fractions of each component, or directly as the amount or flow rate of each component.

$\longrightarrow \iff$ 100 kmol/min $\longrightarrow$

60 kmol N_2/min   0.6 kmol N_2/kmol
40 kmol O_2/min   0.4 kmol O_2/kmol

$\longrightarrow \iff$ 10 lb$_m$ mixture $\longrightarrow$

3.0 lb$_m$ CH$_4$ 0.3 lb$_m$ CH$_4$/lb$_m$
4.0 lb$_m$ C$_2$H$_4$ 0.4 lb$_m$ C$_2$H$_4$/lb$_m$
3.0 lb$_m$ C$_2$H$_6$ 0.3 lb$_m$ C$_2$H$_6$/lb$_m$

Once you have labeled a stream one way, it is easy to calculate the quantities that correspond to the alternative way of labeling. (Verify this for the two examples just given.)

2. *Assign algebraic symbols to unknown stream variables* [such as $\dot{m}$ (kg solution/min), x (lb$_m$ N_2/lb$_m$), and n (kmol C_3H_8)] *and write these variable names and their associated units on the chart.* For example, if you did not know the flow rate of the stream described in the first illustration of step 1, you might label the stream

$\dot{n}$(mol/h)
$\longrightarrow$

0.21 mol O_2/mol
0.79 mol N_2/mol
T = 320°C, P = 1.4 atm

while if the flow rate were known and the mole fractions were not, the stream might be labeled

400 mol/h
$\longrightarrow$

y(mol O_2/mol)
$(1 - y)$(mol N_2/mol)
T = 320°C, P = 1.4 atm

You may ultimately have to derive and solve an equation for each unknown that appears on the chart, and it is therefore to your advantage to keep the number of labeled unknowns to a minimum. When labeling component mass or mole fractions of a stream, for example, variable names need only be assigned to all but one fraction, since the last one must be 1 minus the sum of the others. If you are given that the mass of stream 1 is half that of stream 2, label the masses of these streams m and $2m$ rather than m_1 and m_2; if you know that there is three times as much nitrogen (by mass) in a stream as oxygen, label the mass fractions of O_2 and N_2 y(g O_2/g) and $3y$(g N_2/g) rather than y_1 and y_2.

If a volumetric flow rate of a stream is given, it is generally useful to label the mass or molar flow rate of this stream or to calculate it directly, since balances are not normally written on volumetric qualities.

Note on Notation: Although any symbol may be used to represent any variable, having a consistent notation can aid understanding. In this text, we will generally use m for mass, $\dot{m}$ for mass flow rate, n for moles, $\dot{n}$ for molar flow rate, V for volume, and $\dot{V}$ for volumetric flow rate. Also, we will use x for component fractions (mass or mole) in liquid streams and y for fractions in gas streams.

EXAMPLE 4.3-1 *Flowchart of an Air Humidification and Oxygenation Process*

An experiment on the growth rate of certain organisms requires an environment of humid air enriched in oxygen. Three input streams are fed into an evaporation chamber to produce an output stream with the desired composition.

A: Liquid water, fed at a rate of 20.0 cm³/min
B: Air (21 mole% O_2, the balance N_2)
C: Pure oxygen, with a molar flow rate one-fifth of the molar flow rate of stream B

The output gas is analyzed and is found to contain 1.5 mole% water. Draw and label a flowchart of the process, and calculate all unknown stream variables.

SOLUTION

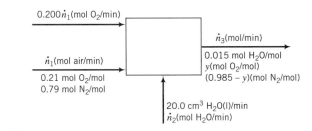

0.200 $\dot{n}_1$ (mol O_2/min)

$\dot{n}_1$ (mol air/min)
0.21 mol O_2/mol
0.79 mol N_2/mol

$\dot{n}_3$ (mol/min)
0.015 mol H_2O/mol
y (mol O_2/mol)
(0.985 − y)(mol N_2/mol)

20.0 cm³ H_2O(l)/min
$\dot{n}_2$ (mol H_2O/min)

Notes on the Labeling:

1. Since the one known flow rate (20 cm³ H_2O/min) is given on a per minute basis, it is most convenient to label all stream flow rates on this basis.
2. Once the variable name ($\dot{n}_1$) is chosen for the air flow rate, the given information about the ratio of the air and O_2 flow rates may be used to label the O_2 flow rate 0.200 $\dot{n}_1$.
3. The mole fractions of the components of any stream must add up to 1. Since the mole fraction of H_2O in the outlet stream is known to be 0.015, once the mole fraction of O_2 is labeled y, that of N_2 must be $1 - (y + 0.015) = (0.985 - y)$ (mol N_2/mol).

The quantity $\dot{n}_2$ may be calculated from the given volumetric flow rate and the density of liquid water:

$$\dot{n}_2 = \frac{20.0 \text{ cm}^3 \text{ H}_2\text{O}}{\text{min}} \left| \frac{1.00 \text{ g H}_2\text{O}}{\text{cm}^3} \right| \frac{1 \text{ mol}}{18.02 \text{ g}} \implies \boxed{\dot{n}_2 = 1.11 \frac{\text{mol H}_2\text{O}}{\text{min}}}$$

The three remaining unknowns ($\dot{n}_1$, $\dot{n}_3$, and y) may be determined from balances, all of which have the simple form *input = output* for this nonreactive steady-state process. The balances are easily written by referring to the flowchart.

H_2O Balance
$$\dot{n}_2 \left(\frac{\text{mol H}_2\text{O}}{\text{min}} \right) = \dot{n}_3 \frac{(\text{mol})}{(\text{min})} \left| \frac{0.015 \text{ mol H}_2\text{O}}{\text{mol}} \right.$$

$$\Downarrow \dot{n}_2 = 1.11 \text{ mol/min}$$

$$\boxed{\dot{n}_3 = 74.1 \frac{\text{mol}}{\text{min}}}$$

Total Mole Balance $0.200\dot{n}_1 + \dot{n}_1 + \dot{n}_2 = \dot{n}_3$

$$\Bigg\Downarrow \begin{array}{l} \dot{n}_2 = 1.11 \text{ mol/min} \\ \dot{n}_3 = 74.1 \text{ mol/min} \end{array}$$

$$\boxed{\dot{n}_1 = 60.8 \; \frac{\text{mol}}{\text{min}}}$$

N_2 Balance

$$\frac{\dot{n}_1(\text{mol})}{(\text{min})} \left| \frac{0.79 \text{ mol N}_2}{\text{mol}} \right. = \frac{\dot{n}_3(\text{mol})}{(\text{min})} \left| \frac{(0.985 - y)(\text{mol N}_2)}{(\text{mol})} \right.$$

$$\Downarrow$$

$$0.79\dot{n}_1 = \dot{n}_3(0.985 - y)$$

$$\Bigg\Downarrow \begin{array}{l} \dot{n}_1 = 60.8 \text{ mol/min} \\ \dot{n}_3 = 74.1 \text{ mol/min} \end{array}$$

$$\boxed{y = 0.337 \text{ mol O}_2/\text{mol}}$$

TEST YOURSELF

Several labeled process streams are shown below. Calculate the indicated quantities in terms of the labeled stream variables. The solution of the first problem is given as an illustration.

1. 100 lb-mole[3]

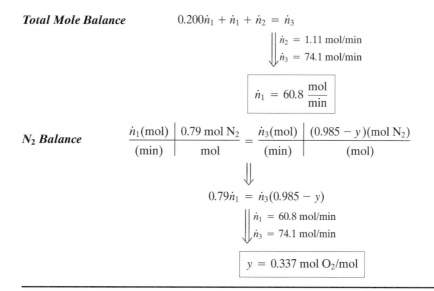

0.300 lb-mole CH$_4$/lb-mole
0.400 lb-mole C$_2$H$_4$/lb-mole
0.300 lb-mole C$_2$H$_6$/lb-mole

Calculate n (lb-mole CH$_4$)
    m (lb$_m$ C$_2$H$_4$)

SOLUTION

$$n = (0.300)(100) \text{ lb-mole CH}_4 = 30.0 \text{ lb-mole CH}_4$$

$$m = \frac{(0.400)(100) \text{ lb-mole C}_2\text{H}_4 \mid 28.0 \text{ lb}_m \text{ C}_2\text{H}_4}{\mid \text{lb-mole C}_2\text{H}_4} = 1120 \text{ lb}_m \text{ C}_2\text{H}_4$$

2. 250 kg/h

x(kg C$_6$H$_6$/kg)
$(1 - x)$(kg C$_7$H$_8$/kg)

Calculate $\dot{m}_T$ (kg C$_7$H$_8$/min)
in terms of x

3. 75 ml CCl$_4$ (liquid)

Calculate n (mol CCl$_4$)

4.
50 kg H$_2$O/s
$\begin{cases} \dot{m}_{dg} \text{ kg dry gas/s} \\ 0.25 \text{ kg CO/kg dry gas} \\ 0.75 \text{ kg CO}_2\text{/kg dry gas} \end{cases}$

Calculate $\dot{m}$ (kg total/s), $\dot{m}_{CO}$ (kg CO/s),
and y (kg CO$_2$/kg total) in terms of $\dot{m}_{dg}$

4.3b   Flowchart Scaling and Basis of Calculation

Suppose a kilogram of benzene is mixed with a kilogram of toluene. The output from this simple process is obviously 2 kg of a mixture that is 50% benzene by mass.

---

[3]Whenever we give a round quantity like 100 lb-moles, assume that it is a basis of calculation and exact, so that it has an infinite number of significant figures.

The process depicted by this flowchart is said to be **balanced**, since material balances on both system components—C_6H_6 and C_7H_8—are satisfied. [1 kg in $= (2 \times 0.5)$ kg out in both cases.]

Observe now that the masses (*but not the mass fractions*) of all streams could be multiplied by a common factor and the process would remain balanced; moreover, the stream masses could be changed to mass flow rates, and the mass units of all stream variables (including the mass fractions) could be changed from kg to g or lb_m or any other mass unit, and the process would still be balanced.

The procedure of changing the values of all stream amounts or flow rates by a proportional amount while leaving the stream compositions unchanged is referred to as **scaling** the flowchart—**scaling up** if the final stream quantities are larger than the original quantities, **scaling down** if they are smaller.

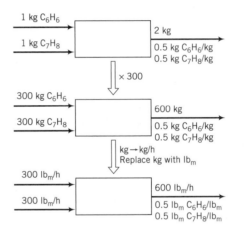

Suppose you have balanced a process and the amount or flow rate of one of the process streams is n_1. You can scale the flowchart to make the amount or flow rate of this stream n_2 by multiplying all stream amounts or flow rates by the ratio n_2/n_1. You cannot, however, scale masses or mass flow rates to molar quantities or vice versa by simple multiplication; conversions of this type must be carried out using the methods of Section 3.3b.

EXAMPLE 4.3-2   *Scale-up of a Separation Process Flowchart*

A 60–40 mixture (by moles) of A and B is separated into two fractions. A flowchart of the process is shown here.

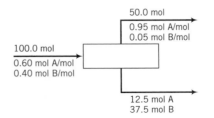

It is desired to achieve the same separation with a continuous feed of 1250 lb-moles/h. Scale the flowchart accordingly.

SOLUTION

The scale factor is

$$\frac{1250 \text{ lb-moles/h}}{100 \text{ mol}} = 12.5 \frac{\text{lb-moles/h}}{\text{mol}}$$

The masses of all streams in the batch process are converted to flow rates as follows:

Feed: $\dfrac{100 \text{ mol} \mid 12.5 \text{ lb-moles/h}}{\text{mol}} = 1250 \dfrac{\text{lb-moles}}{\text{h}}$ (as specified)

Top product stream: $(50.0)(12.5) = 625 \text{ lb-moles/h}$

Bottom product stream: $(12.5)(12.5) = 156 \text{ lb-moles A/h}$

$(37.5)(12.5) = 469 \text{ lb-moles B/h}$

The units of the mole fractions in the top product stream may be changed from mol/mol to lb-mole/lb-mole, but their values remain the same. The flowchart for the scaled-up process follows.

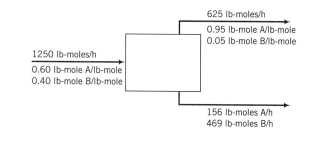

Since a balanced process can always be scaled, material balance calculations can be performed on the basis of any convenient set of stream amounts or flow rates and the results can afterward be scaled to any desired extent. A **basis of calculation** is an amount (mass or moles) or flow rate (mass or molar) of one stream or stream component in a process. The first step in balancing a process is to choose a basis of calculation; all unknown variables are then determined to be consistent with this basis.

If a stream amount or flow rate is given in a problem statement, it is usually most convenient to use this quantity as a basis of calculation. If no stream amounts or flow rates are known, assume one, preferably that of a stream with a known composition. If mass fractions are known, choose a total mass or mass flow rate of that stream (e.g., 100 kg or 100 kg/h) as a basis; if mole fractions are known, choose a total number of moles or a molar flow rate.

TEST YOURSELF

1. What is a balanced process? How do you scale a flowchart? What is a basis of calculation?
2. The processes shown below were balanced using the indicate bases of calculation. Scale as directed, and draw flowcharts for the scaled processes.
   (a) Mix C_2H_6 with air. *Basis of calculation: 100 mol C_2H_6.*

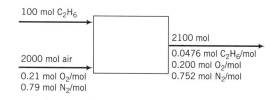

Scale up to a feed of 1000 kmol C_2H_6/h.

(b) Distill a benzene–toluene mixture. *Basis: 1 g feed.*

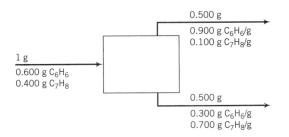

Scale up to a flow rate of 100 lb$_m$/min of the upper product stream.

4.3c Balancing a Process

Suppose 3.0 kg/min of benzene and 1.0 kg/min of toluene are mixed. The process flowchart might be drawn and labeled as follows:

There are two unknown quantities—$\dot{m}$ and x—associated with the process, so two equations are needed to calculate them.

Material balance equations for this nonreactive process all have the simple form input = output. Three possible balances can be written—on total mass, benzene, and toluene—any two of which provide the equations needed to determine $\dot{m}$ and x. For example,

Total Mass Balance: $\qquad 3.0 \dfrac{kg}{min} + 1.0 \dfrac{kg}{min} = \dot{m} \implies \boxed{\dot{m} = 4.0 \text{ kg/min}}$

Benzene Balance: $\qquad 3.0 \dfrac{kg \ C_6H_6}{min} = \dfrac{\dot{m}(kg)}{(min)} \Bigg| \dfrac{x\,(kg \ C_6H_6)}{(kg)}$

$$\Downarrow \dot{m} = 4.0 \text{ kg/min}$$

$$\boxed{x = 0.75 \text{ kg } C_6H_6/kg}$$

A logical question to ask at this point is how far you can go with this procedure: If one of the input flow rates had also been unknown, for example, could another balance (on toluene) have been written to solve for it? Other points to consider are which balances to use when a choice exists and the order in which these balances should be written.

The answers to these questions are not at all obvious when chemical reactions are involved in the process, and we will temporarily postpone consideration of this subject. The following rules apply to nonreactive processes.

1. ***The maximum number of independent equations that can be derived by writing balances on a nonreactive system equals the number of chemical species in the input and output streams.***

In the given example, two substances—benzene and toluene—make up the input and output streams of the process; you can write mass or mole balances on benzene and toluene and a total mass or mole balance, but only two of these three equations are independent—writing the third accomplishes nothing. (If you wrote all three equations in an effort to determine three unknown variables, you would probably go through an elaborate algebraic exercise to prove that $1 = 1$ or something equally uninformative.)

2. *Write balances first that involve the fewest unknown variables.*

In the example, a total mass balance involves only one unknown, $\dot{m}$, while benzene and toluene balances each involve both $\dot{m}$ and x. By writing first a total balance and then a benzene balance, we were able to solve first one equation in one unknown, then a second equation, also in one unknown. If we had instead written benzene and toluene balances, we would have had to solve two simultaneous equations in two unknowns; the same answers would have been obtained, but with greater effort.

EXAMPLE 4.3-3   *Balances on a Mixing Unit*

An aqueous solution of sodium hydroxide contains 20.0% NaOH by mass. It is desired to produce an 8.0% NaOH solution by diluting a stream of the 20% solution with a stream of pure water. Calculate the ratios (liters H_2O/kg feed solution) and (kg product solution/kg feed solution).

SOLUTION

- **Choose a basis of calculation**—an amount or flow rate of one of the feed or product streams—**and then draw and label the flowchart.**

  We will arbitrarily choose a basis of 100 kg of the 20% feed solution. (We could also have chosen a flow rate of 100 lb_m/min of the 8% product solution or 10 tons of diluent water. The final results do not depend on the basis chosen since we are only asked to find ratios of stream amounts.) The flowchart appears as follows:

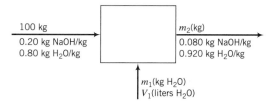

- **Express what the problem asks you to determine in terms of the labeled variables on the flowchart.**

  The desired quantities are $V_1/100$ (liters H_2O/kg feed solution) and $m_2/100$ (kg product solution/kg feed solution). Our task is therefore to calculate the variables V_1 and m_2.

- **Count unknown variables and equations relating them.**

  If the number of unknowns equals the number of independent equations relating them, you will be able to solve the problem; otherwise, either you have forgotten some relations or the problem is not well defined. In the latter case, there is no point wasting your time by jumping into time-consuming calculations.

  (a) Unknowns. Examining the flowchart, we see three unknown variables— m_1, m_2, and V_1.

  (b) Equations. *For a nonreactive process that involves N species, up to N independent material balance equations may be written.* Since there are two species in our process (sodium hydroxide and water), we can write two balances. We could write them on sodium hydroxide, water, total mass, atomic sodium, atomic hydrogen, and so on; the point is that once we have written any two, we can obtain no new information by writing a third one.

  Since we may only write two material balances, we will need a third equation to solve for our three unknowns (m_1, m_2, and V). Fortunately, we have one: the mass and volume of the diluent water, m_1 and V_1, are related by the density of liquid water, which we know. We thus have three equations in three unknowns and therefore a solvable problem.

- **Outline the solution procedure.**

  All balances for this system have the form *input = output*. For example, a total mass balance is 100 kg $+ m_1 = m_2$. Looking at the flowchart, we can see that balances on total mass and water each involve two unknowns (m_1 and m_2), a sodium hydroxide balance only involves one unknown (m_2), and the water density relationship involves two unknowns (m_1 and V_1). We would therefore begin the solution by writing and solving the NaOH balance for m_2, then writing and solving a total mass or water balance for m_1, and finally determining V_1 from m_1 and the density.

- **NaOH balance** (input = output).

$$(0.20 \text{ kg NaOH/kg})(100 \text{ kg}) = (0.080 \text{ kg NaOH/kg})m_2 \implies m_2 = 250 \text{ kg NaOH}$$

It is a good practice to write calculated variable values on the flowchart as soon as they are known for ease of use in later calculations. At this point, 250 would therefore be written in place of m_2 on the chart.

- **Total mass balance** (input = output).

$$100 \text{ kg} + m_1 = m_2 \xrightarrow{\quad m_2 = 250 \text{ kg} \quad} m_1 = 150 \text{ kg H}_2\text{O}$$

- **Diluent water volume.** Although we are not given the temperature or pressure at which the mixing is done, the density of liquid water is approximately constant at 1.00 kg/liter (see Equation 3.1-2). We may therefore calculate

$$V_1 = \frac{150 \text{ kg}}{} \frac{1.00 \text{ liter}}{\text{kg}} = 150 \text{ liters}$$

- **Ratios requested in problem statement.**

$$\frac{V_1}{100 \text{ kg}} = \boxed{1.50 \text{ liters H}_2\text{O/kg feed solution}}$$

$$\frac{m_2}{100 \text{ kg}} = \boxed{2.50 \text{ kg product solution/kg feed solution}}$$

Exercise: Prove to yourself that you would get the same results for a different basis of calculation.

TEST YOURSELF

1. Prove that the following flowchart depicts a balanced process by writing three balances.

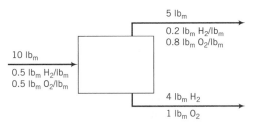

2. Indicate the balances you would write and the order in which you would write them to solve for the unknown stream variables in the following process:

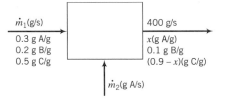

4.3d   Degree-of-Freedom Analysis

Everyone who has done material balance calculations has had the frustrating experience of spending a long time deriving and attempting to solve equations for unknown process variables, only to discover that not enough information is available. Before you do any lengthy calculations, you can use a properly drawn and labeled flowchart to determine whether you have enough information to solve a given problem. The procedure for doing so is referred to as **degree-of-freedom analysis**.

To perform a degree-of-freedom analysis, draw and *completely* label a flowchart, count the unknown variables on the chart, then count the *independent* equations relating them,[4] and subtract the second number from the first. The result is the number of *degrees of freedom* of the process, n_{df} ($= n_{unknowns} - n_{indep\ eqns}$). There are three possibilities:

1. If $n_{df} = 0$, there are n independent equations in n unknowns and the problem can in principle be solved.
2. If $n_{df} > 0$, there are more unknowns than independent equations relating them, and at least n_{df} additional variable values must be specified before the remaining variable values can be determined. Either relations have been overlooked or the problem is *underspecified* and has infinitely many solutions; in either case, plunging into calculations is likely to be a waste of time.[5]
3. If $n_{df} < 0$, there are more independent equations than unknowns. Either the flowchart is incompletely labeled or the problem is *overspecified* with redundant and possibly inconsistent relations. Again there is little point wasting time trying to solve it until the equations and unknowns are brought into balance.

Sources of equations relating unknown process stream variables include the following:

1. ***Material balances.*** For a nonreactive process, no more than n_{ms} independent material balances may be written, where n_{ms} is the number of molecular species (e.g., CH_4, O_2) involved in the process. For example, if benzene and toluene are the species in the streams entering and leaving a distillation column, you could write balances on benzene, toluene, total mass, atomic carbon, atomic hydrogen, and so on, but at most two of those balances would be independent. If additional balances are written, they will not be independent of the first ones and so will provide no new information.

   For a reactive process, the procedure becomes more complicated. We will defer further discussion of this point to Section 4.7.
2. ***An energy balance*** (Chapters 7 through 9). If the amount of energy exchanged between the system and its surroundings is specified or if it is one of the unknown process variables, an energy balance provides a relationship between inlet and outlet material flows and temperatures.
3. ***Process specifications.*** The problem statement may specify how several process variables are related. For example, you may be told that of the acetone fed to a condenser [flow rate = $\dot{m}_1$ (kg acetone/s)], 40% appears in a condensate stream [flow rate = $\dot{m}_2$ (kg acetone/s)]. A system equation would then be $\dot{m}_2 = 0.40\,\dot{m}_1$.
4. ***Physical properties and laws.*** Two of the unknown variables may be the mass and volume of a stream material, in which case a tabulated specific gravity for liquids and solids or an equation of state for gases (Chapter 5) would provide an equation relating the variables. In other instances, saturation or equilibrium conditions for one or more of the process streams (Chapter 6) may provide needed relations.
5. ***Physical constraints.*** For example, if the mole fractions of the three components of a stream are labeled x_A, x_B, and x_C, then a relation among these variables is $x_A + x_B + x_C = 1$. (If instead of x_C you label the last fraction $1 - x_A - x_B$, then you will have one less variable and one less equation to worry about.)

---

[4]Equations are independent if you cannot derive one by adding and subtracting combinations of the others. For example, only two of the three equations $x = 3$, $y = 2$, and $x + y = 5$ are independent; any one of them can be obtained from the other two by addition or subtraction.

[5]When a proposed process has a positive number of degrees of freedom, n_{df}, it is possible to perform a *process optimization*. The engineer chooses n_{df} *design variables* and assigns values to them, calculates values of the remaining system variables from the system equations, and calculates an *objective function* from the full set of system variable values. The objective function may be a cost, a profit, or a rate of return on investment. The goal is to find the set of design variable values that yields the maximum or minimum value of the objective function.

6. **Stoichiometric relations.** If chemical reactions occur in a system, the stoichiometric equations of the reactions (e.g., $2H_2 + O_2 \longrightarrow 2H_2O$) provide relationships between the quantities of the reactants consumed and of the products generated. We will consider how to incorporate these relationships into a degree-of-freedom analysis in Section 4.7.

EXAMPLE 4.3-4   *Degree-of-Freedom Analysis*

A stream of humid air enters a condenser in which 95% of the water vapor in the air is condensed. The flow rate of the condensate (the liquid leaving the condenser) is measured and found to be 225 L/h. Dry air may be taken to contain 21 mole% oxygen, with the balance nitrogen. Calculate the flow rate of the gas stream leaving the condenser and the mole fractions of oxygen, nitrogen, and water in this stream.

SOLUTION   **Basis: 225 L/h Condensate**

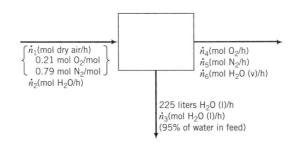

We first do the degree-of-freedom analysis. There are six unknowns on the chart—$\dot{n}_1$ through $\dot{n}_6$. We are allowed up to three material balances—one for each species. We must therefore find three additional relations to solve for all unknowns. One is the relationship between the volumetric and molar flow rates of the condensate: we can determine $\dot{n}_3$ from the given volumetric flow rate and the known specific gravity and molecular weight of liquid water. A second is the given fact that 95% of the water is condensed. This specification provides a relationship between $\dot{n}_3$ and $\dot{n}_2$ ($\dot{n}_3 = 0.95\dot{n}_2$).

However, no information in the problem statement provides a basis for establishing a sixth relation, so that there is one degree of freedom. The problem is therefore underspecified, and there is no point in attempting to solve it. Without the flowchart, it would have been difficult to see this and a great deal of time could have been wasted in a futile effort.

Suppose now that we had been given an additional piece of information—for example, that the entering air contains 10.0 mole% water. The flowchart would then appear as follows:

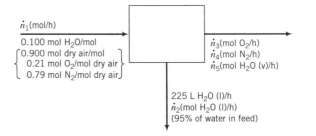

The degree-of-freedom analysis tells us that there are five unknowns and that we have five equations to solve for them [three mole balances, the density relationship between $\dot{V}_2$ ($= 225$ L/h) and $\dot{n}_2$, and the fractional condensation], hence zero degrees of freedom. The problem is therefore solvable in principle. We may now lay out the solution—still before proceeding to any algebraic or numerical calculations—by writing out the equations in an efficient solution order (equations involving only one unknown first, then simultaneous pairs of equations, etc.) and circling the variables for which we would solve each equation or set of simultaneous equations. In this problem, a solution procedure can be found that does not involve simultaneous equations. (Verify that the units are correct in each equation.)

- *Density relationship.*

$$\widehat{\dot{n}_2}\left(\frac{\text{mol } H_2O(l)}{h}\right) = \frac{225 \text{ L } H_2O(l)}{h} \; \frac{1.00 \text{ kg } H_2O(l)}{L} \; \frac{1 \text{ mol } H_2O}{18.0 \times 10^{-3} \text{ kg}}$$

- *95% Condensation.* $\dot{n}_2 = 0.95\,(0.100\,\widehat{\dot{n}_1})$
- *O_2 Balance.* $\dot{n}_1(0.900)(0.21) = \widehat{\dot{n}_3}$
- *N_2 Balance.* $\dot{n}_1(0.900)(0.79) = \widehat{\dot{n}_4}$
- *H_2O Balance.* $\dot{n}_1(0.100) = \dot{n}_2 + \widehat{\dot{n}_5}$
- *Total outlet gas flow rate.* $\widehat{\dot{n}_{\text{total}}} = \dot{n}_3 + \dot{n}_4 + \dot{n}_5$
- *Outlet gas composition.* $\widehat{y_{O_2}} = \dot{n}_3/\dot{n}_{\text{total}}, \; \widehat{y_{N_2}} = \dot{n}_4/\dot{n}_{\text{total}}, \; \widehat{y_{H_2O}} = \dot{n}_5/\dot{n}_{\text{total}}$

The algebra and arithmetic are left as an exercise.

4.3e General Procedure for Single-Unit Process Material Balance Calculations

The problem-solving approach introduced in the previous sections and several additional procedural suggestions are summarized below. Given a description of a process, the values of several process variables, and a list of quantities to be determined:

1. *Choose as a basis of calculation an amount or flow rate of one of the process streams.*
   - If an amount or flow rate of a stream is given in the problem statement, it is usually convenient to use it as the basis of calculation. Subsequently calculated quantities will then be correctly scaled.
   - If several stream amounts or flow rates are given, always use them collectively as the basis.
   - If no stream amount or flow rate is specified in the problem statement, take as a basis an arbitrary amount or flow rate of a stream with a known composition (e.g., 100 kg or 100 kg/h if all mass fractions are known, or 100 mol or 100 mol/h if all mole fractions are known).

2. *Draw a flowchart and fill in all known variable values, including the basis of calculation. Then label unknown stream variables on the chart.*
   - *The flowchart is completely labeled if you can express the mass or mass flow rate (or moles or molar flow rate) of each component of each stream in terms of labeled quantities.* Labeled variables for each process stream should therefore include *either*
     - **(a)** the total mass [e.g., $m_1(\text{kg})$] or mass flow rate [$\dot{m}_1(\text{kg/s})$] and the mass fractions of all stream components [e.g., y_{CH_4} (kg CH_4/kg)], *or*
     - **(b)** the total moles [e.g., $n_1(\text{kmol})$] or molar flow rate [$\dot{n}_1(\text{kmol/s})$] and the mole fractions of all stream components [e.g., $y_{CH_4}(\text{kmol } CH_4/\text{kmol})$], *or*
     - **(c)** for each stream component, the mass [e.g., $m_{H_2}(\text{kg } H_2)$], mass flow rate [$\dot{m}_{H_2}$ (kg SO_2/s)], moles [$n_{CO}(\text{kmol CO})$], or molar flow rate [$\dot{n}_{CO}(\text{kmol CO/s})$].
   - If you are given (or you can easily determine) either the amount or flow rate or any of the component fractions for a stream, label the total stream quantity or flow rate and the component fractions (categories (a) and (b) in the preceding list). If you only know the species present but have no quantitative information, label component quantities or flow rates (category (c) in the preceding list). Any labeling system will work for any stream, but the algebra tends to be simpler if you follow these rules of thumb.
   - Try to incorporate given relationships between unknown quantities in the labeling. For example, if you know that the molar flow rate of Stream 2 is double that of Stream 1, label the flow rates $\dot{n}_1$ and $2\dot{n}_1$ rather than $\dot{n}_1$ and $\dot{n}_2$.
   - *Label volumetric quantities only if they are given in the problem statement or you are asked to calculate them.* You will write mass or mole balances, but not volume balances.

3. *Express what the problem statement asks you to determine in terms of the labeled variables.* You will then know which unknowns you have to determine in order to solve the problem.

4. *If you are given mixed mass and mole units for a stream* (such as a total mass flow rate and component mole fractions or vice versa), *convert all quantities to one basis or the other using the methods of Section 3.3.*

5. *Do the degree-of-freedom analysis.* Count unknowns and identify equations that relate them. The equations may be any of the six types listed in Section 4.3d: material balances, an energy balance, process specifications, physical property relationships and laws, physical constraints, and stoichiometric relations. If you count more unknown variables than equations or vice versa, figure out what's wrong (e.g., the flowchart is not completely labeled, or an additional relation exists that was not counted, or one or more of your equations are not independent of the others, or the problem is underspecified or overspecified). If the number of unknowns does not equal the number of equations, there is no point wasting time trying to solve the problem.

6. *If the number of unknowns equals the number of equations relating them* (i.e., if the system has zero degrees of freedom), *write the equations in an efficient order (minimizing simultaneous equations) and circle the variables for which you will solve* (as in Example 4.3-4). Start with equations that only involve one unknown variable, then pairs of simultaneous equations containing two unknown variables, and so on. *Do no algebra or arithmetic in this step.*

7. *Solve the equations,* either manually or using equation-solving software. Manual solution should go smoothly since you have already worked out an efficient solution procedure.

8. *Calculate the quantities requested in the problem statement if they have not already been calculated.*

9. *If a stream quantity or flow rate n_g was given in the problem statement and another value n_c was either chosen as a basis or calculated for this stream, scale the balanced process by the ratio n_g/n_c to obtain the final result.*

The following example illustrates this procedure.

EXAMPLE 4.3-5 *Material Balances on a Distillation Column*

A liquid mixture containing 45.0% benzene (B) and 55.0% toluene (T) by mass is fed to a distillation column. A product stream leaving the top of the column (the *overhead product*) contains 95.0 mole% B, and a bottom product stream contains 8.0% of the benzene fed to the column (meaning that 92% of the benzene leaves with the overhead product). The volumetric flow rate of the feed stream is 2000 L/h and the specific gravity of the feed mixture is 0.872. Determine the mass flow rate of the overhead product stream and the mass flow rate and composition (mass fractions) of the bottom product stream.

SOLUTION We will explicitly illustrate the implementation of the steps of the procedure just outlined.

1. *Choose a basis.* Having no reason to do otherwise, we choose the given feed stream flow rate (2000 L/h) as the basis of calculation.

2. *Draw and label the flowchart.*

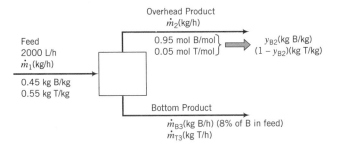

Note several points about the flowchart labeling:

- A volumetric flow rate is given for the feed stream, but mass flow rates and fractions will be needed for balances. The mass flow rate of the stream should therefore be considered an unknown process variable and labeled as such on the chart. Its value will be determined from the known volumetric flow rate and density of the feed stream.
- Since mass balances will be written, the given component mole fractions in the overhead product stream will have to be converted to mass fractions. The mass fractions are accordingly labeled as unknowns.
- We could have labeled the mass flow rate and mass fractions of the bottom stream as we did the overhead. However, since we have no information about either the flow rate or composition of this stream, we have instead labeled the component flow rates (following the rule of thumb given in Step 2 of the general procedure).
- Every component mass flow rate in every process stream can be expressed in terms of labeled quantities and variables. (Verify this statement.) For example, the flow rates of toluene (kg T/h) in the feed, overhead, and bottom streams are, respectively, $0.55\dot{m}_1$, $\dot{m}_2(1 - y_{B2})$, and $\dot{m}_{T3}$. The flowchart is therefore labeled completely.
- The 8%–92% benzene split between the product streams is not a stream flow rate or composition variable; nevertheless, we write it on the chart to remind ourselves that it is an additional relation among the stream variables and so should be included in the degree-of-freedom analysis.

3. **Write expressions for the quantities requested in the problem statement.** In terms of the quantities labeled on the flowchart, the quantities to be determined are $\dot{m}_2$ (the overhead product mass flow rate), $\dot{m}_3 = \dot{m}_{B3} + \dot{m}_{T3}$ (the bottom product mass flow rate), $x_B = \dot{m}_{B3}/\dot{m}_3$ (the benzene mass fraction in the bottom product), and $x_T = 1 - x_B$ (the toluene mass fraction). Once we determine $\dot{m}_2$, $\dot{m}_{B3}$, and $\dot{m}_{T3}$, the problem is essentially solved.

4. **Convert mixed units in overhead product stream** (see procedure preceding Example 3.3-3).

   Basis: 100 kmol overhead $\Longrightarrow$ 95.0 kmol B, 5.00 kmol T

   $\Longrightarrow$ (95.0 kmol B) $\times$ (78.11 kg B/kmol B) = 7420 kg B,   (5.00 $\times$ 92.13) = 461 kg T

   $\Longrightarrow$ (7420 kg B) + (461 kg T) = 7881 kg mixture

   $\Longrightarrow$ $y_{B2} = (7420 \text{ kg B})/(7881 \text{ kg mixture}) = 0.942$ kg B/kg   (write on chart)

   The molecular weights of benzene (78.11) and toluene (92.13) were looked up in Table B.1.

5. **Perform degree-of-freedom analysis.**

   > 4 unknowns ($\dot{m}_1, \dot{m}_2, \dot{m}_{B3}, \dot{m}_{T3}$)
   > -2 material balances (since there are two molecular species in this nonreactive process)
   > -1 density relationship (relating the mass flow rate to the given volumetric flow rate of the feed)
   > -1 specified benzene split (8% in bottom–92% in overhead)
   >
   > **0 degrees of freedom**

   The problem is therefore solvable.

6. **Write system equations and outline a solution procedure.** The variables for which each equation will be solved are circled.

   - **Volumetric flow rate conversion.** From the given specific gravity, the density of the feed stream is 0.872 kg/L. (*Verify.*) Therefore,

   $$\boxed{\dot{m}_1} = \left(2000 \frac{\text{L}}{\text{h}}\right)\left(0.872 \frac{\text{kg}}{\text{L}}\right)$$

   - **Benzene split fraction.** The benzene in the bottom product stream is 8% of the benzene in the feed stream. This statement translates directly into the equation

   $$\boxed{\dot{m}_{B3}} = 0.08(0.45\dot{m}_1)$$

   There are two unknowns remaining on the flowchart ($\dot{m}_2$ and $\dot{m}_{T3}$), and we are allowed to write two balances. Balances on total mass and on toluene each involve both unknowns, but a benzene balance only involves $\dot{m}_2$ (convince yourself, remembering that $\dot{m}_{B3}$ is now known), so we begin with that one.

- **Benzene balance** $0.45\dot{m}_1 = (\dot{m}_2)y_{B2} + \dot{m}_{B3}$

- **Toluene balance** $0.55\dot{m}_1 = (1 - y_{B2})\dot{m}_2 + (\dot{m}_{T3})$

7. **Do the algebra.** The four equations may be solved manually or with equation-solving software. If a manual solution is performed, each newly calculated variable value should be written on the flowchart for ease of reference in the remainder of the solution. The results are $\dot{m}_1 = 1744$ kg/h, $\dot{m}_{B3} = 62.8$ kg benzene/h, $\boxed{\dot{m}_2 = 766 \text{ kg/h}}$, and $\dot{m}_{T3} = 915$ kg toluene/h. (*Verify.*) A total mass balance (which is the sum of the benzene and toluene balances) may be written as a check on this solution:

$$\dot{m}_1 = \dot{m}_2 + \dot{m}_{B3} + \dot{m}_{T3} \implies 1744 \text{ kg/h} = (766 + 62.8 + 915) \text{ kg/h} = 1744 \text{ kg/h}$$

8. **Calculate additional quantities requested in the problem statement.**

$$\dot{m}_3 = \dot{m}_{B3} + \dot{m}_{T3} = 62.8 \text{ kg/h} + 915 \text{ kg/h} = \boxed{978 \text{ kg/h}}$$

$$y_{B3} = \frac{\dot{m}_{B3}}{\dot{m}_3} = \frac{62.8 \text{ kg B}}{978 \text{ kg/h}} = \boxed{0.064 \text{ kg B/kg}}$$

$$y_{T3} = 1 - y_{B3} = \boxed{0.936 \text{ kg T/kg}}$$

If we had chosen a basis of calculation other than an actual stream amount or flow rate, we would now scale the process from the calculated basis value to the actual value of this variable. Since in this case our basis was the actual feed stream flow rate, the solution is now complete.

4.4 BALANCES ON MULTIPLE-UNIT PROCESSES

In the preceding sections, we referred rather loosely to "the system," as in the statement "At steady state the rate at which benzene enters the system equals the rate at which it leaves." Not much was said about what "the system" was. Up to now, however, not much needed to be said, since we have considered only processes involving a single unit—a mixer, or a distillation column, or a reactor—and that unit necessarily constituted the system.

Industrial chemical processes rarely involve just one process unit. One or more chemical reactors are often present, as are units for mixing reactants, blending products, heating and cooling process streams, separating products from each other and from unconsumed reactants, and removing potentially hazardous pollutants from streams prior to discharging the streams to the plant environment. Before we analyze such processes, we must take a closer look at what we mean by a system.

In general terms, a "system" is any portion of a process that can be enclosed within a hypothetical box (boundary). It may be the entire process, an interconnected combination of some of the process units, a single unit, or a point at which two or more process streams come together or one stream splits into branches. The inputs and outputs to a system are the process streams that intersect the system boundary.

A flowchart for a two-unit process is shown in Figure 4.4-1. Five boundaries drawn about portions of the process define systems on which balances may be written.

Boundary Ⓐ encloses the entire process; the system defined by this boundary has as inputs Feed Streams 1, 2, and 3 and Product Streams 1, 2, and 3. (Convince yourself.) Balances on this system are referred to as **overall balances**. The stream that connects Units 1 and 2 is internal to this system and so would not enter into overall system balances.

Boundary Ⓑ encloses a feed stream mixing point. Feed Streams 1 and 2 are inputs to this system and the stream flowing to Unit 1 is an output. Boundary Ⓒ encloses Unit 1 (one input stream and two output streams), Boundary Ⓓ encloses a stream splitting point (one input stream and two output streams), and Boundary Ⓔ encloses Unit 2 (two input streams and one output stream).

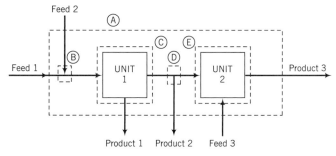

Figure 4.4-1 Flowchart of a two-unit process. Dashed lines denote boundaries of systems about which balances can be written.

The procedure for material balance calculations on multiple-unit processes is basically the same as that outlined in Section 4.3. The difference is that with multiple-unit processes you may have to isolate and write balances on several subsystems of the process to obtain enough equations to determine all unknown stream variables. When analyzing multiple-unit processes, carry out degree-of-freedom analyses on the overall process and on each subsystem, taking into account only the streams that intersect the boundary of the system under consideration. Do not begin to write and solve equations for a subsystem until you have verified that it has zero degrees of freedom.

EXAMPLE 4.4-1   **Two-Unit Process**

A labeled flowchart of a continuous steady-state two-unit process is shown below. Each stream contains two components, A and B, in different proportions. Three streams whose flow rates and/or compositions are not known are labeled 1, 2, and 3.

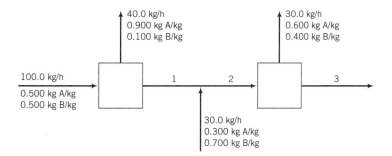

Calculate the unknown flow rates and compositions of streams 1, 2, and 3.

SOLUTION   **Basis—Given Flow Rates**
The systems about which balances might be written are shown on the following representation of the flowchart:

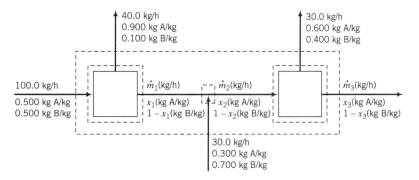

The outer boundary encompasses the entire process and has as input and output streams all of the streams that enter and leave the process. Two of the interior boundaries surround individual process units, and the third encloses a stream junction point.

Degree-of-Freedom Analysis

We first outline the solution procedure by performing degree-of-freedom analyses on different systems. Remember that only variables associated with streams intersecting a system boundary are counted in the analysis of that system.

Overall system (outer dashed boundary):

$$2 \text{ unknowns } (\dot{m}_3, x_3) - 2 \text{ balances (2 species)} = 0 \text{ degrees of freedom}$$
$$\Longrightarrow \underline{\text{Determine } \dot{m}_3 \text{ and } x_3}$$

In subsequent analyses, we may consider these two variables as known. Suppose we decide to consider the intermediate stream mixing point as the next system.

Mixing point:

$$4 \text{ unknowns } (\dot{m}_1, x_1, \dot{m}_2, x_2) - 2 \text{ balances (2 species)} = 2 \text{ degrees of freedom}$$

We have too many unknowns for the number of available equations. Let us try Unit 1.

Unit 1:

$$2 \text{ unknowns } (\dot{m}_1, x_1) - 2 \text{ balances (2 species)} = 0 \text{ degrees of freedom}$$
$$\Longrightarrow \underline{\text{Determine } \dot{m}_1 \text{ and } x_1}$$

We may now analyze either the mixing point or Unit 2, each of which has two unknown variables associated with it.

Mixing point:

$$2 \text{ unknowns } (\dot{m}_2, x_2) - 2 \text{ balances (2 species)} = 0 \text{ degrees of freedom}$$
$$\Longrightarrow \underline{\text{Determine } \dot{m}_2 \text{ and } x_2}$$

The procedure will therefore be to write overall system balances to determine $\dot{m}_3$ and x_3, then balances on Unit 1 to determine $\dot{m}_1$ and x_1, and finally balances on the intermediate mixing point to determine $\dot{m}_2$ and x_2.

The calculations are straightforward. Note that all balances on this steady-state nonreactive process have the form *input = output*, and also note that the balances are written in an order that does not require solution of simultaneous equations (each equation involves only one unknown variable).

Calculations

Overall Mass Balance:

$$(100.0 + 30.0) \frac{\text{kg}}{\text{h}} = (40.0 + 30.0) \frac{\text{kg}}{\text{h}} + \dot{m}_3 \Longrightarrow \boxed{\dot{m}_3 = 60.0 \text{ kg/h}}$$

Overall Balance on A: (Verify that each additive term has the units kg A/h.)

$$(0.500)(100.0) + (0.300)(30.0) = (0.900)(40.0) + (0.600)(30.0) + x_3(60.0)$$
$$\Longrightarrow \boxed{x_3 = 0.0833 \text{ kg A/kg}}$$

Mass Balance on Unit 1: (each term has the units kg/h)

$$100 = 40 + \dot{m}_1 \Longrightarrow \boxed{\dot{m}_1 = 60.0 \text{ kg/h}}$$

A Balance on Unit 1: (each additive term has the units kg A/h)

$$(0.500)(100.0) = (0.900)(40.0) + x_1(60.0) \Longrightarrow \boxed{x_1 = 0.233 \text{ kg A/kg}}$$

Mass Balance on Stream Mixing Point: (Each term has the units kg/h.)

$$\dot{m}_1 + 30.0 = \dot{m}_2 \xrightarrow{\dot{m}_1 = 60.0 \text{ kg/h}} \boxed{\dot{m}_2 = 90.0 \text{ kg/h}}$$

A Balance on Stream Mixing Point: (Each additive term has the units kg A/h.)

$$x_1 \dot{m}_1 + (0.300)(30.0) = x_2 \dot{m}_2$$

$$\Big\Downarrow \begin{array}{l} \dot{m}_1 = 60.0 \text{ kg/h} \\ x_1 = 0.233 \text{ kg/kg} \\ \dot{m}_2 = 90.0 \text{ kg/h} \end{array}$$

$$\boxed{x_2 = 0.255 \text{ kg A/kg}}$$

The situation becomes still more complicated when three or more process units are involved. In such cases, balances may be written not only for the overall process and individual process units, but also for combinations of units. Finding the right combinations can lead to a considerable gain in computational efficiency.

EXAMPLE 4.4-2 **An Extraction–Distillation Process**

A mixture containing 50.0 wt% acetone and 50.0 wt% water is to be separated into two streams—one enriched in acetone, the other in water. The separation process consists of extraction of the acetone from the water into methyl isobutyl ketone (MIBK), which dissolves acetone but is nearly immiscible with water. The description that follows introduces some of the terms commonly used in reference to liquid extraction processes. The process is shown schematically below.

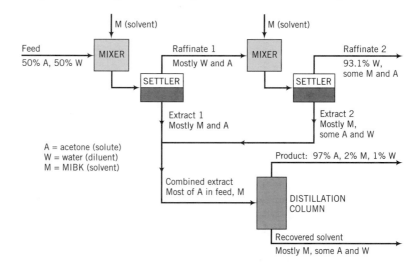

The acetone (**solute**)–water (**diluent**) mixture is first contacted with the MIBK (**solvent**) in a mixer that provides good contact between the two liquid phases. A portion of the acetone in the feed transfers from the aqueous (water) phase to the organic (MIBK) phase in this step. The mixture passes into a settling tank, where the phases separate and are separately withdrawn. The phase rich in the diluent (water, in this process) is referred to as the **raffinate**, and the phase rich in the solvent (MIBK) is the **extract**. The mixer–settler combination is the first **stage** of this separation process.

The raffinate passes to a second extraction stage where it is contacted with a second stream of pure MIBK, leading to the transfer of more acetone. The two phases are allowed to separate in a second settler, and the raffinate from this stage is discarded. The extracts from the two mixer–settler stages are combined and fed to a distillation column. The overhead effluent is rich in acetone and is the process product. The bottom effluent is rich in MIBK and in a real process would be treated further and recycled back to the first extraction stage, but we will not consider recycle in this example.

In a pilot–plant study, for every 100 kg of acetone–water fed to the first extraction stage, 100 kg of MIBK is fed to the first stage and 75 kg is fed to the second stage. The extract from the first stage is found to contain 27.5 wt% acetone. (All percentages in the remainder of this paragraph are weight percents.) The second-stage raffinate has a mass of 43.1 kg and contains 5.3% acetone, 1.6% MIBK, and 93.1% water, and the second-stage extract contains 9.0% acetone, 88.0% MIBK, and 3.0% water. The overhead product from the distillation column contains 2.0% MIBK, 1.0% water, and the balance acetone.

Taking a basis of calculation of 100 kg acetone–water feed, calculate the masses and compositions (component weight percentages) of the Stage 1 raffinate and extract, the Stage 2 extract, the combined extract, and the distillation overhead and bottoms products.

SOLUTION

This is a "trick" problem in that there is not enough information to calculate all of the requested quantities. We will show how degree-of-freedom analysis allows a fairly quick determination of which variables can be determined and how to determine them efficiently, and also helps avoid wasting a great deal of time trying to solve a problem with insufficient information.

As always, we begin by drawing and labeling the flowchart. For simplicity, we will treat each mixer–settler combination as a single "extractor" unit.

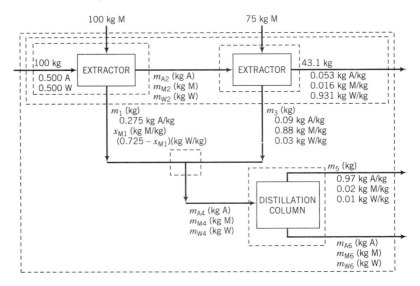

Balances may be written for any of the systems shown on the chart, including the overall process, the individual extractors, the two-extractor combination, the point where the two extract streams combine, and the distillation column. A degree-of-freedom analysis for the overall process indicates that there are four unknown variables ($m_5, m_{A6}, m_{M6}, m_{W6}$) and only three equations that relate them (one material balance for each of the three independent species involved in the process), leaving one degree of freedom. Similarly, each extractor has one degree of freedom, the extract mixing point has three, and the distillation column has four. (Verify these figures.) However, the system composed of the two extraction units involves only three unknowns (m_1, x_{M1}, m_3) and three balances relating them, hence zero degrees of freedom. The solution procedure is as follows:

- **Analyze two-extractor subsystem.** Write a total mass balance and an acetone balance; solve simultaneously to determine m_1 and m_3. Write an MIBK balance to determine x_{M1}.
- **Analyze extract mixing point.** Write acetone, MIBK, and water balances; solve to determine m_{A4}, m_{M4}, and m_{W4}, respectively.
- **Analyze first (or second) extractor.** Write acetone, MIBK, and water balances; solve to determine m_{A2}, m_{M2}, and m_{W2}.

At this point, we may quickly determine that we can go no farther. There are four remaining unknowns—m_5, m_{A6}, m_{M6}, and m_{W6}. Whether we choose the overall process or the distillation column as our system, we will only have three independent equations and hence one degree of freedom, and so we will be unable to solve the problem. Moreover, since acetone, MIBK, and water all appear in both outlet streams, we cannot solve for any one of the individual unknowns. (If there were no water in the overhead product from the distillation column, for example, we could deduce that $m_{W6} = m_{W4}$). The problem is thus underspecified; unless another piece of information is furnished, the amounts and compositions of the distillation column products are indeterminate.

The calculations that can be performed are shown below. All balances have the form *input = output* (why?), and every additive term of each balance has units of kilograms of the balanced species.

Balances Around Two-Extractor Subsystem

$$\textit{Total mass: } (100 + 100 + 75)\text{kg} = 43.1 \text{ kg} + m_1 + m_3$$

$$\text{A: } 100(0.500) \text{ kg A} = (43.1)(0.053) \text{ kg A} + m_1(0.275) + m_3(0.09)$$

⇓ Solve simultaneously

$$\boxed{m_1 = 145 \text{ kg}, m_3 = 86.8 \text{ kg}}$$

$$\text{M: } (100 + 75) \text{ kg M} = (43.1)(0.016) \text{ kg M} + m_1 x_{M1} + m_3(0.88)$$

⇓ $m_1 = 145$ kg, $m_3 = 86.8$ kg

$$\boxed{x_{M1} = 0.675 \text{ kg MIBK/kg}}$$

Balances Around Extract Mixing Point

$$\text{A: } m_1(0.275) + m_3(0.09) = m_{A4}$$

⇓ $m_1 = 145$ kg, $m_3 = 86.8$ kg

$$\boxed{m_{A4} = 47.7 \text{ kg acetone}}$$

$$\text{M: } m_1 x_{M1} + m_3(0.88) = m_{M4}$$

⇓ $m_1 = 145$ kg, $m_3 = 87$ kg, $x_{M1} = 0.675$ kg M/kg

$$\boxed{m_{M4} = 174 \text{ kg MIBK}}$$

$$\text{W: } m_1(0.725 - x_{M1}) + m_3(0.03) = m_{W4}$$

⇓ $m_1 = 145$ kg, $m_3 = 86.8$ kg, $x_{M1} = 0.675$ kg M/kg

$$\boxed{m_{W4} = 9.9 \text{ kg water}}$$

Balances Around First Extractor

$$\text{A: } 100(0.500) \text{ kg A} = m_{A2} + m_1(0.275)$$

⇓ $m_1 = 145$ kg

$$\boxed{m_{A2} = 10.1 \text{ kg acetone}}$$

$$\text{M: } 100 \text{ kg M} = m_{M2} + m_1 x_{M1}$$

⇓ $m_1 = 145$ kg, $x_{M1} = 0.675$ kg M/kg

$$\boxed{m_{M2} = 2.3 \text{ kg MIBK}}$$

$$\text{W: } (100)(0.500) = m_{W2} + m_1(0.725 - x_{M1})$$

⇓ $m_1 = 145$ kg, $x_{M1} = 0.675$ kg M/kg

$$\boxed{m_{W2} = 42.6 \text{ kg water}}$$

If we knew (or could independently determine) any one of the variables m_5, m_{A6}, m_{M6}, or m_{W6}, we could calculate the remaining three. Since we do not, we must terminate the calculations at this point.

4.5 RECYCLE AND BYPASS

It is rare that a chemical reaction A $\rightarrow$ B proceeds to completion in a reactor. No matter how little A is present in the feed or how long the reaction mixture remains in the reactor, some A is normally found in the product.

Unfortunately, you have to pay for *all* the reactant fed to a process, not just the fraction that reacts, and any A that leaves with the product therefore represents wasted resources. Suppose, however, you could find a way to separate most or all of the unconsumed reactant from the product stream. You could then sell the resulting relatively pure product and **recycle** the unconsumed reactant back to the reactor. You would, of course, have to pay for the separation and recycle equipment, but you would compensate for this cost by having to purchase less fresh reactant and being able to sell the purified product at a higher price.

A labeled flowchart of a chemical process involving reaction, product separation, and recycle is shown in Figure 4.5-1. Note the distinction between the fresh feed to the process and the feed to the reactor, which is the sum of the fresh feed and recycle stream. If some of the stream variables shown in Figure 4.5-1 were unknown, you could determine them by writing balances on the overall process and about the reactor, separator, and mixing point.

Students frequently have difficulty when they first encounter the concept of recycle because they find it hard to understand that material can circulate in a system without an accumulation of mass. If you have this difficulty, you might find it helpful to refer back to the flowchart of Figure 4.5-1. Observe that even though there is material circulating within the system, there is no net accumulation: 110 kg of material enters the system each minute, and the same mass leaves each minute. Within the system there is a net circulation rate of 120 kg/min, but the circulation has no effect on the overall process material balance.

EXAMPLE 4.5-1 *Material and Energy Balances on an Air Conditioner*

Fresh air containing 4.00 mole% water vapor is to be cooled and dehumidified to a water content of 1.70 mole% H_2O. A stream of fresh air is combined with a recycle stream of previously dehumidified air and passed through the cooler. The blended stream entering the unit contains 2.30 mole% H_2O. In the air conditioner, some of the water in the feed stream is condensed and removed as liquid. A fraction of the dehumidified air leaving the cooler is recycled and the remainder is delivered to a

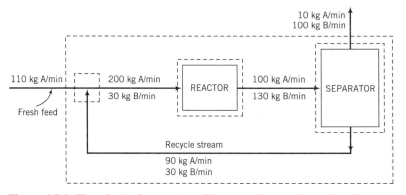

Figure 4.5-1 Flowchart of a reactor with separation and recycle of unconsumed reactant.

room. Taking 100 mol of dehumidified air delivered to the room as a basis of calculation, calculate the moles of fresh feed, moles of water condensed, and moles of dehumidified air recycled.

SOLUTION

The labeled flowchart for this process, including the assumed basis of calculation, is shown below. Dashed lines depict the four subsystems about which balances might be written—the overall process, the recycle–fresh feed mixing point, the air conditioner, and the recycle–product gas splitting point. The quantities to be determined are n_1, n_3, and n_5.

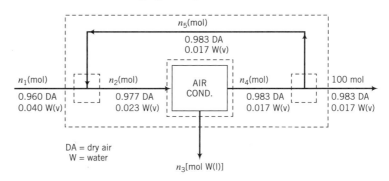

Students first attempting recycle problems are tempted to jump in and start writing balances immediately, and when they do they find themselves spending a *long* time on the problems, often failing to get solutions. If you first take a few minutes to go through a degree-of-freedom analysis you should be able to work out an efficient solution method before doing any calculations, ultimately saving yourself much more time than it took to do the analysis. Observe the procedure.

Overall System

2 variables (n_1, n_3)	(Only consider streams that intersect the system boundary.)
−2 balance equations	(Two species—dry air and water—are in the streams.)
0 degrees of freedom	

We can therefore determine n_1 and n_3 from overall system balances. We would write a dry air balance first since it only involves one unknown (n_1), while total mole and water balances involve both unknowns. Once n_1 has been determined, the second balance yields n_3. No more can be done with the overall system, so we move on to other subsystems.

Mixing Point	**Cooler**	**Splitting Point**
2 variables (n_2, n_5)	2 variables (n_2, n_4)	2 variables (n_4, n_5)
−2 balances	−2 balances	−1 balance (see below)
0 degrees of freedom	0 degrees of freedom	1 degree of freedom

Only one independent balance can be written for the splitting point because the streams entering and leaving this subsystem are labeled as having identical compositions, so that the dry air/water mixture in all three streams behaves like a single species. (Convince yourself: write a total mole balance and a dry air or water balance around the splitting point and observe that you get the same equation each time.)

At this point we could either write balances around the mixing point to determine n_2 and n_5 or around the cooler to determine n_2 and n_4 (but not around the splitting point, which has one degree of freedom). The mixing point is the logical subsystem to attack, since the problem statement asks for n_5 but not n_4. Writing and solving balances around the mixing point will therefore complete the solution.

Now, finally, we do the calculations. All balances have the form *input = output*, and each additive term in each equation has the units (mol of the balanced quantity).

Overall dry air balance: $0.960 n_1 = 0.983(100 \text{ mol}) \implies$ $\boxed{n_1 = 102.4 \text{ mol fresh feed}}$

Overall mole balance: $n_1 = n_3 + 100 \text{ mol} \xrightarrow{\;n_1 = 102.4 \text{ mol}\;}$ $\boxed{n_3 = 2.4 \text{ mol H}_2\text{O condensed}}$

Mole balance on mixing point: $n_1 + n_5 = n_2$

Water balance on mixing point: $0.04n_1 + 0.017n_5 = 0.023n_2$

$$n_1 = 102.4 \text{ mol}$$
Solve simultaneously

$$n_2 = 392.5 \text{ mol}$$

$$\boxed{n_5 = 290 \text{ mol recycled}}$$

Almost three moles are recycled for every mole of air delivered to the room.

There are several reasons for using recycle in a chemical process besides the one given previously (recovering and reusing unconsumed reactants), including the following:

1. ***Recovery of catalyst.*** Many reactors use catalysts to increase the rate of the reaction. Catalysts are usually expensive, and the processes generally include provisions for recovering them from the product stream and recycling them to the reactor. They may be recovered with the unconsumed reactants or recovered separately in special facilities designed for this purpose.

2. ***Dilution of a process stream.*** Suppose a slurry (a suspension of solids in a liquid) is fed to a filter. If the concentration of solids in the slurry is too high, the slurry is difficult to handle and the filter will not operate properly. Rather than diluting the feed with fresh liquid, a portion of the filtrate can be recycled to dilute the feed to the desired solids concentration.

3. ***Control of a process variable.*** Suppose a reaction releases an extremely large amount of heat, making the reactor difficult and expensive to control. The rate of heat generation can be reduced by lowering the reactant concentration, which can in turn be accomplished by recycling a portion of the reactor effluent to the inlet. In addition to acting as a diluent for the reactants, the recycled material also serves as a capacitance for the heat released: the greater the mass of the reaction mixture, the lower the temperature to which that mass will be raised by a fixed amount of heat.

4. ***Circulation of a working fluid.*** The most common example of this application is the refrigeration cycle used in household refrigerators and air conditioners. In these devices, a single material is reused indefinitely, with only small makeup quantities being added to the system to replenish working fluid that may be lost through leaks.

Part 1 of the next example presents a detailed material balance calculation for a separation process that involves recycle. Part 2 of the problem shows what would happen if the recycle were omitted and, in doing so, illustrates one of the reasons for recycling.

EXAMPLE 4.5-2   ***An Evaporative Crystallization Process***

The flowchart of a steady-state process to recover crystalline potassium chromate (K_2CrO_4) from an aqueous solution of this salt is shown below.

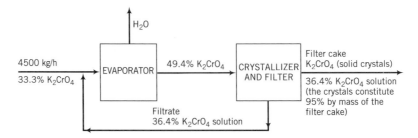

Forty-five hundred kilograms per hour of a solution that is one-third K_2CrO_4 by mass is joined by a recycle stream containing 36.4% K_2CrO_4, and the combined stream is fed into an evaporator. The concentrated stream leaving the evaporator contains 49.4% K_2CrO_4; this stream is fed into a crystallizer in which it is cooled (causing crystals of K_2CrO_4 to come out of solution) and then filtered. The filter cake consists of K_2CrO_4 crystals and a solution that contains 36.4% K_2CrO_4 by mass; the crystals account for 95% of the total mass of the filter cake. The solution that passes through the filter, also 36.4% K_2CrO_4, is the recycle stream.

1. Calculate the rate of evaporation, the rate of production of crystalline K_2CrO_4, the feed rates that the evaporator and the crystallizer must be designed to handle, and the *recycle ratio* (mass of recycle)/(mass of fresh feed).
2. Suppose that the filtrate were discarded instead of being recycled. Calculate the production rate of crystals. What are the benefits and costs of the recycling?

SOLUTION

1. *Basis: 4500 kg/h Fresh Feed.*

Let K denote K_2CrO_4 and W denote water. The flowchart is shown below; included on it are dashed boxes denoting the overall system and subsystems about which balances may be written.

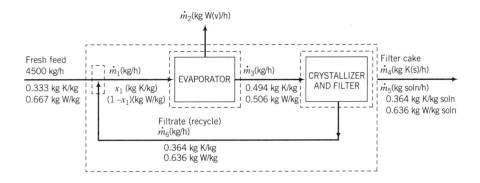

In terms of labeled variables, the quantities requested in the problem statement are $\dot{m}_2$ (kg W evaporated/h), $\dot{m}_4$ [kg K(s)/h], $\dot{m}_1$ (kg/h fed to evaporator), $\dot{m}_3$ (kg/h fed to crystallizer), and $(\dot{m}_6/4500)$ (kg recycle/kg fresh feed).

Notice how the product stream is labeled to take maximum advantage of what we know about its composition. The filter cake is a mixture of solid K crystals and a liquid solution with known K and W mass fractions. Instead of labeling as usual the mass flow rate and component mass fractions or the individual mass flow rates of K and W, we label the flow rates of the crystals and the solution, along with the solution component mass fractions. To confirm that the stream is fully labeled, we must verify that all component flow rates may be expressed in terms of labeled quantities. In this case, the total flow rate of potassium chromate is $\dot{m}_4 + 0.364\dot{m}_5$ (kg K/h) and the flow rate of water is $0.636\dot{m}_5$ (kg W/h), so that the labeling is complete.

Degree-of-Freedom Analysis

The degree-of-freedom analysis begins with the overall system and proceeds as follows:

- ***Overall system***

   3 unknown variables ($\dot{m}_2$, $\dot{m}_4$, $\dot{m}_5$)
   −2 balances (2 species involved)
   −1 additional relation ($\dot{m}_4$ = 95% of the total filter cake mass)
   ───────────────────────────────
   0 degrees of freedom

We will therefore be able to determine $\dot{m}_2$, $\dot{m}_4$, and $\dot{m}_5$ by analyzing the overall system.

- *Recycle–fresh food mixing point*

    3 unknown variables ($\dot{m}_6$, $\dot{m}_1$, x_1)

    −2 balances

    _____

    1 degree of freedom

Since we do not have enough equations to solve for the unknowns associated with this subsystem, we proceed to the next one.

- *Evaporator*

    3 unknown variables ($\dot{m}_1$, x_1, $\dot{m}_3$)

    −2 balances

    _____

    1 degree of freedom

Again, no luck. We have one last hope.

- *Crystallizer/filter*

    2 unknown variables ($\dot{m}_3$, $\dot{m}_6$)

    −2 balances

    _____

    0 degrees of freedom

We can therefore determine $\dot{m}_3$ and $\dot{m}_6$ by writing and solving balances on the crystallizer/filter and then analyze either the mixing point or the evaporator to determine the two remaining unknowns ($\dot{m}_1$ and x_1), thereby completing the solution. (*Suggestion:* Try to reproduce this analysis on your own before reading any farther.) The calculations follow, beginning with the analysis of the overall system.

We are told that the solids constitute 95% by mass of the filter cake. This information translates directly to the following equation:

$$\dot{m}_4 = 0.95(\dot{m}_4 + \dot{m}_5)$$

$$\Downarrow$$

$$\dot{m}_5 = 0.05263\,\dot{m}_4 \tag{1}$$

We next write the two allowed overall system balances. The questions are, which ones and in what order?

- A total mass balance involves all three system variables—$\dot{m}_2$, $\dot{m}_4$, and $\dot{m}_5$.
- A K balance involves $\dot{m}_4$ and $\dot{m}_5$—the same two variables that occur in Equation 1.
- A W balance involves $\dot{m}_2$ and $\dot{m}_5$.

The procedure is therefore to write a K_2CrO_4 balance, solve it simultaneously with Equation 1 to determine $\dot{m}_4$ and $\dot{m}_5$, and then write a total mass balance to determine $\dot{m}_2$. The additive terms in each equation have units of kg/h of the balanced species.

Overall K_2CrO_4 Balance

$$(0.333)(4500)\ \text{kg K/h} = \dot{m}_4 + 0.364\dot{m}_5$$

$$\Downarrow \text{Solve simultaneously with Equation 1}$$

$$\dot{m}_4 = \boxed{1470\ K_2CrO_4\ \text{crystals/h}}$$

$$\dot{m}_5 = 77.5\ \text{kg entrained solution/h}$$

Overall Total Mass Balance

$$4500\ \text{kg/h} = \dot{m}_2 + \dot{m}_4 + \dot{m}_5$$

$$\Downarrow \dot{m}_4 = 1470\ \text{kg/h},\ \dot{m}_5 = 77.5\ \text{kg/h}$$

$$\dot{m}_2 = \boxed{2950\ \text{kg } H_2O\ \text{evaporated/h}}$$

Mass Balance Around Crystallizer

$$\dot{m}_3 = \dot{m}_4 + \dot{m}_5 + \dot{m}_6$$

$$\Downarrow \dot{m}_4 = 1470 \text{ kg/h}, \dot{m}_5 = 77.5 \text{ kg/h}$$

$$\dot{m}_3 = 1550 \text{ kg/h} + \dot{m}_6 \qquad\qquad \textbf{(2)}$$

Water Balance Around Crystallizer

$$0.506\dot{m}_3 = 0.636\dot{m}_5 + 0.636\dot{m}_6$$

$$\Downarrow \dot{m}_5 = 77.5 \text{ kg/h}$$

$$\dot{m}_3 = 97.4 \text{ kg/h} + 1.257\dot{m}_6 \qquad\qquad \textbf{(3)}$$

Solving Equations 2 and 3 simultaneously yields

$$\boxed{\dot{m}_3 = 7200 \text{ kg/h fed to crystallizer}}$$

$$\dot{m}_6 = 5650 \text{ kg/h}$$

and hence

$$\frac{\dot{m}_6 \text{ (kg recycle/h)}}{4500 \text{ kg fresh feed/h}} = \frac{5650}{4500} = \boxed{1.26 \ \frac{\text{kg recycle}}{\text{kg fresh feed}}}$$

Mass Balance Around Recycle-Fresh Feed Mixing Point

$$4500 \text{ kg/h} + \dot{m}_6 = \dot{m}_1$$

$$\Downarrow \dot{m}_6 = 5650 \text{ kg/h}$$

$$\dot{m}_1 = \boxed{10,150 \text{ kg/h feed to evaporator}}$$

Check: A mass balance about the evaporator yields

$$\dot{m}_1 = \dot{m}_2 + \dot{m}_3$$

$\Longrightarrow$ 10,150 kg/h entering evaporator = (2950 + 7200) kg/h = 10,150 kg/h leaving evaporator.

2. Basis: 4500 kg/h Fresh Feed.

The flowchart for the system without recycle appears as shown below.

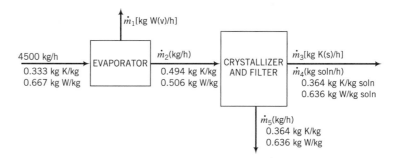

We will not go through the detailed solution but will simply summarize. The degree-of-freedom analysis leads to the results that the overall system has one degree of freedom, the evaporator has zero, and the crystallizer–filter has one. (Verify these statements.) The strategy is therefore to begin with the evaporator and solve the balance equations for $\dot{m}_1$ and $\dot{m}_2$. Once $\dot{m}_2$ is known, the crystallizer has zero degrees of freedom and its three equations may be solved for $\dot{m}_3$, $\dot{m}_4$, and $\dot{m}_5$. The rate of production of crystals is

$$\boxed{\dot{m}_3 = 622 \text{ kg K(s)/h}}$$

With recycle it was 1470 kg/h, a dramatic difference. The mass flow rate of the discarded filtrate is

$$\dot{m}_5 = 2380 \text{ kg/h}$$

The filtrate (which is discarded) contains $0.364 \times 2380 = 866$ kg/h of potassium chromate, more than the filter cake contains. Recycling the filtrate enables us to recover most of this salt. The obvious benefit of recycling is the revenue from selling the additional potassium chromate. The costs include the purchase and installation costs for the recycle piping and pump and the cost of power consumed by the pump. It would probably not take long for the benefit to equal the cost, and thereafter the recycling would continue to increase the process profitability.

A procedure that has several features in common with recycle is **bypass**, in which a fraction of the feed to a process unit is diverted around the unit and combined with the output stream from the unit. (For an example of a process that uses bypass, see Problems 4.32 and 4.33 at the end of this chapter.) By varying the fraction of the feed that is bypassed, we can vary the composition and properties of the product.

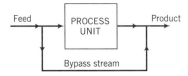

Bypass and recycle calculations are approached in exactly the same manner: the flowchart is drawn and labeled, and overall balances and balances around the process unit or the stream mixing point following the process unit are used to determine unknown variables.

4.6 CHEMICAL REACTION STOICHIOMETRY

The occurrence of a chemical reaction in a process brings several complications into the material balance procedures described in the previous sections. The stoichiometric equation of the reaction imposes constraints on the relative amounts of reactants and products in the input and output streams (if A $\rightarrow$ B, for example, you cannot start with 1 mol of pure A and end with 2 mol of B). In addition, a material balance on a reactive substance does not have the simple form input = output, but must include a generation and/or consumption term.

In this section we review reaction terminology and outline procedures for carrying out material balance calculations on reactive systems.

4.6a Stoichiometry

Stoichiometry is the theory of the proportions in which chemical species combine with one another. The **stoichiometric equation** of a chemical reaction is a statement of the relative number of molecules or moles of reactants and products that participate in the reaction. For example, the stoichiometric equation

$$2 \text{ SO}_2 + \text{O}_2 \rightarrow 2 \text{ SO}_3$$

indicates that for every two molecules (g-moles, lb-moles) of SO_2 that react, one molecule (g-mole, lb-mole) of O_2 reacts to produce two molecules (g-moles, lb-moles) of SO_3. The numbers that precede the formulas for each species are the stoichiometric coefficients of the reaction components.

A valid stoichiometric equation must be *balanced;* that is, the number of atoms of each atomic species must be the same on both sides of the equation, since atoms can neither be created nor destroyed in chemical reactions. The equation

$$SO_2 + O_2 \rightarrow SO_3$$

cannot be valid, for example, since it indicates that three atoms of atomic oxygen (O) are produced for every four atoms that enter into the reaction, for a net loss of one atom, but

$$SO_2 + \tfrac{1}{2}O_2 \rightarrow SO_3 \qquad \begin{pmatrix} 1\,S & \rightarrow & 1\,S \\ 3\,O & \rightarrow & 3\,O \end{pmatrix}$$

and

$$2SO_2 + O_2 \rightarrow 2\,SO_3 \qquad \begin{pmatrix} 2\,S & \rightarrow & 2\,S \\ 6\,O & \rightarrow & 6\,O \end{pmatrix}$$

are balanced.

The **stoichiometric ratio** of two molecular species participating in a reaction is the ratio of their stoichiometric coefficients in the balanced reaction equation. This ratio can be used as a conversion factor to calculate the amount of a particular reactant (or product) that was consumed (or produced), given a quantity of another reactant of product that participated in the reaction. For the reaction

$$2\,SO_2 + O_2 \rightarrow 2\,SO_3$$

you can write the stoichiometric ratios

$$\frac{2 \text{ mol } SO_3 \text{ generated}}{1 \text{ mol } O_2 \text{ consumed}}, \quad \frac{2 \text{ lb-moles } SO_2 \text{ consumed}}{2 \text{ lb-moles } SO_3 \text{ generated}}$$

and so on. If you know, for example, that 1600 kg/h of SO_3 is to be produced, you can calculate the amount of oxygen required as

$$\frac{1600 \text{ kg } SO_3 \text{ generated}}{h} \left| \frac{1 \text{ kmol } SO_3}{80 \text{ kg } SO_3} \right| \frac{1 \text{ kmol } O_2 \text{ consumed}}{2 \text{ kmol } SO_3 \text{ generated}} = 10 \; \frac{\text{kmol } O_2}{h}$$

$$\Longrightarrow 10 \; \frac{\text{kmol } O_2}{h} \left| \frac{32 \text{ kg } O_2}{1 \text{ kmol } O_2} \right. = 320 \text{ kg } O_2/h$$

It is a good practice to include the terms "consumed" and "generated" when performing conversions of this sort: simply writing 1 mol O_2/2 mol SO_3 could be taken to mean that 2 mol of SO_3 contains 1 mol of O_2, which is not true.

TEST YOURSELF

Consider the reaction

$$C_4H_8 + 6\,O_2 \rightarrow 4\,CO_2 + 4\,H_2O$$

1. Is the stoichiometric equation balanced?
2. What is the stoichiometric coefficient of CO_2?
3. What is the stoichiometric ratio of H_2O to O_2? (Include units.)
4. How many lb-moles of O_2 react to form 400 lb-moles of CO_2? (Use a dimensional equation.)
5. One hundred mol/min of C_4H_8 is fed into a reactor, and 50% reacts. At what rate is water formed?

4.6b   Limiting and Excess Reactants, Fractional Conversion, and Extent of Reaction

Two reactants, A and B, are said to be present in **stoichiometric proportion** if the ratio (moles A present)/(mole B present) equals the stoichiometric ratio obtained from the balanced reaction equation. For the reactants in the reaction

$$2 SO_2 + O_2 \rightarrow 2 SO_3$$

to be present in stoichiometric proportion, there must be 2 moles of SO_2 for every mole of O_2 (so that $n_{SO_2}/n_{O_2} = 2:1$) present in the feed to the reactor.

If reactants are fed to a chemical reactor in stoichiometric proportion and the reaction proceeds to completion, all of the reactants are consumed. In the above reaction, for example, if 200 mol of SO_2 and 100 mol of O_2 are initially present and the reaction proceeds to completion, the SO_2 and O_2 would disappear at the same instant. It follows that if you start with 100 mol of O_2 and less than 200 mol of SO_2 (i.e., if the SO_2 is present in less than its stoichiometric proportion), the SO_2 would run out first, while if more than 200 mol of SO_2 is initially present the O_2 would run out.

The reactant that would run out if a reaction proceeded to completion is called the **limiting reactant**, and the other reactants are termed **excess reactants**. *A reactant is limiting if it is present in less than its stoichiometric proportion relative to every other reactant.* If all reactants are present in stoichiometric proportion, then no reactant is limiting (or they all are, depending on how you choose to look at it).

Suppose $(n_A)_{feed}$ is the number of moles of an excess reactant, A, present in the feed to a reactor and that $(n_A)_{stoich}$ is the **stoichiometric requirement** of A, or the amount needed to react completely with the limiting reactant. Then $(n_A)_{feed} - (n_A)_{stoich}$ is the amount by which the A in the feed exceeds the amount needed to react completely if the reaction goes to completion. The **fractional excess** of the reactant is the ratio of the excess to the stoichiometric requirement:

$$\text{fractional excess of A} = \frac{(n_A)_{feed} - (n_A)_{stoich}}{(n_A)_{stoich}} \tag{4.6-1}$$

The **percentage excess of A** is 100 times the fractional excess.

Consider, for example, the hydrogenation of acetylene to form ethane:

$$C_2H_2 + 2H_2 \rightarrow C_2H_6$$

and suppose that 20.0 kmol/h of acetylene and 50.0 kmol/h of hydrogen are fed to a reactor. The stoichiometric ratio of hydrogen to acetylene is 2:1 (the ratio of coefficients in the stoichiometric equation), and since the ratio of H_2 to C_2H_2 in the feed is 2.5:1 (50:20), hydrogen is fed in a greater-than-stoichiometric proportion to acetylene. Acetylene is therefore the limiting reactant. (Convince yourself.) Since it would take 40.0 kmol H_2/h to react completely with all the acetylene fed to the reactor, $(n_{H_2})_{stoich} = 40.0$ kmol/h, and from Equation 4.6-1,

$$\text{fractional excess of } H_2 = \frac{(50.0 - 40.0)\text{kmol/h}}{40.0 \text{ kmol/h}} = 0.25$$

We say that there is *25% excess hydrogen* in the feed.

Chemical reactions do not take place instantaneously, and indeed often proceed rather slowly. In such cases, it is not practical to design the reactor for complete conversion of the limiting reactant; instead, the reactor effluent emerges with some of the limiting reactant still present and is then usually subjected to a separation process to remove the unconverted reactant from the product. The separated reactant is then recycled to the reactor inlet. The **fractional conversion** of a reactant is the ratio

$$f = \frac{\text{moles reacted}}{\text{moles fed}} \tag{4.6-2}$$

The fraction unreacted is accordingly $1 - f$. If 100 moles of a reactant are fed and 90 moles react, the fractional conversion is 0.90 (the **percentage conversion** is 90%) and the fraction unreacted is 0.10. If 20 mol/min of a reactant is fed and the percentage conversion is 80%, then $(20)(0.80) = 16$ mol/min has reacted and $(20)(1 - 0.80) = 4$ mol/min remains unreacted.

Considering the reaction discussed above ($C_2H_2 + 2H_2 \rightarrow C_2H_6$), suppose 20.0 kmol of acetylene, 50.0 kmol of hydrogen, and 50.0 kmol of ethane are charged into a batch reactor. Furthermore, suppose that after some time 30.0 kmol of hydrogen has reacted. How much of each species will be present in the reactor at the moment?

Clearly, if you start with 50.0 kmol of H_2 and 30.0 kmol reacts, you will be left with $\boxed{20.0\ \text{kmol } H_2}$. Also, if 30.0 kmol of H_2 reacts, 15.0 kmol of C_2H_2 also reacts (Why?), leaving you with $(20.0 - 15.0)$ kmol $C_2H_2 = \boxed{5.0\ \text{kmol } C_2H_2}$. Finally, the 30.0 kmol of H_2 that reacts forms 15.0 kmol of C_2H_6, which when added to the 50.0 kmol you started with gives you $\boxed{65.0\ \text{kmol } C_2H_6}$.

More generally, if ξ (kmol of H_2) reacts, we may follow the same reasoning and write

$$n_{H_2} = (n_{H_2})_0 - \xi$$
$$n_{C_2H_2} = (n_{C_2H_2})_0 - \tfrac{1}{2}\xi$$
$$n_{C_2H_6} = (n_{C_2H_6})_0 + \tfrac{1}{2}\xi$$

Once you know how much hydrogen (or acetylene) reacts or how much ethane is formed, you can determine ξ from one of these equations and then determine the remaining component amounts from the other two equations.

Next, we may generalize this result to any chemical reaction. To begin with, *let us define a quantity ν_i to be the stoichiometric coefficient of the ith species in a chemical reaction, making it negative for reactants and positive for products.* For example,

$$C_2H_2 + 2H_2 \rightarrow C_2H_6 \implies \nu_{C_2H_2} = -1,\ \nu_{H_2} = -2,\ \nu_{C_2H_6} = +1$$

Then if n_{i0} (or $\dot{n}_{i0}$) is the moles (batch) or molar flow rate (continuous) of species i in the feed to a batch or continuous steady-state process,

$$n_i = n_{i0} + \nu_i\xi \quad \text{or} \quad \dot{n}_i = \dot{n}_{i0} + \nu_i\dot{\xi} \tag{4.6-3}$$

(Verify that this expression yields the previously given formulas for the acetylene reaction.) The quantity ξ (or $\dot{\xi}$), which has the same units as n (or $\dot{n}$), is called the **extent of reaction**. If you know the feed quantities (n_{i0} for all i) and you also know any one of the n_i values, you can calculate ξ by applying Equation 4.6-3 to the component for which n_i is known. You can then calculate the remaining n_i values by applying Equation 4.6-3 to the other species, substituting the known values of ξ.

For example, consider the ammonia formation reaction:

$$N_2 + 3H_2 = 2NH_3$$

Suppose the feed to a continuous reactor consists of 100 mol/s of nitrogen, 300 mol/s of hydrogen, and 1 mol/s of argon (an inert gas). From Equation 4.6-3, we may write for the reactor outlet flow rates

$$\dot{n}_{N_2} = 100\ \text{mol/s} - \dot{\xi}$$
$$\dot{n}_{H_2} = 300\ \text{mol/s} - 3\dot{\xi}$$
$$\dot{n}_{NH_3} = 2\dot{\xi}$$
$$\dot{n}_{A} = 1\ \text{mol/s}$$

If you know the exit flow rate of any component or the fractional conversion of nitrogen or hydrogen, you can calculate $\dot{\xi}$ and then the other two unknown outlet flow rates. *Try it:* For a fractional hydrogen conversion of 0.60, calculate the outlet flow rate of hydrogen, the extent of reaction, and the outlet flow rates of nitrogen and ammonia. The last value should be 120 mol/s.

TEST YOURSELF

The oxidation of ethylene to produce ethylene oxide proceeds according to the equation

$$2\ C_2H_4 + O_2 \rightarrow 2\ C_2H_4O$$

The feed to a reactor contains 100 kmol C_2H_4 and 100 kmol O_2.

1. Which reactant is limiting?
2. What is the percentage excess of the other reactant?
3. If the reaction proceeds to completion, how much of the excess reactant will be left; how much C_2H_4O will be formed; and what is the extent of reaction?
4. If the reaction proceeds to a point where the fractional conversion of the limiting reactant is 50%, how much of each reactant and product is present at the end, and what is the extent of reaction?
5. If the reaction proceeds to a point where 60 kmol of O_2 is left, what is the fractional conversion of C_2H_4? The fractional conversion of O_2? The extent of reaction?

CREATIVITY EXERCISE

A single chemical reaction A → 2B, takes place in a batch reactor, with known initial quantities of A and B. Think of as many process variables as you can that might vary with the extent of reaction; then suggest means by which each of these variables might be measured in an experiment to determine the extent of reaction versus time. For example, the refractive index of the reaction mixture might vary with the mixture composition; the experiment would then be to pass a beam of light through the reaction vessel and measure the angle of refraction as a function of time.

EXAMPLE 4.6-1   *Reaction Stoichiometry*

Acrylonitrile is produced in the reaction of propylene, ammonia, and oxygen:

$$C_3H_6 + NH_3 + \tfrac{3}{2}O_2 \rightarrow C_3H_3N + 3H_2O$$

The feed contains 10.0 mole% propylene, 12.0% ammonia, and 78.0% air. A fractional conversion of 30.0% of the limiting reactant is achieved. Taking 100 mol of feed as a basis, determine which reactant is limiting, the percentage by which each of the other reactants is in excess, and the molar amounts of all product gas constituents for a 30% conversion of the limiting reactant.

SOLUTION   **Basis: 100 mol Feed**

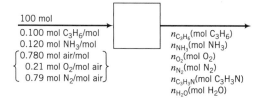

The feed to the reactor contains

$$(n_{C_3H_6})_0 = 10.0 \text{ mol}$$

$$(n_{NH_3})_0 = 12.0 \text{ mol}$$

$$(n_{O_2})_0 = \frac{78.0 \text{ mol air} \mid 0.210 \text{ mol } O_2}{\mid \text{mol air}} = 16.4 \text{ mol}$$

$$\Downarrow$$

$$\left. \begin{array}{l} (n_{NH_3}/n_{C_3H_6})_0 = 12.0/10.0 = 1.20 \\ (n_{NH_3}/n_{C_3H_6})_{stoich} = 1/1 = 1 \end{array} \right\} \Longrightarrow NH_3 \text{ is in excess } (1.20 > 1)$$

$$\left. \begin{array}{l} (n_{O_2}/n_{C_3H_6})_0 = 16.4/10.0 = 1.64 \\ (n_{O_2}/n_{C_3H_6})_{stoich} = 1.5/1 = 1.5 \end{array} \right\} \Longrightarrow O_2 \text{ is in excess } (1.64 > 1.5)$$

Since propylene is fed in less than stoichiometric proportion relative to the two other reactants, *propylene is the limiting reactant.*

To determine the percentages by which ammonia and oxygen are in excess, we must first determine the stoichiometric amounts of these reactants corresponding to the amount of propylene in the feed (10 mol) and then apply Equation 4.6-1.

$$(n_{NH_3})_{stoich} = \frac{10.0 \text{ mol } C_3H_6 \mid 1 \text{ mol } NH_3}{1 \text{ mol } C_3H_6} = 10.0 \text{ mol } NH_3$$

$$(n_{O_2})_{stoich} = \frac{10.0 \text{ mol } C_3H_6 \mid 1.5 \text{ mol } O_2}{1 \text{ mol } C_3H_6} = 15.0 \text{ mol } O_2$$

$$(\% \text{ excess})_{NH_3} = \frac{(NH_3)_0 - (NH_3)_{stoich}}{(NH_3)_{stoich}} \times 100\%$$

$$= (12.0 - 10.0)/10.0 \times 100\% = \boxed{20\% \text{ excess } NH_3}$$

$$(\% \text{ excess})_{O_2} = (16.4 - 15.0)/15.0 \times 100\% = \boxed{9.3\% \text{ excess } O_2}$$

If the fractional conversion of C_3H_6 is 30%, then

$$(n_{C_3H_6})_{out} = 0.700(n_{C_3H_6})_0 = \boxed{7.0 \text{ mol } C_3H_6}$$

But from Equation 4.6-3, $n_{C_3H_6} = 10.0 \text{ mol } C_3H_6 - \xi$. The extent of reaction is therefore $\xi = 3.0$ mol. Then, also from equation 4.6-3,

$$n_{NH_3} = 12.0 \text{ mol } NH_3 - \xi = \boxed{9.0 \text{ mol } NH_3}$$
$$n_{O_2} = 16.4 \text{ mol } O_2 - 1.5\xi = \boxed{11.9 \text{ mol } O_2}$$
$$n_{C_3H_3N} = \xi = \boxed{3.00 \text{ mol } C_3H_3N}$$
$$n_{N_2} = (n_{N_2})_0 = \boxed{61.6 \text{ mol } N_2}$$
$$n_{H_2O} = 3\xi = \boxed{9.0 \text{ mol } H_2O}$$

4.6c   Chemical Equilibrium

Two of the fundamental questions of chemical reaction engineering are, given a set of reactive species and reaction conditions, (a) what will be the final (equilibrium) composition of the reaction mixture, and (b) how long will the system take to reach a specified state short of equilibrium? The field of **chemical equilibrium thermodynamics** concerns itself with the first question, and **chemical kinetics** deals with the second.

Some reactions are essentially **irreversible**; that is, the reaction proceeds only in a single direction (from reactants to products) and the concentration of the limiting reactant eventually approaches zero (although "eventually" could mean seconds for some reactions and years for others). The equilibrium composition for such a reaction is therefore the composition corresponding to complete consumption of the limiting reactant.

Other reactions (or the same reactions at different conditions) are **reversible**; reactants form products and products undergo the reverse reactions to reform the reactants. For example, consider the reaction in which ethylene is *hydrolyzed* to ethanol:

$$C_2H_4 + H_2O \rightleftharpoons C_2H_5OH$$

If you start with ethylene and water, the forward reaction occurs; then once ethanol is present, the reverse reaction begins to take place. As the concentrations of C_2H_4 and H_2O decrease, the rate of the forward reaction decreases, and as the C_2H_5OH concentration increases, the rate of the reverse reaction increases. Eventually a point is reached at which the rates of the forward and reverse reactions are equal. At this point no further composition change takes place and the reaction mixture is in chemical equilibrium.

A full discussion of the relations that can be used to determine equilibrium compositions of reactive mixtures is beyond the scope of this text; however, at this point you have enough knowledge to be able to calculate equilibrium compositions if the relations are given to you. The next example illustrates such a calculation.

EXAMPLE 4.6-2 *Calculation of an Equilibrium Composition*

If the water–gas shift reaction,

$$CO(g) + H_2O(g) \rightleftharpoons CO_2(g) + H_2(g)$$

proceeds to equilibrium at a temperature T (K), the mole fractions of the four reactive species satisfy the relation

$$\frac{y_{CO_2} y_{H_2}}{y_{CO} y_{H_2O}} = K(T)$$

where $K(T)$ is the reaction **equilibrium constant**. At $T = 1105$ K, $K = 1.00$.

Suppose the feed to a reactor contains 1.00 mol of CO, 2.00 mol of H_2O, and no CO_2 or H_2, and the reaction mixture comes to equilibrium at 1105 K. Calculate the equilibrium composition and the fractional conversion of the limiting reactant.

SOLUTION

The strategy is to express all mole fractions in terms of a single variable (ξ_e, the extent of reaction at equilibrium), substitute in the equilibrium relation, solve for ξ_e, and back-substitute to calculate the mole fractions and any other desired quantity.

From Equation 4.6-3,

$$n_{CO} = 1.00 \text{ mol} - \xi_e \quad \text{(number of gram-moles of CO present at equilibrium)}$$
$$n_{H_2O} = 2.00 \text{ mol} - \xi_e$$
$$n_{CO_2} = \xi_e$$
$$\underline{n_{H_2} = \xi_e}$$
$$n_{total} = 3.00 \text{ mol}$$

from which

$$y_{CO} = (1.00 \text{ mol} - \xi_e)/3.00 \text{ mol}$$
$$y_{H_2O} = (2.00 \text{ mol} - \xi_e)/3.00 \text{ mol}$$
$$y_{CO_2} = \xi_e/3.00 \text{ mol}$$
$$y_{H_2} = \xi_e/3.00 \text{ mol}$$

Substitution of these expressions into the equilibrium relation (with $K = 1.00$) yields

$$\frac{y_{CO_2} y_{H_2}}{y_{CO} y_{H_2O}} = \frac{\xi_e^2}{(1.00 \text{ mol} - \xi_e)(2.00 \text{ mol} - \xi_e)} = 1.00$$

This may be rewritten as a standard quadratic equation (*verify*) and solved to yield $\xi_e = 0.667$ mol. This quantity may in turn be substituted back into the expression for y_i to yield

$$\boxed{y_{CO} = 0.111, \quad y_{H_2O} = 0.444, \quad y_{CO_2} = 0.222, \quad y_{H_2} = 0.222}$$

The limiting reactant in this case is CO (*verify*). At equilibrium,

$$n_{CO} = (1.00 - 0.667) \text{ mol} = 0.333 \text{ mol}$$

The fractional conversion of CO at equilibrium is therefore

$$f_{CO} = (1.00 - 0.333) \text{ mol CO reacted}/(1.00 \text{ mol CO fed}) = \boxed{0.667}$$

4.6d   Multiple Reactions, Yield, and Selectivity

In most chemical processes, reactants are brought together with the object of producing a desired product in a single reaction. Unfortunately, reactants can usually combine in more than one way, and the product once formed may react to yield something less desirable. The result of these side reactions is an economic loss: less of the desired product is obtained for a given quantity of raw materials, or a greater quantity of raw materials must be fed to the reactor to obtain a specified product yield.

For example, ethylene can be produced by the dehydrogenation of ethane:

$$C_2H_6 \rightarrow C_2H_4 + H_2$$

Once some hydrogen is produced, it can react with ethane to produce methane:

$$C_2H_6 + H_2 \rightarrow 2\,CH_4$$

Moreover, ethylene can react with ethane to form propylene and methane:

$$C_2H_4 + C_2H_6 \rightarrow C_3H_6 + CH_4$$

Since the object of the process is to produce ethylene, only the first of these reactions may be regarded as desirable; the second one consumes the reactant without yielding the desired product and the third consumes both the reactant and the desired product. The engineer designing the reactor and specifying operating conditions for it must consider not only how to maximize the production of the desired product (C_2H_4), but also how to minimize the production of undesired by-products (CH_4, C_3H_6).

The terms **yield** and **selectivity** are used to describe the degree to which a desired reaction predominates over competing side reactions.

$$\textit{Yield:} \qquad \frac{\text{moles of desired product formed}}{\substack{\text{moles that would have been formed if there were} \\ \text{no side reactions and the limiting reactant had} \\ \text{reacted completely}}} \qquad \textbf{(4.6-4)}$$

$$\textit{Selectivity:} \qquad \frac{\text{moles of desired product formed}}{\text{moles of undesired product formed}} \qquad \textbf{(4.6-5)}$$

The yield defined by Equation 4.6-4 is always a fraction; it may also be expressed as a percentage by multiplying by 100%. If A is the desired product and B is an undesired product, one then refers to the *selectivity of A relative to B*. High values of the yield and selectivity signify that the undesired side reactions have been successfully suppressed relative to the desired reaction.

Besides being defined by Equation 4.6-4, yield is also sometimes defined as moles of desired product divided by either moles of reactant fed or moles of reactant consumed in the reactor. For given feed and product compositions, the yields defined in these three manners may have completely different values, so that when you see a reference to a yield it is important to be aware of the working definition.

The concept of extent of reaction can be extended to multiple reactions, only now each independent reaction has its own extent. If a set of reactions takes place in a batch or continuous steady-state reactor and ν_{ij} is the stoichiometric coefficient of substance i in reaction j (negative for reactants, positive for products), we may then write

$$n_i = n_{i0} + \sum_j \nu_{ij}\xi_j \qquad \textbf{(4.6-6)}$$

For a single reaction, this equation reduces to Equation 4.6-3.

For example, consider the pair of reactions in which ethylene is oxidized either to ethylene oxide (desired) or to carbon dioxide (undesired):

$$C_2H_4 + \tfrac{1}{2}O_2 \rightarrow C_2H_4O$$

$$C_2H_4 + 3O_2 \rightarrow 2\,CO_2 + 2\,H_2O$$

The moles (or molar flow rates) of each of the five species involved in these reactions can be expressed in terms of the feed values and extents of reaction using Equation 4.6-6:

$$(n_{C_2H_4})_{out} = (n_{C_2H_4})_0 - \xi_1 - \xi_2$$
$$(n_{O_2})_{out} = (n_{O_2})_0 - 0.5\xi_1 - 3\xi_2$$
$$(n_{C_2H_4O})_{out} = (n_{C_2H_4O})_0 + \xi_1$$
$$(n_{CO_2})_{out} = (n_{CO_2})_0 + 2\xi_2$$
$$(n_{H_2O})_{out} = (n_{H_2O})_0 + 2\xi_2$$

If values of any two outlet amounts are given, the values of ξ_1 and ξ_2 may be determined from the corresponding two equations, and the remaining amounts may in turn be calculated from the remaining three equations. The next example illustrates this procedure.

EXAMPLE 4.6-3   *Yield and Selectivity in a Dehydrogenation Reactor*

The reactions

$$C_2H_6 \rightarrow C_2H_4 + H_2$$
$$C_2H_6 + H_2 \rightarrow 2\,CH_4$$

take place in a continuous reactor at steady state. The feed contains 85.0 mole% ethane (C_2H_6) and the balance inerts (I). The fractional conversion of ethane is 0.501, and the fractional yield of ethylene is 0.471. Calculate the molar composition of the product gas and the selectivity of ethylene to methane production.

SOLUTION   *Basis: 100 mol Feed*

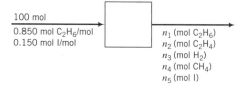

From Equation 4.6-6, the outlet component amounts in terms of extents of reaction are as follows:

$$n_1(\text{mol } C_2H_6) = 85.0 \text{ mol } C_2H_6 - \xi_1 - \xi_2$$
$$n_2(\text{mol } C_2H_4) = \xi_1$$
$$n_3(\text{mol } H_2) = \xi_1 - \xi_2$$
$$n_4(\text{mol } CH_4) = 2\xi_2$$
$$n_5(\text{mol I}) = 15.0 \text{ mol I}$$

Ethane Conversion

If the fractional conversion of ethane is 0.501, the fraction *un*converted (and hence leaving the reactor) must be $(1 - 0.501)$.

$$n_1 = \frac{(1 - 0.501) \text{ mol } C_2H_6 \text{ unreacted}}{\text{mol } C_2H_6 \text{ fed}} \left| 85.0 \text{ mol } C_2H_6 \text{ fed} \right.$$

$$= 42.4 \text{ mol } C_2H_6 = 85.0 \text{ mol } C_2H_6 - \xi_1 - \xi_2 \tag{1}$$

Ethylene Yield

$$\text{maximum possible ethylene formed} = \frac{85.0 \text{ mol } C_2H_6 \text{ fed}}{} \left| \frac{1 \text{ mol } C_2H_4}{1 \text{ mol } C_2H_6} = 85.0 \text{ mol} \right.$$

$$\Downarrow$$

$$n_2 = 0.471(85.0 \text{ mol } C_2H_6) = 40.0 \text{ mol } C_2H_4 = \xi_1$$

Substituting 40.0 mol for ξ_1 in Equation 1 yields $\xi_2 = 2.6$ mol. Then

$$n_3 = \xi_1 - \xi_2 = 37.4 \text{ mol H}_2$$

$$n_4 = 2\xi_2 = 5.2 \text{ mol CH}_4$$

$$n_5 = 15.0 \text{ mol I}$$

$$n_{\text{tot}} = (42.4 + 40.0 + 37.4 + 5.2 + 15.0) \text{ mol} = 140.0 \text{ mol}$$

$$\Downarrow$$

Product: $\boxed{30.3\% \text{ C}_2\text{H}_6, \quad 28.6\% \text{ C}_2\text{H}_4, \quad 26.7\% \text{ H}_2, \quad 3.7\% \text{ CH}_4, \quad 10.7\% \text{ I}}$

$$\text{selectivity} = (40.0 \text{ mol C}_2\text{H}_4)/(5.2 \text{ mol CH}_4)$$

$$= \boxed{7.7 \frac{\text{mol C}_2\text{H}_4}{\text{mol CH}_4}}$$

TEST YOURSELF

Consider the following pair of reactions:

$$A \rightarrow 2B \text{ (desired)}$$

$$A \rightarrow C \text{ (undesired)}$$

Suppose 100 mol of A is fed to a batch reactor and the final product contains 10 mol of A, 160 mol of B, and 10 mol of C. Calculate

1. The fractional conversion of A.
2. The percentage yield of B.
3. The selectivity of B relative to C.
4. The extents of the first and second reactions.

4.7   BALANCES ON REACTIVE PROCESSES

4.7a   Balances on Molecular and Atomic Species

Figure 4.7-1 shows a flowchart for the dehydrogenation of ethane in a steady-state continuous reactor. The reaction is

$$C_2H_6 \longrightarrow C_2H_4 + H_2$$

One hundred kmol/min of ethane is fed to the reactor. The molar flow rate of H_2 in the product stream is 40 kmol/min.

A number of different balances could be written on this process, including balances on total mass, C_2H_6, C_2H_4, and H_2. Of these, only the first has the simple form *input = output:* since the three given species all participate in the reaction, the balance equation for each of them must include a generation term (for C_2H_4 and H_2) or a consumption term (for C_2H_6).

Notice, however, that balance equations may also be written for atomic carbon and atomic hydrogen, regardless of the molecular species in which the carbon and hydrogen atoms happen to be found. Balances on atomic species *can* be written *input = output,* since atoms can neither be created (generation = 0) nor destroyed (consumption = 0) in a chemical reaction.

Before illustrating these balances, let us resolve an ambiguity. When we speak of a hydrogen balance, it could mean two completely different things: a balance on molecular

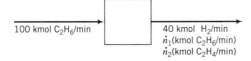

Figure 4.7-1   Dehydrogenation of ethane.

hydrogen (H_2), existing as an independent species, or a balance on total hydrogen (H), bound and unbound. We will henceforth use the terms **molecular hydrogen balance** and **atomic hydrogen balance** to represent these two types of balances.

Some of the balances that can be written for the process shown in Figure 4.7-1 are as follows. (Recall that the general balance equation for a steady-state process is input + generation = output + consumption.)

Molecular H_2 Balance: generation = output

$$\text{Gen}_{H_2}\left(\frac{\text{mol } H_2 \text{ generated}}{\text{min}}\right) = 40 \text{ mol } H_2/\text{min}$$

C_2H_6 Balance: input = output + consumption

$$\frac{100 \text{ mol } C_2H_6}{\text{min}} = \dot{n}_1\left(\frac{\text{mol } C_2H_6}{\text{min}}\right) + \text{Cons}_{C_2H_6}\left(\frac{\text{mol } C_2H_6 \text{ consumed}}{\text{min}}\right)$$

C_2H_4 Balance: generation = output

$$\text{Gen}_{C_2H_4}\left(\frac{\text{mol } C_2H_4 \text{ generated}}{\text{min}}\right) = \dot{n}_2\left(\frac{\text{mol } C_2H_4}{\text{min}}\right)$$

Atomic C Balance: input = output

$$\frac{100 \text{ mol } C_2H_6}{\text{min}} \, \bigg| \, \frac{2 \text{ mol C}}{1 \text{ mol } C_2H_6} = \dot{n}_1 \frac{\text{mol } C_2H_6}{\text{min}} \, \bigg| \, \frac{2 \text{ mol C}}{1 \text{ mol } C_2H_6} + \dot{n}_2 \frac{\text{mol } C_2H_4}{\text{min}} \, \bigg| \, \frac{2 \text{ mol C}}{1 \text{ mol } C_2H_4}$$

$$\Downarrow$$

$$100 \text{ mol C/min} = \dot{n}_1 + \dot{n}_2$$

Atomic H Balance: input = output

$$\frac{100 \text{ mol } C_2H_6}{\text{min}} \, \bigg| \, \frac{6 \text{ mol H}}{1 \text{ mol } C_2H_6} = \frac{40 \text{ mol } H_2}{\text{min}} \, \bigg| \, \frac{2 \text{ mol H}}{1 \text{ mol } H_2}$$

$$+ \dot{n}_1 \frac{\text{mol } C_2H_6}{\text{min}} \, \bigg| \, \frac{6 \text{ mol H}}{1 \text{ mol } C_2H_6} + \dot{n}_2 \frac{\text{mol } C_2H_4}{\text{min}} \, \bigg| \, \frac{4 \text{ mol H}}{1 \text{ mol } C_2H_4}$$

$$\Downarrow$$

$$600 \text{ mol H/min} = 80 \text{ mol H/min} + 6\dot{n}_1 + 4\dot{n}_2$$

You could solve for $\dot{n}_1$ and $\dot{n}_2$ either directly from the two atomic balances or by using the three molecular species balances in conjunction with the stoichiometric equation for the reaction.

In general, systems that involve chemical reactions may be analyzed using (a) molecular species balances (the approach always used for nonreactive systems), (b) atomic species balances, and (c) extents of reaction. Each approach leads to the same results, but any one of them may be more convenient for a given calculation so it is a good idea to become comfortable with all three.

To carry out degree-of-freedom analyses of reactive systems you must first understand the concepts of *independent equations, independent species,* and *independent chemical reactions.* We explain these concepts in the next section, following which we outline and illustrate the three approaches to reactive system balance calculations.

4.7b Independent Equations, Independent Species, and Independent Reactions

When we first described degree-of-freedom analysis in Section 4.3d, we said that the maximum number of material balances you can write for a nonreactive process equals the number of *independent species* involved in the process. It is time to take a closer look at what that means and to see how to extend the analysis to reactive processes.

The key concept is that of *independent equations*. Algebraic equations are independent if you cannot obtain any one of them by adding and subtracting multiples of any of the others. For example, the equations

$$[1] \quad x + 2y = 4$$
$$[2] \quad 3x + 6y = 12$$

are not independent, because [2] = 3 × [1]. In effect, they are really the same equation. (Convince yourself by seeing what happens if you try to solve the two equations for x and y.) Similarly, the equations

$$[1] \quad x + 2y = 4$$
$$[2] \quad 2x - z = 2$$
$$[3] \quad 4y + z = 6$$

are not independent, because [3] = 2 × [1] − [2]. (Prove it.)

If two molecular species are in the same ratio to each other wherever they appear in a process and this ratio is incorporated in the flowchart labeling, balances on those species will not be independent equations. Similarly, if two atomic species occur in the same ratio wherever they appear in a process, balances on those species will not be independent equations.

For example, consider a process in which a stream of liquid carbon tetrachloride is vaporized into a stream of air.

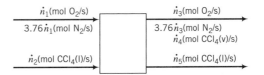

Since nitrogen and oxygen are shown as being in the same ratio wherever they appear on the flowchart (3.76 mol N_2/mol O_2), you cannot count them as two independent species and so you may count only two independent molecular species balances in a degree-of-freedom analysis—one for either O_2 or N_2 and one for CCl_4. (Try writing separate O_2 and N_2 balances and see what you get.)

Similarly, atomic nitrogen (N) and atomic oxygen (O) are always in the same proportion to each other in the process (again 3.76:1) as are atomic chlorine and atomic carbon (4 mol Cl/1 mol C). Consequently, even though four atomic species are involved in this process, you may count only two independent atomic species balances in the degree-of-freedom analysis—one for either O or N and one for either C or Cl. (Again, convince yourself that the O and N balances yield the same equation, as do the C and Cl balances.)

Finally, when you are using either molecular species balances or extents of reaction to analyze a reactive system, the degree-of-freedom analysis must account for the number of independent chemical reactions among the species entering and leaving the system. *Chemical reactions are independent if the stoichiometric equation of any one of them cannot be obtained by adding and subtracting multiples of the stoichiometric equations of the others.*

For example, consider the reactions

$$[1] \quad A \longrightarrow 2B$$
$$[2] \quad B \longrightarrow C$$
$$[3] \quad A \longrightarrow 2C$$

These three reactions are not all independent, since [3] = [1] + 2 × [2].

$$[1]: \quad A \longrightarrow 2B$$
$$2 \times [2]: \quad 2B \longrightarrow 2C$$
$$[1] + 2 \times [2]: \quad A + 2B \longrightarrow 2B + 2C \Longrightarrow A \longrightarrow 2C \ (= [3])$$

However, any two of the reactions are independent. (They must be, since each one involves at least one species that does not appear in the other one.)

TEST
YOURSELF

1. A mixture of ethylene and nitrogen is fed to a reactor in which some of the ethylene is dimerized to butene.

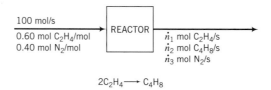

$$2C_2H_4 \longrightarrow C_4H_8$$

How many independent molecular species are involved in the process? How many independent atomic species are involved? Prove the latter claim by writing balances on C, H, and N.

2. Write the stoichiometric equations for the combustion of methane with oxygen to form (a) CO_2 and H_2O and (b) CO and H_2O; and for the combustion of ethane with oxygen to form (c) CO_2 and H_2O and (d) CO and H_2O. Then prove that only three of these four reactions are independent.

4.7c   Molecular Species Balances

If molecular species balances are used to determine unknown stream variables for a reactive process, the balances on reactive species must contain generation and/or consumption terms. The degree-of-freedom analysis is as follows:

      No. unknown labeled variables

+ No. independent chemical reactions (as defined in Section 4.7b)

− No. independent molecular species balances (as defined in Section 4.7b)

− No. other equations relating unknown variables

= No. degrees of freedom

Once a generation or consumption term has been calculated for a species in a given reaction, the generation and consumption terms for all other species in that reaction may be determined directly from the stoichiometric equation. (We will shortly illustrate this determination.) One generation or consumption term must therefore either be specified or calculated for each independent reaction, which is why each reaction adds a degree of freedom to the system.

We will use the dehydrogenation of ethane (Figure 4.7-1) to illustrate the required procedures. The flowchart is shown again here for ease of reference.

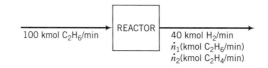

Degree-of-Freedom Analysis

      2 unknown labeled variables ($\dot{n}_1, \dot{n}_2$)

+ 1 independent chemical reaction

− 3 independent molecular species balances (C_2H_6, C_2H_4, and H_2)

− 0 other equations relating unknown variables

= 0 degrees of freedom

The hydrogen balance will be used to determine the rate of hydrogen generation, and the outlet flow rates of ethane and ethylene will be calculated from the balances on those species. Notice how the ethane consumption and ethylene generation terms are determined directly from the hydrogen generation term.

H_2 Balance: generation = output

$$\text{Gen}_{H_2} \left(\frac{\text{kmol } H_2 \text{ generated}}{\text{min}} \right) = 40 \text{ kmol } H_2/\text{min}$$

C_2H_6 Balance: input = output + consumption

$$100 \, \frac{\text{kmol } C_2H_6}{\text{min}} = \dot{n}_1 \left(\frac{\text{kmol } C_2H_6}{\text{min}} \right)$$

$$+ \frac{40 \text{ kmol } H_2 \text{ generated}}{\text{min}} \left| \frac{1 \text{ kmol } C_2H_6 \text{ consumed}}{1 \text{ kmol } H_2 \text{ generated}} \right. \implies \boxed{\dot{n}_1 = 60 \text{ kmol } C_2H_6/\text{min}}$$

C_2H_4 Balance: generation = output

$$\frac{40 \text{ kmol } H_2 \text{ generated}}{\text{min}} \left| \frac{1 \text{ kmol } C_2H_4 \text{ generated}}{1 \text{ kmol } H_2 \text{ generated}} \right. = \dot{n}_2 \left(\frac{\text{kmol } C_2H_4}{\text{min}} \right)$$

$$\implies \boxed{\dot{n}_2 = 40 \text{ kmol } C_2H_4/\text{min}}$$

4.7d   Atomic Species Balances

All balances on atomic species (C, H, O, etc.) take the form "input = output," since atomic species can neither be generated nor consumed in chemical reactions (as opposed to nuclear reactions). The number of degrees of freedom is determined directly by subtracting equations from labeled unknowns: no additional degrees of freedom are contributed by the reactions.

> No. unknown labeled variables
> − No. independent atomic species balances (as defined in Section 4.7b)
> − No. molecular balances on independent nonreactive species
> − No. other equations relating unknown variables
> = No. degrees of freedom

In the ethane dehydrogenation process, the two unknown flow rates will be determined from balances on atomic carbon and atomic hydrogen.

Degree-of-Freedom Analysis

> 2 unknown labeled variables
> − 2 independent atomic species balances (C and H)
> − 0 molecular balances on independent nonreactive species
> − 0 other equations relating unknown variables
> = 0 degrees of freedom

C Balance: input = output

$$\frac{100 \text{ kmol } C_2H_6}{\text{min}} \left| \frac{2 \text{ kmol C}}{1 \text{ kmol } C_2H_6} \right.$$

$$= \frac{\dot{n}_1 (\text{kmol } C_2H_6)}{(\text{min})} \left| \frac{2 \text{ kmol C}}{1 \text{ kmol } C_2H_6} \right. + \frac{\dot{n}_2 (\text{kmol } C_2H_4)}{(\text{min})} \left| \frac{2 \text{ kmol C}}{1 \text{ kmol } C_2H_4} \right.$$

$$\Downarrow$$

$$100 = \dot{n}_1 + \dot{n}_2 \qquad \qquad \textbf{(1)}$$

H Balance: input = output

$$\frac{100 \text{ kmol C}_2\text{H}_6}{\text{min}} \left| \frac{6 \text{ kmol H}}{1 \text{ kmol C}_2\text{H}_6} \right. = \frac{40 \text{ kmol H}_2}{\text{min}} \left| \frac{2 \text{ kmol H}}{1 \text{ kmol H}_2} \right.$$

$$+ \frac{\dot{n}_1 (\text{kmol C}_2\text{H}_6)}{(\text{min})} \left| \frac{6 \text{ kmol H}}{1 \text{ kmol C}_2\text{H}_6} \right. + \frac{\dot{n}_2 (\text{kmol C}_2\text{H}_4)}{(\text{min})} \left| \frac{4 \text{ kmol H}}{1 \text{ kmol C}_2\text{H}_4} \right.$$

$$\Downarrow$$

$$600 \text{ mol H/min} = 80 \text{ mol H/min} + 6\dot{n}_1 + 4\dot{n}_2 \qquad (2)$$

Solving Equations (1) and (2) simultaneously yields the same solutions obtained with atomic species balances:

$$\boxed{\dot{n}_1 = 60 \text{ kmol C}_2\text{H}_6/\text{min}}$$

$$\boxed{\dot{n}_2 = 40 \text{ kmol C}_2\text{H}_4/\text{min}}$$

4.7e Extent of Reaction

The third way to determine unknown molar flow rates for a reactive process is to write expressions for each product species flow rate (or molar amount) in terms of extents of reaction using Equation 4.6-3 (or Equation 4.6-6 for multiple reactions), substitute known feed and product flow rates, and solve for the extents of reaction and the remaining reactive species flow rates. The degree-of-freedom analysis follows:

No. unknown labeled variables

+ No. independent reactions (one extent of reaction for each)

− No. independent reactive species
  (one equation for each species in terms of extents of reaction)

− No. independent nonreactive species (one balance equation for each)

− No. other equations relating unknown variables

= No. degrees of freedom

In the dehydrogenation process (refer one final time to the flowchart), DF = 2 unknown variables $(\dot{n}_1, \dot{n}_2)$ + 1 independent reaction − 3 independent reactive species (C_2H_6, C_2H_4, H_2) = 0. For the same process, Equation 4.6-2 $(\dot{n}_i = \dot{n}_{io} + \nu_i \xi)$ for the three species in the process becomes

$$H_2(\nu = 1): 40 \text{ kmol H}_2/\text{min} = \xi \Longrightarrow \xi = 40 \text{ kmol/min}$$

$$C_2H_6(\nu = -1): \dot{n}_1 = 100 \text{ kmol C}_2\text{H}_6/\text{min} - \xi \xrightarrow{\xi = 40 \text{ kmol/min}} \boxed{\dot{n}_1 = 60 \text{ kmol C}_2\text{H}_6/\text{min}}$$

$$C_2H_4(\nu = 1): \dot{n}_2 = \xi \xrightarrow{\xi = 40 \text{ kmol/min}} \boxed{\dot{n}_2 = 40 \text{ kmol C}_2\text{H}_4/\text{min}}$$

Given that all three methods of carrying out material balances on reactive systems—molecular species balances, atomic species balances, and extents of reaction—necessarily yield the same results, the question is which one to use for a given process. There are no hard and fast rules but we suggest the following guidelines:

- *Atomic species balances generally lead to the most straightforward solution procedure, especially when more than one reaction is involved.*
- *Extents of reaction are convenient for chemical equilibrium problems and when equation-solving software is to be used.*

- *Molecular species balances require more complex calculations than either of the other two approaches and should be used only for simple systems involving one reaction.*

The next example illustrates all three approaches for a system involving two reactions.

EXAMPLE 4.7-1   *Incomplete Combustion of Methane*

Methane is burned with air in a continuous steady-state combustion reactor to yield a mixture of carbon monoxide, carbon dioxide, and water. The reactions taking place are

$$CH_4 + \tfrac{3}{2} O_2 \longrightarrow CO + 2H_2O \tag{1}$$

$$CH_4 + 2O_2 \longrightarrow CO_2 + 2H_2O \tag{2}$$

The feed to the reactor contains 7.80 mole% CH_4, 19.4% O_2, and 72.8% N_2. The percentage conversion of methane is 90.0%, and the gas leaving the reactor contains 8 mol CO_2/mol CO. Carry out a degree-of-freedom analysis on the process. Then calculate the molar composition of the product stream using molecular species balances, atomic species balances, and extents of reaction.

SOLUTION   *Basis = 100 mol Feed*

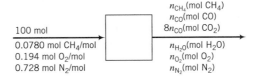

Degree-of-Freedom Analysis

The analysis can be based on any of the three solution methods:

- *Molecular species balances* (Section 4.7c). 5 unknown variables + 2 independent reactions − 6 independent molecular species balances (CH_4, O_2, N_2, CO, CO_2, H_2O) − 1 specified methane conversion = 0 degrees of freedom.
- *Atomic species balances* (Section 4.7d). 5 unknown variables − 3 independent atomic species balances (C, H, O) − 1 nonreactive molecular species balance (N_2) − 1 specified methane conversion = 0 degrees of freedom.
- *Extents of reaction* (Section 4.7e). 5 unknown labeled variables + 2 independent reactions − 5 expressions for $n_i(\xi)$ (i = CH_4, O_2, CO, CO_2, H_2O) − 1 nonreactive molecular species balance (N_2) − 1 specified methane conversion = 0 degrees of freedom.

Before balances are written, the specified methane conversion can be used to determine n_{CH_4}.

90% CH_4 Conversion: (10% remains unconverted)

$$\boxed{n_{CH_4} = 0.100(7.80 \text{ mol } CH_4 \text{ fed}) = 0.780 \text{ mol } CH_4}$$

All three solution methods involve writing a balance on nitrogen (the nonreactive species in the process), so we may as well do that now as well.

N_2 Balance: input = output

$$\boxed{n_{N_2} = 72.8 \text{ mol } N_2}$$

It remains to determine n_{CO}, n_{H_2O}, and n_{O_2}. We will proceed by each of the indicated methods.

Molecular Species Balances

As we mentioned before, this method is the most cumbersome when multiple reactions are involved and we strongly recommend against using it. We present it here only for illustrative purposes (mainly to illustrate why we recommend against using it).

Each balance on a reactive species will contain a generation or consumption term. We will use the notation $C_{CH_4,1}$(mol CH_4) to denote the consumption of methane in Reaction 1, $G_{H_2O,2}$ (mol H_2O) to denote the generation of water in Reaction 2, and so on. Note that any G and C term for a specified reaction may be expressed in terms of any other G or C term for the same reaction directly from the stoichiometric equation. For example, the generation of water in Reaction 1 may be expressed in terms of the consumption of oxygen in that reaction as

$G_{H_2O,1}$ (mol H_2O generated in Reaction 2)

$$= C_{O_2,1} \text{ (mol } O_2 \text{ consumed in Reaction 2)} \times \left(\frac{2 \text{ mol } H_2O \text{ generated}}{1.5 \text{ mol } O_2 \text{ consumed}} \right) \quad (\textit{Verify!})$$

Since the balances on CO and CO_2 each involve the same unknown molar amount (n_{CO}), we will begin with those balances. Make sure you understand the form of each balance (like the CO balance, which simplifies to "output = generation").

CO Balance: output = generation

$$n_{CO} = G_{CO,1} \tag{3}$$

CO₂ Balance: output = generation

$$8n_{CO} = G_{CO_2,2} \tag{4}$$

Since we know both the feed and output amounts of methane, a methane balance should involve only the two methane consumption terms (one for each reaction) as unknowns. Since $C_{CH_4,1}$ can be expressed in terms of $G_{CO,1}$ and $C_{CH_4,2}$ can be expressed in terms of $G_{CO_2,2}$, the CO, CO_2, and CH_4 balances will yield three equations in three unknowns—$n_{CO}, G_{CO,1}$, and $G_{CO_2,2}$.

CH₄ Balance: input = output + consumption

$$7.80 \text{ mol } CH_4 = 0.780 \text{ mol } CH_4 + C_{CH_4,1} + C_{CH_4,2}$$

$$\left\Vert \begin{array}{l} C_{CH_4,1} = G_{CO,1} \times (1 \text{ mol } CH_4 \text{ consumed/1 mol CO generated}) \\ C_{CH_4,2} = G_{CO_2,2} \times (1 \text{ mol } CH_4 \text{ consumed/1 mol } CO_2 \text{ generated}) \end{array} \right.$$

$$7.02 \text{ mol } CH_4 = G_{CO,1} + G_{CO_2,2}$$

$$\Vert \text{Equations 3 and 4}$$

$$7.02 \text{ mol } CH_4 = n_{CO} + 8n_{CO} = 9n_{CO}$$

$$\Downarrow$$

$$\boxed{n_{CO} = 0.780 \text{ mol CO}}$$

$$\boxed{n_{CO_2} = (8 \times 0.780) \text{ mol } CO_2 = 6.24 \text{ mol } CO_2}$$

Equations 3 and 4 now yield

$$G_{CO,1} = n_{CO} = 0.780 \text{ mol CO generated} \tag{5}$$

$$G_{CO_2,2} = 8n_{CO} = 6.24 \text{ mol } CO_2 \text{ generated} \tag{6}$$

Water and oxygen balances complete the calculation of the unknown flow rates.

H₂O Balance: output = generation

$$n_{H_2O} = G_{H_2O,1} + G_{H_2O,2}$$

$$= G_{CO,1} \left(\frac{2 \text{ mol } H_2O \text{ generated}}{1 \text{ mol CO generated}} \right) + G_{CO_2,2} \left(\frac{2 \text{ mol } H_2O \text{ generated}}{1 \text{ mol } CO_2 \text{ generated}} \right)$$

$$\Vert G_{CO,1} = 0.780 \text{ mol CO generated}, \ G_{CO_2,2} = 6.24 \text{ mol } CO_2 \text{ generated}$$

$$\boxed{n_{H_2O} = 14.0 \text{ mol } H_2O}$$

O₂ Balance: output = input − consumption

$$n_{O_2} = 19.4 \text{ mol } O_2 - C_{O_2,1} - C_{O_2,2}$$

$$= 19.4 \text{ mol } O_2 - G_{CO,1}\left(\frac{1.5 \text{ mol } O_2 \text{ consumed}}{1 \text{ mol CO generated}}\right) - G_{CO_2,2}\left(\frac{2 \text{ mol } O_2 \text{ consumed}}{1 \text{ mol } CO_2 \text{ generated}}\right)$$

$\Downarrow G_{CO,1} = 0.780 \text{ mol CO generated}, G_{CO_2,2} = 6.24 \text{ mol } CO_2 \text{ generated}$

$$\boxed{n_{O_2} = 5.75 \text{ mol } O_2}$$

In summary, the stack gas contains 0.780 mol CH_4, 0.780 mol CO, 6.24 mol CO_2, 14.0 mol H_2O, 5.75 mol O_2, and 72.8 mol N_2. The molar composition of the gas is therefore

$$\boxed{0.78\% \text{ } CH_4, 0.78\% \text{ CO}, 6.2\% \text{ } CO_2, 14.0\% \text{ } H_2O, 5.7\% \text{ } O_2, \text{ and } 72.5\% \text{ } N_2}$$

Atomic Species Balances

Referring to the flowchart, we see that a balance on atomic carbon involves only one unknown (n_{CO}) and a balance on atomic hydrogen also involves one unknown (n_{H_2O}), but a balance on atomic oxygen involves three unknowns. We will therefore write the C and H balances first, and then the O balance to determine the remaining unknown variable, n_{O_2}. All atomic balances have the form *input = output*. We will just determine the component amounts; calculation of the mole fractions then follows as in the previous part.

C Balance

$$\frac{7.8 \text{ mol } CH_4 \mid 1 \text{ mol C}}{\mid 1 \text{ mol } CH_4} = \frac{0.78 \text{ mol } CH_4 \mid 1 \text{ mol C}}{\mid 1 \text{ mol } CH_4}$$

$$+ \frac{n_{CO}(\text{mol CO}) \mid 1 \text{ mol C}}{\mid 1 \text{ mol CO}} + \frac{8n_{CO}(\text{mol } CO_2) \mid 1 \text{ mol C}}{\mid 1 \text{ mol } CO_2}$$

$\Downarrow$ Solve for n_{CO}

$$\boxed{n_{CO} = 0.780 \text{ mol CO}}$$

$$\boxed{n_{CO_2} = 8n_{CO} = (8 \times 0.780) \text{ mol } CO_2 = 6.24 \text{ mol } CO_2}$$

H Balance

$$\frac{7.8 \text{ mol } CH_4 \mid 4 \text{ mol H}}{\mid 1 \text{ mol } CH_4} = \frac{0.78 \text{ mol } CH_4 \mid 4 \text{ mol H}}{\mid 1 \text{ mol } CH_4}$$

$$+ \frac{n_{H_2O}(\text{mol } H_2O) \mid 2 \text{ mol H}}{\mid 1 \text{ mol } H_2O} \Longrightarrow \boxed{n_{H_2O} = 14.0 \text{ mol } H_2O}$$

O Balance

$$\frac{19.4 \text{ mol } O_2 \mid 2 \text{ mol O}}{\mid 1 \text{ mol } O_2} = \frac{n_{O_2}(\text{mol } O_2) \mid 2 \text{ mol O}}{\mid 1 \text{ mol } O_2} + \frac{0.78 \text{ mol CO} \mid 1 \text{ mol O}}{\mid 1 \text{ mol CO}}$$

$$+ \frac{6.24 \text{ mol } CO_2 \mid 2 \text{ mol O}}{\mid 1 \text{ mol } CO_2} + \frac{14.0 \text{ mol } H_2O \mid 1 \text{ mol O}}{\mid 1 \text{ mol } H_2O}$$

$$\Longrightarrow \boxed{n_{O_2} = 5.75 \text{ mol } O_2}$$

The flow rates are the same as those calculated with molecular balances (as they must be), only this calculation involves much less effort.

Extents of Reaction

For the reactions

$$CH_4 + \tfrac{3}{2} O_2 \longrightarrow CO + 2H_2O \tag{1}$$

$$CH_4 + 2O_2 \longrightarrow CO_2 + 2H_2O \tag{2}$$

Equation 4.6-6 ($n_i = n_{i0} + \Sigma \nu_{ij} \xi_j$) for the reactive species involved in the process yields the following five equations in five unknowns ($\xi_1, \xi_2, n_{CO}, n_{H_2O}, n_{O_2}$):

$$n_{CH_4}(= 0.78 \text{ mol}) = 7.80 \text{ mol} - \xi_1 - \xi_2 \implies 7.02 \text{ mol} = \xi_1 + \xi_2 \tag{3}$$

$$n_{CO} = \xi_1 \tag{4}$$

$$n_{CO_2}(= 8n_{CO}) = \xi_2 \tag{5}$$

$$n_{H_2O} = 2\xi_1 + 2\xi_2 \tag{6}$$

$$n_{O_2} = 19.4 \text{ mol} - \tfrac{3}{2}\xi_1 - 2\xi_2 \tag{7}$$

From Equation 4, $\xi_1 = n_{CO}$, and from Equation 5, $\xi_2 = 8n_{CO}$. If these two expressions are substituted into Equation 3, the resulting equation may be solved to yield

$$\boxed{n_{CO} = 0.78 \text{ mol CO}} \implies \boxed{n_{CO_2} = 8n_{CO} = (8 \times 0.780) \text{ mol CO}_2 = 6.24 \text{ mol CO}_2}$$

$$\Downarrow \text{Equations 4 and 5}$$

$$\xi_1 = n_{CO} = 0.78 \text{ mol}$$

$$\xi_2 = n_{CO_2} = 6.24 \text{ mol}$$

$$\Downarrow \text{Substitute for } \xi_1 \text{ and } \xi_2 \text{ in Equations 6 and 7}$$

$$\boxed{n_{H_2O} = 14.0 \text{ mol H}_2O}$$

$$\boxed{n_{O_2} = 5.75 \text{ mol O}_2}$$

Once again the same flow rates have been calculated, so that the molar composition of the product gas must therefore also be the same. For this problem, atomic species balances provide the least cumbersome solution.

TEST YOURSELF

Methane is burned to form carbon dioxide and water in a batch reactor:

$$CH_4 + 2O_2 \rightarrow CO_2 + 2H_2O$$

The feed to the reactor and the products obtained are shown in the following flowchart:

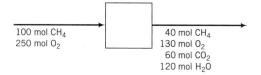

100 mol CH₄
250 mol O₂

40 mol CH₄
130 mol O₂
60 mol CO₂
120 mol H₂O

1. How much methane was consumed? What is the fractional conversion of methane?
2. How much oxygen was consumed? What is the fractional conversion of oxygen?
3. Write the extent of reaction equation (4.6-3) for methane, oxygen, and CO₂. Use each equation to determine the extent of reaction, ξ, substituting inlet and outlet values from the flowchart.

4. How many independent molecular species balances can be written? How many independent atomic species balances can be written?

5. Write the following balances and verify that they are all satisfied. The solution of the first one is given as an example.
  (a) Methane. (I = O + C. 100 mol CH_4 in = 40 mol CH_4 out + 60 mol CH_4 consumed)
  (b) Atomic oxygen (O).
  (c) Molecular oxygen (O_2).
  (d) Water.
  (e) Atomic hydrogen.

4.7f Product Separation and Recycle

Two definitions of reactant conversion are used in the analysis of chemical reactors with product separation and recycle of unconsumed reactants:

Overall Conversion:
$$\frac{\text{reactant input to process} - \text{reactant output from process}}{\text{reactant input to process}} \qquad \textbf{(4.7-1)}$$

Single-Pass Conversion:
$$\frac{\text{reactant input to reactor} - \text{reactant output from reactor}}{\text{reactant input to reactor}} \qquad \textbf{(4.7-2)}$$

As usual, the corresponding percentage conversions are obtained by multiplying these quantities by 100%.

For example, consider the following labeled flowchart for a simple chemical process based on the reaction A → B:

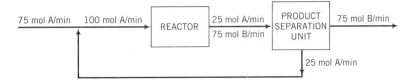

The overall conversion of A is from Equation 4.7-1:

$$\frac{(75 \text{ mol A/min})_{\text{in}} - (0 \text{ mol/min})_{\text{out}}}{(75 \text{ mol A/min})_{\text{in}}} \times 100\% = 100\%$$

The single-pass conversion is from Equation 4.7-2:

$$\frac{(100 \text{ mol A/min})_{\text{in}} - (25 \text{ mol A/min})_{\text{out}}}{(100 \text{ mol A/min})_{\text{in}}} \times 100\% = 75\%$$

This example provides another illustration of the object of recycle. We have achieved complete use of the reactant for which we are paying—the fresh feed—even though only 75% of the reactant entering the reactor is consumed before emerging. The reason the overall conversion is 100% is that perfect separation was assumed: any A that does not react gets sent back to the reactor. If a less-than-perfect separation were achieved and some A left with the product stream, the overall conversion would be less than 100%, although it would always be greater than the single-pass conversion.

TEST YOURSELF

What are the overall and single-pass conversions for the process shown in Figure 4.5-1 on p. 110?

EXAMPLE 4.7-2 *Dehydrogenation of Propane*

Propane is dehydrogenated to form propylene in a catalytic reactor:

$$C_3H_8 \rightarrow C_3H_6 + H_2$$

The process is to be designed for a 95% overall conversion of propane. The reaction products are separated into two streams: the first, which contains H_2, C_3H_6, and 0.555% of the propane that leaves the reactor, is taken off as product; the second stream, which contains the balance of the unreacted propane and 5% of the propylene in the first stream, is recycled to the reactor. Calculate the composition of the product, the ratio (moles recycled)/(mole fresh feed), and the single-pass conversion.

SOLUTION

Basis: 100 mol Fresh Feed

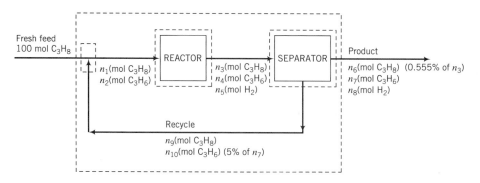

In terms of the labeled variables, the quantities to be calculated are the mole fractions of product stream components $[n_6/(n_6 + n_7 + n_8)]$, ..., the recycle ratio $[(n_9 + n_{10})/100 \text{ mol}]$, and the single-pass conversion $[100\% \times (n_1 - n_3)/n_1]$. We must therefore calculate n_1, n_3, and n_6 through n_{10}. As usual, we begin with the degree-of-freedom analysis to determine whether the problem is properly specified (i.e., whether it is possible to solve it).

Degree-of-Freedom Analysis

When analyzing subsystems in which reactions occur (the overall system and the reactor), we will count atomic balances; for nonreactive subsystems (the recycle mixing point and the separation unit) we will count molecular species balances.

- **Overall system** (the outer dashed box on the flowchart). 3 unknown variables (n_6, n_7, n_8) − 2 independent atomic balances (C and H) − 1 additional relation (95% overall propane conversion) $\Longrightarrow$ 0 degrees of freedom. We will therefore be able to determine n_6, n_7, and n_8 by analyzing the overall system. Let us count these three variables as known at this point.
- **Recycle–fresh feed mixing point.** 4 unknown variables (n_9, n_{10}, n_1, n_2) − 2 balances (C_3H_8, C_3H_6) $\Longrightarrow$ 2 degrees of freedom. Since we do not have enough equations to solve for the unknowns associated with this subsystem, we proceed to the next one.
- **Reactor.** 5 unknown variables (n_1 through n_5) − 2 atomic balances (C and H) $\Longrightarrow$ 3 degrees of freedom. No help here. Let us consider the remaining unit.
- **Separator.** 5 unknown variables (n_3, n_4, n_5, n_9, n_{10}) (n_6 through n_8 are known from the overall system analysis) − 3 balances (C_3H_8, C_3H_6, H_2) − 2 additional relations ($n_6 = 0.00555n_3$, $n_{10} = 0.05n_7$) $\Longrightarrow$ 0 degrees of freedom.

We can therefore determine the five given variables associated with the separator and then return to analyze either the mixing point or the reactor; in either case we can write two atomic balances to solve for the two remaining unknowns (n_1 and n_2), thereby completing the solution. (In fact, not all of the system variables are required by the problem statement so we will be able to stop well short of the full analysis.) The calculations follow, beginning with the analysis of the overall system.

95% Overall Propane Conversion ($\Longrightarrow$ 5% unconverted)

$$n_6 = 0.05(100 \text{ mol}) = 5 \text{ mol } C_3H_8$$

We are left with two overall system atomic balances to write. An H balance involves both remaining unknowns (n_7 and n_8) but a C balance involves only n_7; we therefore begin with the latter balance.

Overall C Balance

$$(100 \text{ mol } C_3H_8)(3 \text{ mol C/mol } C_3H_8) = [n_6(\text{mol } C_3H_8)](3 \text{ mol C/mol } C_3H_8)$$
$$+ [n_7(\text{mol } C_3H_6)](3 \text{ mol C/mol } C_3H_6)$$

$$\xRightarrow{n_6 = 5 \text{ mol}} \boxed{n_7 = 95 \text{ mol } C_3H_6}$$

Overall H Balance (Fill in units.)

$$(100)(8) = n_6(8) + n_7(6) + n_8(2) \xRightarrow{n_6 = 5 \text{ mol},\ n_7 = 95 \text{ mol}} \boxed{n_8 = 95 \text{ mol } H_2}$$

The product therefore contains

$$
\begin{array}{l}
5 \text{ mol } C_3H_8 \\
95 \text{ mol } C_3H_6 \\
95 \text{ mol } H_2
\end{array}
\Longrightarrow
\boxed{
\begin{array}{l}
2.6 \text{ mole\% } C_3H_8 \\
48.7 \text{ mole\% } C_3H_6 \\
48.7 \text{ mole\% } H_2
\end{array}
}
$$

Given Relations Among Separator Variables

$$n_6 = 0.00555 n_3 \xRightarrow{n_6 = 5 \text{ mol}} \boxed{n_3 = 900 \text{ mol } C_3H_8}$$

$$n_{10} = 0.0500 n_7 \xRightarrow{n_7 = 95 \text{ mol}} \boxed{n_{10} = 4.75 \text{ mol } C_3H_6}$$

Propane Balance About Separation Unit

$$n_3 = n_6 + n_9 \xRightarrow{n_3 = 900 \text{ mol},\ n_6 = 5 \text{ mol}} \boxed{n_9 = 895 \text{ mol } C_3H_8}$$

We could continue writing balances about the separation unit to determine the values of n_4 and n_5 but there is no reason to do so, since these values were not requested in the problem statement. The only value still to be determined is that of n_1, which can be calculated from a propane balance about the mixing point.

Propane Balance About Mixing Point

$$100 \text{ mol } + n_9 = n_1 \xRightarrow{n_9 = 895 \text{ mol}} \boxed{n_1 = 995 \text{ mol } C_3H_8}$$

We now have all the variable values we need. The desired quantities are

$$Recycle\ ratio = \frac{(n_9 + n_{10}) \text{ mol recycle}}{100 \text{ mol fresh feed}} \xRightarrow{n_9 = 895 \text{ mol},\ n_{10} = 4.75 \text{ mol}} \boxed{9.00 \ \frac{\text{mol recycle}}{\text{mol fresh feed}}}$$

$$Single\text{-}pass\ conversion = \frac{n_1 - n_3}{n_1} \times 100\% \xRightarrow{n_1 = 995 \text{ mol},\ n_3 = 900 \text{ mol}} \boxed{9.6\%}$$

Consider what is happening in the process just analyzed. Only about 10% of the propane entering the *reactor* is converted to propylene in a single pass; however, over 99% of the unconsumed propane in the reactor effluent is recovered in the separation unit and recycled back to the reactor, where it gets another chance to react. The net result is that 95% of the propane entering the *process* is converted and 5% leaves with the final product.

In general, high overall conversions can be achieved in two ways: (a) design the reactor to yield a high single-pass conversion, or (b) design the reactor to yield a low single-pass conversion (e.g., 10%, as in the preceding example), and follow it with a separation unit to recover and recycle unconsumed reactant. If the second scheme is used, the reactor must handle a larger throughput, but it takes a *much* larger reaction volume to achieve a 95% conversion than a 10% conversion in a single pass. The lower single-pass conversion consequently leads to a decrease in the cost of the reactor. On the other hand, the savings may be offset by the cost of

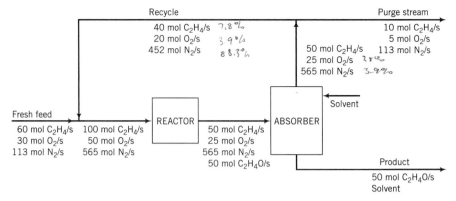

Figure 4.7-2 Process with recycle and purge.

the separation process unit and the pump, pipes, and fittings in the recycle line. The final design would be based on a detailed economic analysis of the alternatives.

4.7g Purging

A problem may arise in processes that involve recycling. Suppose a material that enters with the fresh feed or is produced in a reaction remains entirely in a recycle stream, rather than being carried out in a process product. If nothing were done about this situation, the substance would continuously enter the process and would have no way of leaving; it would therefore steadily accumulate, making the attainment of steady state impossible. To prevent this buildup, a portion of the recycle stream must be withdrawn as a **purge stream** to rid the process of the substance in question.

The flowchart shown in Figure 4.7-2 for the production of ethylene oxide from ethylene illustrates this situation. The reaction is $2\,C_2H_4 + O_2 \rightarrow 2\,C_2H_4O$. A mixture of ethylene and air constitutes the fresh feed to the process. The effluent from the reactor passes to an absorber and is contacted with a liquid solvent. All of the ethylene oxide is absorbed into the solvent. The gas stream leaving the absorber, which contains nitrogen and unreacted ethylene and oxygen, is recycled to the reactor.

If there were no nitrogen (or any other inert and insoluble substance) in the feed, there would be no need for a purge stream. The recycle would contain only ethylene and oxygen; the fresh feed would contain just enough of these substances to make up for the amount lost in the reaction, and the system would be at steady state. However, there is nitrogen. It enters the system at a rate of 113 mol/s and leaves the system at the same rate in the purge stream. If the system were not purged, nitrogen would accumulate at this rate until something—probably unpleasant—occurred to shut down the process.

Material balance calculations on systems involving recycle and purge follow the procedures given in previous sections. When labeling the flowchart, note that the purge stream and the recycle stream before and after the purge takeoff all have the same composition.

**TEST
YOURSELF**

A reaction with stoichiometry A → B takes place in a process with the following flowchart:

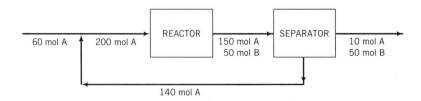

1. What is the overall conversion of A for this process? What is the single-pass conversion?
2. The separation unit and recycle pump and piping are expensive. Why not eliminate them and sell the effluent from the reactor as is? Alternatively, why not keep the separator but discard the bottom stream instead of recycling it?
3. Suppose a trace amount (say, 0.1%) of an inert material C is contained in the fresh feed and all of it stays in the bottom effluent stream from the separation unit (and so is recycled). Why would the process eventually shut down? What would you have to do to get it to work?
4. Why not design the reactor to produce 10 mol A and 50 mol B from 60 mol A in a single pass, thereby eliminating the need for the separation and recycle?

EXAMPLE 4.7-3   *Recycle and Purge in the Synthesis of Methanol*

Methanol is produced in the reaction of carbon dioxide and hydrogen:

$$CO_2 + 3\,H_2 \longrightarrow CH_3OH + H_2O$$

The fresh feed to the process contains hydrogen, carbon dioxide, and 0.400 mole% inerts (I). The reactor effluent passes to a condenser that removes essentially all of the methanol and water formed and none of the reactants or inerts. The latter substances are recycled to the reactor. To avoid buildup of the inerts in the system, a purge stream is withdrawn from the recycle.

The feed to the *reactor* (not the fresh feed to the process) contains 28.0 mole% CO_2, 70.0 mole% H_2, and 2.00 mole% inerts. The single-pass conversion of hydrogen is 60.0%. Calculate the molar flow rates and molar compositions of the fresh feed, the total feed to the reactor, the recycle stream, and the purge stream for a methanol production rate of 155 kmol CH_3OH/h.

SOLUTION   *Basis: 100 mol Combined Feed to the Reactor*

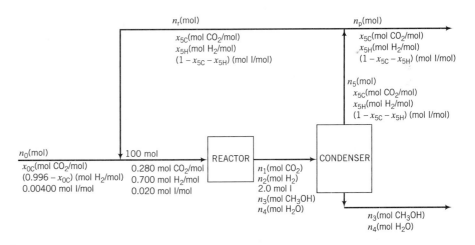

As a general rule, the combined feed to the reactor is a convenient stream to use as a basis of calculation for recycle problems when the stream composition is known. We will therefore temporarily ignore the specified methanol production rate, balance the flowchart for the assumed basis, and then scale the process to the required extent. In terms of the labeled variables, the problem statement will be solved by determining n_0, x_{0C}, n_3, x_{5C}, x_{5H}, n_p, and n_r for the assumed basis, then scaling up n_0, 100 mol (fed to reactor), n_p, and n_r by the factor (155 kmol CH_3OH/h)/n_3.

Degree-of-Freedom Analysis

In the analysis that follows, we will count molecular species balances for all systems. (We could equally well use atomic species balances or the extent of reaction.) Note that the reaction occurs within the overall system and the reactor subsystem and so must be included in the degree-of-freedom analyses for both systems.

- *Overall system.* 7 unknowns ($n_0, x_{0C}, n_3, n_4, n_p, x_{5C}, x_{5H}$) + 1 reaction − 5 independent balances ($CO_2, H_2, I, CH_3OH, H_2O$) $\Longrightarrow$ 3 degrees of freedom. Since we do not have enough equations to solve for the number of unknowns in the overall system, we check subsystems to see if one can be found with zero degrees of freedom.
- *Recycle–fresh feed mixing point.* 5 unknowns ($n_0, x_{0C}, n_r, x_{5C}, x_{5H}$) − 3 independent balances (CO_2, H_2, I) $\Longrightarrow$ 2 degrees of freedom.
- *Reactor.* 4 unknowns (n_1, n_2, n_3, n_4) + 1 reaction − 4 independent balances (CO_2, H_2, CH_3OH, H_2O) − 1 single-pass conversion $\Longrightarrow$ 0 degrees of freedom. We will therefore be able to determine n_1, n_2, n_3, and n_4 and proceed from there.

  Notice that we only subtracted four balances and not one for each of the five species. The reason is that when we labeled the outlet flow of I as 2.0 mol, we implicitly used the balance on I (*input = output*) and so can no longer count it in the degree-of-freedom analysis. We will use the same reasoning in the analysis of the condenser.
- *Condenser.* 3 unknowns (n_5, x_{5C}, x_{5H}) − 3 independent balances (CO_2, H_2, I) $\Longrightarrow$ 0 degrees of freedom. We may now presume that n_5, x_{5C}, and x_{5H} are known.

  In this analysis we presumed that we knew n_1, n_2, n_3, and n_4 from the reactor analysis, and since we used the methanol and water balances when we labeled the bottom product stream we only counted three available balances in the degree-of-freedom analysis.
- *Purge–recycle splitting point.* 2 unknowns (n_r, n_p) − 1 independent balance $\Longrightarrow$ 1 degree of freedom. Since the labeled component mole fractions are the same in all three streams in this subsystem, balances on all three species reduce to the same equation (try it and see).
- *Recycle–fresh feed mixing point* (revisited). 3 unknowns (n_0, x_{0C}, n_r) − 3 independent balances $\Longrightarrow$ 0 degrees of freedom. We can now determine (n_0, x_{0C}, and n_r).
- *Purge–recycle splitting point* (revisited). 1 unknown (n_p) − 1 independent balance $\Longrightarrow$ 0 degrees of freedom. The final unknown variable can now be calculated.

The solution procedure will therefore be to write balances on the reactor, then the condenser, then the fresh feed–recycle mixing point, and finally the purge–recycle splitting point. The flowchart may then be scaled up by the required amount to obtain a methanol production rate of 155 kmol/h. The calculations follow.

Reactor Analysis

We will use molecular balances. Recall that the stoichiometric reaction is

$$CO_2 + 3H_2 \longrightarrow CH_3OH + H_2O$$

60% Single-Pass H_2 Conversion: ($\Longrightarrow$ 40% is unconverted and emerges at the reactor outlet)

$$n_2 = 0.40(70.0 \text{ mol } H_2 \text{ fed}) = 28.0 \text{ mol } H_2$$

H_2 Balance: consumption = input − output

$$\text{Cons}_{H_2} = (70.0 - 28.0) \text{ mol } H_2 = 42.0 \text{ mol } H_2 \text{ consumed}$$

CO_2 Balance: output = input − consumption

$$n_1 = 28.0 \text{ mol } CO_2 - \frac{42.0 \text{ mol } H_2 \text{ consumed}}{} \left| \frac{1 \text{ mol } CO_2 \text{ consumed}}{3 \text{ mol } H_2 \text{ consumed}} \right. = 14.0 \text{ mol } CO_2$$

CH_3OH Balance: output = generation

$$n_3 = \frac{42.0 \text{ mol } H_2 \text{ consumed}}{} \left| \frac{1 \text{ mol } CH_3OH \text{ generated}}{3 \text{ mol } H_2 \text{ consumed}} \right. = 14.0 \text{ mol } CH_3OH$$

H_2O Balance: output = generation

$$n_4 = \frac{42.0 \text{ mol } H_2 \text{ consumed}}{} \left| \frac{1 \text{ mol } H_2O \text{ generated}}{3 \text{ mol } H_2 \text{ consumed}} \right. = 14.0 \text{ mol } H_2O$$

Condenser Analysis

Total Mole Balance: input = output

$$n_1 + n_2 + n_3 + n_4 + 2.0 \text{ mol} = n_3 + n_4 + n_5$$

$$\Downarrow n_2 = 28.0 \text{ mol}, \ n_1 = n_3 = n_4 = 14.0 \text{ mol}$$

$$n_5 = 44.0 \text{ mol}$$

CO_2 Balance: input = output

$$n_1 = n_5 x_{5C}$$

$$\Downarrow n_1 = 14.0 \text{ mol}, \ n_5 = 44.0 \text{ mol}$$

$$x_{5C} = 0.3182 \text{ mol } CO_2/\text{mol}$$

H_2 Balance: input = output

$$n_2 = n_5 x_{5H}$$

$$\Downarrow n_2 = 28.0 \text{ mol}, \ n_5 = 44.0 \text{ mol}$$

$$x_{5H} = 0.6364 \text{ mol } CO_2/\text{mol}$$

$$\Downarrow$$

$$x_I = 1 - x_{5C} - x_{5H} = 0.04545 \text{ mol I/mol}$$

Fresh Feed–Recycle Mixing Point Analysis

Total Mole Balance: input = output

$$n_0 + n_r = 100 \text{ mol}$$

I Balance: input = output

$$n_0(0.00400) + n_r(0.04545) = 2.0 \text{ mol I}$$

Solving these two equations simultaneously yields

$$n_0 = 61.4 \text{ mol fresh feed}, \ n_r = 38.6 \text{ mol recycle}$$

CO_2 Balance: input = output

$$n_0 x_{0C} + n_r x_{5C} = 28.0 \text{ mol } CO_2$$

$$\Downarrow n_0 = 61.4 \text{ mol}, \ n_r = 38.6 \text{ mol}, \ x_{5C} = 0.3182 \text{ mol } CO_2/\text{mol}$$

$$x_{0C} = 0.256 \text{ mol } CO_2/\text{mol}$$

$$\Downarrow$$

$$x_{0H} = (1 - x_{0C} - x_{0I}) = 0.740 \text{ mol } H_2/\text{mol}$$

Recycle–Purge Splitting Point Analysis

Total Mole Balance: input = output

$$n_5 = n_r + n_p$$

$$\Downarrow n_5 = 44.0 \text{ mol}, \ n_r = 38.6 \text{ mol}$$

$$n_p = 5.4 \text{ mol purge}$$

Flowchart Scaling

For the assumed basis of 100 mol feed to the reactor, the production rate of methanol is $n_3 = 14.0$ mol CH_3OH. To scale the process to a methanol production rate of 155 kmol CH_3OH/h, we multiply each total and component molar flow rate by the factor

$$\left(\frac{155 \text{ kmol } CH_3OH/\text{h}}{14.0 \text{ mol } CH_3OH} \right) = \frac{11.1 \text{ kmol/h}}{\text{mol}}$$

The mole fractions remain unchanged by the scaling. The results follow.

Variable	Basis Value	Scaled Value
Fresh feed	61.4 mol	681 kmol/h
	25.6 mole% CO_2	25.6 mole% CO_2
	74.0 mole% H_2	74.0 mole% H_2
	0.400 mole% I	0.400 mole% I
Feed to reactor	100 mol	1110 kmol/h
	28.0 mole% CO_2	28.0 mole% CO_2
	70.0 mole% H_2	70.0 mole% H_2
	2.0 mole% I	2.0 mole% I
Recycle	38.6 mol	428 kmol/h
	31.8 mole% CO_2	31.8 mole% CO_2
	63.6 mole% H_2	63.6 mole% H_2
	4.6 mole% I	4.6 mole% I
Purge	5.4 mol	59.9 kmol/h
	31.8 mole% CO_2	31.8 mole% CO_2
	63.6 mole% H_2	63.6 mole% H_2
	4.6 mole% I	4.6 mole% I

4.8 COMBUSTION REACTIONS

Combustion—the rapid reaction of a fuel with oxygen—is perhaps more important than any other class of industrial chemical reactions, despite the fact that combustion products (CO_2, H_2O, and possibly CO and SO_2) are worth much less than the fuels burned to obtain them. The significance of these reactions lies in the tremendous quantities of energy they release— energy that is used to boil water to produce steam, which is then used to drive the turbines that generate most of the world's electrical power.

The job of designing power generation equipment usually falls to mechanical engineers, but the analysis of combustion reactions and reactors and the abatement and control of environmental pollution caused by combustion products like CO, CO_2, and SO_2 are problems with which chemical engineers are heavily involved. In Chapter 14, for example, we present a case study involving the generation of electricity from the combustion of coal and removal of SO_2 (a pollutant) from combustion products.

In the sections that follow, we introduce terminology commonly used in the analysis of combustion reactors and discuss material balance calculations for such reactors. Methods of determining the energy that can be obtained from combustion reactions are given in Chapter 9.

4.8a Combustion Chemistry

Most of the fuel used in power plant combustion furnaces is either coal (carbon, some hydrogen and sulfur, and various noncombustible materials), fuel oil (mostly high molecular weight hydrocarbons, some sulfur), gaseous fuel (such as *natural gas,* which is primarily methane), or *liquefied petroleum gas,* which is usually propane and/or butane.

When a fuel is burned, carbon in the fuel reacts to form either CO_2 or CO, hydrogen forms H_2O, and sulfur forms SO_2. At temperatures greater than approximately 1800°C, some of the nitrogen in the air reacts to form nitric acid (NO). A combustion reaction in which CO is formed from a hydrocarbon is referred to as **partial combustion** or **incomplete combustion** of the hydrocarbon.

Examples:

$$C + O_2 \rightarrow CO_2 \qquad \text{Complete combustion of carbon}$$

$$C_3H_8 + 5\,O_2 \rightarrow 3\,CO_2 + 4\,H_2O \qquad \text{Complete combustion of propane}$$

$$C_3H_8 + \tfrac{7}{2}O_2 \rightarrow 3\,CO + 4\,H_2O \qquad \text{Partial combustion of propane}$$

$$CS_2 + 3\,O_2 \rightarrow CO_2 + 2\,SO_2 \qquad \text{Complete combustion of carbon disulfide}$$

Perry's Chemical Engineer's Handbook provides a thorough discussion of fuels and combustion.[6]

For obvious economic reasons, air is the source of oxygen in most combustion reactors. Dry air has the following average molar composition:

N_2	78.03%	
O_2	20.99%	
Ar	0.94%	Average molecular weight = 29.0
CO_2	0.03%	
H_2, He, Ne, Kr, Xe	0.01%	
	100.00%	

In most combustion calculations, it is acceptable to simplify this composition to 79% N_2, 21% $O_2 \Longrightarrow$ 79 moles N_2/21 moles O_2 = 3.76 moles N_2/mole O_2.

The term **composition on a wet basis** is commonly used to denote the component mole fractions of a gas that contains water, and **composition on a dry basis** signifies the component mole fractions of the same gas without the water. For example, a gas that contains 33.3 mole% CO_2, 33.3% N_2, and 33.3% H_2O (wet basis) contains 50% CO_2 and 50% N_2 on a dry basis.

The product gas that leaves a combustion furnace is referred to as the **stack gas** or **flue gas**. When the flow rate of a gas in a stack is measured, it is the total flow rate of the gas including water; on the other hand, common techniques for analyzing stack gases provide compositions on a dry basis. You must therefore be able to convert a composition on a dry basis to its corresponding composition on a wet basis before writing material balances on the combustion reactor. The procedure to convert a composition from one basis to another is similar to the one used to convert mass fractions to mole fractions and vice versa given in Chapter 3: assume an amount of the stack gas (e.g., 100 mol of wet gas if composition on a wet basis is known or 100 mol of dry gas if a dry basis composition is known), calculate how much of each component is present, and use this information to calculate mole fractions on the desired basis.

EXAMPLE 4.8-1   *Composition on Wet and Dry Bases*

   1. *Wet Basis $\Longrightarrow$ Dry Basis.*

A stack gas contains 60.0 mole% N_2, 15.0% CO_2, 10.0% O_2, and the balance H_2O. Calculate the molar composition of the gas on a dry basis.

---

[6]R. H. Perry and D. W. Green, Eds., *Perry's Chemical Engineers' Handbook*, 7th Edition, McGraw-Hill, New York, 1997, pp. 27-3 through 27-51.

SOLUTION   ***Basis: 100 mol Wet Gas***

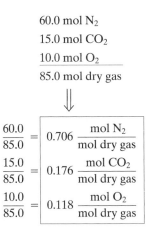

$$\begin{array}{l} 60.0 \text{ mol N}_2 \\ 15.0 \text{ mol CO}_2 \\ \underline{10.0 \text{ mol O}_2} \\ 85.0 \text{ mol dry gas} \end{array}$$

$$\frac{60.0}{85.0} = \left| \; 0.706 \; \frac{\text{mol N}_2}{\text{mol dry gas}} \right.$$

$$\frac{15.0}{85.0} = \left| \; 0.176 \; \frac{\text{mol CO}_2}{\text{mol dry gas}} \right.$$

$$\frac{10.0}{85.0} = \left| \; 0.118 \; \frac{\text{mol O}_2}{\text{mol dry gas}} \right.$$

2. *Dry Basis* $\Longrightarrow$ *Wet Basis.*

An **Orsat analysis** (a technique for stack analysis) yields the following dry basis composition:

$$\begin{array}{ll} \text{N}_2 & 65\% \\ \text{CO}_2 & 14\% \\ \text{CO} & 11\% \\ \text{O}_2 & 10\% \end{array}$$

A humidity measurement shows that the mole fraction of H_2O in the stack gas is 0.0700. Calculate the stack gas composition on a wet basis.

SOLUTION   ***Basis: 100 lb-moles Dry Gas***

$$0.0700 \; \frac{\text{lb-mole H}_2\text{O}}{\text{lb-mole wet gas}} \iff 0.930 \; \frac{\text{lb-mole dry gas}}{\text{lb-mole wet gas}}$$

$$\frac{0.0700 \text{ lb-mole H}_2\text{O/lb-mole wet gas}}{0.930 \text{ lb-mole dry gas/lb-mole wet gas}} = 0.0753 \; \frac{\text{lb-mole H}_2\text{O}}{\text{lb-mole dry gas}}$$

Hence the gas in the assumed basis contains

$$\frac{100 \text{ lb-moles dry gas} \; \left| \; 0.0753 \text{ lb-mole H}_2\text{O} \right.}{\text{lb-mole dry gas}} = 7.53 \text{ lb-moles H}_2\text{O}$$

$$\frac{100 \text{ lb-moles dry gas} \; \left| \; 0.650 \text{ lb-mole N}_2 \right.}{\text{lb-mole dry gas}} = 65.0 \text{ lb-moles N}_2$$

$$\begin{array}{r} (100)(0.140) \text{ lb-moles CO}_2 = 14.0 \text{ lb-moles CO}_2 \\ (100)(0.110) \text{ lb-moles CO} = 11.0 \text{ lb-moles CO} \\ (100)(0.100) \text{ lb-moles O}_2 = \underline{10.0 \text{ lb-moles O}_2} \\ 107.5 \text{ lb-moles wet gas} \end{array}$$

The mole fractions of each stack gas component may now easily be calculated:

$$y_{\text{H}_2\text{O}} = \frac{7.53}{107.5} \; \frac{\text{lb-moles H}_2\text{O}}{\text{lb-moles wet gas}} = 0.070 \; \frac{\text{lb-mole H}_2\text{O}}{\text{lb-mole wet gas}}, \dots$$

TEST YOURSELF

1. What is the approximate molar composition of air? What is the approximate molar ratio of N_2 to O_2 in air?
2. A gas contains 1 mol H_2, 1 mol O_2, and 2 mol H_2O. What is the molar composition of this gas on a wet basis? On a dry basis?
3. A flue gas contains 5 mole% H_2O. Calculate the ratios
   (a) kmol flue gas/kmol H_2O.
   (b) kmol dry flue gas/kmol flue gas.
   (c) kmol H_2O/kmol dry flue gas.

4.8b   Theoretical and Excess Air

If two reactants participate in a reaction and one is considerably more expensive than the other, the usual practice is to feed the less expensive reactant in excess of the valuable one. This has the effect of increasing the conversion of the valuable reactant at the expense of the cost of the excess reactant and additional pumping costs.

The extreme case of an inexpensive reactant is air, which is free. Combustion reactions are therefore invariably run with more air than is needed to supply oxygen in stoichiometric proportion to the fuel. The following terms are commonly used to describe the quantities of fuel and air fed to a reactor.

Theoretical Oxygen: The moles (batch) or molar flow rate (continuous) of O_2 needed for complete combustion of all the fuel fed to the reactor, assuming that all carbon in the fuel is oxidized to CO_2 and all the hydrogen is oxidized to H_2O.

Theoretical Air: The quantity of air that contains the theoretical oxygen.

Excess Air: The amount by which the air fed to the reactor exceeds the theoretical air.

Percent Excess Air:
$$\frac{(\text{moles air})_{\text{fed}} - (\text{moles air})_{\text{theoretical}}}{(\text{moles air})_{\text{theoretical}}} \times 100\% \qquad \textbf{(4.8-1)}$$

If you know the fuel feed rate and the stoichiometric equation(s) for complete combustion of the fuel, you can calculate the theoretical O_2 and air feed rates. If in addition you know the actual feed rate of air, you can calculate the percent excess air from Equation 4.8-1. It is also easy to calculate the air feed rate from the theoretical air and a given value of the percentage excess: if 50% excess air is supplied, for example, then

$$(\text{moles air})_{\text{fed}} = 1.5\,(\text{moles air})_{\text{theoretical}}$$

EXAMPLE 4.8-2   *Theoretical and Excess Air*

One hundred mol/h of butane (C_4H_{10}) and 5000 mol/h of air are fed into a combustion reactor. Calculate the percent excess air.

SOLUTION

First, calculate the theoretical air from the feed rate of fuel and the stoichiometric equation for complete combustion of butane:

$$C_4H_{10} + \tfrac{13}{2}O_2 \rightarrow 4\,CO_2 + 5\,H_2O$$

$$(\dot{n}_{O_2})_{\text{theoretical}} = \frac{100 \text{ mol } C_4H_{10}}{h} \left| \frac{6.5 \text{ mol } O_2 \text{ required}}{\text{mol } C_4H_{10}} \right.$$

$$= 650 \,\frac{\text{mol } O_2}{h}$$

$$(\dot{n}_{\text{air}})_{\text{theoretical}} = \frac{650 \text{ mol } O_2}{h} \left| \frac{4.76 \text{ mol air}}{\text{mol } O_2} \right. = 3094 \,\frac{\text{mol air}}{h}$$

Hence

$$\% \text{ excess air} = \frac{(\dot{n}_{air})_{fed} - (\dot{n}_{air})_{theoretical}}{(\dot{n}_{air})_{theoretical}} \times 100\% = \frac{5000 - 3094}{3094} \times 100\% = \boxed{61.6\%}$$

If instead you had been *given* 61.6% excess air, you could have calculated the feed rate of air as $(\dot{n}_{air})_{fed} = 1.616(\dot{n}_{air})_{theoretical} = 1.616(3094 \text{ mol/h}) = 5000 \text{ mol/h}$.

Two points of confusion often arise in the calculation of theoretical and excess air, both of which are caused by ignoring the definitions of these terms.

1. *The theoretical air required to burn a given quantity of fuel does not depend on how much is actually burned.* The fuel may not react completely, and it may react to form both CO and CO_2, but the theoretical air is still that which would be required to react with *all* of the fuel to form CO_2 only.
2. *The value of the percent excess air depends only on the theoretical air and the air feed rate, and not on how much O_2 is consumed in the reactor or whether combustion is complete or partial.*

TEST YOURSELF

Methane burns in the reactions

$$CH_4 + 2\,O_2 \rightarrow CO_2 + 2\,H_2O$$
$$CH_4 + \tfrac{3}{2}O_2 \rightarrow CO + 2\,H_2O$$

One hundred mol/h of methane is fed to a reactor.

1. What is the theoretical O_2 flow rate if complete combustion occurs in the reactor?
2. What is the theoretical O_2 flow rate assuming that only 70% of the methane reacts? (Careful!)
3. What is the theoretical air flow rate?
4. If 100% excess air is supplied, what is the flow rate of air entering the reactor?
5. If the actual flow rate of air is such that 300 mol O_2/h enters the reactor, what is the percent excess air?

CREATIVITY EXERCISES

1. Years ago it was common to operate boiler furnaces with air fed in 20% excess or more, while today improved boiler designs enable the use of 5–10% excess air. Cite as many possible negative consequences as you can think of for the air-to-fuel feed ratio being (a) too low and (b) too high.
2. The costs of petroleum and natural gas have increased dramatically since the early 1970s, and there is some question about their continued long-term availability. List as many alternative energy sources as you can think of, being as creative as you can, and then go back and suggest possible drawbacks to each one.

4.8c Material Balances on Combustion Reactors

The procedure for writing and solving material balances for a combustion reactor is the same as that for any other reactive system. Bear in mind these points, however:

1. When you draw and label the flowchart, be sure the outlet stream (the stack gas) includes (a) unreacted fuel unless you are told that all the fuel is consumed, (b) unreacted oxygen, (c) water and carbon dioxide, as well as carbon monoxide if the problem statement says any is present, and (d) nitrogen if the fuel is burned with air and not pure oxygen.

2. To calculate the oxygen feed rate from a specified percent excess oxygen or percent excess air (both percentages have the same value, so it doesn't matter which one is stated), first calculate the theoretical O_2 from the fuel feed rate and the reaction stoichiometry for complete combustion, then calculate the oxygen feed rate by multiplying the theoretical oxygen by (1 + fractional excess oxygen).

3. If only one reaction is involved, all three balance methods (molecular species balances, atomic species balances, extent of reaction) are equally convenient. If several reactions occur simultaneously, however—such as combustion of a fuel to form both CO and CO_2—atomic species balances are usually most convenient.

EXAMPLE 4.8-3 *Combustion of Ethane*

Ethane is burned with 50% excess air. The percentage conversion of the ethane is 90%; of the ethane burned, 25% reacts to form CO and the balance reacts to form CO_2. Calculate the molar composition of the stack gas on a dry basis and the mole ratio of water to dry stack gas.

SOLUTION   *Basis: 100 mol C_2H_6 Fed*

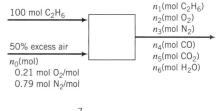

$$C_2H_6 + \frac{7}{2}O_2 \longrightarrow 2CO_2 + 3H_2O$$

$$C_2H_6 + \frac{5}{2}O_2 \longrightarrow 2CO + 3H_2O$$

Notes

1. Since no product stream mole fractions are known, subsequent calculations are easier if individual component amounts rather than a total amount and mole fractions are labeled.
2. The composition of air is taken to be approximately 21 mole% O_2, 79 mole% N_2.
3. If the ethane reacted completely, n_1 would be omitted. Since excess air is supplied, O_2 *must* appear in the product stream.
4. In material balance calculations on combustion processes it is reasonable to assume that nitrogen is inert—that is, to neglect the trace amounts of NO, NO_2, and N_2O_4 (collectively referred to as NO_x) that might form in the burner. On the other hand, in environmental impact studies NO_x may not automatically be neglected; trace amounts of nitrogen oxides may have little impact on the nitrogen balance but may have a significant polluting effect if they are released into the atmosphere.

Degree-of-Freedom Analysis

$$7 \text{ unknowns } (n_0, n_1, \ldots, n_6)$$
$$- 3 \text{ atomic balances (C, H, O)}$$
$$- 1 \text{ N}_2 \text{ balance}$$
$$- 1 \text{ excess air specification (relates } n_0 \text{ to the quantity of fuel fed)}$$
$$- 1 \text{ ethane conversion specification}$$
$$- 1 \text{ CO/CO}_2 \text{ ratio specification}$$
$$= 0 \text{ degrees of freedom}$$

50% Excess Air

$$(n_{O_2})_{\text{theoretical}} = \frac{100 \text{ mol } C_2H_6 \mid 3.50 \text{ mol } O_2}{1 \text{ mol } C_2H_6} = 350 \text{ mol } O_2$$

$$\Downarrow 50\% \text{ excess air}$$

$$0.21n_0 = 1.50(350 \text{ mol } O_2) \Longrightarrow n_0 = 2500 \text{ mol air fed}$$

90% Ethane Conversion: ($\Longrightarrow$ 10% unreacted)

$$n_1 = 0.100(100 \text{ mol } C_2H_6 \text{ fed}) = \boxed{10.0 \text{ mol } C_2H_6}$$

$$0.900(100 \text{ mol } C_2H_6 \text{ fed}) = 90.0 \text{ mol } C_2H_6 \text{ react}$$

25% Conversion to CO

$$n_4 = \frac{(0.25 \times 90.0) \text{ mol } C_2H_6 \text{ react to form CO} \mid 2 \text{ mol CO generated}}{1 \text{ mol } C_2H_6 \text{ react}} = \boxed{45.0 \text{ mol CO}}$$

Nitrogen Balance: output = input

$$n_3 = 0.79(2500 \text{ mol}) = \boxed{1975 \text{ mol } N_2}$$

Atomic Carbon Balance: input = output

$$\frac{100 \text{ mol } C_2H_6 \mid 2 \text{ mol C}}{1 \text{ mol } C_2H_6} = \frac{n_1(\text{mol } C_2H_6) \mid 2 \text{ mol C}}{1 \text{ mol } C_2H_6} + \frac{n_4(\text{mol CO}) \mid 1 \text{ mol C}}{1 \text{ mol CO}}$$

$$+ \frac{n_5(\text{mol } CO_2) \mid 1 \text{ mol C}}{1 \text{ mol } CO_2}$$

$$\Downarrow \begin{array}{l} n_1 = 10 \text{ mol} \\ n_4 = 45 \text{ mol} \end{array}$$

$$\boxed{n_5 = 135 \text{ mol } CO_2}$$

Atomic Hydrogen Balance: input = output

$$\frac{100 \text{ mol } C_2H_6 \mid 6 \text{ mol H}}{1 \text{ mol } C_2H_6} = \frac{10 \text{ mol } C_2H_6 \mid 6 \text{ mol H}}{1 \text{ mol } C_2H_6} + \frac{n_6(\text{mol } H_2O) \mid 2 \text{ mol H}}{1 \text{ mol } H_2O}$$

$$\Downarrow$$

$$\boxed{n_6 = 270 \text{ mol } H_2O}$$

Atomic Oxygen Balance: input = output

$$\frac{525 \text{ mol } O_2 \mid 2 \text{ mol O}}{1 \text{ mol } O_2} = \frac{n_2(\text{mol } O_2) \mid 2 \text{ mol O}}{1 \text{ mol } O_2} + \frac{45 \text{ mol CO} \mid 1 \text{ mol O}}{1 \text{ mol CO}}$$

$$+ \frac{135 \text{ mol } CO_2 \mid 2 \text{ mol O}}{1 \text{ mol } CO_2} + \frac{270 \text{ mol } H_2O \mid 1 \text{ mol O}}{1 \text{ mol } H_2O}$$

$$\Downarrow$$

$$\boxed{n_2 = 232 \text{ mol } O_2}$$

The analysis of the stack gas is now complete. Summarizing:

$$
\begin{aligned}
n_1 &= 10 \text{ mol } C_2H_6 \\
n_2 &= 232 \text{ mol } O_2 \\
n_3 &= 1974 \text{ mol } N_2 \\
n_4 &= 45 \text{ mol } CO \\
n_5 &= \underline{135 \text{ mol } CO_2} \\
& 2396 \text{ mol dry gas} \\[4pt]
+n_6 &= \underline{270 \text{ mol } H_2O} \\
& 2666 \text{ mol total}
\end{aligned}
$$

Hence the stack gas composition on a dry basis is

$$
\begin{aligned}
y_1 &= \frac{10 \text{ mol } C_2H_6}{2396 \text{ mol dry gas}} = 0.00417 \; \frac{\text{mol } C_2H_6}{\text{mol}} \\[8pt]
y_2 &= \frac{232 \text{ mol } O_2}{2396 \text{ mol dry gas}} = 0.0970 \; \frac{\text{mol } O_2}{\text{mol}} \\[8pt]
y_3 &= \frac{1974 \text{ mol } N_2}{2396 \text{ mol dry gas}} = 0.824 \; \frac{\text{mol } N_2}{\text{mol}} \\[8pt]
y_4 &= \frac{45 \text{ mol } CO}{2396 \text{ mol dry gas}} = 0.019 \; \frac{\text{mol } CO}{\text{mol}} \\[8pt]
y_5 &= \frac{135 \text{ mol } CO_2}{2396 \text{ mol dry gas}} = 0.0563 \; \frac{\text{mol } CO_2}{\text{mol}}
\end{aligned}
$$

and the mole ratio of water to dry stack gas is

$$
\frac{270 \text{ mol } H_2O}{2396 \text{ mol dry stack gas}} = \boxed{0.113 \; \frac{\text{mol } H_2O}{\text{mol dry stack gas}}}
$$

If a fuel of unknown composition is burned, you may be able to deduce something about its composition by analyzing the combustion products and writing and solving atomic species balances. The procedure is illustrated in the next example.

EXAMPLE 4.8-4   *Combustion of a Hydrocarbon Fuel of Unknown Composition*

A hydrocarbon gas is burned with air. The dry-basis product gas composition is 1.5 mole% CO, 6.0% CO_2, 8.2% O_2, and 84.3% N_2. There is no atomic oxygen in the fuel. Calculate the ratio of hydrogen to carbon in the fuel gas and speculate on what the fuel might be. Then calculate the percent excess air fed to the reactor.

SOLUTION   **Basis: 100 mol Product Gas**

Since the molecular composition of the fuel is unknown, we label its atomic species composition. We also recognize that since the fuel is a hydrocarbon, water must be one of the combustion products.

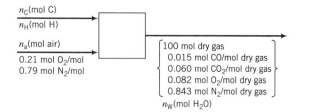

$$C + O_2 \longrightarrow CO_2$$
$$2C + O_2 \longrightarrow 2CO$$
$$4H + O_2 \longrightarrow 2H_2O$$

Degree-of-Freedom Analysis

$$4 \text{ unknowns } (n_H, n_C, n_a, n_w)$$
$$-3 \text{ independent atomic balances (C, H, O)}$$
$$\underline{-1 \text{ } N_2 \text{ balance}}$$
$$= 0 \text{ degrees of freedom}$$

A solution procedure that does not require solving simultaneous equations is as follows:

N_2 Balance $\quad 0.79n_a = (100)(0.843) \text{ mol } N_2 \implies n_a = 106.7 \text{ mol air}$

Atomic C Balance $\quad n_C = \dfrac{100 \text{ mol}}{} \left| \dfrac{0.015 \text{ mol CO}}{\text{mol}} \right| \dfrac{1 \text{ mol C}}{1 \text{ mol CO}} + (100)(0.060)(1) \text{ mol C}$

$$\implies n_C = 7.5 \text{ mol C}$$

Atomic O Balance $\quad 0.21n_a(2) = n_w(1) + 100[\overbrace{(0.015)(1)}^{CO} + \overbrace{(.060)(2)}^{CO_2} + \overbrace{(0.082)(2)}^{O_2}] \text{ mol O}$

$$\xrightarrow{\quad n_a = 106.7 \text{ mol} \quad} n_w = 14.9 \text{ mol } H_2O$$

Atomic H Balance $\quad n_H = \overbrace{n_w(2)}^{H_2O} \xrightarrow{\quad n_w = 14.9 \text{ mol} \quad} n_H = 29.8 \text{ mol H}$

C/H Ratio in the Fuel $\quad \dfrac{n_H}{n_C} = \dfrac{29.8 \text{ mol H}}{7.5 \text{ mol C}} = \boxed{3.97 \text{ mol H/mol C}}$

The fuel composition may therefore be described by the formula $(CH_{3.97})_N$.

Since there is only one hydrocarbon for which the ratio of H to C is close to 3.97—that is, CH_4—we may conclude in this case that the fuel is essentially pure methane, perhaps with trace amounts of other hydrocarbons. [If we had obtained, say, $n_H/n_C \approx 2$, we could have gone no further than to label the fuel as $(CH_2)_n$; from the information given, there would have been no way to distinguish between C_2H_4, C_3H_6, a mixture of CH_4 and C_2H_2, and so on.]

Percent Excess Air

We must first determine the theoretical oxygen required to consume the carbon and hydrogen in the fuel. The reactions may be written as

$$C + O_2 \longrightarrow CO_2$$
$$4H + O_2 \longrightarrow 2H_2O$$
$$\Downarrow$$

$$(n_{O_2})_{\text{theoretical}} = \dfrac{7.5 \text{ mol C}}{} \left| \dfrac{1 \text{ mol } O_2}{1 \text{ mol C}} + \dfrac{29.8 \text{ mol H}}{} \right| \dfrac{1 \text{ mol } O_2}{4 \text{ mol H}} = 14.95 \text{ mol } O_2$$

$$(n_{O_2})_{\text{fed}} = 0.21(106.7 \text{ mol air}) = 22.4 \text{ mol } O_2$$

$$\% \text{ excess air} = \dfrac{(n_{O_2})_{\text{fed}} - (n_{O_2})_{\text{theoretical}}}{(n_{O_2})_{\text{theoretical}}} \times 100\% = \dfrac{(22.4 - 14.95) \text{ mol } O_2}{14.95 \text{ mol } O_2} \times 100\%$$

$$= \boxed{49.8\% \text{ excess air}}$$

4.9   SOME ADDITIONAL CONSIDERATIONS ABOUT CHEMICAL PROCESSES

The methods we present in this chapter and the rest of the text are applied universally in the chemical process industry. There are several features of industrial processes that tend not to show up in textbooks, however. You will find out about them as soon as you go to work as a chemical engineer, but it could be helpful to know about them in advance.

- Textbook processes always work the way they are designed to work. In practice, unexpected occurrences are common, especially when processes are first run.
- Textbook process variables are measured with relatively high precision. In practice, every measurement introduces an error.
- The invisible people who run textbook processes never do anything wrong. Real process operators and managers, being human, sometimes make mistakes.
- In textbooks you always have exactly the data you need to determine what you want to know, no matter how complex the problem may be. In practice, you may not have all the data you need and may find it necessary to use approximate correlations and make assumptions based on common sense and experience.
- In textbooks, the *closure* of every steady-state material balance [defined as (output/input) $\times$ 100%] is 100%. In practice, measurement imprecision and inaccurate assumptions may lead to closures that differ—possibly significantly—from 100%. Also, in practice there is no such thing as true steady state: variable values *always* fluctuate or drift to some extent.
- Textbook problems usually have one and only one correct answer, and your job is to follow prescribed procedures to determine it. In practice you may have trouble even defining what the real problem is, and once you have defined it you can usually find a variety of solutions, each of which has advantages and disadvantages. Making the choice involves considerations of technological capability, short-range profit, long-range profit, safety, environmental protection, and ethics. Textbooks provide little help with this type of problem.

Since our purpose in this text is to help you to develop skills in certain basic methods of chemical process analysis, we have intentionally omitted most of the complicating factors that can make industrial processes so difficult to manage. Once you have mastered the basics, you can start learning how to adjust for the complications.

EXAMPLE 4.9-1   *Material Balances in Process Design and Process Operation*

Methyl ethyl ketone (MEK) is to be recovered from a gas mixture containing 20.0 mole% MEK and 80.0 mole% N_2 at 85°C and 3.5 atm. In a proposed process design, a stream of this mixture is fed to a condenser at a rate of 500 L/s and is cooled at constant pressure, causing most of the MEK to condense.

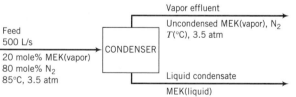

The design engineer (a) converts the volumetric flow rate of the feed stream to a molar flow rate using the *ideal gas equation of state*, an approximate relationship between the pressure, temperature, volumetric flow rate, and molar flow rate of a gas (Chapter 5); (b) specifies a condenser temperature of 15°C; (c) calculates the mole fraction of MEK in the vapor product using *Raoult's law*—an approximate relationship between the compositions of liquid and vapor phases in equilibrium with each other at a specified temperature and pressure (Chapter 6); and (d) calculates the molar flow rates of the vapor and liquid products from nitrogen and MEK balances (input = output). The results follow.

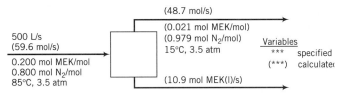

Values without parentheses are specified quantities and those with parentheses are calculated.

A condenser is then installed and run at the design temperature and pressure. The volumetric flow rates of the feed stream and the vapor and liquid product streams are measured with rotameters (see p. 46), and the MEK mole fractions in the feed and vapor effluent streams are measured with a gas chromatograph. The feed stream flow rate is set to 500 liters/s and enough time is allowed to pass for the product stream rotameter readings to reach steady levels. The feed and product gas flow rates are then converted to molar flow rates using the ideal gas equation of state, and the product liquid flow rate is converted to a molar flow rate using a tabulated MEK density and the molecular weight of MEK. Here are the results.

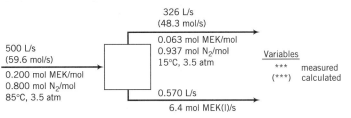

1. Calculate the MEK balance closures for the condenser design and the experimental condenser.
2. List possible reasons for the differences between the design predictions and the experimental values of the output stream variables and for the failure of the experimental system balance to close.

SOLUTION

1. *Material balance closures.*

Design

$$\text{MEK in} = (59.6 \text{ mol/s})(0.200 \text{ mol MEK/mol}) = 11.9 \text{ mol MEK/s}$$

$$\text{MEK out} = (48.7 \text{ mol/s})(0.021 \text{ mol MEK/mol}) + 10.9 \text{ mol MEK/s} = 11.9 \text{ mol MEK/s}$$

$$\Downarrow$$

$$\text{Closure} = \frac{\text{MEK out}}{\text{MEK in}} \times 100\% = \frac{11.9 \text{ mol/s}}{11.9 \text{ mol/s}} \times 100\% = \boxed{100\% \text{ closure}}$$

The nitrogen balance closure is also 100% (*verify*).

Experiment

$$\text{MEK in} = (59.6 \text{ mol/s})(0.200 \text{ mol MEK/mol}) = 11.9 \text{ mol MEK/s}$$

$$\text{MEK out} = (48.3 \text{ mol/s})(0.063 \text{ mol MEK/mol}) + 6.4 \text{ mol MEK/s} = 9.44 \text{ mol MEK/s}$$

$$\Downarrow$$

$$\text{Closure} = \frac{\text{MEK out}}{\text{MEK in}} \times 100\% = \frac{9.44 \text{ mol/s}}{11.9 \text{ mol/s}} \times 100\% = \boxed{79\% \text{ closure}}$$

The nitrogen balance closure is 95% (*verify*).

2. *Possible reasons for differences between design values and experimental values.*

- *Human errors, instrument errors, and random data scatter.* Plant or laboratory personnel are responsible for setting and maintaining the process operating conditions, reading the feed and product stream flowmeters, and drawing samples of the product gas and analyzing them. Any error that any of these individuals makes could lead to errors in measured variable values and values calculated from them. Moreover, any measured value (e.g., an input or output stream volumetric flow rate, the mole fraction of MEK in the feed or vapor product stream, any stream temperature or pressure) is subject to errors due to a faulty instrument (e.g., a malfunctioning or poorly calibrated flowmeter or gas chromatograph) or random data scatter.

- ***Impurities in the feed.*** The design calculations were based on an assumption that the feed contains only MEK vapor and nitrogen. Impurities present in the feed could react with the MEK, or they could condense and affect the vapor–liquid equilibrium distribution of MEK in the products.
- ***Incorrect assumption of steady state.*** Closure should be expected only after the system reaches steady state, so that input = output. In the experimental run, steady state was declared when the operator could no longer see changes in the outlet stream rotameter readings. It is possible that the flow rates were still changing but the rotameter was not sensitive enough to show the changes. It is also possible that MEK was still accumulating in the system—for example, by adsorbing on the container walls—and much more time would be required for the buildup to be complete.
- ***Incorrect assumption that MEK is not reactive.*** If MEK undergoes a reaction in the system—decomposition, for example, or reaction with something on the reactor wall—then input = output + consumption. The output would then necessarily be less than the input and the balance would not close.
- ***Errors due to approximations in the experimental data analysis.*** Several potential errors were introduced when the measured volumetric flow rates were converted to molar flow rates. Volumetric gas flow rates were converted using the ideal gas equation of state, which is approximate, and the volumetric liquid flow rate was converted using a tabulated density that may not have been measured at the system temperature. Also, the fact that a physical property value has been published is no guarantee that it is correct.
- ***Approximations in the design analysis.*** Like the ideal gas equation of state, Raoult's law is an approximation that may be excellent or seriously in error, depending on the experimental process conditions.

There are other possibilities, but you get the idea. The point is that no matter how carefully you design a process, you cannot predict exactly what the real process will do. Approximations and assumptions must be made for every process design; closures on real process material balances are never exactly 100%; nothing can be measured with complete accuracy; and everyone sometimes makes mistakes.

Experienced design engineers know these things and account for them with *overdesign factors*. If they calculate that they need a 2500-liter reactor, they might order a 3000-liter or 3500-liter reactor to make sure they have enough reactor capacity to meet both current and anticipated product demands. The more uncertainties in the design or the projected product demand, the greater the overdesign. A large part of what engineers do involves reducing the uncertainties and thus lowering the required overdesign, resulting in major reductions in equipment purchase and maintenance costs.

4.10 SUMMARY

Every chemical process analysis involves writing and solving material balances to account for all process species in feed and product streams. This chapter outlines and illustrates a systematic approach to material balance calculations. The procedure is to draw and label a flowchart, perform a degree-of-freedom analysis to verify that enough equations can be written to solve for all unknown process variables, and write and solve the equations.

- The **general balance equation** is

$$input + generation - output - consumption = accumulation$$

A **differential balance** applies to an instant of time and each term is a rate (mass/time or moles/time). An **integral balance** applies to a time interval and each term is an amount (mass or moles). Balances may be applied to total mass, individual species, or energy. (They may also be applied to momentum, but we will not consider momentum balances in this text.)

- For a differential balance on a continuous process (material flows in and out throughout the process) at steady-state (no process variables change with time), the accumulation term in the balance (the rate of buildup or depletion of the balanced species) equals zero. For an integral balance on a batch process (no material flows in or out during the process), the input and output terms equal zero and accumulation = initial input − final output. In both cases, the balance simplifies to

$$input + generation = output + consumption$$

If the balance is on total mass or on a nonreactive species, the equation simplifies further to

$$input = output$$

- A process stream on a flowchart is *completely labeled* if values or variable names are assigned to one of the following sets of stream variables: (a) total mass flow rate or total mass and component mass fractions; (b) mass flow rates or masses of each stream component; (c) total molar flow rate or total moles and component mole fractions; and (d) molar flow rates or moles of each stream component. *If a total amount or flow rate or one or more component fractions are known for a stream, use (a) or (c) to incorporate the known values into the labeling. If neither the total nor any fractions are known, using (b) or (d) (component amounts or flow rates) often leads to easier algebra.* Volumetric quantities should be labeled only if they are either given or requested in the problem statement. A flowchart is completely labeled if every stream is completely labeled.

- A **basis of calculation** for a process is an amount or flow rate of one of the process streams. If two or more stream flow rates or amounts are given in the problem statement, they constitute the basis of calculation. If one is given, it may be assumed as a basis but it may also be convenient to assume another basis and then scale the flowchart to the specified value. If no flow rates or amounts are given, assume one as a basis, preferably an amount of a stream with known composition.

- To perform a **degree-of-freedom analysis on a single-unit nonreactive process**, count unknown variables on the flowchart, then subtract independent relations among them. The difference, which equals the number of degrees of freedom for the process, must equal zero for a unique solution of the problem to be determinable. Relations include material balances (as many as there are independent species in the feed and product streams), process specifications, density relations between labeled masses and volumes, and physical constraints (e.g., the sum of the component mass or mole fractions of a stream must add up to 1.)

- To perform a **degree-of-freedom analysis on a multiple-unit process**, perform separate analyses on the overall process, each process unit, each stream mixing or stream splitting point, and, if necessary, on combinations of process units. When you find a system with zero degrees of freedom, assume that you can solve for the unknown variables in the feed and output streams for that system; then, considering those variables as known, try to find another system with zero degrees of freedom. This procedure helps you to find an efficient solution procedure before you undertake time-consuming calculations.

- Once you have written the system equations for a process, you may solve them either manually or using an equation-solving computer program. *If you solve system equations manually, write them in an order that minimizes the number that must be solved simultaneously, starting with equations that only involve one unknown variable.*

- **Recycle** is a common feature of chemical processes. Its most common use is to send unused raw materials emerging from a process unit back to the unit. Overall system balances are usually (but not always) convenient starting points for analyzing process with recycle. A **purge** stream is withdrawn from a process when a species enters in the process feed and is completely recycled. If this species were not removed in the purge, it would keep accumulating in the process system and eventually lead to shutdown.

- The **limiting reactant** in a reactive process is the one that would be completely consumed if the reaction proceeded to completion. All other reactants must either be fed in **stoichiometric**

- **proportion** to the limiting reactant (the feed rates are in the ratio of the stoichiometric coefficients) or in **excess** of the limiting reactant (in greater than stoichiometric proportion to it).

- The **theoretical requirement** for an excess reactant is the amount required to react completely with the limiting reactant. The **percentage excess** of the reactant is

$$\% \text{ excess} = \frac{\text{amount fed} - \text{amount theoretically required}}{\text{amount theoretically required}}$$

  The percentage excess depends only on the feed rates of the excess and limiting reactants and on their stoichiometric coefficients; it does *not* depend on how much actually reacts or on anything else that happens in the reactor.

- The **fractional conversion** of a reactant is the ratio of amount reacted to amount fed. The fractional conversions of different reactants are generally different unless the reactants are fed in stoichiometric proportion.

- The **extent of reaction**, ξ (or $\dot{\xi}$ for a continuous process), is a species-independent quantity that satisfies the equation

$$n_i = n_{i0} + \nu_i \xi \quad \text{or} \quad \dot{n}_i = \dot{n}_{i0} + \nu_i \dot{\xi}$$

  where n_{i0} ($\dot{n}_{i0}$) is the number of moles (molar flow rate) of species i in the feed to the reactor, n_i ($\dot{n}_i$) is the number of moles (molar flow rate) of species i in the stream leaving the reactor, and ν_i is the stoichiometric coefficient of species i (negative for reactants, positive for products, and zero for nonreactive species). The units of ξ ($\dot{\xi}$) are the same as those of n ($\dot{n}$). If you know the inlet and outlet amounts or flow rates of any reactive species, you can determine ξ or $\dot{\xi}$ by applying this equation to that species. You may then substitute the calculated value into the equations for the other species in the stream leaving the reactor to determine the amounts or flow rates of those species.

- You may analyze reactive processes using (a) *molecular species balances* (the only method used for nonreactive processes), (b) *atomic species balances,* or (c) *extents of reaction.* Molecular species balances on reactive processes are often cumbersome: they must include generation and consumption terms for each species, and one degree of freedom must be added for each independent reaction. Atomic species balances have the simple form *input = output* and are usually more straightforward than either of the other two methods. Extents of reaction are particularly convenient for reaction equilibrium calculations.

- **Combustion** is a rapid reaction between a fuel and oxygen. The carbon in the fuel is oxidized to CO_2 (**complete combustion**) or CO (**partial combustion**) and the hydrogen in the fuel is oxidized to water. Other species in the fuel like sulfur and nitrogen may be partially or completely converted to their oxides. Combustion reactions are carried out commercially either to generate heat or to consume waste products.

PROBLEMS

4.1. Water enters a 2.00-m³ tank at a rate of 6.00 kg/s and is withdrawn at a rate of 3.00 kg/s. The tank is initially half full.
  (a) Is this process continuous, batch, or semibatch? Is it transient or steady state?
  (b) Write a mass balance for the process (see Example 4.2-1). Identify the terms of the general balance equation (Equation 4.2-1) present in your equation and state the reason for omitting any terms.
  (c) How long will the tank take to overflow?

4.2. A liquid-phase chemical reaction A $\rightarrow$ B takes place in a well-stirred tank. The concentration of A in the feed is C_{A0} (mol/m³), and that in the tank and outlet stream is C_A (mol/m³). Neither concentration varies with time. The volume of the tank contents is V (m³) and the volumetric flow rate of the inlet and outlet streams is $\dot{v}$ (m³/s). The **reaction rate** (the rate at which A is consumed by reaction in the tank) is given by the expression

$$r(\text{mol A consumed/s}) = kVC_A$$

where k is a constant.

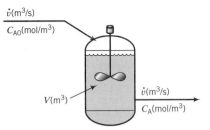

(a) Is this process continuous, batch, or semibatch? Is it transient or steady-state?

(b) What would you expect the reactant concentration C_A to equal if $k = 0$ (no reaction)? What should it approach if $k \to \infty$ (infinitely rapid reaction)?

(c) Write a differential balance on A, stating which terms in the general balance equation (*accumulation = input + generation − output − consumption*) you discarded and why you discarded them. Use the balance to derive the following relation between the inlet and outlet reactant concentrations:

$$C_A = \frac{C_{A0}}{1 + kV/\dot{v}}$$

Verify that this relation predicts the results in part (b).

4.3. A liquid mixture of benzene and toluene contains 55.0% benzene by mass. The mixture is to be partially evaporated to yield a vapor containing 85.0% benzene and a residual liquid containing 10.6% benzene by mass.

(a) Suppose the process is to be carried out continuously and at steady state, with a feed rate of 100.0 kg/h of the 55% mixture. Let $\dot{m}_v$(kg/h) and $\dot{m}_l$(kg/h) be the mass flow rates of the vapor and liquid product streams, respectively. Draw and label a process flowchart, then write and solve balances on total mass and on benzene to determine the expected values of $\dot{m}_v$ and $\dot{m}_l$. For each balance, state which terms of the general balance equation (*accumulation = input + generation − output − consumption*) you discarded and why you discarded them. (See Example 4.2-2.)

(b) Next, suppose the process is to be carried out in a closed container that initially contains 100.0 kg of the liquid mixture. Let m_v(kg) and m_l(kg) be the masses of the final vapor and liquid phases, respectively. Draw and label a process flowchart, then write and solve integral balances on total mass and on benzene to determine m_v and m_l. For each balance, state which terms of the general balance equation (*accumulation = input + generation − output − consumption*) you discarded and why you discarded them.

(c) Returning to the continuous process, suppose the evaporator is built and started up and the product stream flow rates and compositions are measured. The measured percentage of benzene in the vapor stream is 85% and the product stream flow rates have the values calculated in part (a), but the liquid product stream is found to contain 7% benzene instead of 10.6%. One possible explanation is that a mistake was made in the measurement. Give at least five others. [Think about assumptions you made in obtaining the solution of part (a).]

4.4. Draw and label the given streams and derive expressions for the indicated quantities in terms of labeled variables. The solution of part (a) is given as an illustration.

(a) A continuous stream contains 40.0 mole% benzene and the balance toluene. Write expressions for the molar and mass flow rates of benzene, $\dot{n}_B$(mol C_6H_6/s) and $\dot{m}_B$(kg C_6H_6/s), in terms of the total molar flow rate of the stream, $\dot{n}$(mol/s).

Solution

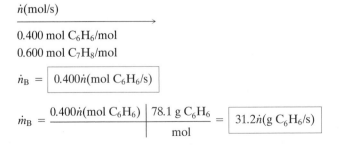

$\dot{n}$(mol/s)

0.400 mol C_6H_6/mol
0.600 mol C_7H_8/mol

$\dot{n}_B = \boxed{0.400\dot{n}(\text{mol } C_6H_6/s)}$

$\dot{m}_B = \dfrac{0.400\dot{n}(\text{mol } C_6H_6)}{} \Bigg| \dfrac{78.1 \text{ g } C_6H_6}{\text{mol}} = \boxed{31.2\dot{n}(\text{g } C_6H_6/s)}$

(b) The feed to a batch process contains equimolar quantities of nitrogen and methane. Write an expression for the kilograms of nitrogen in terms of the total moles n(mol) of this mixture.

(c) A stream containing ethane, propane, and butane has a mass flow rate of 100.0 g/s. Write an expression for the molar flow rate of ethane, $\dot{n}_E$ (lb-mole C_2H_6/h), in terms of the mass fraction of this species, x_E.

(d) A continuous stream of humid air contains water vapor and dry air, the latter containing approximately 21 mole% O_2 and 79% N_2. Write expressions for the molar flow rate of O_2 and for the mole fractions of H_2O and O_2 in the gas in terms of $\dot{n}_1$ (lb-mole H_2O/s) and $\dot{n}_2$ (lb-mole dry air/s).

(e) The product from a batch reactor contains NO, NO_2, and N_2O_4. The mole fraction of NO is 0.400. Write an expression for the gram-moles of N_2O_4 in terms of n(mol mixture) and y_{NO_2} (mol NO_2/mol).

4.5. (a) Draw a flowchart for the catalytic dehydrogenation of propane from the description of this process that begins Section 4.3a. Label all feed, product, and connecting streams between units.

(b) Write clear statements of the overall objective of the process and the functions of each of the process units (the preheater, the reactor, the absorption and stripping towers, and the distillation column).

4.6. A *distillation column* is a process unit in which a feed mixture is separated by multiple partial vaporizations and condensations to form two or more product streams. The overhead product stream is rich in the most *volatile* components of the feed mixture (the ones that vaporize most readily), and the bottom product stream is rich in the least volatile components.

   The following flowchart shows a distillation column with two feed streams and three product streams:

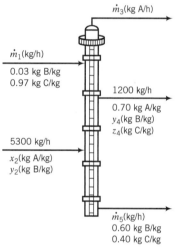

(a) How many independent material balances may be written for this system?

(b) How many of the unknown flow rates and/or mole fractions must be specified before the others may be calculated? (See Example 4.3-4. Also, remember what you know about the component mole fractions of a mixture—for example, the relationship between x_2 and y_2.) Briefly explain your answer.

(c) Suppose values are given for $\dot{m}_1$ and x_2. Give a series of equations, each involving only a single unknown, for the remaining variables. Circle the variable for which you would solve. (Once a variable has been calculated in one of these equations, it may appear in subsequent equations without being counted as an unknown.)

4.7. *Liquid extraction* is an operation used to separate the components of a liquid mixture of two or more species. In the simplest case, the mixture contains two components: a solute (A) and a liquid solvent (B). The mixture is contacted in an agitated vessel with a second liquid solvent (C) that has two key properties: A dissolves in it, and B is immiscible or nearly immiscible with it. (For example, B may be water, C a hydrocarbon oil, and A a species that dissolves in both water and oil.) Some of the A transfers from B to C, and then the B-rich phase (the raffinate) and the C-rich phase (the extract) separate from each other in a settling tank. If the raffinate is then contacted with fresh C in another stage, more A will be transferred from it. This process can be repeated until essentially all of the A has been extracted from the B.

Shown below is a flowchart of a process in which acetic acid (A) is extracted from a mixture of acetic acid and water (B) into 1-hexanol (C), a liquid immiscible with water.

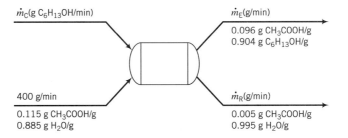

$\dot{m}_C$(g C_6H_{13}OH/min)

$\dot{m}_E$(g/min)
0.096 g CH_3COOH/g
0.904 g C_6H_{13}OH/g

400 g/min
0.115 g CH_3COOH/g
0.885 g H_2O/g

$\dot{m}_R$(g/min)
0.005 g CH_3COOH/g
0.995 g H_2O/g

(a) What is the maximum number of independent material balances that can be written for this process?

(b) Calculate $\dot{m}_C$, $\dot{m}_E$, and $\dot{m}_R$, using the given mixture feed rate as a basis and writing balances in an order such that you never have an equation that involves more than one unknown variable.

(c) Calculate the difference between the amount of acetic acid in the feed mixture and that in the 0.5% mixture, and show that it equals the amount that leaves in the 9.6% mixture.

(d) Acetic acid is relatively difficult to separate completely from water by distillation (see Problem 4.6) and relatively easy to separate from hexanol by distillation. Sketch a flowchart of a two-unit process that might be used to recover nearly pure acetic acid from an acetic acid–water mixture.

4.8. Eggs are sorted into two sizes (large and extra large) at the Cheerful Chicken Dairy. Unfortunately, business has not been good lately, and since the Cheerful Chicken's 40-year-old egg-sorting machine finally gave up the ghost there have been no funds available to replace it. Instead, Old Fred, one of the firm's sharper-eyed employees, has been equipped with a "Large" rubber stamp in his right hand and an "X-large" stamp in his left and assigned to stamp each egg with the appropriate label as it goes by on the conveyor belt. Down the line, another employee puts the eggs into either of two hoppers, each egg according to its stamp. The system works reasonably well, all things considered, except that Old Fred has a heavy hand and on the average breaks 30% of the 120 eggs that pass by him each minute. At the same time, a check of the "X-large" stream reveals a flow rate of 70 eggs/min, of which 25 eggs/min are broken.

(a) Draw and label a flowchart for this process.

(b) Write and solve balances about the egg sorter on total eggs and broken eggs.

(c) How many "large" eggs leave the plant each minute, and what fraction of them are broken?

(d) Is Old Fred right- or left-handed?

4.9. Strawberries contain about 15 wt% solids and 85 wt% water. To make strawberry jam, crushed strawberries and sugar are mixed in a 45:55 mass ratio, and the mixture is heated to evaporate water until the residue contains one-third water by mass.

(a) Draw and label a flowchart of this process.

(b) Do the degree-of-freedom analysis and show that the system has zero degrees of freedom (i.e., the number of unknown process variables equals the number of equations relating them). If you have too many unknowns, think about what you might have forgotten to do.

(c) Calculate how many pounds of strawberries are needed to make a pound of jam.

4.10. Three hundred gallons of a mixture containing 75.0 wt% ethanol (ethyl alcohol) and 25% water (mixture specific gravity = 0.877) and a quantity of a 40.0 wt% ethanol–60% water mixture (SG = 0.952) are blended to produce a mixture containing 60.0 wt% ethanol. The object of this problem is to determine V_{40}, the required volume of the 40% mixture.

(a) Draw and label a flowchart of the mixing process and do the degree-of-freedom analysis.

(b) Calculate V_{40}.

4.11. If the percentage of fuel in a fuel–air mixture falls below a certain value called the *lower flammability limit (LFL)*, the mixture cannot be ignited. For example, the LFL of propane in air is 2.05 mole% C_3H_8. If the percentage of propane in a propane–air mixture is greater than 2.05 mole%, the gas mixture can ignite if it is exposed to a flame or spark; if the percentage is lower than the LFL, the mixture will not ignite. (There is also an *upper flammability limit*, which for propane in air is 11.4%.)

A mixture of propane in air containing 4.03 mole% C_3H_8 (*fuel gas*) is the feed to a combustion furnace. If there is a problem in the furnace, a stream of pure air (*dilution air*) is added to the fuel mixture prior to the furnace inlet to make sure that ignition is not possible.

(a) Draw and label a flowchart of the fuel gas-dilution air mixing unit, presuming that the gas entering the furnace contains propane at the LFL, and do the degree-of-freedom analysis.

(b) If propane flows at a rate of 150 mol C_3H_8/s in the original fuel–air mixture, what is the minimum molar flow rate of the dilution air?

(c) How would the actual dilution air feed rate probably compare with the value calculated in part (b)? ($>, <, =$) Explain.

4.12. One thousand kilograms per hour of a mixture containing equal parts by mass of methanol and water is distilled. Product streams leave the top and the bottom of the distillation column. The flow rate of the bottom stream is measured and found to be 673 kg/h, and the overhead stream is analyzed and found to contain 96.0 wt% methanol.

(a) Draw and label a flowchart of the process and do the degree-of-freedom analysis.

(b) Calculate the mass and mole fractions of methanol and the molar flow rates of methanol and water in the bottom product stream.

(c) Suppose the bottom product stream is analyzed and the mole fraction of methanol is found to be significantly higher than the value calculated in part (b). List as many possible reasons for the discrepancy as you can think of. Include in your list possible violations of assumptions made in part (b).

4.13. A pharmaceutical product, P, is made in a batch reactor. The reactor effluent goes through a purification process to yield a final product stream and a waste stream. The initial charge (feed) to the reactor and the final product are each weighed, and the reactor effluent, final product, and waste stream are each analyzed for P. The analyzer calibration is a series of meter readings, R, corresponding to known mass fractions of P, x_P.

x_P	0.08	0.16	0.25	0.45
R	105	160	245	360

(a) Plot the analyzer calibration data on logarithmic axes and determine an expression for $x_P(R)$.

(b) The data sheet for one run is shown below:

Batch #: 23601   Date: 10/4
Mass charged to reactor: 2253 kg
Mass of purified product: 1239 kg
Reactor effluent analysis: $R = 388$
Final product analysis: $R = 583$
Waste stream analysis: $R = 140$

Calculate the mass fractions of P in all three streams. Then calculate the percentage yield of the purification process,

$$Y_P = \frac{\text{kg P in final product}}{\text{kg P in reactor effluent}} \times 100\%$$

(c) You are the engineer in charge of the process. You review the given run sheet and the calculations of part (b), perform additional balance calculations, and realize that all of the recorded run data cannot possibly be correct. State how you know, itemize possible causes of the problem, state which cause is most likely, and suggest a step to correct it.

4.14. A stream of humid air containing 1.00 mole% $H_2O(v)$ and the balance dry air is to be humidified to a water content of 10.0 mole% H_2O. For this purpose, liquid water is fed through a flowmeter and evaporated into the air stream. The flowmeter reading, R, is 95. The only available calibration data for the flowmeter are two points scribbled on a sheet of paper, indicating that readings $R = 15$ and $R = 50$ correspond to flow rates $\dot{V} = 40.0$ ft^3/h and $\dot{V} = 96.9$ ft^3/h, respectively.

(a) Assuming that the process is working as intended, draw and label the flowchart, do the degree-of-freedom analysis, and estimate the molar flow rate (lb-mole/h) of the humidified (outlet) air.

(b) Suppose the outlet air is analyzed and found to contain only 7% water instead of the desired 10%. List as many possible reasons as you can think of for the discrepancy, concentrating on assumptions made in the calculation of part (a) that might be violated in the real process.

4.15. A liquid mixture contains 60.0 wt% ethanol (E), 5.0 wt% of a dissolved solute (S), and the balance water. A stream of this mixture is fed to a continuous distillation column operating at steady state. Product streams emerge at the top and bottom of the column. The column design calls for the product streams to have equal mass flow rates and for the top stream to contain 90.0 wt% ethanol and no S.

(a) Assume a basis of calculation, draw and fully label a process flowchart, do the degree-of-freedom analysis, and verify that all unknown stream flows and compositions can be calculated. (Don't do any calculations yet.)

(b) Calculate (i) the mass fraction of S in the bottom stream and (ii) the fraction of the ethanol in the feed that leaves in the bottom product stream (i.e., kg E in bottom stream/kg E in feed) if the process operates as designed.

(c) An analyzer is available to determine the composition of ethanol–water mixtures. The calibration curve for the analyzer is a straight line on a plot *on logarithmic axes* of mass fraction of ethanol, x(kg E/kg mixture), versus analyzer reading, R. The line passes through the points ($R = 15$, $x = 0.100$) and ($R = 38$, $x = 0.400$). Derive an expression for x as a function of $R(x = \cdots)$ based on the calibration, and use it to determine the value of R that should be obtained if the *top* product stream from the distillation column is analyzed.

(d) Suppose a sample of the top stream is taken and analyzed and the reading obtained is not the one calculated in part (c). Assume that the calculation in part (c) is correct and that the plant operator followed the correct procedure in doing the analysis. Give five significantly different possible causes for the deviation between R_{measured} and $R_{\text{predicted}}$, including several assumptions made when writing the balances of part (c). For each one, suggest something that the operator could do to check whether it is in fact the problem.

4.16. Two aqueous sulfuric acid solutions containing 20.0 wt% H_2SO_4 (SG = 1.139) and 60.0 wt% H_2SO_4 (SG = 1.498) are mixed to form a 4.00 molar solution (SG = 1.213).

(a) Calculate the mass fraction of sulfuric acid in the product solution.

(b) Taking 100 kg of the 20% feed solution as a basis, draw and label a flowchart of this process, labeling both masses and volumes, and do the degree-of-freedom analysis. Calculate the feed ratio (liters 20% solution/liter 60% solution).

(c) What feed rate of the 60% solution (L/h) would be required to produce 1250 kg/h of the product?

4.17. A paint mixture containing 25.0% of a pigment and the balance water sells for $18.00/kg, and a mixture containing 12.0% pigment sells for $10.00/kg. If a paint retailer produces a blend containing 17.0% pigment, for how much ($/kg) should it be sold to yield a 10% profit?

4.18. Wet sugar that contains one-fifth water by mass is conveyed through an evaporator in which 85.0% of the entering water is vaporized.

(a) Taking a basis of 100 kg feed, calculate (i) x_W, the mass fraction of water in the wet sugar leaving the evaporator, and (ii) the ratio (kg H_2O vaporized/kg wet sugar leaving the evaporator).

(b) If 1000 tons/day of wet sugar is fed to the evaporator, how much additional water must be removed from the outlet sugar to dry it completely, and what annual revenue can be expected if dry sugar sells for $0.15/lb_m$?

(c) The evaporator is built to achieve the production rate of part (b), installed, and started up, and the water content of the partially dried sugar is measured on successive days of operation. The results follow.

Day	1	2	3	4	5	6	7	8	9	10
x_W	.0513	.0486	.0500	.0507	.0541	.0498	.0512	.0474	.0511	.0494

In subsequent runs, the evaporator is to be shut down for maintenance if x_W falls more than three standard deviations from the mean of this series of runs. Calculate the endpoints of this range.

(d) Considering the results of parts (a) and (c) together, what can you conclude about the recently installed evaporator?

4.19. A sedimentation process is to be used to separate pulverized coal from slate. A suspension of finely divided particles of galena (lead sulfide, SG $= 7.44$) in water is prepared. The overall specific gravity of the suspension is 1.48.

  (a) Four hundred kilograms of galena and a quantity of water are loaded into a tank and stirred to obtain a uniform suspension with the required specific gravity. Draw and label the flowchart (label both the masses and volumes of the galena and water), do the degree-of-freedom analysis, and calculate how much water (m^3) must be fed to the tank.

  (b) A mixture of coal and slate is placed in the suspension. The coal rises to the top and is skimmed off, and the slate sinks. What can you conclude about the specific gravities of coal and slate?

  (c) The separation process works well for several hours, but then a region of clear liquid begins to form at the top of the cloudy suspension and the coal sinks to the bottom of this region, making skimming more difficult. What might be happening to cause this behavior and what corrective action might be taken? Now what can you say about the specific gravity of coal?

4.20. Wet air containing 4.0 mole% water vapor is passed through a column of calcium chloride pellets. The pellets adsorb 97.0% of the water and none of the other constituents of the air. The column packing was initially dry and had a mass of 3.40 kg. Following 5.0 hours of operation, the pellets are reweighed and found to have a mass of 3.54 kg.

  (a) Calculate the molar flow rate (mol/h) of the feed gas and the mole fraction of water vapor in the product gas.

  (b) The mole fraction of water in the product gas is monitored and found to have the value calculated in part (a) for the first 10 hours of operation, but then it begins to increase. What is the most likely cause of the increase? If the process continues to run, what will the mole fraction of water in the product gas eventually be?

4.21. A dilute aqueous solution of H_2SO_4 (Solution A) is to be mixed with a solution containing 90.0 wt% H_2SO_4 (Solution B) to produce a 75.0 wt% solution (Solution C).

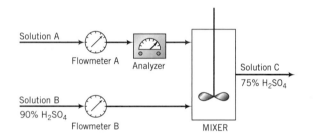

The flow rate and concentration of Solution A change periodically, so that it is necessary to adjust the flow rate of Solution B to keep the product H_2SO_4 concentration constant.

    Flowmeters A and B have linear calibration plots of mass flow rate ($\dot{m}$) versus meter reading (R), which pass through the following points:

$$\text{Flowmeter A:}\quad \dot{m}_A = 150\ \text{lb}_m/\text{h},\quad R_A = 25$$
$$\dot{m}_A = 500\ \text{lb}_m/\text{h},\quad R_A = 70$$
$$\text{Flowmeter B:}\quad \dot{m}_B = 200\ \text{lb}_m/\text{h},\quad R_B = 20$$
$$\dot{m}_B = 800\ \text{lb}_m/\text{h},\quad R_B = 60$$

The analyzer calibration is a straight line on a semilog plot of %$H_2SO_4(x)$ on a logarithmic scale versus meter reading (R_x) on a linear scale. The line passes through the points ($x = 20\%, R_x = 4.0$) and ($x = 100\%, R_x = 10.0$).

  (a) Calculate the flow rate of Solution B needed to process 300 lb$_m$/h of 55% H_2SO_4 (Solution A), and the resulting flow rate of Solution C. (The calibration data are not needed for this part.)

  (b) Derive the calibration equations for $\dot{m}_A(R_A)$, $\dot{m}_B(R_B)$, and $x(R_x)$. Calculate the values of R_A, R_B, and R_x corresponding to the flow rates and concentrations of part (a).

  (c) The process technician's job is to read Flowmeter A and the analyzer periodically, and then to adjust the flow rate of Solution B to its required value. Derive a formula that the technician can use for R_B in terms of R_A and R_x, and then check it by substituting the values of part (a).

*4.22. Gas streams containing hydrogen and nitrogen in different proportions are produced on request by blending gases from two feed tanks: Tank A (hydrogen mole fraction $= x_A$) and Tank B (hydrogen mole fraction $= x_B$). The requests specify the desired hydrogen mole fraction, x_P, and *mass* flow rate of the product stream, $\dot{m}_P$(kg/h).

(a) Suppose the feed tank compositions are $x_A = 0.10$ mol H_2/mol and $x_B = 0.50$ mol H_2/mol, and the desired blend–stream mole fraction and mass flow rate are $x_P = 0.20$ mol H_2/mol and $\dot{m}_P = 100$ kg/h. Draw and label a flowchart and calculate the required *molar* flow rates of the feed mixtures, $\dot{n}_A$(kmol/h) and $\dot{n}_B$(kmol/h).

(b) Derive a series of formulas for $\dot{n}_A$ and $\dot{n}_B$ in terms of x_A, x_B, x_P, and $\dot{m}_P$. Test them using the values in part (a).

(c) Write a spreadsheet that has column headings x_A, x_B, x_P, $\dot{m}_P$, $\dot{n}_A$, and $\dot{n}_B$. The spreadsheet should calculate entries in the last two columns corresponding to data in the first four. In the first six data rows of the spreadsheet, do the calculations for $x_A = 0.10$, $x_B = 0.50$, and $x_P = 0.10, 0.20$, 0.30, 0.40, 0.50, and 0.60, all for $\dot{m}_P = 100$ kg/h. Then in the next six rows repeat the calculations for the same values of x_A, x_B, and x_P for $\dot{m}_P = 250$ kg/h. *Explain any of your results that appear strange or impossible.*

(d) Enter the formulas of part (b) into an equation-solving program. Run the program to determine $\dot{n}_A$ and $\dot{n}_B$ for the 12 sets of input variable values given in part (c) and explain any physically impossible results.

4.23. An **artificial kidney** is a device that removes water and waste metabolites from blood. In one such device, the **hollow fiber hemodialyzer**, blood flows from an artery through the insides of a bundle of hollow cellulose acetate fibers, and *dialyzing fluid,* which consists of water and various dissolved salts, flows on the outside of the fibers. Water and waste metabolites—principally urea, creatinine, uric acid, and phosphate ions—pass through the fiber walls into the dialyzing fluid, and the purified blood is returned to a vein.

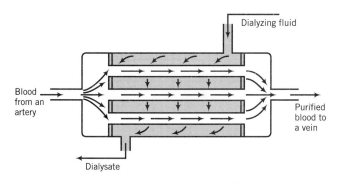

At some time during a dialysis the arterial and venous blood conditions are as follows:

	Arterial (entering) Blood	Venous (exiting) Blood
Flow Rate	200.0 mL/min	195.0 mL/min
Urea (H_2NCONH_2) Concentration	1.90 mg/mL	1.75 mg/mL

(a) Calculate the rates at which urea and water are being removed from the blood.

(b) If the dialyzing fluid enters at a rate of 1500 mL/min and the exiting solution (*dialysate*) leaves at approximately the same rate, calculate the concentration of urea in the dialysate.

(c) Suppose we want to reduce the patient's urea level from an initial value of 2.7 mg/mL to a final value of 1.1 mg/mL. If the total blood volume is 5.0 liters and the average rate of urea removal is that calculated in part (a), how long must the patient be dialyzed? (Neglect the loss in total blood volume due to the removal of water in the dialyzer.)

---

*Computer problem.

4.24. The *indicator dilution method* is a technique used to determine flow rates of fluids in channels for which devices like rotameters and orifice meters cannot be used (e.g., rivers, blood vessels, and large-diameter pipelines). A stream of an easily measured substance (the *tracer*) is injected into the channel at a known rate and the tracer concentration is measured at a point far enough downstream of the injection point for the tracer to be completely mixed with the flowing fluid. The larger the flow rate of the fluid, the lower the tracer concentration at the measurement point.

A gas stream that contains 1.50 mole% CO_2 flows through a pipeline. Twenty (20.0) kilograms of CO_2 per minute is injected into the line. A sample of the gas is drawn from a point in the line 150 meters downstream of the injection point and found to contain 2.3 mole% CO_2.

(a) Estimate the gas flow rate (kmol/min) upstream of the injection point.

(b) Eighteen seconds elapses from the instant the additional CO_2 is first injected to the time the CO_2 concentration at the measurement point begins to rise. Assuming that the tracer travels at the average velocity of the gas in the pipeline (i.e., neglecting diffusion of CO_2), estimate the average velocity (m/s). If the molar gas density is 0.123 kmol/m^3, what is the pipe diameter?

4.25. A variation of the indicator dilution method (see preceding problem) is used to measure total blood volume. A known amount of a tracer is injected into the bloodstream and disperses uniformly throughout the circulatory system. A blood sample is then withdrawn, the tracer concentration in the sample is measured, and the measured concentration [which equals (tracer injected)/(total blood volume) if no tracer is lost through blood vessel walls] is used to determine the total blood volume.

In one such experiment, 0.60 cm^3 of a solution containing 5.00 mg/L of a dye is injected into an artery of a grown man. About 10 minutes later, after the tracer has had time to distribute itself uniformly throughout the bloodstream, a blood sample is withdrawn and placed in the sample chamber of a spectrophotometer. A beam of light passes through the chamber, and the spectrophotometer measures the intensity of the transmitted beam and displays the value of the solution absorbance (a quantity that increases with the amount of light absorbed by the sample). The value displayed is 0.18. A calibration curve of absorbance A versus tracer concentration C (micrograms dye/liter blood) is a straight line through the origin and the point ($A = 0.9$, $C = 3$ μg/L). Estimate the patient's total blood volume from these data.

4.26. *Gas absorption* or *gas scrubbing* is a commonly used method for removing environmentally undesirable species from waste gases in chemical manufacturing and combustion processes. The waste gas is contacted with a liquid solvent in which the potential pollutants are highly soluble and the other species in the waste gas are relatively insoluble. Most of the pollutants go into solution and emerge with the liquid effluent from the scrubber, and the cleaned gas is discharged to the atmosphere. The liquid effluent may be discharged to a waste lagoon or subjected to further treatment to recover the solvent and/or to convert the pollutant to a species that can be released safely to the environment.

A waste gas containing SO_2 (a precursor of acid rain) and several other species (collectively designated as A) is fed to a scrubbing tower where it contacts a solvent (B) that absorbs SO_2. The solvent feed rate to the tower is 1000 L/min. The specific gravity of the solvent is 1.30. Absorption of A and evaporation of B in the scrubber may be neglected.

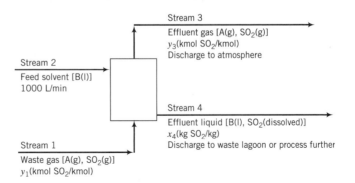

The gas in the scrubber rises through a series of *trays* (metal plates perforated with many small holes), and the solvent flows over the trays and through *downcomers* to the trays below. Gas bubbles emerge from the holes in each tray and rise through the covering liquid, and SO_2 diffuses out of the bubbles and into solution.

The volumetric flow rate of the feed gas is determined with an orifice meter, with a differential mercury manometer being used to measure the pressure drop across the orifice. Calibration data for this meter are tabulated here:

h(mm)	$\dot{V}$(m^3/min)
100	142
200	204
300	247
400	290

The molar density of the feed gas may be determined from the formula

$$\rho\left(\frac{\text{mol}}{\text{liter}}\right) = \frac{12.2P(\text{atm})}{T(\text{K})}$$

where P and T are the absolute pressure and temperature of the gas. An electrochemical detector is used to measure the SO_2 concentration in the inlet and outlet gas streams: SO_2 in the sampled gas is absorbed in a solution across which a fixed voltage is applied, and the mole fraction of SO_2 in the gas is determined from the resulting current. The calibration curve for the analyzer is a straight line on a semilog plot of y(mol SO_2/mol total) versus R (analyzer reading), which passes through the following points:

y (log scale)	R (rectangular scale)
0.00166	20
0.1107	90

The following data are taken:

$$\left. \begin{array}{l} T = 75°\text{F} \\ P = 150 \text{ psig} \\ h(\text{orifice meter}) = 210 \text{ mm} \\ R(SO_2 \text{ analyzer}) = 82.4 \end{array} \right\} \quad \text{(feed gas)}$$

$R(SO_2 \text{ analyzer}) = 11.6$ (outlet gas)

(a) Draw and completely label a process flowchart. Include in the labeling the molar flow rates and SO_2 mole fractions of the gas streams and the mass flow rates and SO_2 mass fractions of the liquid streams. Show that the scrubber has zero degrees of freedom.

(b) Determine (i) the orifice meter calibration formula by plotting $\dot{V}$ versus h on logarithmic axes and (ii) the SO_2 analyzer calibration formula.

(c) Calculate (i) the mass fraction of SO_2 in the liquid effluent stream and (ii) the rate at which SO_2 is removed from the feed gas (kg SO_2/min).

(d) The scrubber column trays commonly have diameters on the order of 1–5 meters and perforation holes on the order of 4–12 mm in diameter, leading to the formation of many tiny bubbles in the liquid on each tray. Speculate on the advantages of making the bubbles as small as possible.

*4.27. The SO_2 scrubbing tower described in Problem 4.26 is to be used to reduce the SO_2 mole fraction in the waste gas to a level that meets air quality control regulations. The solvent feed rate must be large enough to keep the SO_2 mass fraction in the exit liquid below a specified maximum value.

(a) Draw and label a flowchart of the tower. Along with the molar flow rates and SO_2 mole fractions of the four process streams, label the feed gas temperature and pressure [T_1(°F), P_1(psig)], the feed gas orifice meter reading [h_1(mm)], the SO_2 analyzer reading for the feed gas (R_1), the volumetric feed rate of solvent to the column [$\dot{V}_2$(m^3/min)], and the SO_2 analyzer reading for

---

*Computer problem. It is advisable to solve Problem 4.26 before attempting this problem.

the exit gas (R_3). (The temperature and pressure units are taken from the calibration curves for the gauges used to measure these variables.)

(b) Derive a series of equations relating all of the labeled variables on the flowchart. The equations should include the calibration formulas calculated in part (b) of Problem 4.26. Determine the number of degrees of freedom for the process system.

In the remainder of the problem, you will be given values of measured feed gas stream variables [T_1(°F), P_1(psig), h_1(mm), and R_1], the maximum allowed SO_2 mole fraction in the exit solution [x_4], and the specified SO_2 mole fraction in the exit gas (y_3), and asked to calculate the expected SO_2 analyzer reading for the exit gas (R_3) and the minimum volumetric flow rate of solvent ($\dot{V}_2$). Part (c) involves using a spreadsheet for the calculation, and part (d) calls for an independent calculation using an equation-solving program.

(c) Create a spreadsheet to store input values of T_1, P_1, h_1, R_1, x_4, and y_3 and to calculate R_3 and $\dot{V}_2$. In the first five rows, insert the values $T_1 = 75$, $P_1 = 150$, $h_1 = 210$, $R_1 = 82.4$, $x_4 = 0.10$, and $y_3 = 0.05, 0.025, 0.01, 0.005,$ and 0.001. In the next five rows, insert the same input values except make $x_4 = 0.02$. On a single graph, draw plots of $\dot{V}_2$ versus y_3 for each of the two given values of x_4 (preferably using the spreadsheet program to generate the graph). Briefly explain the shape of the curves and their position relative to each other.

(d) Enter the equations of part (b) into an equation-solving program. Use the program to calculate R_3 and $\dot{V}_2$ corresponding to $T_1 = 75$, $P_1 = 150$, $h_1 = 210$, $R_1 = 82.4$, $x_4 = 0.10$, and $y_3 = 0.05,$ $0.025, 0.01, 0.005,$ and 0.001, and then $T_1 = 75$, $P_1 = 150$, $h_1 = 210$, $R_1 = 82.4$, $x_4 = 0.02$, and $y_3 = 0.05, 0.025, 0.01, 0.005,$ and 0.001. If you did not already do so in part (c), draw on a single graph plots of $\dot{V}_2$ versus y_3 for each of the two given values of x_4 and briefly explain the shape of the curves and their position relative to each other.

4.28. Following is a labeled flowchart for a steady-state two-unit process, with boundaries shown to denote subsystems about which balances can be taken. State the maximum number of balances that can be written for each subsystem and the order in which you would write balances to determine the unknown process variables. (See Example 4.4-1.)

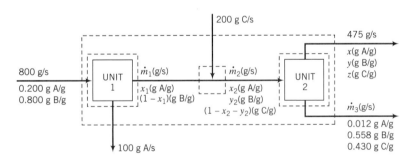

4.29. A liquid mixture containing 30.0 mole% benzene (B), 25.0% toluene (T), and the balance xylene (X) is fed to a distillation column. The bottoms product contains 98.0 mole% X and no B, and 96.0% of the X in the feed is recovered in this stream. The overhead product is fed to a second column. The overhead product from the second column contains 97.0% of the B in the feed to this column. The composition of this stream is 94.0 mole% B and the balance T.

(a) Draw and label a flowchart of this process and do the degree-of-freedom analysis to prove that for an assumed basis of calculation, molar flow rates and compositions of all process streams can be calculated from the given information. Write in order the equations you would solve to calculate unknown process variables. In each equation (or pair of simultaneous equations), circle the variable(s) for which you would solve. Do not do the calculations.

(b) Calculate (i) the percentage of the benzene in the process feed (i.e., the feed to the first column) that emerges in the overhead product from the second column and (ii) the percentage of toluene in the process feed that emerges in the bottom product from the second column.

4.30. Seawater containing 3.50 wt% salt passes through a series of 10 evaporators. Roughly equal quantities of water are vaporized in each of the 10 units and then condensed and combined to obtain a product stream of fresh water. The brine leaving each evaporator but the tenth is fed to the next evaporator. The brine leaving the tenth evaporator contains 5.00 wt% salt.

(a) Draw a flowchart of the process showing the first, fourth, and tenth evaporators. Label all the streams entering and leaving these three evaporators.

(b) Write in order the set of equations you would solve to determine the fractional yield of fresh water from the process (kg H_2O recovered/kg H_2O in process feed) and the weight percent of salt in the solution leaving the fourth evaporator. Each equation you write should contain no more than one previously undetermined variable. In each equation, circle the variable for which you would solve. Do not do the calculations.

(c) Solve the equations derived in part (b) for the two specified quantities.

4.31. An equimolar liquid mixture of benzene and toluene is separated into two product streams by distillation. A process flowchart and a somewhat oversimplified description of what happens in the process follow:

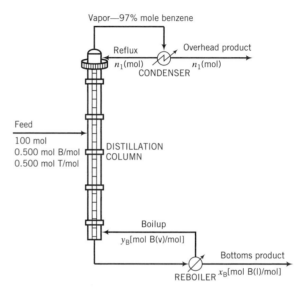

Inside the column a liquid stream flows downward and a vapor stream rises. At each point in the column some of the liquid vaporizes and some of the vapor condenses. The vapor leaving the top of the column, which contains 97 mole% benzene, is completely condensed and split into two equal fractions: one is taken off as the overhead product stream, and the other (the **reflux**) is recycled to the top of the column. The overhead product stream contains 89.2% of the benzene fed to the column. The liquid leaving the bottom of the column is fed to a partial reboiler in which 45% of it is vaporized. The vapor generated in the reboiler (the **boilup**) is recycled to become the rising vapor stream in the column, and the residual reboiler liquid is taken off as the bottom product stream. The compositions of the streams leaving the reboiler are governed by the relation

$$\frac{y_B/(1 - y_B)}{x_B/(1 - x_B)} = 2.25$$

where y_B and x_B are the mole fractions of benzene in the vapor and liquid streams, respectively.

(a) Take a basis of 100 mol fed to the column. Draw and completely label a flowchart, and for each of four systems (overall process, column, condenser, and reboiler), do the degree-of-freedom analysis and identify a system with which the process analysis might appropriately begin (one with zero degrees of freedom).

(b) Write in order the equations you would solve to determine all unknown variables on the flowchart, circling the variable for which you would solve in each equation. Do not do the calculations in this part.

(c) Calculate the molar amounts of the overhead and bottoms products, the mole fraction of benzene in the bottoms product, and the percentage recovery of toluene in the bottoms product (100 × moles toluene in bottoms/mole toluene in feed).

4.32. Fresh orange juice contains 12.0 wt% solids and the balance water, and concentrated orange juice contains 42.0 wt% solids. Initially a single evaporation process was used for the concentration, but volatile constituents of the juice escaped with the water, leaving the concentrate with a flat taste. The current process overcomes this problem by bypassing the evaporator with a fraction of the fresh juice. The juice that enters the evaporator is concentrated to 58 wt% solids, and the evaporator product stream is mixed with the bypassed fresh juice to achieve the desired final concentration.

   (a) Draw and label a flowchart of this process, neglecting the vaporization of everything in the juice but water. First prove that the subsystem containing the point where the bypass stream splits off from the evaporator feed has one degree of freedom. (If you think it has zero degrees, try determining the unknown variables associated with this system.) Then perform the degree-of-freedom analysis for the overall system, the evaporator, and the bypass–evaporator product mixing point, and write in order the equations you would solve to determine all unknown stream variables. In each equation, circle the variable for which you would solve, but don't do any calculations.

   (b) Calculate the amount of product (42% concentrate) produced per 100 kg fresh juice fed to the process and the fraction of the feed that bypasses the evaporator.

   (c) Most of the volatile ingredients that provide the taste of the concentrate are contained in the fresh juice that bypasses the evaporator. You could get more of these ingredients in the final product by evaporating to (say) 90% solids instead of 58%; you could then bypass a greater fraction of the fresh juice and thereby obtain an even better tasting product. Suggest possible drawbacks to this proposal.

4.33. A stream containing 5.15 wt% chromium, Cr, is contained in the wastewater from a metal finishing plant. The wastewater stream is fed to a treatment unit that removes 95% of the chromium in the feed and recycles it to the plant. The residual liquid stream leaving the treatment unit is sent to a waste lagoon. The treatment unit has a maximum capacity of 4500 kg wastewater/h. If wastewater leaves the finishing plant at a rate higher than the capacity of the treatment unit, the excess (anything above 4500 kg/h) bypasses the unit and combines with the residual liquid leaving the unit, and the combined stream goes to the waste lagoon.

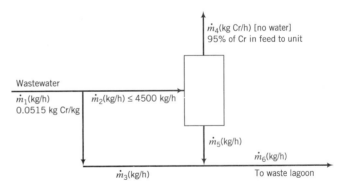

   (a) Without assuming a basis of calculation, draw and label a flowchart of the process.

   (b) Wastewater leaves the finishing plant at a rate $\dot{m}_1 = 6000$ kg/h. Calculate the flow rate of liquid to the waste lagoon, $\dot{m}_6$(kg/h), and the mass fraction of Cr in this liquid, x_6(kg Cr/kg).

   (c) Calculate the flow rate of liquid to the waste lagoon and the mass fraction of Cr in this liquid for $\dot{m}_1$ varying from 1000 kg/h to 10,000 kg/h in 1000 kg/h increments. Generate a plot of x_6 versus $\dot{m}_1$. (*Suggestion:* Use a spreadsheet for these calculations.)

   (d) The company has hired you as a consultant to help them determine whether or not to add capacity to the treatment unit to increase the recovery of chromium. What would you need to know to make this determination?

4.34. An evaporation–crystallization process of the type described in Example 4.5-2 is used to obtain solid potassium sulfate from an aqueous solution of this salt. The fresh feed to the process contains 19.6 wt% K_2SO_4. The wet filter cake consists of solid K_2SO_4 crystals and a 40.0 wt% K_2SO_4 solution, in a ratio 10 kg crystals/kg solution. The filtrate, also a 40.0% solution, is recycled to join the fresh feed. Of the water fed to the evaporator, 45.0% is evaporated. The evaporator has a maximum capacity of 175 kg water evaporated/s.

(a) Assume the process is operating at maximum capacity. Draw and label a flowchart and do the degree-of-freedom analysis for the overall system, the recycle–fresh feed mixing point, the evaporator, and the crystallizer. Then write in an efficient order (minimizing simultaneous equations) the equations you would solve to determine all unknown stream variables. In each equation, circle the variable for which you would solve, but don't do the calculations.

(b) Calculate the maximum production rate of solid K_2SO_4, the rate at which fresh feed must be supplied to achieve this production rate, and the ratio kg recycle/kg fresh feed.

(c) Calculate the composition and feed rate of the stream entering the crystallizer if the process is scaled to 75% of its maximum capacity.

(d) The wet filter cake is subjected to another operation after leaving the filter. Suggest what it might be. Also, list what you think the principal operating costs for this process might be.

*(e) Use an equation-solving computer program to solve the equations derived in part (a). Verify that you get the same solutions determined in part (b).

4.35. In an **absorption tower** (or **absorber**), a gas is contacted with a liquid under conditions such that one or more species in the gas dissolve in the liquid. A **stripping tower** (or **stripper**) also involves a gas contacting a liquid, but under conditions such that one or more components of the feed liquid come out of solution and exit in the gas leaving the tower.

A process consisting of an absorption tower and a stripping tower is used to separate the components of a gas containing 30.0 mole% carbon dioxide and the balance methane. A stream of this gas is fed to the bottom of the absorber. A liquid containing 0.500 mole% dissolved CO_2 and the balance methanol is recycled from the bottom of the stripper and fed to the top of the absorber. The product gas leaving the top of the absorber contains 1.00 mole% CO_2 and essentially all of the methane fed to the unit. The CO_2-rich liquid solvent leaving the bottom of the absorber is fed to the top of the stripper and a stream of nitrogen gas is fed to the bottom. Ninety percent of the CO_2 in the liquid feed to the stripper comes out of solution in the column, and the nitrogen/CO_2 stream leaving the column passes out to the atmosphere through a stack. The liquid stream leaving the stripping tower is the 0.500% CO_2 solution recycled to the absorber.

The absorber operates at temperature T_a and pressure P_a and the stripper operates at T_s and P_s. Methanol may be assumed to be nonvolatile—that is, none enters the vapor phase in either column—and N_2 may be assumed insoluble in methanol.

(a) In your own words, explain the overall objective of this two-unit process and the functions of the absorber and stripper in the process.

(b) The streams fed to the tops of each tower have something in common, as do the streams fed to the bottoms of each tower. What are these commonalities and what is the probable reason for them?

(c) Taking a basis of 100 mol/h of gas fed to the absorber, draw and label a flowchart of the process. For the stripper outlet gas, label the component molar flow rates rather than the total flow rate and mole fractions. Do the degree-of-freedom analysis and write in order the equations you would solve to determine all unknown stream variables *except the nitrogen flow rate entering and leaving the stripper.* Circle the variable(s) for which you would solve each equation (or set of simultaneous equations), but don't do any of the calculations yet.

(d) Calculate the fractional CO_2 removal in the absorber (moles absorbed/mole in gas feed) and the molar flow rate and composition of the liquid feed to the stripping tower.

(e) Calculate the molar feed rate of gas to the absorber required to produce an absorber product gas flow rate of 1000 kg/h.

(f) Would you guess that T_s would be higher or lower than T_a? Explain. (*Hint:* Think about what happens when you heat a carbonated soft drink and what you want to happen in the stripper.) What about the relationship of P_s to P_a?

(g) What properties of methanol would you guess make it the solvent of choice for this process? (In more general terms, what would you look for when choosing a solvent for an absorption–stripping process to separate one gas from another?)

4.36. In the production of a bean oil, beans containing 13.0 wt% oil and 87.0% solids are ground and fed to a stirred tank (the *extractor*) along with a recycled stream of liquid *n*-hexane. The feed ratio is 3 kg

---

hexane/kg beans. The ground beans are suspended in the liquid, and essentially all of the oil in the beans is extracted into the hexane. The extractor effluent passes to a filter. The filter cake contains 75.0 wt% bean solids and the balance bean oil and hexane, the latter two in the same ratio in which they emerge from the extractor. The filter cake is discarded and the liquid filtrate is fed to a heated evaporator in which the hexane is vaporized and the oil remains as a liquid. The oil is stored in drums and shipped. The hexane vapor is subsequently cooled and condensed, and the liquid hexane condensate is recycled to the extractor.

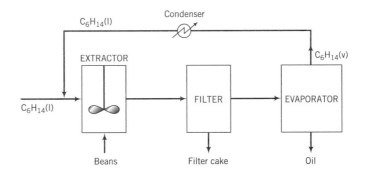

(a) Draw and label a flowchart of the process, do the degree-of-freedom analysis, and write in an efficient order the equations you would solve to determine all unknown stream variables, circling the variables for which you would solve.

(b) Calculate the yield of bean oil product (kg oil/kg beans fed), the required fresh hexane feed (kg C_6H_{14}/kg beans fed), and the recycle to fresh feed ratio (kg hexane recycled/kg fresh feed).

(c) It has been suggested that a *heat exchanger* might be added to the process. This process unit would consist of a bundle of parallel metal tubes contained in an outer shell. The liquid filtrate would pass from the filter through the inside of the tubes and then go on to the evaporator. The hot hexane vapor on its way from the evaporator to the extractor would flow through the shell, passing over the outside of the tubes and heating the filtrate. How might the inclusion of this unit lead to a reduction in the operating cost of the process?

(d) Suggest additional steps that might improve the process economics.

4.37. Following is a diagram of the shirt-cleaning process used by the Floods of Suds One-Day Laundry Services, Inc. The shirts are soaked in an agitated tub containing Whizzo, the Wonder Detergent, and are then wrung out and sent to a rinse stage. The dirty Whizzo is sent to a filter in which most of the dirt is removed, and the cleaned detergent is recycled back to join a stream of pure Whizzo, with the combined stream serving as the feed to the washtub.

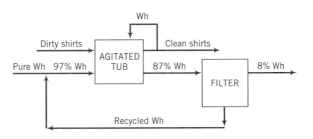

Data:

1. Each 100 lb_m of dirty shirts contains 2 lb_m of dirt.
2. The washing removes 95% of the dirt in the dirty shirts.
3. For each 100 lb_m of dirty shirts, 25 lb_m of Whizzo leaves with the clean shirts, of which 22 lb_m is wrung back into the tub.
4. The detergent that enters the tub contains 97% Whizzo, and that which enters the filter contains 87%. The wet dirt that leaves the filter contains 8% Whizzo.

   (a) How much pure Whizzo must be supplied per 100 lb_m of dirty shirts?
   (b) What is the composition of the recycled stream?

4.38. A drug (D) is produced in a three-stage extraction from the leaves of a tropical plant. About 1000 kg of leaf is required to produce 1 kg of the drug. The extraction solvent (S) is a mixture containing 16.5 wt% ethanol (E) and the balance water (W). The following process is carried out to extract the drug and recover the solvent.

   1. A mixing tank is charged with 3300 kg of S and 620 kg of leaf. The mixer contents are stirred for several hours, during which a portion of the drug contained in the leaf goes into solution. The contents of the mixer are then discharged through a filter. The liquid filtrate, which carries over roughly 1% of the leaf fed to the mixer, is pumped to a holding tank, and the solid cake (spent leaf and entrained liquid) is sent to a second mixer. The entrained liquid has the same composition as the filtrate and a mass equal to 15% of the mass of liquid charged to the mixer. The extracted drug has a negligible effect on the total mass and volume of the spent leaf and the filtrate.

   2. The second mixer is charged with the spent leaf from the first mixer and with the filtrate from the previous batch in the third mixer. The leaf is extracted for several more hours, and the contents of the mixer are then discharged to a second filter. The filtrate, which contains 1% of the leaf fed to the second mixer, is pumped to the same holding tank that received the filtrate from the first mixer, and the solid cake—spent leaf and entrained liquid—is sent to the third mixer. The entrained liquid mass is 15% of the mass of liquid charged to the second mixer.

   3. The third mixer is charged with the spent leaf from the second mixer and with 2720 kg of solvent S. The mixer contents are filtered; the filtrate, which contains 1% of the leaf fed to the third mixer, is recycled to the second mixer; and the solid cake is discarded. As before, the mass of the entrained liquid in the solid cake is 15% of the mass of liquid charged to the mixer.

   4. The contents of the filtrate holding tank are filtered to remove the carried-over spent leaf, and the wet cake is pressed to recover entrained liquid, which is combined with the filtrate. A negligible amount of liquid remains in the wet cake. The filtrate, which contains D, E, and W, is pumped to an extraction unit (another mixer).

   5. In the extraction unit, the alcohol–water–drug solution is contacted with another solvent (F), which is almost but not completely immiscible with ethanol and water. Essentially all of the drug (D) is extracted into the second solvent, from which it is eventually separated by a process of no concern in this problem. Some ethanol but no water is also contained in the extract. The solution from which the drug has been extracted (the **raffinate**) contains 13.0 wt% E, 1.5% F, and 85.5% W. It is fed to a stripping column for recovery of the ethanol.

   6. The feeds to the stripping column are the solution just described and steam. The two streams are fed in a ratio such that the overhead product stream from the column contains 20.0 wt% E and 2.6% F, and the bottom product stream contains 1.3 wt% E and the balance W.

Draw and label a flowchart of the process, taking as a basis one batch of leaf processed. Then calculate

   (a) the masses of the components of the filtrate holding tank.
   (b) the masses of the components D and E in the extract stream leaving the extraction unit.
   (c) the mass of steam fed to the stripping column, and the masses of the column overhead and bottoms products.

4.39. Acetylene is hydrogenated to form ethane. The feed to the reactor contains 1.50 mol H_2/mol C_2H_2.
   (a) Calculate the stoichiometric reactant ratio (mol H_2 react/mol C_2H_2 react) and the yield ratio (kmol C_2H_6 formed/kmol H_2 react).
   (b) Determine the limiting reactant and calculate the percentage by which the other reactant is in excess.
   (c) Calculate the mass feed rate of hydrogen (kg/s) required to produce 4×10^6 metric tons of ethane per year, assuming that the reaction goes to completion and that the process operates for 24 hours a day, 300 days a year.
   (d) There is a definite drawback to running with one reactant in excess rather than feeding the reactants in stoichiometric proportion. What is it? [*Hint:* In the process of part (c), what does the reactor effluent consist of and what will probably have to be done before the product ethane can be sold or used?]

4.40. Ammonia is burned to form nitric oxide in the following reaction:

$$4NH_3 + 5O_2 \longrightarrow 4NO + 6H_2O$$

(a) Calculate the ratio (lb-mole O_2 react/lb-mole NO formed).

(b) If ammonia is fed to a continuous reactor at a rate of 100.0 kmol NH_3/h, what oxygen feed rate (kmol/h) would correspond to 40.0% excess O_2?

(c) If 50.0 kg of ammonia and 100.0 kg of oxygen are fed to a batch reactor, determine the limiting reactant, the percentage by which the other reactant is in excess, and the extent of reaction (mol) and mass of NO produced (kg) if the reaction proceeds to completion.

4.41. A stream containing H_2S and inert gases and a second stream of pure SO_2 are fed to a sulfur recovery reactor, where the reaction

$$2H_2S + SO_2 \longrightarrow 3S + 2H_2O$$

takes place. The feed rates are adjusted so that the ratio of H_2S to SO_2 in the combined feed is always stoichiometric.

In the normal operation of the reactor the flow rate and composition of the H_2S feed stream both fluctuate. In the past, each time either variable changed the required SO_2 feed rate had to be reset by adjusting a valve in the feed line. A control system has been installed to automate this process. The H_2S feed stream passes through an electronic flowmeter that transmits a signal R_f directly proportional to the molar flow rate of the stream, $\dot{n}_f$. When $\dot{n}_f = 100$ kmol/h, the transmitted signal $R_f = 15$ mV. The mole fraction of H_2S in this stream is measured with a thermal conductivity detector, which transmits a signal R_a. Analyzer calibration data are as follows:

R_a(mV)	0	25.4	42.8	58.0	71.9	85.1
x(mol H_2S/mol)	0.00	0.20	0.40	0.60	0.80	1.00

The controller takes as input the transmitted values of R_f and R_a and calculates and transmits a voltage signal R_c to a flow control valve in the SO_2 line, which opens and closes to an extent dependent on the value of R_c. A plot of the SO_2 flow rate, $\dot{n}_c$, versus R_c on rectangular coordinates is a straight line through the points ($R_c = 10.0$ mV, $\dot{n}_c = 25.0$ kmol/h) and ($R_c = 25.0$ mV, $\dot{n}_c = 60.0$ kmol/h).

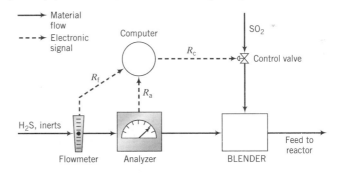

(a) Why would it be important to feed the reactants in stoichiometric proportion? (*Hint:* SO_2 and especially H_2S are serious pollutants.) What are several likely reasons for wanting to automate the SO_2 feed rate adjustment?

(b) If the first stream contains 85.0 mole% H_2S and enters the unit at a rate of $\dot{n}_f = 3.00 \times 10^2$ kmol/h, what must the value of $\dot{n}_c$(kmol SO_2/h) be?

(c) Fit a function to the H_2S analyzer calibration data to derive an expression for x as a function of R_a. Check the fit by plotting both the function and the calibration data on the same graph.

(d) Derive a formula for R_c from specified values of R_f and R_a, using the result of part (c) in the derivation. (This formula would be built into the controller.) Test the formula using the flow rate and composition data of part (a).

(e) The system has been installed and made operational, and at some point the concentration of H_2S in the feed stream suddenly changes. A sample of the blended gas is collected and analyzed a short time later and the mole ratio of H_2S to SO_2 is not the required 2:1. List as many possible reasons as you can think of for this apparent failure of the control system.

4.42. The reaction between ethylene and hydrogen bromide to form ethyl bromide is carried out in a continuous reactor. The product stream is analyzed and found to contain 51.7 mole% C_2H_5Br and 17.3% HBr. The feed to the reactor contains only ethylene and hydrogen bromide. Calculate the fractional conversion of the limiting reactant and the percentage by which the other reactant is in excess. If the molar flow rate of the feed stream is 165 mol/s, what is the extent of reaction? (Give its numerical value and its units.)

4.43. In the Deacon process for the manufacture of chlorine, HCl and O_2 react to form Cl_2 and H_2O. Sufficient air (21 mole% O_2, 79% N_2) is fed to provide 35% excess oxygen and the fractional conversion of HCl is 85%.

(a) Calculate the mole fractions of the product stream components, using atomic species balances in your calculation.

(b) Again calculate the mole fractions of the product stream components, only this time use the extent of reaction in the calculation.

(c) An alternative to using air as the oxygen source would be to feed pure oxygen to the reactor. Running with oxygen imposes a significant extra process cost relative to running with air, but also offers the potential for considerable savings. Speculate on what the cost and savings might be. What would determine which way the process should be run?

4.44. Titanium dioxide (TiO_2) is used extensively as a white pigment. It is produced from an ore that contains ilmenite ($FeTiO_3$) and ferric oxide (Fe_2O_3). The ore is digested with an aqueous sulfuric acid solution to produce an aqueous solution of titanyl sulfate [$(TiO)SO_4$] and ferrous sulfate ($FeSO_4$). Water is added to hydrolyze the titanyl sulfate to H_2TiO_3, which precipitates, and H_2SO_4. The precipitate is then roasted, driving off water and leaving a residue of pure titanium dioxide. (Several steps to remove iron from the intermediate solutions as iron sulfate have been omitted from this description.)

Suppose an ore containing 24.3% Ti by mass is digested with an 80% H_2SO_4 solution, supplied in 50% excess of the amount needed to convert all the ilmenite to titanyl sulfate and all the ferric oxide to ferric sulfate [$Fe_2(SO_4)_3$]. Further suppose that 89% of the ilmenite actually decomposes. Calculate the masses (kg) of ore and 80% sulfuric acid solution that must be fed to produce 1000 kg of pure TiO_2.

4.45. Coal containing 5.0 wt% S is burned at a rate of 1250 lb_m/min in a boiler furnace. All of the sulfur in the coal is oxidized to SO_2. The product gas is sent to a scrubber in which most of the SO_2 is removed, and the scrubbed gas then passes out of a stack. An Environmental Protection Agency regulation requires that the gas in the stack must contain no more than 0.018 lb_m SO_2/lb_m coal burned. To test compliance with this regulation a flowmeter and an SO_2 analyzer are mounted in the stack. The volumetric flow rate of the scrubbed gas is found to be 2867 ft^3/s, and the SO_2 analyzer reading is 37. Calibration data for the analyzer are given in the table below.

SO_2 Analyzer Calibration Data

C(g SO_2/m^3 gas)	Reading (0–100 scale)
0.30	10
0.85	28
2.67	48
7.31	65
18.2	81
30.0	90

(a) Determine the equation that relates SO_2 concentration in lb_m/ft^3 to the analyzer reading.

(b) Is the process in compliance with the EPA regulation?

(c) What percentage of the SO_2 produced in the furnace is removed in the scrubber?

(d) An earlier EPA regulation set a limit on the mole fraction of SO_2 in the gas emerging from the stack (as opposed to the quantity of SO_2 emitted per mass of coal burned), but a way was found to release large quantities of SO_2 from stacks without violating this regulation. Speculate on what the method of getting around the old regulation was. (*Hint:* It involved feeding a second stream to the base of the stack.) Explain why this method no longer worked when the new regulation was imposed.

4.46. The gas-phase reaction between methanol and acetic acid to form methyl acetate and water

$$CH_3OH + CH_3COOH \rightleftharpoons CH_3COOCH_3 + H_2O$$
$$\quad (A) \qquad\qquad (B) \qquad\qquad (C) \qquad (D)$$

takes place in a batch reactor and proceeds to equilibrium. When the reaction mixture comes to equilibrium, the mole fractions of the four reactive species satisfy the relation

$$\frac{y_C y_D}{y_A y_B} = 4.87$$

(a) Suppose the feed to the reactor consists of n_{A0}, n_{B0}, n_{C0}, n_{D0}, and n_{I0} gram-moles of A, B, C, D, and an inert gas, I, respectively. Let ξ(mol) be the extent of reaction. Write expressions for the gram-moles of each reactive species in the final product, $n_A(\xi)$, $n_B(\xi)$, $n_C(\xi)$, and $n_D(\xi)$. Then use these expressions and the given equilibrium relation to derive an equation for ξ_e, the equilibrium extent of reaction, in terms of $n_{A0}, \ldots, n_{I0}$. (see Example 4.6-2.)
(b) If the feed to the reactor contains equimolar quantities of methanol and acetic acid and no other species, calculate the equilibrium fractional conversion.
(c) It is desired to produce 70 mol of methyl acetate starting with 80 mol of acetic acid. If the reaction proceeds to equilibrium, how much methanol must be fed? What is the composition of the final product?
(d) If you wanted to carry out the process of part (b) or (c) commercially, what would you need to know besides the equilibrium composition to determine whether the process would be profitable? (List several things.)

4.47. At low to moderate pressures, the equilibrium state of the water–gas shift reaction

$$CO + H_2O \rightleftharpoons CO_2 + H_2$$

is approximately described by the relation

$$\frac{y_{CO_2}\, y_{H_2}}{y_{CO}\, y_{H_2O}} = K_e(T) = 0.0247\,\exp[4020/T(K)]$$

where T is the reactor temperature, K_e is the reaction equilibrium constant, and y_i is the mole fraction of species i in the reactor contents at equilibrium.

The feed to a batch shift reactor contains 20.0 mole% CO, 10.0% CO_2, 40.0% water, and the balance an inert gas. The reactor is maintained at $T = 1123$ K.

(a) Assume a basis of 1 mol feed and draw and label a flowchart. Carry out a degree-of-freedom analysis of the reactor based on extents of reaction and use it to prove that you have enough information to calculate the composition of the reaction mixture at equilibrium. Do no calculations.
(b) Calculate the total moles of gas in the reactor at equilibrium (if it takes you more than 5 seconds you're missing the point) and then the equilibrium mole fraction of hydrogen in the product. (*Suggestion:* Begin by writing expressions for the moles of each species in the product gas in terms of the extent of reaction, and then write expressions for the species mole fractions.)
(c) Suppose a gas sample is drawn from the reactor and analyzed shortly after startup and the mole fraction of hydrogen is significantly different from the calculated value. Assuming that no calculation mistakes or measurement errors have been made, what is a likely explanation for the discrepancy between the calculated and measured hydrogen yields?
*(d) Write a spreadsheet to take as input the reactor temperature and the feed component mole fractions x_{CO}, x_{H_2O}, and x_{CO_2} (assume no hydrogen is fed) and to calculate the mole fraction y_{H_2} in the product gas when equilibrium is reached. The spreadsheet column headings should be

$$\text{T} \quad \text{x(CO)} \quad \text{x(H2O)} \quad \text{x(CO2)} \quad \text{Ke} \quad \cdots \quad \text{y(H2)}$$

Columns between Ke and y(H2) may contain intermediate quantities in the calculation of y_{H_2}. First test your program for the conditions of part (a) and verify that it is correct. Then try a variety of values of the input variables and draw conclusions about the conditions (reactor temperature and feed composition) that maximize the equilibrium yield of hydrogen.

---

*Computer problem.

4.48. Methanol is formed from carbon monoxide and hydrogen in the gas-phase reaction

$$CO + 2\,H_2 \rightleftharpoons CH_3OH$$
$$\text{(A)} \quad \text{(B)} \qquad \text{(C)}$$

The mole fractions of the reactive species at equilibrium satisfy the relation

$$\frac{y_C}{y_A y_B^2}\frac{1}{P^2} = K_e(T)$$

where P is the total pressure (atm), K_e the reaction equilibrium constant (atm^{-2}), and T the temperature (K). The equilibrium constant K_e equals 10.5 at 373 K, and 2.316×10^{-4} at 573 K. A semilog plot of K_e (logarithmic scale) versus $1/T$ (rectangular scale) is approximately linear between $T = 300$ K and $T = 600$ K.

(a) Derive a formula for $K_e(T)$, and use it to show that $K_e(450\text{K}) = 0.0548$ atm^{-2}.

(b) Write expressions for n_A, n_B, and n_C (gram-moles of each species), and then y_A, y_B, and y_C, in terms of n_{A0}, n_{B0}, n_{C0}, and ξ, the molar extent of reaction. Then derive an equation involving only n_{A0}, n_{B0}, n_{C0}, P, T, and ξ_e, where ξ_e (mol) is the value of the extent of reaction at equilibrium.

(c) Suppose you begin with equimolar quantities of CO and H_2 and no CH_3OH, and the reaction proceeds to equilibrium at 423 K and 2.00 atm. Calculate the molar composition of the product (y_A, y_B, and y_C) and the fractional conversion of CO.

*(d) Write a set of equations for y_A, y_B, y_C, and f_A (the fractional conversion of CO) in terms of y_{A0}, y_{B0}, T, and P (the reactor temperature and pressure at equilibrium). Enter the equations in an equation-solving program. Check the program by running it for the conditions of part (c), then use it to determine the effects on f_A (increase, decrease, or no effect) of separately increasing (i) the fraction of CO in the feed, (ii) the fraction of CH_3OH in the feed, (iii) temperature, and (iv) pressure.

*(e) Write a computer program to take as input y_{A0}, y_{B0}, T, and P (the reactor temperature and pressure at equilibrium) and to calculate and print out y_A, y_B, y_C, and f_A (the fractional conversion of CO). Test your program with the data of part (c). [*Suggestion:* Write the equilibrium relations derived in part (b) as a cubic equation in ξ_e and use Newton's rule—Appendix A.2—to obtain the solution.]

4.49. Methane and oxygen react in the presence of a catalyst to form formaldehyde. In a parallel reaction, methane is oxidized to carbon dioxide and water:

$$CH_4 + O_2 \longrightarrow HCHO + H_2O$$
$$CH_4 + 2\,O_2 \longrightarrow CO_2 + 2\,H_2O$$

The feed to the reactor contains equimolar amounts of methane and oxygen. Assume a basis of 100 mol feed/s.

(a) Draw and label a flowchart. Use a degree-of-freedom analysis based on extents of reaction to determine how many process variable values must be specified for the remaining variable values to be calculated.

(b) Use Equation 4.6-6 to derive expressions for the product stream component flow rates in terms of the two extents of reaction, ξ_1 and ξ_2.

(c) The fractional conversion of methane is 0.900 and the fractional yield of formaldehyde is 0.855. Calculate the molar composition of the reactor output stream and the selectivity of formaldehyde production relative to carbon dioxide production.

4.50. Ethane is chlorinated in a continuous reactor:

$$C_2H_6 + Cl_2 \longrightarrow C_2H_5Cl + HCl$$

Some of the product monochloroethane is further chlorinated in an undesired side reaction:

$$C_2H_5Cl + Cl_2 \longrightarrow C_2H_4Cl_2 + HCl$$

(a) Suppose your principal objective is to maximize the selectivity of monochloroethane production relative to dichloroethane production. Would you design the reactor for a high or low conversion of ethane? Explain your answer. (*Hint:* If the reactor contents remained in the reactor long

---

*Computer problem.

enough for most of the ethane in the feed to be consumed, what would the main product constituent probably be?) What additional processing steps would almost certainly be carried out to make the process economically sound?

(b) Take a basis of 100 mol C_2H_5Cl produced. Assume that the feed contains only ethane and chlorine and that all of the chlorine is consumed and carry out a degree-of-freedom analysis based on atomic species balances.

(c) The reactor is designed to yield a 15% conversion of ethane and a selectivity of 14 mol C_2H_5Cl/mol $C_2H_4Cl_2$, with a negligible amount of chlorine in the product gas. Calculate the feed ratio (mol Cl_2/mol C_2H_6) and the fractional yield of monochloroethane.

(d) Suppose the reactor is built and started up and the conversion is only 14%. Chromatographic analysis shows that there is no Cl_2 in the product but another species with a molecular weight higher than that of dichloroethane is present. Offer a likely explanation for these results.

4.51. Ethanol is produced commercially by the hydration of ethylene:

$$C_2H_4 + H_2O \longrightarrow C_2H_5OH$$

Some of the product is converted to diethyl ether in the side reaction

$$2\,C_2H_5OH \longrightarrow (C_2H_5)_2O + H_2O$$

The feed to the reactor contains ethylene, steam, and an inert gas. A sample of the reactor effluent gas is analyzed and found to contain 43.3 mole% ethylene, 2.5% ethanol, 0.14% ether, 9.3% inerts, and the balance water.

(a) Take as a basis 100 mol of effluent gas, draw and label a flowchart, and do a degree-of-freedom analysis based on atomic species to prove that the system has zero degrees of freedom.[7]

(b) Calculate the molar composition of the reactor feed, the percentage conversion of ethylene, the fractional yield of ethanol, and the selectivity of ethanol production relative to ether production.

(c) The percentage conversion of ethylene you calculated should be very low. Why do you think the reactor would be designed to consume so little of the reactant? (*Hint:* If the reaction mixture remained in the reactor long enough to use up most of the ethylene, what would the main product constituent probably be?) What additional processing steps are likely to take place downstream from the reactor?

4.52. Solid calcium fluoride (CaF_2) reacts with sulfuric acid to form solid calcium sulfate and gaseous hydrogen fluoride. The HF is then dissolved in water to form hydrofluoric acid. A source of calcium fluoride is fluorite ore containing 96.0 wt% CaF_2 and 4.0% SiO_2.

In a typical hydrofluoric acid manufacturing process, fluorite ore is reacted with 93 wt% aqueous sulfuric acid, supplied 15% in excess of the stoichiometric amount. Ninety-five percent of the ore dissolves in the acid. Some of the HF formed reacts with the dissolved silica in the reaction

$$6\,HF + SiO_2(aq) \longrightarrow H_2SiF_6(s) + 2\,H_2O(l)$$

The hydrogen fluoride exiting from the reactor is subsequently dissolved in enough water to produce 60.0 wt% hydrofluoric acid. Calculate the quantity of fluorite ore needed to produce a metric ton of acid. Note: Some of the given data are not needed to solve the problem.

4.53. Chlorobenzene (C_6H_5Cl), an important solvent and intermediate in the production of many other chemicals, is produced by bubbling chlorine gas through liquid benzene in the presence of ferric chloride catalyst. In an undesired side reaction, the product is further chlorinated to dichlorobenzene, and in a third reaction the dichlorobenzene is chlorinated to trichlorobenzene.

The feed to a chlorination reactor consists of essentially pure benzene and a technical grade of chlorine gas (98 wt% Cl_2, the balance gaseous impurities with an average molecular weight of 25.0). The liquid output from the reactor contains 65.0 wt% C_6H_6, 32.0 wt% C_6H_5Cl, 2.5 wt% $C_6H_4Cl_2$, and 0.5 wt% $C_6H_3Cl_3$. The gaseous output contains only HCl and the impurities that entered with the chlorine.

(a) You wish to determine (i) the percentage by which benzene is fed in excess, (ii) the fractional conversion of benzene, (iii) the fractional yield of monochlorobenzene, and (iv) the mass ratio

---

[7]Your first attempt is likely to yield D.F. $= -1$, meaning that either (i) you counted one too many independent balances, (ii) you forgot to label one unknown variable, or (iii) the problem statement contains one redundant—and possibly inconsistent—process variable specification. Prove that (i) is actually the case. (Review the definition of independent balances in Section 4.7b.)

of the gas feed to the liquid feed. Without doing any calculations, prove that you have enough information about the process to determine these quantities.

(b) Perform the calculations.

(c) Why would benzene be fed in excess and the fractional conversion kept low?

(d) What might be done with the gaseous effluent?

(e) It is possible to use 99.9% pure ("reagent-grade") chlorine instead of the technical grade actually used in the process. Why is this probably not done? Under what conditions might extremely pure reactants be called for in a commercial process? (*Hint:* Think about possible problems associated with the impurities in technical grade chemicals.)

*4.54. The following two reactions take place in a gas-phase reactor:

$$2 CO_2 \rightleftharpoons 2CO + O_2$$
$$\text{(A)} \quad \text{(B)} \quad \text{(C)}$$

$$O_2 + N_2 \rightleftharpoons 2NO$$
$$\text{(C)} \quad \text{(D)} \quad \text{(E)}$$

If the system comes to equilibrium at 3000 K and 1 atm, the product gas mole fractions satisfy the relations

$$\frac{y_B^2 y_C}{y_A^2} = 0.1071 \qquad \frac{y_E^2}{y_C y_D} = 0.01493$$

(a) Let $n_{A0}, \ldots, n_{E0}$ be the initial number of gram-moles of each species and ξ_{e1} and ξ_{e2} be the extents of reactions 1 and 2, respectively, at equilibrium (see Equation 4.6-6). Derive expressions for the mole fractions $y_A, y_B, \ldots, y_E$ in terms of $n_{A0}, n_{B0}, \ldots, n_{E0}, \xi_{e1}$, and ξ_{e2}. Then substitute in the equilibrium relations to derive two simultaneous equations for the two extents of reaction.

(b) One-third of a gram-mole each of CO_2, O_2, and N_2 are charged into a batch reactor and the reactor contents equilibrate at 3000 K and 1 atm. Without doing any calculations, prove that you have enough information to calculate the component mole fractions of the reactor contents at equilibrium.

(c) Perform the calculation of part (b), using either (i) an equation-solving program or (ii) a spreadsheet that implements the Newton–Raphson method outlined in Section A.2i of Appendix A. If you use the spreadsheet, guess initial values of 0.1 for both ξ_{e1} and ξ_{e2} and iterate until successive estimates of these values differ by less than 0.1%.

(d) Write a computer program to implement the Newton–Raphson procedure of part (c) for an arbitrary starting composition. The program should take input values of $n_{A0}, n_{B0}, n_{C0}, n_{D0}$, and n_{E0} and calculate the total moles and mole fractions of each species at equilibrium, stopping when the values of ξ_{e1} and ξ_{e2} each change by less than 0.001% from one iteration to the next. Run the program for the following feed mixtures:

n_{A0}	n_{B0}	n_{C0}	n_{D0}	n_{E0}
1/3	0	1/3	1/3	0
0	1/3	1/3	1/3	0
1/2	0	0	0	1/2
1/5	1/5	1/5	1/5	1/5

4.55. An intermediate (A) in a continuous pharmaceutical manufacturing process contains up to 10 wt% of an ingredient R [mass fraction x_{RA}(g R/g A)]. The intermediate is subjected to a reaction step in which the conversion of R is 99%. The final product (P) must contain 0.75 wt% R. To achieve this goal, a fraction of the A fed to the process bypasses the reactor and combines with the product stream leaving the reactor to produce P. A second feed stream (B), which contains no R, enters the reactor along with the A not bypassed. The mass flow rates of A and B entering the *reactor* (not the process) are equal.

---

*Computer problem.

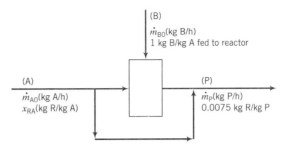

(a) Without taking a basis of calculation, draw and label a flowchart and prove that the process has two degrees of freedom. (Be careful when counting allowable balances on the pre-reactor stream splitting point.)

(b) Suppose the production rate of the final product, $\dot{m}_P$, and the mass fraction of R in the process feed, x_{RA}, are known. Write a set of equations that may be solved for the rates, $\dot{m}_{A0}$(kg/h) and $\dot{m}_{B0}$(kg/h), at which A and B must be fed to the *process* (not the reactor), and for the fraction of the process feed, f(kg bypass/kg fresh feed), that must be bypassed to achieve the desired mass fraction of R in the final product (0.0075 kg R/kg P). Do not perform any numerical calculations.

(c) Suppose $\dot{m}_P$ = 4850 kg P/h and x_{RA} = 0.0500 kg R/kg A. Calculate $\dot{m}_{A0}$, $\dot{m}_{B0}$, and f.

*(d) The desired production rate ($\dot{m}_P$) and the mass fraction of R in the feed (x_{RA}) both vary from day to day. Use an equation-solving program such as E-Z Solve to define. Use it to generate a plot of f versus x_{RA} for $\dot{m}_P$ = 4850 kg/h and x_{RA} varying between 0.02 and 0.10. Prove that you would get the same curve regardless of the value of $\dot{m}_P$.

4.56. A catalytic reactor is used to produce formaldehyde from methanol in the reaction

$$CH_3OH \longrightarrow HCHO + H_2$$

A single-pass conversion of 60.0% is achieved in the reactor. The methanol in the reactor product is separated from the formaldehyde and hydrogen in a multiple-unit process. The production rate of formaldehyde is 900.0 kg/h.

(a) Calculate the required feed rate of methanol to the process (kmol/h) if there is no recycle.

(b) Suppose the recovered methanol is recycled to the reactor and the single-pass conversion remains 60%. Without doing any calculations, prove that you have enough information to determine the required fresh feed rate of methanol (kmol/h) and the rates (kmol/h) at which methanol enters and leaves the reactor. Then perform the calculations.

(c) The single-pass conversion in the reactor, X_{sp}, affects the costs of the reactor (C_r) and the separation process and recycle line (C_s). What effect would you expect an increased X_{sp} would have on each of these costs for a fixed formaldehyde production rate? (*Hint:* To get a 100% single-pass conversion you would need an infinitely large reactor, and lowering the single-pass conversion leads to a need to process greater amounts of fluid through both process units and the recycle line.) What would you expect a plot of ($C_r + C_s$) versus X_{sp} to look like? What does the design specification X_{sp} = 60% probably represent?

4.57. Methanol is produced by reacting carbon monoxide and hydrogen. A fresh feed stream containing CO and H_2 joins a recycle stream and the combined stream is fed to a reactor. The reactor outlet stream flows at a rate of 350 mol/min and contains 10.6 wt% H_2, 64.0 wt% CO, and 25.4 wt% CH_3OH. (Notice that those are percentages by mass, not mole percents.) This stream enters a cooler in which most of the methanol is condensed. The liquid methanol condensate is withdrawn as a product, and the gas stream leaving the condenser—which contains CO, H_2, and 0.40 mole% uncondensed CH_3OH vapor—is the recycle stream that combines with the fresh feed.

(a) Without doing any calculations, prove that you have enough information to determine (i) the molar flow rates of CO and H_2 in the fresh feed, (ii) the production rate of liquid methanol, and (iii) the single-pass and overall conversions of carbon monoxide. Then perform the calculations.

*Computer problem.

(b) After several months of operation, the flow rate of liquid methanol leaving the condenser begins to decrease. List at least three possible explanations of this behavior and state how you might check the validity of each one. (What would you measure and what would you expect to find if the explanation is valid?)

4.58. Methane reacts with chlorine to produce methyl chloride and hydrogen chloride. Once formed, the methyl chloride may undergo further chlorination to form methylene chloride (CH_2Cl_2), chloroform, and carbon tetrachloride.

A methyl chloride production process consists of a reactor, a condenser, a distillation column, and an absorption column. A gas stream containing 80.0 mole% methane and the balance chlorine is fed to the reactor. In the reactor a single-pass chlorine conversion of essentially 100% is attained, the mole ratio of methyl chloride to methylene chloride in the product is 5:1, and negligible amounts of chloroform and carbon tetrachloride are formed. The product stream flows to the condenser. Two streams emerge from the condenser: the liquid condensate, which contains essentially all of the methyl chloride and methylene chloride in the reactor effluent, and a gas containing the methane and hydrogen chloride. The condensate goes to the distillation column in which the two component species are separated. The gas leaving the condenser flows to the absorption column where it contacts an aqueous solution. The solution absorbs essentially all of the HCl and none of the CH_4 in the feed. The liquid leaving the absorber is pumped elsewhere in the plant for further processing, and the methane is recycled to join the fresh feed to the process (a mixture of methane and chlorine). The combined stream is the feed to the reactor.

(a) Choose a quantity of the reactor feed as a basis of calculation, draw and label a flowchart, and determine the degrees of freedom for the overall process and each single unit and stream mixing point. Then write in order the equations you would use to calculate the molar flow rate and molar composition of the fresh feed, the rate at which HCl must be removed in the absorber, the methyl chloride production rate, and the molar flow rate of the recycle stream. Do no calculations.

(b) Calculate the quantities specified in part (a), either manually or with an equation-solving program.

(c) What molar flow rates and compositions of the fresh feed and the recycle stream are required to achieve a methyl chloride production rate of 1000 kg/h?

4.59. Ethylene oxide is produced by the catalytic oxidation of ethylene:

$$2C_2H_4 + O_2 \longrightarrow 2C_2H_4O$$

An undesired competing reaction is the combustion of ethylene:

$$C_2H_4 + 3O_2 \longrightarrow 2CO_2 + 2H_2O$$

The feed to the reactor (*not* the fresh feed to the process) contains 3 moles of ethylene per mole of oxygen. The single-pass conversion of ethylene is 20%, and for every 100 moles of ethylene consumed in the reactor, 90 moles of ethylene oxide emerges in the reactor products. A multiple-unit process is used to separate the products: ethylene and oxygen are recycled to the reactor, ethylene oxide is sold as a product, and carbon dioxide and water are discarded.

(a) Assume a quantity of the reactor feed stream as a basis of calculation, draw and label the flowchart, perform a degree-of-freedom analysis, and write the equations you would use to calculate (i) the molar flow rates of ethylene and oxygen in the fresh feed, (ii) the production rate of ethylene oxide, and (iii) the overall conversion of ethylene. Do no calculations.

(b) Calculate the quantities specified in part (a), either manually or with an equation-solving program.

(c) Calculate the molar flow rates of ethylene and oxygen in the fresh feed needed to produce 1 ton per hour of ethylene oxide.

4.60. Methanol is synthesized from carbon monoxide and hydrogen in a catalytic reactor. The fresh feed to the process contains 32.0 mole% CO, 64.0% H_2, and 4.0% N_2. This stream is mixed with a recycle stream in a ratio 5 mol recycle/1 mol fresh feed to produce the feed to the reactor, which contains 13.0 mole% N_2. A low single-pass conversion is attained in the reactor. The reactor effluent goes to a condenser from which two streams emerge: a liquid product stream containing essentially all the methanol formed in the reactor, and a gas stream containing all the CO, H_2, and N_2 leaving the reactor. The gas stream is split into two fractions: one is removed from the process as a purge stream, and the other is the recycle stream that combines with the fresh feed to the reactor.

(a) For a basis of 100 mol fresh feed/h, calculate the production rate of methanol (mol/h), the molar flow rate and composition of the purge gas, and the overall and single-pass conversions.

(b) Briefly explain in your own words the reasons for including (i) the recycle stream and (ii) the purge stream in the process design.

4.61. The fresh feed to an ammonia production process contains nitrogen and hydrogen in stoichiometric proportion, along with an inert gas (I). The feed is combined with a recycle stream containing the same three species, and the combined stream is fed to a reactor in which a low single-pass conversion of nitrogen is achieved. The reactor effluent flows to a condenser. A liquid stream containing essentially all of the ammonia formed in the reactor and a gas stream containing all the inerts and the unreacted nitrogen and hydrogen leave the condenser. The gas stream is split into two fractions with the same composition: one is removed from the process as a purge stream, and the other is the recycle stream combined with the fresh feed. *In every stream containing nitrogen and hydrogen, the two species are in stoichiometric proportion.*

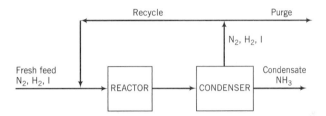

(a) Let x_{I0} be the mole fraction of inerts in the fresh feed, f_{sp} the single-pass conversion of nitrogen (and of hydrogen) in the reactor, and y_p the fraction of the gas leaving the condenser that is purged (mol purged/mol total). Taking a basis of 1 mol fresh feed, draw and fully label a process flowchart, incorporating x_{I0}, f_{sp}, and y_p in the labeling to the greatest possible extent. Then, assuming that the values of these three variables are given, write a set of equations for the total moles fed to the reactor (n_r), moles of ammonia produced (n_p), and overall nitrogen conversion (f_{ov}). Each equation should involve only one unknown variable, which should be circled.

(b) Solve the equations of part (a) for $x_{I0} = 0.01$, $f_{sp} = 0.20$, and $y_p = 0.10$.

(c) Briefly explain in your own words the reasons for including (i) the recycle stream and (ii) the purge stream in the process design.

*(d) Write a spreadsheet program to perform the calculations of part (a) for given values of x_{I0}, f_{sp}, and y_p. Test it with the values in part (b). Then in successive rows of the spreadsheet, vary each of the three input variables two or three times, holding the other two constant. The first six columns and first five rows of the spreadsheet should appear as follows:

xio	fsp	yp	nr	np	fov
0.01	0.20	0.10			
0.05	0.20	0.10			
0.10	0.20	0.10			
0.01	0.30	0.10			

Summarize the effects on ammonia production (n_p) and reactor throughput (n_r) of changing each of the three input variables.

4.62. Iso-octane is produced in the reaction of isobutane and butylene in an emulsion with concentrated sulfuric acid:

$$i\text{-}C_4H_{10} + C_4H_8 \longrightarrow i\text{-}C_8H_{18}$$

---

*Computer problem.

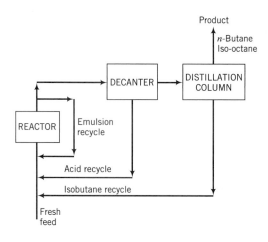

The fresh feed to the process flows at a rate of 60,000 kg/h and contains 25.0 mole% isobutane, 25.0% butylene, and 50.0% n-butane, which is chemically inert in this process. The fresh feed combines with three separate recycle streams, as shown in the flowchart, and the combined stream enters the reactor. Essentially all of the butylene fed to the reactor is consumed. A portion of the reactor effluent is recycled to the reactor inlet and the remainder passes to a decanter, in which the aqueous (sulfuric acid) and hydrocarbon phases are allowed to separate. The acid is recycled to the reactor, and the hydrocarbons pass to a distillation column. The overhead from the column contains iso-octane and n-butane, and the bottoms product, which is recycled to the reactor, contains only isobutane. The stream entering the reactor contains 200 moles of isobutane per mole of butylene, and 2 kg of 91 wt% H_2SO_4(aq) per kg of hydrocarbon. The stream obtained by combining the fresh feed and isobutane recycle contains 5.0 moles of isobutane per mole of butylene.

You wish to determine the molar flow rates (kmol/h) of each component of the fresh feed, the product stream, and the emulsion, isobutane, and acid recycle streams.

(a) Draw and completely label a flowchart of the process, perform degree-of-freedom analyses on the overall process and subprocesses, and write the equations you would use to determine the required flow rates. (*Suggestion:* Begin by calculating the total molar flow rate of the fresh feed stream.)

(b) Perform the calculations.

(c) List the assumptions in this problem that are not likely to be fully satisfied in practice.

*4.63. Ethyl acetate (A) undergoes a reaction with sodium hydroxide (B) to form sodium acetate and ethyl alcohol:

$$CH_3COOC_2H_5 + NaOH \longrightarrow CH_3COONa + C_2H_5OH$$
$$\quad\quad (A) \quad\quad\quad\quad (B)$$

The reaction is carried out at steady state in a series of stirred-tank reactors. The output from the i th reactor is the input to the $(i + 1)$st reactor. The volumetric flow rate between the reactors is constant at $\dot{v}$(L/min), and the volume of each tank is V(L).

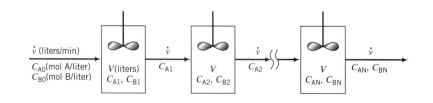

The concentrations of A and B in the feed to the first tank are C_{A0} and C_{B0} (mol/L). The tanks are stirred sufficiently for their contents to be uniform throughout, so that C_A and C_B in a tank equal

---

*Computer problem.

C_A and C_B in the stream leaving that tank. The rate of reaction is given by the expression

$$r\left(\frac{\text{mol A or B reacting}}{\text{min} \cdot \text{L}}\right) = kC_A C_B$$

where $k\,[\text{L}/(\text{mol}\cdot\text{min})]$ is the *reaction rate constant*.

(a) Write a material balance on A in the ith tank, and show that it yields

$$C_{A,i-1} = C_{Ai} + k\tau C_{Ai} C_{Bi}$$

where $\tau = V/\dot{v}$ is the *mean residence time* in each tank. Then write a balance on B in the ith tank and subtract the two balances, using the result to prove that

$$C_{Bi} - C_{Ai} = C_{B0} - C_{A0}, \quad \text{all } i$$

(b) Use the equations derived in part (a) to prove that

$$C_{A,i-1} = C_{Ai} + k\tau C_{Ai}(C_{Ai} + C_{B0} - C_{A0})$$

and from this relation derive an equation of the form

$$\alpha C_{Ai}^2 + \beta C_{Ai} + \gamma = 0$$

where α, β, and γ are functions of k, C_{A0}, C_{B0}, $C_{A,i-1}$, and τ. Then write the solution of this equation for C_{Ai}.

(c) Write a spreadsheet or computer program to calculate N, the number of tanks needed to achieve a fractional conversion $x_{AN} \geq x_{Af}$ at the outlet of the final reactor. Your program should implement the following procedure:

(i) Take as input values of k, $\dot{v}$, V, C_{A0} (mol/L), C_{B0}(mol/L), and x_{Af}.

(ii) Use the equation for C_{Ai} derived in part (b) to calculate C_{A1}; then calculate the corresponding fractional conversion x_{A1}.

(iii) Repeat the procedure to calculate C_{A2} and x_{A2}, then C_{A3} and x_{A3}, continuing until $x_{Ai} \geq x_{Af}$.

Test the program supposing that the reaction is to be carried out at a temperature at which $k = 36.2$ L/(mol·min), and that the other process variables have the following values:

$$\text{Feed concentration:} \quad C_{A0} = 5.0 \text{ g/L } (= ??? \text{ mol/L})$$
$$C_{B0} = 0.10 \text{ molar } (= ??? \text{ mol/L})$$
$$\text{Throughput:} \quad \dot{v} = 5000 \text{ L/min}$$
$$\text{Tank volume:} \quad V = 2000 \text{ L}$$

Use the program to calculate the required number of tanks and the final fractional conversion for the following values of the desired minimum final fractional conversion, x_{Af}: 0.50, 0.80, 0.90, 0.95, 0.99, 0.999. Briefly describe the nature of the relationship between N and x_{Af} and what probably happens to the process cost as the required final fractional conversion approaches 1.0. *Hint:* If you write a spreadsheet, it might appear in part as follows:

Spreadsheet for Problem 4-63						
k =	36.2		N	gamma	CA(N)	xA(N)
v =	5000		1	−5.670E−02	2.791E−02	0.5077
V =	2000		2	−2.791E−02	1.512E−02	0.7333
CA0 =	0.0567		3	⋮	⋮	⋮
CB0 =	0.1000		⋮	⋮	⋮	⋮
alpha =	14.48		⋮	⋮	⋮	⋮
beta =	1.6270		⋮	⋮	⋮	⋮
			⋮	⋮	⋮	⋮

(d) Suppose a 95% conversion is desired. Use your program to determine how the required number of tanks varies as you increase (i) the rate constant, k; (ii) the throughput, $\dot{v}$; and (iii) the individual reactor volume, V. Then briefly explain why the results you obtain make sense physically.

4.64. A gas contains 80.0 wt% propane, 15.0 wt% n-butane, and the balance water.

(a) Calculate the *molar* composition of this gas on both a wet and a dry basis and the ratio (mol H_2O/mol dry gas).

(b) If 100 kg/h of this fuel is to be burned with 30% excess air, what is the required air feed rate (kmol/h)? How would the answer change if the combustion were only 75% complete?

4.65. Five liters of liquid n-hexane and 4 liters of liquid n-heptane are mixed and burned with 4000 gram-moles of air. Not all of the hydrocarbons are burned in the furnace, and both CO and CO_2 are formed. If it is possible to do so without additional information, calculate the percent excess air supplied to the furnace; if more information is needed, state what it is and outline the calculation of the percent excess air.

4.66. A fuel gas produced by gasifying coal is to be burned with 20% excess air. The gas contains 50.0 mole% nitrogen and the balance carbon monoxide and hydrogen. A sample of the gas is passed through an infrared spectrometer, which registers a signal R that depends on the mole fraction of carbon monoxide in the sample, and a reading $R = 38.3$ is recorded.

Analyzer calibration data are as follows:

x(mol CO/mol)	0.05	0.10	0.40	0.80	1.00
R	10.0	17.0	49.4	73.6	99.7

A power law ($x = aR^b$) should be suitable for fitting the calibration data. Derive the equation relating x and R (use a graphical method), and then calculate the molar flow rate of air required for a fuel feed rate of 175 kmol/h, assuming that CO and H_2 are oxidized but N_2 is not.

4.67. Natural gas containing a mixture of methane, ethane, propane, and butane is burned in a furnace with excess air.

(a) One hundred kmol/h of a gas containing 94.4 mole% methane, 3.40% ethane, 0.60% propane, and 0.50% butane is to be burned with 17% excess air. Calculate the required molar flow rate of the air.

(b) Let

$$\dot{n}_f(\text{kmol/h}) = \text{molar flow of the fuel gas}$$
$$x_1, x_2, x_3, x_4, = \text{mole fractions of methane, ethane, propane,}$$
$$\text{and butane, respectively, in the fuel}$$
$$P_{xs} = \text{percent excess air}$$
$$\dot{n}_a(\text{kmol/h}) = \text{molar flow rate of the air fed to the furnace}$$

Derive an expression for $\dot{n}_a$ in terms of the other variables. Check your formula with the results of part (a).

*(c) Suppose the feed rate and composition of the fuel gas are subject to periodic variations, and a process control computer is to be used to adjust the flow rate of air to maintain a constant percentage excess. A calibrated electronic flowmeter in the fuel gas line transmits a signal R_f that is directly proportional to the flow rate ($\dot{n}_f = aR_f$), with a flow rate of 75.0 kmol/h yielding a signal $R_f = 60$. The fuel gas composition is obtained with an on-line gas chromatograph. A sample of the gas is injected into the gas chromatograph (GC), and signals A_1, A_2, A_3, and A_4, which are directly proportional to the moles of methane, ethane, propane, and butane, respectively, in the sample, are transmitted. (Assume the same proportionality constant for all species.) The control computer processes these data to determine the required air flow rate and then sends a signal R_a to a control valve in the air line. The relationship between R_a and the resulting air flow rate, $\dot{n}_a$, is another direct proportionality, with a signal $R_a = 25$ leading to an air flow rate of 550 kmol/h.

---

*Computer problem.

Write a spreadsheet or computer program to perform the following tasks:

(i) Take as input the desired percentage excess and values of R_f, A_1, A_2, A_3, and A_4.

(ii) Calculate and print out $\dot{n}_f$, x_1, x_2, x_3, x_4, $\dot{n}_a$, and R_a.

Test your program on the data given below, assuming that 15% excess air is required in all cases. Then explore the effects of variations in P_{xs} and R_f on $\dot{n}_a$ for the values of $A_1 - A_4$ given on the third line of the data table. Briefly explain your results.

R_f	A_1	A_2	A_3	A_4
62	248.7	19.74	6.35	1.48
83	305.3	14.57	2.56	0.70
108	294.2	16.61	4.78	2.11

(d) Finally, suppose that when the system is operating as described, stack gas analysis indicates that the air feed rate is consistently too high to achieve the specified percentage excess. Give several possible explanations.

4.68. Butane is burned with air. No carbon monoxide is present in the combustion products.

(a) Use a degree-of-freedom analysis to prove that if the percentage excess air and the percentage conversion of butane are specified, the molar composition of the product gas can be determined.

(b) Calculate the molar composition of the product gas for each of the following three cases: (i) theoretical air supplied, 100% conversion of butane; (ii) 20% excess air, 100% conversion of butane; and (iii) 20% excess air, 90% conversion of butane.

4.69. A mixture of 75 mole% propane and 25 mole% hydrogen is burned with 25% excess air. Fractional conversions of 90% of the propane and 85% of the hydrogen are achieved; of the propane that reacts, 95% reacts to form CO_2 and the balance reacts to form CO. The hot combustion product gas passes through a boiler in which heat transferred from the gas converts boiler feedwater into steam.

(a) Calculate the concentration of CO (ppm) in the stack gas.

(b) The CO in the stack gas is a pollutant. Its concentration can be decreased by increasing the percent excess air fed to the furnace. Think of at least two costs of doing so. (*Hint:* The heat released by the combustion goes into heating the combustion products, and the higher the combustion product temperature, the more steam is produced.)

4.70. *n*-Pentane is burned with excess air in a continuous combustion chamber.

(a) A technician runs an analysis and reports that the product gas contains 0.270 mole% pentane, 5.3% oxygen, 9.1% carbon dioxide, and the balance nitrogen *on a dry basis.* Assume 100 mol of dry product gas as a basis of calculation, draw and label a flowchart, perform a degree-of-freedom analysis based on atomic species balances, and show that the system has -1 degrees of freedom. Interpret this result.

(b) Use balances to prove that the reported percentages could not possibly be correct.

(c) The technician reruns the analysis and reports new values of 0.304 mole% pentane, 5.9% oxygen, 10.2% carbon dioxide, and the balance nitrogen. Verify that this result *could* be correct and, assuming that it is, calculate the percent excess air fed to the reactor and the fractional conversion of pentane.

4.71. Liquid methanol is fed to a space heater at a rate of 12.0 L/h and burned with excess air. The product gas is analyzed and the following dry-basis mole percentages are determined: $CH_3OH = 0.45\%$, $CO_2 = 9.03\%$, and $CO = 1.81\%$.

(a) Draw and label a flowchart and verify that the system has zero degrees of freedom.

(b) Calculate the fractional conversion of methanol, the percentage excess air fed, and the mole fraction of water in the product gas.

(c) Suppose the combustion products are released directly into a room. What potential problems do you see and what remedies can you suggest?

4.72. A gas containing methane, ethane, and carbon dioxide is analyzed with a gas chromatograph (GC) and flame ionization detector (FID): the GC separates the components of the gas, and the FID registers signals proportional to the amount of each hydrocarbon (but not CO_2) in its sample chamber.

The output of the FID is as follows:

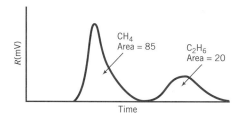

The area under each peak is proportional to the number of carbon atoms in the sample, so that 1 mol of ethane would yield a peak with twice the area of a peak corresponding to 1 mol of methane.

This fuel is being burned with air in a continuous combustion chamber. The molar feed ratio of air to fuel was supposed to be 7:1, but you suspect the air flowmeter is not functioning properly. To check it, you take a 0.50-mol sample of the product gas and pass it through a condenser, which condenses essentially all of the water in the sample. The condensate (which can be assumed to be pure water) is weighed and found to have a mass of 1.134 g. The dry gas leaving the condenser is analyzed and found to contain no hydrocarbons, no CO, and 11.9% CO_2.

(a) Calculate the molar composition (component mole fractions) of the fuel gas and the desired percent excess air.

(b) Calculate the actual molar feed ratio of air to fuel and the actual percent excess air.

4.73. A mixture of propane and butane is burned with pure oxygen. The combustion products contain 47.4 mole% H_2O. After all the water is removed from the products, the residual gas contains 69.4 mole% CO_2 and the balance O_2.

(a) What is the mole percent of propane in the fuel?

(b) It now turns out that the fuel mixture may contain not only propane and butane but also other hydrocarbons. All that is certain is that there is no oxygen in the fuel. Use atomic balances to calculate the elemental molar composition of the fuel from the given combustion product analysis (i.e., what mole percent is C and what percent is H). Prove that your solution is consistent with the result of part (a).

4.74. A fuel oil is analyzed and found to contain 85.0 wt% carbon, 12.0% elemental hydrogen (H), 1.7% sulfur, and the remainder noncombustible matter. The oil is burned with 20.0% excess air, based on complete combustion of the carbon to CO_2, the hydrogen to H_2O, and the sulfur to SO_2. The oil is burned completely, but 8% of the carbon forms CO. Calculate the molar composition of the stack gas.

4.75. The analysis of a coal indicates 75 wt% C, 17% H, 2% S, and the balance noncombustible ash. The coal is burned at a rate of 5000 kg/h, and the feed rate of air to the furnace is 50 kmol/min. All of the ash and 6% of the carbon in the fuel leave the furnace as a molten slag; the remainder of the carbon leaves in the stack gas as CO and CO_2; the hydrogen in the coal is oxidized to water, and the sulfur emerges as SO_2. The selectivity of CO_2 to CO production is 10:1.

(a) Calculate the percent excess air fed to the reactor.

(b) Calculate the mole fractions of the gaseous pollutants—CO and SO_2—in the stack gas.

(c) Emitted sulfur dioxide by itself is a health hazard, but it is a more serious threat to the environment as a precursor to **acid rain**. Under the catalytic action of sunlight, the sulfur dioxide is oxidized to sulfur trioxide, which in turn combines with water vapor to form sulfuric acid, which eventually returns to earth as rainfall. Acid rain formed in this manner has caused extensive damage to forests, fields, and lakes in many parts of the world. For the furnace described above, calculate the rate of formation of sulfuric acid (kg/h) if all the emitted SO_2 is converted in the indicated manner.

4.76. The composition of a coal is determined by a **proximate analysis**. The coal is first finely ground and air-dried. Samples of the dried coal are then subjected to several operations, with the sample weights being recorded before and after each operation. **Moisture content** is determined as the weight loss when a sample is held at 105°C in an oxygen-free atmosphere for roughly 2 h, added to the weight loss in the initial drying step. **Volatile matter** (primarily organic tars) is determined by holding a sample at 925°C in an oxygen-free atmosphere for 7 min and subtracting the moisture loss from the total weight loss. **Ash** (or **mineral matter**—oxides and sulfates of silicon, aluminum, iron, calcium, sulfur, and trace minerals) is the residue that remains after a sample has been heated to 800°C in an oxygen-containing atmosphere until all the organic matter has been burned away. **Fixed carbon** is what is present in coal besides moisture, volatile matter, and ash.

(a) Use the following proximate analysis data to determine the percentages by mass of moisture, fixed carbon, volatile matter, and ash in a coal:

$$1.207 \text{ g} \xrightarrow[25°C, 12 \text{ h}]{\text{air-dry}} 1.147 \text{ g}$$

The remaining tests are performed on air-dried samples.

$$1.234 \text{ g} \xrightarrow[2 \text{ h}]{105°C, \text{ N}_2} 1.204 \text{ g}$$

$$1.347 \text{ g} \xrightarrow[7 \text{ min}]{925°C, \text{ N}_2} 0.811 \text{ g}$$

$$1.175 \text{ g} \xrightarrow[1 \text{ h}]{800°C, \text{ N}_2} 0.111 \text{ g}$$

(b) If the mass ratio of C to H in the volatile matter is 6:1, calculate the gram-moles of air theoretically required to burn 1 metric ton of this coal.

4.77. The product gas from a solid fuel combustion reaction has the following dry-basis molar composition: 72.0% CO_2, 2.57% CO, 0.0592% SO_2, and 25.4% O_2. Pure oxygen is fed to the furnace in 20% excess of that required to burn the fuel completely. There is no oxygen in the fuel. Calculate the elemental composition (mole% of the various elements) of the fuel, *stating any assumptions you have to make to arrive at your answer.*

4.78. A fuel oil is fed to a furnace and burned with 25% excess air. The oil contains 87.0 wt% C, 10.0% H, and 3.0% S. Analysis of the furnace exhaust gas shows only N_2, O_2, CO_2, SO_2, and H_2O. The sulfur dioxide emission rate is to be controlled by passing the exhaust gas through a scrubber, in which most of the SO_2 is absorbed in an alkaline solution. The gases leaving the scrubber (all of the N_2, O_2, and CO_2, and some of the H_2O and SO_2 entering the unit) pass out to a stack. The scrubber has a limited capacity, however, so that a fraction of the furnace exhaust gas must be bypassed directly to the stack.

At one point during the operation of the process, the scrubber removes 90% of the SO_2 in the gas fed to it, and the combined stack gas contains 612.5 ppm (parts per million) SO_2 on a dry basis; that is, every million moles of dry stack gas contains 612.5 moles of SO_2. Calculate the fraction of the exhaust bypassing the scrubber at this moment.

4.79. You have been sent by the Environmental Protection Agency to measure SO_2 emissions from a small industrial power plant. You withdraw and analyze a gas sample from the boiler stack and obtain the following composition: 75.66% N_2, 10.24% CO_2, 8.27% H_2O, 5.75% O_2, and 0.0825% SO_2. You show these figures to the plant superintendent next day, and she insists they must be wrong, since the fuel was a natural gas containing methane and ethane, and no sulfur. You ask if they ever burn another fuel, and the superintendent replies that they sometimes use a fuel oil but the plant log shows that they were not doing so when the measurements were made. You do some calculations and prove that the oil and not the gas must have been the fuel; the superintendent checks further and discovers that the plant log is in error and you are right.

(a) Calculate the mole ratio of carbon to hydrogen in the fuel, and use the result to prove that the fuel could not have been the natural gas.

(b) Calculate the *mass* ratio of carbon to hydrogen and the weight% of sulfur in the fuel, assuming that C, H, and S are the only elements present. Then use the results in conjunction with the ultimate fuel oil analyses in Table 27-6 on p. 27-10 of *Perry's Chemical Engineers' Handbook,* 7th Edition, to deduce the most probable classification of the fuel oil.

4.80. Fuel oils contain primarily organic compounds and sulfur. The molar composition of the organic fraction of a fuel oil may be represented by the formula $C_pH_qO_r$; the mass fraction of sulfur in the fuel is x_S (kg S/kg fuel); and the percentage excess air, P_{xs}, is defined in terms of the theoretical air required to burn only the carbon and hydrogen in the fuel.

(a) For a certain high-sulfur No. 6 fuel oil, $p = 0.71$, $q = 1.1$, $r = 0.003$, and $x_S = 0.02$. Calculate the composition of the stack gas on a dry basis if this fuel is burned with 18% excess air, assuming complete combustion of the fuel to form CO_2, SO_2, and H_2O and expressing the SO_2 fraction as ppm (mol $SO_2/10^6$ mol dry gas).

*(b) Create a spreadsheet to calculate the mole fractions of the stack gas components on a dry basis for specified values of p, q, r, x_S, and P_{xs}. The output should appear as follows:

Solution to Problem 4-80			
Run	1	2	...
p	0.71	0.71	...
q	1.1	1.1	...
r	0.003	0.003	...
xS	0.02	0.02	...
Pxs	18%	36%	...
y(CO2)	13.4%	...	...
y(O2)	...	...	...
y(N2)	...	...	...
ppm SO2	1165	...	...

(Rows below the last one shown can be used to calculate intermediate quantities.) Execute enough runs (including the two shown above) to determine the effect on the stack gas composition of each of the five input parameters. Then for the values of p, q, r, and x_S given in part (a), find the minimum percentage excess air needed to keep the dry-basis SO_2 composition below 700 ppm. (Make this the last run in the output table.)

You should find that for a given fuel oil composition, increasing the percentage excess air decreases the SO_2 concentration in the stack gas. Explain why this should be the case.

(c) Someone has proposed using the relationship between P_{xs} and ppm SO_2 as the basis of a pollution control strategy. The idea is to determine the minimum acceptable concentration of SO_2 in the stack gas, then run with the percentage excess air high enough to achieve this value. Give several reasons why this is a poor idea.

---

*Computer problem.

Chapter 5

Single-Phase Systems

Most of the material balance problems in Chapter 4 could be solved entirely from information given in the problem statements. As you will come to discover, problems in process analysis are rarely so conveniently self-contained; before you can carry out a complete material balance on a process, you usually must determine various physical properties of the process materials and use these properties to derive additional relations among the system variables. The following methods can be used to determine a physical property of a process material:

Look It Up. When you need a value for a physical property of a substance—whether it be a density, vapor pressure, solubility, or heat capacity—there is a good chance that someone, somewhere has measured this property and published the result. Since experiments are usually costly and time consuming, a reliable source of physical property data is an invaluable asset in process analysis. Four excellent sources of data are the following:

Perry's Chemical Engineers' Handbook, 7th Edition, R. H. Perry and D. W. Green, Eds., McGraw-Hill, New York, 1997.

CRC Handbook of Chemistry and Physics, 79th Edition, D. Lide, Ed., Chemical Rubber Company, Boca Raton, FL, 1998.

TRC Databases in Chemistry and Engineering. TRC Thermodynamic Tables Version 1.0. Thermodynamic Research Center, Texas A&M University, College Station, Texas, 1994. This is a continuation of the American Petroleum Institute Project 44: "Selected Values of the Properties of Hydrocarbons and Related Compounds."

T. E. Daubert and R. P. Danner, *Physical and Thermodynamic Properties of Pure Chemicals: Data Compilation,* Hemisphere Publishing Corporation, New York, 1991. This is a hard-copy version of the Design Institute of Physical Properties Research (DIPPR) Data Compilation.

If the desired information cannot be found in these references, the species in question can be looked up in the index of *Chemical Abstracts* in an effort to locate data in the open literature.

Estimate It. There are a relatively small number of atomic elements and a much larger but still countable number of molecular species of interest to chemical engineers. Chemical species can be combined into mixtures in an infinite number of ways, however, and it is clearly impossible to tabulate physical property data for even a small fraction of the possible combinations. Moreover, even when data are found they are likely to have been determined for conditions other than the ones for which you need them. Reid, Prausnitz, and Poling[1] summarize a large number of empirical correlations that express physical properties of a mixture in terms of the

---

[1]R. C. Reid, J. M. Prausnitz, and B. E. Poling, *The Properties of Gases and Liquids,* 4th Edition, McGraw-Hill, New York, 1986.

pure-component properties and the mixture composition. These correlations can be used to estimate physical properties when no data are available and to extrapolate available data to conditions other than the ones at which the data were obtained.

Measure It. When no information on a particular physical property of a substance can be found in the literature or when the property must be known with a precision greater than that provided by general estimation formulas, the only recourse is to determine the property experimentally. Information on experimental techniques for measuring physical properties can be found in any of numerous texts on experimental physical, organic, and analytical chemistry.

Density is a frequently needed physical property of a process fluid. For example, engineers often know volumetric flow rates ($\dot{V}$) of process streams from flowmeter readings but need to know mass flow rates ($\dot{m}$) or molar flow rates ($\dot{n}$) for material balance calculations. The factor needed to calculate $\dot{m}$ or $\dot{n}$ from $\dot{V}$ is the density of the stream. This chapter illustrates the uses of both tabulated data and estimation formulas for calculating densities. Section 5.1 concerns solids and liquids; Section 5.2 covers *ideal gases,* gases for which the ideal gas equation of state ($PV = nRT$) is a good approximation; and Section 5.3 extends the discussion to nonideal gases.

5.0 INSTRUCTIONAL OBJECTIVES

After completing this chapter, you should be able to do the following:

- Explain in your own words and without the use of jargon (a) the three ways of obtaining values of physical properties; (b) why some fluids are referred to as incompressible; (c) the "liquid volume additivity assumption" and the species for which it is most likely to be valid; (d) the term "equation of state"; (e) what it means to assume ideal gas behavior; (f) what it means to say that the specific volume of an ideal gas at standard temperature and pressure is 22.4 L/mol; (g) the meaning of partial pressure; (h) why volume fraction and mole fraction for ideal gases are identical; (i) what the compressibility factor, z, represents, and what its value indicates about the validity of the ideal gas equation of state; (j) why certain equations of state are referred to as cubic; and (k) the physical meaning of critical temperature and pressure (explain them in terms of what happens when a vapor either below or above its critical temperature is compressed).
- For a mixture of liquids with known composition, determine V (or $\dot{V}$) from a known m ($\dot{m}$) or vice versa using (a) tabulated density data for the mixture and (b) pure-component densities and an assumption of volume additivity. Derive the density estimation formula for the second case (Equation 5.1-1).
- Given any three of the quantities P, V (or $\dot{V}$), n (or $\dot{n}$), and T for an ideal gas, (a) calculate the fourth one either directly from the ideal gas equation of state or by conversion from standard conditions; (b) calculate the density of the gas; and (c) test the assumption of ideality either by using a rule of thumb about the specific volume or by estimating a compressibility factor and seeing how much it differs from 1.
- Explain the meaning of "37.5 SCFH" (37.5 standard cubic feet per hour) and what it means to say that the flow rate of a gas stream at 120°F and 2.8 atm is 37.5 SCFH. (Why doesn't this statement specify the impossible condition that the gas is at two sets of temperatures and pressures simultaneously?) Calculate the true volumetric flow rate of that gas.
- Given the component partial pressures of an ideal gas mixture and the total gas pressure, determine the mixture composition expressed in either mole fractions (or mole percents), volume fractions (or % v/v), or mass fractions (or % w/w).
- Carry out *PVT* calculations for a gas using (a) the truncated virial equation of state, (b) the van der Waals equation of state, (c) the SRK equation of state, and (d) the compressibility

factor equation of state with either tabulated compressibility factors or a generalized compressibility chart for a single species and Kay's rule for a nonideal mixture of gases.

- Given a description of a process system in which a volumetric flow rate is either specified or requested for any process stream, (a) carry out the degree-of-freedom analysis, including density estimates for liquid and solid streams and equations of state for gas streams; (b) write the system equations and outline the procedure you would use to solve for all requested quantities; (c) carry out the calculations; (d) list all your assumptions (e.g., volume additivity for liquids or ideal gas behavior) and state whether or not they are reasonable for the given process conditions.

5.1   LIQUID AND SOLID DENSITIES

Specific gravities of solids and liquids were discussed in Sections 3.1 and 3.2. Values for several substances at a single temperature are listed in Table B.1 in Appendix B, and more extensive tabulations can be found in *Perry's Chemical Engineers' Handbook*[2] on pp. 2-7 through 2-47 and 2-91 through 2-120.

When you heat a liquid or a solid it normally expands (i.e., its density decreases). In most process applications, however, it can be assumed with little error that solid and liquid densities are independent of temperature. Similarly, changes in pressure do not cause significant changes in liquid or solid densities; these substances are therefore termed **incompressible**.

Perry's Chemical Engineers' Handbook (pp. 2-91 through 2-93) lists densities of liquid water and mercury at different temperatures and gives expressions on pp. 2-128 through 2-131 that may be used to calculate densities for many other substances at different temperatures. Reid, Prausnitz, and Poling (see footnote 1) present a number of methods to estimate the density of a liquid for which tabulated densities are not available. Some of these estimation formulas are also given in *Perry's Chemical Engineers' Handbook* on pp. 2-358 and 2-361.

The most accurate way to determine the density of a mixture of liquids or a solution of a solid in a liquid is from experimental data. *Perry's Chemical Engineers' Handbook* provides data for mixtures and solutions of a number of substances on pp. 2-99 through 2-118 and lists additional sources of data on p. 2-99.

In the absence of data, the density $\bar{\rho}$ of a mixture of n liquids $(A_1, A_2, \ldots, A_n)$ can be estimated from the component mass fractions $[x_i]$ and pure-component densities $[\rho_i]$ in two ways. First, we might assume *volume additivity*—that is, if 2 mL of liquid A and 3 mL of liquid B are mixed, the resulting volume would be exactly 5 mL. Making this assumption and recognizing that component masses are always additive leads to the formula

$$\frac{1}{\bar{\rho}} = \sum_{i=1}^{n} \frac{x_i}{\rho_i} \tag{5.1-1}$$

Second, we might simply average the pure-component densities, weighting each one by the mass fraction of the component:

$$\bar{\rho} = \sum_{i=1}^{n} x_i \rho_i \tag{5.1-2}$$

(Equation 5.1-1 calculates the inverse of the mixture density, or the *specific volume* of the mixture, as the weighted average of the pure-component specific volumes.)

One of these estimation formulas might work better for some species and the other might work better for other species. For example, Figure 5.1-1 shows densities of both methanol–water and sulfuric acid–water mixtures at 20°C. The densities for each pair of components

---

[2]R. H. Perry and D. W. Green, Eds., *Perry's Chemical Engineers' Handbook*, 7th Edition, McGraw-Hill, New York, 1997.

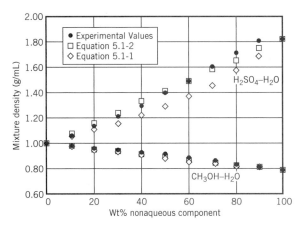

Figure 5.1-1 Experimental and estimated mixture densities. Experimental values from *Perry's Chemical Engineers' Handbook*, p. 2-107 for sulfuric acid–water and p. 2-111 for methanol–water, both at 20°C.

are obtained in three ways: from experimental data in *Perry's Chemical Engineers' Handbook* (pp. 2-107 and 2-111), using Equation 5.1-1, and using Equation 5.1-2. Equation 5.1-1 provides a slightly better estimate (i.e., values closer to the experimental data) for methanol and water, and Equation 5.1-2 provides a much better estimate for sulfuric acid and water.

Which method should you use? Equation 5.1-1 and the assumption of volume additivity work best for mixtures of liquid species with similar molecular structures (e.g., all straight-chain hydrocarbons of nearly equal molecular weight, such as *n*-pentane, *n*-hexane, and *n*-heptane). There are no general rules for when Equation 5.1-2 works better—all we can do is rely on empirical (experimental) findings.

TEST YOURSELF

1. The specific gravity of water is 1.0000 at 4.0°C. A stream of water at 4°C has a mass flow rate of 255 g/s. What is its volumetric flow rate? If the stream temperature is raised to 75°C, does the mass flow rate change? How would you expect the volumetric flow rate to change? How would you calculate the volumetric flow rate at 75°C without actually measuring it?

2. The pressure equivalence 14.696 $lb_f/in.^2 \iff$ 760 mm Hg is not complete—it should be stated as

$$14.696 \text{ lb}_f/\text{in.}^2 \iff 760 \text{ mm Hg at } 0°\text{C}$$

   Why is it technically necessary to state a temperature? Why is omitting the temperature not a serious mistake?

3. Suppose you mix $m_1(g)$ of liquid A_1 with density $\rho_1(g/cm^3)$, $m_2(g)$ of liquid A_2 with density $\rho_2, \ldots$, and $m_n(g)$ of liquid A_n with density ρ_n. Assuming that the volumes are additive, show that the density of the mixture is given by Equation 5.1-1.

EXAMPLE 5.1-1 **Determination of a Solution Density**

Determine the density in g/cm³ of a 50 wt% aqueous solution of H_2SO_4 at 20°C, both by (1) looking up a tabulated value and (2) assuming volume additivity of the solution components.

SOLUTION

1. **Look It Up.** *Perry's Chemical Engineers' Handbook*, pp. 2-107 and 2-108, tabulates specific gravities of sulfuric acid solutions. From this table,

$$\rho(50\% \text{ H}_2\text{SO}_4, 20°\text{C}) = 1.3951 \text{ g/cm}^3$$

2. *Estimate It.*   The pure-component densities are

$$\rho(H_2O,\ 20°C) = 0.998\ \text{g/cm}^3\ (\textit{Handbook, p. 2-91})$$

$$\rho(H_2SO_4, 18°C) = 1.834\ \text{g/cm}^3\ (\textit{Handbook, p. 2-25})$$

We will neglect the density change for H_2SO_4 between 18°C and 20°C, although *Perry's Chemical Engineers' Handbook,* p. 2-131, gives thermal expansion data for H_2SO_4 that could be used to make this minor correction. We then estimate from Equation 5.1-1:

$$1/\bar\rho = (0.500/0.998 + 0.500/1.834)\ \text{cm}^3/\text{g} = 0.7736\ \text{cm}^3/\text{g}$$

$$\Downarrow$$

$$\bar\rho = 1.29\ \text{g/cm}^3$$

The density estimated assuming volume additivity thus differs from the true density, given in part (1), by $[(1.29 - 1.3951)/1.3951] \times 100\% = -7.3\%$. Alternatively, we could estimate the density from Equation 5.1-2:

$$\bar\rho = (0.500 \times 0.998 + 0.500 \times 1.834)\ \frac{\text{g}}{\text{cm}^3} = 1.42\ \text{g/cm}^3$$

This leads to an estimation error of $[(1.42 - 1.3951)/1.3951] \times 100\% = 1.5\%$. Clearly, the accuracy of Equation 5.1-2 is better in this instance than that of Equation 5.1-1.

5.2 IDEAL GASES

Looking up a density or specific volume at one temperature and pressure and using it at another temperature and pressure usually works well for a solid or a liquid, but not at all for a gas. An expression is needed for gases that relates specific volume to temperature and pressure, so that if any two of these quantities are known the third can be calculated.

Typical problems that require a *PVT* relationship for their solution include the following:

1. Propane at 120°C and 2.3 bars passes through a flowmeter that reads 250 L/min. What is the mass flow rate of the gas?
2. A pure hydrocarbon gas fills a two-liter vessel at 30°C with an absolute pressure of 25 atm. How many gram-moles of gas are contained in the vessel? If the mass of the gas is 60 g, what might the gas be?
3. A 20-ft^3 gas cylinder in your laboratory can hold pressures up to 400 atm. The Bourdon gauge on the cylinder reads 380 atm one morning when the temperature is 55°F. How high can the temperature go before it becomes advisable to move to a different laboratory?

An **equation of state** relates the molar quantity and volume of a gas to temperature and pressure. The simplest and most widely used of these relationships is the **ideal gas equation of state** (the familiar $PV = nRT$), which, while approximate, is adequate for many engineering calculations involving gases at low pressures. However, some gases deviate from ideal behavior at nearly all conditions and all gases deviate substantially at certain conditions (notably at high pressures and/or low temperatures). In such cases it is necessary to use more complex equations of state for *PVT* calculations.

In this section we discuss the ideal gas equation of state and show how it is applied to systems containing single gaseous substances and mixtures of gases. Section 5.3 outlines methods used for a single nonideal gas (by definition, a gas for which the ideal gas equation of state does not work well) and for mixtures of nonideal gases.

5.2a The Ideal Gas Equation of State

The ideal gas equation of state can be derived from the kinetic theory of gases by assuming that gas molecules have a negligible volume, exert no forces on one another, and collide elastically with the walls of their container. The equation usually appears in the form

$$\boxed{PV = nRT} \quad \text{or} \quad \boxed{P\dot{V} = \dot{n}RT} \tag{5.2-1}$$

where

$$
\begin{aligned}
P &= \textit{absolute} \text{ pressure of a gas} \\
V(\dot{V}) &= \text{volume (volumetric flow rate) of the gas} \\
n(\dot{n}) &= \text{number of moles (molar flow rate) of the gas} \\
R &= \text{the } \textit{gas constant,} \text{ whose value depends on the units of } P, V, n, \text{ and } T \\
T &= \textit{absolute} \text{ temperature of the gas}
\end{aligned}
$$

The equation may also be written as

$$P\hat{V} = RT \tag{5.2-2}$$

where $\hat{V} = V/n$ (or $\dot{V}/\dot{n}$) is the **specific molar volume** of the gas.

A gas whose *PVT* behavior is well represented by Equation 5.2-1 is said to behave as an **ideal gas** or a **perfect gas**. The use of this equation does not require a knowledge of the gas species: *1 mol of an ideal gas at 0°C and 1 atm occupies 22.415 liters,* whether the gas is argon, nitrogen, a mixture of propane and air, or any other single species or mixture of gases.

The gas constant R has units of (pressure × volume)/(mole × temperature); moreover, since pressure times volume has units of energy (prove it), R may also be expressed in units of (energy)/(mole × temperature). Values of the gas constant expressed in various units are listed in the table on the inside back cover of this book.

The ideal gas equation of state is an approximation. It works well under some conditions—generally speaking, at temperatures above about 0°C and pressures below about 1 atm—but at other conditions its use may lead to substantial errors. Here is a useful rule of thumb for when it is reasonable to assume ideal gas behavior. Let X_{ideal} be a quantity calculated using the ideal gas equation of state [$X = P$ (absolute), T (absolute), n or V] and ϵ be the error in the estimated value,

$$\epsilon = \frac{X_{\text{ideal}} - X_{\text{true}}}{X_{\text{true}}} \times 100\%$$

An error of no more than about 1% may be expected if the quantity RT/P (the *ideal specific molar volume*) satisfies the following criterion:[3]

$$|\epsilon| < 1\% \text{ if } \hat{V}_{\text{ideal}} = \frac{RT}{P} > 5 \text{ L/mol } (80 \text{ ft}^3/\text{lb-mole}) \quad \text{(diatomic gases)} \tag{5.2-3a}$$

$$> 20 \text{ L/mol } (320 \text{ ft}^3/\text{lb-mole}) \quad \text{(other gases)} \tag{5.2-3b}$$

EXAMPLE 5.2-1   *The Ideal Gas Equation of State*

One hundred grams of nitrogen is stored in a container at 23.0°C and 3.00 psig.

1. Assuming ideal gas behavior, calculate the container volume in liters.
2. Verify that the ideal gas equation of state is a good approximation for the given conditions.

SOLUTION

1. The ideal gas equation of state relates absolute temperature, absolute pressure, and the quantity of a gas in moles. We therefore first calculate

$$n = \frac{100.0 \text{ g}}{28.0 \text{ g/mol}} = 3.57 \text{ mol}$$

$$T = 296 \text{ K}$$

---

[3]O. A. Hougen, K. M. Watson, and R. A. Ragatz, *Chemical Process Principles*. Part I. Material and Energy Balances, 2nd Edition, John Wiley & Sons, New York, 1956, p. 67.

and (assuming $P_{atm} = 14.7$ psia) $P = 17.7$ psia. Then from the ideal gas equation of state

$$V(\text{liters}) = \frac{nRT}{P}$$

$$= \frac{(3.57 \text{ mol})(296 \text{ K})}{17.7 \text{ psia}} \left| \frac{R(\text{liter} \cdot \text{psia})}{(\text{mol} \cdot \text{K})} \right.$$

Unfortunately, the table of gas constants at the back of this book does not list the value of R with this particular set of units. In its absence, we use an available value and carry out the necessary additional unit conversions.

$$V = \frac{(3.57 \text{ mol})(296 \text{ K})}{17.7 \text{ psi}} \left| \frac{0.08206 \text{ liter} \cdot \text{atm}}{\text{mol} \cdot \text{K}} \right| \frac{14.7 \text{ psi}}{\text{atm}} = \boxed{72.0 \text{ liters}}$$

2. To check the assumption of ideal gas behavior for N_2 (a diatomic gas), we apply Criterion 5.3-a. Since we have already determined n and V_{ideal}, we can determine $\hat{V}_{ideal} = V_{ideal}/n$ rather than as RT/P. (Both calculations yield the same value, which you might want to verify for yourself.)

$$\hat{V}_{ideal} = \frac{V}{n} = \frac{72.0 \text{ L}}{3.57 \text{ mol}} = 20.2 \text{ L/mol} > 5 \text{ L/mol}$$

Since the calculated value of $\hat{V}_{ideal}$ exceeds the criterion value of 5 L/mol, the ideal gas equation of state should yield an error of less than 1%.

TEST YOURSELF

1. What is an equation of state? What is the ideal gas equation of state? At what conditions (high or low temperature, high or low pressure) does the ideal gas equation provide the most accurate estimates?
2. Two gas cylinders have identical volumes and contain gases at identical temperatures and pressures. Cylinder A contains hydrogen and cylinder B contains carbon dioxide. Assuming ideal gas behavior, which of the following variables differ for the two gases: (a) number of gram-moles, (b) number of molecules, (c) masses, (d) specific molar volumes (L/mol), (e) mass densities (g/L)? For each of the quantities that differ, which one is bigger and by how much? (Assume ideal gas behavior.)
3. One hundred grams per hour of ethylene (C_2H_4) flows through a pipe at 1.2 atm and 70°C and 100 g/h of butene (C_4H_8) flows through a second pipe at the same temperature and pressure. Which of the following quantities differ for the two gases: (a) volumetric flow rate, (b) specific molar volume (L/mol), (c) mass density (g/L)? For each of the quantities that differ, which one is bigger and by how much? (Assume ideal gas behavior.)
4. A gas is stored at $T = 200$ K and $P = 20$ atm. Show whether or not the ideal gas equation of state would provide an estimate of the specific volume of the gas, $\hat{V}(\text{L/mol})$, within 1% of the true value.

The relationship between the density ρ (mass/volume), temperature, and pressure of an ideal gas can be obtained by first relating the specific molar volume, $\hat{V}$ (volume/mole), to the density. Using a specific set of units for illustration,

$$\hat{V}\left(\frac{\text{liters}}{\text{mol}}\right) = \frac{\overline{M} \text{ (g/mol)}}{\rho \text{ (g/liter)}}$$

where $\overline{M}$ is the average molecular weight of the gas (the molecular weight if the gas is a single species or Equation 3.3-7 for a mixture). Substituting $\hat{V} = \overline{M}/\rho$ into Equation 5.2-2 and solving for ρ yields

$$\rho = \frac{P\overline{M}}{RT} \tag{5.2-4}$$

CREATIVITY EXERCISE

An unlabeled gas cylinder is equipped with a sensitive pressure gauge. Devise several experiments that could be used to estimate the molecular weight of the gas, using only equipment and materials likely to be found in the average home or purchasable at a neighborhood hardware store. (You may have a sensitive balance if you need one, but you may not assume that the average home has its own chemistry laboratory.)

5.2b Standard Temperature and Pressure

Doing *PVT* calculations by substituting given values of variables into the ideal gas equation of state is straightforward, but to use this method you must have on hand either a table of values of R with different units or a good memory. A way to avoid these requirements is to use *conversion from standard conditions.*

For an ideal gas at an arbitrary temperature T and pressure P,

$$PV = nRT \tag{5.2-1}$$

and for the same ideal gas at a specified reference temperature T_s and pressure P_s (referred to as *standard temperature and pressure,* or STP), we may write Equation 5.2-2 as

$$P_s \hat{V}_s = RT_s$$

The first equation divided by the second yields

$$\frac{PV}{P_s \hat{V}_s} = n \frac{T}{T_s} \tag{5.2-5}$$

(For a flowing stream, $\dot{n}$ and $\dot{V}$ would replace n and V in this equation.) Since the standard conditions (P_s, T_s, $\hat{V}_s = RT_s/P_s$) are known, Equation 5.2-5 may be used to determine V for a given value of n or vice versa. Note that when you use this method you do not need a value for R.

The standard conditions most commonly used are shown in Table 5.2-1. Standard temperature ($T_s = 0°C \implies 273$ K) and standard pressure ($P_s = 1$ atm) are easy to remember. You should also commit to memory the following values of the standard specific molar volume:

$$\hat{V}_s = 22.4 \frac{m^3(STP)}{kmol} \iff 22.4 \frac{L(STP)}{mol} \iff 359 \frac{ft^3(STP)}{lb\text{-mole}} \tag{5.2-6}$$

The term **standard cubic meters** (or **SCM**) is often used to denote $m^3(STP)$, and **standard cubic feet** (or **SCF**) denotes $ft^3(STP)$. A volumetric flow rate of 18.2 SCMH means 18.2 m^3/h at 0°C and 1 atm.

Caution: While standard temperature and pressure for most equation of state calculations are 0°C and 1 atm, some specialized industries have adopted different values. If you see a reference to standard temperature and pressure, try to find out which values were assumed. (In this book it is always safe to assume 0°C and 1 atm.)

Table 5.2-1 Standard Conditions for Gases

System	T_s	P_s	$\hat{V}_s$	n_s
SI	273 K	1 atm	0.022415 m^3	1 mol
CGS	273 K	1 atm	22.415 L	1 mol
American Engineering	492°R	1 atm	359.05 ft^3	1 lb-mole

EXAMPLE 5.2-2 *Conversion from Standard Conditions*

Butane (C_4H_{10}) at 360°C and 3.00 atm absolute flows into a reactor at a rate of 1100 kg/h. Calculate the volumetric flow rate of this stream using conversion from standard conditions.

SOLUTION

As always, molar quantities and absolute temperature and pressure must be used.

$$\dot{n} = \frac{1100 \text{ kg/h}}{58.1 \text{ kg/kmol}} = 19.0 \text{ kmol/h}$$

$$T = 633 \text{ K}, \qquad P = 3.00 \text{ atm}$$

From Equation 5.2-5

$$\frac{P\dot{V}}{P_s \hat{V}_s} = \dot{n}\frac{T}{T_s} \implies \dot{V} = \dot{n}\hat{V}_s \frac{T}{T_s}\frac{P_s}{P}$$

$$\Downarrow$$

$$\dot{V} = \frac{19.0 \text{ kmol}}{\text{h}} \left| \frac{22.4 \text{ m}^3(\text{STP})}{\text{kmol}} \right| \frac{633 \text{ K}}{273 \text{ K}} \left| \frac{1.00 \text{ atm}}{3.00 \text{ atm}} \right. = \boxed{329 \frac{\text{m}^3}{\text{h}}}$$

You will frequently encounter problems that involve gases at two different states (sets of conditions)—for example, at the inlet and outlet of a process unit. A convenient way to determine an unknown variable (P, V, n, or T) of the gas at one of the states is to write the gas law for both states and divide one equation by the other. Any variables that are the same at both points will cancel, leaving an equation containing only the variable you wish to determine and known quantities.

EXAMPLE 5.2-3 *Effect of T and P on Volumetric Flow Rates*

Ten cubic feet of air at 70°F and 1.00 atm is heated to 610°F and compressed to 2.50 atm. What volume does the gas occupy in its final state?

SOLUTION

Let 1 denote the initial state of the gas and 2 the final state. Note that $n_1 = n_2$ (the number of moles of the gas does not change). Assume ideal gas behavior.

$$\begin{array}{l} P_2V_2 = nRT_2 \\ P_1V_1 = nRT_1 \end{array} \implies \frac{P_2V_2}{P_1V_1} = \frac{T_2}{T_1}$$

$$\implies V_2 = V_1\left(\frac{P_1}{P_2}\right)\left(\frac{T_2}{T_1}\right) = \frac{10.0 \text{ ft}^3}{} \left| \frac{1.00 \text{ atm}}{2.50 \text{ atm}} \right| \frac{1070°\text{R}}{530°\text{R}} = \boxed{8.08 \text{ ft}^3}$$

You may on occasion be told that the flow rate of a gas stream is, say, 23.8 SCMH [or m³(STP)/h] at 150°C and 2.5 atm. This looks like a contradiction: How can a gas be at standard temperature and pressure (0°C and 1 atm) *and* at 150°C and 2.5 atm?

The answer is that it can't—the gas is *not* at standard temperature and pressure. A flow rate specified in the given manner (23.8 SCMH) is not the true volumetric flow rate of the stream at its actual temperature and pressure (150°C and 2.5 atm) but the flow rate that would be obtained *if* the stream were brought from its actual conditions to standard temperature and pressure. From the given value of 23.8 SCMH, you could (a) calculate the molar flow rate (kmol/h) by dividing 23.8 m^3/h by 22.4 m^3(STP)/kmol, or (b) calculate the true volumetric flow rate (m^3/h) by multiplying 23.8 m^3/h by (323 K/273 K)(1 atm/2.5 atm). (Convince yourself of both of these claims.)

EXAMPLE 5.2-4   *Standard and True Volumetric Flow Rates*

The flow rate of a methane stream at 285°F and 1.30 atm is measured with an orifice meter. The calibration chart for the meter indicates that the flow rate is 3.95×10^5 SCFH. Calculate the molar flow rate and the true volumetric flow rate of the stream.

SOLUTION

Recall that SCFH means ft^3(STP)/h.

$$\dot{n} = \frac{3.95 \times 10^5 \text{ ft}^3\text{(STP)}}{h} \left| \frac{1 \text{ lb-mole}}{359 \text{ ft}^3\text{(STP)}} \right. = \boxed{1.10 \times 10^3 \text{ lb-moles/h}}$$

Note that to calculate the molar flow rate from a standard volumetric flow rate, you don't need to know the actual gas temperature and pressure.

The true volumetric flow rate of the methane is calculated using the method illustrated in Example 5.2-3, only now we are bringing the gas from standard conditions ($T_1 = 492$°R, $P_1 = 1.0$ atm, $\dot{V}_1 = 3.95 \times 10^5$ ft^3/h) to actual conditions ($T_2 = 745$°R, $P_2 = 1.30$ atm, $\dot{V}_2 = ?$). We therefore obtain

$$\dot{V}_2 = \dot{V}_1 \left(\frac{T_2}{T_1} \right) \left(\frac{P_1}{P_2} \right) = (3.95 \times 10^5 \text{ ft}^3\text{/h}) \left(\frac{745°\text{R}}{492°\text{R}} \right) \left(\frac{1.00 \text{ atm}}{1.30 \text{ atm}} \right) = \boxed{4.60 \times 10^5 \text{ ft}^3\text{/h}}$$

TEST YOURSELF

1. What are standard temperature and pressure? What are the values of $\hat{V}_s$ in the SI, CGS, and American Engineering systems?
2. What happens to the volume of an ideal gas when you double the pressure at a fixed temperature? When you double the temperature at a fixed pressure?
3. What happens to the density of a fixed quantity of an ideal gas when you raise the temperature at a fixed pressure? When you raise the temperature at a fixed volume?
4. The volumetric flow rate of an ideal gas is given as 35.8 SCMH. The temperature and pressure of the gas are −15°C and 1.5 atm. Is the actual volumetric flow rate of the gas (a) <35.8 m^3/h, (b) 35.8 m^3/h, (c) >35.8 m^3/h, or (d) indeterminate without additional information?

5.2c Ideal Gas Mixtures

Suppose n_A moles of substance A, n_B moles of B, n_C moles of C, and so on, are contained in a volume V at a temperature T and total pressure P. The **partial pressure** p_A and **pure-component volume** v_A of A in the mixture are defined as follows:

p_A: the pressure that would be exerted by n_A moles of A alone in the same total volume V at the same temperature T.

v_A: the volume that would be occupied by n_A moles of A alone at the total pressure P and temperature T of the mixture.

Suppose next that each of the individual mixture components and the mixture as a whole behave in an ideal manner. (This is the definition of an **ideal gas mixture**.) If there are n moles of all species in the volume V at pressure P and temperature T, then

$$PV = nRT$$

In addition, from the definition of partial pressure,

$$p_A V = n_A RT$$

Dividing the second equation by the first yields

$$\frac{p_A}{P} = \frac{n_A}{n} = y_A \qquad \text{(the mole fraction of A in the gas)}$$

or

$$p_A = y_A P \tag{5.2-7}$$

That is, *the partial pressure of a component in an ideal gas mixture is the mole fraction of that component times the total pressure.*[4] Moreover, since $y_A + y_B + \cdots = 1$,

$$p_A + p_B + \cdots = (y_A + y_B + \cdots)P = P \tag{5.2-8}$$

or, *the partial pressures of the components of an ideal gas mixture add up to the total pressure (Dalton's law).*

A similar series of calculations can be performed for pure-component volumes:

$$Pv_A = n_A RT$$

$$\Downarrow \text{Divide by } PV = nRT$$

$$\frac{v_A}{V} = \frac{n_A}{n} = y_A$$

or

$$v_A = y_A V \tag{5.2-9}$$

and

$$v_A + v_B + \cdots = V \qquad \text{(Amagat's law)}$$

The quantity v_A/V is the **volume fraction** of A in the mixture, and 100 times this quantity is the **percentage by volume** (% v/v) of this component. As shown above, *the volume fraction of a substance in an ideal gas mixture equals the mole fraction of this substance.* Stating, for example, that an ideal gas mixture contains 30% CH_4 and 70% C_2H_6 by volume (or 30% v/v CH_4 and 70% v/v C_2H_6) is equivalent to specifying 30 mole% CH_4 and 70 mole% C_2H_6.

TEST YOURSELF

1. A cylinder containing a mixture of N_2 and O_2 is taken from the basement to the top of a 50-story building. Assuming that the temperature remains constant, which of the following properties of the gas change during the ascent? (a) mass, (b) weight, (c) absolute pressure, (d) gauge pressure, (e) partial pressure of N_2, (f) pure-component volume of O_2, (g) density, (h) specific molar volume.
2. An ideal gas mixture at 10 bar absolute and 200°C in a 100-m^3 tank contains 50 mole% H_2 and 50 mole% N_2. What is the partial pressure of H_2? What is the pure-component volume of H_2? What would happen to p_{H_2} and v_{H_2} if the temperature were raised?
3. An ideal gas mixture at 10 atm absolute contains 50 wt% H_2 and 50 wt% N_2. Is the partial pressure of H_2 less than, equal to, or greater than 5 atm?

EXAMPLE 5.2-5 *Material Balances on an Evaporator-Compressor*

Liquid acetone (C_3H_6O) is fed at a rate of 400 L/min into a heated chamber, where it evaporates into a nitrogen stream. The gas leaving the heater is diluted by another nitrogen stream flowing at a measured rate of 419 m^3(STP)/min. The combined gases are then compressed to a total pressure $P = 6.3$ atm gauge at a temperature of 325°C. The partial pressure of acetone in this stream is $p_a = 501$ mm Hg. Atmospheric pressure is 763 mm Hg.

---

[4]Equation 5.2-7 is often used as the definition of the partial pressure. For an ideal gas mixture, the definition given and Equation 5.2-7 are equivalent; for a nonideal gas the concept of partial pressure has little utility.

1. What is the molar composition of the stream leaving the compressor?
2. What is the volumetric flow rate of the nitrogen entering the evaporator if the temperature and pressure of this stream are 27°C and 475 mm Hg gauge?

SOLUTION

Basis: Given Feed Rates

Assume ideal gas behavior. Let $\dot{n}_1, \dot{n}_2, \ldots$ (mol/min) be the molar flow rates of each stream.

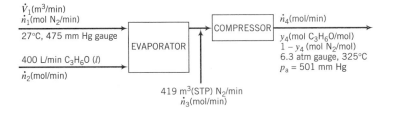

You should be able to examine the flowchart and see exactly how the solution will proceed.

1. Calculate $\dot{n}_2$ (from the given volumetric flow rate and a tabulated density of liquid acetone), $\dot{n}_3$ (from the ideal gas equation of state), and y_4 ($= p_a/P$).
2. Calculate $\dot{n}_4$ (overall acetone balance), $\dot{n}_1$ (overall mass balance), and $\dot{V}_1$ (ideal gas equation of state).

Calculate Molar Flow Rate of Acetone

From Table B.1 in Appendix B, the density of liquid acetone is 0.791 g/cm³ (791 g/L), so that

$$\dot{n}_2 = \frac{400 \text{ L}}{\text{min}} \left|\frac{791 \text{ g}}{\text{L}}\right| \frac{1 \text{ mol}}{58.08 \text{ g}} = 5450 \frac{\text{mol C}_3\text{H}_6\text{O}}{\text{min}}$$

Determine Mole Fractions from Partial Pressures

In the stream leaving the compressor,

$$\frac{p_a}{P} = y_4 \left(\frac{\text{mol C}_3\text{H}_6\text{O}}{\text{mol}}\right)$$

$$P = P_{\text{gauge}} + P_{\text{atm}} = \frac{6.3 \text{ atm}}{} \left|\frac{760 \text{ mm Hg}}{1 \text{ atm}}\right| + 763 \text{ mm Hg} = 5550 \text{ mm Hg}$$

so that

$$\boxed{\begin{aligned} y_4 &= \frac{501 \text{ mm Hg}}{5550 \text{ mm Hg}} = 0.0903 \frac{\text{mol C}_3\text{H}_6\text{O}}{\text{mol}} \\[2mm] 1 - y_4 &= 0.9097 \frac{\text{mol N}_2}{\text{mol}} \end{aligned}}$$

Calculate $\dot{n}_3$ from PVT Information $\dot{n}_3 = \dfrac{419 \text{ m}^3(\text{STP})}{\text{min}} \left|\dfrac{1 \text{ mol}}{0.0224 \text{ m}^3(\text{STP})}\right| = 18{,}700 \dfrac{\text{mol}}{\text{min}}$

Overall Mole Balance on Acetone $\dot{n}_2 = \dot{n}_4 y_4$

$$\Bigg\Downarrow \begin{aligned} \dot{n}_2 &= 5450 \text{ mol/min} \\ y_4 &= 0.0903 \end{aligned}$$

$$\dot{n}_4 = 60{,}400 \text{ mol/min}$$

Overall Mole Balance $\dot{n}_1 + \dot{n}_2 + \dot{n}_3 = \dot{n}_4$

$\dot{n}_2 = 5450$ mol/min
$\dot{n}_3 = 18{,}700$ mol/min
$\dot{n}_4 = 60{,}400$ mol/min

$\dot{n}_1 = 36{,}200$ mol/min

Ideal gas equation of state
$T_1 = 27°C$ (300 K)
$P_1 = 475$ mm Hg gauge (1238 mm Hg)

$$\dot{V}_1 = \dot{n}_1 \frac{V_s}{n_s} \frac{T_1}{T_s} \frac{P_s}{P_1}$$

$$= \frac{36{,}200 \text{ mol/min}}{} \left| \frac{0.0224 \text{ m}^3}{1 \text{ mol}} \right| \frac{300 \text{ K}}{273 \text{ K}} \left| \frac{760 \text{ mm Hg}}{1238 \text{ mm Hg}} \right.$$

$$\boxed{\dot{V}_1 = 550 \text{ m}^3 \text{ N}_2/\text{min}}$$

5.3 EQUATIONS OF STATE FOR NONIDEAL GASES

The ideal gas is the basis of the simplest and most convenient equation of state: solving it is trivial, regardless of which variable is unknown, and the calculation is independent of the species of the gas and is the same for single species and mixtures. Its shortcoming is that it can be seriously inaccurate. At a sufficiently low temperature and/or a sufficiently high pressure, a value of $\hat{V}$ predicted with the ideal gas equation could be off by a factor of two or three or more in either direction. Making things worse, the predicted value for one species at a given temperature and pressure could be much too high, that for another species at the same T and P could be much too low, and that for a third species could be quite close to the true value.

In this section we introduce several more complex but more accurate equations of state for single species: the *virial equation,* the *van der Waals equation,* and the *Soave–Redlich–Kwong equation.* In Section 5.4 we introduce another approach to nonideal gas analysis that makes use of *compressibility factors,* and we describe *Kay's rule,* a method for performing *PVT* calculations on gas mixtures.

5.3a Critical Temperature and Pressure

How well or poorly the ideal gas equation of state fits *PVT* data for a species often depends on the values of the system temperature and pressure relative to two physical properties of the species—the **critical temperature** (T_c) and **critical pressure** (P_c). Values of these *critical constants* can be looked up in Table B.1 and in most standard chemical reference handbooks. Let us first consider their physical significance and then see how they are used in nonideal gas calculations.

Suppose a quantity of water is kept in a closed piston-fitted cylinder.

The cylinder temperature is first set to a specified value with the cylinder pressure low enough for all the water to be vapor; then the water is compressed at constant temperature by lowering the piston until a drop of liquid water appears (i.e., until condensation occurs). The pressure at which condensation begins (P_{cond}) and the densities of the vapor (ρ_v) and of the liquid (ρ_l) at that point are noted, and the experiment is then repeated at several progressively higher temperatures. The following results might be obtained (observe the pattern for the three observed variables as T increases):

Run	$T(°C)$	$P_{cond}(atm)$	$\rho_v(kg/m^3)$	$\rho_l(kg/m^3)$
1	25.0	0.0329	0.0234	997.0
2	100.0	1.00	0.5977	957.9
3	201.4	15.8	8.084	862.8
4	349.8	163	113.3	575.0
5	373.7	217.1	268.1	374.5
6	374.15	218.3	315.5	315.5
7	>374.15	*No condensation occurs!*		

Notice what happens. At 25°C, water condenses at a very low pressure, and the density of the liquid is more than four orders of magnitude greater than that of the vapor. At higher temperatures, the condensation pressure increases and the densities of the vapor and liquid at condensation approach each other. At 374.15°C, the densities of the two phases are virtually equal, and above that temperature no phase separation is observed, no matter how high the pressure is raised.

In general, *the highest temperature at which a species can coexist in two phases (liquid and vapor) is the* **critical temperature** *of that species,* T_c, *and the corresponding pressure is the* **critical pressure**, P_c. A substance at T_c and P_c is said to be at its **critical state**. The above experiment demonstrates and Table B.1 confirms that, for water, $T_c = 374.15°C$ and $P_c = 218.3$ atm.

The terms "gas" and "vapor" are often used interchangeably, but there is a technical difference between them that you are now in a position to understand. A **vapor** is a gaseous species below its critical temperature, and a **gas** is a species above its critical temperature at a pressure low enough for the species to be more like a vapor than a liquid (i.e., a density closer to 1 g/L than 1000 g/L). You can condense a vapor by compressing it isothermally, but while you can make a gas denser and denser by compressing it isothermally you will never achieve a separation into two phases. Substances at temperatures above T_c and pressures above P_c are referred to as **supercritical fluids**.

TEST YOURSELF

The critical temperature and pressure of isopropanol (isopropyl alcohol) are $T_c = 508.8$ K and $P_c = 53.0$ atm.

1. Isopropanol is in a gaseous state at $T = 400$ K and $P = 1$ atm. Would it be classified as a vapor or a gas?

2. Isopropanol is compressed isothermally at 400 K until at pressure P_a a liquid phase forms. The vapor and liquid densities at that point are, respectively, ρ_{va} and ρ_{la}. In a second experiment, isopropanol is compressed at 450 K until condensation occurs, at which point the pressure and vapor and liquid densities are P_b, ρ_{vb}, and ρ_{lb}. What are the relationships ($>$, $=$, $<$, or ?) between (a) P_a and P_b, (b) ρ_{va} and ρ_{vb}, and (c) ρ_{la} and ρ_{lb}?

3. If isopropanol at 550 K and 1 atm is compressed isothermally to 100 atm, will a condensate form? What term might you use to refer to the fluid at its initial condition? At its final condition?

5.3b Virial Equations of State

A **virial equation of state** expresses the quantity $P\hat{V}/RT$ as a power series in the inverse of specific volume:

$$\frac{P\hat{V}}{RT} = 1 + \frac{B}{\hat{V}} + \frac{C}{\hat{V}^2} + \frac{D}{\hat{V}^3} + \cdots \qquad (5.3\text{-}1)$$

where B, C, and D are functions of temperature and are known as the second, third, and fourth virial coefficients, respectively. This equation of state has a theoretical basis in statistical mechanics but procedures for estimating the coefficients are not well developed, especially for those beyond B. Note that the ideal gas equation of state is obtained if $B = C = D = \cdots = 0$.

Truncating the virial equation of state after the second term yields

$$\frac{P\hat{V}}{RT} = 1 + \frac{B}{\hat{V}} \qquad (5.3\text{-}2)$$

Reid et al. (see footnote 1) caution against using this equation for polar compounds (asymmetrical compounds with a nonzero dipole moment, such as water). The following procedure may be used to estimate $\hat{V}$ or P for a given T for a nonpolar species (one with a dipole moment close to zero, such as hydrogen and oxygen and all other molecularly symmetrical compounds).

- Look up the critical temperature and pressure (T_c and P_c) for the species of interest in Table B.1 or elsewhere. Also look up the **Pitzer acentric factor**, ω, a parameter that reflects the geometry and polarity of a molecule. Table 5.3-1 lists values of ω for selected compounds, and a more complete list can be found in Reid et al.

Table 5.3-1 Pitzer Acentric Factors

Compound	Acentric Factor, ω
Ammonia	0.250
Argon	−0.004
Carbon dioxide	0.225
Carbon monoxide	0.049
Chlorine	0.073
Ethane	0.098
Hydrogen sulfide	0.100
Methane	0.008
Methanol	0.559
Nitrogen	0.040
Oxygen	0.021
Propane	0.152
Sulfur dioxide	0.251
Water	0.344

SOURCE: R. C. Reid, J. M. Prausnitz, and B. E. Poling, *The Properties of Gases and Liquids,* 4th Edition, McGraw-Hill, New York, 1986.

- Calculate the **reduced temperature**, $T_r = T/T_c$.
- Estimate B using the following equations:

$$B_0 = 0.083 - \frac{0.422}{T_r^{1.6}} \qquad (5.3\text{-}3)$$

$$B_1 = 0.139 - \frac{0.172}{T_r^{4.2}} \qquad (5.3\text{-}4)$$

$$B = \frac{RT_c}{P_c}(B_0 + \omega B_1) \qquad (5.3\text{-}5)$$

- Substitute into Equation 5.3-2 the values of B and whichever of the variables P and $\hat{V}$ is known and solve for the other variable. Solution for P is straightforward. If $\hat{V}$ is to be determined, the equation can be rearranged into a quadratic and solved using the quadratic formula. Normally one of the two solutions is reasonable and the other is not and should be discarded; if there is any doubt, estimate $\hat{V}$ from the ideal gas equation of state and accept the virial equation solution that comes closest to $\hat{V}_{\text{ideal}}$.

EXAMPLE 5.3-1 *The Truncated Virial Equation*

Two gram-moles of nitrogen is placed in a three-liter tank at $-150.8°C$. Estimate the tank pressure using the ideal gas equation of state and then using the virial equation of state truncated after the second term. Taking the second estimate to be correct, calculate the percentage error that results from the use of the ideal gas equation at the system conditions.

SOLUTION

$T = (-150.8 + 273.2)$ K $= 122.4$ K, and $\hat{V} = 3.00$ L/2.00 mol $= 1.50$ L/mol. From the ideal gas equation of state,

$$P_{\text{ideal}} = \frac{RT}{\hat{V}} = \frac{0.08206 \text{ L·atm}}{\text{mol·K}} \left| \frac{123 \text{ K}}{} \right| \frac{1 \text{ mol}}{1.50 \text{ L}} = \boxed{6.73 \text{ atm}}$$

The virial equation solution procedure is as follows:

- Table B.1 $\Longrightarrow (T_c)_{N_2} = 126.2$ K, $(P_c)_{N_2} = 33.5$ atm

  Table 5.3-1 $\Longrightarrow \omega_{N_2} = 0.040$

- $T_r = \dfrac{T}{T_c} = \dfrac{122.4 \text{ K}}{126.2 \text{ K}} = 0.970$

- Equation 5.3-3 $\Longrightarrow B_0 = 0.083 - \dfrac{0.422}{0.970^{1.6}} = -0.36$

  Equation 5.3-4 $\Longrightarrow B_1 = 0.139 - \dfrac{0.172}{0.970^{4.2}} = -0.056$

  Equation 5.3-5 $\Longrightarrow B = \dfrac{\left(0.08206 \dfrac{\text{L·atm}}{\text{mol·K}}\right)(126.2 \text{ K})}{33.5 \text{ atm}} [-0.36 + 0.040(-0.056)]$

  $= -0.113 \text{ L/mol}$

- Equation 5.3-2 $\Longrightarrow P = \dfrac{RT}{\hat{V}}\left(1 + \dfrac{B}{\hat{V}}\right)$

  $= \dfrac{\left(0.08206 \dfrac{\text{L·atm}}{\text{mol·K}}\right)(122.4 \text{ K})}{1.50 \text{ L/mol}}\left(1 + \dfrac{-0.113 \text{ L/mol}}{1.50 \text{ L/mol}}\right) = \boxed{6.19 \text{ atm}}$

The error in the pressure calculated using the ideal gas equation of state is

$$\epsilon = \frac{P_{\text{ideal}} - P}{P} \times 100\% = \boxed{8.7\% \text{ error}}$$

Other forms of the virial equation of state have been developed for specific compounds and mixtures. For example, the **Benedict–Webb–Rubin** (BWR) equation, which has eight empirical constants, has been used extensively for light hydrocarbons and other nonpolar gases. *Perry's Chemical Engineers' Handbook* (see footnote 2) describes the BWR equation of state and gives the eight constants for a number of gases on pp. 3-270 to 3-272.

5.3c   Cubic Equations of State

A number of analytical *PVT* relationships are referred to as **cubic equations of state** because, when expanded, they yield third-order equations for the specific volume. The **van der Waals equation of state** is the earliest of these expressions, and it remains useful for discussing deviations from ideal behavior.

$$P = \frac{RT}{\hat{V} - b} - \frac{a}{\hat{V}^2} \tag{5.3-6}$$

where

$$a = \frac{27R^2T_c^2}{64P_c} \qquad b = \frac{RT_c}{8P_c}$$

In the van der Waals derivation, the term $a/\hat{V}^2$ accounts for attractive forces between molecules and b is a correction accounting for the volume occupied by the molecules themselves.[5]

Reid, Prausnitz, and Poling (see footnote 1) discuss other important cubic equations of state including the **Redlich–Kwong**, **Soave–Redlich–Kwong (SRK)**, and **Peng–Robinson** equations. These equations are empirical but have proved remarkably robust in describing a wide variety of systems. Here we will use the SRK expression to illustrate the general characteristics of cubic equations of state.

The SRK equation of state is

$$P = \frac{RT}{\hat{V} - b} - \frac{\alpha a}{\hat{V}(\hat{V} + b)} \tag{5.3-7}$$

where the parameters a, b, and α are empirical functions of the critical temperature and pressure (T_c and P_c from Table B.1), the Pitzer acentric factor (ω from Table 5.3-1), and the system temperature. The following correlations are used to estimate these three parameters:

$$a = 0.42747 \frac{(RT_c)^2}{P_c} \tag{5.3-8}$$

$$b = 0.08664 \frac{RT_c}{P_c} \tag{5.3-9}$$

$$m = 0.48508 + 1.55171\omega - 0.1561\omega^2 \tag{5.3-10}$$

$$T_r = T/T_c \tag{5.3-11}$$

$$\alpha = \left[1 + m\left(1 - \sqrt{T_r}\right)\right]^2 \tag{5.3-12}$$

EXAMPLE 5.3-2   *The SRK Equation of State*

A gas cylinder with a volume of 2.50 m³ contains 1.00 kmol of carbon dioxide at $T = 300$ K. Use the SRK equation of state to estimate the gas pressure in atm.

SOLUTION

The specific molar volume is calculated as

$$\hat{V} = \frac{V}{n} = \frac{2.5 \text{ m}^3}{1.00 \text{ kmol}} \left| \frac{10^3 \text{ L}}{1 \text{ m}^3} \right| \frac{1 \text{ kmol}}{10^3 \text{ mol}} = 2.50 \text{ L/mol}$$

---

[5]B. G. Kyle, *Chemical and Process Thermodynamics,* 2nd Edition, Prentice Hall, Englewood Cliffs, 1992, p. 41.

From Table B.1, $T_c = 304.2$ K and $P_c = 72.9$ atm, and from Table 5.3-1, $\omega = 0.225$. The parameters in the SRK equation of state are evaluated using Equations 5.3-7 through 5.3-12:

$$\text{Equation 5.3-8} \implies a = 0.42747 \frac{\{[0.08206 \text{ L} \cdot \text{atm}/(\text{mol} \cdot \text{K})](304.2 \text{ K})\}^2}{72.9 \text{ atm}}$$

$$= 3.654 \text{ L}^2 \cdot \text{atm}/\text{mol}^2$$

$$\text{Equation 5.3-9} \implies b = 0.08664 \frac{[0.08206 \text{ L} \cdot \text{atm}/(\text{mol} \cdot \text{K})](304.2 \text{ K})}{72.9 \text{ atm}}$$

$$= 0.02967 \text{ L/mol}$$

$$\text{Equation 5.3-10} \implies m = 0.8263$$

$$\text{Equation 5.3-11} \implies T_r = 0.986$$

$$\text{Equation 5.3-12} \implies \alpha = 1.0115$$

The SRK equation (5.3-7) can now be solved for the tank pressure:

$$\begin{aligned}
P &= \frac{RT}{\hat{V} - b} - \frac{\alpha a}{\hat{V}(\hat{V} + b)} \\
&= \frac{[0.08206 \text{ L} \cdot \text{atm}/(\text{mol} \cdot \text{K})](300 \text{ K})}{[(2.50 - 0.02967)\text{L/mol}]} - \frac{1.0115(3.654 \text{ L}^2 \cdot \text{atm}/\text{mol}^2)}{(2.50 \text{ L/mol})[(2.50 + 0.02967) \text{ L/mol}]} \\
&= \boxed{9.38 \text{ atm}}
\end{aligned}$$

Use of the ideal gas equation of state leads to an estimated pressure of 9.85 atm (*verify*), a deviation of 5% from the more accurate SRK-determined value.

Evaluation of a system volume from a given temperature and pressure using a cubic equation of state requires a trial-and-error procedure. A spreadsheet is ideally suited to solving problems of this type. The next example illustrates the procedure.

EXAMPLE 5.3-3   *Estimation of Volumes Using the SRK Equation of State*

A stream of propane at temperature $T = 423$ K and pressure P(atm) flows at a rate of 100.0 kmol/h. Use the SRK equation of state to estimate the volumetric flow rate of the stream for $P = 0.7$ atm, 7 atm, and 70 atm. In each case, calculate the percentage differences between the predictions of the SRK equation and the ideal gas equation of state.

SOLUTION   The calculation of $\hat{V}$(L/mol) proceeds as follows: the SRK equation of state is written in the form

$$f(\hat{V}) = P - \frac{RT}{\hat{V} - b} + \frac{\alpha a}{\hat{V}(\hat{V} + b)} = 0$$

the values of T_c, P_c, and ω are looked up; a, b, and α are calculated from the given formulas; specified values of T and P are substituted; and the value of $\hat{V}$ for which $f(\hat{V}) = 0$ is found by trial and error. The percentage difference between $\hat{V}_{\text{SRK}}$ and $\hat{V}_{\text{ideal}}(= RT/P)$ is

$$D(\%) = \frac{\hat{V}_{\text{ideal}} - \hat{V}_{\text{SRK}}}{\hat{V}_{\text{SRK}}} \times 100\%$$

Once $\hat{V}$ is known for a given P, the volumetric flow rate corresponding to a molar flow rate of 100.0 kmol/h is obtained as

$$\dot{V}(\text{m}^3/\text{h}) = \frac{\hat{V} \text{ (L)}}{(\text{mol})} \left| \frac{10^3 \text{ mol}}{1 \text{ kmol}} \right| \frac{1 \text{ m}^3}{10^3 \text{ L}} \left| \frac{100.0 \text{ kmol}}{\text{h}} \right. = 100.0\hat{V}(\text{L/mol})$$

The calculations are performed easily with a spreadsheet. In Figure 5.3-1 we show one of many possible spreadsheet configurations that might be created for this purpose, along with the formulas

	A	B	C	D	E	F
1	**Spreadsheet for Example 5.3-3**					
2						
3	$T_c =$	369.9	$P_c =$	42.0	$\omega =$	0.152
4	$a =$	9.3775	$b =$	0.06262	$m =$	0.7173
5	$T =$	423	$T_r =$	1.14355	$\alpha =$	0.903
6						
7	P	$\hat{V}_{ideal}$	$\hat{V}$	$f(\hat{V})$	D	$\dot{V}$
8	(atm)	(L/mol)	(L/mol)	(atm)	(%)	(m^3/h)
9	0.7	49.59	49.41	1.6E–5	0.37%	4941
10	7	4.959	4.775	9.9E–6	3.9%	478
11	70	0.4959	0.2890	9.2E–5	72%	28.9

[**B4**] = 0.42747*(0.08206*B3)**2/D3
[**D4**] = 0.08664*0.08206 * B3/D3
[**F4**] = 0.48508 + 1.55171*F3 − 0.1561*F3**2
[**D5**] = B5/B3
[**F5**] = (1 + F4*(1–SQRT(D5)))**2
[**B9**] = 0.08206*$B$5/A9
[**C9**] = 49.588
[**D9**] = A9 − 0.08206*$B$5/(C9–$D$4) + $F$5*$B$4/(C9*(C9 + $D$4))
[**E9**] = 100*(B9–C9)/C9
[**F9**] = 100*C9

Figure 5.3-1 Spreadsheet for Example 5.3-3.

entered into selected cells. The critical constants of propane (T_c = 369.9 K and P_c = 42.0 atm) come from Table B.1 and the Pitzer acentric factor (ω = 0.152) comes from Table 5.3-1. In the spreadsheet formulas, a double asterisk signifies exponentiation. (Many spreadsheets use the carat symbol for this purpose.) Notice that the ideal gas equation of state works very well at 0.7 atm and reasonably well at 7 atm, but for 70 atm the differences between the two estimates of $\hat{V}$ are considerable.

When the spreadsheet is constructed, the displayed contents of Rows 1, 3–5, 7, and 8 are entered exactly as shown except for the given formulas in Cells B4, D4, F4, D5, and F5. After the cell contents of Row 9 have been entered they are copied into Rows 10 and 11, and the pressures in Column A are then changed to their desired values. The entries in Cells C9–C11 (the initial guesses for $\hat{V}$) are the values copied from the adjacent cells in Column B (the values obtained using the ideal gas equation of state). The correct values are then obtained by trial and error; for example, the value in Cell C9 would be varied until the value in Cell D9 is sufficiently close to zero, and similarly for Rows 10 and 11. The search is conveniently done using the spreadsheet's *goalseek* tool. If you are not experienced in the use of spreadsheets, it might be useful to construct this one and attempt to reproduce the given results.

Remember that the SRK equation of state (and every other equation of state) is itself an approximation. Referring back to the preceding example, a published study provides experimental data for the *PVT* behavior of propane.[6] The data indicate that at 423 K and 70 atm,

---

[6]R. D. Gray, N. H. Rent, and D. Zudkevitch, *AIChE Journal,* **16**, 991(1970).

the value of $\hat{V}$ is 0.2579 L/mol. The percentage error in the SRK estimate ($\hat{V} = 0.2890$ L/mol) is a far from insignificant 12%, and that in the ideal gas estimate ($\hat{V} = 0.4959$ L/mol) is 92%.

All equations of state have parameters obtained by fitting empirical expressions to experimental PVT data. The fit may be excellent in the temperature and pressure ranges where the data were obtained but may be terrible elsewhere. You should always try to ascertain the region of validity of any equation of state you intend to use. At conditions far removed from this region, you have no assurance of the accuracy of the equation.

TEST YOURSELF

1. Why is the SRK equation of state called a *cubic equation of state?*
2. What physical properties of a species do you have to look up in order to use the SRK equation of state? Where can you find values of these properties in this text?
3. The SRK equation of state is to be used to determine one of the variables T, P, and $\hat{V}$ from given values of the other two. Rank the following problems from easiest to most difficult: (a) given T and P, find $\hat{V}$; (b) given T and $\hat{V}$, find P; and (c) given P and $\hat{V}$, find T.
4. Explain in your own words why estimates obtained using an equation of state may be inaccurate and when you need to be particularly skeptical of them.

5.4 THE COMPRESSIBILITY FACTOR EQUATION OF STATE

The **compressibility factor** of a gaseous species is defined as the ratio

$$z = \frac{P\hat{V}}{RT} \tag{5.4-1}$$

If the gas behaves ideally, $z = 1$. The extent to which z differs from 1 is a measure of the extent to which the gas is behaving nonideally.

Equation 5.4-1 may be rearranged to form the **compressibility factor equation of state**,

$$P\hat{V} = zRT \tag{5.4-2a}$$

or since $\hat{V} = V/n$ for a fixed quantity of gas and $\dot{V}/\dot{n}$ for a flowing stream,

$$PV = znRT \tag{5.4-2b}$$

$$P\dot{V} = z\dot{n}RT \tag{5.4-2c}$$

An alternative to using a nonideal gas equation of state like those described in Section 5.3 is to determine z and substitute it into Equation 5.4-2a, 5.4-2b, or 5.4-2c. The next section describes an application of this method to a system for which tabulated z values are available. The following sections outline methods for estimating z in the absence of tabulated data.

5.4a Compressibility Factor Tables

Perry's Chemical Engineers' Handbook (see footnote 2), pp. 2-140 through 2-150, gives values of $z(T, P)$ for air, argon, CO_2, CO, H_2, CH_4, N_2, O_2, steam, and a limited number of other compounds. Once z is known, it can be substituted in the compressibility factor equation of state, which may in turn be solved for whichever variable is unknown.

EXAMPLE 5.4-1   *Tabulated Compressibility Factors*

Fifty cubic meters per hour of methane flows through a pipeline at 40.0 bar absolute and 300.0 K. Use z from page 2-144 of *Perry's Chemical Engineers' Handbook* to estimate the mass flow rate in kg/h.

SOLUTION   From the given reference, $z = 0.934$ at 40.0 bar and 300.0 K. Rearranging Equation 5.4-2c yields

$$\dot{n} = \frac{P\dot{V}}{zRT} = \frac{(40.0 \text{ bar})(50.0 \text{ m}^3/\text{h})}{(0.934)(300.0 \text{ K})} \left| \frac{\text{kmol} \cdot \text{K}}{8.314 \text{ m}^3 \cdot \text{kPa}} \right| \frac{101.325 \text{ kPa}}{1.01325 \text{ bar}} = 85.9 \text{ kmol/h}$$

$$\Downarrow$$

$$\dot{m} = \frac{85.9 \text{ kmol}}{\text{h}} \left| \frac{16.04 \text{ kg}}{\text{kmol}} \right| = \boxed{1370 \text{ kg/h}}$$

5.4b   The Law of Corresponding States and Compressibility Charts

It would be convenient if the compressibility factor at a single temperature and pressure were the same for all gases, so that a single chart or table of $z(T, P)$ could be used for all PVT calculations. Nature is not that accommodating, unfortunately; for example, z for nitrogen at 0°C and 100 atm is 0.9848 while z for carbon dioxide at the same temperature and pressure is 0.2020. Consequently, to use tabulated z values for all PVT calculations as in the preceding example, you would have to measure compressibilities as functions of temperature and pressure separately for every chemical species. Equations of state such as the van der Waals and the Soave–Redlich–Kwong equations were developed to avoid having to compile the massive volumes of z data that would be involved in such an effort.

An alternative approach is presented in this section. We will show that z can be estimated for a species at a given temperature, T, and pressure, P, with this procedure:

1. Look up (e.g., in Table B.1) the *critical temperature, T_c,* and *critical pressure, P_c,* of the species.
2. Calculate the *reduced temperature, $T_r = T/T_c$,* and *reduced pressure, $P_r = P/P_c$.*
3. Look up the value of z on a **generalized compressibility chart**, which plots z versus P_r for specified values of T_r.

The basis for estimating z in this manner is the empirical **law of corresponding states,** which holds that *the values of certain physical properties of a gas—such as the compressibility factor— depend to great extent on the proximity of the gas to its critical state.* The reduced temperature and pressure provide a measure of this proximity; the closer T_r and P_r are to 1, the closer the gas is to its critical state. This observation suggests that a plot of z versus T_r and P_r should be approximately the same for all substances, which proves to be the case. Such a plot is called the **generalized compressibility chart**.[7]

Figure 5.4-1 shows a generalized compressibility chart for those fluids having a critical compressibility factor of 0.27.[8] Conditions for both gases and liquids are illustrated, although in our discussions here we only consider estimation of z for gases. Note the increasing deviations from ideal gas behavior as pressures approach P_c (i.e., when $P_r \rightarrow 1$).

Figures 5.4-2 through 5.4-4 are expansions of various regions of Figure 5.4-1. The parameter V_r^{ideal} is introduced in these figures to eliminate the need for trial-and-error calculations in problems where either temperature or pressure is unknown. This parameter is defined in terms of the ideal critical volume[9] as

$$V_r^{\text{ideal}} = \frac{\hat{V}}{\hat{V}_c^{\text{ideal}}} = \frac{\hat{V}}{RT_c/P_c} = \frac{P_c\hat{V}}{RT_c} \tag{5.4-3}$$

---

[7]L. C. Nelson and E. F. Obert, *Trans. ASME,* **76,** 1057(1954).

[8]An extension of the generalized charts that provides somewhat greater accuracy also allows for a dependence of $z(T, P)$ on z_c, the compressibility factor at the critical point, which generally varies between 0.25 and 0.29.

[9]$V_c^{\text{ideal}} = RT_c/P_c$ is the specific molar volume that would be calculated by the ideal gas equation of state for temperature T_c and pressure P_c. It has no physical significance for the gas in question, unlike T_c, P_c, and the *critical volume* $\hat{V}_c$, another property of the gas.

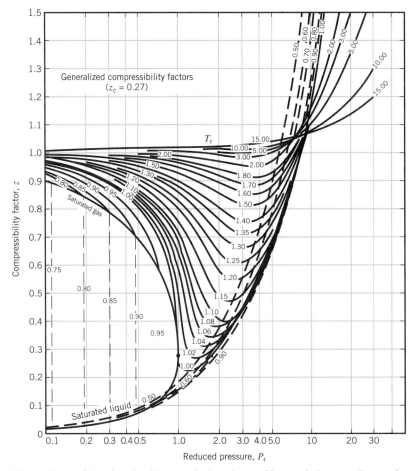

Figure 5.4-1 (Reprinted with permission from *Chemical Process Principles Charts,* 2nd Edition, by O. A. Hougen, K. M. Watson, and R. A. Ragatz, John Wiley & Sons, New York, 1960.)

The procedure for using the generalized compressibility chart for *PVT* calculations is as follows:

1. Look up or estimate the critical temperature, T_c, and pressure, P_c, of the substance of interest (Table B.1).

2. If the gas is either hydrogen or helium, determine *adjusted critical constants* from the empirical formulas

$$T_c^a = T_c + 8 \text{ K} \tag{5.4-4}$$

$$P_c^a = P_c + 8 \text{ atm} \tag{5.4-5}$$

These equations are known as **Newton's corrections**.

3. Calculate reduced values of the two known variables (temperature and pressure, temperature and volume, or pressure and volume) using the definitions

$$T_r = \frac{T}{T_c} \tag{5.4-6}$$

$$P_r = \frac{P}{P_c} \tag{5.4-7}$$

$$V_r^{\text{ideal}} = \frac{P_c \hat{V}}{R T_c} \tag{5.4-8}$$

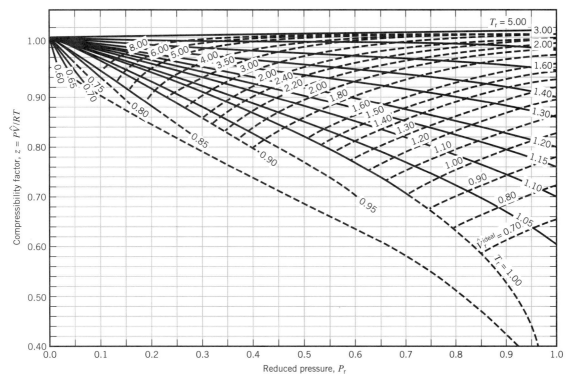

Figure 5.4-2 Generalized compressibility chart, low pressures. (From D. M. Himmelblau, *Basic Principles and Calculations in Chemical Engineering*, 3rd Edition, copyright © 1974, p. 175. Reprinted by permission of Prentice Hall, Inc., Englewood Cliffs, NJ.)

Don't forget to perform whatever unit conversions may be necessary to make the calculated reduced variables dimensionless. If the gas is H_2 or He, substitute for T_c and P_c the adjusted values of these quantities calculated in Step 2. *All temperatures and pressures used in these calculations must be absolute.*

4. Use the compressibility charts to determine the compressibility factor, and then solve for the unknown variable from the compressibility-factor equation of state (Equation 5.4-2).

The compressibility-factor equation of state used in conjunction with the generalized compressibility chart is not generally as accurate as a multiple-constant equation of state for *PVT* calculations under highly nonideal conditions. Furthermore, it lacks precision and cannot readily be adapted to computer calculations. Its advantages include relative computational simplicity and (as will be shown) adaptability to multicomponent gas mixtures.

EXAMPLE 5.4-2   *The Generalized Compressibility Chart*

One hundred gram-moles of nitrogen is contained in a 5-liter vessel at $-20.6°C$. Estimate the pressure in the cylinder.

SOLUTION   From Table B.1, the critical temperature and pressure of nitrogen are

$$T_c = 126.2 \text{ K}, \qquad P_c = 33.5 \text{ atm}$$

The reduced temperature and volume are calculated from Equations 5.4-6 and 5.4-7 as

$$T_r = \frac{T}{T_c} = \frac{(-20.6 + 273.2)\text{ K}}{126.2\text{ K}} = 2.00$$

$$V_r^{\text{ideal}} = \frac{\hat{V} P_c}{R T_c} = \frac{5\text{ L}}{100\text{ mol}} \left| \frac{33.5\text{ atm}}{126.2\text{ K}} \right| \frac{\text{mol·K}}{0.08206\text{ L·atm}} = 0.161$$

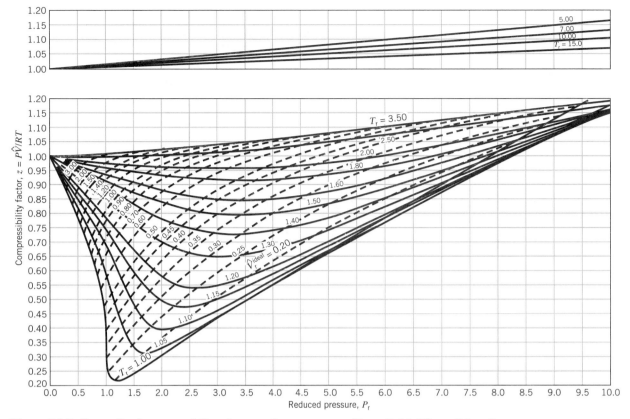

Figure 5.4-3 Generalized compressibility chart, medium pressures. (From D. M. Himmelblau, *Basic Principles and Calculations in Chemical Engineering*, 3rd Edition, copyright © 1974, p. 176. Reprinted by permission of Prentice Hall, Inc., Englewood Cliffs, NJ.)

From Figure 5.4-4, the intersection of $T_r = 2$ and $V_r^{ideal} = 0.161$ occurs at approximately $z = 1.77$. From Equation 5.4-2a, we can now calculate

$$P = \frac{zRT}{\hat{V}} = \frac{1.77}{} \left| \frac{0.08206\ \text{L·atm}}{\text{mol·K}} \right| \frac{252.4\ \text{K}}{0.05\ \text{L/mol}} = \boxed{733\ \text{atm}}$$

Note: You could also read the value of P_r at the intersection and calculate $P = P_r P_c$; however, it is generally more accurate to calculate unknown variables by first determining z and then using the equation of state, as was done above.

TEST YOURSELF

1. Suppose you need to know the volume that would be occupied by 10 kmol of H_2 at $-190°C$ and 300 atm. Would you consider it reasonable to use a value calculated from the ideal gas equation of state? How would you use the generalized compressibility chart for this calculation?

2. Why would a chart such as that in Figure 5.4-1 be useless if T and P were the chart parameters?

3. What is the law of corresponding states, and how does it provide a basis for the generalized compressibility chart?

5.4c Nonideal Gas Mixtures

Whether an analytical or graphical correlation is used to describe nonideal gas behavior, difficulties arise when the gas contains more than one species. Consider, for example, the SRK

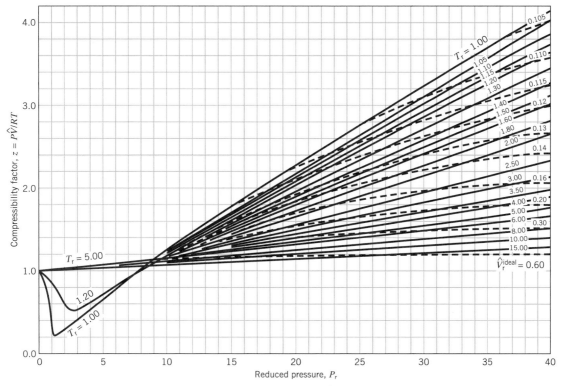

Figure 5.4-4 Generalized compressibility chart, high pressures. (From D. M. Himmelblau, *Basic Principles and Calculations in Chemical Engineering*, 3rd Edition, copyright © 1974, p. 177. Reprinted by permission of Prentice Hall, Inc., Englewood Cliffs, NJ.)

equation of state (Equation 5.3-7): How would you estimate the parameters a, b, and α if the gas consisted of methane, carbon dioxide, and nitrogen? **Mixture rules** developed for such circumstances are summarized by Reid, Prausnitz, and Poling (see footnote 1). We will illustrate *PVT* calculations for mixtures with a simple rule developed by Kay[10] that utilizes the generalized compressibility charts.

Kay's rule estimates *pseudocritical properties* of mixtures as simple averages of pure-component critical constants:[11]

Pseudocritical Temperature: $\qquad T_c' = y_A T_{cA} + y_B T_{cB} + y_C T_{cC} + \cdots$ $\qquad$ **(5.4-9)**

Pseudocritical Pressure: $\qquad P_c' = y_A P_{cA} + y_B P_{cB} + y_C P_{cC} + \cdots$ $\qquad$ **(5.4-10)**

where $y_A, y_B, \ldots$ are mole fractions of species A, B, $\ldots$ in the mixture. Assuming that the system temperature T and pressure P are known, the pseudocritical properties can be used to estimate the *pseudoreduced temperature and pressure* of the mixture:

Pseudoreduced Temperature: $\qquad T_r' = T/T_c'$ $\qquad$ **(5.4-11)**

Pseudoreduced Pressure: $\qquad P_r' = P/P_c'$ $\qquad$ **(5.4-12)**

The compressibility factor for a gas mixture, z_m, can now be estimated from the compressibility charts and the pseudoreduced properties, and $\hat{V}$ for the mixture can be calculated as

$$\hat{V} = \frac{z_m RT}{P} \qquad \textbf{(5.4-13)}$$

---

[10]W. B. Kay, *Ind. Eng. Chem.*, **28**, 1014(1936).

[11]The pseudocritical constants are simply empirical parameters that have been found useful for correlating the physical properties of a mixture. Unlike T_c and P_c for a single component, T_c' and P_c' have no physical significance.

As with single-component gases, if you know $\hat{V}$ and either T or P you can estimate the ideal pseudoreduced volume $\hat{V}_r^{ideal} = \hat{V} P_c' / R T_c'$ and use the other known reduced property to determine the unknown temperature or pressure from the compressibility chart.

Like the theory of corresponding states on which it is based, Kay's rule provides only approximate values of the quantities it is used to calculate. It works best when used for mixtures of nonpolar compounds whose critical temperatures and pressures are within a factor of two of one another. Reid, Prausnitz, and Poling (see footnote 1) provide more complex but more accurate mixing rules for systems that do not fall into this category.

TEST YOURSELF

What is Kay's rule? How would you use it to calculate the specific molar volume of an equimolar mixture of gases at a given temperature and pressure? For what types of gases would you have the greatest confidence in the answer?

EXAMPLE 5.4-3   *Kay's Rule*

A mixture of 75% H_2 and 25% N_2 (molar basis) is contained in a tank at 800 atm and $-70°C$. Estimate the specific volume of the mixture in L/mol using Kay's rule.

SOLUTION

Critical Constants: From Table B.1:

H_2: $T_c = 33\ K$

$\qquad T_c^a = (33 + 8)\ K = 41\ K$ (Newton's correction: Equation 5.4-4)

$\qquad P_c = 12.8\ atm$

$\qquad P_c^a = (12.8 + 8)\ atm = 20.8\ atm$ (Newton's correction: Equation 5.4-5)

N_2: $T_c = 126.2\ K$

$\qquad P_c = 33.5\ atm$

Pseudocritical Constants: From Equations 5.4-9 and 5.4-10:

$$T_c' = y_{H_2}(T_c^a)_{H_2} + y_{N_2}(T_c)_{N_2} = 0.75 \times 41\ K + 0.25 \times 126.2\ K = 62.3\ K$$

$$P_c' = y_{H_2}(P_c^a)_{H_2} + y_{N_2}(P_c)_{N_2} = 0.75 \times 20.8\ atm + 0.25 \times 33.5\ atm = 24.0\ atm$$

Reduced Conditions: $T = (-70 + 273)\ K = 203\ K,\ P = 800\ atm$

$$T_r' = \frac{T}{T_c'} = \frac{203\ K}{62.3\ K} = 3.26$$

$$P_r' = \frac{P}{P_c'} = \frac{800\ atm}{24.0\ atm} = 33.3$$

Mixture Compressibility: From Figure 5.4-4:

$$z_m(T_r' = 3.26, P_r' = 33.3) = 1.86$$

Calculation of Specific Volume $P\hat{V} = z_m RT$

$$\Downarrow$$

$$\hat{V}\left(\frac{L}{mol}\right) = \frac{z_m T(K)}{P(atm)} \times R\left(\frac{L \cdot atm}{mol \cdot K}\right) = \frac{(1.86)(203)(0.08206)}{800}\ \frac{L}{mol}$$

$$= \boxed{0.0387\ \frac{L}{mol}}$$

5.5   SUMMARY

Problems often arise that call for determining the value of one of the four variables P, T, V, and n (or $\dot{V}$ and $\dot{n}$) for a process material from known values of the other three.

- If the material is a solid or liquid and consists of a single species, look up the specific gravity or density in Table B.1 or one of the references on p. 187. As a first approximation, assume that the tabulated value is independent of temperature and pressure. For a more refined estimate, find and apply a correlation for the dependence of the density on temperature.

- If the material is a liquid mixture, either find a table of mixture density as a function of composition or assume volume additivity and estimate the mixture density from either Equation 5.1-1 or Equation 5.1-2. If the material is a dilute liquid solution, either find a table of mixture density as a function of composition or take the density to be that of the pure solvent.

- If the material is a gas, the ideal gas equation of state ($PV = nRT$) may provide a reasonable approximation for PVT calculations. The equation works best at low pressures (on the order of 1 atm or less) and high temperatures (generally not much lower than $0°C$). A rule of thumb is that the ideal gas equation provides reasonable estimates if RT/P is greater than 5 L/mol for diatomic gases and greater than 20 L/mol for other gases.

- *Standard temperature and pressure* (STP) are generally defined as $0°C$ and 1 atm. These values and the corresponding *standard specific volume*, $\hat{V}_s = 22.4$ L(STP)/mol $\implies$ 359 ft^3 – (STP)/lb-mole, can be used in conjunction with Equation 5.2-5 for PVT calculations on ideal gases.

- The *partial pressure* of a component in an ideal gas mixture is $y_i P$, where y_i is the mole fraction of the component and P is the total absolute pressure. The component partial pressures add up to the total pressure.

- The volume percent of a component in an ideal gas mixture (%v/v) is the same as the mole percent of that component. If the gas mixture is nonideal, the volume percent has no useful meaning.

- The *critical temperature* T_c of a species is the highest temperature at which isothermal compression of the species vapor results in the formation of a separate liquid phase, and the *critical pressure* is the pressure at which that phase forms. Isothermal compression of a species above its critical temperature—a *gas* (as opposed to *vapor*) or *supercritical fluid*—results in a fluid of increasing density but not a separate liquid phase.

- If process conditions are such that the ideal gas equation of state is a poor approximation, a more complex species-specific equation of state must be used. Most such equations, including the *Soave-Redlich-Kwong (SRK)* equation of state, contain adjustable parameters that depend on the critical temperature and pressure of the species and possibly other factors that depend on the molecular geometry and polarity of the species.

- An alternative to using quadratic equations of state (such as the truncated virial equation) and cubic equations of state (such as the SRK equation) is to use the *compressibility-factor equation of state: $PV = znRT$*. The compressibility factor z, defined as the ratio $P\hat{V}/RT$, equals 1 if the gas behaves ideally. For some species it may be looked up in a table (e.g., in *Perry's Chemical Engineers' Handbook*) or, more generally, estimated from the generalized compressibility charts (Figures 5.4-1 through 5.4-4).

- The basis of the generalized compressibility charts is the *law of corresponding states,* an empirical rule stating that the compressibility factor of a species at a given temperature and pressure depends primarily on the *reduced temperature* and *reduced pressure, $T_r = T/T_c$* and $P_r = P/P_c$. Once you have determined these quantities, you may use the charts to determine z and then substitute the value in the compressibility-factor equation of state and solve for whichever variable is unknown.

- To perform *PVT* calculations for nonideal gas mixtures, you may use *Kay's rule.* Determine *pseudocritical constants* (temperature and pressure) by weighting the critical constants for each mixture component by the mole fraction of that component in the mixture; then calculate the reduced temperature and pressure and the compressibility factor as before.

- Remember that every equation of state for nonideal gases is an approximation, often based on fitting adjustable constants to experimental *PVT* data. Be skeptical of every value you estimate, especially if you are using an equation of state beyond the range of conditions for which it was derived.

- If a *PVT* calculation is part of a material balance problem and a volume (or volumetric flow rate) is either given or required for a process stream, label both n (or $\dot{n}$) and V (or $\dot{V}$) on the flowchart and count a density relationship (for solids and liquids) or an equation of state (for gases) as an additional relation in the degree-of-freedom analysis.

PROBLEMS

Note: *Unless otherwise specified, all pressures given in these problems are absolute.*

5.1. A liquid mixture containing 40.0 wt% n-octane and the balance n-decane flows into a tank mounted on a balance. The mass in kg indicated by the scale is plotted against time. The data fall on a straight line that passes through the points ($t = 3$ min, $m = 150$ kg) and ($t = 10$ min, $m = 250$ kg).
   (a) Estimate the volumetric flow rate of the liquid mixture.
   (b) What does the empty tank weigh?

5.2. When a liquid or a gas occupies a volume, it may be assumed to fill the volume completely. On the other hand, when solid particles occupy a volume, there are always spaces (voids) among the particles. The **porosity** or **void fraction** of a bed of particles is the ratio (void volume)/(total bed volume). The **bulk density** of the solids is the ratio (mass of solids)/(total bed volume), and the **absolute density** of the solids has the usual definition, (mass of solids)/(volume of solids).

   Suppose 600.0 g of a crushed ore is placed in a graduated cylinder, filling it to the 184 cm³ level. One hundred cm³ of water is then added to the cylinder, whereupon the water level is observed to be at the 233.5 cm³ mark. Calculate the porosity of the dry particle bed, the bulk density of the ore in this bed, and the absolute density of the ore.

5.3. Two liquid streams are flowing at constant rates into a blender. One is benzene, which flows at a measured rate of 20.0 L/min, and the other is toluene. The blended mixture enters a storage tank (inner diameter = 5.5 m) equipped with a sight gauge. During an interval in which no liquid leaves the storage tank, the liquid level in the tank is observed to increase by 0.15 meters over a one-hour period. Calculate the flow rate of toluene into the blender (L/min) and the composition of the tank contents (wt% benzene).

5.4. A slurry contains crystals of copper sulfate pentahydrate [$CuSO_4\cdot5H_2O(s)$, specific gravity = 2.3] suspended in an aqueous copper sulfate solution (liquid SG = 1.2). A sensitive transducer is used to measure the pressure difference, ΔP(Pa), between two points in the sample container separated by a vertical distance of h meters. The reading is in turn used to determine the mass fraction of crystals in the slurry, x_c(kg crystals/kg slurry).

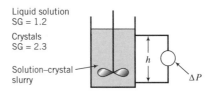

Liquid solution
SG = 1.2

Crystals
SG = 2.3

Solution–crystal slurry

   (a) Derive an expression for the transducer reading, ΔP(Pa), in terms of the overall slurry density, ρ_{sl}(kg/m³), assuming that the pressure head formula of Chapter 3 ($P = P_0 + \rho gh$) is valid for this two-phase system.

(b) Validate the following expression relating the overall slurry density to the liquid and solid crystal densities (ρ_l and ρ_c) and the mass fraction of crystals in the slurry:

$$\frac{1}{\rho_{sl}} = \frac{x_c}{\rho_c} + \frac{(1 - x_c)}{\rho_l}$$

(*Suggestion:* Insert units for all variables.)

(c) Suppose 175 kg of the slurry is placed in the sample container with $h = 0.200$ m and a transducer reading $\Delta P = 2775$ Pa is obtained. Calculate (i) ρ_{sl}, (ii) x_c, (iii) the total slurry volume, (iv) the mass of crystals in the slurry, (v) the mass of anhydrous copper sulfate ($CuSO_4$ without the water of hydration) in the crystals, (vi) the mass of liquid solution, and (vii) the volume of liquid solution.

*(d) Write a spreadsheet program to generate a calibration curve of x_c versus ΔP for this device. Take as inputs $\rho_c(kg/m^3)$, $\rho_l(kg/m^3)$, and $h(m)$, and calculate $\Delta P(Pa)$ for $x_c = 0.0, 0.05, 0.10, \ldots$, 0.60. Run the program for the parameter values in this problem ($\rho_c = 2300$, $\rho_l = 1200$, and $h = 0.200$). Then plot x_c versus ΔP (have the spreadsheet program do it, if possible), and verify that the value of x_c corresponding to $\Delta P = 2775$ Pa on the calibration curve corresponds to the value calculated in part (c).

(e) Derive the expression in part (b). Take a basis of 1 kg of slurry [x_c (kg), V_c (m^3) crystals, $(1 - x_c)$(kg), V_l (m^3) liquid], and use the fact that the volumes of the crystals and liquid are additive.

5.5. Use the ideal gas equation of state to estimate the molar volume in m^3/mol and the density of air in kg/m^3 at 40°C and a gauge pressure of 3.0 atm.

5.6. One gram-mole of methyl chloride vapor is contained in a vessel at 100°C and 10 atm.

    (a) Use the ideal gas equation of state to estimate the system volume.

    (b) Suppose the actual volume of the vessel is 2.8 liters. What percentage error results from assuming ideal gas behavior?

5.7. The pressure gauge on a 20.0 m^3 tank of nitrogen at 25°C reads 10 bar. Estimate the mass of nitrogen in the tank by (a) direct solution of the ideal gas equation of state and (b) conversion from standard conditions. (See Example 5.2-2.)

5.8. From the standard conditions given in Table 5.2-1, calculate the value of the gas constant R in (a) atm·m^3/(kmol·K) and (b) torr·ft^3/(lb-mole·°R).

5.9. The volume of a **dry box** (a closed chamber with dry nitrogen flowing through it) is 2.0 m^3. The dry box is maintained at a slight positive gauge pressure of 10 cm H_2O and room temperature (25°C). If the contents of the box are to be replaced every five minutes, calculate the required mass flow rate of nitrogen in g/min by (a) direct solution of the ideal gas equation of state and (b) conversion from standard conditions.

5.10. A stream of air enters a 7.50-cm ID pipe at a velocity of 60.0 m/s at 27°C and 1.80 bar (gauge). At a point downstream, the air flows through a 5.00 cm ID pipe at 60°C and 1.53 bar (gauge). What is the velocity of the gas at this point?

5.11. The label has come off a cylinder of gas in your laboratory. You know only that one species of gas is contained in the cylinder, but you do not know whether it is hydrogen, oxygen, or nitrogen. To find out, you evacuate a 5-liter flask, seal it and weigh it, then let gas from the cylinder flow into it until the gauge pressure equals 1.00 atm. The flask is reweighed, and the mass of the added gas is found to be 13.0 g. Room temperature is 27°C, and barometric pressure is 1.00 atm. What is the gas?

5.12. A gas cylinder filled with nitrogen at standard temperature and pressure has a mass of 37.289 g. The same container filled with carbon dioxide at STP has a mass of 37.440 g. When filled with an unknown gas at STP, the container mass is 37.062 g. Calculate the molecular weight of the unknown gas, and then state its probable identity.

5.13. A nitrogen rotameter is calibrated by feeding N_2 from a compressor through a pressure regulator, a needle valve, the rotameter, and a **dry test meter**, a device that measures the total volume of gas that passes through it. A water manometer is used to measure the gas pressure at the rotameter outlet. A flow rate is set using the needle valve, the rotameter reading, ϕ, is noted, and the change in the

---

*Computer problem.

dry gas meter reading (ΔV) for a measured running time (Δt) is recorded.

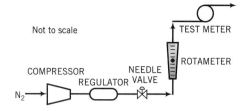

Not to scale

The following calibration data are taken on a day when the temperature is 23°C and barometric pressure is 763 mm Hg.

ϕ	Δt(min)	ΔV(liters)
5.0	10.0	1.50
9.0	10.0	2.90
12.0	5.0	2.00

(a) Prepare a calibration chart of ϕ versus $\dot{V}_{std}$, the flow rate in standard cm³/min equivalent to the actual flow rate at the measurement conditions.

(b) Suppose the rotameter–valve combination is to be used to set the flow rate to 0.010 mol N_2/min. What rotameter reading must be maintained by adjusting the valve?

5.14. The flow rate required to yield a specified reading on an orifice meter varies inversely as the square root of the fluid density; that is, if a fluid with density ρ_1(g/cm³) flowing at a rate $\dot{V}_1$ (cm³/s) yields a meter reading ϕ, then the flow rate of a fluid with density ρ_2 required to yield the same reading is

$$\dot{V}_2 = \dot{V}_1(\rho_1/\rho_2)^{1/2}$$

(a) Suppose an orifice meter calibrated with nitrogen at 25°C and 758 mm Hg is used to measure the flow rate of a process stream of hydrogen at 50°C and 1800 mm Hg, and the value read from the calibration chart is 350 cm³/s. Calculate the true volumetric flow rate of the gas.

(b) Repeat part (a), only suppose the process fluid contains 25 mole% CH_4 and the balance C_3H_8.

5.15. A device has been designed to measure the flow rate of carbon dioxide evolved from a fermentation reactor. The reactor is sealed except for a tube that allows the generated carbon dioxide to bubble through a soap solution and into a vertical glass tube with an internal diameter of 1.2 cm. On leaving the soap solution, the gas forces thin soap films stretched across the tube to traverse the length of the tube. Ambient temperature and pressure are 27°C and 755 mm Hg. It takes the films 7.4 s to traverse the 1.2 m between two calibration marks on the tube.

(a) Sketch the apparatus.

(b) What is the rate of generation of CO_2 in mol/min?

5.16. A stream of air (21 mole% O_2, the rest N_2) flowing at a rate of 10.0 kg/h is mixed with a stream of CO_2. The CO_2 enters the mixer at a rate of 20.0 m³/h at 150°C and 1.5 bar. What is the mole percent of CO_2 in the product stream?

5.17. **Spray drying** is a process in which a liquid containing dissolved or suspended solids is injected into a chamber through a spray nozzle or centrifugal disk atomizer. The resulting mist is contacted with hot air, which evaporates most or all of the liquid, leaving the dried solids to fall to a conveyor belt at the bottom of the chamber.

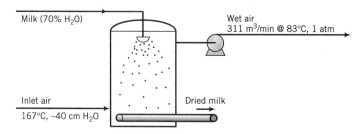

Powdered milk is produced in a spray dryer 6 m in diameter by 6 m high. Air enters at 167°C and −40 cm H_2O. The milk fed to the atomizer contains 70% water by mass, all of which evaporates. The outlet gas contains 12 mole% water and leaves the chamber at 83°C and 1 atm (absolute) at a rate of 311 m^3/min.

(a) Calculate the production rate of dried milk and the volumetric flow rate of the inlet air. Estimate the upward velocity of air (m/s) at the bottom of the dryer.

(b) What problem would you guess would occur if the velocity is too high?

5.18. Many references give the specific gravity of gases with reference to air. For example, the specific gravity of carbon dioxide is 1.52 relative to air at the same temperature and pressure. Show that this value is correct as long as the ideal gas equation of state applies.

5.19. Sax and Lewis[12] describe the hazards of breathing air containing appreciable amounts of an *asphyxiant* (a gas that has no specific toxicity but, when inhaled, excludes oxygen from the lungs). When the mole percent of the asphyxiant in the air reaches 50%, marked symptoms of distress appear, and at 75% death occurs in a matter of minutes.

A small storage room whose dimensions are 2 m × 1.5 m × 3 m contains a number of expensive and dangerous chemicals. To prevent unauthorized entry, the room door is always locked and can be opened with a key from either side. A cylinder of liquid carbon dioxide is stored in the room. The valve on the cylinder is faulty and some of the contents have escaped over the weekend. The room temperature is 25°C.

(a) If the concentration of CO_2 reaches the lethal 75 mole% level, what would be the mole percent of O_2?

(b) How much CO_2 (kg) is present in the room when the lethal concentration is reached? Why would more than that amount have to escape from the cylinder for this concentration to be reached?

(c) Describe a set of events that could result in a fatality in the given situation. Suggest at least two measures that would reduce the hazards associated with storage of this seemingly harmless substance.

5.20. A tank in a room at 19°C is initially open to the atmosphere on a day when the barometric pressure is 102 kPa. A block of dry ice (solid CO_2) with a mass of 15.7 kg is dropped into the tank, which is then sealed. The reading on the tank pressure gauge initially rises very quickly, then much more slowly, eventually reaching a value of 3.27 MPa. Assume $T_{final} = 19°C$.

(a) How many moles of air were in the tank initially? Neglect the volume occupied by CO_2 in the solid state, and assume that a negligible amount of CO_2 escapes prior to the sealing of the tank.

(b) What is the final density (g/liter) of the gas in the tank?

(c) Explain the observed variation of pressure with time. More specifically, what is happening in the tank during the initial rapid pressure increase and during the later slow pressure increase?

5.21. In **froth flotation**, air is bubbled through an aqueous solution or slurry to which a foaming agent (soap) has been added. The air bubbles carry finely dispersed solids and hydrophobic materials such as grease and oil to the surface where they can be skimmed off in the foam.

An ore-containing slurry is to be processed in a froth flotation tank at a rate of 300 tons/h. The slurry consists of 20.0 wt% solids (the ore, SG = 1.2) and the remainder an aqueous solution with a density close to that of water. Air is *sparged* (blown through a nozzle designed to produce small bubbles) into the slurry at a rate of 40.0 ft^3(STP)/1000 gal of slurry. The entry point of the air is 10 ft below the slurry surface. The tank contents are at 75°F and the barometric pressure is 28.3 in. Hg. The sparger design is such that the average bubble diameter on entry is 2.0 mm.

(a) What is the volumetric flow rate of the air at its entering conditions?

(b) By what percentage does the average bubble diameter change between the entry point and the slurry surface?

5.22.[13] Several decades ago benzene was thought to be a harmless chemical with a somewhat pleasant odor and was widely used as a cleaning solvent. It has since been found that chronic exposure to

[12]N. I. Sax and R. J. Lewis, *Hazardous Chemicals Desk Reference,* Van Nostrand Reinhold, New York, 1987, p. 183.

[13]From D. A. Crowl, D. W. Hubbard, and R. M. Felder, *Problem Set: Stoichiometry,* Center for Chemical Process Safety, New York, 1993.

benzene can cause health problems such as anemia and possibly leukemia. Benzene has a *permissible exposure level* (PEL) of 1.0 ppm (part per million on a molar basis, equivalent to a mole fraction of 1.0×10^{-6}) averaged over an 8-hour period.

The safety engineer in a plant wishes to determine whether the benzene concentration in a laboratory exceeds the PEL. One Monday at 9 a.m., 1 p.m., and 5 p.m., she collects samples of room air (33°C, 99 kPa) in evacuated 2-liter stainless steel containers. To collect a sample she opens the container valve, allows room air to enter until the container pressure equals atmospheric pressure, and then charges clean dry helium into the container until the pressure reaches 500 kPa. Next, she takes the containers to an analytical laboratory in which the temperature is 23°C, leaves them there for a day, and then feeds gas from each container to a gas chromatograph (GC) until the pressure in the container is reduced to 400 kPa. In the order in which they were collected, the samples that pass through the GC are found to contain 0.656 µg (microgram), 0.788 µg, and 0.910 µg of benzene, respectively.

(a) What were the concentrations of benzene (ppm on a molar basis) in the original room air at the three collection times? (Assume ideal gas behavior.) Is the average concentration below the PEL?

(b) Why did the engineer add helium to the container after collecting the room air sample? Why did she wait a day before analyzing the container contents?

(c) Why might a finding that the average benzene concentration is below the PEL not necessarily mean that the laboratory is safe insofar as exposure to benzene is concerned? Give several reasons, including possible sources of error in the sampling and analysis procedure. (Among other things, note the day on which the samples were taken.)

5.23. A balloon 20 m in diameter is filled with helium at a gauge pressure of 2.0 atm. A man is standing in a basket suspended from the bottom of the balloon. A restraining cable attached to the basket keeps the balloon from rising. The balloon (not including the gas it contains), the basket, and the man have a combined mass of 150 kg. The temperature is 24°C that day, and the barometer reads 760 mm Hg.

(a) Calculate the mass (kg) and weight (N) of the helium in the balloon.

(b) How much force is exerted on the balloon by the restraining cable? (*Recall:* The buoyant force on a submerged object equals the weight of the fluid—in this case, the air—displaced by the object. Neglect the volume of the basket and its contents.)

(c) Calculate the initial acceleration of the balloon when the restraining cable is released.

(d) Why does the balloon eventually stop rising? What would you need to know to calculate the altitude at which it stops?

(e) Suppose at its point of suspension in midair the balloon is heated, raising the temperature of the helium. What happens and why?

5.24. The Cookenwythe Gas Company pumps propane gas to the nearby Noxious Chemicals, Inc., polypropylene production plant. The gas is metered at the Noxious plant at 400 m³/h at 4.7 atm gauge and 30°C. The pressure at the Cookenwythe plant is 8.5 atm gauge, and the temperature is also 30°C. Noxious pays Cookenwythe at the rate of $0.60/kg C_3H_8.

One dark night Sebastian Goniff, a Noxious engineer who is really a spy for the Rancid Plastics Corporation—Noxious's principal competitor and a bunch of really rotten guys—puts into effect his plan to divert propane from the Cookenwythe–Noxious line into an underground pipe that leads to a secret Rancid tank-truck loading station in the middle of a nearby abandoned garbage dump. To cover the operation, Goniff obtains a broken pressure gauge that is stuck at 4.7 atm and substitutes it for the gauge at the Noxious plant. He adjusts the gas pressure regulator so that the real gauge pressure is 1.8 atm, instructs his associate in the field via walkie-talkie to open the Rancid line gradually, and tells him to hold it when the flowmeter at Noxious reads 400 m³/h. To the unsuspecting Noxious meter reader, the flow rate and pressure would therefore appear to be normal.

The plan goes according to schedule until the associate smells gas, suspects a leak near the valve, and lights a match to see if he can pinpoint the location.

(a) What should the flowmeter read at the Cookenwythe end of the pipeline?

(b) How much does Noxious pay Cookenwythe each month?

(c) What flow rate of propane (kg/h) should the Rancid trucks be prepared to handle?

(d) What happened?

5.25. An ideal gas mixture contains 35% helium, 20% methane, and 45% nitrogen by volume at 2.00 atm absolute and 90°C. Calculate (a) the partial pressure of each component, (b) the mass fraction of methane, (c) the average molecular weight of the gas, and (d) the density of the gas in kg/m^3.

5.26. There is a percentage of fuel in a fuel–air mixture called the *lower flammable limit* (LFL), below which the mixture cannot be ignited. There is also an *upper flammable limit* (UFL), a percentage of fuel above which ignition will not take place. The LFL and UFL are collectively known as the *flammability limits* of the fuel.

　　The LFL and UFL of propane in air at 1 atm are, respectively, 2.05 mole% C_3H_8 and 11.4 mole% C_3H_8. If the mole percent of propane in a propane–air mixture is between 2.05% and 11.4%, the gas mixture will burn explosively if exposed to a flame or spark; if the percentage is outside these limits, the mixture is safe—a match may burn in it but the flame will not spread. If the percentage of propane is below the LFL, the mixture is said to be too *lean* to ignite; if it is above the UFL, the mixture is too *rich* to ignite.

　　(a) Which would be safer to release into the atmosphere—a fuel–air mixture that is too lean or too rich to ignite? Explain.

　　(b) A mixture of propane in air containing 4.03 mole% C_3H_8 is the feed to a combustion furnace. If there is a problem in the furnace, the mixture is diluted with a stream of pure air to make sure that it cannot accidentally ignite. If propane enters the furnace at a rate of 150 mol C_3H_8/s in the original fuel–air mixture, what is the minimum molar flow rate of the diluting air?

　　(c) The actual diluting air molar flow rate is specified to be 130% of the minimum value. Assuming the fuel mixture (4.03 mole% C_3H_8) enters the furnace at the same rate as in part (b) at 125°C and 131 kPa and the diluting air enters at 25°C and 110 kPa, calculate the ratio m^3 diluting air/m^3 fuel gas and the mole percent of propane in the diluted mixture.

　　(d) Give several possible reasons for feeding air at a value greater than the calculated minimum rate.

5.27. An adult takes about 12 breaths per minute, inhaling roughly 500 mL of air with each breath. The molar compositions of the inspired and expired gases are as follows:

Species	Inspired Gas (%)	Expired Gas (%)
O_2	20.6	15.1
CO_2	0.0	3.7
N_2	77.4	75.0
H_2O	2.0	6.2

The inspired gas is at 24°C and 1 atm, and the expired gas is at body temperature and pressure—37°C and 1 atm. Nitrogen is not transported into or out of the blood in the lungs, so that $(N_2)_{in} = (N_2)_{out}$.

　　(a) Calculate the masses of O_2, CO_2, and H_2O transferred from the pulmonary gases to the blood or vice versa (specify which) per minute.

　　(b) Calculate the volume of air exhaled per milliliter inhaled.

　　(c) At what rate (g/min) is this individual losing weight by merely breathing?

5.28. As everyone who has used a fireplace knows, when a fire burns in a furnace a **draft**, or slight vacuum, is induced that causes the hot combustion gases and entrained particulate matter to flow up and out of the stack. The reason is that the hot gas in the stack is less dense than air at ambient temperature, leading to a lower hydrostatic head inside the stack than at the furnace inlet. The **theoretical draft** D (N/m^2) is the difference in these hydrostatic heads; the **actual draft** takes into account pressure losses undergone by the gases flowing in the stack.

　　Let T_s(K) be the average temperature in a stack of height L(m) and T_a the ambient temperature, and let M_s and M_a be the average molecular weights of the gases inside and outside the stack. Assume that the pressures inside and outside the stack are both equal to atmospheric pressure, P_a(N/m^2) (in fact, the pressure inside the stack is normally a little lower).

(a) Use the ideal gas equation of state to prove that the theoretical draft is given by the expression

$$D(\text{N/m}^2) = \frac{P_a L g}{R} \left(\frac{M_a}{T_a} - \frac{M_s}{T_s} \right)$$

(b) Suppose the gas in a 53-m stack has an average temperature of 655 K and contains 18 mole% CO_2, 2% O_2, and 80% N_2 on a day when barometric pressure is 755 mm Hg and the outside temperature is 294 K. Calculate the theoretical draft (cm H_2O) induced in the furnace.

5.29. Phosgene (CCl_2O) is a colorless gas that was used as an agent of chemical warfare in World War I. It has the odor of newly mown hay (which is a good warning if you know the smell of newly mown hay).

Pete Brouillette, an innovative chemical engineering student, came up with what he believed was an effective new process that utilized phosgene as a starting material. He immediately set up a reactor and a system for analyzing the reaction mixture with a gas chromatograph. To calibrate the chromatograph (i.e., to determine its response to a known quantity of phosgene), he evacuated a 15.0 cm length of tubing with an outside diameter of 0.635 cm and a wall thickness of 0.559 mm, and then connected the tube to the outlet valve of a cylinder containing pure phosgene. The idea was to crack the valve, fill the tube with phosgene, close the valve, feed the tube contents into the chromatograph, and observe the instrument response.

What Pete hadn't thought about (among other things) was that the phosgene was stored in the cylinder at a pressure high enough for it to be a liquid. When he opened the cylinder valve, the liquid rapidly flowed into the tube and filled it. Now he was stuck with a tube full of liquid phosgene at a pressure the tube was not designed to support. Within a minute he was reminded of a tractor ride his father had once given him through a hayfield, and he knew that the phosgene was leaking. He quickly ran out of the lab, called campus security, and told them that a toxic leak had occurred and that the building had to be evacuated and the tube removed and disposed. Personnel in air masks shortly appeared, took care of the problem, and then began an investigation that is still continuing.

(a) Show why one of the reasons phosgene was an effective weapon is that it would collect in low spots soldiers often entered for protection.

(b) Pete's intention was to let the tube equilibrate at room temperature (23°C) and atmospheric pressure. How many gram-moles of phosgene would have been contained in the sample fed to the chromatograph if his plan had worked?

(c) The laboratory in which Pete was working had a volume of 2200 ft³, the specific gravity of liquid phosgene is 1.37, and Pete had read somewhere that the maximum "safe" concentration of phosgene in air is 0.1 ppm (0.1×10^{-6} mol CCl_2O/mol air). Would the "safe" concentration have been exceeded if all the liquid phosgene in the tube had leaked into the room and evaporated? Even if the limit would not have been exceeded, give several reasons why the lab would still have been unsafe.

(d) List several things Pete did (or failed to do) that made his experiment unnecessarily hazardous.

5.30. A fuel gas containing 86% methane, 8% ethane, and 6% propane by volume flows to a furnace at a rate of 1450 m³/h at 15°C and 150 kPa (gauge), where it is burned with 8% excess air. Calculate the required flow rate of air in SCMH (standard cubic meters per hour).

*5.31. The flow of air to a gas-fired boiler furnace is regulated by a minicomputer controller. The fuel gases used in the furnace are mixtures of methane (A), ethane (B), propane (C), *n*-butane (D), and isobutane (E). At periodic intervals the temperature, pressure, and volumetric flow rate of the fuel gas are measured, and voltage signals proportional to the values of these variables are transmitted to the computer. Whenever a new feed gas is used, a sample of the gas is analyzed and the mole fractions of each of the five components are determined and read into the computer. The desired percent excess air is then specified, and the computer calculates the required volumetric flow rate of air and transmits the appropriate signal to a flow control valve in the air line.

The linear proportionalities between the input and the output signals and the corresponding process variables may be determined from the following calibration data:

---

*Computer problem.

Fuel Temperature:	$T = 25.0°C,$	$R_T = 14$	
	$T = 35.0°C,$	$R_T = 27$	
Fuel Pressure:	$P_{gauge} = 0$ kPa,	$R_P = 0$	
	$P_{gauge} = 20.0$ kPa,	$R_P = 6$	
Fuel Flow Rate:	$V_f = 0$ m³/h,	$R_f = 0$	
	$V_f = 2.00 \times 10^3$ m³/h,	$R_f = 10$	
Air Flow Rate:	$V_a = 0$ m³(STP)/h,	$R_a = 0$	
	$V_a = 1.0 \times 10^5$ m³(STP)/h,	$R_a = 25$	

(a) Create a spreadsheet or write a program to read in values of R_f, R_T, R_P, the fuel gas component mole fractions x_A, x_B, x_C, x_D, and x_E, and the percent excess air PX, and to calculate and print out the required value of R_A.

(b) Run your program for the following data.

R_f	R_T	R_P	x_A	x_B	x_C	x_D	x_E	PX
7.25	23.1	7.5	0.81	0.08	0.05	0.04	0.02	15%
5.80	7.5	19.3	0.58	0.31	0.06	0.05	0.00	23%
2.45	46.5	15.8	0.00	0.00	0.65	0.25	0.10	33%

5.32. The oxidation of nitric oxide

$$NO + \tfrac{1}{2}O_2 \rightleftharpoons NO_2$$

takes place in an isothermal batch reactor. The reactor is charged with a mixture containing 20.0 volume percent NO and the balance air at an initial pressure of 380 kPa (absolute).

(a) Assuming ideal gas behavior, determine the composition of the mixture (component mole fractions) and the final pressure (kPa) if the conversion of NO is 90%.

(b) Suppose the pressure in the reactor eventually equilibrates (levels out) at 360 kPa. What is the equilibrium percent conversion of NO? Calculate the reaction equilibrium constant at the prevailing temperature, $K_p[(\text{atm})^{-0.5}]$, defined as

$$K_p = \frac{(p_{NO_2})}{(p_{NO})(p_{O_2})^{0.5}}$$

where $p_i(\text{atm})$ is the partial pressure of species i (NO_2, NO, O_2) at equilibrium.

5.33. Monochlorobenzene (M) is produced commercially by the direct catalytic chlorination of benzene (B) at 40°C and 120 kPa absolute. In the process, dichlorobenzene (D) is generated as a coproduct:

$$C_6H_6 + Cl_2 \rightarrow C_6H_5Cl + HCl$$

$$C_6H_5Cl + Cl_2 \rightarrow C_6H_4Cl_2 + HCl$$

Liquid and gas streams leave the reactor. The liquid contains 49.2 wt% M, 29.6% D, and the remainder unreacted B. The gas, which is sent to a treatment facility, contains 92%(v/v) HCl and 8% unreacted chlorine.

(a) What volume of gas leaves the reactor (m³/kg B fed)?

(b) The pipe through which the gas is to flow is sized so that the gas velocity is no greater than 10 m/s. Derive an expression relating pipe diameter d_p (cm) to benzene feed rate $\dot{m}_{B0}$ (kg B/min).

(c) In 1996, the demand for monochlorobenzene was projected to decrease by 6%/year through the year 2000.[14] What factors were contributing to the reduced demand when the projection was made?

---

[14]ChemExpo, Schnell Publishing, September 23, 1996. Web address:
http://www.chemexpo.com/chemexpo2/news/PROFILEsep23.html

5.34.[15] In *chemical vapor deposition* (CVD), a semiconducting or insulating solid material is formed in a reaction between a gaseous species and a species adsorbed on the surface of silicon *wafers* (disks about 10 cm in diameter and 1 mm thick). The coated wafers are subjected to further processing to produce the microelectronic chips in computers and most other electronic devices in use today.

In one such process, silicon dioxide (MW = 60.06, SG = 2.67) is formed in the reaction between gaseous dichlorosilane (DCS) and adsorbed nitrous oxide:

$$SiH_2Cl_2(g) + 2\,N_2O(ads) \longrightarrow SiO_2(s) + 2\,N_2(g) + 2\,HCl(g)$$

A mixture of DCS and N_2O flows through a "boat reactor"—a horizontal pipe in which 50–100 silicon wafers about 12 cm in diameter and 1 mm thick are set upright along the reactor length, with about 20 mm separation between each wafer. A side view of the reactor is shown below:

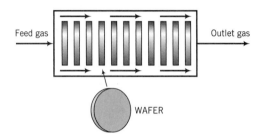

The feed gas enters the reactor at a rate of 3.74 SCMM (standard cubic meters per minute) and contains 22.0 mole% DCS and the balance N_2O. In the reactor, the gas flows around the wafers, DCS and N_2O diffuse into the spaces between the wafers, N_2O is adsorbed on the wafer surfaces, and the adsorbed N_2O reacts with gaseous DCS. The silicon dioxide formed remains on the surface, and the nitrogen and hydrogen chloride go into the gas phase and eventually leave the reactor with the unconsumed reactants. The temperature and absolute pressure in the reactor are constant at 900°C and 604 millitorr.

(a) The percentage conversion of DCS at a certain axial position (distance along the length of the reactor) is 60%. Calculate the volumetric flow rate (m³/min) of gas at this axial position.

(b) The rate of deposition of silicon dioxide per unit area of wafer surface is given by the formula

$$r\left(\frac{\text{mol SiO}_2}{\text{m}^2 \cdot \text{s}}\right) = 3.16 \times 10^{-8}\,p_{DCS}\,p_{N_2O}^{0.65}$$

where p_{DCS} and p_{N_2O} are the partial pressures of DCS and N_2O in millitorr. What is r at the axial position in the reactor where the DCS conversion is 60%?

(c) Consider a wafer located at the axial position determined in part (b). How thick is the silicon dioxide layer on that wafer after two hours of reactor operation, assuming that gas diffusion is rapid enough at the low reactor pressure for the composition of the gas (and hence the component partial pressures) to be uniform over the wafer surface? Express your answer in angstroms, where 1 Å = 1.0×10^{-10} m. (*Hint:* You can calculate the rate of growth of the SiO_2 layer in Å/min from r and properties of SiO_2 given in the problem statement.) Would the thickness be greater or less than this value at an axial position closer to the reactor entrance? Briefly explain your answer.

5.35. A gas turbine power plant receives a shipment of hydrocarbon fuel whose composition is uncertain but may be represented by the expression C_xH_y. The fuel is burned with excess air. An analysis of the product gas gives the following results on a moisture-free basis: 10.5%(v/v) CO_2, 5.3% O_2, and 84.2% N_2.

(a) Determine the molar ratio of hydrogen to carbon in the fuel (r), where $r = y/x$, and the percentage excess air used in the combustion.

[15]Based on a problem in H. S. Fogler, *Elements of Chemical Reaction Engineering,* 2nd Edition, Prentice Hall, Englewood Cliffs, NJ, 1992, p. 323.

(b) What is the air-to-fuel ratio (m³ air/kg of fuel) if the air is fed to the power plant at 30°C and 98 kPa?

5.36. Liquid hydrazine (SG = 0.82) undergoes a family of decomposition reactions that can be represented by the stoichiometric expression

$$3\,N_2H_4 \longrightarrow 6x\,H_2 + (1 + 2x)N_2 + (4 - 4x)NH_3$$

(a) For what range of values of x is this equation physically meaningful?

(b) Plot the volume of product gas $[V\,(L)]$ at 600°C and 10 bars absolute that would be formed from 50 liters of liquid hydrazine as a function of x, covering the range of x values determined in part (b).

(c) Speculate on what makes hydrazine a good propellant.

5.37.[16] Chemicals are stored in a laboratory with volume $V\,(m^3)$. As a consequence of poor laboratory practices, a hazardous species, A, enters the room air (from inside the room) at a constant rate $\dot{m}_A(g\,A/h)$. The room is ventilated with clean air flowing at a constant rate $\dot{V}_{air}(m^3/h)$. The average concentration of A in the room air builds up until it reaches a steady-state value $C_{A,r}(g\,A/m^3)$.

(a) List at least four situations that could lead to A getting into the room air.

(b) Assume that the A is perfectly mixed with the room air and derive the formula

$$m_A = \dot{V}_{air}C_A$$

(c) The assumption of perfect mixing is never justified when the enclosed space is a room (as opposed to, say, a stirred reactor). In practice, the concentration of A varies from one point in the room to another: it is relatively high near the point where A enters the room air and relatively low in regions far from that point, including the ventilator outlet duct. If we say that $C_{A,duct} = kC_A$, where $k < 1$ is a nonideal mixing factor (generally between 0.1 and 0.5, with the lowest value corresponding to the poorest mixing), then the equation of part (b) becomes

$$\dot{m}_A = k\dot{V}_{air}C_A$$

Use this equation and the ideal gas equation of state to derive the following expression for the average mole fraction of A in the room air:

$$y_A = \frac{m_A}{k\dot{V}_{air}} \frac{RT}{M_A P}$$

where M_A is the molecular weight of A.

(d) The *permissible exposure level* (PEL) for styrene ($M = 104.14$) defined by the U.S. Occupational Safety and Health Administration is 50 ppm (molar basis). An open storage tank in a polymerization laboratory contains styrene. The evaporation rate from this tank is estimated to be 9.0 g/h. Room temperature is 20°C. Assuming that the laboratory air is reasonably well mixed (so that $k = 0.5$), calculate the minimum ventilation rate (m³/h) required to keep the average styrene concentration at or below the PEL. Then give several reasons why working in the laboratory might still be hazardous if the calculated minimum ventilation rate is used.

(e) Would the hazard level in the situation described in part (d) increase or decrease if the temperature in the room were to increase? (Increase, decrease, no way to tell.) Explain your answer, citing at least two temperature effects in your explanation.

5.38. Propylene is hydrogenated in a batch reactor:

$$C_3H_6(g) + H_2(g) \rightarrow C_3H_8(g)$$

Equimolar amounts of propylene and hydrogen are fed into the reactor at 25°C and a total absolute pressure of 32.0 atm. The reactor temperature is raised to 235°C and held constant thereafter until the reaction is complete. The propylene conversion at the beginning of the isothermal period is 53.2%. You may assume ideal gas behavior for this problem, although at the high pressures involved this assumption constitutes a crude approximation at best.

---

[16]From D. A. Crowl, D. W. Hubbard, and R. M. Felder, *Problem Set: Stoichiometry*, Center for Chemical Process Safety, New York, 1993.

(a) What is the final reactor pressure?

(b) What is the percentage conversion of propylene when $P = 35.1$ atm?

(c) Construct a graph of pressure versus fractional conversion of propylene covering the isothermal period of operation. Use the graph to confirm the results in parts (a) and (b). (Suggestion: Use a spreadsheet.)

5.39. A natural gas contains 95 wt% CH_4 and the balance C_2H_6. Five hundred cubic meters per hour of this gas at 40°C and 1.1 bar is to be burned with 25% excess air. The air flowmeter is calibrated to read the volumetric flow rate at standard temperature and pressure. What should the meter read (in SCMH) when the flow rate is set to the desired value?

5.40. A stream of hot dry nitrogen flows through a process unit that contains liquid acetone. A substantial portion of the acetone vaporizes and is carried off by the nitrogen. The combined gases leave the recovery unit at 205°C and 1.1 bar and enter a condenser in which a portion of the acetone is lique-fied. The remaining gas leaves the condenser at 10°C and 40 bar. The partial pressure of acetone in the feed to the condenser is 0.100 bar, and that in the effluent gas from the condenser is 0.379 bar. Assume ideal gas behavior.

(a) Calculate for a basis of 1 m^3 of gas fed to the condenser the mass of acetone condensed (kg) and the volume of gas leaving the condenser (m^3).

(b) Suppose the volumetric flow rate of the gas leaving the condenser is 20.0 m^3/h. Calculate the rate (kg/h) at which acetone is vaporized in the solvent recovery unit.

5.41. Ammonia is one of the chemical constituents of industrial waste that must be removed in a treatment plant before the waste can safely be discharged into a river or estuary. Ammonia is normally present in wastewater as aqueous ammonium hydroxide ($NH_4^+OH^-$). A two-part process is frequently car-ried out to accomplish the removal. Lime (CaO) is first added to the wastewater, leading to the reaction

$$CaO + H_2O \longrightarrow Ca^{2+} + 2(OH^-)$$

The hydroxide ions produced in this reaction drive the following reaction to the right, resulting in the conversion of ammonium ions to dissolved ammonia:

$$NH_4^+ + OH^- \rightleftharpoons NH_3(g) + H_2O(l)$$

Air is then contacted with the wastewater, stripping out the ammonia.

(a) One million gallons of alkaline wastewater containing 0.03 mole NH_3/mole H_2O is fed to a stripping tower that operates at 68°F. Air at 68°F and 21.3 psia contacts the wastewater coun-tercurrently as it passes through the tower. The feed ratio is 300 ft^3 air/gal wastewater, and 93% of the ammonia is stripped from the wastewater. Calculate the volumetric flow rate of the gas leaving the tower and the partial pressure of ammonia in this gas.

(b) Briefly explain in terms a first-year chemistry student could understand how this process works. Include the equilibrium constant for the second reaction in your explanation.

5.42. You have purchased a gas cylinder that is supposed to contain 5.0 mole% Cl_2 ($\pm 0.1\%$) and 95% air. The experiments you have been running are not giving reasonable results, and you suspect that the chlorine concentration in the gas cylinder is incorrect.

To check this hypothesis, you bubble gas from the suspicious cylinder through 2.0 L of an aque-ous NaOH solution (12.0 wt% NaOH, SG = 1.13) for exactly one hour. The inlet gas is metered at a gauge pressure of 510 mm H_2O and a temperature of 23°C. Prior to entering the vessel, the gas passes through a flowmeter that indicates a flow rate of 2.00 L/min. At the conclusion of the ex-periment, a sample of the residual NaOH solution is analyzed and the results show that the NaOH content has been reduced by 23%. What is the concentration of Cl_2 in the cylinder gas? (Assume the Cl_2 is completely consumed in the reaction $Cl_2 + 2\,NaOH \rightarrow NaCl + NaOCl + H_2O$.)

5.43. Two humid gas streams are combined in a heated mixing chamber. The first stream contains 23.5 mole% ethane and 76.5% ethylene on a dry basis and enters the chamber at 25°C and 105 kPa at a rate of 125 L/min. The second stream is humid air, which enters at 75°C and 115 kPa at a rate of 355 L/min. The product stream emerges at 65°C and 1.00 atm. A *hygrometer* is used to measure the water contents of the two feed streams and the combined product stream. The calibration curve for the hygrometer is a straight line on a semilog plot of y (mole fraction of water) versus R (hygrometer reading), which passes through the points ($y = 10^{-4}$, $R = 5$) and ($y = 0.2$, $R = 90$). The following readings are recorded.

Hydrocarbon Feed Stream: $R = 86.0$

Air Feed Stream: $R = 12.8$

(a) Derive an expression for $y(R)$.

(b) Calculate the volumetric flow rate of the product stream and the molar composition of the product gas on a dry basis.

(c) Calculate the partial pressure of water in the product gas and the hygrometer reading for this stream.

5.44. Most of the concrete used in the construction of buildings, roads, dams, and bridges is made from **portland cement**, a substance obtained by pulverizing the hard, granular residue (**clinker**) from the roasting of a mixture of clay and limestone and adding other materials to modify the setting properties of the cement and the mechanical properties of the concrete.

The charge to a portland cement rotary kiln contains 17% of a dried building clay (72 wt% SiO_2, 16% Al_2O_3, 7% Fe_2O_3, 1.7% K_2O, 3.3% Na_2O) and 83% limestone (95 wt% $CaCO_3$, 5% impurities). When the solid temperature reaches about 900°C, *calcination* of the limestone to lime (CaO) and carbon dioxide occurs. As the temperature continues to rise to about 1450°C, the lime reacts with the minerals in the clay to form such compounds as 3 $CaO \cdot SiO_2$, 3 $CaO \cdot Al_2O_3$, and 4 $CaO \cdot Al_2O_3 \cdot Fe_2O_3$. The flow rate of CO_2 from the kiln is 1350 m³/h at 1000°C and 1 atm. Calculate the feed rates of clay and limestone (kg/h) and the weight percent of Fe_2O_3 in the final cement.

5.45. The ultimate analysis of a No. 4 fuel oil is 86.47 wt% carbon, 11.65% hydrogen, 1.35% sulfur, and the balance noncombustible inerts. This oil is burned in a steam-generating furnace with 15% excess air. The air is preheated to 175°C and enters the furnace at a gauge pressure of 180 mm Hg. The sulfur and hydrogen in the fuel are completely oxidized to SO_2 and H_2O; 5% of the carbon is oxidized to CO, and the balance forms CO_2.

(a) Calculate the feed ratio (m³ air)/(kg oil).

(b) Calculate the mole fractions (dry basis) and the ppm (parts per million on a wet basis, or moles contained in 10^6 moles of the wet stack gas) of the stack gas species that might be considered environmental hazards.

5.46. A stream of liquid *n*-pentane flows at a rate of 50.4 L/min into a heating chamber, where it evaporates into a stream of air 15% in excess of the amount needed to burn the pentane completely. The temperature and gauge pressure of the entering air are 336 K and 208.6 kPa. The heated gas flows into a combustion furnace in which a fraction of the pentane is burned. The product gas, which contains all of the unreacted pentane and no CO, goes to a condenser in which both the water formed in the furnace and the unreacted pentane are liquefied. The uncondensed gas leaves the condenser at 275 K and 1 atm absolute. The liquid condensate is separated into its components, and the flow rate of the pentane is measured and found to be 3.175 kg/min.

(a) Calculate the fractional conversion of pentane achieved in the furnace and the volumetric flow rates (L/min) of the feed air, the gas leaving the condenser, and the liquid condensate before its components are separated.

(b) Sketch the apparatus that could have been used to separate the pentane and water in the condensate. *Hint:* Remember that pentane is a hydrocarbon and recall what is said about oil (hydrocarbons) and water.

5.47. The feed stream to a *Claus plant* consists of 20.0 mole% H_2S and 80.0% CO_2. One-third of the stream is sent to a furnace where the H_2S is burned completely with a stoichiometric amount of air fed at 1 atm and 25°C. The reaction is

$$H_2S + \tfrac{3}{2}O_2 \longrightarrow SO_2 + H_2O$$

The product gases from this reaction are then mixed with the remaining two-thirds of the feed stream and sent to a reactor in which the following reaction goes to completion:

$$2H_2S + SO_2 \longrightarrow 2 H_2O + 3 S$$

The gases leave the reactor flowing at a rate of 10.0 m³/min at 380°C and 205 kPa absolute. Assuming ideal gas behavior, determine the feed rate of air in kmol/min.

5.48. Sulfuric acid is the chemical produced in the United States with the highest volume of production. In one of the earliest processes used to make it, an ore containing iron pyrites (FeS_2) is roasted (burned) with air. The following reactions take place in the roasting furnace:

$$2\,FeS_2(s) + \tfrac{11}{2}\,O_2(g) \longrightarrow Fe_2O_3(s) + 4\,SO_2(g) \qquad [1]$$

$$2\,FeS_2(s) + \tfrac{15}{2}\,O_2(g) \longrightarrow Fe_2O_3(s) + 4\,SO_3(g) \qquad [2]$$

The gas leaving the reactor goes to a catalytic converter in which most of the SO_2 produced is further oxidized to SO_3:

$$SO_2 + \tfrac{1}{2}O_2 \longrightarrow SO_3 \qquad [3]$$

Finally, the gas leaving the converter passes through an absorption tower, in which the SO_3 is absorbed in water to produce sulfuric acid (H_2SO_4).

(a) An ore containing 82 wt% FeS_2 and 18% inerts is fed to a roasting furnace. Dry air is fed to the furnace in 40% excess of the amount theoretically required to oxidize all of the sulfur in the ore to SO_3. An FeS_2 oxidation of 85% is obtained, with 40% of the FeS_2 converted forming sulfur dioxide and the rest forming sulfur trioxide. Two streams leave the roaster: a gas stream containing SO_2, SO_3, O_2, and N_2, and a solid stream containing unconverted pyrites, ferric oxide, and inert material in the ore. Calculate the required feed rate of air in standard cubic meters per 100 kg of ore roasted and the molar composition and volume (SCM/100 kg of ore) of the gas leaving the roasting oven.

(b) The gas leaving the roasting oven enters the catalytic converter, which operates at 1 atm. The conversion reaction [3] proceeds to an equilibrium point at which the component partial pressures satisfy the relation

$$\frac{p_{SO_3}}{p_{SO_2} p_{O_2}^{1/2}} = K_p(T)$$

The gases are first heated to 600°C, at which $K_p = 9.53$ atm$^{-1/2}$, and are then cooled to 400°C, where $K_p = 397$ atm$^{-1/2}$. The rate of the forward reaction increases sharply with temperature and is several orders of magnitude greater at 600°C than at 400°C. Calculate the equilibrium fractional conversions of sulfur dioxide in the converter when the temperature is 600°C and when it is 400°C. Briefly explain why the converter gases are initially heated and then cooled.

(c) Assuming complete conversion to sulfuric acid of the sulfur trioxide leaving the converter, how many kg of H_2SO_4 will be produced per kg of sulfur in the ore? What would this ratio have been if all of the sulfur in the ore had been converted? Summarize the factors responsible for the second number being larger than the first one.

5.49. You have been assigned the task of measuring the equilibrium constant for the reaction $N_2O_4 \rightleftharpoons 2NO_2$ as a function of temperature. To do so, you obtain a rigid 2-liter vessel equipped with a pressure gauge, evacuate and then fill the vessel with a mixture of NO_2 and N_2O_4, and heat the vessel to $T_0 = 473$ K, a temperature at which you know the gas is essentially pure NO_2. The gauge pressure at this point is noted to be 1.00 atm. You then decrease the temperature in stages, recording the equilibrium gauge pressure at each temperature. The data are as follows:

T (K)	473	350	335	315	300
P_{gauge}(atm)	1.00	0.272	0.111	−0.097	−0.224

$\uparrow$

Pure NO_2

(a) How many gram-moles of NO_2 are in the vessel at 473 K?

(b) The reaction equilibrium constant is

$$K_p = p_{NO_2}^2 / p_{N_2O_4}$$

where p_{NO_2} and $p_{N_2O_4}$ are the equilibrium partial pressures of NO_2 and N_2O_4. Derive an equation or a series of equations for calculating K_p(atm) from specified values of T and P_{gauge}. (*Suggestion*: Begin by defining n_1 and n_2 as the moles of NO_2 and N_2O_4 present at equilibrium.) Then calculate K_p for $T = 350$ K, 335 K, 315 K, and 300 K. (*Suggestion*: Use a spreadsheet program.)

(c) The equilibrium constant should vary with temperature according to the relation

$$K_p = ae^{-b/T}$$

Use the results of part (b) to determine the values of a and b by a graphical curve-fitting procedure. [*Suggestion*: Use the spreadsheet program from part (b).]

5.50. The demand for a particular hydrogenated compound, S, is 5.00 kmol/h. This chemical is synthesized in the gas-phase reaction

$$A + H_2 \rightleftharpoons S$$

The reaction equilibrium constant at the reactor operating temperature is

$$K_p = \frac{p_S}{p_A p_{H_2}} = 0.1 \text{ atm}^{-1}$$

The fresh feed to the process is a mixture of A and hydrogen that is mixed with a recycle stream consisting of the same two species. The resulting mixture, which contains 3 kmol A/kmol H_2, is fed to the reactor, which operates at an absolute pressure of 10.0 atm. The reaction products are in equilibrium. The effluent from the reactor is sent to a separation unit that recovers all of the S in essentially pure form. The A and hydrogen leaving the separation unit form the recycle that is mixed with fresh feed to the process. Calculate the feed rates of hydrogen and A to the process in kmol/h and the recycle stream flow rate in SCMH (standard cubic meters per hour).

5.51. Methanol is synthesized from carbon monoxide and hydrogen in the reaction

$$CO + 2 H_2 \rightleftharpoons CH_3OH$$

A process flowchart is shown below.

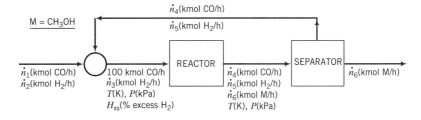

The fresh feed to the system, which contains only CO and H_2, is blended with a recycle stream containing the same species. The combined stream is heated and compressed to a temperature T (K) and a pressure P (kPa) and fed to the reactor. The percentage excess hydrogen in this stream is H_{xs}. The reactor effluent—also at T and P—goes to a separation unit where essentially all of the methanol produced in the reactor is condensed and removed as product. The unreacted CO and H_2 constitute the recycle stream blended with fresh feed.

Provided that the reaction temperature (and hence the rate of reaction) is high enough and the ideal gas equation of state is a reasonable approximation at the reactor outlet conditions (a questionable assumption), the ratio

$$K_{pc} = \frac{p_{CH_3OH}}{p_{CO} p_{H_2}^2}$$

approaches the equilibrium value

$$K_p(T) = 1.390 \times 10^{-4} \exp\left(21.225 + \frac{9143.6}{T} - 7.492 \ln T + 4.076 \times 10^{-3} T - 7.161 \times 10^{-8} T^2\right)$$

In these equations, p_i is the partial pressure of species i in kilopascals (i = CH_3OH, CO, H_2) and T is in Kelvin.

(a) Suppose $P = 5000$ kPa, $T = 500$ K, and $H_{xs} = 5.0\%$. Calculate $\dot{n}_4$, $\dot{n}_5$, and $\dot{n}_6$, the component flow rates (kmol/h) in the reactor effluent. [*Suggestion:* Use the known value of H_{xs}, atomic balances around the reactor, and the equilibrium relationship, $K_{pc} = K_p(T)$, to write four equations in the four variables $\dot{n}_3$ to $\dot{n}_6$; use algebra to eliminate all but $\dot{n}_6$; and use trial and error to solve the remaining nonlinear equation for $\dot{n}_6$.] Then calculate component fresh feed rates ($\dot{n}_1$ and $\dot{n}_2$) and the flow rate (SCMH) of the recycle stream.

*(b) Write a spreadsheet program to perform the calculations of part (a) for the same basis of calculation (100 kmol CO/h fed to the reactor) and different specified values of P (kPa), T (K), and

―――――
*Computer problem.

$H_{xs}(\%)$. The spreadsheet should have the following columns:

A. $P(kPa)$
B. $T(K)$
C. $H_{xs}(\%)$
D. $K_p(T) \times 10^8$. (The given function of T multiplied by 10^8. When $T = 500$ K, the value in this column should be 91.113.)
E. $K_p P^2$
F. $\dot{n}_3$. The rate (kmol/h) at which H_2 enters the reactor.
G. $\dot{n}_4$. The rate (kmol/h) at which CO leaves the reactor.
H. $\dot{n}_5$. The rate (kmol/h) at which H_2 leaves the reactor.
I. $\dot{n}_6$. The rate (kmol/h) at which methanol leaves the reactor.
J. $\dot{n}_{tot}$. The total molar flow rate (kmol/h) of the reactor effluent.
K. $K_{pc} \times 10^8$. The ratio $y_M/(y_{CO} y_{H_2}^2)$ multiplied by 10^8. When the correct solution has been attained, this value should equal the one in Column E.
L. $K_p P^2 - K_{pc} P^2$. Column E − Column K, which equals zero for the correct solution.
M. $\dot{n}_1$. The molar flow rate (kmol/h) of CO in the fresh feed.
N. $\dot{n}_2$. The molar flow rate (kmol/h) of H_2 in the fresh feed.
O. $\dot{V}_{rec}(SCMH)$. The flow rate of the recycle stream in $m^3(STP)/h$.

When the correct formulas have been entered, the value in Column I should be varied until the value in Column L equals 0.
   Run the program for the following nine conditions (three of which are the same):

- $T = 500$ K, $H_{xs} = 5\%$, and $P = 1000$ kPa, 5000 kPa, and 10,000 kPa.
- $P = 5000$ kPa, $H_{xs} = 5\%$, and $T = 400$ K, 500 K, and 600 K.
- $T = 500$ K, $P = 5000$ kPa, and $H_{xs} = 0\%, 5\%$, and 10%.

Summarize the effects of reactor pressure, reactor temperature, and excess hydrogen on the yield of methanol (kmol M produced per 100 kmol CO fed to the reactor).

(c) You should find that the methanol yield increases with increasing pressure and decreasing temperature. What cost is associated with increasing the pressure?

(d) Why might the yield be much lower than the calculated value if the temperature is too low?

(e) If you actually ran the reaction at the given conditions and analyzed the reactor effluent, why might the spreadsheet values in Columns F–M be significantly different from the measured values of these quantities? (Give several reasons, including assumptions made in obtaining the spreadsheet values.)

5.52. One gram-mole each of CO_2, O_2, and N_2 are fed to a batch reactor and heated to 3000 K and 5.0 atm. The two reactions given here proceed to equilibrium (also shown are the equilibrium constants at 3000 K).

$$CO_2 \rightleftharpoons CO + \tfrac{1}{2}O_2$$

$$K_1 = (p_{CO} p_{O_2}^{1/2})/p_{CO_2} = 0.3272 \text{ atm}^{1/2}$$

$$\tfrac{1}{2}O_2 + \tfrac{1}{2}N_2 \rightleftharpoons NO$$

$$K_2 = p_{NO}/(p_{O_2} p_{N_2})^{1/2} = 0.1222$$

Calculate the equilibrium composition (component mole fractions) of the reactor contents. [*Suggestion:* Express K_1 and K_2 in terms of the extents of the two reactions, ξ_1 and ξ_2. (See Section 4.6d.) Then use an equation-solving program or a trial-and-error procedure, such as the Newton–Raphson method (Appendix A.2), to solve for ξ_1 and ξ_2, and use the results to determine the equilibrium mole fractions.]

5.53. Terephthalic acid (TPA), a raw material in the manufacture of polyester fiber, film, and soft drink bottles, is synthesized from p-xylene (PX) in the process shown below.

$$\underbrace{C_8H_{10}}_{PX} + \tfrac{3}{2}O_2 \longrightarrow \underbrace{C_8H_6O}_{TPA} + 2H_2O$$

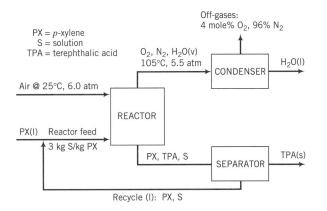

A fresh feed of pure liquid PX combines with a recycle stream containing PX and a solution (S) of a catalyst (a cobalt salt) in a solvent (methanol). The combined stream, which contains S and PX in a 3:1 mass ratio, is fed to a reactor in which 90% of the PX is converted to TPA. A stream of air at 25°C and 6.0 atm absolute is also fed to the reactor. The air bubbles through the liquid and the reaction given above takes place under the influence of the catalyst. A liquid stream containing unreacted PX, dissolved TPA, and all the S that entered the reactor goes to a separator in which solid TPA crystals are formed and filtered out of the solution. The filtrate, which contains all the S and PX leaving the reactor, is the recycle stream. A gas stream containing unreacted oxygen, nitrogen, and the water formed in the reaction leaves the reactor at 105°C and 5.5 atm absolute and goes through a condenser in which essentially all the water is condensed. The uncondensed gas contains 4.0 mole% O_2.

(a) Taking 100 kmol TPA produced/h as a basis of calculation, draw and label a flowchart for the process.

(b) What is the required fresh feed rate (kmol PX/h)?

(c) What are the volumetric flow rates (m^3/h) of the air fed to the reactor, the gas leaving the reactor, and the liquid water leaving the condenser? Assume ideal gas behavior for the two gas streams.

(d) What is the mass flow rate (kg/h) of the recycle stream?

(e) Briefly explain in your own words the functions of the oxygen, nitrogen, catalyst, and solvent in the process.

(f) In the actual process, the liquid condensate stream contains both water and PX. Speculate on what might be done with the latter stream to improve the economics of the process. [*Hint:* Note that PX is expensive, and recall what is said about oil (hydrocarbons) and water.]

5.54. A flowchart of a methanol synthesis process is shown below.

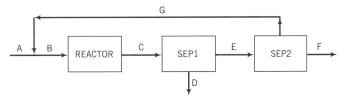

The following specifications apply to the labeled streams and process units:

A. Fresh feed—a mixture of CO, H_2, N_2, and CO_2

B. Feed to the reactor—30.0 mole% CO, 63.0% H_2, 2.0% N_2, and 5.0% CO_2.

Reactor. Two reactions occur and proceed to equilibrium at 200°C and 4925 kPa absolute:

$$CO + 2\,H_2 \rightleftharpoons CH_3OH(M), \qquad (K_p)_1 = \frac{p_M}{p_{CO}\,p_{H_2}^2} = 3.49 \times 10^{-6}\ kPa^{-2}$$

$$CO_2 + 3\,H_2 \rightleftharpoons CH_3OH + H_2O \qquad (K_p)_2 = \frac{p_M\,p_{H_2O}}{p_{CO_2}\,p_{H_2}^3} = 5.19 \times 10^{-8}\ kPa^{-2}$$

C. Reactor effluent—contains all feed and product species at the reactor temperature and pressure. Species partial pressures satisfy the two given equations.

Sep1. Condenses all methanol and water in reactor effluent.

D. Liquid methanol and water. (These species will be separated by distillation in a unit not shown.)

E. Gas containing N_2 and unreacted CO, H_2, and CO_2.

Sep2. Multiple-unit separation process.

F. All of the nitrogen and some of the hydrogen in Stream E.

G. Recycle stream—CO, CO_2, and 10% of the hydrogen fed to Sep2.

(a) Taking 100 kmol/h of Stream B as a basis of calculation, calculate the molar flow rates (kmol/h) and molar compositions of the remaining six labeled streams.

(b) The process is to be used to provide 237 kmol/h of methanol. Scale up the flowchart of part (a) to calculate the required fresh feed rate (SCMH), the flow rate of the reactor effluent (SCMH), and the actual volumetric flow rate of the reactor effluent (m^3/h), assuming ideal gas behavior.

(c) Use the rule of thumb for a diatomic gas given on p. 192 to test the ideal gas assumption at the reactor outlet. If the assumption is invalid, which of the values calculated in part (b) are in error?

5.55. The measured volumetric flow rate of ethane at 10.0 atm absolute and 35°C is 1.00×10^3 L/h. Using an estimated value of the second virial coefficient in the truncated virial equation (Equation 5.3-4), (a) calculate $\hat{V}$ (L/mol); (b) estimate the compressibility factor, z; and (c) determine the mass flow rate of ethane in kg/h.

5.56. Methanol is to be delivered to a process unit at a rate of 15.0 kmol/h by a stream that is 30.0 mole% methanol and 70.0 mole% propane. Estimate the volumetric flow rate of this stream at 10.0 atm and 100.0°C using the truncated virial equation and the following mixing rule:

$$B_{mix} = \sum_i \sum_j y_i y_j B_{ij}$$

where the virial coefficients for the pure species, B_{ii} and B_{jj}, are determined from Equation 5.3-5 and $B_{ij} \approx 0.5(B_{ii} + B_{jj})$.

5.57. The van der Waals equation of state (Equation 5.3-6) is to be used to estimate the specific molar volume $\hat{V}$(L/mol) of air at specified values of T (K) and P (atm). The van der Waals constants for air are $a = 1.33$ atm·L^2/mol^2 and $b = 0.0366$ L/mol.

(a) Show why the van der Waals equation is classified as a cubic equation of state by expressing it in the form

$$f(\hat{V}) = c_3 \hat{V}^3 + c_2 \hat{V}^2 + c_1 \hat{V} + c_0 = 0$$

where the coefficients c_3, c_2, c_1, and c_0 involve P, R, T, a, and b. Calculate the values of these coefficients for air at 223 K and 50.0 atm. (Include the proper units when giving the values.)

(b) What would the value of $\hat{V}$ be if the ideal gas equation of state were used for the calculation? Use this value as an initial estimate of $\hat{V}$ for air at 223 K and 50.0 atm and solve the van der Waals equation by trial and error to obtain a better estimate. What percentage error results from the use of the ideal gas equation of state, taking the van der Waals estimate to be correct?

*(c) Set up a spreadsheet to carry out the calculations of part (b) for air at 223 K and several pressures. The spreadsheet should appear as follows:

T(K)	P(atm)	c3	c2	c1	c0	V(ideal) (L/mol)	V (L/mol)	f(V)	% error
223	1.0	...	...	...	...	...	...	...	...
223	10.0	...	...	...	...	...	...	...	...
223	50.0	...	...	...	...	...	...	...	...
223	100.0	...	...	...	...	...	...	...	...
223	200.0	...	...	...	...	...	...	...	...

*Computer problem.

The polynomial expression for $\hat{V}$ ($f = c_3\hat{V}^3 + c_2\hat{V}^2 + \cdots$) should be entered in the $f(V)$ column, and the value in the V column should be varied until $f(V)$ is essentially zero. Use goal-seeking if the spreadsheet program includes this feature.

(d) Do the calculation for 223 K and 50.0 atm using the Newton-Raphson method (Appendix A.2).

5.58. A 5.0-m^3 tank is charged with 75.0 kg of propane gas at 25°C. Use the SRK equation of state to estimate the pressure in the tank; then calculate the percentage error that would result from the use of the ideal gas equation of state for the calculation.

5.59. The absolute pressure within a 35.0-liter gas cylinder should not exceed 51.0 atm. Suppose the cylinder contains 50.0 mol of a gas. Use the SRK equation of state to calculate the maximum permissible cylinder temperature if the gas is (a) carbon dioxide and (b) argon. Finally, calculate the values that would be predicted by the ideal gas equation of state.

5.60. A stream of oxygen at −65°C and 8.3 atm flows at a rate of 250 kg/h. Use the SRK equation of state to estimate the volumetric flow rate of this stream. (See Example 5.3-3.)

5.61. An innovative engineer has invented a device to replace the hydraulic jacks found at many service stations. A movable piston with a diameter of 0.15 m is fitted into a cylinder. Cars are raised by opening a small door near the base of the cylinder, inserting a block of dry ice (solid CO_2), closing and sealing the door, and vaporizing the dry ice by applying just enough heat to raise the cylinder contents to ambient temperature (25°C). The car is subsequently lowered by opening a valve and venting the cylinder gas.

The device is tested by raising a car a vertical distance of 1.5 m. The combined mass of the piston and the car is 5500 kg. Before the piston rises, the cylinder contains 0.030 m^3 of CO_2 at ambient temperature and pressure (1 atm). Neglect the volume of the dry ice.

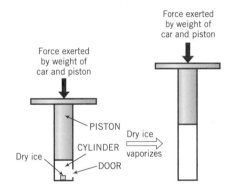

(a) Calculate the pressure in the cylinder when the piston comes to rest at the desired elevation.

(b) How much dry ice (kg) must be placed in the cylinder? Use the SRK equation of state for this calculation.

(c) Outline how you would calculate the minimum piston diameter required for any elevation of the car to occur if the calculated amount of dry ice is added. (Just give formulas and describe the procedure you would follow—no numerical calculations required.)

5.62.[17] An oxygen tank with a volume of 2.5 ft^3 is kept in a room at 50°F. An engineer has used the ideal gas equation of state to determine that if the tank is first evacuated and then charged with 35.3 lb$_m$ of pure oxygen, its rated *maximum allowable working pressure* (MAWP) will be attained. Operation at pressures above this value is considered unsafe.

(a) What is the maximum allowable working pressure (psig) of the tank?

(b) You suspect that at the conditions of the fully charged tank, ideal gas behavior may not be a good assumption. Use the SRK equation of state to obtain a better estimate of the maximum mass of oxygen that may be charged into the tank. Did the ideal gas assumption lead to a conservative

[17]From D. A. Crowl, D. W. Hubbard, and R. M. Felder, *Problem Set: Stoichiometry,* Center for Chemical Process Safety, New York, 1993.

estimate (on the safe side) or a nonconservative estimate of the amount of oxygen that could be charged?

(c) Suppose the tank is charged and ruptures before the amount of oxygen calculated in part (b) enters it. (It should have been able to withstand pressures up to four times the MAWP.) Think of at least five possible explanations for the failure of the tank below its rated pressure limit.

*5.63. Using the SRK equation of state (or any other cubic equation of state) to determine a specific volume from a specified temperature and pressure requires a trial-and-error calculation. Three computer-based approaches to solving this problem may be used: (1) spreadsheeting; (2) mathematical packages such as Mathcad, Mathematica, Maple, and E-Z Solve; and (3) programming languages such as Fortran and C^{++}. The goal of this problem is to use each approach to determine $\hat{V}$(L/mol) for CO_2 at (i) 200 K and 6.8 atm; (ii) 250 K and 12.3 atm; (iii) 300 K and 6.8 atm; (iv) 300 K and 21.5 atm; and (v) 300 K and 50.0 atm.

(a) Starting with Equation 5.3-7, derive the following equivalent expression for the SRK equation of state:

$$f(\hat{V}) = P\hat{V}^3 - RT\hat{V}^2 + (a\alpha - b^2P - bRT)\hat{V} - a\alpha b = 0$$

(b) Write a spreadsheet program to take as inputs a species identifier (such as CO_2), the critical temperature, critical pressure, and Pitzer acentric factor, and the temperatures and pressures for which $\hat{V}$ is to be calculated, and to calculate $\hat{V}$ using Equations 5.3-9 to 5.3-13 for each of the specified conditions. The spreadsheet should have the following structure:

PROBLEM 5.63—SRK EQUATION SPREADSHEET					
Species	CO2				
Tc(K)	304.2				
Pc(atm)	72.9				
w	0.225				
a	*.****				
b	*.****				
m	*.****				
T(K)	P(atm)	alpha	V(ideal)	V(SRK)	f(V)
200	6.8	1.3370	2.4135	2.1125	*.**E−**
250	12.3	*.****	*.****	*.****	*.**E−**
300	6.8	*.****	*.****	*.****	*.**E−**
300	21.5	*.****	*.****	*.****	*.**E−**
300	50.0	*.****	*.****	*.****	*.**E−**

Single digits should appear in place of each asterisk shown on the table. Formulas should be entered into the row for 200 K and 6.8 atm and copied into the next four rows. The goalseek tool should be used to determine each $\hat{V}$(SRK), starting with the ideal gas value and varying the cell value to make $f(\hat{V})$ as close as possible to zero.

---

*Computer problem.

(c) Use a root-finding procedure in a mathematical software package to determine $\hat{V}$ for each of the five conditions.

(d) Write a program (in Fortran or another programming language) to determine $\hat{V}$ for each of the five conditions using Newton's rule (Appendix A.2d). The program should

    (i) Read in values of the species formula (CO_2), the critical temperature and pressure, and the Pitzer acentric factor.

    (ii) Calculate a, b, and m.

    (iii) Read values of T and P for which $\hat{V}$ is to be calculated. Terminate if a negative value of T is input.

    (iv) Use the ideal gas equation of state to generate the initial value of $\hat{V}$.

    (v) Calculate and print successive estimates of $\hat{V}$ using Equation A.2-2, stopping when the fractional change in $\hat{V}$ from one iteration to the next (ϵ of Equation A.2-8) is less than 1.0×10^{-5}. Build in an upper limit of 15 iterations for each process condition; if convergence is not achieved within that limit, print an error message and quit.

    (vi) Go back to Step (iii).

(e) Briefly summarize the advantages and disadvantages of the three problem-solving approaches.

5.64. Use the generalized compressibility chart to estimate z for (a) nitrogen at 40°C and 40 MPa and (b) helium at −200°C and 350 atm. (Don't forget Newton's corrections.)

5.65. A certain gas has a molecular weight of 30.0, a critical temperature of 310 K, and a critical pressure of 4.5 MPa. Calculate the density in kg/m³ of this gas at 465 K and 9.0 MPa (a) if the gas is ideal and (b) if the gas obeys the law of corresponding states.

5.66. One hundred pounds of CO_2 is contained in a 10.0-ft³ tank. The safety limit of the tank is 1600 psig. Use the compressibility chart to estimate the maximum permissible gas temperature.

5.67. A stream of oxygen enters a compressor at 298 K and 1.00 atm at a rate of 127 m³/h and is compressed to 358 K and 1000 atm. Estimate the volumetric flow rate of compressed O_2, using the compressibility-factor equation of state.

5.68. A 10-liter cylinder containing oxygen at 175 atm absolute is used to supply O_2 to an oxygen tent. The cylinder can be used until its absolute pressure drops to 1.1 atm. Assuming a constant temperature of 27°C, calculate the gram-moles of O_2 that can be obtained from the cylinder, using the compressibility-factor equation of state when appropriate.

5.69. Dry ice (solid CO_2) has been used as a mine explosive in the following manner. A hole is drilled into the mine wall, filled with dry ice plus a small charge of gunpowder, and then plugged. The gunpowder is lit with a fuse, vaporizing the CO_2 and building up an explosively high pressure within the hole. Use each of the following correlations to estimate the pressure that will develop if 5.00 g of dry ice is placed in a 50.0-mL hole and heated to 1000 K: (a) the ideal gas equation of state, (b) the compressibility-factor equation of state, and (c) the SRK equation of state.

5.70. The concentration of oxygen in a 5000-liter tank containing air at 1 atm is to be reduced by pressure purging prior to charging a fuel into the tank. The tank is charged with nitrogen up to a high pressure and then vented back down to atmospheric pressure. The process is repeated as many times as required to bring the oxygen concentration below 10 ppm (i.e., to bring the mole fraction of O_2 below 10.0×10^{-6}). Assume that the temperature is 25°C at the beginning and end of each charging cycle.

When doing *PVT* calculations in parts (b) and (c), use the generalized compressibility chart if possible for the fully charged tank and assume that the tank contains pure nitrogen.

(a) Speculate on why the tank is being purged.

(b) Estimate the gauge pressure (atm) to which the tank must be charged if the purge is to be done in one charge–vent cycle. Then estimate the mass of nitrogen (kg) used in the process. (For this part, if you can't find the tank condition on the compressibility chart, assume ideal gas behavior and state whether the resulting estimate of the pressure is too high or too low.)

(c) Suppose nitrogen at 700 kPa gauge is used for the charging. Calculate the number of charge–vent cycles required and the total mass of nitrogen used.

(d) Use your results to explain why multiple cycles at a lower gas pressure are preferable to a single cycle. What is a probable disadvantage of multiple cycles?

5.71. A stream of propane at an average temperature $T = 566°R$ and absolute pressure $P = 6.8$ atm flows from a hydrocarbon processing plant to a nearby customer's production facility. A technician at the processing plant periodically measures the volumetric flow rate of the stream, $\dot{V}(\text{ft}^3/\text{h})$, and reports the value to the billing office. The customer is charged for the propane at a rate

$$C(\$/h) = 60.4\frac{SP\dot{V}}{T}$$

where $S(\$/\text{lb}_m)$ is the unit cost of propane.
(a) Derive the given formula, assuming ideal gas behavior.
(b) One day a recent chemical engineering graduate working at the processing plant comes across the formula used to calculate the charge for the propane stream. She figures out where the formula came from and uses the generalized compressibility chart to derive an improved formula. What is her result? Calculate the percentage by which the plant was overcharging or undercharging the customer (state which) when it used the old formula.

5.72. Approximately 150 SCFM (standard cubic feet per minute) of nitrogen is required by a process facility. As shown in the diagram below, plans call for supplying the facility from a tank of liquid nitrogen (SG = 0.81) at its normal boiling point ($-350°F$) and 1 atm. Nitrogen vapor leaves the tank and is compressed and heated to obtain the desired conditions, 150°F and 600 psia.
(a) Using the generalized compressibility charts, determine the volumetric flow rate of nitrogen delivered from the heater.
(b) What would the required minimum tank size be if deliveries are made to the site no more frequently than every two weeks?

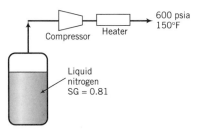

5.73. A 150-liter cylinder of carbon monoxide is stored in a 30.7-m³ room. The pressure gauge on the tank reads 2500 psi when the tank is delivered. Sixty hours later the gauge reads 2245 psi, indicating a leak. The *Ceiling Threshold Limit Value* (TLV-C) molar concentration of CO—that is, the concentration considered unsafe for even instantaneous human exposure—is 200 ppm (200×10^{-6} mol CO/mol room air). The temperature of the room is constant at 27°C.
(a) Estimate the average leak rate (mol CO/h). Do not assume that the gas in the cylinder behaves ideally.
(b) Calculate $t_{\min}$(h), the minimum time from delivery at which the average concentration of CO in the room could have reached the TLV-C concentration. Explain why the actual time to reach this concentration would be greater.
(c) Why could it be disastrous to enter the room at any time without proper personal protective equipment on, even at a time $t < t_{\min}$? (Think of at least three possible reasons.)

5.74. A gas consists of 20.0 mole% CH_4, 30.0% C_2H_6, and 50.0% C_2H_4. Ten kilograms of this gas is to be compressed to a pressure of 200 bar at 90°C. Using Kay's rule, estimate the final volume of the gas.

5.75. A 30-liter cylinder is evacuated and filled with 5.00 kg of a gas containing 10.0 mole% N_2O and the balance N_2. The gas temperature is 24°C. Use the compressibility chart to solve the following problems.
(a) What is the gauge pressure (atm) of the cylinder gas after the tank is filled?
(b) A fire breaks out in the plant where the cylinder is kept, and the cylinder valve ruptures when the gas gauge pressure reaches 273 atm. What was the gas temperature (°C) at the moment before the rupture occurred?

5.76. The product gas from a coal gasification plant consists of 60.0 mole% CO and the balance H_2; it leaves the plant at 150°C and 2000 psia. The gas expands through a turbine, and the outlet gas from the turbine is fed to a boiler furnace at 100°C and 1 atm at a rate of 15,000 ft³/min. Estimate the inlet flow rate to the turbine in ft³/min, using Kay's rule. What percentage error would result from the use of the ideal gas equation of state at the turbine inlet?

5.77. A 30.0-liter cylinder of a gas containing 97.0 mole% CO and 3.0% CO_2 is delivered to your plant. You sign the receipt for it, noting that the gauge on the tank reads 2000 psi. Several days later you notice that the gauge reads 1875 psi, indicating a leak. The storage room in which the cylinder is kept has a volume of 24.2 m³ and is very poorly ventilated. Calculate the maximum mole% of CO in the room at the time the leak is discovered, assuming that the leaking gas spreads uniformly throughout the room and that the room temperature is constant at 30°C, using Kay's rule when appropriate.

5.78. Methanol is produced by reacting carbon monoxide and hydrogen at 644 K over a ZnO–Cr_2O_3 catalyst. A mixture of CO and H_2 in a ratio 2 mol H_2/mol CO is compressed and fed to the catalyst bed at 644 K and 34.5 MPa absolute. A single-pass conversion of 25% is obtained. The **space velocity**, or ratio of the volumetric flow rate of the feed gas to the volume of the catalyst bed, is (25,000 m³/h)/(1 m³ catalyst bed). The product gases are passed through a condenser, in which the methanol is liquefied.

    (a) You are designing a reactor to produce 54.5 kmol CH_3OH/h. Estimate the volumetric flow rate that the compressor must be capable of delivering if no gases are recycled, and the required volume of the catalyst bed. (Use Kay's rule for pressure–volume calculations.)

    (b) If (as is done in practice) the gases from the condenser are recycled to the reactor, the compressor is then required to deliver only the fresh feed. What volumetric flow rate must it deliver assuming that the methanol produced is completely recovered in the condenser? (In practice it is not; moreover, a purge stream must be taken off to prevent the buildup of impurities in the system.)

5.79. A process stream flowing at 35 kmol/h contains 15 mole% hydrogen and the remainder 1-butene. The stream pressure is 10.0 atm absolute, the temperature is 50°C, and the velocity is 150 m/min. Determine the diameter (in cm) of the pipe transporting this stream, using Kay's rule in your calculations.

5.80. A gas mixture consisting of 15.0 mole% methane, 60.0% ethylene, and 25.0% ethane is compressed to a pressure of 175 bar at 90°C. It flows through a process line in which the velocity should be no greater than 10 m/s. What flow rate (kmol/min) of the mixture can be handled by a 2-cm internal diameter pipe?

5.81.[18] A system has been devised to store acetonitrile safely at high pressures and temperatures. The acetonitrile is contained in a 0.2-ft³ tank maintained at 4500 psia and 550°F. This tank is placed inside a second tank whose volume, excluding the volume of the first tank, is 2 ft³. Nitrogen at 10.0 atm and 550°F is contained in the second tank. Use the compressibility chart to estimate the final system pressure (atm) if the first tank ruptures and the final system temperature is 550°F. The critical temperature and pressure of acetonitrile are 548 K and 47.7 atm, respectively.

5.82. A solid carbohydrate ($C_aH_bO_c$) with a specific gravity of 1.59 is placed in a 1.000-liter combustion chamber. The chamber is evacuated and then charged with pure oxygen. Complete combustion of the carbohydrate takes place. A sample of the product gas is cooled to condense all the water formed in the combustion, and the remaining gas is analyzed by gas chromatography. The following data are obtained:

Mass of Carbohydrate Charged: 3.42 g

Chamber Conditions Prior to Combustion: $T = 26.8°C$, $P = 499.9$ kPa

Chamber Conditions Following Combustion: $T = 483.4°C$, $P = 1950.0$ kPa

Product Gas Analysis: 38.7 mole% CO_2, 25.8% O_2, 35.5% H_2O

Assume (i) none of the carbohydrate is lost when the chamber is evacuated and (ii) the vapor pressure of the carbohydrate at 27°C is negligible. Do *not* neglect the volume of the carbohydrate and do not assume ideal gas behavior.

---

[18]This problem is adapted from *Professional Engineering Examinations,* Volume 1, National Council of Engineering Examiners, 1972, p. 347.

     (a) Determine at least two possible molecular formulas for the carbohydrate (i.e., sets of integer values of a, b, and c) consistent with the given data.

     (b) If the molecular weight of the carbohydrate is independently determined to be in the range 300 to 350, what is its molecular formula?

5.83. The **adiabatic flame temperature** of a fuel is the temperature achieved if the fuel is burned completely in a perfectly insulated container.

     You are performing an experiment to measure the adiabatic flame temperature of cyclopentane. You put 10.0 mL of liquid cyclopentane into a well-insulated steel vessel with a volume of 11.2 L and pressurize the vessel with air to achieve a stoichiometric ratio of oxygen to cyclopentane. You then ignite the fuel, planning to record the final temperature. The vessel is equipped with a thermocouple and a pressure gauge.

     (a) If room temperature is 27°C and barometric pressure is 1.00 bar, what should the pressure gauge read prior to ignition?

     (b) Suppose you discover after the combustion has been carried out that the thermocouple is not functioning properly. Use the final pressure gauge reading of 75.3 bar to estimate the adiabatic flame temperature of cyclopentane. Do not assume ideal gas behavior.

Chapter 6

Multiphase Systems

Virtually all commercial chemical processes involve operations in which material is transferred from one phase (gas, liquid, or solid) into another. These multiphase operations include all **phase-change operations** on a single species, such as freezing, melting, evaporation, and condensation, and most **separation** and **purification** processes, which are designed to separate components of mixtures from one another. Most separations are accomplished by feeding a mixture of species A and B into a two-phase system under conditions such that most of the A remains in its original phase and most of the B transfers into a second phase. The two phases then either separate themselves under the influence of gravity—as when gases and liquids or two immiscible liquids separate—or are separated with the aid of a device such as a filter or a skimmer.

Here are some examples of multiphase separation processes.

- *Brewing a cup of coffee.* Hot liquid water and solid ground coffee beans are contacted. Soluble constituents of the beans are transferred from the solid phase to a liquid solution (coffee), and then the residual solids (grounds) are filtered from the solution. The operation of dissolving a component of a solid phase in a liquid solvent is referred to as **leaching**.
- *Removal of sulfur dioxide from a gas stream.* If a fuel that contains sulfur is burned, the product gas contains sulfur dioxide. If the gas is released directly into the atmosphere, the SO_2 combines with atmospheric oxygen to form sulfur trioxide. The SO_3 in turn combines with water vapor in the atmosphere to form sulfuric acid (H_2SO_4), which eventually precipitates as *acid rain*. To prevent this occurrence, the combustion product gas is contacted with a liquid solution in an **absorption** or **scrubbing** process. The SO_2 dissolves in the solvent and the clean gas that remains is released to the atmosphere.
- *Recovery of methanol from an aqueous solution.* After being used as a reactant or solvent, methanol (methyl alcohol) often leaves a process in an aqueous mixture (combined with water). Methanol has a higher *vapor pressure* than water, meaning that it has a greater tendency to vaporize when a mixture of the two species is heated. The separation process **distillation** exploits this difference by partially vaporizing a liquid mixture, yielding a vapor relatively rich in methanol and a residual liquid relatively rich in water. Subsequent partial condensations and vaporizations can be used to recover almost pure methanol. The recovered methanol can be recycled and reused, resulting in considerable savings in raw material costs.
- *Separation of paraffinic and aromatic hydrocarbons.* Liquid paraffinic hydrocarbons (such as pentane, hexane, and heptane) and liquid aromatic hydrocarbons (such as benzene, toluene, and xylene) have different chemical characteristics; for example, the paraffinic compounds are almost completely immiscible with liquid ethylene glycol, while aromatic compounds and ethylene glycol readily form homogeneous liquid mixtures. Paraffinics and aromatics may therefore be separated from each other by blending a mixture of the two

237

with ethylene glycol. When allowed to settle, the aromatic compounds distribute between a paraffin-rich phase and a glycol phase. This process is known as **liquid extraction**. Subsequent processing separates the aromatics from the glycol, recovering the glycol for recycle and reuse in the extraction process.

- **Separation of an isomeric mixture.** *Para*-xylene, a constituent in the synthesis of polyesters, must be separated from two of its isomers, *ortho*- and *meta*-xylene.

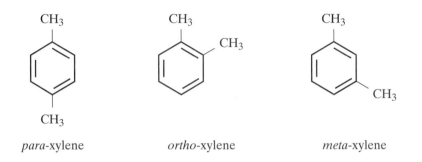

para-xylene   *ortho*-xylene   *meta*-xylene

Two alternative commercial operations have been developed to perform the separation. In one, a mixture of the isomers is contacted with a *molecular sieve* that has pores large enough to accommodate *para*-xylene but not the *meta* and *ortho* isomers. This operation is referred to as **adsorption**. In another process, the difference in freezing points of the three isomers (*para*-xylene freezes at 13.3°C, *ortho* at −25.2°C, and *meta* at −47.9°C) forms the basis of a **crystallization** operation. The mixture is cooled to a temperature at which *para* crystallizes and can then be separated physically from the remaining *ortho* and *meta* liquid.

When a species transfers from one phase to another, the transfer rate generally decreases with time until the second phase is **saturated** with the species, holding as much as it can hold at the prevailing process conditions. When the concentrations of all species in each phase no longer change with time, the phases are said to be in **phase equilibrium**. The effectiveness of any of the separation processes described above depends both on how species are distributed between the phases at equilibrium and on the rate at which the system approaches equilibrium from its initial state.

TEST YOURSELF

Suggest a method that might be suitable to achieve each of the following separations.

1. Separate crude oil into volatile low molecular weight compounds (naphthas used to make gasoline and light chemicals), intermediate molecular weight compounds (used for heating oils), and nonvolatile high molecular weight compounds (used for lubricating oils).
2. Remove water from an aqueous slurry of bleached wood pulp.
3. Obtain fresh water from seawater.
4. Separate NH_3 from a mixture of N_2, H_2, and NH_3. Ammonia is highly soluble in water; also, it condenses at −33.4°C.
5. Concentrate O_2 for breathing-impaired patients.

CREATIVITY EXERCISE

A gas contains two species, A and B. Suggest as many methods as you can think of, both conventional and unconventional, for separating the two species. Briefly indicate the conditions required for each method to work. (For example, find a third substance, C, that reacts with A to form a solid, and introduce C into the mixture. The A will react and the product will deposit as a solid, leaving B in the gas phase.)

6.0   INSTRUCTIONAL OBJECTIVES

After completing this chapter, you should be able to do the following:

- Explain in your own words the terms *separation process, distillation, absorption, scrubbing, liquid extraction, crystallization, adsorption,* and *leaching.* (What are they and how do they work?)
- Sketch a phase diagram (*P* versus *T*) for a single species and label the regions (solid, liquid, vapor, gas). Explain the difference between a vapor and a gas. Use the phase diagram to define (a) the vapor pressure at a specified temperature, (b) the boiling point at a specified pressure, (c) the normal boiling point, (d) the melting point at a specified pressure, (e) the sublimation point at a specified pressure, (f) the triple point, and (h) the critical temperature and pressure. Explain how the melting and boiling point temperatures of water vary with pressure and how *P* and *T* vary (increase, decrease, or remain constant) as a specified path on the diagram is followed.
- Estimate the vapor pressure of a pure substance at a specified temperature or the boiling point at a specified pressure using (a) the Antoine equation, (b) the Cox chart, (c) the Clausius–Clapeyron equation and known vapor pressures at two specified temperatures, or (d) Table B.3 for water.
- Distinguish between intensive and extensive variables, giving examples of each. Use the Gibbs phase rule to determine the number of degrees of freedom for a multicomponent multiphase system at equilibrium, and state the meaning of the value you calculate in terms of the system's intensive variables. Specify a feasible set of intensive variables that will enable the remaining intensive variables to be calculated.
- In the context of a system containing a single condensable species and noncondensable gases, explain in your own words the terms *saturated vapor, superheated vapor, dew point, degrees of superheat,* and *relative saturation.* Explain the following statement from a weather report in terms a first-year engineering student could understand: *The temperature is 75°F, barometric pressure is 29.87 inches of mercury and falling, the relative humidity is 50%, and the dew point is 54°F.*
- Given an equilibrated gas–liquid system containing only a single condensable component A, a correlation for $p_A^*(T)$, and any two of the variables y_A (mole fraction of A in the gas phase), temperature, and total pressure, calculate the third variable using Raoult's law.
- Given a mixture of a single condensable vapor, A, and one or more noncondensable gases, a correlation for $p_A^*(T)$, and any two of the variables y_A (mole fraction of A), temperature, total pressure, dew point, degrees of superheat, and relative, molal, absolute, and percentage saturation (or humidity if A is water and the noncondensable gas is air), use Raoult's law for a single condensable species to calculate the remaining variables.
- For a process system that involves a single condensable component, a vapor-liquid phase change, and specified or requested values of feed or product stream properties (temperature, pressure, dew point, relative saturation or humidity, degrees of superheat, etc.), draw and label the flowchart, carry out the degree-of-freedom analysis, and perform the required calculations.
- Explain the meaning of the term *ideal solution behavior* applied to a liquid mixture of volatile species. Write and clearly explain the formulas for Raoult's law and Henry's law, state the conditions for which each relationship is most likely to be accurate, and apply the appropriate one to determine any of the variables T, P, x_A, or y_A (temperature, pressure, and mole fractions of A in the liquid and gas phases) from given values of the other three.
- Explain in your own words the terms *bubble point, boiling point,* and *dew point* of a mixture of condensable species, and the difference between *vaporization* and *boiling.* Use Raoult's law to determine (a) the bubble-point temperature (or pressure) of a liquid mixture of known composition at a specified pressure (or temperature) and the composition of the first bubble

that forms; (b) the dew-point temperature (or pressure) of a vapor mixture of known composition at a specified pressure (or temperature) and the composition of the first liquid drop that forms; (c) whether a mixture of known amount (moles) and composition (component mole fractions) at a given temperature and pressure is a liquid, a gas, or a gas–liquid mixture and, if the latter, the amounts and compositions of each phase; and (d) the boiling point temperature of liquid mixture of known composition at a specified total pressure.

- Use a *Txy* or *Pxy* diagram to determine bubble- and dew-point temperatures and pressures, compositions and relative amounts of each phase in a two-phase mixture, and the effects of varying temperature and pressure on bubble points, dew points, and phase amounts and compositions. Outline how the diagrams are constructed for mixtures of components that obey Raoult's law.

- For a process system that involves liquid and gas streams in equilibrium and vapor–liquid equilibrium relations for all distributed components, draw and label the flowchart, carry out the degree-of-freedom analysis, and perform the required calculations.

- Explain in your own words the terms *solubility* of a solid in a liquid, *saturated solution,* and *hydrated salt.* Given solubility data, determine the saturation temperature of a feed solution of given composition and the quantity of solid crystals that form if the solution is cooled to a specified temperature below the saturation point.

- Given a liquid solution of a nonvolatile solute, estimate the solvent vapor-pressure lowering, the boiling-point elevation, and the freezing-point depression, and list the assumptions required for your estimate to be accurate.

- Explain the term *distribution coefficient* (or *partition ratio*) for a solute distributed between two nearly immiscible liquids. Given feed-stream flow rates and compositions for a liquid extraction process and either solute distribution coefficient data or a triangular phase diagram, calculate the product stream flow rates and compositions.

- Explain the term *adsorption isotherm.* Given adsorption equilibrium data or an expression for an adsorption isotherm, calculate the maximum quantity of adsorbate that can be removed from a gas by a specified quantity of adsorbent or, conversely, the minimum quantity of adsorbent needed to remove a specified quantity of adsorbate.

6.1 SINGLE-COMPONENT PHASE EQUILIBRIUM

6.1a Phase Diagrams

At most temperatures and pressures, a single pure substance at equilibrium exists entirely as a solid, liquid, or gas; but at certain temperatures and pressures, two and even all three phases may coexist. Pure water is a gas at 130°C and 100 mm Hg, for example, and a solid at −40°C and 10 atm, but at 100°C and 1 atm it may be a gas, a liquid, or a mixture of both, and at approximately 0.0098°C and 4.58 mm Hg it may be a solid, a liquid, a gas, or any combination of the three.

A **phase diagram** of a pure substance is a plot of one system variable against another that shows the conditions at which the substance exists as a solid, a liquid, and a gas. The most common of these diagrams plots pressure on the vertical axis versus temperature on the horizontal axis. The boundaries between the single-phase regions represent the pressures and temperatures at which two phases may coexist. The phase diagrams of water and carbon dioxide are shown in Figure 6.1-1.

What the phase diagram means and what can be done with it are illustrated by a hypothetical experiment in which pure water is placed in a leakproof evacuated cylinder fitted with a movable piston, as shown in the diagram below. Heat can be added to or withdrawn from the cylinder, so that the temperature in the chamber can be adjusted to any desired value, and the absolute pressure of the cylinder contents [which equals $(F + W)/A$, where W is the weight of the piston] can similarly be adjusted by varying the force F on the piston.

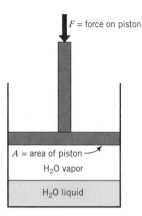

Suppose the system is initially at 20°C, and the force is set at a value such that the system pressure is 3 mm Hg. As the phase diagram shows, water can only exist as a vapor at these conditions, so any liquid that may initially have been in the chamber evaporates, until finally the chamber contains only water vapor at 20°C and 3 mm Hg (point A on Figure 6.1-1a).

Now suppose the force on the piston is slowly increased with the system temperature held constant at 20°C until the pressure in the cylinder reaches 760 mm Hg, and thereafter heat is added to the system with the pressure remaining constant until the temperature reaches 130°C. The state of the water throughout this process can be determined by following path A → B → C → D → E on Figure 6.1-1a. The conditions of the system at various stages of the process are shown in the diagram on the next page.

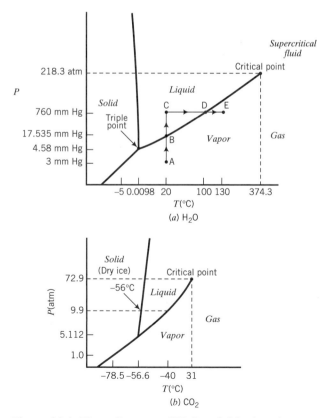

Figure 6.1-1 Phase diagrams of H_2O and CO_2 (not drawn to scale).

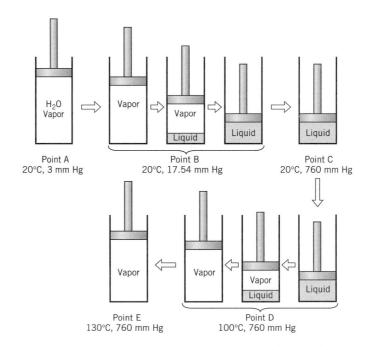

Point A
20°C, 3 mm Hg

Point B
20°C, 17.54 mm Hg

Point C
20°C, 760 mm Hg

Point E
130°C, 760 mm Hg

Point D
100°C, 760 mm Hg

Notice that the phase transitions—condensation at point B and evaporation at point D—take place at boundaries on the phase diagram; the system cannot move off these boundaries until the transitions are complete.

Several familiar terms may be defined with reference to the phase diagram.

1. If T and P correspond to a point on the vapor–liquid equilibrium curve for a substance, P is the **vapor pressure** of the substance at temperature T, and T is the **boiling point** (more precisely, the **boiling point temperature**) of the substance at pressure P.

2. The boiling point of a substance at $P = 1$ atm is the **normal boiling point** of that substance.

3. If (T, P) falls on the solid–liquid equilibrium curve, then T is the **melting point** or **freezing point** at pressure P.

4. If (T, P) falls on the solid–vapor equilibrium curve, then P is the vapor pressure of the solid at temperature T, and T is the **sublimation point** at pressure P.

5. The point (T, P) at which solid, liquid, and vapor phases can all coexist is called the **triple point** of the substance.

6. The vapor–liquid equilibrium curve terminates at the **critical temperature** and **critical pressure** (T_c and P_c). Above and to the right of the critical point, two separate phases never coexist.

As you can see from Figure 6.1-1, the freezing point of water decreases with increasing pressure. (*Verify.*) This behavior is extremely unusual; most substances, including carbon dioxide, exhibit the opposite behavior. Note also, however, that the changes are very slight; the solid–liquid equilibrium curves are in fact nearly vertical. Their slopes have been exaggerated in Figure 6.1-1 for illustrative purposes.

Normal boiling and melting points for many substances are given in Table B.1 of Appendix B and for many more substances on pp. 2-7 through 2-47 of *Perry's Chemical Engineers' Handbook,*[1] and vapor pressures are tabulated on pp. 2-48 through 2-75 of the *Handbook.* Techniques for estimating vapor pressures are presented in Section 6.1b.

---

[1]R. H. Perry and D. W. Green, Eds., *Perry's Chemical Engineers' Handbook,* 7th Edition, McGraw-Hill, New York, 1997

TEST YOURSELF

(Referring to Figure 6.1-1)

1. What is the sublimation point of H_2O at 3 mm Hg? What is the vapor pressure of ice at $-5°C$?
2. What is the triple point of CO_2?
3. Describe what happens when the pressure of pure CO_2 is raised from 1 atm to 9.9 atm at $-78.5°C$, and then the temperature is raised from $-78.5°C$ to $0°C$ at 9.9 atm.
4. What is the vapor pressure of CO_2 at $-78.5°C$? At $-40°C$?
5. What is the sublimation point of CO_2 at 1 atm? The melting point at 9.9 atm? The boiling point at the latter pressure?
6. Does the state of water at point E in Figure 6.1-1a depend on the path followed in changing the temperature and pressure at point A to the values at point E?

6.1b   Estimation of Vapor Pressures

The **volatility** of a species is the degree to which the species tends to transfer from the liquid (or solid) state to the vapor state. At a given temperature and pressure, a highly volatile substance is much more likely to be found as a vapor than is a substance with low volatility, which is more likely to be found in a condensed phase (liquid or solid).

Separation processes such as distillation are used to separate more volatile species from less volatile species by partially vaporizing liquid mixtures. The vapor product is relatively rich in the more volatile feed components and the residual liquid is rich in the components with lower volatility. *The vapor pressure of a species is a measure of its volatility:* the higher the vapor pressure at a given temperature, the greater the volatility of the species at that temperature. Engineers who design and analyze separation processes therefore need to know the vapor pressures of process species as functions of temperature.

It often happens that tabulated vapor pressure data are not available at temperatures of interest, or they may not be available at all for a given species. One solution to this problem is to measure p^* at the desired temperatures. Doing so is not always convenient, however, especially if a highly precise value is not required. An alternative is to *estimate* the vapor pressure using an empirical correlation for $p^*(T)$. Reid, Prausnitz, and Poling[2] summarize and compare vapor pressure estimation methods, several of which are given in the paragraphs that follow.

A relationship between p^*, the vapor pressure of a pure substance, and T, the absolute temperature, is the **Clapeyron equation**

$$\frac{dp^*}{dT} = \frac{\Delta \hat{H}_v}{T(\hat{V}_g - \hat{V}_l)} \tag{6.1-1}$$

where T is *absolute* temperature; $\hat{V}_g$ and $\hat{V}_l$ are the specific molar volumes (volume/mole) of gas (vapor) and liquid, respectively; and $\Delta \hat{H}_v$ is the **latent heat of vaporization**, or the energy required to vaporize one mole of the liquid (to be defined more precisely in Chapter 8).

Unless the pressure is extremely high, the specific volume of the liquid is negligible relative to that of the vapor (i.e. $\hat{V}_g - \hat{V}_l \approx \hat{V}_g$). If we assume that this is the case, apply the ideal gas equation of state to the vapor (so that $\hat{V}_g$ is replaced with RT/p^* in Equation 6.1-1) and rearrange the resulting equation with the aid of elementary calculus. We obtain

$$\frac{d(\ln p^*)}{d(1/T)} = -\frac{\Delta \hat{H}_v}{R} \tag{6.1-2}$$

---

[2]R. C. Reid, J. H. Prausnitz, and B. E. Poling, *The Properties of Gases and Liquids,* 4th Edition, McGraw-Hill, New York, 1986.

(Convince yourself that Equation 6.1-2 is correct by working backward to derive Equation 6.1-1.)

If the vapor pressure of a substance is measured at several temperatures and $\ln p^*$ is plotted versus $1/T$ (or p^* is plotted versus $1/T$ on semilog axes), then from Equation 6.1-2, the slope of the resulting curve at a given temperature equals $-\Delta \hat{H}_v/R$. This is the method most commonly used to determine heats of vaporization experimentally.

Suppose now that the heat of vaporization of a substance is independent of temperature (or nearly so) in the temperature range over which vapor pressures are available. Equation 6.1-2 may then be integrated to yield the **Clausius–Clapeyron equation**

$$\ln p^* = -\frac{\Delta \hat{H}_v}{RT} + B \qquad \text{(6.1-3)}$$

where B is a constant that varies from one substance to another. According to this equation, a plot of $\ln p^*$ versus $1/T$ (or a semilog plot of p^* versus $1/T$) should be a straight line with slope $-\Delta \hat{H}_v/R$ and intercept B.

If you know $\Delta \hat{H}_v$ and p^* at a single temperature T_0, you can solve the Clausius–Clapeyron equation for B and thereafter use this equation to estimate p^* at any temperature close to T_0. If you have p^* versus T data, you can plot $\ln p^*$ versus $1/T$ and determine $\Delta \hat{H}_v/R$ and B graphically or by the method of least squares (see Appendix A.1).

EXAMPLE 6.1-1   *Vapor Pressure Estimation Using the Clausius–Clapeyron Equation*

The vapor pressure of benzene is measured at two temperatures, with the following results:

$$T_1 = 7.6°C, \qquad p_1^* = 40 \text{ mm Hg}$$

$$T_2 = 15.4°C, \qquad p_2^* = 60 \text{ mm Hg}$$

Calculate the latent heat of vaporization and the parameter B in the Clausius–Clapeyron equation and then estimate p^* at 42.2°C using this equation.

SOLUTION

p^*(mm Hg)	T(°C)	T(K)
40	7.6	280.8
60	15.4	288.6

The slope of the line through the two data points on a plot of $\ln p^*$ versus $1/T$ is

$$-\frac{\Delta \hat{H}_v}{R} = \frac{\ln(p_2^*/p_1^*)}{[(1/T_2) - (1/T_1)]} = \frac{T_1 T_2 \ln(p_2^*/p_1^*)}{(T_1 - T_2)}$$

$$= \frac{(280.8 \text{ K})(288.6 \text{ K})\ \ln(60 \text{ mm Hg}/40 \text{ mm Hg})}{(280.8 - 288.6) \text{ K}} = -4213 \text{ K}$$

The intercept B is obtained from Equation 6.1-3 as

$$B = \ln p_1^* + \frac{\Delta \hat{H}_v}{RT_1}$$

$$= \ln 40 + (4213/280.8) = 18.69$$

The Clausius–Clapeyron equation is therefore

$$\boxed{\ln p^* = -\frac{4213 \text{ K}}{T(\text{K})} + 18.69} \quad p^* \text{ in mm Hg}$$

Check: T = 15.4°C $\Longrightarrow$ 288.6 K

$$\ln p^* = -\frac{4213}{288.6} + 18.69 = 4.093$$

$$\Downarrow$$

$$p^* = \exp(4.093) = 60 \text{ mm Hg} \quad \checkmark$$

Finally, at $T = 42.2°C = 315.4$ K

$$\ln p^* = -\frac{4213}{315.4} + 18.69 = 5.334$$

$$\Downarrow$$

$$p^* = \exp(5.334) = \boxed{207 \text{ mm Hg}}$$

Perry's Chemical Engineers' Handbook, p. 2-61, lists the vapor pressure of benzene at 42.2°C as 200 mm Hg, so that the use of the Clausius–Clapeyron equation results in an estimation error of approximately 3.5%.

The heat of vaporization of benzene $\Delta\hat{H}_v$ may be estimated from the slope of the Clausius–Clapeyron plot $(-\Delta\hat{H}_v/R)$ as

$$\Delta\hat{H}_v = (\Delta\hat{H}_v/R)(R)$$

$$= \frac{4213 \text{ K}}{} \frac{8.314 \text{ J}}{\text{mol·K}} = \boxed{35{,}030 \text{ J/mol}}$$

(The true value is approximately 31,000 J/mol.)

Chemical engineers often encounter a need to know the vapor pressure of a species at a specified temperature. Tables of p^* at different temperatures exist for many species, but it is difficult to interpolate between tabulated values since p^* varies sharply with T. Plots of p^* versus T for different species would not be particularly useful since many closely spaced data points would be required to generate each plot, and the curvature of the plots would make it awkward to show data for many different species on a single graph.

Fortunately, while the temperature dependence of vapor pressure may be highly nonlinear, the logarithm of p^* varies with T in roughly the same way for a large number of species. A log plot of the vapor pressure of a species at a given temperature versus the vapor pressure of a reference species at the same temperature consequently tends to be linear. This observation gives rise to the **equal-temperature reference-substance plot**. If you have values of p^* for a species at two or more temperatures, you can look up the vapor pressures of a reference substance (usually water) at the same temperature and plot $p^*(T)$ versus $p_{ref}^*(T)$ on log axes. You can draw a straight line through the points plotted in this manner, and then, with fair accuracy, use the plot to estimate p^* for any temperature from the known value of p_{ref}^* at the same temperature, as shown in Figure 6.1-2.

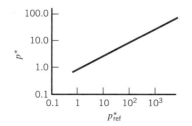

Figure 6.1-2 Reference-substance plot for vapor pressure correlation.

An additional step is normally taken in the construction of plots of this type. The usual use of the plot is to determine p^* for a given T; p_{ref}^* is only an intermediate quantity used to relate these variables. The necessity of looking up p_{ref}^* for each given temperature can be avoided if values of $T(p_{\text{ref}}^*)$ are shown on a second abscissa scale (see Figure 6.1-3). Now to find $p^*(T)$ you need only find T on the new abscissa scale; the value of $p_{\text{ref}}^*(T)$ will be located at the same abscissa value, and the curve of p^* versus p_{ref}^* may then be used to determine $p^*(T)$.

Notice, however, that there is no longer a need for the p_{ref}^* scale, since once you find T on the abscissa you can proceed directly to the curve. The p_{ref}^* scale can therefore be omitted, leaving what has come to be called a **Cox chart**—log p^* on the ordinate and the temperature scale on the abscissa, as shown in Figure 6.1-3. Special graph paper—sometimes called **vapor pressure paper**—that has these coordinate axes is available commercially. Cox chart plots for various substances are shown in Figure 6.1-4.

It is also possible to get straight-line correlations with **Duhring plots**, which are plots of the temperature at which a substance has a certain vapor pressure versus the temperature at which a reference substance has the same vapor pressure. The principles of the preparation and use of these charts are identical to those of the Cox chart.

A relatively simple empirical equation that correlates vapor pressure–temperature data extremely well is the **Antoine equation**

$$\log_{10} p^* = A - \frac{B}{T + C} \tag{6.1-4}$$

Values of A, B, and C for several compounds are listed in Table B.4. Observe the units of p^* and T (mm Hg and °C for the constants in Table B.4) and the logarithm base (10 in the case of Equation 6.1-4 and Table B.4). Gathering parameters from multiple sources increases the need for care relative to the units used.

TEST YOURSELF

1. You need to know the vapor pressure of n-hexane at 87°C. Name two ways you could estimate this quantity using material readily available to you.
2. Suppose you are given the vapor pressure p^* of a substance at three closely spaced temperatures T_1, T_2, and T_3, and you wish to determine p^* at a fourth temperature T_4 far removed from the other three. If you use the Clausius–Clapeyron equation to correlate p^* and T, how would you plot the data and extrapolate it to T_4?
3. Why would it be preferable to use a Cox chart to plot and extrapolate the data of question 2 rather than the Clausius–Clapeyron equation?

CREATIVITY EXERCISE

Give as many reasons as you can think of, however farfetched, for wanting to know the vapor pressure of a substance at a given temperature. (*Example:* You want to know whether you can leave it overnight in an unstoppered flask without most of it evaporating.)

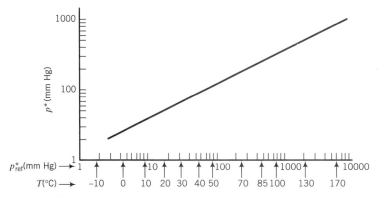

Figure 6.1-3 Reference-substance plot: temperature scale for water on abscissa.

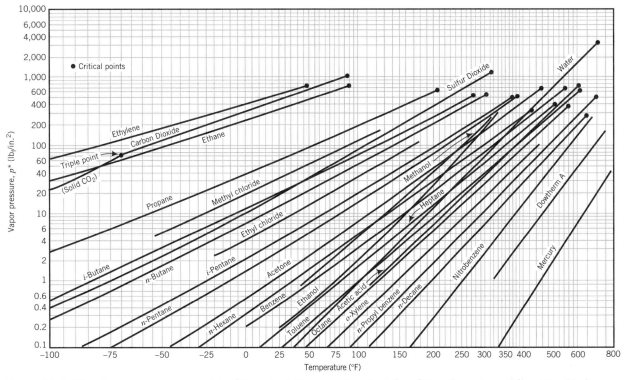

Figure 6.1-4   Cox chart vapor pressure plots. (From A. S. Foust et al., *Principles of Unit Operations,* Wiley, New York, 1960, p. 550.)

6.2   THE GIBBS PHASE RULE

When two phases are brought into contact with each other, a redistribution of the components of each phase normally takes place—species evaporate, condense, dissolve, or precipitate until a state of equilibrium is reached in which the temperatures and pressures of both phases are the same and the composition of each phase no longer changes with time.

Suppose you have a closed vessel containing three components A, B, and C distributed between gas and liquid phases, and you wish to describe this system to someone else in sufficient detail for that person to duplicate it exactly. Specifying the system temperature and pressure, the masses of each phase, and two mass or mole fractions for each phase would certainly be sufficient; however, these variables are not all independent—once some of them are specified, others are fixed by nature and, in some cases, may be calculated from physical properties of the system components.

The variables that describe the condition of a process system fall into two categories: **extensive variables**, which depend on the size of the system, and **intensive variables**, which do not. Mass and volume are examples of extensive variables; intensive variables include temperature, pressure, density and specific volume, and mass and mole fractions of individual system components in each phase.

The number of intensive variables that can be specified independently for a system at equilibrium is called the **degrees of freedom** of the system. Let

Π = number of phases in a system at equilibrium

c = number of chemical species

DF = degrees of freedom

The relationship among DF, Π, and c is given by the **Gibbs phase rule.** If no reactions occur among the system components, the phase rule is

$$DF = 2 + c - \Pi \qquad \textbf{(6.2-1)}$$

If r independent reactions occur among the system components and the reactions proceed to equilibrium, then the right-hand side of this equation must be reduced by r. [*Note: Perry's Chemical Engineers' Handbook* (see footnote 1), p. 4-24, presents a proof of the phase rule and outlines a method for determining how many independent reactions may occur among the components of a system.]

The meaning of the term *degrees of freedom* in the Gibbs phase rule is similar to its meaning in the degree-of-freedom analyses you have been performing since Chapter 4. In those analyses, the degrees of freedom represents the number of process variables that must be specified for a process system before the remaining variables can be calculated. In the Gibbs phase rule, the degrees of freedom equals the number of *intensive* variables that must be specified for a system *at equilibrium* before the remaining intensive variables can be calculated.

The example that follows illustrates the application of the Gibbs phase rule to several simple systems. The remainder of the chapter presents the equilibrium relationships that are used to determine the remaining intensive system variables once the allowed number of these variables has been specified.

EXAMPLE 6.2-1   *The Gibbs Phase Rule*

Determine the degrees of freedom for each of the following systems at equilibrium. Specify a feasible set of independent variables for each system.

1. *Pure liquid water*

$$\text{One phase } (\Pi = 1), \text{ one component } (c = 1)$$
$$\Downarrow$$
$$DF = 2 + 1 - 1 = 2$$

Two intensive variables must be specified to fix the state of the system, for example, T and P. Once these variables have been specified, other intensive variables such as density and viscosity may be determined.

2. *A mixture of liquid, solid, and vapor water*

$$\text{Three phases } (\Pi = 3), \text{ one component } (c = 1)$$
$$\Downarrow$$
$$DF = 2 + 1 - 3 = 0$$

No further information about the system may be specified and all intensive variables are fixed. Note from Figure 6.1-1a that three phases coexist at equilibrium at only one temperature and pressure.

3. *A vapor–liquid mixture of acetone and methyl ethyl ketone*

$$\text{Two phases } (\Pi = 2), \text{ two components } (c = 2)$$
$$\Downarrow$$
$$DF = 2 + 2 - 2 = 2$$

Two variables must be specified to fix the state of the system. For example, setting T and P fixes the acetone and MEK mole fractions in both the vapor and liquid phases. Alternatively, T and the acetone mole fraction in the vapor may be specified, and P and the acetone mole fraction in the liquid are then fixed.

TEST YOURSELF

1. Define and give examples of extensive and intensive variables. Define "degrees of freedom of a system." What is the Gibbs phase rule?

2. Use the phase rule to determine the degrees of freedom of each of the following equilibrium systems and give a possible set of variables that may be specified.

   (a) NaCl crystals suspended in an aqueous NaCl solution.

   (b) Humid air in equilibrium with condensed water (dry air may be considered a single species).

   (c) A vapor–liquid mixture of four hydrocarbons.

   (d) A gaseous mixture of H_2, Br_2, and HBr, given that the single reaction

$$H_2 + Br_2 \rightleftharpoons 2\,HBr$$

   has proceeded to equilibrium. (See the sentence following Equation 6.2-1.)

6.3 GAS–LIQUID SYSTEMS: ONE CONDENSABLE COMPONENT

Systems containing several components, of which only one is capable of existing as a liquid at the process conditions, are common in industrial processes. Separation processes that involve such systems include **evaporation**, **drying**, and **humidification**—all of which involve transfer of liquid into the gas phase—and **condensation** and **dehumidification**, which involve transfer of the condensable species from the gas to the liquid phase.

Suppose liquid water is introduced into a chamber that initially contains dry air and that the temperature and pressure in the system are kept constant at 75°C and 760 mm Hg. Initially the gas phase contains no water ($p_{H_2O} = 0$), and water molecules consequently begin to evaporate. The mole fraction of water in the gas phase, y_{H_2O}, increases, and hence so does the partial pressure of water, $p_{H_2O} = y_{H_2O}P$. Eventually, however, the amount of water in the gas phase is such that the rate at which water molecules enter the gas phase approaches zero, and thereafter no change occurs in the amount or composition of either phase. The gas phase is then said to be **saturated** with water—it contains all the water it can hold at the system temperature and pressure—and the water in the gas phase is referred to as a **saturated vapor**.

Let us apply the Gibbs phase rule to this equilibrium system. Since there are two phases and two components,

$$DF = 2 + c - \Pi = 2$$

It follows that only two out of the three intensive variables T, P, and y_{H_2O} can be specified, and that some relationship must exist that uniquely determines the value of the third variable once the first two have been specified.[3]

A law that describes the behavior of gas–liquid systems over a wide range of conditions provides the desired relationship. *If a gas at temperature T and pressure P contains a saturated vapor whose mole fraction is y_i (mol vapor/mol total gas), and if this vapor is the only species that would condense if the temperature were slightly lowered, then the partial pressure of the vapor in the gas equals the pure-component vapor pressure $p_i^*(T)$ at the system temperature.*

Raoult's Law, Single Condensable Species: $\quad p_i = \boxed{y_i P = p_i^*(T)}$ **(6.3-1)**

Equation 6.3-1 is a limiting case of **Raoult's law**, which will be introduced in a more general context in Section 6.4. It is the fundamental relation used in the analysis of equilibrated gas–liquid systems containing one condensable component. A wide variety of problems occur in

---

[3]It could be argued that setting the mole fraction of air in the liquid water to zero uses up a degree of freedom. In fact, however, we are not fixing a precise value of this variable; all we are saying is that the mole fraction of air in the liquid water approaches zero and the value is so small that it does not affect liquid-phase behavior and mass balances on the system.

connection with such systems, but they all eventually involve knowing two of the variables y_i, P, and T and having to determine the third using Equation 6.3-1.

EXAMPLE 6.3-1 *Composition of a Saturated Gas–Vapor System*

Air and liquid water are contained at equilibrium in a closed chamber at 75°C and 760 mm Hg. Calculate the molar composition of the gas phase.

SOLUTION

Since the gas and liquid are in equilibrium, the air must be saturated with water vapor (if it were not, more water would evaporate), so that Raoult's law may be applied:

$$y_{H_2O} = p^*_{H_2O}(75°C)/P$$

From Table B.3 in Appendix B, $p^*_{H_2O}(75°C) = 289$ mm Hg. Consequently,

$$y_{H_2O} = \frac{289 \text{ mm Hg}}{760 \text{ mm Hg}} = \boxed{0.380 \ \frac{\text{mol } H_2O}{\text{mol}}}$$

$$y_{\text{dry air}} = 1 - y_{H_2O} = \boxed{0.620 \ \frac{\text{mol dry air}}{\text{mol}}}$$

Several important points concerning the behavior of gas–liquid systems and several terms used to describe the state of such systems are summarized here.

1. A gas in equilibrium with a liquid must be saturated with the volatile components of that liquid.

2. The partial pressure of a vapor at equilibrium in a gas mixture containing a single condensable component cannot exceed the vapor pressure of the pure component at the system temperature. If $p_i = p^*_i$, the vapor is saturated; any attempt to increase p_i—either by adding more vapor to the gas phase or by increasing the total pressure at constant temperature—must instead lead to condensation.

3. A vapor present in a gas in less than its saturation amount is referred to as a **superheated vapor**. For such a vapor,

$$p_i = \boxed{y_i P < p^*_i(T)} \tag{6.3-2}$$

Since only a saturated vapor can condense (why?), to achieve condensation in a system containing a superheated vapor one or more of the variables of Equation 6.3-2 must be changed so that the inequality becomes the equality of Raoult's law. This can be done in several ways, such as by increasing the pressure at constant temperature (the left side increases, while the right side stays constant) or by decreasing the temperature at constant pressure (the left side remains constant, and the right side decreases).

4. If a gas containing a single superheated vapor is cooled at constant pressure, the temperature at which the vapor becomes saturated is referred to as the **dew point** of the gas. From Raoult's law (Equation 6.3-1),

$$p_i = \boxed{y_i P = p^*_i(T_{dp})} \tag{6.3-3}$$

The difference between the temperature and the dew point of a gas is called the **degrees of superheat** of the gas. If any two of the quantities y_i, P, and T_{dp} (or, equivalently, the temperature of the gas and the degrees of superheat) are known, the third quantity may be determined from Equation 6.3-3 and a table, graph, or equation relating p^*_i and T.

| EXAMPLE 6.3-2 | *Material Balances Around a Condenser* |

A stream of air at 100°C and 5260 mm Hg contains 10.0% water by volume.

1. Calculate the dew point and degrees of superheat of the air.
2. Calculate the percentage of the vapor that condenses and the final composition of the gas phase if the air is cooled to 80°C at constant pressure.
3. Calculate the percentage condensation and the final gas-phase composition if, instead of being cooled, the air is compressed isothermally to 8500 mm Hg.
4. Suppose the process of part 2 is run, the product gas is analyzed, and the mole fraction of water differs considerably from the calculated value. What could be responsible for the disparity between calculated and measured values? (List several possibilities.)

SOLUTION

1. $p_{H_2O} = y_{H_2O}P = (0.100)(5260 \text{ mm Hg}) = 526 \text{ mm Hg}$
$p_{H_2O}^*(100°C) = 760 \text{ mm Hg} > p_{H_2O} \implies$ the vapor is superheated (see Inequality 6.3-2)

From Equation 6.3-3

$$p_{H_2O} = p_{H_2O}^*(T_{dp}) = 526 \text{ mm Hg}$$

$\Downarrow$ Table B.3

$$\boxed{T_{dp} = 90°C}$$

and the air has $100°C - 90°C = \boxed{10°C \text{ of superheat}}$

2. Since the air becomes saturated at 90°C, further cooling must lead to condensation. Since the products are liquid water in equilibrium with a gas phase, the water vapor in the gas must remain saturated.

On the following flowchart, the symbol BDA stands for **bone-dry air**, a term used to signify the water–free component of an air-water vapor mixture.

Basis: 100 mol Feed Gas

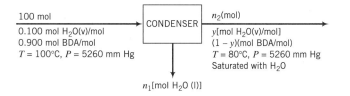

Let us first do the degree-of-freedom analysis. Three unknown variables appear on the chart—n_1, n_2, and y. Since only two species are involved in the process, we can only write two independent material balances, leaving us one equation short. If we fail to observe that the gas at the condenser outlet is saturated with water, solution of the problem would be impossible; however, the saturation condition supplies the needed third equation, Raoult's law.

The solution outline is as follows: apply Raoult's law at the outlet to determine y, the mole fraction of water in the outlet gas; then use a dry air balance to determine n_2 and a total mole balance or a water balance to determine the final unknown, n_1.

Raoult's Law at Outlet $\qquad yP = p_{H_2O}^*(T)$

$\Downarrow$

$$y = \frac{p_{H_2O}^*(80°C)}{P} = \frac{355 \text{ mm Hg}}{5260 \text{ mm Hg}} = \boxed{0.0675 \frac{\text{mol } H_2O}{\text{mol}}}$$

Balance on Dry Air
$$\frac{100\ \text{mol}}{}\ \left|\ \frac{0.900\ \text{mol BDA}}{\text{mol}}\right. = n_2(1-y)$$

$$\Downarrow\ y = 0.0675$$

$$n_2 = 96.5\ \text{mol}$$

Total Mole Balance $\quad 100\ \text{mol} = n_1 + n_2$

$$\Downarrow\ n_2 = 96.5\ \text{mol}$$

$$n_1 = 3.5\ \text{mol H}_2\text{O condensed}$$

Percentage Condensation $\quad \dfrac{3.5\ \text{mol H}_2\text{O condensed}}{(0.100 \times 100)\ \text{mol H}_2\text{O fed}} \times 100\% = \boxed{35\%}$

3. Initially $y_{\text{H}_2\text{O}}P < p^*_{\text{H}_2\text{O}}(100°\text{C})$. Saturation occurs when P is high enough for the inequality to become an equality, or

$$P_{\text{saturation}} = \frac{p^*_{\text{H}_2\text{O}}(100°\text{C})}{y_{\text{H}_2\text{O}}} = \frac{760\ \text{mm}}{0.100} = 7600\ \text{mm Hg}$$

Any increase in P above 7600 mm Hg must cause condensation, so that the products from the compression to 8500 mm Hg must include a liquid stream.

Basis: 100 mol Feed Gas

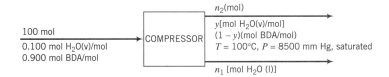

Before going through the solution, try to outline it as was done following the flowchart of part 2.

Raoult's Law $\qquad y = \dfrac{p^*_{\text{H}_2\text{O}}(100°\text{C})}{P} = \dfrac{760\ \text{mm Hg}}{8500\ \text{mm Hg}} = \boxed{0.0894\ \dfrac{\text{mol H}_2\text{O}}{\text{mol}}}$

Balance on Dry Air $\qquad (100\ \text{mol})(0.900) = n_2(1-y)$

$$\Downarrow\ y = 0.0894$$

$$n_2 = 98.8\ \text{mol}$$

Total Mole Balance $\qquad 100\ \text{mol} = n_1 + n_2$

$$\Downarrow\ n_2 = 98.8\ \text{mol}$$

$$n_1 = 1.2\ \text{mol H}_2\text{O condensed}$$

Percentage Condensation $\qquad \dfrac{1.2\ \text{mol H}_2\text{O condensed}}{(0.100 \times 100)\ \text{mol H}_2\text{O fed}} \times 100\% = \boxed{12\%}$

4. (a) Experimental error (you should be able to list many possibilities). (b) The condenser was not at steady state when the measurements were made, possibly because the system had not yet leveled out following startup or because water vapor was adsorbing on the walls of the condenser. (c) The emerging gas and liquid streams were not at equilibrium (e.g., condensation actually occurred at a temperature below 100°C and the product streams were separated and reheated before emerging). (d) Raoult's law does not apply (this is not a likely explanation for the air–water system at the given conditions).

The mechanism of evaporation of a liquid depends on the relative values of the vapor pressure of the liquid and the total system pressure. If evaporation takes place at a temperature such that $p^* < P$, the process involves transfer of molecules from the surface of the liquid to the gas above the surface, while if $p^* = P$, vapor bubbles form throughout the entire liquid, but predominantly at the heated container walls: that is, the liquid *boils*. The temperature at which $p^* = P$ is the **boiling point** of the liquid at the given pressure.

TEST YOURSELF

1. If water vapor is in equilibrium with liquid water, must the vapor be saturated? Can a vapor be saturated if there is no liquid present in the system?

2. The vapor pressure of acetone is 200 mm Hg at 22.7°C. Liquid acetone is kept in a sealed flask at 22.7°C, and the gas above the liquid contains air and acetone vapor at a pressure of 960 mm Hg. What is (a) the partial pressure of acetone in the gas, (b) the partial pressure of N_2, and (c) the mole fraction of acetone in the gas? What assumption have you made in answering the preceding question? How would you determine the boiling point of the acetone, assuming a constant total pressure of 960 mm Hg?

3. Suppose you have a curve of $p^*_{H_2O}$ versus T, and you are given the temperature and pressure (T_0 and P_0) of a mixture of water and noncondensable gases.

   a. Define the dew point of the gas. Would the vapor be saturated or superheated if $T_0 > T_{dp}$? If $T_0 = T_{dp}$?
   b. If you are told that the gas is saturated, how would you calculate the mole fraction of water in it? What would happen to the vapor if you (i) heated the gas isobarically (at constant pressure), (ii) cooled the gas isobarically, (iii) compressed the gas isothermally (at constant temperature), and (iv) expanded the gas isothermally?
   c. If you are given the mole fraction of water in the gas, how would you calculate the dew point of the gas?
   d. If you are given the degrees of superheat of the gas, how would you calculate the mole fraction of water in the gas?

Several quantities besides those introduced in the previous section are commonly used to describe the state and composition of a gas containing a single condensable vapor. *In the definitions to be given, the term "saturation" refers to any gas–vapor combination, while "humidity" refers specifically to an air–water system.*

Suppose a gas at temperature T and pressure P contains a vapor whose partial pressure is p_i and whose vapor pressure is $p_i^*(T)$.

Relative Saturation (Relative Humidity) $s_r(h_r) = \dfrac{p_i}{p_i^*(T)} \times 100\%$ **(6.3-4)**

A relative humidity of 40%, for example, signifies that the partial pressure of water vapor equals $\frac{4}{10}$ of the vapor pressure of water at the system temperature.

Molal Saturation (Molal Humidity)

$$s_m(h_m) = \frac{p_i}{P - p_i} = \frac{\text{moles of vapor}}{\text{moles of vapor-free (dry) gas}} \qquad \textbf{(6.3-5)}$$

(Can you prove the second equality?)

Absolute Saturation (Absolute Humidity)

$$s_a(h_a) = \frac{p_i M_i}{(P - p_i)M_{dry}} = \frac{\text{mass of vapor}}{\text{mass of dry gas}} \qquad \textbf{(6.3-6)}$$

where M_i is the molecular weight of the vapor and M_{dry} is the average molecular weight of the dry (vapor-free) gas.

Percentage Saturation (Percentage Humidity)

$$s_\text{p}(h_\text{p}) = \frac{s_\text{m}}{s_\text{m}^*} \times 100\% = \frac{p_i/(P - p_i)}{p_i^*/(P - p_i^*)} \times 100\% \tag{6.3-7}$$

If you are given any of these quantities for a gas at a given temperature and pressure, you can solve the defining equation to calculate the partial pressure or mole fraction of the vapor in the gas; thereafter, you can use the formulas given previously to calculate the dew point and degrees of superheat.

TEST YOURSELF

The vapor pressure of styrene is 100 mm Hg at 82°C and 200 mm Hg at 100°C. A gas that consists of 10 mole% styrene and 90 mole% noncondensables is contained in a tank at 100°C and 1000 mm Hg. Calculate:

1. The dew point of the gas.
2. The relative saturation.
3. The molal saturation and percentage saturation.

CREATIVITY EXERCISES

1. Suppose you know the temperature and barometric pressure on a given day. List as many ways as you can think of to determine—exactly or approximately—the mole fraction of water vapor in the air.
2. Repeat question 1, only this time limit yourself to methods in which a stuffed bear plays a part in the determination. (*Example:* Saturate the bear with water, and measure the rate at which it loses weight due to evaporation.)

EXAMPLE 6.3-3

Humid air at 75°C, 1.1 bar, and 30% relative humidity is fed into a process unit at a rate of 1000 m³/h. Determine (1) the molar flow rates of water, dry air, and oxygen entering the process unit, (2) the molal humidity, absolute humidity, and percentage humidity of the air, and (3) the dew point.

SOLUTION

1.
$$h_\text{r}(\%) = 100 p_{H_2O}/p_{H_2O}^*(75°C)$$

$$\Downarrow \quad \begin{array}{l} h_\text{r} = 30\% \\ p_{H_2O}^*(75°C) = 289 \text{ mm Hg (from Table B.3)} \end{array}$$

$$p_{H_2O} = (0.3)(289 \text{ mm Hg}) = 86.7 \text{ mm Hg}$$

$$\Downarrow \quad \begin{array}{l} y_{H_2O} = p_{H_2O}/P \\ P = 1.1 \text{ bar} \Longrightarrow 825 \text{ mm Hg} \end{array}$$

$$y_{H_2O} = (86.7 \text{ mm Hg})/(825 \text{ mm Hg}) = 0.105 \text{ mol } H_2O/\text{mol}$$

The molar flow rate of wet air is given by the ideal gas equation of state as

$$\dot{n} = P\dot{V}/RT = \frac{1000 \text{ m}^3}{\text{h}} \left| \frac{1.1 \text{ bar}}{348 \text{ K}} \right| \frac{\text{kmol·K}}{0.0831 \text{ m}^3\text{·bar}} = 38.0 \frac{\text{kmol}}{\text{h}}$$

Consequently,

$$\dot{n}_{H_2O} = \frac{38.0 \text{ kmol}}{\text{h}} \left| \frac{0.105 \text{ kmol } H_2O}{\text{kmol}} \right. = \boxed{3.99 \frac{\text{kmol } H_2O}{\text{h}}}$$

$$\dot{n}_{BDA} = \frac{38.0 \text{ kmol}}{\text{h}} \left| \frac{(1 - 0.105) \text{ kmol BDA}}{\text{kmol}} \right. = \boxed{34.0 \frac{\text{kmol BDA}}{\text{h}}}$$

$$\dot{n}_{O_2} = \frac{34.0 \text{ kmol BDA}}{\text{h}} \left| \frac{0.21 \text{ kmol } O_2}{\text{kmol BDA}} \right. = \boxed{7.14 \frac{\text{kmol } O_2}{\text{h}}}$$

2.
$$h_m = \frac{p_{H_2O}}{P - p_{H_2O}} = \frac{86.7 \text{ mm Hg}}{(825 - 86.7) \text{ mm Hg}} = \boxed{0.117 \, \frac{\text{mol } H_2O}{\text{mol BDA}}}$$

The same result could have been obtained from the results of part 1 as (3.99 kmol H_2O/h)/ (34.0 kmol BDA/h).

$$h_a = \frac{0.117 \text{ kmol } H_2O}{\text{kmol BDA}} \, \bigg| \, \frac{18.0 \text{ kg } H_2O}{\text{kmol } H_2O} \, \bigg| \, \frac{1 \text{ kmol BDA}}{29.0 \text{ kg BDA}} = \boxed{0.0726 \, \frac{\text{kg } H_2O}{\text{kg BDA}}}$$

$$h_m^* = \frac{p_{H_2O}^*}{P - p_{H_2O}^*} = \frac{289 \text{ mm Hg}}{(825 - 289) \text{ mm Hg}} = 0.539 \, \frac{\text{kmol } H_2O}{\text{kmol BDA}}$$

$$h_p = 100 h_m / h_m^* = (100)(0.117)/(0.539) = \boxed{21.7\%}$$

3. $p_{H_2O} = 86.7 \text{ mm Hg} = p_{H_2O}^*(T_{dp})$

$\Downarrow$ Table B.3

$$\boxed{T_{dp} = 48.7°C}$$

6.4 MULTICOMPONENT GAS–LIQUID SYSTEMS

Gas–liquid processes that involve several components in each phase include many chemical reactions, distillation, and transfer of one or more species from a gas to a liquid (**absorption** or **scrubbing**) or vice versa (**stripping**).

When multicomponent gas and liquid phases are in equilibrium, a limited number of intensive system variables may be specified arbitrarily (the number is given by the Gibbs phase rule), and the remaining variables can then be determined using equilibrium relationships for the distribution of components between the two phases. In this section we define several such relationships and illustrate how they are used in the solution of material balance problems.

6.4a Vapor–Liquid Equilibrium Data

The best way to evaluate equilibrium compositions is from tabulated data. *Perry's Chemical Engineers' Handbook* (see footnote 1), pp. 2-76 through 2-89, gives partial pressures of vapors over various liquid solutions. Example 6.4-1 illustrates the use of such data.

EXAMPLE 6.4-1 | *Absorption of SO$_2$*

A gas stream consisting of 100 lb-mole/h of an SO$_2$–air mixture containing 45 mole% SO$_2$ is contacted with liquid water in a continuous absorber at 30°C. The liquid leaving the absorber is analyzed and found to contain 2.00 g of SO$_2$ per 100 g of H$_2$O. Assuming that the gas and liquid streams leaving the absorber are in equilibrium at 30°C and 1 atm, calculate the fraction of the entering SO$_2$ absorbed in the water and the required water feed rate.

SOLUTION | *Basis: Given Feed Rate of Gas*

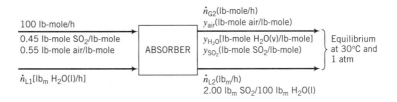

From Table 3-12 on p. 3-65 of the 6th Edition of *Perry's Chemical Engineers' Handbook*,[4] the equilibrium partial pressures of H_2O and SO_2 over a solution of the indicated composition are

$$p_{H_2O} = 31.6 \text{ mm Hg}$$

$$p_{SO_2} = 176 \text{ mm Hg}$$

so that the composition of the exit gas stream is

$$y_{H_2O} = \frac{31.6 \text{ mm Hg}}{760 \text{ mm Hg}} = 0.0416 \frac{\text{lb-mole } H_2O}{\text{lb-mole}}$$

$$y_{SO_2} = \frac{176 \text{ mm Hg}}{760 \text{ mm Hg}} = 0.232 \frac{\text{lb-mole } SO_2}{\text{lb-mole}}$$

$$y_{air} = 1 - y_{H_2O} - y_{SO_2} = 0.727 \frac{\text{lb-mole air}}{\text{lb-mole}}$$

Three unknown process variables remain—$\dot{n}_{L_1}$, $\dot{n}_{G_2}$, and $\dot{n}_{L_2}$—and since a total of three independent balances can be written, the system is determinate.

Air Balance

$$(0.55 \times 100)\frac{\text{lb-mole air}}{\text{h}} = y_{air}\dot{n}_{G_2}$$

$$\Downarrow y_{air} = 0.727 \text{ lb-mole air/lb-mole}$$

$$\dot{n}_{G_2} = 75.7 \text{ lb-mole/h}$$

To write the remaining two balances, it is necessary to determine the mass fractions of SO_2 and H_2O in the liquid effluent.

$$\frac{2.00 \text{ lb}_m \ SO_2}{100 \text{ lb}_m \ H_2O} \Longrightarrow \frac{2.00 \text{ lb}_m \ SO_2}{102 \text{ lb}_m \text{ total}} \Longrightarrow x_{SO_2} = 0.0196 \text{ lb}_m \ SO_2/\text{lb}_m$$

$$\Downarrow x_{SO_2} + x_{H_2O} = 1$$

$$x_{H_2O} = 0.9804 \text{ lb}_m \ H_2O/\text{lb}_m$$

SO₂ Balance

$$\frac{100 \text{ lb-mole}}{\text{h}} \left| \frac{0.45 \text{ lb-mole } SO_2}{\text{lb-mole}} \right. = \dot{n}_{G_2}y_{SO_2} + \frac{\dot{L}_2 \ (\text{lb}_m)}{(\text{h})} \left| \frac{x_{SO_2} \ (\text{lb}_m \ SO_2)}{(\text{lb}_m)} \right| \frac{\text{lb-mole}}{64 \text{ lb}_m \ SO_2}$$

$$\left\Downarrow \begin{array}{l} \dot{n}_{G_2} = 75.7 \text{ lb-mole/h} \\ y_{SO_2} = 0.232 \\ x_{SO_2} = 0.0196 \end{array}\right.$$

$$\dot{n}_{L_2} = 89,600 \text{ lb}_m/\text{h}$$

H₂O Balance

$$\dot{n}_{L_1}(\text{lb}_m \ H_2O/\text{h}) = \frac{\dot{n}_{G_2} \ (\text{lb-mole})}{(\text{h})} \left| \frac{y_{H_2O} \ (\text{lb-mole } H_2O)}{(\text{lb-mole})} \right| \frac{18 \text{ lb}_m \ H_2O}{\text{lb-mole}} + \dot{n}_{L_2}x_{H_2O}$$

$$\left\Downarrow \begin{array}{l} \dot{n}_{G_2} = 75.7 \text{ lb-mole/h} \\ y_{H_2O} = 0.0416 \text{ lb-mole } H_2O/\text{lb-mole} \\ \dot{n}_{L_2} = 89,600 \text{ lb}_m/\text{h} \\ x_{H_2O} = 0.9804 \text{ lb}_m \ H_2O/\text{lb}_m \end{array}\right.$$

$$\boxed{\dot{n}_{L_1} = 87,900 \text{ lb}_m \ H_2O/\text{h}} \quad \text{(feed to absorber)}$$

---

[4]R. H. Perry and D. W. Green, Eds., *Perry's Chemical Engineers' Handbook,* 6th Edition, McGraw-Hill, New York, 1984. (*Note:* These data do not appear in the 7th edition.)

Fraction SO₂ Absorbed

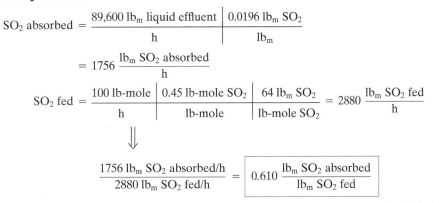

$$SO_2 \text{ absorbed} = \frac{89{,}600 \text{ lb}_m \text{ liquid effluent}}{h} \; \bigg| \; \frac{0.0196 \text{ lb}_m \text{ SO}_2}{\text{lb}_m}$$

$$= 1756 \frac{\text{lb}_m \text{ SO}_2 \text{ absorbed}}{h}$$

$$SO_2 \text{ fed} = \frac{100 \text{ lb-mole}}{h} \; \bigg| \; \frac{0.45 \text{ lb-mole SO}_2}{\text{lb-mole}} \; \bigg| \; \frac{64 \text{ lb}_m \text{ SO}_2}{\text{lb-mole SO}_2} = 2880 \frac{\text{lb}_m \text{ SO}_2 \text{ fed}}{h}$$

$$\Downarrow$$

$$\frac{1756 \text{ lb}_m \text{ SO}_2 \text{ absorbed/h}}{2880 \text{ lb}_m \text{ SO}_2 \text{ fed/h}} = \boxed{0.610 \frac{\text{lb}_m \text{ SO}_2 \text{ absorbed}}{\text{lb}_m \text{ SO}_2 \text{ fed}}}$$

6.4b   Raoult's Law and Henry's Law

If you apply the Gibbs phase rule to a multicomponent gas–liquid system at equilibrium, you will discover that the compositions of the two phases at a given temperature and pressure are not independent. Once the composition of one of the phases is specified (in terms of mole fractions, mass fractions, concentrations, or, for the vapor phase, partial pressures), the composition of the other phase is fixed and, in principle, can be determined from physical properties of the system components.

Relationships governing the distribution of a substance between gas and liquid phases are the subject matter of **phase-equilibrium thermodynamics** and, for the most part, fall beyond the scope of this text. However, we will cover several simple approximate relationships that provide reasonably accurate results over a wide range of conditions. Such relationships form the bases of more precise methods that must be used when system conditions require them.

Suppose A is a substance contained in a gas–liquid system in equilibrium at temperature T and pressure P. Two simple expressions—**Raoult's law** and **Henry's law**—provide relationships between p_A, the partial pressure of A in the gas phase, and x_A, the mole fraction of A in the liquid phase.

Raoult's Law: $\qquad\qquad p_A \equiv \boxed{y_A P = x_A p_A^*(T)}$ **(6.4-1)**

where p_A^* is the vapor pressure of pure liquid A at temperature T and y_A is the mole fraction of A in the gas phase.

Raoult's law is an approximation that is generally valid when x_A is close to 1—that is, when the liquid is almost pure A. It is also sometimes valid over the entire range of compositions for mixtures of similar substances, such as paraffinic hydrocarbons of similar molecular weights.

Note: When $x_A = 1$—that is, when the liquid is pure A—Raoult's law reduces to the expression $p_A = p_A^*(T)$ given previously for systems with only one condensable component.

Henry's Law: $\qquad\qquad p_A \equiv \boxed{y_A P = x_A H_A(T)}$ **(6.4-2)**

where $H_A(T)$ is the **Henry's law constant** for A in a specific solvent.

Henry's law is generally valid for solutions in which x_A is close to 0 (dilute solutions of A) provided that A does not dissociate, ionize, or react in the liquid phase. The law is often applied to solutions of noncondensable gases. Values of Henry's law constants (or closely related quantities) are given for several gases in water on pp. 2-125 through 2-128 of *Perry's Chemical Engineers' Handbook* (see footnote 1), and references to other data sources are given on p. 14-4 of the *Handbook*.

A gas–liquid system in which the vapor–liquid equilibrium relationship for every volatile species is either Raoult's law or Henry's law is said to exhibit **ideal solution** behavior. An ideal liquid solution is a mixture of liquids that exhibits ideal solution behavior at equilibrium.

TEST YOURSELF

1. What is Raoult's law, and when is it most likely to be valid?
2. What is Henry's law, and when is it most likely to be valid?
3. What is an ideal solution?
4. A gas containing CO_2 is in equilibrium with liquid water containing a small amount of dissolved CO_2 at 30°C and 3 atm. Would you use Raoult's law or Henry's law to estimate the relationship between (a) x_{CO_2} and p_{CO_2}, (b) x_{H_2O} and p_{H_2O}, where x denotes mole fraction in the liquid and p denotes partial pressure in the gas? In each case, what would you look up and where would you look for it? Would you expect to observe ideal solution behavior for this system?

EXAMPLE 6.4-2 *Raoult's Law and Henry's Law*

Use either Raoult's law or Henry's law to solve the following problems.

1. A gas containing 1.00 mole% ethane is in contact with water at 20.0°C and 20.0 atm. Estimate the mole fraction of dissolved ethane.
2. An equimolar liquid mixture of benzene (B) and toluene (T) is in equilibrium with its vapor at 30.0°C. What is the system pressure and the composition of the vapor?

SOLUTION

1. Hydrocarbons normally are relatively insoluble in water, so that the solution of ethane is probably extremely dilute. Let us therefore apply Henry's law. Page 2-126 of *Perry's Chemical Engineers' Handbook* (see footnote 1) gives the Henry's law constant for ethane in water at 20°C as 2.63×10^4 atm/mole fraction. From Equation 6.4-2

$$x_{C_2H_6} = \frac{y_{C_2H_6}P}{H_{C_2H_6}} = \frac{(0.0100)(20.0 \text{ atm})}{2.63 \times 10^4 \text{ atm/mole fraction}} = \boxed{7.60 \times 10^{-6} \frac{\text{mol } C_2H_6}{\text{mol}}}$$

2. Since benzene and toluene are structurally similar compounds, we may apply Raoult's law. From Table B.4,

$$\log_{10} p_B^* = 6.906 - \frac{1211}{T + 220.8} \xrightarrow{T = 30°C} p_B^* = 119 \text{ mm Hg}$$

$$\log_{10} p_T^* = 6.9533 - \frac{1343.9}{T + 219.38} \xrightarrow{T = 30°C} p_T^* = 36.7 \text{ mm Hg}$$

Using Equation 6.4-1,

$$p_B = x_B p_B^* = (0.500)(119 \text{ mm Hg}) = 59.5 \text{ mm Hg}$$

$$p_T = x_T p_T^* = (0.500)(36.7 \text{ mm Hg}) = 18.35 \text{ mm Hg}$$

$$P = p_B + p_T = \boxed{77.9 \text{ mm Hg}}$$

$$y_B = p_B/P = \boxed{0.764 \text{ mole benzene/mole}}$$

$$y_T = p_T/P = \boxed{0.236 \text{ mole toluene/mole}}$$

CREATIVITY EXERCISE

Give as many cases as you can think of where it would be useful or necessary to know the Henry's law constant for a gas in a liquid. (*Example:* You wish to calculate the pressure needed to achieve a specified level of carbonation in bottled soda.) In your list, include several examples with environmental relevance.

6.4c Vapor–Liquid Equilibrium Calculations for Ideal Solutions

Suppose heat is added slowly to a closed vessel that contains a liquid and that the pressure in the vessel is held constant. We have already examined what happens in a situation of this sort if the liquid is a single species: the temperature increases until the boiling point of the liquid is reached, and thereafter the liquid vaporizes at a constant temperature. Once the vaporization is complete, further addition of heat raises the temperature of the vapor.

Consider now what happens if the liquid is a mixture of several components. As heat is added, the liquid temperature rises until a temperature is reached at which the first bubble of vapor forms. Up to this point, the process looks like that for a single component. However, if the liquid is a mixture, the vapor generated generally will have a composition different from that of the liquid. *As vaporization proceeds, the composition of the remaining liquid continuously changes, and hence so does its vaporization temperature.* A similar phenomenon occurs if a mixture of vapors is subjected to a condensation process at constant pressure: at some temperature the first droplet of liquid forms, and thereafter the composition of the vapor and the condensation temperature both change.

To design or control an evaporation or condensation process, you must know the conditions at which the transition from liquid to vapor or vapor to liquid takes place. Design or control of other separation processes such as distillation, absorption, and stripping also requires information on the conditions at which phase transitions occur and on the compositions of the resulting phases. This section outlines the required calculations for a relatively simple class of mixtures.

When a liquid is heated slowly at constant pressure, the temperature at which the first vapor bubble forms is the **bubble-point temperature** of the liquid at the given pressure. When a gas (vapor) is cooled slowly at constant pressure, the temperature at which the first liquid droplet forms is the **dew-point temperature** at the given pressure. Calculating bubble-point and dew-point temperatures can be a complex task for an arbitrary mixture of components. However, if the liquid behaves as an **ideal solution** (one for which Raoult's or Henry's law is obeyed for all components) and the gas phase can also be considered ideal, the calculations are relatively straightforward.

Suppose an ideal liquid solution follows Raoult's law and contains species A, B, C, ... with known mole fractions x_A, x_B, x_C, ... If the mixture is heated at a constant pressure P to its bubble-point temperature T_{bp}, the further addition of a slight amount of heat will lead to the formation of a vapor phase. Since the vapor is in equilibrium with the liquid, and we now assume that the vapor is ideal (follows the ideal gas equation of state), the partial pressures of the components are given by Raoult's law, Equation 6.4-1.

$$p_i = x_i p_i^*(T_{bp}), \quad i = A, B, \dots \tag{6.4-3}$$

where p_i^* is the vapor pressure of component i at the bubble-point temperature. Moreover, since we have assumed that only A, B, C, ... are present in the system, the sum of the partial pressures must be the total system pressure, P; hence,

$$P = x_A p_A^*(T_{bp}) + x_B p_B^*(T_{bp}) + \cdots \tag{6.4-4}$$

The bubble-point temperature may be calculated by trial and error as the value of T_{bp} that satisfies this equation; all that is needed is a set of relationships for $p_i^*(T)$, such as the Antoine equation or vapor-pressure charts or tables. Once T_{bp} is known, the composition of the vapor phase can easily be determined by evaluating the partial pressures of each component from Equation 6.4-3 and determining each vapor–phase mole fraction as $y_i = p_i / P$.

The pressure at which the first vapor forms when a liquid is decompressed at a constant temperature is the **bubble-point pressure** of the liquid at the given temperature. Equation 6.4-4 can be used to determine such a pressure for an ideal liquid solution at a specific temperature, and the mole fractions in the vapor in equilibrium with the liquid can then be determined as

$$y_i = \frac{p_i}{P_{bp}} = \frac{x_i p_i^*(T)}{P_{bp}} \tag{6.4-5}$$

The dew-point temperature of a gas (vapor) may be found using a method similar to that for bubble-point temperature estimation. Again, suppose a gas phase contains the condensable components A, B, C, ... and a noncondensable component G at a fixed pressure P. Let y_i be the mole fraction of component i in the gas. If the gas mixture is cooled slowly to its dew point, T_{dp}, it will be in equilibrium with the first liquid that forms. Assuming that Raoult's law applies, the liquid-phase mole fractions may be calculated as

$$x_i = \frac{y_i P}{p_i^*(T_{dp})}, \qquad i = A, B, C, \ldots \quad \text{excluding } G \tag{6.4-6}$$

At the dew point of the gas mixture, the mole fractions of the liquid components (those that are condensable) must sum to 1:

$$x_A + x_B + x_C + \cdots = 1$$

$$\Downarrow \quad \text{Equation 6.4-6}$$

$$\frac{y_A P}{p_A^*(T_{dp})} + \frac{y_B P}{p_B^*(T_{dp})} + \cdots = 1 \tag{6.4-7}$$

The value of T_{dp} can be found by trial and error once expressions for $p_i^*(T)$ have been substituted. The composition of the liquid phase may then be determined from Equation 6.4-6.

The **dew-point pressure**, which relates to condensation brought about by increasing system pressure at constant temperature, can be determined by solving Equation 6.4-7 for P:

$$P_{dp} = \frac{1}{\dfrac{y_A}{p_A^*(T)} + \dfrac{y_B}{p_B^*(T)} + \dfrac{y_C}{p_C^*(T)} + \cdots} \tag{6.4-8}$$

Liquid mole fractions may then be calculated from Equation 6.4-6 with T_{dp} replaced by the system temperature, T.

EXAMPLE 6.4-3 **Bubble- and Dew-Point Calculations**

1. Calculate the temperature and composition of a vapor in equilibrium with a liquid that is 40.0 mole% benzene–60.0 mole% toluene at 1 atm. Is the calculated temperature a bubble-point or dew-point temperature?
2. Calculate the temperature and composition of a liquid in equilibrium with a gas mixture containing 10.0 mole% benzene, 10.0 mole% toluene, and the balance nitrogen (which may be considered noncondensable) at 1 atm. Is the calculated temperature a bubble-point or dew-point temperature?
3. A gas mixture consisting of 15.0 mole% benzene, 10.0 mole% toluene, and 75.0 mole% nitrogen is compressed isothermally at 80°C until condensation occurs. At what pressure will condensation begin? What will be the composition of the initial condensate?

SOLUTION Let A = benzene and B = toluene.

1. Equation 6.4-4 may be written in the form

$$f(T_{bp}) = 0.400 p_A^*(T_{bp}) + 0.600 p_B^*(T_{bp}) - 760 \text{ mm Hg} = 0$$

The solution procedure is to choose a temperature, evaluate p_A^* and p_B^* for that temperature from the Antoine equation using constants from Table B.4, evaluate $f(T_{bp})$ from the above equation, and repeat the calculations until a temperature is found for which $f(T_{bp})$ is sufficiently close to 0.

Trial-and-error searches of this sort are easy to do using spreadsheet programs. Enter a guessed value of T_{bp} in one cell and the formula for $f(T_{bp})$ (including the Antoine equation for each vapor pressure) in an adjacent cell, then vary the value in the first cell until the value in the second cell is sufficiently close to zero. If the spreadsheet program has a *goalseek* tool, the calculation is even easier: simply instruct the program to vary the value in the first cell to drive the value in the second cell to zero. (See Appendix A.2c.)

Several numerical techniques for accelerating trial-and-error searches of this sort are discussed in Appendix A.2. One of them, the *regula-falsi* method, is used by goalseek tools in many spreadsheet programs. This procedure has been used to generate the trial temperatures shown here for the third and subsequent trials.

$T\,(^\circ\mathrm{C})$	$p_A^*\,(\mathrm{mm\ Hg})$	$p_B^*\,(\mathrm{mm\ Hg})$	$f(T_{\mathrm{bp}})$	$T_{\mathrm{new}}(^\circ\mathrm{C})$
80	757.66	291.21	−282.21	
100	1350.49	556.32	113.99	94.25
94.25	1152.23	465.71	−19.68	95.09
95.09	1179.91	478.26	−1.08	95.14
95.14	1181.43	478.95	−0.06	95.14

The solution is taken to be $\boxed{T_{\mathrm{bp}} = 95.1^\circ\mathrm{C}}$. At this temperature, Equation 6.4-1 yields

$$p_A = 0.400\,(1181\ \mathrm{mm\ Hg}) = 472.5\ \mathrm{mm\ Hg}$$

$$p_B = 0.600\,(479\ \mathrm{mm\ Hg}) = 287.5\ \mathrm{mm\ Hg}$$

$$\Downarrow$$

$$P = (472.5 + 287.5)\ \mathrm{mm\ Hg} = 760\ \mathrm{mm\ Hg}$$

Furthermore, from Equation 6.4-5,

$$y_A = \frac{472.5}{760.0} = \boxed{0.622\ \mathrm{mol\ benzene/mol}}$$

$$y_B = 1 - y_A = \boxed{0.378\ \mathrm{mol\ toluene/mol}}$$

Since the composition of the liquid was given, this was a $\boxed{\text{bubble-point}}$ calculation.

2. Equation 6.4-7 may be written as

$$f(T_{\mathrm{dp}}) = \frac{(0.100)(760\ \mathrm{mm\ Hg})}{p_A^*(T_{\mathrm{dp}})} + \frac{(0.100)(760\ \mathrm{mm\ Hg})}{p_B^*(T_{\mathrm{dp}})} - 1.00 = 0$$

A trial and error procedure similar to that in part (1) leads to the result $\boxed{T_{\mathrm{dp}} = 52.4^\circ\mathrm{C}}$, at which temperature $p_A^* = 297.4\ \mathrm{mm\ Hg}$ and $p_B^* = 102.1\ \mathrm{mm\ Hg}$. Then, from Equation 6.4-6,

$$x_A = \frac{0.100(760\ \mathrm{mm\ Hg})}{p_A^*(52.4^\circ\mathrm{C})} = \boxed{0.256\ \mathrm{mol\ benzene/mol}}$$

$$x_B = 1 - x_A = \boxed{0.744\ \mathrm{mol\ toluene/mol}}$$

The composition of the vapor was given and that of the liquid was calculated; therefore, this was a $\boxed{\text{dew-point}}$ calculation.

3. The vapor pressures of benzene and toluene at 80°C are determined from the Antoine equation to be 757.7 mm Hg and 291.2 mm Hg, respectively. Assuming that nitrogen is insoluble in the condensate, Equation 6.4-8 gives

$$P = \frac{1}{(0.150/757.7\ \mathrm{mm\ Hg}) + (0.100/291.2\ \mathrm{mm\ Hg})} = \boxed{1847\ \mathrm{mm\ Hg}}$$

$$x_A = \frac{y_A P}{p_A^*} = \frac{0.150(1847\ \mathrm{mm\ Hg})}{757.7\ \mathrm{mm\ Hg}} = \boxed{0.366\ \mathrm{mol\ benzene/mol}}$$

$$x_B = 1 - x_A = \boxed{0.634\ \mathrm{mol\ toluene/mol}}$$

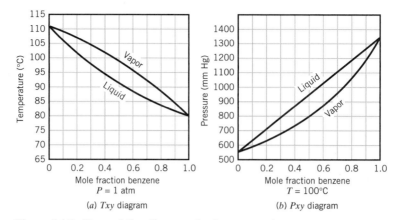

Figure 6.4-1 *Txy* and *Pxy* diagrams for benzene–toluene system.

6.4d   Graphical Representations of Vapor–Liquid Equilibrium

Vapor–liquid equilibrium calculations for binary (two-component) systems can be simplified considerably using a ***Txy* diagram**. Suppose the bubble-point temperature T of a solution of two components—A and B—is determined at a fixed pressure P and several values of x_A, and the equilibrium vapor composition y_A is determined for each liquid composition. The *Txy* diagram is a plot of the equilibrium temperature versus the mole fraction of one of the components—usually the more volatile one—with curves being drawn for both the liquid phase (T versus x_A) and the vapor phase (T versus y_A). A plot of this type is shown in Figure 6.4-1a for the benzene–toluene system at $P = 1$ atm. Alternatively, equilibrium pressure can be plotted against a mole fraction for a fixed temperature to generate a ***Pxy* diagram** (Figure 6.4-1b).

Once you have a *Txy* diagram like that of Figure 6.4-1, bubble- and dew-point calculations become trivial. To determine a bubble-point temperature for a given liquid composition, go to the liquid curve on the *Txy* diagram for the system pressure and read the desired temperature from the ordinate scale. (If you are not sure why this works, go back and consider again how the curve was generated.) You can then move horizontally to the vapor curve to determine the composition of the vapor in equilibrium with the given liquid at that temperature.

The dew point of a vapor mixture of A and B at P can be determined from the *Txy* diagram if there are no species other than A and B in the gas phase. Look up the specified mole fraction of A in the vapor phase, read the dew-point temperature from the corresponding ordinate value of the vapor curve, and move horizontally to the liquid curve and down to read the composition of the liquid in equilibrium with the vapor. If a noncondensable species is present in the gas phase, however, you must use Equation 6.4-6 to find the dew point, as in the previous example.

What happens as a liquid mixture is continuously vaporized is easily seen from the *Txy* diagram. Consider the benzene–toluene system again, and suppose heat is added to a 55 mole% benzene and 45 mole% toluene liquid mixture at a fixed pressure of 1 atm. As Figure 6.4-1a shows, the mixture will begin to boil at 90°C, and the vapor generated will contain 77% benzene. However, once a small amount of liquid has been vaporized, the remainder no longer contains 55% benzene; it contains less, since the vapor produced is relatively rich in this component. Consequently, the temperature of the system steadily rises as more and more liquid is vaporized, and the compositions of both phases change continuously during the process.

EXAMPLE 6.4-4   *Bubble- and Dew-Point Calculations Using Txy Diagrams*

1. Using the *Txy* diagram, estimate the bubble-point temperature and the equilibrium vapor composition associated with a 40 mole% benzene–60 mole% toluene liquid mixture at 1 atm. If the mixture is steadily vaporized until the remaining liquid contains 25% benzene, what is the final temperature?

2. Using the Txy diagram, estimate the dew-point temperature and the equilibrium liquid composition associated with a vapor mixture of benzene and toluene containing 40 mole% benzene at 1 atm. If condensation proceeds until the remaining vapor contains 60% benzene, what is the final temperature?

SOLUTION

1. From Figure 6.4-1a, for $x_B = 0.40$, $\boxed{T_{bp} \approx 95°C}$ and $\boxed{y_B \approx 0.62}$. (This matches the result of the much lengthier solution of the previous example.) When $x_B = 0.25$, $\boxed{T_{bp} \approx 100°C}$. The temperature thus rises by 5°C as the vaporization proceeds.

2. From Figure 6.4-1a, for $y_B = 0.40$, $\boxed{T_{dp} \approx 102°C}$ and $\boxed{x_B \approx 0.20}$. When $y_B = 0.60$, $\boxed{T_{bp} \approx 96°C}$.

Note: The precision associated with graphical calculations is less than that of numerical calculations, as reflected in this example by the use of the $\approx$. However, the simplicity and clarity in following the process path make graphical calculations quite useful.

The term "boiling" is sometimes incorrectly used to describe any process involving the transition from liquid to vapor. In fact, boiling refers to a specific type of vaporization process in which vapor bubbles form at a heated surface and escape from the liquid; it does *not* refer to molecular evaporation of liquid from a gas–liquid interface, which may occur at temperatures below the boiling point. (Recall your experience with leaving a container of liquid open to the atmosphere and finding the liquid partially or completely evaporated on your return.)

In our discussion of single-component systems, we considered the case of a liquid being heated in a container exposed to the atmosphere and observed that the liquid boils at a temperature at which the vapor pressure of the liquid equals the total pressure of the atmosphere above it. A similar phenomenon occurs for liquid mixtures. If a mixture is heated slowly in an open container, vapor bubbles will form at the heated surface and emerge into the gas phase when the vapor pressure of the liquid equals the pressure above the liquid.[5] Some reflection should convince you that the temperature at which this occurs is the bubble point of the liquid at that pressure. For an ideal liquid solution, the boiling point may therefore be determined *approximately* from Equation 6.4-9.

$$x_A p_A^*(T_{bp}) + x_B p_B^*(T_{bp}) + \cdots = P \tag{6.4-9}$$

EXAMPLE 6.4-5   *Boiling Point of a Mixture*

A mixture that is 70 mole% benzene and 30 mole% toluene is to be distilled in a batch distillation column. The column startup procedure calls for charging the reboiler at the base of the column and slowly adding heat until boiling begins. Estimate the temperature at which boiling begins and the initial composition of the vapor generated, assuming the system pressure is 760 mm Hg.

SOLUTION

From the Txy diagram, the mixture will boil at approximately $\boxed{87°C}$. The initial vapor composition is approximately $\boxed{88 \text{ mole}\% \text{ benzene and } 12 \text{ mole}\% \text{ toluene}}$.

We conclude this discussion with one final reminder. The vapor–liquid equilibrium calculations we have shown in Section 6.4c are based on the ideal-solution assumption and the corresponding use of Raoult's law. Many commercially important systems involve nonideal solutions, or systems of immiscible or partially miscible liquids, for which Raoult's law is inapplicable and the Txy diagram looks nothing like the one shown for benzene and toluene.

---

[5]This is only an approximation, albeit generally a good one. In fact, the vapor pressure must be slightly greater than the gas-phase pressure to overcome the effects of liquid surface tension and the hydrostatic head of liquid at the heated surface.

Calculations for systems of this sort are considered in texts on phase-equilibrium thermodynamics.

TEST
YOURSELF

1. What is the bubble point of a liquid mixture at a given pressure? What is the dew point of a vapor mixture at a given pressure?
2. At what temperature will an equimolar liquid mixture of benzene and toluene begin to boil at 1 atm? What is the mole fraction of benzene in the first bubble?
3. At what temperature will an equimolar vapor mixture of benzene and toluene at 1 atm begin to condense? What is the mole fraction of benzene in the first drop? What happens to the system temperature as the condensation proceeds?
4. Would you expect the bubble-point temperature of a liquid mixture to increase, decrease, or remain the same as the pressure increases? What about the dew-point temperature of a vapor mixture?
5. When you raise a liquid to its boiling point, the pressure beneath the liquid surface where the bubbles form is different from the gas-phase pressure (why?), so that the boiling does not take place at precisely the calculated temperature. Explain. If the bottom of a vat of water is heated on a day when atmospheric pressure is 1 atm and the height of the water is 5 ft, how would you estimate the boiling temperature?
6. Why does determination of T_{bp} from Equation 6.4-4 or T_{dp} from Equation 6.4-8 involve a trial-and-error calculation?

6.5   SOLUTIONS OF SOLIDS IN LIQUIDS

6.5a   Solubility and Saturation

The **solubility** of a solid in a liquid is the maximum amount of that substance that can be dissolved in a specified amount of the liquid at equilibrium. This physical property varies considerably from one solute–solvent pair to another: for example, 100 g of water at 20°C can dissolve 222 g of $AgNO_3$, 0.003 g of $AgCO_3$, and 0.00002 g of $AgBr$. The limit may also depend strongly on temperature: the solubility of $AgNO_3$ in 100 g of water increases from 222 g at 20°C to 952 g at 100°C. *Perry's Chemical Engineers' Handbook* (see footnote 1) on pp. 2-7 through 2-47 and 2-121 through 2-124 gives the solubilities of many substances in water, ethyl alcohol, and diethyl ether at specified temperatures.

A solution that contains as much of a dissolved species as it can *at equilibrium* is said to be **saturated** with that species. *A solution in equilibrium with solid solute must be saturated with that solute; if it were not, more solute would dissolve.*

If a saturated solution is cooled, the solubility of the solute generally decreases; in order for the cooled solution to return to equilibrium, some solute must come out of solution as solid crystals. The crystallization rate may be slow, however, so that a metastable condition can exist in which the concentration of the solute is higher than the equilibrium value at the solution temperature. Under such conditions, the solution is said to be **supersaturated** and the difference between actual and equilibrium concentrations is referred to as **supersaturation**. All problems involving solid–liquid separations in this text assume that equilibrium exists between the solid and liquid phases, so that supersaturation need not be considered.

EXAMPLE 6.5-1   *Crystallization and Filtration*

One hundred fifty kilograms of a saturated aqueous solution of $AgNO_3$ at 100°C is cooled to 20°C, thereby forming $AgNO_3$ crystals, which are filtered from the remaining solution. The wet filter cake, which contains 80% solid crystals and 20% saturated solution by mass, passes to a dryer in which the remaining water is vaporized. Calculate the fraction of the $AgNO_3$ in the feed stream eventually recovered as dry crystals and the amount of water that must be removed in the drying stage.

SOLUTION   *Basis: 150 kg Feed*

Both the filtrate and the liquid retained in the filter cake are in equilibrium with solid $AgNO_3$ crystals and must therefore be saturated with $AgNO_3$ at 20°C. The compositions of saturated solutions of silver nitrate at 100°C and 20°C are given at the beginning of this section and are used in the flowchart.

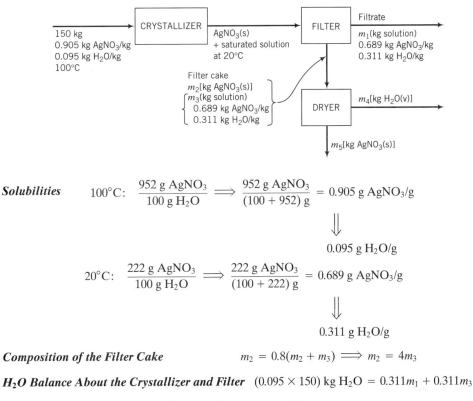

Solubilities   100°C: $\dfrac{952 \text{ g AgNO}_3}{100 \text{ g H}_2\text{O}} \Longrightarrow \dfrac{952 \text{ g AgNO}_3}{(100 + 952) \text{ g}} = 0.905 \text{ g AgNO}_3/\text{g}$

$$\Downarrow$$

$$0.095 \text{ g H}_2\text{O/g}$$

20°C: $\dfrac{222 \text{ g AgNO}_3}{100 \text{ g H}_2\text{O}} \Longrightarrow \dfrac{222 \text{ g AgNO}_3}{(100 + 222) \text{ g}} = 0.689 \text{ g AgNO}_3/\text{g}$

$$\Downarrow$$

$$0.311 \text{ g H}_2\text{O/g}$$

Composition of the Filter Cake $m_2 = 0.8(m_2 + m_3) \Longrightarrow m_2 = 4m_3$

H_2O Balance About the Crystallizer and Filter $(0.095 \times 150) \text{ kg H}_2\text{O} = 0.311m_1 + 0.311m_3$

Mass Balance About the Crystallizer and Filter $150 \text{ kg} = m_1 + m_2 + m_3$

The simultaneous solution of these three equations yields

$$m_1 = 20 \text{ kg}$$
$$m_2 = 104 \text{ kg}$$
$$m_3 = 26 \text{ kg}$$

Overall $AgNO_3$ Balance $(0.905 \times 150) \text{ kg AgNO}_3 = 0.689m_1 + m_5$

$$\Downarrow m_1 = 20 \text{ kg}$$

$$m_5 = 122 \text{ kg AgNO}_3 \text{ crystals recovered}$$

Percentage Recovery $\dfrac{122 \text{ kg AgNO}_3 \text{ recovered}}{(0.905 \times 150) \text{ kg AgNO}_3 \text{ fed}} \times 100\% = \boxed{89.9\%}$

Overall Mass Balance $150 \text{ kg} = m_1 + m_4 + m_5$

$$\Downarrow \begin{array}{l} m_1 = 20 \text{ kg} \\ m_5 = 122 \text{ kg} \end{array}$$

$$m_4 = \boxed{8 \text{ kg H}_2\text{O removed in the dryer}}$$

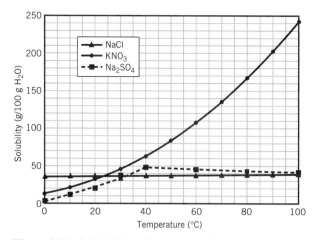

Figure 6.5-1 Solubilities of inorganic solutes.

6.5b Solid Solubilities and Hydrated Salts

The Gibbs phase rule shows that specifying temperature and pressure for a two-component system at equilibrium containing a solid solute and a liquid solution fixes the values of all other intensive variables. (Verify this statement.) Furthermore, because the properties of liquids and solids are only slightly affected by pressure, a single plot of solubility (an intensive variable) versus temperature may be applicable over a wide pressure range.

The solubility plots shown in Figure 6.5-1 illustrate how the effect of temperature on solubility can vary from system to system. Increasing the temperature from 0°C to 100°C barely changes NaCl solubility but increases the solubility of KNO_3 by more than a factor of 10. For Na_2SO_4, solubility increases up to about 40°C and decreases thereafter.

EXAMPLE 6.5-2 *Material Balances on a Crystallizer*

An aqueous potassium nitrate solution containing 60.0 wt% KNO_3 at 80°C is fed to a cooling crystallizer in which the temperature is reduced to 40°C. Determine the temperature at which the solution reaches saturation and the percentage of the potassium nitrate in the feed that forms crystals.

SOLUTION

The feed concentration must be converted to a solute/solvent ratio to use Figure 6.5-1. Since 100.0 g of the solution contains 60.0 g KNO_3 and 40.0 g H_2O, the desired ratio is

$$\frac{60.0 \text{ g } KNO_3}{40.0 \text{ g } H_2O} = 1.50 \frac{\text{g } KNO_3}{\text{g } H_2O} = \frac{150 \text{ g } KNO_3}{100 \text{ g } H_2O}$$

From Figure 6.5-1, the saturation temperature of this solution is $\boxed{74° \text{ C}}$.

Here is a flowchart of the process for an assumed basis of 100 kg feed.

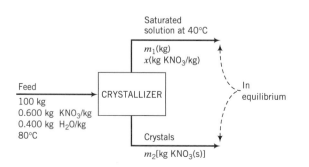

There are three unknowns on the chart (m_1, x, m_2). We will assume that the solution leaving the crystallizer is saturated at 40°C. Accordingly, the value of x may be determined from the known solubility of KNO_3 at that temperature, and the remaining two variables may be determined from material balances. From Figure 6.5-1, the solubility at 40°C is roughly 63 kg KNO_3/100 kg H_2O. The calculations follow.

$$x = \frac{63 \text{ kg } KNO_3}{(63 + 100)\text{kg solution}} = 0.386 \text{ kg } KNO_3/\text{kg}$$

H₂O Balance $\dfrac{100 \text{ kg}}{} \left| \dfrac{0.400 \text{ kg } H_2O}{\text{kg}} = \dfrac{m_1(\text{kg})}{} \right| \dfrac{(1 - 0.386) \text{ kg } H_2O}{\text{kg}} \implies m_1 = 65.1 \text{ kg}$

Mass Balance $100 \text{ kg} = m_1 + m_2 \xrightarrow{\quad m_1 = 65.1 \text{ kg} \quad} m_2 = 34.9 \text{ kg } KNO_3(s)$

The percentage of the potassium nitrate in the feed that crystallizes is therefore

$$\frac{34.9 \text{ kg } KNO_3 \text{ crystallized}}{60.0 \text{ kg } KNO_3 \text{ fed}} \times 100\% = \boxed{58.2\%}$$

The solid crystals that formed in the preceding example consisted of *anhydrous* (water-free) potassium nitrate. When certain solutes crystallize from aqueous solutions, the crystals are **hydrated salts**, containing water molecules bonded to solute molecules (**water of hydration**). The number of water molecules associated with each solute molecule may vary with the crystallization temperature.

For example, when sodium sulfate crystallizes from aqueous solution below 40°C the crystals that form are anhydrous Na_2SO_4, while above 40°C each molecule of Na_2SO_4 that crystallizes has 10 molecules of water associated with it. The hydrated salt, $Na_2SO_4 \cdot 10H_2O(s)$, is called *sodium sulfate decahydrate*. The change from the anhydrous to the hydrated form of the solid at 40°C is responsible for the discontinuity in the plot of Figure 6.5-1. Another solute that forms hydrated salts is magnesium sulfate, which can exist in five different forms in different temperature ranges. (See Table 6.5-1.)

Table 6.5-1 Hydrated $MgSO_4$ Salts

Form	Name	Wt% $MgSO_4$	Conditions
$MgSO_4$	Anhydrous magnesium sulfate	100.0	$> 100°C$
$MgSO_4 \cdot H_2O$	Magnesium sulfate monohydrate	87.0	67 to 100°C
$MgSO_4 \cdot 6 H_2O$	Magnesium sulfate hexahydrate	52.7	48 to 67°C
$MgSO_4 \cdot 7 H_2O$	Magnesium sulfate heptahydrate	48.8	2 to 48°C
$MgSO_4 \cdot 12 H_2O$	Magnesium sulfate dodecahydrate	35.8	-4 to 2°C

EXAMPLE 6.5-3 *Production of a Hydrated Salt*

An aqueous solution of magnesium sulfate at 104°C containing 30.1 wt% $MgSO_4$ is fed to a cooling crystallizer that operates at 10°C. The stream leaving the crystallizer is a slurry of solid magnesium sulfate heptahydrate particles [$MgSO_4 \cdot 7H_2O(s)$] suspended in a liquid solution. Tabulated solubility data for magnesium sulfate [*Perry's Chemical Engineers' Handbook* (see footnote 1), p. 18-35] show that a saturated solution at 10°C contains 23.2 wt% $MgSO_4$. Determine the rate at which solution must be fed to the crystallizer to produce 1 metric ton (1 tonne, 1000 kg) of magnesium sulfate heptahydrate per hour.

SOLUTION *Basis: 1 tonne MgSO₄·7H₂O(s) Produced/h*

We assume that the solution leaving the crystallizer is in equilibrium with the solid crystals and is therefore saturated with $MgSO_4$. A flowchart of the crystallizer follows:

$\dot{m}_1$(tonne /h) → CRYSTALLIZER → 1 tonne $MgSO_4 \cdot 7H_2O(s)$/h

0.301 tonne $MgSO_4$/tonne
0.699 tonne H_2O/tonne
104°C

$\dot{m}_2$(tonne solution/h)
0.232 tonne $MgSO_4$/tonne
0.768 tonne H_2O/tonne
10°C

There are two unknowns on the chart ($\dot{m}_1$ and $\dot{m}_2$) and two independent molecular species on which balances may be written ($MgSO_4$ and H_2O), so that the problem can be solved. The atomic weights on the inside back cover may be used to show that the molecular weight of anhydrous magnesium sulfate is 120.4 and that of the heptahydrate salt is 246.4. The balances follow.

Total Mass Balance $\dot{m}_1 = 1 \text{ tonne/h} + \dot{m}_2$

$MgSO_4$ Balance

$$0.301\dot{m}_1 \left(\frac{\text{tonne } MgSO_4}{\text{h}} \right) = \frac{1 \text{ tonne } MgSO_4 \cdot 7H_2O}{\text{h}} \left| \frac{120.4 \text{ tonne } MgSO_4}{246.4 \text{ tonne } MgSO_4 \cdot 7H_2O} \right.$$

$$+ \frac{\dot{m}_2(\text{tonne solution/h})}{} \left| \frac{0.232 \text{ tonne } MgSO_4}{\text{tonne solution}} \right.$$

Solving these two equations simultaneously yields $\boxed{\dot{m}_1 = 3.71 \text{ tonne/h}}$ and $\dot{m}_2 = 2.71$ tonne/h.

TEST YOURSELF

1. Solid crystals of sodium chloride are slowly added to 1000 kg of water at 60°C. After each small addition, the mixture is stirred until the salt dissolves, and then more salt is added. How much salt can be dissolved if the temperature is maintained at 60°C? What will happen if more than this quantity of salt is added? (Refer to Figure 6.5-1.)

2. An aqueous solution contains 50.0 wt% KNO_3 at 80°C. To what temperature would this solution have to be cooled before solid crystals begin to form? What happens as the solution is cooled to progressively lower temperatures?

3. What do the terms *hydrated salt, water of hydration,* and *anhydrous salt* mean? What would $MgSO_4 \cdot 4H_2O(s)$ probably be called if this species were found in nature? (*Hint:* Think about CCl_4.)

4. Given that the molecular weight of $MgSO_4$ is 120.4, what is the mass fraction of $MgSO_4$ in magnesium sulfate monohydrate?

5. Why is there a slope discontinuity at 40°C in the Na_2SO_4 solubility curve of Figure 6.5-1?

6.5c Colligative Solution Properties

Physical properties of a solution generally differ from the same properties of the pure solvent. Under certain conditions, the changes in the values of several properties—such as vapor pressure, boiling point, and freezing point—depend only on the concentration of solute in the solution, and not on what the solute and solvent are. Such properties are referred to as **colligative solution properties**. (A fourth colligative property—osmotic pressure—will not concern us in this text.)

An understanding of colligative properties is important in the determination of the operating conditions of certain processes. For example, a process can be designed to recover pure water by evaporation or freezing of seawater. In the first instance, pure water is recovered by condensation of vapor from the evaporator, while in the second it is recovered by separating and melting ice from the freezer. An engineer who intends to design or operate an evaporator or an ice crystallizer clearly must know the temperature at which the phase transition takes place—the boiling point in the first case, the freezing point in the second. In addition, measured values of colligative solution properties are often used to deduce properties of either the solvent or the solute, such as molecular weight, which cannot easily be determined by more direct means.

This section presents an introductory view of colligative properties, considering only the simple case of a solution in which the solute is nonvolatile (i.e., has a negligible vapor pressure at the solution temperature) and the dissolved solute neither dissociates (which rules out ionizing acids, bases, and salts) nor reacts with the solvent. Discussions of more complex systems can be found in most physical chemistry texts.

Consider a solution in which the solute mole fraction is x and the vapor pressure of the pure solvent at the solution temperature is p_s^*. Applying Raoult's law (Equation 6.4-1) to the solution, we obtain for the partial pressure of the solvent

$$p_s(T) = (1 - x)p_s^*(T) \tag{6.5-1}$$

If the liquid is pure solvent ($x = 0$), this equation predicts that the partial pressure of the solvent vapor equals the vapor pressure of the solvent, as we would expect. Since the solute is nonvolatile, the solvent is the only component of the liquid solution that is also in the vapor. The pressure exerted by this vapor is referred to as the *effective* solvent vapor pressure:

$$(p_s^*)_e = p_s = (1 - x)p_s^* \tag{6.5-2}$$

Since x—and hence $(1 - x)$—is less than one, the effect of the solute is to lower the effective solvent vapor pressure. The **vapor pressure lowering**, defined as the difference between the vapor pressure of the pure component and the effective vapor pressure of the solvent, is

$$\Delta p_s^* = p_s^* - (p_s^*)_e = xp_s^* \tag{6.5-3}$$

The simplicity and generality of Equation 6.5-3 are surprising. According to the equation, if a solution contains 20 mole% solute, then the solvent partial pressure is 80% of the vapor pressure of pure solvent at the system temperature, regardless of the temperature, the pressure, *and what the solute and solvent are.* (Hence, vapor pressure lowering is a colligative property, by definition.) The only stipulations are that Raoult's law holds and the solute is nonvolatile, nonreactive, and nondissociative.

The lowering of solvent vapor pressure has two important consequences. *The solvent in a solution at a given pressure boils at a higher temperature and freezes at a lower temperature than does the pure solvent at the same pressure.* The validity of these statements can be seen by referring to Figure 6.5-2, a phase diagram for an arbitrary solute–solvent system. Shown on this figure are the vapor–liquid and solid–liquid equilibrium curves for a pure solvent (solid curves) and for a solution of a fixed solute concentration (dashed curves). The solution vapor–liquid and solid–liquid equilibrium curves lie below the solvent curves, reflecting the fact that the effective vapor pressure at a given temperature and freezing point at a given pressure for the solution are lower than those of the pure solvent. The higher the solute concentration, the greater is the separation between the pure solvent and the solution curves.

The effect of the solute on the solution boiling point is easy to see from the diagram. Recall that the boiling point of a liquid at a given pressure is the intersection of a horizontal line at that pressure with the vapor–liquid equilibrium curve. At pressure P_0, the pure solvent boils at temperature T_{b0}, while the solution boils at a higher temperature, T_{bs}.

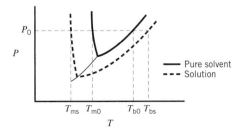

Figure 6.5-2 Phase-equilibrium curves for pure solvent and solution.

The change in the solvent freezing point is a little less obvious. First, consider the triple point—the intersection of the solid–vapor and the liquid–vapor equilibrium curves. It is clear from Figure 6.5-2 that the effect of the vapor pressure lowering is to lower the triple point of the solution relative to pure solvent. If in addition the solid–liquid equilibrium curve for the solution is (like that for the pure solvent) almost vertical, then the freezing point at an arbitrary pressure P_0 also drops—on the diagram, from T_{m0} for the pure solvent to T_{ms} for the solution.

Since we know how the solution vapor pressure varies with concentration (the relationship being given by Equation 6.5-2) and temperature (through the Clausius–Clapeyron equation, Equation 6.1-3), we can determine the relationships between concentration and both boiling point elevation and freezing point depression. The relationships are particularly simple for dilute solutions ($x \rightarrow 0$, where x is solute mole fraction).

$$\Delta T_b = T_{bs} - T_{b0} = \frac{RT_{b0}^2}{\Delta \hat{H}_v} x \tag{6.5-4}$$

$$\Delta T_m = T_{m0} - T_{ms} = \frac{RT_{m0}^2}{\Delta \hat{H}_m} x \tag{6.5-5}$$

In these equations, $\Delta \hat{H}_v$ refers to the heat of vaporization of the pure solvent at its boiling point T_{b0}, and $\Delta \hat{H}_m$ denotes the heat of fusion of the pure solvent at its melting point T_{m0}. These properties of the solvent can be looked up in data tables, such as Table B.1 of this text. Their physical significance is discussed in Chapter 8. The derivation of Equation 6.5-4 is dealt with in Problem 6.87 at the end of this chapter.

Since the coefficients of x in these two equations are constant, it follows that for dilute solutions of nonvolatile, nonreactive, nondissociative solutes, both boiling point elevation and freezing point depression vary linearly with solute mole fraction.

The next example shows applications of Equations 6.5-2 through 6.5-5 to the determination of a vapor pressure and phase-transition temperatures for a known solution concentration, and to the calculation of a solution composition and solute molecular weight from a measured colligative property.

TEST YOURSELF

1. What is a colligative solution property? Name three of them.
2. The vapor pressure of a solvent at 120°C is 1000 mm Hg. A solution contains 15 mole% of a solute in this solvent at 120°C. If the behavior described in this section is followed, what is the effective vapor pressure of the solvent? What conditions must be met for your answer to be valid?
3. The solution described in question 2 is heated to a temperature at which boiling occurs at a total pressure of 1000 mm Hg. Is the boiling temperature greater than, less than, or equal to 120°C? What is the vapor pressure of the pure solvent at the solution boiling point?
4. Explain why salt is scattered on roads and sidewalks on a snowy day.
5. Explain why antifreeze (which you may think of as a nonvolatile solute) is a useful addition to an automobile radiator in both the cold of winter and the heat of summer.

EXAMPLE 6.5-4 *Colligative Property Calculations*

A solution of 5.000 g of a solute in 100.0 g of water is heated slowly at a constant pressure of 1.00 atm and is observed to boil at 100.421°C. Estimate the molecular weight of the solute, the effective solvent vapor pressure at 25°C, and the solution freezing point at 1 atm. The necessary properties of water can be found in Table B.1.

SOLUTION

If the values of the normal boiling point and heat of vaporization of pure water (from Table B.1) and the gas constant are substituted into Equation 6.5-4, the result is

$$\Delta T_b(\text{K}) = \frac{[8.314 \text{ J/(mol·K)}](373.16 \text{ K})^2 x}{40,656 \text{ J/mol}} = 28.5x$$

From the measured boiling point elevation, $\Delta T_b = 0.421$ K, we may deduce that the mole fraction of the solute in the solution is $x = 0.421/28.5 = 0.0148$. But since the solution is known to contain $(5.000/M_s)$ mol of solute, where M_s is the solute molecular weight, and 100.0 g/18.016 g/mol $= 5.551$ mol of water, we may write

$$0.0148 = (5.000 \text{ g}/M_s)/(5.000 \text{ g}/M_s + 5.551 \text{ mol})$$

$$\Downarrow$$

$$\boxed{M_s = 60.1 \text{ g/mol}}$$

From Equation 6.5-2 the effective solvent vapor pressure at 25°C is determined from the vapor pressure of pure water at this temperature (found in Table B.3) as

$$(p_s^*)_e = (1.000 - 0.0148)(23.756 \text{ mm Hg}) = \boxed{23.40 \text{ mm Hg}}$$

Finally, substituting values of the melting point and heat of fusion of water (from Table B.1) and the gas constant into Equation 6.5-5, we obtain

$$\Delta T_m = \frac{[8.314 \text{ J/(mol·K)}](273.16 \text{ K})^2(0.0148)}{(6009.5 \text{ J/mol})} = 1.53 \text{ K} = 1.53°C$$

$$\Downarrow$$

$$T_{ms} = (0.000 - 1.53)°C = \boxed{-1.53°C}$$

CREATIVITY EXERCISE

A solution contains an unknown amount of table salt dissolved in water. List as many ways as you can think of to measure or estimate the concentration of salt in the solution without leaving the kitchen of your home. The only instruments you are allowed to bring home from work are a thermometer that covers the range $-10°C$ to $120°C$ and a small laboratory balance. (*Example:* Make up several solutions with known salt concentrations, and compare their tastes with that of the unknown solution.)

6.6   EQUILIBRIUM BETWEEN TWO LIQUID PHASES

6.6a   Miscibility and Distribution Coefficients

If water and methyl isobutyl ketone (MIBK) are mixed at 25°C, a single phase results if the mixture contains more than either 98% water or 97.7% MIBK by mass; otherwise, the mixture separates into two liquid phases, one of which contains 98% H_2O and 2% MIBK and the other 97.7% MIBK and 2.3% H_2O. Water and MIBK are examples of **partially miscible** liquids; they would be termed **immiscible** if one phase contained a negligible amount of water and the other a negligible amount of MIBK.

If a third substance is added to a two-phase liquid mixture, it distributes itself according to its relative solubility in each phase. For example, acetone is soluble in both water and chloroform—two nearly immiscible liquids—but much more so in chloroform. If a mixture of acetone and water is contacted with chloroform, a substantial portion of the acetone enters the chloroform-rich phase. Separation of the acetone and water may then be accomplished easily by allowing the mixture to settle and separating the two phases. This example illustrates the separation process of **liquid extraction**.

Suppose A and S are two nearly immiscible liquids and B is a solute distributed between the phases of an A–S mixture. The **distribution coefficient** (also known as **partition ratio**) of component B is the ratio of the mass fraction of B in the S phase to that in the A phase. *Perry's*

Chemical Engineers' Handbook (see footnote 1) on pp. 15-10 through 15-14 lists distribution coefficients for a number of ternary (three-component) liquid systems. Example 6.6-1 illustrates the use of this physical property in a material balance calculation.

EXAMPLE 6.6-1 *Extraction of Acetone from Water*

Two hundred cubic centimeters of an acetone–water mixture that contains 10.0 wt% acetone is mixed with 400.0 cm^3 of chloroform at 25°C, and the phases are then allowed to settle. What percentage of the acetone is transferred from the water to the chloroform?

SOLUTION **Basis: Given Quantities**

The densities of the pure substances are given in Table B.1:

Acetone (A)	0.792 g/cm^3
Chloroform (C)	1.489 g/cm^3
Water (W)	1.000 g/cm^3

Since density data for the acetone–water system are not readily available, let us use Equation 5.1-1 to estimate the density of the feed solution:

$$\frac{1}{\rho} = \frac{x_A}{\rho_A} + \frac{x_W}{\rho_W} = \left(\frac{0.100}{0.792} + \frac{0.900}{1.000}\right)\frac{\text{cm}^3}{\text{g}} = 1.026\ \frac{\text{cm}^3}{\text{g}}$$

$$\Downarrow$$

$$\bar{\rho} = 0.974\ \text{g/cm}^3$$

The mass of the solution fed is therefore

$$\frac{200.0\ \text{cm}^3}{}\ \left|\ \frac{0.974\ \text{g}}{\text{cm}^3}\right. = 195\ \text{g}$$

and that of the chloroform is

$$\frac{400.0\ \text{cm}^3}{}\ \left|\ \frac{1.489\ \text{g}}{\text{cm}^3}\right. = 596\ \text{g}$$

Let us next assume that chloroform and water are immiscible. (How to deal with partial miscibility is discussed in Section 6.6b.) In drawing the flowchart for this problem, it is advisable to label the quantities of each component in each of the two output streams, rather than the total stream masses and component mass fractions. (Labeling in the latter manner results in having to solve four simultaneous equations in four unknowns.)

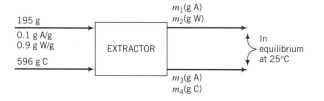

The distribution coefficient for the A-C-W system is given on p. 15-12 of *Perry's Chemical Engineers' Handbook* (see footnote 1) as 1.72. If x denotes mass fraction of acetone,

$$K = \frac{(x)_{\text{C phase}}}{(x)_{\text{W phase}}} = \frac{m_3/(m_3 + m_4)}{m_1/(m_1 + m_2)} = 1.72 \tag{6.6-1}$$

This provides one equation in the four unknowns m_1, m_2, m_3, and m_4. The others are provided by material balances.

C Balance	$596 \text{ g} = m_4$
W Balance	$(0.900)(195 \text{ g}) = m_2 \implies m_2 = 175.5 \text{ g}$
A Balance	$(0.100)(195 \text{ g}) = m_1 + m_3$

Substituting the known values of m_2 and m_4 into the first equation yields (with the acetone balance) two equations in two unknowns, which may be solved to obtain

$$m_1 = 2.7 \text{ g A in water phase}$$
$$m_3 = 16.8 \text{ g A in chloroform phase}$$

The percentage of acetone transferred is therefore

$$\frac{16.8 \text{ g acetone in chloroform phase}}{(0.100 \times 195) \text{ g acetone fed}} \times 100\% = \boxed{86.1\%}$$

In practice, extraction is often carried out in several consecutive stages, with the solution leaving each stage being contacted with additional solvent in the next stage. If enough stages are used, almost complete transfer of the solute can be achieved. Problem 6.91 at the end of this chapter illustrates this method of operation.

TEST YOURSELF

1. What is a distribution coefficient? What is liquid extraction?
2. The distribution coefficient for the system water–acetic acid–vinyl acetate is

$$\frac{\text{mass fraction of acetic acid in vinyl acetate}}{\text{mass fraction of acetic acid in water}} = 0.294$$

Is acetic acid more or less soluble in vinyl acetate than in water? If you use vinyl acetate to extract a large portion of the acetic acid in an aqueous solution, how would the relative masses of the two phases compare ($m_{VA} \ll m_W$, $m_{VA} \approx m_W$, or $m_{VA} \gg m_A$)?

6.6b   Phase Diagrams for Ternary Systems

The behavior of partially miscible ternary (three-component) systems may be represented on a **triangular phase diagram**, which may take the form of an equilateral triangle (as shown in Figure 6.6-1 for H_2O–MIBK–acetone at 25°C) or a right triangle. The latter is easier to construct on rectangular axes, but the two forms are equally easy to use. In both cases, each apex of the triangle represents a single component and edges represent binary solutions. For example, edge b on Figure 6.6-1 represents solutions of H_2O and acetone. Point K represents a mixture that is 20.0 wt% MIBK, 65.0% acetone, and 15.0% water. Any mixture whose composition falls in region A, such as at point K, is a single-phase liquid, whereas any mixture whose *overall* composition falls in region B separates into two phases.

The lines shown within region B—called **tie lines**—connect compositions of the two liquid phases in equilibrium with each other. For example, if MIBK, water, and acetone are blended so that a mixture with overall composition at point M (55.0 wt% water, 15.0% acetone, 30.0% MIBK) results, the mixture separates into phases having compositions given by points L (85 wt% water, 12% acetone, 3% MIBK) and N (4 wt% water, 20% acetone, 76% MIBK). When a mixture does not fall on a tie line, interpolation between the lines is necessary to determine the composition of each phase.

The Gibbs phase rule demonstrates that a mixture of three components that forms two liquid phases at equilibrium has three degrees of freedom. (*Verify.*) If pressure (which in any case has little effect on liquid properties) and temperature are fixed, one degree of freedom remains. Specifying the mole fraction of one of the components in one of the phases is therefore sufficient to determine the compositions of both phases. For example (referring to Figure 6.6-1), specifying that the weight fraction of acetone is 0.25 in the MIBK-rich phase fixes

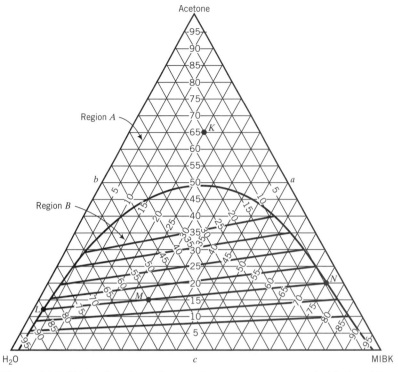

Figure 6.6-1 Triangular phase diagram for water–acetone–methyl isobutyl ketone (composition in wt%) at 25°C. (From D. F. Othmer, R. E. White, and E. Trueger, *Ind. Eng. Chem.* **33**: 1240, 1941.)

both the composition of that phase and the composition of the water-rich phase. (Verify this statement by determining both compositions.)

TEST YOURSELF

1. What is a tie line on a triangular phase diagram?
2. Show that a mixture with a composition of 4% acetone, 51% MIBK, and 45% H_2O separates into two phases. What is the composition of each phase? Calculate the ratio of the mass of the MIBK-rich phase to the mass of the H_2O-rich phase.

EXAMPLE 6.6-2 *Extraction of Acetone from Water: Use of the Phase Diagram*

One thousand kilograms of a 30.0 wt% solution of acetone in water and a second stream of pure methyl isobutyl ketone (MIBK) is fed to a mixer. The mixture is then fed to a settler where two phases form and are withdrawn separately at 25°C. How much MIBK must be fed to the process to reduce the acetone concentration in the water-rich phase to 5 wt%, assuming that the fluids remain in the settler long enough for equilibrium to be achieved?

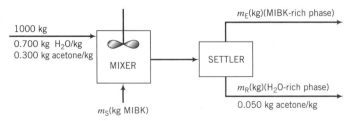

SOLUTION Since the two product streams are in equilibrium, their compositions must lie on the phase envelope and must be connected by a tie line. Accordingly, the composition of m_R is 5% acetone, 93% H_2O, and 2% MIBK and that of m_E is 10% acetone, 87% MIBK, and 3% H_2O. Overall balances can now be used to determine m_E, m_R, and m_S.

Mass Balance $m_S + 1000 \text{ kg} = m_E + m_R$

Acetone Balance $(0.30)(1000 \text{ kg}) = 0.10m_E + 0.05m_R$

H_2O Balance $(0.70)(1000 \text{ kg}) = 0.03m_E + 0.93m_R$

Solving these three equations yields (before rounding for significant figures)

$$m_E = 2667 \text{ kg}$$
$$m_R = 667 \text{ kg}$$

and

$$m_S = 2334 \text{ kg MIBK}$$

6.7 ADSORPTION ON SOLID SURFACES

The attraction of chemical species in gases and liquids to the surfaces of solids is the basis of a number of separation processes. For example, baking soda or charcoal may be placed in a refrigerator to remove unacceptable odors, and compressed air may be dried and purified by passing it through a bed of calcium chloride to remove water vapor and then through a bed of activated carbon to separate hydrocarbons taken up by the air during compression. Each of these operations utilizes a solid with an extremely high surface area (e.g., about 320 m^2/g of activated carbon) and takes advantage of the affinity of specific components in a fluid for the surface of a solid. The solid is known as an **adsorbent** and the component attracted to the solid surface is an **adsorbate**.

Adsorbate equilibrium data on a specific adsorbent are often taken at a specific temperature and are referred to as **adsorption isotherms**. These functions or plots relate X_i^*, the maximum mass of adsorbate i that can be held by a unit mass of the adsorbent, to c_i or p_i, the concentration or partial pressure of adsorbate i in the fluid contacting the solid.

Consider how an isotherm might be determined for the system carbon tetrachloride and activated carbon.

- Place a known mass of activated carbon in a chamber whose temperature is controlled at a specified value.
- Draw a vacuum in the chamber and then admit carbon tetrachloride vapor until a desired pressure is reached.
- Allow the system to come to equilibrium, read the equilibrium pressure, and determine the mass of carbon tetrachloride adsorbed by weighing the solid.
- Admit more carbon tetrachloride into the system and repeat the procedure.

Data resulting from a series of such experiments might appear as shown in Table 6.7-1.

Chapter 16 of *Perry's Chemical Engineers' Handbook* (see footnote 1) gives physical properties of several important adsorbents and several different expressions for adsorption isotherms. Equilibrium data for specific adsorbent–adsorbate systems may be found in published articles, adsorbent manufacturers' specification sheets, or company records. If no data can be found, isotherms must be obtained experimentally.

Table 6.7-1 Equilibrium Data for CCl_4 Adsorbed on Activated Carbon at 34°C

p (mm Hg)	0	1.69	3.38	6.76	8.45	11.8	20.7	32.1	40.0	84.5	104	123	133
X^*(g CCl_4/g carbon)	0	0.07	0.14	0.27	0.34	0.48	0.57	0.63	0.68	0.70	0.71	0.71	0.71

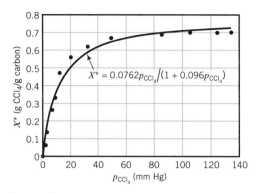

Figure 6.7-1 Langmuir adsorption isotherm for carbon tetrachloride on activated carbon at 34°C.

At low adsorbate partial pressures, isotherms may be linear:

$$X_i^* = Kc_i \qquad \text{or} \qquad X_i^* = K'p_i \tag{6.7-1}$$

The **Langmuir isotherm** is a more complex expression that is valid for some systems over a wider range of adsorbate partial pressures or concentrations.

$$X_i^* = \frac{aK_L p_i}{1 + K_L p_i} \qquad \text{or} \qquad X_i^* = \frac{aK_L' c_i}{1 + K_L' c_i} \tag{6.7-2}$$

In these equations, a, K_L and K_L' are parameters determined by fitting the equations to equilibrium data. Figure 6.7-1 shows the fit of the Langmuir isotherm (Equation 6.7-2) to the adsorption data in Table 6.7-1. The fitted parameter values are $a = 0.794$ g CCl$_4$/g carbon and $K_L = 0.096$ (mm Hg)$^{-1}$.

EXAMPLE 6.7-1   *Balances on an Adsorption Process*

A 50.0-liter tank contains an air–carbon tetrachloride mixture at 1 atm absolute, 34°C, and 30.0% relative saturation. Activated carbon is placed in the tank to adsorb CCl$_4$. The temperature of the tank contents is maintained at 34°C, and clean air is continuously supplied to the tank throughout the process to maintain the total pressure at 1.00 atm. The process may be shown schematically as follows:

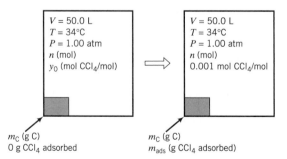

Calculate the minimum amount of activated carbon needed to reduce the CCl$_4$ mole fraction in the gas to 0.001. Neglect the volume of the activated carbon and the adsorbed CCl$_4$. Why would the actual amount placed in the tank be larger than the calculated value?

SOLUTION

The minimum amount of activated carbon is required if adsorption equilibrium is achieved in the final state, so that the adsorbent holds as much CCl$_4$ as it can. The strategy will be to determine

1. n from the ideal gas equation of state.
2. y_0 from the specified initial relative saturation.

3. p_{CCl_4} (the final partial pressure of CCl_4) = 0.001P.
4. $X^*_{CCl_4}$ (the mass ratio of adsorbed CCl_4 to carbon at equilibrium) from the Langmuir isotherm (Equation 6.7-2).
5. The mass of CCl_4 adsorbed (m_{ads}) as the difference between the mass initially present in the gas ($= y_0 n M_{CCl_4}$) and the mass present at the end ($= 0.001 n M_{CCl_4}$).
6. The mass of carbon from $X^*_{CCl_4}$ and m_{ads} ($m_C = m_{ads} X^*_{CCl_4}$).

Ideal Gas Equation of State
$$n = \frac{PV}{RT} = \frac{(1.00\ \text{atm})(50.0\ \text{L})}{\left(0.08206\ \dfrac{\text{L·atm}}{\text{mol·K}}\right)(307\ \text{K})} = 1.98\ \text{mol}$$

Initial Relative Saturation = 0.300

From the Antoine equation (Table B.4), the vapor pressure of carbon tetrachloride at 34°C is $p^*_{CCl_4} = 169$ mm Hg. Consequently,

$$\frac{p_{CCl_4}}{p^*_{CCl_4}(34°C)} = \frac{y_0 P}{169\ \text{mm Hg}} = 0.300 \xrightarrow{\ P\ =\ 760\ \text{mm Hg}\ } y_0 = 0.0667\ \text{mol}\ CCl_4/\text{mol}$$

Langmuir Isotherm

The final partial pressure of carbon tetrachloride is
$$p_{CCl_4} = y_0 P = 0.001(760\ \text{mm Hg}) = 0.760\ \text{mm Hg}$$

From Equation 6.7-2,

$$X^*_{CCl_4} = \frac{a K_L p_{CCl_4}}{1 + K_L p_{CCl_4}}$$

$$\left\Vert\begin{array}{l} a = 0.794\ \text{g}\ CCl_4/\text{g C} \\ K_L = 0.096\ (\text{mm Hg})^{-1} \\ p_{CCl_4} = 0.760\ \text{mm Hg} \end{array}\right.$$

$$X^*_{CCl_4} = 0.0540\ \frac{\text{g}\ CCl_4\ \text{ads}}{\text{g C}}$$

Mass of CCl_4 Adsorbed

$$m_{ads} = \left(\frac{0.0667\ \text{mol}\ CCl_4}{\text{mol}}\ \bigg|\ \frac{1.98\ \text{mol}}{} - \frac{0.001\ \text{mol}\ CCl_4}{\text{mol}}\ \bigg|\ \frac{1.98\ \text{mol}}{}\right)\left(\frac{154\ \text{g}\ CCl_4}{\text{mol}\ CCl_4}\right)$$

$$= 20.0\ \text{g}\ CCl_4\ \text{adsorbed}$$

Mass of Carbon Required
$$m_C = \frac{20.0\ \text{g}\ CCl_4\ \text{ads}}{0.0540\ \text{g}\ CCl_4\ \text{ads/g C}} = \boxed{370\ \text{g carbon}}$$

More activated carbon than this would be put into the container, for several reasons. First, since the rate of adsorption approaches zero as the adsorbent approaches saturation, it would take an infinite amount of time for the mole fraction of CCl_4 in the gas phase to reach 0.001. If more carbon is present, the target mole fraction would be reached in a finite amount of time (before the carbon becomes saturated). Second, the Langmuir isotherm is an approximate correlation with parameters obtained by fitting scattered experimental data, and so the estimated adsorption capacity of the adsorbent (X^*) could be too high. Third, we have assumed that nothing but CCl_4 is adsorbed on the carbon. If any oxygen, nitrogen, or other species that may be present in the gas is adsorbed, it could lower the amount of carbon tetrachloride adsorbed.

In this section we have assumed that the adsorption isotherm of an adsorbate is unaffected by the presence of constituents other than the adsorbate in the fluid mixture. If such ideality is assumed for the Langmuir isotherm developed in the previous example, you could use the derived expression for any gaseous system containing carbon tetrachloride and the same activated carbon. In reality, however, the presence of other solutes that have an affinity for the carbon surface alters the CCl_4 equilibrium behavior. An accurate system representation would require data or models for the complete multicomponent mixture.

1. What is the difference between adsorption and absorption?
2. What is the difference between an adsorbate and an adsorbent?
3. Why is it possible to use either molar concentration or partial pressure as the independent variable in the Langmuir isotherm without changing the form of the expression?
4. An **air-purifying respirator**[6]—often incorrectly called a gas mask—is a device that allows its wearer to breathe in an environment that contains low levels of a toxic substance. The inspired air passes through a filter containing an adsorbent such as activated carbon. Give a brief explanation of how the device works. How would using a nonactivated carbon affect the performance of the respirator?

6.8 SUMMARY

Two phases in contact with each other are said to be in *equilibrium* when the temperature, pressure, composition, and all other variables that characterize each phase do not change with time. Many chemical process operations—particularly separation processes such as distillation, absorption, crystallization, liquid extraction, and adsorption—work by distributing mixture components between two phases and then separating the phases. An essential step in analyzing such processes is determining how the feed mixture components distribute themselves between the two phases at equilibrium. This chapter summarizes common procedures for making this determination.

- The **phase diagram** of a pure species is a plot of pressure versus temperature that shows regions where the species exists as a solid, a liquid, or a gas; curves that bound these regions where pairs of phases can coexist in equilibrium; and a point (the *triple point*) where all three phases can coexist.

- The temperature coordinate of a point on the vapor–liquid equilibrium curve (the curve separating the liquid and vapor regions on a phase diagram) is the **boiling point** of the species at the corresponding pressure, and the pressure coordinate is the **vapor pressure** of the species at the corresponding temperature. The **normal boiling point** is the boiling point at $P = 1$ atm. Normal boiling points (and normal melting points) of selected species can be found in Table B.1. Vapor pressures at specified temperatures can be estimated using the Antoine equation (Table B.4), the Cox chart (Figure 6.1-4), and Table B.3 for water.

- The vapor pressure of a species is a measure of its **volatility**, or tendency to vaporize. Heating a liquid mixture tends to form a vapor enriched in more volatile components (those with higher vapor pressures) and to leave the residual liquid enriched in components with lower volatility. The separation process **distillation** is based on this principle. Similarly, if a gas mixture contains one or more components with relatively low volatilities, cooling the mixture in a **condensation** operation can be used to recover a liquid enriched in these components.

- The **Gibbs phase rule** gives the *degrees of freedom* of a multiphase system in equilibrium, or the number of intensive (size-independent) system variables that must be specified before the others can be determined.

- For a gas at temperature T and pressure P containing a single condensable vapor A with mole fraction y_A and vapor pressure $p_A^*(T)$, **Raoult's law** $[y_A P = p_A^*(T)]$ provides the basis for a number of definitions. If Raoult's law is satisfied, the vapor is **saturated** (or equivalently, the gas is saturated with A); if $y_A P < p_A^*(T)$, the vapor is **superheated**. If A is saturated and either the temperature is decreased or the pressure is increased, A will begin to condense. If liquid A is in contact with a gas phase and the system is at equilibrium, the A vapor in the gas must be saturated.

- If a gas containing a single superheated vapor A is cooled at constant pressure, the temperature at which the vapor becomes saturated is the **dew point** of the gas. The dew point may be determined from Raoult's law, $y_A P = p_A^*(T_{dp})$. The **degrees of superheat** is the difference

[6]For a concise discussion of both air-purifying and atmosphere-supplying respirators, see N. I. Sax and R. J. Lewis, Sr., *Hazardous Chemicals Desk Reference,* Van Nostrand Reinhold, New York, 1987, pp. 22–42.

between the actual temperature and the dew point of the gas. The **relative saturation** of the gas (or **relative humidity** for an air–water system) is the ratio of the partial pressure of the vapor to the vapor pressure at the system temperature, expressed as a percentage: $[y_A P / p_A^*(T)] \times 100\%$. If you are given the temperature, pressure, and either the dew point, the degrees of superheat, the relative saturation, or a related quantity (molal or absolute or percentage saturation), you can use Raoult's law to calculate the mole fraction of A in the gas.

- If pure liquid A is placed in an open container at pressure P and a temperature for which $p_A^*(T) < P$ and $p_A^*(T) > p_A$, the liquid **evaporates**: molecules of A transfer from the liquid surface to the surrounding gas. If the container is heated to a temperature such that $p_A^*(T) = P$, the liquid **boils**: vapor bubbles form at the heated surface and rise through the liquid into the surrounding gas. The liquid temperature remains constant as boiling continues.

- If the volatile components of a liquid mixture are all structurally similar compounds (e.g., all paraffins), the general form of **Raoult's law** may be a good approximation for all species: $y_i P = x_i p_i^*(T)$, where x_i and y_i are the mole fractions of species i in the liquid and gas phases, respectively. If the liquid is nearly pure A $(x_A \approx 1)$, Raoult's law might apply only to species A.

- In the separation process **absorption**, a gas mixture contacts a liquid solvent and one or more mixture components dissolve in the solvent. If a liquid solution contains only small amounts of a dissolved solute, A $(x_A \approx 0)$, **Henry's law** may apply to A: $y_A P = x_A H_A(T)$, where H_A is the **Henry's law constant**.

- An **ideal liquid solution** is one for which all volatile components are distributed between liquid and gas phases at equilibrium according to either Raoult's law or Henry's law.

- The **bubble-point temperature** of a liquid mixture is the temperature at which the first vapor bubble forms if the mixture is heated at constant pressure. Contrary to what many students mistakenly assume, the bubble point is *not* the boiling temperature of the most volatile species in the liquid; it is always higher than this temperature for an ideal liquid solution. The **dew-point temperature** of a vapor mixture is the temperature at which the first liquid droplet forms if the mixture is cooled at constant pressure. If Raoult's law applies to all species, either of these temperatures can be determined by trial and error using Equation 6.4-4 (for the bubble point) or Equation 6.4-7 (for the dew point).

- If a liquid mixture is heated above its bubble point, the vapor generated is rich in the more volatile mixture components. As vaporization continues, the system temperature steadily increases (unlike the case for a single-component system, in which T remains constant). Similarly, if a vapor mixture is cooled below its dew point, the liquid that condenses is rich in the less volatile components and the temperature progressively decreases.

- The **solubility** of a solid (the *solute*) in a liquid (the *solvent*) is the maximum amount of that solute that can dissolve in a specified amount of the liquid at equilibrium. A solution that contains all the dissolved solute it can hold is **saturated** with that solute. If additional solute is added, it will not dissolve unless the temperature is changed in a way that increases the solubility.

- In the separation process **crystallization**, a solution of a solute is cooled below its saturation temperature, causing solid crystals of solute to form; alternatively, solvent may be evaporated to cause solute crystallization. For aqueous solutions of some solutes in certain temperature ranges, the crystals that form are **hydrated salts**, containing **water of hydration** molecules bonded to solute molecules in specific ratios. For example, if magnesium sulfate crystallizes at a temperature above 100°C, the crystals contain **anhydrous** (water-free) $MgSO_4$, while if the crystallization takes place between 48°C and 67°C, the crystals consist of $MgSO_4 \cdot 6H_2O$ (magnesium sulfate hexahydrate).

- Provided that the solute in a solution is nonvolatile and does not react with the solvent, the vapor pressure of the solution at a given temperature is lower than that of the pure solvent, the boiling point at a given pressure is higher, and the freezing point at a given pressure is lower. Vapor pressure lowering, boiling point elevation, and freezing point depression are examples of **colligative solution properties**; formulas for them are given in Section 6.5c.

- **Liquid extraction** is a separation process in which a liquid feed solution is combined with a second solvent that is immiscible or nearly immiscible with the feed solvent, causing some (and ideally most) of the solute to transfer to the phase containing the second solvent. The **distribution coefficient** is the ratio of the solute mass fractions in the two phases at equilibrium. Its value determines how much solvent must be added to the feed solution to achieve a specified solute transfer. When the two solvents are partially miscible, a **triangular phase diagram** like that in Figure 6.6-1 simplifies balance calculations on extraction processes.

- **Adsorption** is a process in which a species in a fluid (liquid or gas) mixture adheres to the surface of a solid with which the fluid is in contact. (This process should not be confused with *ab*sorption, in which a component of a gas mixture dissolves in a liquid solvent.) The solid is the **adsorbent**, and the species that adheres to the surface is the **adsorbate**. Good adsorbents such as activated carbon have extremely high specific surface areas (m^2 surface/g solid), enabling small quantities of adsorbent to remove large quantities of adsorbate from fluid mixtures. An **adsorption isotherm** is a plot or equation that relates the equilibrium amount of an adsorbate held by a given mass of adsorbent to the adsorbate partial pressure or concentration in the surrounding gas at a specified temperature.

- Material balance calculations on separation processes follow the same procedures used in Chapters 4 and 5. If the product streams leaving a unit include two phases in equilibrium, an equilibrium relationship for each species distributed between the phases should be counted in the degree-of-freedom analysis and included in the calculations. If a species is distributed between gas and liquid phases (as in distillation, absorption, and condensation), use tabulated vapor–liquid equilibrium data, Raoult's law, or Henry's law. If a solid solute is in equilibrium with a liquid solution, use tabulated solubility data. If a solute is distributed between two immiscible liquid phases, use a tabulated distribution coefficient or equilibrium data. If an adsorbate is distributed between a solid surface and a gas phase, use an adsorption isotherm.

PROBLEMS

Unless otherwise instructed, use physical property data in this text when solving these problems.

6.1. Ten mL of pure liquid water in a cylinder with a movable piston is heated at a constant pressure of 1 atm from an initial temperature of 80°C. The temperature of the system is monitored, and the following behavior is observed:

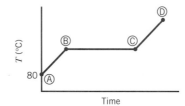

(a) What is happening in steps AB, BC, and CD? What is the temperature corresponding to the horizontal portion of the curve?

(b) Estimate the volume occupied by the water at points B and C. (Assume the vapor follows the ideal gas equation of state.)

6.2. A quantity of liquid chloroform is placed in an open, transparent, three-liter flask and boiled long enough to purge all air from the vapor space. The flask is then sealed and allowed to equilibrate at 30°C, at which temperature chloroform has a vapor pressure of 243 mm Hg. Visual inspection shows 10 mL of liquid chloroform present.

(a) What is the pressure in the flask at equilibrium? Explain your reasoning.

(b) What is the total mass (grams) of chloroform in the flask? What fraction is in the vapor phase at equilibrium?

6.3. Ethyl acetate has a vapor pressure of 118.3 mm Hg at 29.5°C and a normal boiling point of 77.0°C. Estimate the vapor pressure at 45°C using (a) the Antoine equation and constants from Table B.4; (b) the Clausius–Clapeyron equation and the two given data points; and (c) linear interpolation between the two given points. Taking the first estimate to be correct, calculate the percentage error associated with the second and third estimates.

6.4. The vapor pressure of ethylene glycol at several temperatures is given below:

T (°C)	79.7	105.8	120.0	141.8	178.5	197.3
p^* (mm Hg)	5.0	20.0	40.0	100.0	400.0	760.0

Use a semilog plot based on the Clausius–Clapeyron equation to derive an equation for p^* (mm Hg) as a function of T (°C). From the plot, estimate the heat of vaporization of ethylene glycol in kJ/mol. (Remember to use absolute temperatures in the Clausius–Clapeyron equation.)

*6.5. You are given vapor pressure data in the form of $[T\,(°C), p^*\,(mm\ Hg)]$ pairs. Construct a spreadsheet or write a computer program to perform the following tasks:

(a) Read in $T_1, p_1^*, T_2, p_2^*, \ldots, T_N, p_N^*$.

(b) Fit the Clausius–Clapeyron equation to the data using the method of least squares (Appendix A.1) or a fitting routine embedded in the spreadsheet. In completing this task you should find the values of a and b in the formula $y = ax + b$, where $y = \ln p^*$ and $x = 1/(T + 273.2)$. Print out the values of a and b.

Test your program by fitting the data for ethylene glycol given in Problem 6.4. Then use your formula to estimate the vapor pressures of this substance at 50°C, 80°C, and 110°C, and the boiling points at 760 mm Hg and 2000 mm Hg. In which of the last two values would you have the least confidence? Explain your reasoning.

6.6. The apparatus shown here is used to measure the vapor pressure of ethylene diamine.

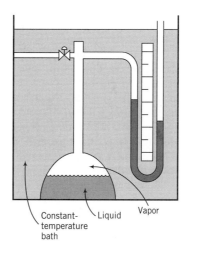

Constant-temperature bath   Liquid   Vapor

The system is loaded with pure ethylene diamine and the bath is adjusted to each of several known temperatures. The following readings are taken on a day when the atmospheric pressure is 758.9 mm Hg:

	Mercury Level	
T (°C)	Right Arm (mm)	Left Arm (mm)
42.7	138	862
58.9	160	840
68.3	182	818
77.9	213	787
88.6	262	738
98.3	323	677
105.8	383	617

---

*Computer problem.

(a) Calculate p^* for ethylene diamine at each temperature.

(b) Use a semilog plot of p^* versus $1/T$ to estimate the normal boiling point and the heat of vaporization of ethylene diamine.

(c) Does the Clausius–Clapeyron equation appear to be justified for ethylene diamine in the temperature range covered by the data? Explain.

6.7. Estimate the vapor pressure of acetone (mm Hg) at 50°C (a) from data in *Perry's Chemical Engineers' Handbook* and the Clausius–Clapeyron equation, (b) from the Cox chart (Figure 6.1-4), and (c) from the Antoine equation using parameters from Table B.4.

6.8. The vapor pressure of an organic solvent is 50 mm Hg at 25°C and 200 mm Hg at 45°C. The solvent is the only species in a closed flask at 35°C and is present in both liquid and vapor states. The volume of gas above the liquid is 150 mL. Estimate the amount of the solvent (mol) contained in the gas phase.

6.9. Liquid methyl ethyl ketone (MEK) is introduced into a vessel containing air. The system temperature is increased to 55°C, and the vessel contents reach equilibrium with some MEK remaining in the liquid state. The equilibrium pressure is 1200 mm Hg.

(a) Use the Gibbs phase rule to determine how many degrees of freedom exist for the system at equilibrium. State the meaning of your result in your own words.

(b) Mixtures of MEK vapor and air that contain between 1.8 mole% MEK and 11.5 mole% MEK can ignite and burn explosively if exposed to a flame or spark. Determine whether or not the given vessel constitutes an explosion hazard.

6.10. When a flammable liquid (such as gasoline or charcoal lighter fluid) ignites, it is not the liquid itself that burns: what actually happens is that the liquid vaporizes and the resulting air–vapor mixture burns. If the temperature is such that the percentage of the vapor in the mixture is below a certain level (the *lower flammable limit*), the liquid will not ignite if exposed to a spark or other ignition source. A match may burn in the mixture, but the flame will not spread.

(a) The *flash point* of a liquid is the lowest temperature at which the liquid vaporizes sufficiently to form an ignitable mixture with air. For example, the flash point of octane at 1 atm is 13°C (55°F), which means that dropping a match into an open container of octane is likely to start a fire on a warm summer day but not on a cold winter day. (Please do not try it!)

Suppose you are keeping two solvents in your laboratory—one with a flash point of 15°C and the other with a flash point of 75°C. How do these solvents differ from the standpoint of safety? How might you treat them differently?

(b) The lower flammable limit (LFL) of methanol in air is 6.0 mole%. Calculate the temperature at which the equilibrium percentage of methanol vapor in a saturated methanol–air mixture would equal the LFL. (This temperature is a rough estimate of the flash point.)

(c) Suppose an open container of methanol is kept at a temperature below the temperature calculated in part (b). Why would it still be unsafe to expose the container to a flame?

6.11. A gas mixture contains 10.0 mole% $H_2O(v)$ and 90.0 mole% N_2. The gas temperature and absolute pressure at the start of each of the three parts of this problem are 50°C and 500 mm Hg. Ideal gas behavior may be assumed in every part of this problem.

(a) If some of the gas mixture is put in a cylinder and slowly cooled at constant pressure, at what temperature would the first drop of liquid form?

(b) If a 30.0-liter flask is filled with some of the gas mixture and sealed and the water vapor in the flask is completely condensed, what volume (cm^3) would be occupied by the liquid water?

(c) If the gas mixture is stored in a rigid-walled cylinder and a low-pressure weather front moves in and the barometric (atmospheric) pressure drops, which of the following would change: (i) the gas density, (ii) the absolute pressure of the gas, (iii) the partial pressure of water in the gas, (iv) the gauge pressure of the gas, (v) the mole fraction of water in the gas, (vi) the dew-point temperature of the mixture?

6.12. Pure chlorobenzene is contained in a flask attached to an open-end mercury manometer. When the flask contents are at 58.3°C, the height of the mercury in the arm of the manometer connected to the flask is 747 mm and that in the arm open to the atmosphere is 52 mm. At 110°C, the mercury level is 577 mm in the arm connected to the flask and 222 mm in the other arm. Atmospheric pressure is 755 mm Hg.

(a) Extrapolate the data using the Clausius–Clapeyron equation to estimate the vapor pressure of chlorobenzene at 130°C.

(b) Air saturated with chlorobenzene at 130°C and 101.3 kPa is cooled to 58.3°C at constant pressure. Estimate the percentage of the chlorobenzene originally in the vapor that condenses. (See Example 6.3-2.)

(c) Summarize the assumptions you made in doing the calculation of part (b).

6.13. The latest weather report includes the following statement: "The temperature is 78°F, barometric pressure is 29.9 inches, and the relative humidity is 87%." From this information, estimate the mole fraction of water in the air and the dew point (°F), molal humidity, absolute humidity, and percentage humidity of the air.

6.14. It has been suggested that Atlanta–Fulton County Stadium earned the nickname "The Launching Pad" because baseballs carried farther than normal in the region's hot, humid atmosphere. With this suggestion in mind, examine the effect of temperature and humidity on the buoyant force exerted on a baseball by calculating the density (g/L) of air at the following conditions and a pressure of 1 atm:

Condition	Temperature (°F)	Relative Humidity
I	70	50%
II	70	80%
III	90	80%

Explain why your results make sense. Then argue for or against the suggestion in the first sentence.

6.15. Air at 50% relative humidity is cooled isobarically at 1 atm absolute from 90°C to 25°C.

(a) Estimate the dew point and degrees of superheat of the air at 90°C.

(b) How much water condenses (mol) per cubic meter of feed gas? (See Example 6.3-2.)

(c) Suppose a sample of the 90°C air is put in a closed variable-volume chamber containing a mirror and the pressure is raised at constant temperature until a mist forms on the mirror. At what pressure (atm) would the mist form? (Assume ideal gas behavior.)

6.16. In a device to produce potable water, humid air at 90°F, 29.7 in Hg, and 95% relative humidity is cooled to 40°F at constant pressure. What volumetric flow rate of air into the cooler (ft³/min) is required to provide 10.0 gal/min of condensed water?

6.17. Air containing 20.0 mole% water vapor at an initial pressure of 1 atm absolute is cooled in a 1-liter sealed vessel from 200°C to 15°C.

(a) What is the pressure in the vessel at the end of the process? (*Hint:* The partial pressure of air in the system can be determined from the expression $p_{air} = n_{air}RT/V$ and $P = p_{air} + p_{H_2O}$. You may neglect the volume of the liquid water condensed, but you must show that condensation occurs.)

(b) What is the mole fraction of water in the gas phase at the end of the process?

(c) How much water (grams) condenses?

6.18. Air at 90°C and 1.00 atm (absolute) contains 10.0 mole% water. A continuous stream of this air enters a compressor–condenser, in which the temperature is lowered to 15.6°C and the pressure is raised to 3.00 atm. The air leaving the condenser is then heated isobarically to 100°C. Calculate the fraction of water that is condensed from the air, the relative humidity of the air at 100°C, and the ratio m³ outlet air @ 100°C/m³ feed air @ 90°C.

6.19. Dry air is bubbled through 25.0 liters of water at a rate of 15.0 liters (STP)/min. The air leaving the liquid is saturated with water at 25°C and 1.5 atm. How long will it take for all of the water to vaporize?

6.20. A storage tank for liquid *n*-octane has a diameter of 30 ft and a height of 20 ft. During a typical 24-h period the level of liquid octane falls from 18 ft to 8 ft, after which fresh octane is pumped into the tank to return the level to 18 ft. As the level in the tank falls, nitrogen is fed into the free space to maintain the pressure at 16 psia; when the tank is being refilled, the pressure is maintained at 16 psia by discharging gas from the vapor space to the environment. The nitrogen in the tank may be considered saturated with octane vapor at all times. The average tank temperature is 90°F.

(a) What is the daily rate, in gallons and lb$_m$, at which octane is used?

(b) What is the variation in absolute pressure at the bottom of the tank in inches of mercury?

(c) How much octane is lost to the environment during a 24-h period?

(d) Why is nitrogen used in the vapor space of the tank when air would be cheaper?

6.21. A 1000-gallon tank currently contains 100.0 gallons of liquid toluene and a gas saturated with toluene vapor at 85°F and 1 atm.

    (a) What quantity of toluene (lb_m) will enter the atmosphere when the tank is filled and the gas displaced?

    (b) Suppose that 90% of the displaced toluene is to be recovered by compressing the displaced gas to a total pressure of 5 atm and then cooling it isobarically to a temperature T (°F). Calculate T.

6.22. A gas mixture containing 85.0 mole% N_2 and the balance n-hexane flows through a pipe at a rate of 100.0 m^3/h. The pressure is 2.00 atm absolute and the temperature is 100°C.

    (a) What is the molar flow rate of the gas in kmol/h?

    (b) Is the gas saturated? If not, to what temperature (°C) would it have to be cooled at constant pressure in order to begin condensing hexane?

    (c) To what temperature (°C) would the gas have to be cooled at constant pressure in order to condense 80% of the hexane?

6.23. Solids soaked with liquid hexane are dried by being contacted with nitrogen at an elevated temperature. The gas stream leaving the dryer is at 80°C, 1 atm absolute, and 50% relative saturation.

    (a) One of several possibilities for recovering the hexane from the gas is to send the stream to a cooling condenser. The gas stream leaving the condenser would contain 5.00 mole% hexane, and hexane condensate would be recovered at a rate of 1.50 kmol/min. The condenser would be operated at a pressure of 1 atm absolute.

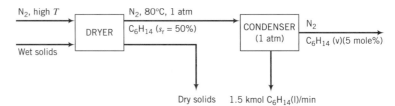

    Calculate the temperature to which the gas must be cooled and the required flow rate of fresh nitrogen to the dryer in standard cubic meters per minute (SCMM).

    (b) In an alternative arrangement, the gas leaving the dryer would be compressed to 10.0 atm and the temperature would simultaneously be increased so that the relative saturation remains at 50%. The gas then would be cooled at constant pressure to produce a stream containing 5.00 mole% hexane. Calculate the final gas temperature and the ratio of volumetric flow rates of the gas streams leaving and entering the condenser. State any assumptions you make.

    (c) What would you need to know to determine which of processes (a) and (b) is more cost-effective?

6.24. A 20,000-liter storage tank was taken out of service to repair and reattach a feed line damaged in a collision with a tanker. The tank was drained and then opened several days later for a welder to enter and perform the required work. No one realized, however, that 15 liters of liquid nonane (C_9H_{20}) remained in a collection sump at the bottom of the tank after the draining had been completed.

    (a) Nonane has a lower explosion limit of 0.80 mole% and an upper explosion limit of 2.9 mole%[7] (i.e., nonane–air mixtures at 1 atm can explode when exposed to a spark or flame if the nonane mole fraction is between the two given values). Assume any liquid nonane that evaporates spreads uniformly throughout the tank. Is it possible for the average gas-phase composition in the tank to be within the explosion limits at any time? Even when the average composition falls outside those limits, why is an explosion still a possibility? (*Hint:* Think about that assumption.)

    (b) Nonane has a vapor pressure of 5.00 mm Hg at 25.8°C and 40.0 mm Hg at 66.0°C. Use the Clausius–Clapeyron equation (6.1-3) to derive an expression for $p^*(T)$. Then calculate the temperature at which the system would have to equilibrate in order for the gas in the tank to be at the lower explosion limit.

    (c) Fortunately, a safety inspector examined the system before the welder began work and immediately canceled the work order. The welder was cited and fined for violating established safety

----

[7]N. I. Sax and R. J. Lewis, *Hazardous Chemicals Desk Reference,* Van Nostrand Reinhold, New York, 1987, p. 681.

procedures. One requirement was for the tank to be purged thoroughly with steam after being drained. What is the purpose of this requirement? (Why purge, and why with steam rather than air?) What other precautions should be taken to be sure that the welder is in no danger?

6.25. An adult takes roughly 12 breaths each minute, inhaling approximately 500 mL with each breath. Oxygen and carbon dioxide are exchanged in the lungs. The amount of nitrogen exhaled equals the amount inhaled, and the mole fraction of nitrogen in the exhaled air is 0.75. The exhaled air is saturated with water vapor at body temperature, 37°C. Estimate the increase in the rate of water loss (g/day) when a person breathing air at 23°C and a relative humidity of 50% enters an airplane in which the temperature is also 23°C but the relative humidity is 10%.

6.26. Recovering and reusing organic solvents (rather than discharging the solvents in waste streams) is an important part of the operation of most chemical plants. The magnitude of these recovery efforts can be staggering: in recent years the Eastman Chemical Company used 3.6 billion pounds of solvents and recovered 3.5 billion pounds (97%). Eastman's installation of a $26 million acetone-recovery system reduced acetone emissions by 50% in the division that had been responsible for most of these emissions.[8]

In an acetone-recovery process, a gas stream containing 20.0 mole% acetone and the remainder nitrogen leaves a chemical plant at 90°C and 1 atm. The stream is cooled at constant pressure in a condenser, enabling some of the acetone vapor to be recovered as a liquid. The nitrogen and uncondensed acetone are discharged to the atmosphere.

(a) Give two major benefits of recovering the acetone.

(b) Two cooling fluids are available—cooling-tower water at 20°C and a refrigerant at −35°C. (At the latter temperature, the vapor pressure of acetone is essentially zero.) For each fluid, calculate the percentage acetone recovery [(mol acetone condensed/mol acetone fed to condenser) × 100%], assuming that the condenser temperature equals the coolant temperature.

(c) What more would you need to know to decide which coolant to use?

(d) In a real system, the condenser temperature could never be as low as the initial cooling fluid temperature. Why not? (*Hint:* In a condenser, heat is transferred from the process fluid to the cooling fluid.) Explain how this fact would affect the percentage solvent recovery.

6.27. On a hot summer day the temperature is 35°C, barometric pressure is 103 kPa, and the relative humidity is 90%. An air conditioner draws in outside air, cools it to 20°C, and delivers it at a rate of 12,500 L/h. Calculate the rate of moisture condensation (kg/h) and the volumetric flow rate of the air drawn from the outside.

6.28. An air conditioner is designed to bring 10,000 ft³/min of outside air (90°F, 29.8 in Hg, 88% relative humidity) to 40°F, thereby condensing a portion of the water vapor, and then to reheat the air, releasing it into a room at 65°F. Calculate the rate of condensation (gallons H_2O/min) and the volumetric flow rate of the air delivered to the room. (*Suggestion:* On the flowchart, treat the cooling–condensation and the reheating as separate process steps.)

6.29. The air in a building is to be maintained at 25°C and 55% relative humidity by passing outside air through a water spray. The air enters the spray chamber at 32°C and 70% relative humidity, leaves the chamber cooled and saturated with water vapor, and is then reheated to 25°C. Estimate the temperature of the air leaving the spray chamber and the water (kg) added to or removed from (specify which) each kilogram of dry air processed.

6.30. A **hygrometer** is used to measure the moisture content of humid air. Calibration of the instrument leads to a straight line on a semilog plot of y, the mole fraction of water in air (logarithmic scale), versus H, the instrument reading (linear scale).

Room air is charged into the hygrometer sample chamber on a day when the temperature is 22°C, barometric pressure is 1.00 atm, and the relative humidity is 40%. The resulting meter reading is $H = 5.0$. A second measurement is then made by heating water to 50°C in a sealed flask containing air. The system is allowed to equilibrate at a pressure of 839 mm Hg with liquid still present in the flask, and a sample of the air above the liquid is withdrawn and injected into the sample chamber (which is heated to prevent condensation). The meter reading in this case is $H = 48$.

---

[8]A Pollution Prevention Report from the Chemical Manufacturers Association, A Chemical Industry Progress Report (1988–1992), Responsible Care—A Public Commitment.

(a) Determine the expression for y as a function of H.

(b) Suppose you wish to condition air at 35°C and 1 atm to produce air at 22°C, 1 atm, and 40% relative humidity. The air conditioner first cools the air, condensing the necessary amount of water, and then reheats the remaining air to 22°C. A sample of the outside air is injected into the hygrometer chamber, and the resulting reading is $H = 30$. Calculate the temperature to which the air must be cooled before it is reheated and determine the amount of water condensed in kg/m³ of delivered conditioned air.

*6.31. Recovery of a solvent vapor from a gas stream by condensation can be achieved by cooling the gas, by compressing it, or by a combination of these operations. The greater the compression, the less cooling is needed.

(a) A gas mixture at a pressure P_0 and temperature T_0 is the feed to a recovery process. A single condensable vapor and several noncondensable gases are present in the mixture, giving the feed a dew point of T_{d0}. A fraction f of the vapor is to be condensed. The vapor pressure $p^*(T)$ of the condensable component may be expressed as a function of temperature with the Antoine equation. For a gas feed rate of $\dot{n}_0$, draw and label a flowchart. Then derive the following relationship for the final condenser pressure in terms of the final temperature T_f and the specified feed conditions and fractional solvent recovery:

$$P_f = \frac{p^*(T_f)[1 - f p^*(T_{d0})/P_0]}{(1 - f)p^*(T_{d0})/P_0}$$

(b) The cost of refrigeration equipment and the compressor can be estimated using the empirical formulas[9]

$$C_{refr}(\$/\text{kmol feed gas}) = 2000 + 27(\Delta T)^2$$
$$C_{comp}(\$/\text{kmol feed gas}) = 4500 + 5.58(\Delta P)$$

where $\Delta T(°C) = T_f - T_0$ and $\Delta P(\text{mm Hg}) = P_f - P_0$. Your task is to prepare a spreadsheet to estimate the operating cost of a process in which ethylbenzene is recovered from an ethylbenzene–nitrogen gas mixture. The spreadsheet should have the following form:

Condensation of ethylbenzene from nitrogen											
Antoine constants for ethylbenzene											
A =	6.95719										
B =	1424.26										
C =	213.206										
Run	T0	P0	Td0	f	Tf	p*(Td0)	p*(Tf)	Pf	Crefr	Ccomp	Ctot
1	50	765	40	0.95	45	21.493	27.62	19137	2675	107013	109688
2	50	765	40	0.95	40						
3	50	765	40	0.95	35						
4	50	765	40	0.95	45						

Enter the values in the first six columns of the first row of the 12-column table $(1, 50, \ldots, 45)$ and enter formulas in the next six columns (including $C_{tot} = C_{refr} + C_{comp}$).

---

*Computer problem.

[9]These formulas are fictitious. Real cost-estimation formulas can be found in a number of texts including M. S. Peters and K. D. Timmerhaus, *Plant Design and Economics for Chemical Engineers*, 4th Edition, McGraw-Hill, New York, 1991; W. D. Seider, J. D. Seader, and D. R. Lewin, *Process Design Principles*, John Wiley & Sons, New York, 1999; and G. D. Ulrich, *A Guide to Chemical Engineering Process Design and Economics*, John Wiley & Sons, New York, 1984.

The row shown above for Run 1 contains results for a feed gas at 50°C and 765 mm Hg with a dew point of 40°C, from which 95% of the ethylbenzene is to be recovered by cooling the mixture to 45°C. The output shows that the mixture must be compressed to 19,137 mm Hg to achieve the desired recovery, and that the costs of refrigeration and compression and the total cost ($/kmol feed gas) are, respectively, $2675, $107,013, and $109,688.

When you have constructed the spreadsheet and duplicated the results just described for Run 1, (i) copy that row into the next three rows and change the values in the first six columns to duplicate those shown above; (ii) let Runs 2 and 3 stand; and (iii) in Run 4, vary the value of T_f to find the most cost-effective final temperature and pressure for the given feed conditions and fractional recovery, noting what happens to P_f, C_{refr}, C_{comp}, and C_{tot} as you carry out your search.

(c) Use the results for Runs 1–3 to deduce the effect of lowering the final temperature on the pressure required to achieve a specified fractional recovery of ethylbenzene. Explain why this result makes sense.

(d) Summarize the effect of T_f on the refrigeration and compression costs and explain why the total cost has a minimum.

6.32. A gas stream containing 40.0 mole% hydrogen, 35.0% carbon monoxide, 20.0% carbon dioxide, and 5.0% methane is cooled from 1000°C to 10°C at a constant absolute pressure of 35.0 atm. Gas enters the cooler at 120 m^3/min and upon leaving the cooler is fed to an absorber, where it is contacted with refrigerated liquid methanol. The methanol is fed to the absorber at a molar flow rate 1.2 times that of the inlet gas and absorbs essentially all of the CO_2, 98% of the methane, and none of the other components of the feed gas. The gas leaving the absorber, which is saturated with methanol at −12°C, is fed to a cross-country pipeline.

(a) Calculate the volumetric flow rate of methanol entering the absorber (m^3/min) and the molar flow rate of methanol in the gas leaving the absorber. *Do not assume ideal gas behavior when doing PVT calculations.*

(b) What is a possible intended use of the product gas? Why is it desirable to remove the CO_2 from the gas prior to feeding it to the pipeline?

6.33. A hot-air dryer is used to reduce the moisture content of 1500 kg/min of wet wood pulp from 0.75 kg H_2O/kg dry pulp to 0.15 wt% H_2O. Air is drawn from the atmosphere at 28°C, 760 mm Hg, and 50% relative humidity, sent through a blower-heater, and then fed to the dryer. The air leaves the dryer at 80°C and 10 mm Hg (gauge). A sample of the exit air is drawn into a chamber containing a mirror and cooled slowly, keeping the gauge pressure at 10 mm Hg. A mist is observed to form on the mirror at a temperature of 40.0°C. Calculate the mass of water removed from the pulp (kg/min) and the volumetric flow rate of air entering the system (m^3/min).

6.34. Wet leather containing 61% water enters a continuous tunnel dryer and leaves at a rate of 485 lb$_m$/h containing 6.0% water. Dry air enters the dryer at 140°F and 1 atm, and the outlet air is at 130°F and 1 atm with a relative humidity of 50%. Calculate the rate at which wet leather enters the dryer and the volumetric flow rate of the inlet air (ft^3/h).

6.35. *n*-Hexane is used to extract oil from soybeans. The solid residue from the extraction unit, which contains 0.78 kg liquid hexane/kg dry solids, is contacted in a dryer with nitrogen that enters at 85°C. The solids leave the dryer containing 0.05 kg liquid hexane/kg dry solids, and the gas leaves the dryer at 80°C and 1.0 atm with a relative saturation of 70%. The gas is then fed to a condenser in which it is compressed to 5.0 atm and cooled to 28°C, enabling some of the hexane to be recovered as condensate.

(a) Calculate the fractional recovery of hexane (kg condensed/kg fed in wet solids).

(b) A proposal has been made to split the gas stream leaving the condenser, combining 90% of it with fresh makeup nitrogen, heating the combined stream to 85°C, and recycling the heated stream to the dryer inlet. What fraction of the fresh nitrogen required in the process of part (a) would be saved by introducing the recycle? What costs would be incurred by introducing the recycle?

6.36. In the final stage of the manufacturing process for a solid organic product, the product is cleaned with liquid toluene and then dried in a process whose flowchart is shown on the next page.[10]

---

[10]Adapted from *Professional Engineering Examinations,* Vol. 1 (1965–1971), National Council of Engineering Examiners, p. 60.

The wet product enters the dryer at a rate of 300 lb_m/h containing 0.200 lb_m toluene/lb_m dry solids. A stream of nitrogen at 200°F, 1.2 atm, and containing a small amount of toluene vapor also enters the dryer. (A higher temperature would cause the product to soften and degrade.) Heat is transferred in the dryer from the gas to the wet solids, causing most of the toluene to evaporate. The final product contains 0.020 lb_m toluene/lb_m dry solids. Gas leaves the dryer at 150°F and 1.2 atm with a relative saturation of 70% and passes through a water-cooled condenser. Gas and liquid streams leave the condenser in equilibrium at 90°F and 1 atm. The gas is reheated to 200°F and reenters the dryer.

(a) Briefly explain this process in your own words. In your explanation, include the purposes of the condenser and the nitrogen reheater and a likely reason that nitrogen rather than air is used as the recirculating gas. What do you suppose happens to the liquid toluene leaving the condenser?

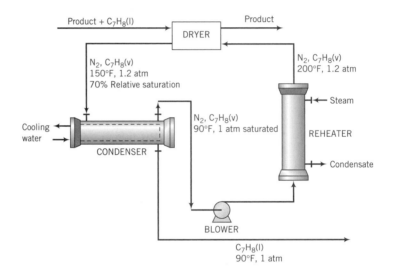

(b) Calculate the compositions (component mole fractions) of the gas streams entering and leaving the dryer, the circulation rate of dry nitrogen (lb_m/h), and the volumetric flow rate of gas entering the dryer (ft^3/h).

6.37. n-Hexane is burned with excess air. An analysis of the product gas yields the following *dry-basis* molar composition: 82.1% N_2, 6.9% CO_2, 2.1% CO, 8.6% O_2, and 0.265% C_6H_{14}. The stack gas emerges at 760 mm Hg. Calculate the percentage conversion of hexane, the percentage excess air fed to the burner, and the dew point of the stack gas, taking water to be the only condensable species.

6.38. A fuel gas containing methane and ethane is burned with air in a furnace, producing a stack gas at 300°C and 105 kPa (absolute). The stack gas contains CO_2 at a partial pressure of 80 mm Hg and no CO, O_2, methane, or ethane. Calculate the mole fraction of methane in the fuel and the dew-point temperature of the stack gas.

6.39. A mixture of propane and butane is burned with air. Partial analysis of the stack gas produces the following dry-basis volume percentages: 0.0527% C_3H_8, 0.0527% C_4H_{10}, 1.48% CO, and 7.12% CO_2. The stack gas is at an absolute pressure of 780 mm Hg and the dew point of the gas is 46.5°C. Calculate the molar composition of the fuel.

6.40. An important parameter in the design of gas absorbers is the ratio of the flow rate of the feed liquid to that of the feed gas. The lower the value of this ratio, the lower the cost of the solvent required to process a given quantity of gas but the taller the absorber must be to achieve a specified separation.

Propane is recovered from a 7 mole% propane–93% nitrogen mixture by contacting the mixture with liquid n-decane. An insignificant amount of decane is vaporized in the process, and 98.5% of the propane entering the unit is absorbed.

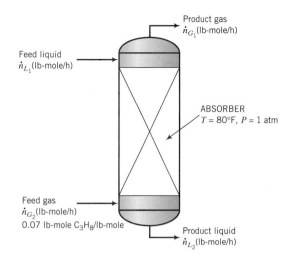

(a) The highest possible propane mole fraction in the exiting liquid would be that in equilibrium with the propane mole fraction in the entering gas (a condition requiring an infinitely tall column). Using the Cox chart (Figure 6.1-4) and Raoult's law to relate the mole fractions of propane in the entering gas and exiting liquid, calculate the ratio $(\dot{n}_{L_1}/\dot{n}_{G_2})$ corresponding to this limiting condition.

(b) Suppose the actual feed ratio $(\dot{n}_{L_1}/\dot{n}_{G_2})$ is 1.2 times the value calculated in part (a) and the percentage of the entering propane absorbed is the same (98.5%). Calculate the mole fraction of propane in the exiting liquid.

(c) What are the costs and benefits associated with increasing $(\dot{n}_{L_1}/\dot{n}_{G_2})$ from its minimum value [the value calculated in part (a)]? What would you have to know to determine the most cost-effective value of this ratio?

6.41. A liquid stream consisting of 12.5 mole% *n*-butane and the balance a heavy nonvolatile hydrocarbon is fed to the top of a stripping column, where it is contacted with an upward-flowing stream of nitrogen. The residual liquid leaves the bottom of the column containing all of the heavy hydrocarbon, 5% of the butane entering the column, and a negligible amount of dissolved nitrogen.

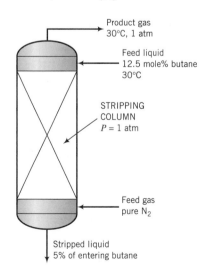

(a) The highest possible butane mole fraction in the exiting gas would be that in equilibrium with the butane in the entering liquid (a condition that would require an infinitely tall column to achieve). Using Raoult's law to relate the mole fractions of butane in the entering liquid and exiting gas, calculate the molar feed stream ratio (mol gas fed/mol liquid fed) corresponding to this limiting condition.

(b) Suppose the actual mole fraction of butane in the exit gas is 80% of its theoretical maximum value and the percentage stripped (95%) is the same as in part (a). Calculate the ratio (mol gas fed/mol liquid fed) for this case.

(c) Increasing the nitrogen feed rate for a given liquid feed rate and butane recovery decreases the cost of the process in one way and increases it in another. Explain. What would you have to know to determine the most cost-effective value of the gas/liquid feed ratio?

6.42. Nitric acid is used extensively for the production of inorganic and organic nitrates, for metal treatments of various kinds, and for photoengraving. It is produced by oxidizing ammonia to nitric oxide over a platinum–rhodium catalyst, oxidizing the nitric oxide to nitrogen dioxide, and dissolving the NO_2 in water:

$$4\,NH_3(g) + 5\,O_2(g) \rightarrow 4\,NO(g) + 6\,H_2O(g)$$
$$2\,NO(g) + O_2(g) \rightarrow 2\,NO_2(g)$$
$$3\,NO_2(g) + H_2O(l) \rightarrow 2\,HNO_3(aq) + NO(g)$$

A side reaction that lowers the product yield is the oxidation of ammonia to nitrogen and water vapor:

$$4\,NH_3(g) + 3\,O_2(g) \rightarrow 2\,N_2(g) + 6\,H_2O(g)$$

Saturated ammonia vapor, produced by vaporizing pure liquid ammonia at 820 kPa absolute, is mixed with a stoichiometric quantity of air, and the combined stream enters a converter. Prior to being mixed with the ammonia, the air is compressed and passed through a preheater. It enters the compressor at 30°C and 1 atm with a relative humidity of 50%, and it exchanges heat in the preheater with the gases emerging from the converter. The quantity of oxygen in the feed is the amount theoretically required to convert all of the ammonia to HNO_3.

In the converter, the ammonia reacts completely, with 97% forming NO and the balance forming N_2. In the short time in which the reaction mixture is in the presence of the catalyst (less than 0.001 s), a negligible amount of NO_2 is formed. The product gas is subjected to a series of cooling and hydration steps in which the NO is completely oxidized to NO_2, which in turn combines with water (some of which is present in the product gas, the rest of which is added) to form a 55 wt% aqueous nitric acid solution. The NO formed in the latter reaction is reoxidized and the added NO_2 is hydrated to form still more HNO_3. The product gas from the process may be taken to contain only N_2 and O_2. A simplified flowchart of the process follows.

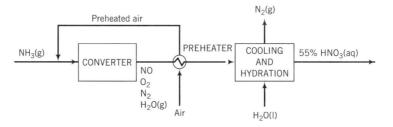

(a) Taking a basis of 100 mol of ammonia fed to the process, calculate (i) the volume (m³) of the ammonia vapor and of the air fed to the process, using the compressibility factor equation of state for the ammonia calculation; (ii) the moles and molar composition of the gas leaving the converter; and (iii) the required feed of liquid water (m³) to the cooling and hydration step.

(b) Scale up the results calculated in part (a) to a new basis of 1000 metric tons of 55% nitric acid solution produced.

6.43. A dry gas containing 10.0% NH_3 by volume is contacted with water at 10°C and 1 atm in a single-stage bubble contactor. The effluent liquid and gas streams may be considered to be in equilibrium with each other. A small slip stream taken from the effluent liquid is fed to a continuous densitometer, which indicates that the liquid density is 0.9534 g/mL.

(a) Using tabulated data from *Perry's Chemical Engineers' Handbook* (pp. 2-85, 2-87, and 2-99),[11] estimate the percentage of the ammonia in the feed that is removed in the contactor.

---

[11]R. H. Perry and D. W. Green, Eds., *Perry's Chemical Engineers' Handbook,* 7th Edition, McGraw-Hill, New York, 1997.

(b) Why is it important to maintain the slip stream and densitometer chamber at a known temperature at or below the temperature of the contactor?

6.44. Sulfur trioxide (SO_3) dissolves in and reacts with water to form an aqueous solution of sulfuric acid (H_2SO_4). The vapor in equilibrium with the solution contains both SO_3 and H_2O. If enough SO_3 is added, all of the water reacts and the solution becomes pure H_2SO_4. If still more SO_3 is added, it dissolves to form a solution of SO_3 in H_2SO_4, called **oleum** or **fuming sulfuric acid**. The vapor in equilibrium with oleum is pure SO_3. A 20% oleum by definition contains 20 kg of dissolved SO_3 and 80 kg of H_2SO_4 per hundred kilograms of solution. Alternatively, the oleum composition can be expressed as % SO_3 by mass, with the constituents of the oleum considered to be SO_3 and H_2O.
   (a) Prove that a 15.0% oleum contains 84.4% SO_3.
   (b) Suppose a gas stream at 40°C and 1.2 atm containing 90 mole% SO_3 and 10% N_2 contacts a liquid stream of 98 wt% H_2SO_4 (aq), producing 15% oleum at the tower outlet. Tabulated equilibrium data indicate that the partial pressure of SO_3 in equilibrium with this oleum is 1.15 mm Hg. Calculate (i) the mole fraction of SO_3 in the outlet gas if this gas is in equilibrium with the liquid product at 40°C and 1 atm, and (ii) the ratio (m^3 gas feed)/(kg liquid feed).

6.45. State whether you would use Raoult's law or Henry's law to perform vapor–liquid equilibrium calculations for each component in the following liquid mixtures: (a) water and dissolved nitrogen; (b) hexane, octane, and decane; and (c) club soda or any other carbonated beverage.

6.46. A gas containing nitrogen, benzene, and toluene is in equilibrium with a 40 mole% benzene–60 mole% toluene liquid mixture at 100°C and 10 atm. Estimate the gas-phase composition (mole fractions) using Raoult's law.

6.47. Using Raoult's law or Henry's law for each substance (whichever one you think appropriate), calculate the pressure and gas-phase composition (mole fractions) in a system containing a liquid that is 0.3 mole% N_2 and 99.7 mole% water in equilibrium with nitrogen gas and water vapor at 80°C.

6.48. The pressure in a vessel containing methane and water at 70°C is 10 atm. At the given temperature, the Henry's law constant for methane is 6.66×10^4 atm/mole fraction. Estimate the mole fraction of methane in the liquid.

6.49. When air ($\approx$21 mole% O_2, 79 mole% N_2) is placed in contact with 1000 cm^3 of liquid water at body temperature, 36.9°C, and 1 atm absolute, approximately 14.1 standard cubic centimeters [cm^3(STP)] of gas are absorbed in the water at equilibrium. Subsequent analysis of the liquid reveals that 33.4 mole% of the dissolved gas is oxygen and the balance is nitrogen.
   (a) Estimate the Henry's law coefficients (atm/mole fraction) of oxygen and nitrogen at 36.9°C.
   (b) An adult absorbs approximately 0.4 g O_2/min in the blood flowing though the lungs. Assuming that blood behaves like water and that it enters the lungs free of oxygen, estimate the flow rate of blood into the lungs in L/min.
   (c) The actual flow rate of blood into the lungs is roughly 5 L/min. Identify the assumptions made in the calculation of part (b) that are likely causes of the discrepancy between the calculated and actual blood flows.

6.50. The **solubility coefficient** of a gas may be defined as the number of cubic centimeters (STP) of the gas that dissolves in 1 cm^3 of a solvent under a partial pressure of 1 atm. The solubility coefficient of CO_2 in water at 20°C is 0.0901 cm^3 CO_2(STP)/cm^3 H_2O(l).
   (a) Calculate the Henry's law constant in atm/mole fraction for CO_2 in H_2O at 20°C from the given solubility coefficient.
   (b) How many grams of CO_2 can be dissolved in a 12-oz bottle of soda at 20°C if the gas above the soda is pure CO_2 at a gauge pressure of 2.5 atm (1 liter = 33.8 fluid ounces)? Assume the liquid properties are those of water.
   (c) What volume would the dissolved CO_2 occupy if it were released from solution at body temperature and pressure—37°C and 1 atm?

6.51. The sulfur dioxide content of a stack gas is monitored by passing a sample stream of the gas through an SO_2 analyzer. The analyzer reading is 1000 ppm SO_2 (parts per million on a molar basis). The sample gas leaves the analyzer at a rate of 1.50 L/min at 30°C and 10.0 mm Hg gauge and is bubbled through a tank containing 140 liters of initially pure water. In the bubbler, SO_2 is absorbed and water evaporates. The gas leaving the bubbler is in equilibrium with the liquid in the bubbler at 30°C and 1 atm absolute. The SO_2 content of the gas leaving the bubbler is periodically monitored with the

SO$_2$ analyzer, and when it reaches 100 ppm SO$_2$ the water in the bubbler is replaced with 140 liters of fresh water.

(a) Speculate on why the sample gas is not just discharged directly into the atmosphere after leaving the analyzer. Assuming that the equilibrium between SO$_2$ in the gas and dissolved SO$_2$ is described by Henry's law, explain why the SO$_2$ content of the gas leaving the bubbler increases with time. What value would it approach if the water were never replaced? Explain. (The word "solubility" should appear in your explanation.)

(b) Use the following data for aqueous solutions of SO$_2$ at 30°C[12] to estimate the Henry's law constant in units of mm Hg/mole fraction:

g SO$_2$ dissolved/100 g H$_2$O(l)	0.0	0.5	1.0	1.5	2.0
p_{SO_2} (mm Hg)	0.0	42	85	129	176

(c) Estimate the SO$_2$ concentration of the bubbler solution (mol SO$_2$/liter), the total moles of SO$_2$ dissolved, and the molar composition of the gas leaving the bubbler (mole fractions of air, SO$_2$, and water vapor) at the moment when the bubbler solution must be changed. Make the following assumptions:

- The feed and outlet streams behave as ideal gases.
- Dissolved SO$_2$ is uniformly distributed throughout the liquid.
- The liquid volume remains essentially constant at 140 liters.
- The water lost by evaporation is small enough for the total moles of water in the tank to be considered constant.
- The distribution of SO$_2$ between the exiting gas and the liquid in the vessel at any instant of time is governed by Henry's law, and the distribution of water is governed by Raoult's law (assume $x_{H_2O} \approx 1$).

(d) Suggest changes in both scrubbing conditions and the scrubbing solution that might lead to an increased removal of SO$_2$ from the feed gas.

6.52. A vapor stream that is 65 mole% styrene and 35 mole% toluene is in equilibrium with a liquid mixture of the same two species. The pressure in the system is 150 mm Hg absolute. Use Raoult's law to estimate the composition of the liquid and the system temperature.

6.53. A gas containing nitrogen, benzene, and toluene is in equilibrium with a liquid consisting of 35 mole% benzene and 65 mole% toluene at 85°C and 10 atm. Estimate the gas composition (mole fractions) using Raoult's law and assuming ideal gas behavior.

6.54. A liquid mixture containing 50 mole% propane, 30% n-butane, and 20% isobutane is stored in a rigid container at 77°F. The container has a maximum allowable working pressure of 200 psig. The head space above the liquid contains only vapors of the three hydrocarbons.

(a) Show that the container is currently safe, using Raoult's law and the Cox chart (Figure 6.1-4) in your calculations.

(b) Obtain a rough estimate of the temperature above which the maximum allowable pressure would be exceeded. Comment on the suitability of the container to store the given mixture.

6.55. A closed system contains an equimolar mixture of n-pentane and isopentane.

(a) Suppose the system is initially all liquid at 120°C and a high pressure, and the pressure is gradually reduced at a constant temperature. Estimate the pressures at which the first bubble of vapor forms and at which the last drop of liquid evaporates. Also calculate the liquid and vapor compositions (mole fractions) at those two conditions. (*Suggestion:* Use a spreadsheet.)

(b) Now suppose the system starts as a vapor at 1200 mm Hg gauge and a high temperature, and the temperature is gradually reduced at constant pressure. Estimate the temperatures at which the first drop of liquid forms and at which the last bubble of vapor condenses. Also calculate the liquid and vapor compositions (mole fractions) at those two conditions.

6.56. Nitrogen is bubbled through a liquid mixture that initially contains equimolar amounts of benzene and toluene. The system pressure is 3 atm and the temperature is 80°C. The nitrogen flow rate is

---

[12]R. H. Perry and D. W. Green, Eds., *Perry's Chemical Engineers' Handbook,* 7th Edition, McGraw-Hill, New York, 1997, p. 2-77.

10.0 standard liters per minute. The gas leaving the bubbler is saturated with benzene and toluene vapors.

(a) Estimate the initial rates (mol/min) at which benzene and toluene leave the bubbler.

(b) How will the mole fractions of benzene and toluene in the liquid change with time (increase, decrease, or remain constant)? Explain your answer.

(c) How will the mole fractions of benzene and toluene in the exiting gas change with time (increase, decrease, or remain constant)? Explain your answer.

6.57. Calculate the following:

(a) The bubble-point temperature of an equimolar mixture of liquid n-hexane and n-heptane at 1.0 atm and the composition (mole fractions) of the vapor in equilibrium with this mixture.

(b) The dew-point temperature of a gas mixture with a molar composition of 30% n-hexane, 30% n-heptane, and 40% air at 1 atm and the composition (mole fractions) of the liquid in equilibrium with this mixture.

*6.58. A liquid mixture contains N components (N may be any number from 2 to 10) at pressure P(mm Hg). The mole fraction of the ith component is x_i ($i = 1, 2, \ldots, N$), and the vapor pressure of that component is given by the Antoine equation (see Table B.4) with constants A_i, B_i, and C_i. Raoult's law may be applied to each component.

(a) Write the equations you would use to calculate the bubble-point temperature of the mixture, ending with an equation of the form $f(T) = 0$. (The value of T that satisfies this equation is the bubble-point temperature.) Then write the equations for the component mole fractions (y_1, $y_2, \ldots, y_N$) in the first bubble that forms, assuming that the temperature is now known.

(b) Prepare a spreadsheet to perform the calculations of part (a). The spreadsheet should include a title line and two tables: the first table should contain the Antoine-equation constants and the total pressure, and the second should contain columns for the liquid-phase mole fractions, guessed values of the bubble-point temperature, any intermediate quantities generated in the bubble-point calculation (such as vapor pressures at the guessed temperatures), the function $f(T)$, and the values of the vapor-phase mole fractions. Enter the values of A_i, B_i, C_i, P, and x_i for each species in the mixture, assume a value of T, and enter formulas for the other variables in the spreadsheet including f. Then determine the bubble-point temperature by using the goalseek tool (or simple trial and error) to find the value of T for which $f = 0$. Test your program by calculating the bubble-point temperatures and vapor compositions for liquids at 760 mm Hg containing (i) 22.6 mole% benzene, 44.3 mole% ethylbenzene, and the balance toluene; (ii) 44.3 mole% benzene, 22.6% ethylbenzene, and the balance toluene; and (iii) 22.6 mole% benzene, 22.6% ethylbenzene, and the balance toluene. Briefly explain why the variations in bubble-point temperature for these three cases make sense.

(c) Write a computer program to perform the calculations of part (b), and test it using the same three cases. To calculate the bubble-point temperature, evaluate f for the first guessed value of T, and then vary T in increments of $\pm 5°C$ until the value of f changes sign from its initial value. Use the two values of T for which the corresponding values of f bracket 0 as the starting point for a *regula-falsi* method calculation (Appendix A.2c), stopping when $|f(T)| < 1.0 \times 10^{-4}$.

6.59. A vapor mixture of n-butane (B) and n-hexane (H) contains 50.0 mole% butane at 120°C and 1.0 atm. A stream of this mixture flowing at a rate of 150.0 L/s is cooled and compressed, causing some but not all of the vapor to condense. (Treat this process as a single-unit operation.) Liquid and vapor product streams emerge from the process in equilibrium at T(°C) and 1100 mm Hg. The vapor product contains 60.0 mole% butane.

(a) Draw and label a flowchart. Perform a degree-of-freedom analysis to show that you have enough information to determine the required final temperature (T), the composition of the liquid product (component mole fractions), and the molar flow rates of the liquid and vapor products from the given information and Antoine expressions for the vapor pressures $p_B^*(T)$ and $p_H^*(T)$. *Just identify the equations—for example, mole balance on butane or Raoult's law for hexane—but don't write them yet.*

(b) Write in order the equations that you would use to determine the quantities listed in part (a) and also the fractional condensation of *hexane* (mol H condensed/mol H fed). In each equation, circle the variable for which you would solve. Do no algebra or calculations.

---

*Computer problem.

(c) Complete the calculations either manually or with an equation-solving program.

(d) State three assumptions you made that could lead to errors in the calculated quantities.

6.60. The feed to a distillation column is a 45.0 mole% *n*-pentane–55.0 mole% *n*-hexane liquid mixture. The vapor stream leaving the top of the column, which contains 98.0 mole% pentane and the balance hexane, goes to a total condenser (in which all the vapor is condensed). Half of the liquid condensate is returned to the top of the column as *reflux* and the rest is withdrawn as overhead product (*distillate*) at a rate of 85.0 kmol/h. The distillate contains 95.0% of the pentane fed to the column. The liquid stream leaving the bottom of the column goes to a *reboiler.* Part of the stream is vaporized; the vapor is recycled to the bottom of the column as *boilup,* and the residual liquid is withdrawn as *bottoms product.*

(a) Calculate the molar flow rate of the feed stream and the molar flow rate and composition of the bottoms product stream.

(b) Estimate the temperature of the vapor entering the condenser, assuming that it is saturated (at its dew point) at an absolute pressure of 1 atm and that Raoult's law applies to both pentane and hexane. Then estimate the volumetric flow rates of the vapor stream leaving the column and of the liquid distillate product. State any assumptions you make.

(c) Estimate the temperature of the reboiler and the composition of the vapor boilup, again assuming operation at 1 atm.

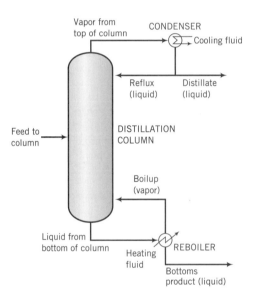

(d) Calculate the minimum diameter of the pipe connecting the column and the condenser if the maximum allowable velocity in the pipe is 10 m/s. Then list all the assumptions underlying the calculation of that number.

6.61. The vapor leaving the top of a distillation column goes to a condenser in which either total or partial condensation takes place. If a total condenser is used, a portion of the condensate is returned to the top of the column as *reflux* and the remaining liquid is taken off as the *overhead product* (or *distillate*). (See Problem 6.60.) If a partial condenser is used, the liquid condensate is returned as reflux and the uncondensed vapor is taken off as the overhead product.

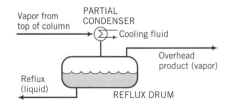

The overhead product from an *n*-butane–*n*-pentane distillation column is 96 mole% butane. The temperature of the cooling fluid limits the condenser temperature to 40°C or higher.

(a) Using Raoult's law, estimate the minimum pressure at which the condenser can operate as a partial condenser (i.e., at which it can produce liquid for reflux) and the minimum pressure at which it can operate as a total condenser. In terms of dew point and bubble point, what do each of these pressures represent for the given temperature?

(b) Suppose the condenser operates as a total condenser at 40°C, the production rate of overhead product is 75 kmol/h, and the mole ratio of reflux to overhead product is 1.5:1. Calculate the molar flow rates and compositions of the reflux stream and the vapor feed to the condenser.

(c) Suppose now that a partial condenser is used, with the reflux and overhead product in equilibrium at 40°C and the overhead product flow rate and reflux-to-overhead product ratio having the values given in part (b). Calculate the operating pressure of the condenser and the compositions of the reflux and vapor feed to the condenser.

6.62. Vapor–liquid equilibrium calculations can sometimes be simplified through the use of a quantity called the **relative volatility**, which may be defined in terms of the following depiction of vapor and liquid phases in equilibrium:

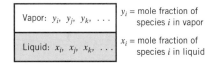

The relative volatility of species i to species j is

$$\alpha_{ij} = \frac{y_i/x_i}{y_j/x_j}$$

If α_{ij} is much greater than 1, species i is much more volatile than species j (i.e., it has a much greater tendency to vaporize at the system temperature and pressure); conversely, if $\alpha_{ij} \ll 1$, species i is much less volatile than species j. The closer α_{ij} is to 1, the more difficult it is to separate species i from species j by a process such as distillation or partial condensation of a vapor mixture.

(a) Show that the relative volatility of species A to species B, α_{AB}, equals the ratio of vapor pressures at the system temperature, p_A^*/p_B^*, if both species obey Raoult's law and follow ideal gas behavior.

(b) Determine the relative volatility of styrene to ethylbenzene at 85°C and the relative volatility of benzene to ethylbenzene at the same temperature. Which pair would you classify as more difficult to separate by distillation?

(c) Show that for a binary mixture of i and j

$$y_i = \frac{\alpha_{ij}x_i}{1 + (\alpha_{ij} - 1)x_i}$$

(d) Apply the equation from part (c) to a benzene–ethylbenzene system at 85°C, using it to estimate the mole fractions of benzene in the vapor phase in equilibrium with liquids having benzene mole fractions of 0.0, 0.2, 0.4, 0.6, 0.8, and 1.0. Then calculate the total system pressure for each of these six conditions.

6.63. A **stage** of a separation process is defined as an operation in which components of one or more feed streams divide themselves between two phases, and the phases are taken off separately. In an **ideal stage** or **equilibrium stage**, the effluent (exiting) streams are in equilibrium with each other.

Distillation columns often consist of a series of vertically distributed stages. Vapor flows upward and liquid flows downward between adjacent stages; some of the liquid fed to each stage vaporizes, and some of the vapor fed to each stage condenses. A representation of the top section of a distillation column is shown on the next page. (See Problem 4.26 for a more realistic representation.)

Consider a distillation column operating at 0.4 atm absolute in which benzene and styrene are being separated. A vapor stream containing 65 mole% benzene and 35 mole% styrene enters stage 1 at a rate of 200 mol/h, and liquid containing 55 mole% benzene and 45 mole% styrene leaves this stage at a rate of 150 mol/h. You may assume (1) the stages are ideal, (2) Raoult's law can be used to relate the compositions of the streams leaving each stage, and (3) the total vapor and liquid molar flow rates do not change by a significant amount from one stage to the next.

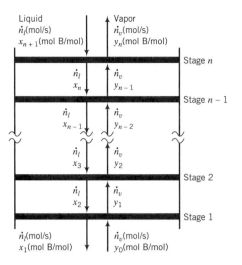

(a) How would you expect the mole fraction of benzene in the liquid to vary from one stage to another, beginning with stage 1 and moving up the column? In light of your answer and considering that the pressure remains essentially constant from one stage to another, how would you then expect the temperature to vary at progressively higher stages? Briefly explain.

(b) Estimate the temperature at stage 1 and the compositions of the vapor stream leaving this stage and the liquid stream entering it. Then repeat these calculations for stage 2.

(c) Describe how you would calculate the number of ideal stages required to reduce the styrene content of the vapor to less than 5 mole%.

*6.64. The following diagram shows a staged absorption column in which n-hexane (H) is absorbed from a gas into a heavy oil.

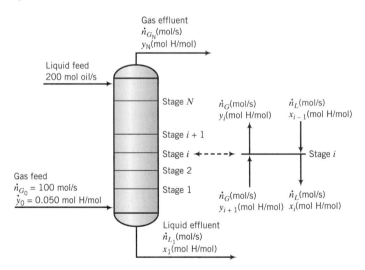

A gas feed stream containing 5.0 mole% hexane vapor and the balance nitrogen enters at the bottom of an absorption column at a basis rate of 100 mol/s, and a nonvolatile oil enters the top of the column in a ratio 2 mol oil fed/mol gas fed. The absorber consists of a series of ideal stages (see Problem 6.63), arranged so that gas flows upward and liquid flows downward. The liquid and gas streams leaving each stage are in equilibrium with each other (by the definition of an ideal stage), with compositions related by Raoult's law. The absorber operates at an approximately constant temperature T (°C) and pressure P(mm Hg). Of the hexane entering the column, 99.5% is absorbed and leaves in the

---

*Computer problem.

liquid column effluent. At the given conditions it may be assumed that N_2 is insoluble in the oil and that none of the oil vaporizes.

(a) Calculate the molar flow rates and mole fractions of hexane in the gas and liquid streams leaving the column. Then calculate the average values of the liquid and gas molar flow rates in the column, $\dot{n}_L$(mol/s) and $\dot{n}_G$(mol/s). For simplicity, in subsequent calculations use these values as the molar flow rates of the liquid and gas streams leaving each stage.

(b) Estimate the mole fraction of hexane in the gas leaving the bottom stage of the column (y_1) and in the liquid entering this stage (x_2).

(c) Suppose that x_i and y_i are the mole fractions of hexane in the liquid and gas streams leaving stage i. Derive the following formulas and verify that they yield the answers you calculated in part (b):

$$y_i = x_i p_i^*(T)/P \tag{1}$$

$$x_{i+1} = x_i + \frac{\dot{n}_G}{\dot{n}_L}(y_i - y_{i-1}) \tag{2}$$

(d) Create a spreadsheet to determine the number of stages (N) required to reduce the mole fraction of hexane to its required final value [calculated in part (a)] or less for $P = 760$ torr and temperatures of 30°C, 50°C, and 70°C. The spreadsheet should have the following structure (some calculated values are shown):

Hexane Absorption								
P =	760	PR =	1					
y0 =	0.05	x1 =		ye =	2.63E-04			
nGN =		nL1 =		nG =		nL =		
A =	6.8776	B =	1172	C =	224.366			
T	p*(T)		T	p*(T)		T	p*(T)	
30	187.1		50			70		
i	x(i)	y(i)	i	x(i)	y(i)	i	x(i)	y(i)
0		5.00E-02	0		5.00E-02	0		5.00E-02
1	2.43E-02	5.98E-03	1			1		
2	3.07E-03	7.56E-04	2			2		
3	5.57E-04	1.37E-04	3			3		

Enter the values of x_i, y_N, $\dot{n}_{G_N}$, $\dot{n}_{L_1}$, and the average flow rates $\dot{n}_G$ and $\dot{n}_L$ calculated in parts (a) and (b). Then in the appropriate cells for the calculation at 30°C, enter the Antoine formula for the vapor pressure, the value of x_1, the formula for y_1 (Equation 1), and the formulas for x_2 and y_2 (Equations 2 and 1). Then copy the formulas into successive rows, proceeding until the value of y_i is less than or equal to the calculated effluent value (y_N). The results (which should match the ones shown) indicate that three stages are required to achieve the specified hexane recovery at 30°C. Repeat the calculations for the other two temperatures. (You should be able to do so entirely by copying cells from one location to another on the spreadsheet.) *Do not go beyond 25 stages for any temperature, whether or not you achieve the required separation.*

(e) You should have found that at 70°C and 760 mm Hg the hexane mole fraction in the vapor levels out at a value above the target value, which means that the specified separation cannot be achieved at those conditions. Explain this result. Then use your spreadsheet to determine the minimum pressure at which the target absorption can be achieved at that temperature.

6.65. A vapor mixture containing 30 mole% benzene and 70% toluene at 1 atm is cooled isobarically in a closed container from an initial temperature of 115°C. Use the *Txy* diagram of Figure 6.4-1 to answer the following questions.

(a) At what temperature does the first drop of condensate form? What is its composition?

(b) At one point during the process the system temperature is 100°C. Determine the mole fractions of benzene in the vapor and liquid phases and the ratio (total moles in vapor/total moles in liquid) at this point.

(c) At what temperature does the last bubble of vapor condense? What is its composition?

6.66. Three gram-moles of benzene and 7 gram-moles of toluene are placed in a closed cylinder equipped with a piston. The cylinder is immersed in a boiling-water bath that maintains the temperature at 100°C. The force exerted on the piston can be varied to adjust the cylinder pressure to any desired value. The pressure is initially 1000 mm Hg and is gradually lowered to 600 mm Hg. Use the *Pxy* diagram of Figure 6.4-1 to convince yourself that the cylinder initially contains only liquid benzene and toluene and to answer the following questions.

(a) At what pressure does the first vapor bubble form? What is its composition?

(b) At what pressure does the last droplet of liquid evaporate? What is its composition?

(c) What are the liquid and vapor compositions in equilibrium with each other when the pressure is 750 mm Hg? What is the ratio (moles vapor/mole liquid) at this point?

(d) Estimate the volume of the cylinder contents when the pressure is (i) 1000 mm Hg, (ii) 750 mm Hg, and (iii) 600 mm Hg.

6.67. A methanol–water feed stream is introduced to a vaporizer in which a molar fraction f of the feed is vaporized. The feed has a methanol mole fraction of $x_F = 0.4$, and the vaporizer operates at a pressure of 1 atm absolute and 80°C. Vapor and liquid leaving the device are in equilibrium at the temperature and pressure of the system and have methanol mole fractions of y and x, respectively.

A *Txy* diagram for methanol–water mixtures at 1 atm absolute is shown below. The feed to the vaporizer and the liquid and vapor product streams are shown as points B, A, and C, respectively.

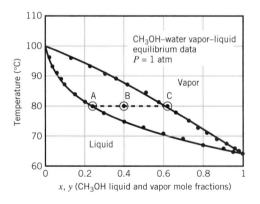

(a) Prove that f can be determined from the equation

$$f = \frac{\text{moles of vapor}}{\text{moles of liquid}} = \frac{x_F - x}{y - x}$$

Use this result to determine f for the specific conditions cited above ($x_F = 0.4$, $T = 80°C$).

(b) Use the *Txy* diagram to estimate the minimum and maximum temperatures at which the given feed stream could be separated into vapor and liquid fractions at 1 atm. In each case, what fraction of the feed would be vaporized?

6.68. Vapor–liquid equilibrium data for mixtures of acetone (A) and ethanol at 1 atm are given in the following table:

$T\,(°C)$	78.3	67.3	65.9	63.6	61.8	60.4	59.1	58.0	57.0	56.1
x_A	0.000	0.250	0.300	0.400	0.500	0.600	0.700	0.800	0.900	1.000
y_A	0.000	0.478	0.524	0.605	0.674	0.739	0.802	0.865	0.929	1.00

(a) Use the given data to construct a *Txy* diagram for this system.

(b) A thermocouple inserted into a two-phase mixture of acetone and ethanol at equilibrium reads 62.1°C. The system pressure is 1 atm. Use the *Txy* diagram to estimate the mole fractions of acetone in the liquid and vapor phases.

(c) An equimolar mixture of acetone and ethanol is fed to an evacuated vessel and allowed to come to equilibrium at 65°C and 1.00 atm absolute. Estimate (i) the molar compositions of each phase, (ii) the percentage of the total moles in the vessel that are in the vapor phase, and (iii) the percentage of the vessel volume occupied by the vapor phase.

(d) A liquid mixture containing 40.0 mole% acetone and 60.0 mole% ethanol is fed to a continuous flash evaporator. Vapor and product streams leave the unit in equilibrium at 1.00 atm. The molar flow rate of the vapor product stream is 20% of the molar flow rate of the feed stream. Estimate the operating temperature of the evaporator and the compositions of the liquid and vapor product streams.

(e) Use Raoult's law to estimate the bubble-point temperature and vapor composition in equilibrium with an equimolar liquid mixture of acetone and ethanol. Calculate the percentage errors in the estimated values of T_{bp} and y. Propose a reason why Raoult's law produces poor estimates for this system. (*Suggestion:* Consider the molecular structure of the two components.)

*6.69. Your task in this problem will be to use a spreadsheet to generate a *Txy* diagram for a two-component system, using Raoult's law to express the vapor–liquid equilibrium distribution of each species. The spreadsheet will be constructed for the chloroform–benzene system at 1 atm (for which Raoult's law is not a very good approximation), but it can then be used for any other system by substituting different Antoine equation constants.

(a) Look up the normal boiling points of chloroform and benzene and sketch the expected shape of a *Txy* diagram for these two species at 1 atm. Do not perform any calculations.

(b) Create a spreadsheet that contains a title line in Row 1 (**Txy Diagram for an Ideal Binary Solution**), the names of the two species (**Chloroform** and **Benzene**), and the labels (**A, B, C**) and values of their Antoine equation constants (Rows 3 and 4), and a label **P(mm Hg)** = and in the adjacent cell the pressure for which the diagram is to be generated (760) (Row 5). Then insert column headings **x, T, y, p1, p2, p1+p2** in Row 7. These headings denote the mole fraction of the first species in the liquid phase, the equilibrium temperature, the mole fraction of the first species in the vapor phase, the partial pressures of the first and second species in the vapor phase, and the sum of the partial pressures of the two species. In the first column (under the **x** heading) enter 0.0, 0.05, 0.10, … , 0.95, 1.0. (You should be able to enter a formula in the second cell and then copy it into the remaining cells of the column rather than entering each number individually.) Then carry out the following procedure for each x value.

- Enter a temperature (for all but $x = 0$ and $x = 1$ you will have to guess a value).
- Enter formulas for the two partial pressures (use Raoult's law) and for their sum, and then enter a formula for y.
- Vary the value of T to determine the temperature at which the sum of the component partial pressures equals the specified total pressure of the system (760 mm Hg). This calculation can easily be done using the goalseek tool on most spreadsheet programs, or it can be done by manual trial and error. The row now contains the correct x and y values for the given temperature.

Once you have done the calculation for the first x value, you should be able to copy formulas into subsequent rows rather than having to enter them again. When the calculation has been

---

completed for all rows of the table, draw the *Txy* diagram (using the graphing facility of your spreadsheet program if possible, otherwise by hand).

(c) Explain in your own words exactly what you are doing in the bulleted sequence of steps in part (b) and give the relevant formulas. The phrase "bubble point" should appear in your explanation.

(d) The following vapor–liquid equilibrium data have been obtained for mixtures of chloroform (C) and benzene (B) at 1 atm.

$T(°C)$	80.6	79.8	79.0	77.3	75.3	71.9	68.9	61.4
x_C	0.00	0.08	0.15	0.29	0.44	0.66	0.79	1.00
y_C	0.00	0.10	0.20	0.40	0.60	0.80	0.90	1.00

Plot these data on the graph generated in part (b). Estimate the percentage errors in the Raoult's law values of the bubble-point temperature and vapor mole fraction for $x_C = 0.44$, taking the tabulated values to be correct. Why does Raoult's law give poor estimates for this system?

6.70. A liquid mixture containing 40.0 mole% methanol and 60.0 mole% 1-propanol is placed in an open vessel and heated slowly. Estimate the temperature at which the mixture begins to boil. List assumptions made in your calculations. If heat is supplied continuously, how will the liquid temperature and composition change with time?

6.71. Acetaldehyde is synthesized by the catalytic dehydrogenation of ethanol:

$$C_2H_5OH \rightarrow CH_3CHO + H_2$$

Fresh feed (pure ethanol) is blended with a recycle stream (95 mole% ethanol and 5% acetaldehyde), and the combined stream is heated and vaporized, entering the reactor at 280°C. Gases leaving the reactor are cooled to $-40°C$ to condense the acetaldehyde and unreacted ethanol. Off-gas from the condenser is sent to a scrubber, where the uncondensed organic compounds are removed and hydrogen is recovered as a by-product. The condensate from the condenser, which is 45 mole% ethanol, is sent to a distillation column that produces a distillate containing 97 mole% acetaldehyde and a bottoms product that constitutes the recycle blended with fresh feed to the process. The production rate of the distillate is 1000 kg/h. The pressure throughout the process may be taken as 1 atm absolute.

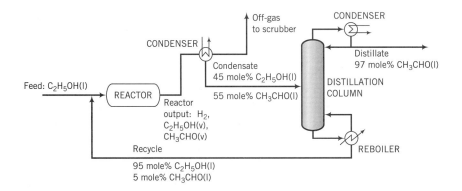

(a) Calculate the molar flow rates (kmol/h) of the fresh feed, the recycle stream, and the hydrogen in the off-gas. Also determine the volumetric flow rate (m³/h) of the feed to the reactor. (*Suggestion:* Use Raoult's law in the analysis of the condenser.)

(b) Estimate (i) the overall and single-pass conversions of ethanol and (ii) the rates (kmol/h) at which ethanol and acetaldehyde are sent to the scrubber.

6.72. Dehydration of natural gas is necessary to prevent the formation of gas hydrates, which can plug valves and other components of a gas pipeline, and also to reduce potential corrosion problems. Water removal can be accomplished as shown in the following schematic diagram:

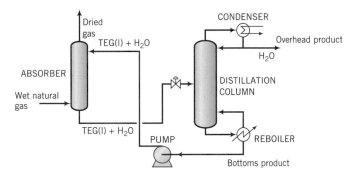

Natural gas containing 80 lb_m $H_2O/10^6$ SCF gas [SCF = ft^3 (STP)] enters the bottom of an ab-sorber at a rate of 4.0×10^6 SCF/day. A liquid stream containing triethylene glycol (TEG, molecular weight = 150.2) and a small amount of water is fed to the top of the absorber. The absorber oper-ates at 500 psia and 90°F. The dried gas leaving the absorber contains 10 lb_m $H_2O/10^6$ SCF gas. The solvent leaving the absorber, which contains all the TEG–water mixture fed to the column plus all the water absorbed from the natural gas, goes to a distillation column. The overhead product stream from the distillation column contains only liquid water. The bottoms product stream, which contains TEG and water, is the stream recycled to the absorber.

(a) Draw and completely label a flowchart of this process. Calculate the mass flow rate (lb_m/day) and volumetric flow rate (ft^3/day) of the overhead product from the distillation column.

(b) The greatest possible amount of dehydration is achieved if the gas leaving the absorption column is in equilibrium with the solvent entering the column. If the Henry's law constant for water in TEG at 90°F is 0.398 psia/mol fraction, what is the maximum allowable mole fraction of water in the solvent fed to the absorber?

(c) A column of infinite height would be required to achieve equilibrium between the gas and liquid at the top of the absorber. For the desired separation to be achieved in practice, the mole fraction of water in the entering solvent must be less than the value calculated in part (b). Suppose it is 80% of that value and the flow rate of TEG in the recirculating solvent is 37 lb_m TEG/lb_m water absorbed in the column. Calculate the flow rate (lb_m/day) of the solvent stream entering the absorber and the mole fraction of water in the solvent stream leaving the absorber.

(d) What is the purpose of the distillation column in the process? (*Hint:* Think about how the process would operate without it.)

6.73. A two-stage process is used to separate H_2S from a gas containing 96% H_2 and 4% H_2S by volume. The H_2S is absorbed in a solvent, which is then regenerated by air in a stripping column. The Henry's law constant for the absorption of H_2S in the solvent at 0°C is 22 atm/mole fraction.

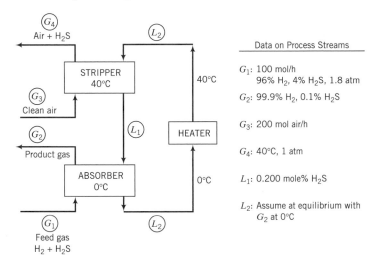

    (a) Briefly explain in your own words the functions of the three process units. Include in your explanation the purpose of the air in the stripper and the reason the stripper operates at a higher temperature than the absorber.

    (b) Calculate the molar flow rate of pure solvent and the volumetric flow rate of the gas at G_4, neglecting evaporation of solvent in both columns. (See flowchart.)

6.74. The solubility of sodium bicarbonate in water is 11.1 g $NHCO_3$/100 g H_2O at 30°C and 16.4 g $NHCO_3$/100 g H_2O at 60°C. If a saturated solution of $NaHCO_3$ at 60°C is cooled and comes to equilibrium at 30°C, what percentage of the dissolved salt crystallizes?

6.75. An aqueous solution of potassium hydroxide is fed at a rate of 875 kg/h to an evaporative crystallizer operating at 10°C, producing crystals of $KOH\cdot2H_2O$. A 5 g aliquot of the feed solution is titrated to neutrality with 22.4 mL of 0.85 molar H_2SO_4. The solubility of KOH at 10°C is 103 kg KOH/100 kg H_2O. At what rate must water be evaporated to crystallize 60% of the KOH in the feed?

6.76. A salt A is soluble in a solvent S. A conductivity meter used to measure the solute concentration in A–S solutions is calibrated by dissolving a known quantity of A in S, adding more S to bring the solution volume to a fixed value, and noting the conductivity meter reading. The data given below are taken at 30°C:

Solute Dissolved (g)	Solution Volume (mL)	Meter Reading R
0	100.0	0
20.0	100.0	30
30.0	100.0	45

The following experiment is performed. One hundred sixty grams of A is dissolved in S at 30°C. S is added until a final solution volume of 500 mL is obtained. The solution is cooled slowly to 0°C while being stirred and is maintained at this temperature long enough for crystallization to be complete. The concentration of A in the supernatant liquid is then measured with the conductivity meter, yielding $R = 17.5$. The solution is next reheated in small temperature increments. The last crystal is observed to dissolve at 10.2°C. A specific gravity of 1.10 may be assumed for all A–S solutions.

    (a) Derive an expression for C(g A/mL solution) in terms of R.

    (b) Calculate the solubilities (g A/100 g S) at 10.2°C and 0°C and the mass of solid crystals in the beaker at 0°C.

    (c) If half the solvent in the flask were to evaporate at 0°C, how much more A would come out of solution?

6.77. A saturated $MgSO_4$ solution at 130°F is fed to a crystallizer operating at 50°F. The solution leaving the crystallizer is saturated. Magnesium sulfate solubilities are 35 wt% $MgSO_4$ at 130°F and 23 wt% $MgSO_4$ at 50°F.

    (a) Write the molecular formula for the crystalline product that forms. (See Table 6.5-1.)

    (b) A production rate of 1000 kg/h of crystalline material is desired. Calculate (i) the required feed rate to the crystallizer (kg/h), and (ii) the rate (kg/h) at which anhydrous $MgSO_4$ could be recovered from the crystals.

6.78. A solution containing 100 lb_m KNO_3/100 lb_m H_2O at 80°C is fed to a cooling crystallizer operated at 25°C. Slurry from the crystallizer (KNO_3 crystals suspended in saturated solution) is fed to a filter, where the crystals are separated from the solution. Use the solubility data in Figure 6.5-1 to determine the production rate of crystals (lb_m/lb_m feed) and the solid-to-liquid mass ratio (lb_m crystals/lb_m liquid) in the slurry leaving the crystallizer.

6.79. A 10.0 wt% aqueous solution of sodium chloride is fed to an evaporative crystallizer operated under a partial vacuum. Evaporation of water concentrates the remaining solution beyond its saturation point at the crystallizer temperature and causes crystallization of NaCl. The crystallizer product is a slurry of solute crystals suspended in a saturated solution at 80°C. The unit is to produce 1000 kg NaCl(s)/h. The solubility of NaCl in water is given by Figure 6.5-1.

    (a) Derive expressions for the required rate of evaporation of water (kg/h) and the mass flow rate of solution in the exit slurry in terms of the mass flow rate of the feed stream to the crystallizer.

Determine the minimum possible feed rate (explain why it is the minimum rate) and the corresponding values of the evaporation rate and exit solution flow rate.

(b) The pump that conveys the exit slurry from the crystallizer to a downstream filter cannot handle material containing more than 40 wt% solids. Determine the maximum feed rate to the crystallizer and the corresponding evaporation rate.

6.80. Potassium dichromate ($K_2Cr_2O_7$) is to be recovered from a 21 wt% aqueous solution in a continuous crystallization operation. The solution is joined by a recycle stream and fed to a vacuum evaporator where water is removed and the remaining solution is cooled to 30°C, at which temperature the solubility of the salt is 0.20 kg $K_2Cr_2O_7$/kg H_2O. The solution and suspended potassium dichromate crystals flow to a centrifuge. The crystals and 5.0% of the solution constitute the solid effluent from the centrifuge, and the remaining solution is recycled to the evaporator. The solid effluent, which contains 90 wt% crystals and 10% entrained solution, is fed to a dryer, where it is contacted with hot air. The remaining water in the effluent is evaporated, leaving pure potassium dichromate crystals. The air leaves the dryer at 90°C and 1 atm with a dew point of 39.2°C. For a production rate of 1000 kg solid $K_2Cr_2O_7$/h, calculate the required feed rate (kg/h) of 21% solution, the rate of evaporation of water in the evaporator (kg/h), the flow rate (kg/h) of the recycle stream, and the feed rate of air (standard liters/h).

6.81. Sodium bicarbonate is synthesized by reacting sodium carbonate with carbon dioxide and water at 70°C and 2.0 atm gauge:

$$Na_2CO_3 + CO_2 + H_2O \rightarrow 2\,NaHCO_3$$

An aqueous solution containing 7.00 wt% sodium carbonate and a gas stream containing 70.0 mole% CO_2 and the balance air are fed to the reactor. All of the sodium carbonate and some of the carbon dioxide in the feed react. The gas leaving the reactor, which contains the air and unreacted CO_2, is saturated with water vapor at the reactor conditions. A liquid–solid slurry of sodium bicarbonate crystals in a saturated aqueous solution containing 2.4 wt% dissolved sodium bicarbonate and no dissolved CO_2 leaves the reactor and is pumped to a filter. The wet filter cake contains 86 wt% sodium bicarbonate crystals and the balance saturated solution, and the filtrate is also saturated solution. The production rate of solid crystals is 500 kg/h.

Suggestion: Although the problems to be given can be solved in terms of the product flow rate of 500 kg NaHCO₃(s)/h, it might be easier to assume a different basis and then scale the process to the desired production rate of crystals.

(a) Calculate the composition (component mole fractions) and volumetric flow rate (m³/min) of the gas stream leaving the reactor.

(b) Calculate the feed rate of gas to the process in standard cubic meters/min [m³(STP)/min].

(c) Calculate the flow rate (kg/h) of the liquid feed to the process. What more would you need to know to calculate the volumetric flow rate of this stream?

(d) The filtrate was assumed to leave the filter as a saturated solution at 70°C. What would be the effect on your calculations if the temperature of the filtrate actually dropped to 50°C as it passed through the filter?

(e) The reactor pressure of 2 atm gauge was arrived at in an optimization study. What benefit do you suppose would result from increasing the pressure? What penalty would be associated with this increase? The term "Henry's law" should appear in your explanation. (*Hint:* The reaction occurs in the liquid phase and the CO_2 enters the reactor as a gas. What step must precede the reaction?)

6.82. An ore containing 90 wt% $MgSO_4 \cdot 7H_2O$ and the balance insoluble minerals is fed to a dissolution tank at a rate of 60,000 lb_m/h along with fresh water and a recycle stream. The tank contents are heated to 110°F, causing all of the magnesium sulfate heptahydrate in the ore to dissolve, forming a saturated solution. The resulting slurry of the insoluble minerals in saturated $MgSO_4$ solution is pumped to a heated filter, where a wet filter cake is separated from a solids-free filtrate. The filter cake retains 5 lb_m of saturated solution per 100 lb_m of minerals. The filtrate is sent to a crystallizer in which the temperature is reduced to 50°F, producing a slurry of $MgSO_4 \cdot 7H_2O$ crystals in a saturated solution that is sent to another filter. The product filter cake contains all of the crystals and entrained solution, again in a ratio of 5 lb_m solution per 100 lb_m crystals. The filtrate from this filter is returned to the dissolution tank as the recycle stream.

Solubility data: Saturated magnesium sulfate solutions at 110°F and 50°F contain 32 wt% $MgSO_4$ and 23 wt% $MgSO_4$, respectively.

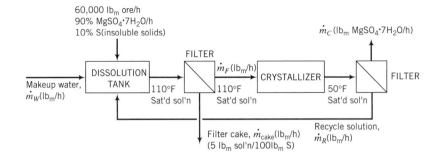

(a) Explain why the solution is first heated (in the dissolution tank) and filtered and then cooled (in the crystallizer) and filtered.

(b) Calculate the production rate of crystals and the required feed rate of fresh water to the dissolution tank. (*Note:* Don't forget to include water of hydration when you write a mass balance on water.)

(c) Calculate the ratio lb_m recycle/lb_m makeup water.

6.83. An aqueous waste stream leaving a process contains 10.0 wt% sulfuric acid and 1 kg nitric acid per kg sulfuric acid. The flow rate of sulfuric acid in the waste stream is 1000 kg/h. The acids are neutralized before being sent to a wastewater treatment facility by combining the waste stream with an aqueous slurry of solid calcium carbonate that contains 2 kg of recycled liquid per kg solid calcium carbonate. (The source of the recycled liquid will be given later in the process description.)

The following neutralization reactions occur in the reactor:

$$CaCO_3 + H_2SO_4 \rightarrow CaSO_4 + H_2O + CO_2$$
$$CaCO_3 + 2\,HNO_3 \rightarrow Ca(NO_3)_2 + H_2O + CO_2$$

The sulfuric and nitric acids and calcium carbonate fed to the reactor are completely consumed. The carbon dioxide leaving the reactor is compressed to 30 atm absolute and 40°C and sent elsewhere in the plant. The remaining reactor effluents are sent to a crystallizer operating at 30°C, at which temperature the solubility of calcium sulfate is 2.0 g $CaSO_4$/1000 g H_2O. Calcium sulfate crystals form in the crystallizer and all other species remain in solution.

The slurry leaving the crystallizer is filtered to produce (i) a filter cake containing 96% calcium sulfate crystals and the remainder entrained saturated calcium sulfate solution, and (ii) a filtrate solution saturated with $CaSO_4$ at 30°C that also contains dissolved calcium nitrate. The filtrate is split, with a portion being recycled to mix with the solid calcium carbonate to form the slurry fed to the reactor, and the remainder being sent to the wastewater treatment facility.

(a) Draw and completely label a flowchart for this process.

(b) Speculate on why the acids must be neutralized before being sent to the wastewater treatment facility.

(c) Calculate the mass flow rates (kg/h) of the calcium carbonate fed to the process and of the filter cake; also determine the mass flow rates and compositions of the solution sent to the wastewater facility and of the recycle stream. (*Caution:* If you write a water balance around the reactor or the overall system, remember that water is a reaction product and not just an inert solvent.)

(d) Calculate the volumetric flow rate (L/h) of the carbon dioxide leaving the process at 30 atm absolute and 40°C. Do not assume ideal gas behavior.

(e) The solubility of $Ca(NO_3)_2$ at 30°C is 152.6 kg $Ca(NO_3)_2$ per 100 kg H_2O. What is the maximum ratio of nitric acid to sulfuric acid in the feed that can be tolerated without encountering difficulties associated with contamination of the calcium sulfate by-product by $Ca(NO_3)_2$?

6.84. A solution of diphenyl (MW = 154.2) in benzene is formed by mixing 56.0 g diphenyl with 550.0 mL of benzene. Estimate the effective vapor pressure of the solution at 30°C and the melting and boiling points of the solution at 1 atm.

6.85. An aqueous solution of urea (MW = 60.06) freezes at −4.6°C and 1 atm. Estimate the normal boiling point of the solution; then calculate the mass of urea (grams) that would have to be added to 1.00 kg of water to raise the normal boiling point by 3°C.

6.86. A solution is prepared by dissolving 0.5150 g of a solute (MW = 110.1) in 100.0 g of an organic solvent (MW = 94.10). The solution is observed to have a freezing point 0.41°C below that of the pure solvent. A second solution is prepared by dissolving 0.4460 g of a solute having an unknown molecular weight in 95.60 g of the original solvent. A freezing point depression of 0.49°C is observed. Determine the molecular weight of the second solute and the heat of fusion (kJ/mol) of the solvent. The melting point of the pure solvent is −5.000°C.

6.87. Derive Equation 6.5-4 for the boiling-point elevation of a dilute solution of a nonvolatile solute with mole fraction x in a solvent that has a pure-component vapor pressure $p_s^*(T)$. To do so, suppose that when the pressure is P_0, the pure solvent boils at temperature T_{b0} [so that $P_0 = p_s^*(T_{b0})$] and the solvent in the solution boils at $T_{bs} > T_{b0}$. Further suppose that at temperature T_{b0} the effective vapor pressure of the solvent is $P_s = (p_s^*)_e(T_{b0}) < P_0$. (See diagram.)

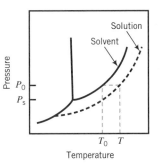

The procedure is as follows.

(a) Write the Clausius–Clapeyron equation (Equation 6.1-3) for P_s (the effective solvent vapor pressure at T_{b0}) and then for P_0 (the effective solvent vapor pressure at T_{bs}), assuming that at the low solute concentrations in question the heat of vaporization is the same at both temperatures. Subtract the two equations. Simplify the equation algebraically, assuming that T_{b0} and T_{bs} are close enough together to say that $T_{b0}T_{bs} \approx T_{b0}^2$.

(b) Substitute the Raoult's law expression (Equation 6.5-2) for $P_s = (p_s^*)_e(T_{b0})$. Observe that if $x \ll 1$ (which it is for highly dilute solutions), then $\ln(1 - x) \approx -x$. The desired result follows.

6.88. The distribution coefficient of styrene distributed between ethylbenzene and ethylene glycol at 25°C is 0.19 mass fraction in the ethylene glycol phase per mass fraction in the ethylbenzene phase. One hundred grams of pure ethylene glycol is added to 120 g of a mixture containing containing 75 wt% ethylbenzene and 25% styrene, and the resulting blend is allowed to equilibrate. How much styrene transfers to the ethylene glycol phase, assuming that ethylene gylcol and ethylbenzene are immiscible?

6.89. A stream of 5.00 wt% oleic acid in cottonseed oil enters an extraction unit at a rate of 100.0 kg/h. The unit operates as an equilibrium stage (the streams leaving the unit are in equilibrium) at 85°C. At this temperature, propane and cottonseed oil are essentially immiscible, and the distribution coefficient (oleic acid mass fraction in propane/oleic acid mass fraction in cottonseed oil) is 0.15.

(a) Calculate the rate at which liquid propane must be fed to the unit to extract 90% of the oleic acid.

(b) Estimate the minimum operating pressure of the extraction unit (i.e., the pressure required to keep the propane liquid at 85°C).

(c) High-pressure operation is costly and introduces potential safety hazards. Suggest two possible reasons for using propane as the solvent when other less volatile hydrocarbons are equally good solvents for oleic acid.

6.90. Benzene and hexane are being considered as solvents to extract acetic acid from aqueous mixtures. At 30°C, distribution coefficients for the two solvents are $K_B = 0.098$ mass fraction acetic acid in benzene/mass fraction acetic acid in water and $K_H = 0.017$ mass fraction acetic acid in hexane/mass fraction acetic acid in water.

(a) Based on the distribution coefficients only, which of the two solvents would you use and why? Demonstrate the logic of your decision by comparing the quantities of the two solvents required to reduce the acetic acid content in 100 kg of an aqueous solution from 30 wt% to 10 wt%.

(b) What other factors may be important in choosing between benzene and cyclohexane?

6.91. Acetone is to be extracted with *n*-hexane from a 40.0 wt% acetone–60.0 wt% water mixture at 25°C. The acetone distribution coefficient (mass fraction acetone in the hexane-rich phase/mass fraction acetone in the water-rich phase) is 0.343.[13] Water and hexane may be considered immiscible. Three different processing alternatives are to be considered: a two-stage process and two single-stage processes.

(a) In the first stage of the proposed two-stage process, equal masses of the feed mixture and pure hexane are blended vigorously and then allowed to settle. The organic phase is withdrawn and the aqueous phase is mixed with 75% of the amount of hexane added in the first stage. The mixture is allowed to settle and the two phases are separated. What percentage of the acetone in the original feed solution remains in the water at the end of the process?

(b) Suppose all of the hexane added in the two-stage process of part (a) is instead added to the feed mixture and the process is carried out in a single equilibrium stage. What percentage of the acetone in the feed solution remains in the water at the end of the process?

(c) Finally, suppose a single-stage process is used but it is desired to reduce the acetone content of the water to the final value of part (a). How much hexane must be added to the feed solution?

(d) Under what circumstances would each of the three processes be the most cost-effective? What additional information would you need to make the choice?

6.92. Penicillin is produced by fermentation and recovered from the resulting aqueous broth by extraction with butyl acetate. The penicillin distribution coefficient K (mass fraction of penicillin in the butyl acetate phase/mass fraction of penicillin in the water phase) depends strongly on the pH in the aqueous phase:

pH	2.1	4.4	5.8
K	25.0	1.38	0.10

This dependence provides the basis for the process to be described. Water and butyl acetate may be considered immiscibile.

The extraction is performed in the following three-unit process:

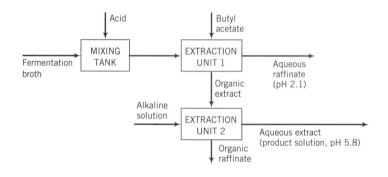

- Broth from a fermentor containing dissolved penicillin, other dissolved impurities, and water is acidified in a mixing tank. The acidified broth, which contains 1.5 wt% penicillin, is contacted with liquid butyl acetate in an extraction unit consisting of a mixer, in which the aqueous and organic phases are brought into intimate contact with each other, followed by a settling tank, in which the two phases separate under the influence of gravity. The pH of the aqueous phase in the extraction unit equals 2.1. In the mixer 90% of the penicillin in the feed broth transfers from the aqueous phase to the organic phase.

- The two streams leaving the settler are in equilibrium with each other—that is, the ratio of the penicillin mass fractions in the two phases equals the value of K corresponding to the pH of the aqueous phase (= 2.1 in Unit 1). The impurities in the feed broth remain in the aqueous phase. The *raffinate* (by definition, the product stream containing the feed-solution solvent) leav-

---

[13] *Perry's Chemical Engineers' Handbook*, 7th Edition, McGraw-Hill, New York, 1997.

ing Extraction Unit 1 is sent elsewhere for further processing, and the organic *extract* (the product stream containing the extracting solvent) is sent to a second mixer–settler unit.

- In the second unit, the organic solution fed to the mixing stage is contacted with an alkaline aqueous solution that adjusts the pH of the aqueous phase in the unit to 5.8. In the mixer, 90% of the penicillin entering in the organic feed solution transfers to the aqueous phase. Once again, the two streams emerging from the settler are in equilibrium. The aqueous extract is the process product.

(a) Taking a basis of 100 kg of acidified broth fed to the first extraction unit, draw and completely label a flowchart of this process and carry out the degree-of-freedom analysis to show that all labeled variables can be determined. (*Suggestion:* Consider the combination of water, impurities, and acid as a single species and the alkaline solution as a second single species, since the components of these "pseudospecies" always stay together in the process.)

(b) Calculate the ratios (kg butyl acetate required/kg acidified broth) and (kg alkaline solution required/kg acidified broth) and the mass fraction of penicillin in the product solution.

(c) Briefly explain the following:

  (i) What is the likely reason for transferring most of the penicillin from an aqueous phase to an organic phase and then transferring most of it back to an aqueous phase, when each transfer leads to a loss of some of the drug?

  (ii) What is the purpose of acidifying the broth prior to the first extraction stage, and why is the extracting solution added to the second unit a base?

  (iii) Why are the two "raffinates" in the process the aqueous phase leaving the first unit and the organic phase leaving the second unit, and vice versa for the "extracts"? (Look again at the definitions of these terms.)

6.93. A mixture of 20 wt% water, 33% acetone, and the remainder methyl isobutyl ketone is brought to equilibrium at 25°C. If the total mass of the system is 1.2 kg, use the data in Figure 6.6-1 to estimate the composition and mass of each phase of the mixture.

6.94. Five kilograms of a 30 wt% acetone–70% water mixture is added to 3.5 kg of a 20 wt% acetone–80% MIBK mixture at 25°C. Use Figure 6.6-1 to estimate the mass and composition of each phase of the resulting mixture.

6.95. An aqueous acetone solution is fed at a rate of 32.0 lb_m/h to a stirred tank. A stream of pure methyl isobutyl ketone is also fed to the tank, and the resulting mixture is sent to a settler operating at 25°C. One of the phases formed has a flow rate of 41.0 lb_m/h and contains 70 wt% MIBK. Use Figure 6.6-1 to determine the flow rate and composition of the second product stream and the rate at which MIBK is fed to the unit.

6.96. Two systems contain water, acetone, and methyl isobutyl ketone in equilibrium at 25°C. The first system contains equal masses of the three species, and the second one contains 9.0% acetone, 21.0% water, and 70.0% MIBK by mass. Let $x_{a,aq}$ and $x_{a,org}$, respectively, denote the mass fractions of acetone in the aqueous phase (the phase that contains most of the water in the system) and the organic phase (the phase that contains most of the MIBK), and let $x_{w,aq}$ and $x_{w,org}$ denote the mass fractions of water in the two phases.

(a) Use Figure 6.6-1 to estimate the mass and composition (component mass fractions) of each phase of the mixtures in System 1 and in System 2.

(b) Determine the distribution coefficient of acetone in the organic phase relative to the aqueous phase in each system, $K_a = x_{a,org}/x_{a,aq}$. If a process is being designed to extract acetone from one of the two solvents (water and MIBK) to the other one, when would a high value of K_a be desirable and when would a low value be desirable?

(c) Determine the *selectivity,* β_{aw}, of acetone relative to water in the two systems, where

$$\beta_{aw} = \frac{(\text{mass fraction acetone/mass fraction water})_{\text{extract phase}}}{(\text{mass fraction acetone/mass fraction water})_{\text{raffinate phase}}}$$

What would be the value of β_{aw} if water and MIBK were completely immiscible?

(d) Express the selectivity, β_{aw}, in terms of the distribution coefficients of acetone and water, K_a and K_w. [Start with the formula given in part (c).] If MIBK is being used to extract acetone from an aqueous phase, under what circumstances might it be important to have a very high value of β_{aw}, even if it means that less acetone is being extracted?

6.97. Water is used to extract acetone from a 30 wt% acetone–70% MIBK mixture flowing at a rate of 200 kg/h. Two equilibrium stages at 25°C are used as shown in the following diagram. If 300 kg H_2O/h is fed to each extraction unit, what fraction of the acetone in the feed solution would be removed and what would be the composition of the combined extract?

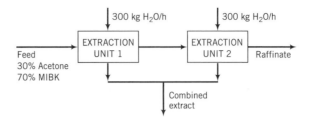

6.98. Air at 25°C and 1 atm with a relative humidity of 25% is to be dehumidified in an adsorption column packed with silica gel. The equilibrium adsorptivity of water on silica gel is given by the expression[14]

$$X^*(\text{kg water/100 kg silica gel}) = 12.5 \, \frac{p_{H_2O}}{p^*_{H_2O}}$$

where p_{H_2O} is the partial pressure of water in the gas contacting the silica gel and $p^*_{H_2O}$ is the vapor pressure of water at the system temperature. Air is fed to the column at a rate of 1.50 L/min until the silica gel is saturated (i.e., until it reaches equilibrium with the feed air), at which point the flow is stopped and the silica gel replaced.

(a) Calculate the minimum amount of silica gel needed in the column if replacement is to take place no more frequently than every two hours. State any assumptions you make.

(b) Briefly describe this process in terms that a high school student would have no trouble understanding. (What is the process designed to do, what happens within the column, and why is replacement of the column packing necessary?)

6.99. A 50.0-L tank contains an air–carbon tetrachloride gas mixture at an absolute pressure of 1 atm, a temperature of 34°C, and a relative saturation of 30%. Activated carbon is added to the tank to remove the CCl_4 from the gas by adsorption and the tank is then sealed. The volume of added activated carbon may be assumed negligible in comparison to the tank volume.

(a) Calculate p_{CCl_4} at the moment the tank is sealed, assuming ideal gas behavior and neglecting adsorption that occurs prior to sealing.

(b) Calculate the total pressure in the tank and the partial pressure of carbon tetrachloride at a point when half of the CCl_4 initially in the tank has been adsorbed. *Note:* It was shown in Example 6.7-1 that at 34°C

$$X^*\left(\frac{\text{g } CCl_4 \text{ adsorbed}}{\text{g carbon}}\right) = \frac{0.0762 p_{CCl_4}}{1 + 0.096 p_{CCl_4}}$$

where p_{CCl_4} is the partial pressure (mm Hg) of carbon tetrachloride in the gas contacting the carbon.

(c) How much activated carbon must be added to the tank to reduce the mole fraction of CCl_4 in the gas to 0.001?

6.100. The following equilibrium data[15] have been obtained for the adsorption of nitrogen dioxide, NO_2, on silica gel at 25°C and 1 atm:

p_{NO_2} (mm Hg)	0	2	4	6	8	10	12
X^* (kg NO_2/100 kg silica gel)	0	0.4	0.9	1.65	2.60	3.65	4.85

(a) Confirm that these data are reasonably correlated by the **Freundlich isotherm**

$$X^* = K_F p_{NO_2}^{\beta}$$

[14]R. Yang, *Gas Separation by Adsorption Processes*, Butterworths, London, 1987, p. 13.

[15]Adapted from R. E. Treybal, *Mass-Transfer Operations*, 3rd Edition, McGraw-Hill, New York, 1980, p. 653.

and determine the values of K_F and β that provide the best correlation. (Use one of the graphical methods introduced in Section 2.7c.)

(b) The adsorption column shown in the figure below has an internal diameter of 10.0 cm and a bed height of 1.00 m. The bed of silica gel has a bulk density of 0.75 kg/L. The adsorber is to remove NO_2 from a stream containing 1.0 mole% NO_2 and the balance air that enters the adsorber at 8.00 kg/h. The pressure and temperature are maintained at 1 atm and 25°C. Past experience with this system has shown that a plot of the partial pressure ratio $[(p_{NO_2})_{outlet}/(p_{NO_2})_{inlet}]$ versus time produces a **breakthrough curve** with the following appearance.

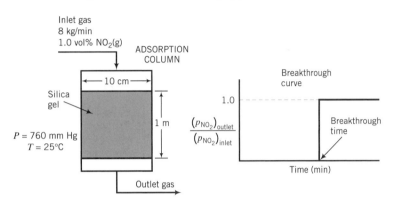

Using the isotherm derived in part (a), determine the time (in min) required for breakthrough of the NO_2.

(c) Silica gel in the column can be *regenerated* (i.e., adsorbed NO_2 can be removed so that the silica gel column can be reused) by elevating the bed temperature and/or purging the bed with clean air. Suppose such a regeneration process requires 1.5 hours to accomplish. Process shutdowns can be avoided by installing several silica gel columns in parallel, using one to carry out the purification while the others are being regenerated. What is the minimum number of columns required to achieve continuous process operation?

6.101. Various amounts of activated carbon were added to a fixed amount of raw cane sugar solution (48 wt% sucrose in water) at 80°C. A colorimeter was used to measure the color of the solutions, R, which is proportional to the concentration of trace unknown impurities in the solution. The following data were obtained (See footnote 15, p. 654.)

kg carbon/kg dry sucrose	0	0.005	0.010	0.015	0.020	0.030
R (color units/kg sucrose)	20.0	10.6	6.0	3.4	2.0	1.0

The reduction in color units is a measure of the mass of impurities (the adsorbate) adsorbed on the carbon (the adsorbent).

(a) The general form of the **Freundlich isotherm** is

$$X_i^* = K_F c_i^{\beta}$$

where X_i^* is the mass of i adsorbed/mass of adsorbent and c_i is the concentration of i in solution. Demonstrate that the Freundlich isotherm may be formulated for the system described above as

$$\vartheta = K_F' R^{\beta}$$

where ϑ is the % removal of color / [mass of carbon / mass of dissolved sucrose]. Then determine K_F' and β by fitting this expression to the given data, using one of the graphical methods in Section 2.7.

(b) Calculate the amount of carbon that would have to be added to a vat containing 1000 kg of the 48 wt% sugar solution at 80°C for a reduction in color content to 2.5% of the original value.

Part Three

Energy Balances

Chapter 7

Energy and Energy Balances

Energy is expensive. We have not yet learned how to use efficiently the seemingly endless supply of "free" energy provided by the sun, winds, and tides; nuclear power generation is feasible, but the need for safe disposal of radioactive wastes from nuclear reactors is a serious unresolved problem; and there are not nearly enough waterfalls and dams to provide sufficient hydroelectric power to meet the world's energy needs. This leaves us with fuel combustion—burning a combustible gas, oil, or solid fuel, and using the energy released as a source of thermal or (indirectly) electrical energy.

Process industries have always recognized that wasting energy leads to reduced profits, but throughout most of this century the cost of energy was often an insignificant part of the overall process cost, and gross operational inefficiencies were tolerated. In the 1970s, a sharp increase in the price of natural gas and petroleum raised the cost of energy severalfold and intensified the need to eliminate unnecessary energy consumption. If a plant uses more energy than its competitors, its product could be priced out of the marketplace.

As an engineer designing a process, one of your principal jobs would be to account carefully for the energy that flows into and out of each process unit and to determine the overall energy requirement for the process. You would do this by writing **energy balances** on the process, in much the same way that you write material balances to account for the mass flows to and from the process and its units. Typical problems that may be solved using energy balances include the following:

1. How much power (energy/time) is required to pump 1250 m³/h of water from a storage vessel to a process unit? (The answer determines the size of the required pump motor.)
2. How much energy is required to convert 2000 kg of water at 30°C to steam at 180°C?
3. A hydrocarbon mixture is distilled, producing a liquid and a vapor stream, each with a known or calculable flow rate and composition. The energy input to the distillation column is provided by condensing saturated steam at a pressure of 15 bar. At what rate must steam be supplied to process 2000 mol/h of the feed mixture?
4. A highly exothermic chemical reaction A → B takes place in a continuous reactor. If a 75% conversion of A is to be achieved, at what rate must energy be transferred from the reactor to keep the contents at a constant temperature?
5. How much coal must be burned each day to produce enough energy to generate the steam to run the turbines to produce enough electricity to meet the daily power requirements of a city of 500,000 people?
6. A chemical process consists of four reactors, 25 pumps, and a number of compressors, distillation columns, mixing tanks, evaporators, filter presses, and other materials handling and separation units. Each individual unit either requires or releases energy.

(a) How can the process operation be designed to minimize the total energy requirement? (For example, can the energy released by an energy-emitting process unit be transferred to an energy-absorbing process unit?)

(b) What is the total energy requirement for the process as it is finally designed, and how much will it cost to provide this energy? (The answer could determine whether or not the process is economically feasible.)

In this chapter we show how energy balances are formulated and applied. Section 7.1 defines the types of energy a process system can possess and the ways in which energy can be transferred to and from a system. Section 7.2 reviews the procedure for calculating the kinetic energy and gravitational potential energy of a process stream. Sections 7.3 and 7.4 derive the general energy balance equation for closed (batch) systems and open (semibatch and continuous) systems, and various applications of these equations are illustrated in Sections 7.5 through 7.7.

7.0 INSTRUCTIONAL OBJECTIVES

After completing this chapter, you should be able to do the following:

- List and define in your own words the three components of the total energy of a process system and the two forms of energy transfer between a system and its surroundings. State the conditions under which heat and work are positive. Convert an energy or power (energy/time) expressed in any unit (e.g., J, dyne·cm, Btu, ft·lb$_f$/h, kW, hp) to its equivalent in any other dimensionally consistent units.

- Calculate the kinetic energy of a body of mass m moving with velocity u or the rate of transport of kinetic energy by a stream moving with mass flow rate $\dot{m}$ and velocity u. Calculate the gravitational potential energy of a body of mass m at elevation z or the rate of transport of gravitational potential energy by a stream moving with mass flow rate $\dot{m}$ at elevation z, where z is height above a reference plane at which potential energy is defined to equal zero.

- Define the terms **closed process system, open process system, isothermal process,** and **adiabatic process.** Write the first law of thermodynamics (the energy balance equation) for a closed process system and state the conditions under which each of the five terms in the balance can be neglected. Given a description of a closed process system, simplify the energy balance and solve it for whichever term is not specified in the process description.

- Define the terms **flow work, shaft work, specific internal energy, specific volume,** and **specific enthalpy.** Write the energy balance for an open process system in terms of enthalpy and shaft work and state the conditions under which each of the five terms can be neglected. Given a description of an open process system, simplify the energy balance and solve it for whichever term is not specified in the process description.

- State why the actual values of $\hat{U}$ and $\hat{H}$ can never be known for a given species in a specified **state** (temperature, pressure, and phase) and define the concept of a **reference state.** Explain in your own terms the statement, "The specific enthalpy of CO(g) at 100°C and 0.5 atm relative to CO(g) at 500°C and 1 atm is $-12{,}141$ J/mol." (Your explanation should involve a process in which carbon monoxide gas goes from one state to another.)

- Explain why the reference state used to generate a table of specific internal energies or enthalpies is irrelevant if one is only interested in calculating ΔU or ΔH for a process. (The term "state property" should appear in your explanation.)

- Given a process in which a specified mass m of a species goes from one state to another and tabulated values of $\hat{U}$ or $\hat{H}$ for the species at the initial and final states are available, calculate ΔU or ΔH. Given values of $\hat{V}$ at each state, calculate ΔH from the previously calculated ΔU or vice versa. Carry out the corresponding calculations to determine $\Delta \dot{U}$ and $\Delta \dot{H}$ for a stream with mass flow rate $\dot{m}$ going from one state to another.

- Use the saturated and superheated steam tables (Tables B.5, B.6, and B.7) to determine (a) whether water at a specified temperature and pressure is liquid, saturated vapor, or superheated vapor; (b) the specific volume, specific internal energy, and specific enthalpy of liquid water or steam at a specified temperature and pressure; (c) the vapor pressure of water at a specified temperature; (d) the boiling point of water at a specified pressure; and (e) the dew point of superheated steam at a specified pressure.

- Explain the significance of the specific internal energies and enthalpies tabulated in the steam tables (B.5, B.6, and B.7), remembering that we can never know the true values of either variable in a given state. Given any process in which a specified mass (or mass flow rate) of water changes from one state to another, use the steam tables to calculate ΔU (or $\Delta \dot{U}$) and/or ΔH (or $\Delta \dot{H}$).

- Given a description of any nonreactive process for which tabulated specific internal energies or specific enthalpies are available at all input and output states for all process species, (a) draw and completely label a flowchart, including Q and W (or $\dot{Q}$ and $\dot{W}_s$ for an open system) if their values are either specified or called for in a problem statement; (b) perform a degree-of-freedom analysis; and (c) write the necessary equations (including the appropriately simplified energy balance) to determine all requested variables.

- Starting with the open system balance equation, derive the steady-state mechanical energy balance equation (Equation 7.7-2) for an incompressible fluid and simplify the equation further to derive the Bernoulli equation. List all the assumptions made in the derivation of the latter equation.

- Given fluid conditions (pressure, flow rate, velocity, elevation) at the inlet and outlet of an open system and values of friction loss and shaft work within the system, substitute known quantities into the mechanical energy balance (or the Bernoulli equation if friction loss and shaft work can be neglected) and solve the equation for whichever variable is unknown.

7.1   FORMS OF ENERGY: THE FIRST LAW OF THERMODYNAMICS

The total energy of a system has three components:

1. **Kinetic energy:** Energy due to the translational motion of the system as a whole relative to some frame of reference (usually the earth's surface) or to rotation of the system about some axis. In this text, we will deal only with translational kinetic energy.

2. **Potential energy:** Energy due to the position of the system in a potential field (such as a gravitational or electromagnetic field). In this text, we will deal only with gravitational potential energy.

3. **Internal energy:** All energy possessed by a system other than kinetic and potential energy, such as energy due to the motion of molecules relative to the center of mass of the system, to the rotational and vibrational motion and the electromagnetic interactions of the molecules, and to the motion and interactions of the atomic and subatomic constituents of the molecules.

Suppose a process system is **closed**, meaning that no mass is transferred across its boundaries while the process is taking place. Energy may be transferred between such a system and its surroundings in two ways:

1. As **heat**, or energy that flows as a result of temperature difference between a system and its surroundings. The direction of flow is always from a higher temperature to a lower one. *Heat is defined as positive when it is transferred to the system from the surroundings.*

2. As **work**, or energy that flows in response to any driving force other a temperature difference, such as a force, a torque, or a voltage. For example, if a gas in a cylinder expands and moves a piston against a restraining force, the gas does work on the piston (energy is transferred as work from the gas to its surroundings, which include the piston). *In this*

text, work is defined as positive when it is done by the system on the surroundings. (*Note:* The opposite sign convention is sometimes used. The choice is arbitrary, as long as it is used consistently; however, to avoid confusion when reading thermodynamics references, you should be sure which convention has been adopted.)

The terms "work" and "heat" refer only to energy that is being transferred: you can speak of the heat or work added to a system or given off by a system, but it is meaningless to speak of the heat or work possessed by or contained within a system.

Energy, like work, has units of force times distance: for example, joules (N·m), ergs (dyne·cm), and ft·lb$_f$. It is also common to use energy units defined in terms of the amount of heat that must be transferred to a specified mass of water to raise the temperature of the water by a specified temperature interval at a constant pressure of 1 atm. The most common of these units are tabulated here.

Unit	Symbol	Mass of Water	Temperature Interval
Kilogram-calorie or kilocalorie	kcal	1 kg	15°C to 16°C
Gram-calorie or calorie	cal	1 g	15°C to 16°C
British thermal unit	Btu	1 lb$_m$	60°F to 61°F

Conversion between these and other energy units may be performed using the conversion factors in the table on the inside front cover.

The principle that underlies all energy balances is the law of conservation of energy, which states that energy can neither be created nor destroyed. This law is also called the **first law of thermodynamics**. In its most general form, the first law states that the rate at which energy (kinetic + potential + internal) is carried into a system by the input streams, plus the rate at which it enters as heat, minus the rate at which it is transported out of the system by the output streams, minus the rate at which it leaves as work, equals the rate of accumulation of energy in the system. (That is, accumulation = input − output, as would be expected.)

Instead of presenting the equation in its most general form at this point, we will build up to it in stages. The next section reviews how to evaluate the kinetic and potential energies of an object and shows how the calculation can readily be extended to determine the rates at which kinetic and potential energies are transported by a flowing stream. Section 7.3 presents an integrated form of the transient balance equation that describes the behavior of a system between an initial state and a final state. This form of the equation is particularly useful for analyzing batch process systems. In Section 7.4 the first law is developed for a continuous steady-state process.

Most process systems are conveniently analyzed using one of the two forms of the energy balance equation presented in Sections 7.3 and 7.4. To perform energy balance calculations on other types of processes, such as semibatch processes or continuous processes that are being started up or shut down, the full transient energy balance equation is required. This equation is discussed in an introductory fashion in Chapter 11. A more thorough treatment of the full equation is deferred to thermodynamics courses and texts.

TEST YOURSELF

1. What forms of energy may a system possess? In what forms may energy be transferred to and from a closed system?
2. Why is it meaningless to speak of the heat possessed by a system?
3. Suppose the initial energy of a system (internal + kinetic + potential) is E_i, the final energy is E_f, an amount of energy Q is transferred from the environment to the system as heat, and an amount W is transferred from the system to the environment as work. According to the first law of thermodynamics, how must E_i, E_f, Q, and W be related?

7.2 KINETIC AND POTENTIAL ENERGY

The kinetic energy, E_k(J), of an object of mass m(kg) moving with velocity u(m/s) relative to the surface of the earth is

$$E_k = \tfrac{1}{2}mu^2 \tag{7.2-1a}$$

If a fluid enters a system with a mass flow rate $\dot{m}$(kg/s) and uniform velocity u(m/s), then

$$\dot{E}_k = \tfrac{1}{2}\dot{m}u^2 \tag{7.2-1b}$$

$\dot{E}_k$(J/s) may be thought of as the rate at which kinetic energy is transported into the system by the fluid.

EXAMPLE 7.2-1   *Kinetic Energy Transported by a Flowing Stream*

Water flows into a process unit through a 2-cm ID pipe at a rate of 2.00 m³/h. Calculate $\dot{E}_k$ for this stream in joules/second.

SOLUTION

First calculate the linear velocity (which equals the volumetric flow rate divided by the cross-sectional area of the pipe) and the mass flow rate of the fluid:

$$u = \frac{2.00\ \text{m}^3}{\text{h}} \left| \frac{100^2\ \text{cm}^2}{1^2\ \text{m}^2} \right| \frac{1}{\pi(1)^2\ \text{cm}^2} \left| \frac{1\ \text{h}}{3600\ \text{s}} \right. = 1.77\ \text{m/s}$$

$$\dot{m} = \frac{2.00\ \text{m}^3}{\text{h}} \left| \frac{1000\ \text{kg}}{\text{m}^3} \right| \frac{1\ \text{h}}{3600\ \text{s}} = 0.556\ \text{kg/s}$$

Then, from Equation 7.2-1b

$$\dot{E}_k = \frac{0.556\ \text{kg/s}}{2} \left| \frac{(1.77)^2\ \text{m}^2}{\text{s}^2} \right| \frac{1\ \text{N}}{1\ \text{kg·m/s}^2} = 0.870\ \text{N·m/s} = \boxed{0.870\ \text{J/s}}$$

The gravitational potential energy of an object of mass m is

$$E_p = mgz \tag{7.2-2a}$$

where g is the acceleration of gravity and z is the height of the object above a reference plane at which E_p is arbitrarily defined to be zero. If a fluid enters a system with a mass flow rate $\dot{m}$ and an elevation z relative to the potential energy reference plane, then

$$\dot{E}_p = \dot{m}gz \tag{7.2-2b}$$

$\dot{E}_p$(J/s) may be thought of as the rate at which gravitational potential energy is transported into the system by the fluid. Since we are normally interested in the *change* in potential energy when a body or fluid moves from one elevation to another [$\dot{E}_{p2} - \dot{E}_{p1} = \dot{m}g(z_2 - z_1)$], the elevation chosen as the reference plane does not matter.

EXAMPLE 7.2-2   *Potential Energy Increase of a Flowing Fluid*

Crude oil is pumped at a rate of 15.0 kg/s from a point 220 meters below the earth's surface to a point 20 meters above ground level. Calculate the attendant rate of increase of potential energy.

SOLUTION

Let subscripts 1 and 2 denote the first and second points, respectively:

$$\Delta\dot{E}_p = \dot{E}_{p2} - \dot{E}_{p1} = \dot{m}g(z_2 - z_1)$$

$$= \frac{15.0 \text{ kg}}{\text{s}} \left| \frac{9.81 \text{ m}}{\text{s}^2} \right| \frac{1 \text{ N}}{1 \text{ kg·m/s}^2} \left| [20 - (-220)] \text{ m} \right.$$

$$= 35{,}300 \text{ N·m/s} = \boxed{35{,}300 \text{ J/s}}$$

The answer could also have been expressed as 35,300 W or 35.3 kW. A pump would have to deliver at least this much power to raise the oil at the given rate.

TEST YOURSELF

A gas flows through a long pipe of constant diameter. The outlet of the pipe is higher than the inlet, and the pressure of the gas at the outlet is less than the inlet pressure. The gas temperature is constant throughout the pipe and the system is at steady state.

1. How do the mass flow rates at the inlet and outlet compare? The densities? The volumetric flow rates? (Assume ideal gas behavior.)
2. Is the change in potential energy of the gas from inlet to outlet positive, negative, or zero? What about the change in kinetic energy?

7.3 ENERGY BALANCES ON CLOSED SYSTEMS

A system is termed **open** or **closed** according to whether or not mass crosses the system boundary during the period of time covered by the energy balance. A batch process system is, by definition, closed, and semibatch and continuous systems are open.

An integral energy balance may be derived for a closed system between two instants of time. Since energy can neither be created nor destroyed, the generation and consumption terms of the general balance (4.2-1) drop out, leaving

$$\text{accumulation} = \text{input} - \text{output} \tag{7.3-1}$$

In deriving the integral mass balance for a closed system in Section 4.2c we eliminated the input and output terms, since by definition no mass crosses the boundaries of a closed system. It is possible, however, for energy to be transferred across the boundaries as heat or work, so that the right side of Equation 7.3-1 may not be eliminated automatically. As with mass balances, however, the accumulation term equals the final value of the balanced quantity (in this case, the system energy) minus the initial value of this quantity. Equation 7.3-1 may therefore be written

$$\begin{matrix} \text{final system} \\ \text{energy} \end{matrix} - \begin{matrix} \text{initial system} \\ \text{energy} \end{matrix} = \begin{matrix} \text{net energy transferred to} \\ \text{the system (in} - \text{out)} \end{matrix} \tag{7.3-2}$$

Now

$$\text{intitial system energy} = U_i + E_{ki} + E_{pi}$$

$$\text{final system energy} = U_f + E_{kf} + E_{pf}$$

$$\text{energy transferred} = Q - W$$

where the subscripts i and f refer to the initial and final states of the system and U, E_k, E_p, Q, and W represent internal energy, kinetic energy, potential energy, heat transferred to the system from its surroundings, and work done by the system on its surroundings. Equation 7.3-2 then becomes

$$(U_f - U_i) + (E_{kf} - E_{ki}) + (E_{pf} - E_{pi}) = Q - W \tag{7.3-3}$$

or, if the symbol Δ is used to signify (final − initial),

$$\Delta U + \Delta E_k + \Delta E_p = Q - W \qquad \text{(7.3-4)}$$

Equation 7.3-4 is the basic form of the first law of thermodynamics for a closed system. When applying this equation to a given process, you should be aware of the following points:

1. The internal energy of a system depends almost entirely on the chemical composition, state of aggregation (solid, liquid, or gas), and temperature of the system materials. It is independent of pressure for ideal gases and nearly independent of pressure for liquids and solids. *If no temperature changes, phase changes, or chemical reactions occur in a closed system and if pressure changes are less than a few atmospheres, then $\Delta U \approx 0$.*
2. *If a system is not accelerating, then $\Delta E_k = 0$. If a system is not rising or falling, then $\Delta E_p = 0$.*
3. *If a system and its surroundings are at the same temperature or the system is perfectly insulated, then $Q = 0$.* The process is then termed **adiabatic.**
4. Work done on or by a closed system is accomplished by movement of the system boundary against a resisting force or the passage of an electrical current or radiation across the system boundary. Examples of the first type of work are motion of a piston or rotation of a shaft that projects through the system boundary. *If there are no moving parts or electrical currents or radiation at the system boundary, then $W = 0$.*

TEST YOURSELF

1. What do the terms closed system and open system mean? What is an adiabatic process?
2. If 250 J is added to a system as heat, what is the value of Q in the energy balance equation? If 250 J of work is done to the system, what is the value of W?
3. If a closed system has an internal energy of 100 kcal at the beginning of a process and 50 kcal at the end, what is ΔU?
4. Under what circumstances might U be considered independent of pressure for a pure substance?

EXAMPLE 7.3-1 *Energy Balance on a Closed System*

A gas is contained in a cylinder fitted with a movable piston.

The initial gas temperature is 25°C.

The cylinder is placed in boiling water with the piston held in a fixed position. Heat in the amount of 2.00 kcal is transferred to the gas, which equilibrates at 100°C (and a higher pressure). The piston is then released, and the gas does 100 J of work in moving the piston to its new equilibrium position. The final gas temperature is 100°C.

Write the energy balance equation for each of the two stages of this process, and in each case solve for the unknown energy term in the equation. In solving this problem, consider the gas in the cylinder to be the system, neglect the change in potential energy of the gas as the piston moves vertically, and assume the gas behaves ideally. Express all energies in joules.

SOLUTION

1.

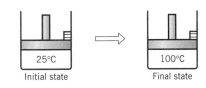

$$\Delta U + \Delta E_k + \Delta E_p = Q - W \quad \text{(Equation 7.3-4)}$$

$$\left. \begin{array}{l} \Delta E_k = 0 \quad \text{(the system is stationary)} \\ \Delta E_p = 0 \quad \text{(no vertical displacement)} \\ W = 0 \quad \text{(no moving boundaries)} \end{array} \right.$$

$$\Delta U = Q$$

$$\left. \right. Q = 2.00 \text{ kcal}$$

$$\Delta U = \frac{2.00 \text{ kcal}}{} \left| \frac{10^3 \text{ cal}}{\text{kcal}} \right| \frac{1 \text{ J}}{0.23901 \text{ cal}} = \boxed{8370 \text{ J} = \Delta U}$$

The gas thus gains 8370 J of internal energy in going from 25 to 100°C.

2.

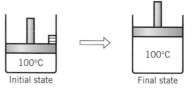

$$\Delta U + \Delta E_k + \Delta E_p = Q - W$$

$$\left. \begin{array}{l} \Delta E_k = 0 \quad \text{(the system is stationary at the initial and final states)} \\ \Delta E_p = 0 \quad \text{(assumed negligible by hypothesis)} \\ \Delta U = 0 \quad (U \text{ depends only on } T \text{ for an ideal gas, and } T \text{ does not change)} \end{array} \right.$$

$$0 = Q - W$$

$$\left. \right. W = +100 \text{ J} \quad \text{(Why is it positive?)}$$

$$\boxed{Q = 100 \text{ J}}$$

Thus an additional 100 J of heat is transferred to the gas as it expands and reequilibrates at 100°C.

7.4   ENERGY BALANCES ON OPEN SYSTEMS AT STEADY STATE

An open process system by definition has mass crossing its boundaries as the process occurs. Work must be done on such a system to push mass in, and work is done on the surroundings by mass that emerges. Both work terms must be included in the energy balance.

In Section 7.4a we outline the calculation of the work (or more precisely, the rate of energy transferred as work) required to move fluid through a continuous process system, and in Section 7.4b we review the concepts of intensive and extensive variables introduced in Chapter 6 and introduce the concept of specific properties of a substance. Section 7.4c uses the results of the two preceding sections to derive the energy balance for an open system at steady state.

7.4a   Flow Work and Shaft Work

The net rate of work done by an open system on its surroundings may be written as

$$\dot{W} = \dot{W}_s + \dot{W}_{fl} \tag{7.4-1}$$

where

$\dot{W}_s$ = **shaft work**, or rate of work done by the process fluid on a moving part within the system (e.g., a pump rotor)

$\dot{W}_{fl}$ = **flow work**, or rate of work done by the fluid at the system outlet minus the rate of work done on the fluid at the system inlet

To derive an expression for $\dot{W}_{fl}$, we initially consider the single-inlet–single-outlet system shown here.

$$\frac{\dot{V}_{in}(m^3/s)}{P_{in}(N/m^2)} \rightarrow \boxed{\begin{array}{c} \text{PROCESS} \\ \text{UNIT} \end{array}} \frac{\dot{V}_{out}(m^3/s)}{P_{out}(N/m^2)} \rightarrow$$

Fluid at a pressure $P_{in}(N/m^2)$ enters a pipe at a volumetric flow rate $\dot{V}_{in}(m^3/s)$ and exits at a pressure $P_{out}(N/m^2)$ and volumetric flow rate $\dot{V}_{out}(m^3/s)$. The fluid that enters the system has work done on it by the fluid just behind it at a rate

$$\dot{W}_{in}(N \cdot m/s) = P_{in}(N/m^2)\dot{V}_{in}(m^3/s) \tag{7.4-2}$$

while the fluid leaving the system performs work on the surroundings at a rate

$$\dot{W}_{out} = P_{out}\dot{V}_{out} \tag{7.4-3}$$

The net rate at which work is done by the system at the inlet and outlet is therefore

$$\dot{W}_{fl} = P_{out}\dot{V}_{out} - P_{in}\dot{V}_{in} \tag{7.4-4}$$

If several input and output streams enter and leave the system, the $P\dot{V}$ products for each stream must be summed to determine $\dot{W}_{fl}$.

TEST YOURSELF

An incompressible liquid flows through a straight horizontal pipe. Friction of the fluid within the pipe causes a small amount of heat to be transferred from the fluid; to compensate, flow work must be done on the fluid to move it through the system (so that $\dot{W}_{fl}$ is less than zero).

1. How are $\dot{V}_{in}$ and $\dot{V}_{out}$ related, where $\dot{V}$ is the volumetric flow rate of the liquid? (Remember, the fluid is incompressible.)
2. How must the pressures P_{in} and P_{out} be related? ($P_{in} > P_{out}$, $P_{in} = P_{out}$, or $P_{in} < P_{out}$?)

7.4b Specific Properties and Enthalpy

As we observed in Section 6.2, the properties of a process material are either extensive (proportional to the quantity of the material) or intensive (independent of the quantity). Mass, number of moles, and volume (or mass flow rate, molar flow rate, and volumetric flow rate for a continuous stream), and kinetic energy, potential energy, and internal energy (or the rates of transport of these quantities by a continuous stream) are extensive properties, while temperature, pressure, and density are intensive.

A **specific property** is an intensive quantity obtained by dividing an extensive property (or its flow rate) by the total amount (or flow rate) of the process material. Thus, if the volume of a fluid is 200 cm^3 and the mass of the fluid is 200 g, the **specific volume** of the fluid is 1 cm^3/g. Similarly, if the mass flow rate of a stream is 100 kg/min and the volumetric flow rate is 150 L/min, the specific volume of the stream material is $(150 \text{ L/min} / 100 \text{ kg/min}) = 1.5$ L/kg; if the rate at which kinetic energy is transported by this stream is 300 J/min, then the **specific kinetic energy** of the stream material is $(300 \text{ J/min}) / (100 \text{ kg/min}) = 3$ J/kg. *We will use the symbol ^ to denote a specific property:* $\hat{V}$ will denote specific volume, $\hat{U}$ specific internal energy, and so on.

If the temperature and pressure of a process material are such that the specific internal energy of the material is $\hat{U}(\text{J/kg})$, then a mass $m(\text{kg})$ of this material has a total internal energy

$$U(\text{J}) = m(\text{kg})\hat{U}(\text{J/kg}) \tag{7.4-5}$$

Similarly, a continuous stream of this material with a mass flow rate $\dot{m}(\text{kg/s})$ transports internal energy at a rate

$$\dot{U}(\text{J/s}) = \dot{m}(\text{kg/s})\hat{U}(\text{J/kg}) \tag{7.4-6}$$

A property that occurs in the energy balance equation for open systems (Section 7.4c) is the **specific enthalpy**, defined as

$$\hat{H} \equiv \hat{U} + P\hat{V} \tag{7.4-7}$$

where P is total pressure and $\hat{U}$ and $\hat{V}$ are specific internal energy and specific volume. The gas constants tabulated on the inside back cover provide a convenient source for the conversion factors needed to evaluate $\hat{H}$ from Equation 7.4-7, as the following example shows.

EXAMPLE 7.4-1   *Calculation of Enthalpy*

The specific internal energy of helium at 300 K and 1 atm is 3800 J/mol, and the specific molar volume at the same temperature and pressure is 24.63 L/mol. Calculate the specific enthalpy of helium at this temperature and pressure, and the rate at which enthalpy is transported by a stream of helium at 300 K and 1 atm with a molar flow rate of 250 kmol/h.

SOLUTION

$$\hat{H} = \hat{U} + P\hat{V} = 3800 \text{ J/mol} + (1 \text{ atm})(24.63 \text{ L/mol})$$

To convert the second term to joules we need the factor J/(L·atm). From the gas constant table on the inside back cover,

$$0.08206 \text{ L·atm/(mol·K)} = 8.314 \text{ J/(mol·K)}$$

Dividing the right side by the left side yields the desired factor:

$$\frac{8.314 \text{ J/mol·K}}{0.08206 \text{ L·atm/(mol·K)}} = 101.3 \text{ J/(L·atm)}$$

Therefore,

$$\hat{H} = 3800 \text{ J/mol} + \frac{24.63 \text{ L·atm}}{\text{mol}} \left| \frac{101.3 \text{ J}}{1 \text{ L·atm}} \right. = \boxed{6295 \text{ J/mol}}$$

If $\dot{n} = 250$ kmol/h

$$\dot{H} = \dot{n}\hat{H} = \frac{250 \text{ kmol}}{\text{h}} \left| \frac{10^3 \text{ mol}}{\text{kmol}} \right| \frac{6295 \text{ J}}{\text{mol}} = \boxed{1.57 \times 10^9 \text{ J/h}}$$

The enthalpy function is important in the analysis of open systems, as we will show in the next section. It can also be shown, however, that *if a closed system expands (or contracts) against a constant external pressure,* ΔE_k *and* ΔE_p *are negligible, and the only work done by or on the system is the work of the expansion, then the energy balance equation reduces to* $Q = \Delta H$. A proof of this statement is required in Problem 7.15.

TEST YOURSELF

The specific internal energy of a fluid is 200 cal/g.

1. What is the internal energy of 30 g of this fluid?
2. If the fluid leaves a system at a flow rate of 5 g/min, at what rate does it transport internal energy out of the system?
3. What would you need to know to calculate the specific enthalpy of this fluid?

7.4c   The Steady-State Open-System Energy Balance

The first law of thermodynamics for an open system at steady state has the form

$$\text{input} = \text{output} \tag{7.4-8}$$

(Why do the accumulation, generation, and consumption terms of the general balance equation vanish?) "Input" here signifies the total rate of transport of kinetic energy, potential energy, and internal energy by all process input streams plus the rate at which energy is transferred in as heat, and "output" is the total rate of energy transport by the output streams plus the rate at which energy is transferred out as work.

If $\dot{E}_j$ denotes the total rate of energy transport by the jth input or output stream of a process, and $\dot{Q}$ and $\dot{W}$ are again defined as the rates of flow of heat into and work out of the process, then Equation 7.4-8 may be written

$$\dot{Q} + \sum_{\substack{\text{input} \\ \text{streams}}} \dot{E}_j = \sum_{\substack{\text{output} \\ \text{streams}}} \dot{E}_j + \dot{W}$$

$$\Downarrow$$

$$\sum_{\substack{\text{output} \\ \text{streams}}} \dot{E}_j - \sum_{\substack{\text{input} \\ \text{streams}}} \dot{E}_j = \dot{Q} - \dot{W} \qquad \textbf{(7.4-9)}$$

If $\dot{m}_j$, $\dot{E}_{kj}$, $\dot{E}_{pj}$, and $\dot{U}_j$ are the flow rates of mass, kinetic energy, potential energy, and internal energy for the jth process stream, then the total rate at which energy is transported into or out of the system by this stream is

$$\dot{E}_j = \dot{U}_j + \dot{E}_{kj} + \dot{E}_{pj}$$

$$\Downarrow \left\| \begin{array}{l} \dot{U}_j = \dot{m}_j \hat{U}_j \\ \dot{E}_{kj} = \dot{m}_j u_j^2/2 \\ \dot{E}_{pj} = \dot{m}_j g z_j \end{array} \right.$$

$$\dot{E}_j = \dot{m}_j \left(\hat{U}_j + \frac{u_j^2}{2} + g z_j \right) \qquad \textbf{(7.4-10)}$$

where u_j is the velocity of the jth stream and z_j is the height of this stream relative to a reference plane at which $E_p = 0$.

The total work $\dot{W}$ done by the system on its surroundings equals the shaft work $\dot{W}_s$ plus the flow work $\dot{W}_{fl}$ (Equation 7.4-1). If $\dot{V}_j$ is the volumetric flow rate of the jth stream and P_j is the pressure of this stream as it crosses the system boundary, then as was shown in Section 7.4a,

$$\dot{W}_{fl} = \sum_{\substack{\text{output} \\ \text{streams}}} P_j \dot{V}_j - \sum_{\substack{\text{input} \\ \text{streams}}} P_j \dot{V}_j$$

$$\Downarrow \dot{V}_j = \dot{m}_j \hat{V}_j$$

$$\dot{W} = \dot{W}_s + \sum_{\substack{\text{output} \\ \text{streams}}} \dot{m}_j P_j \hat{V}_j - \sum_{\substack{\text{input} \\ \text{streams}}} \dot{m}_j P_j \hat{V}_j \qquad \textbf{(7.4-11)}$$

Substituting the expression for $\dot{E}_j$ of Equation 7.4-10 and that for $\dot{W}$ of Equation 7.4-11 into Equation 7.4-9 and bringing the $P\hat{V}$ terms to the left side yields

$$\sum_{\substack{\text{output} \\ \text{streams}}} \dot{m}_j \left[\hat{U}_j + P_j \hat{V}_j + \frac{u_j^2}{2} + g z_j \right] - \sum_{\substack{\text{input} \\ \text{streams}}} \dot{m}_j \left[\hat{U}_j + P_j \hat{V}_j + \frac{u_j^2}{2} + g z_j \right] = \dot{Q} - \dot{W}_s \qquad \textbf{(7.4-12)}$$

Equation 7.4-12 could be used for all steady-state open system energy balance problems. As a rule, however, the term $\hat{U}_j + P_j \hat{V}_j$ is combined and written as $\hat{H}_j$, the variable previously defined as the specific enthalpy. In terms of this variable, Equation 7.4-12 becomes

$$\sum_{\substack{\text{output} \\ \text{streams}}} \dot{m}_j \left(\hat{H}_j + \frac{u_j^2}{2} + g z_j \right) - \sum_{\substack{\text{input} \\ \text{streams}}} \dot{m}_j \left(\hat{H}_j + \frac{u_j^2}{2} + g z_j \right) = \dot{Q} - \dot{W}_s \qquad \textbf{(7.4-13)}$$

Finally, let us use the symbol Δ to denote total output minus total input, so that

$$\Delta \dot{H} = \sum_{\substack{\text{output} \\ \text{streams}}} \dot{m}_j \hat{H}_j - \sum_{\substack{\text{input} \\ \text{streams}}} \dot{m}_j \hat{H}_j \qquad \textbf{(7.4-14a)}$$

$$\Delta \dot{E}_k = \sum_{\substack{\text{output} \\ \text{streams}}} m_j u_j^2/2 - \sum_{\substack{\text{input} \\ \text{streams}}} m_j u_j^2/2 \qquad \textbf{(7.4-14b)}$$

$$\Delta \dot{E}_p = \sum_{\substack{\text{output} \\ \text{streams}}} m_j g z_j - \sum_{\substack{\text{input} \\ \text{streams}}} m_j g z_j \qquad \textbf{(7.4-14c)}$$

In terms of these quantities, Equation 7.4-13 becomes

$$\boxed{\Delta \dot{H} + \Delta \dot{E}_k + \Delta \dot{E}_p = \dot{Q} - \dot{W}_s} \qquad \textbf{(7.4-15)}$$

Equation 7.4-15 states that the net rate at which energy is transferred to a system as heat and/or shaft work ($\dot{Q} - \dot{W}_s$) equals the difference between the rates at which the quantity (enthalpy + kinetic energy + potential energy) is transported into and out of the system ($\Delta \dot{H} + \Delta \dot{E}_k + \Delta \dot{E}_p$). We will use this equation as the starting point for most energy balance calculations on open systems at steady state.

Notice that if a process has a single input stream and a single output stream and there is no accumulation of mass in the system (so that $\dot{m}_{in} = \dot{m}_{out} = \dot{m}$), the expression for $\Delta \dot{H}$ of Equation 7.4-14a simplifies to

$$\Delta \dot{H} = \dot{m}(\hat{H}_{out} - \hat{H}_{in}) = \dot{m}\Delta\hat{H} \qquad \textbf{(7.4-16)}$$

Also notice that if a specific variable has the same value for all input and output streams, the corresponding term of Equation 7.4-15 drops out. For example, if $\hat{H}_j$ is the same for all streams, then from Equation 7.4-14a

$$\Delta \dot{H} = \hat{H} \left[\sum_{\substack{\text{output} \\ \text{streams}}} \dot{m}_j - \sum_{\substack{\text{input} \\ \text{streams}}} \dot{m}_j \right] \qquad \textbf{(7.4-17)}$$

But from a total mass balance the quantity in brackets (which is simply total mass in minus total mass out) equals zero, and hence $\Delta \dot{H} = 0$, as claimed.

TEST YOURSELF

How would you simplify Equation 7.4-15 in each of the following cases?

1. There are no moving parts in the system.
2. The system and its surroundings are at the same temperature.
3. The linear velocities of all streams are the same.
4. All streams enter and leave the process at a single height.

EXAMPLE 7.4-2 *Energy Balance on a Turbine*

Five hundred kilograms per hour of steam drives a turbine. The steam enters the turbine at 44 atm and 450°C at a linear velocity of 60 m/s and leaves at a point 5 m below the turbine inlet at atmospheric pressure and a velocity of 360 m/s. The turbine delivers shaft work at a rate of 70 kW, and the heat loss from the turbine is estimated to be 10^4 kcal/h. Calculate the specific enthalpy change associated with the process.

SOLUTION

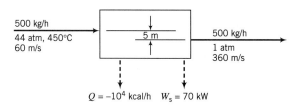

From Equation 7.4-15

$$\Delta \dot{H} = \dot{Q} - \dot{W}_s - \Delta \dot{E}_k - \Delta \dot{E}_p$$

Normally, heat, work, and kinetic and potential energy terms are determined in different units. To evaluate $\Delta \dot{H}$, we will convert each term to kW (kJ/s) using conversion factors given on the inside front cover, first noting that $\dot{m} = (500 \text{ kg/h}/3600 \text{ s/h}) = 0.139$ kg/s.

$$\Delta \dot{E}_k = \frac{\dot{m}}{2}(u_2^2 - u_1^2) = \frac{0.139 \text{ kg/s}}{2} \left| \frac{1 \text{ N}}{1 \text{ kg·m/s}^2} \right| \frac{(360^2 - 60^2) \text{ m}^2}{\text{s}^2} \left| \frac{1 \text{ W}}{1 \text{ N·m/s}} \right| \frac{1 \text{ kW}}{10^3 \text{ W}}$$

$$= 8.75 \text{ kW}$$

$$\Delta \dot{E}_p = \dot{m}g(z_2 - z_1) = \frac{0.139 \text{ kg/s}}{} \left| \frac{9.81 \text{ N}}{\text{kg}} \right| \frac{(-5) \text{ m}}{} \left| \frac{1 \text{ kW}}{10^3 \text{ N·m/s}} \right. = -6.81 \times 10^{-3} \text{ kW}$$

$$\dot{Q} = \frac{-10^4 \text{ kcal}}{\text{h}} \left| \frac{1 \text{ J}}{0.239 \times 10^{-3} \text{ kcal}} \right| \frac{1 \text{ h}}{3600 \text{ s}} \left| \frac{1 \text{ kW}}{10^3 \text{ J/s}} \right. = -11.6 \text{ kW}$$

$$\dot{W}_s = 70 \text{ kW}$$

$$\Downarrow$$

$$\Delta \dot{H} = \dot{Q} - \dot{W}_s - \Delta \dot{E}_k - \Delta \dot{E}_p = -90.3 \text{ kW}$$

But

$$\Delta \dot{H} = \dot{m}(\hat{H}_2 - \hat{H}_1) \quad \text{(from Equation 7.4-16)}$$

$$\Downarrow$$

$$\hat{H}_2 - \hat{H}_1 = \Delta \dot{H} / \dot{m}$$

$$= \frac{-90.3 \text{ kJ/s}}{0.139 \text{ kg/s}} = \boxed{-650 \text{ kJ/kg}}$$

7.5   TABLES OF THERMODYNAMIC DATA

7.5a   Reference States and State Properties

It is not possible to know the absolute value of $\hat{U}$ or $\hat{H}$ for a process material, but you can determine the *change* in $\hat{U}(\Delta\hat{U})$ or in $\hat{H}(\Delta\hat{H})$ corresponding to a specified change of state (temperature, pressure, and phase). This may be done, for example, by bringing a known mass m of a substance through the specified change of state in such a way that all terms of the energy balance except ΔU (i.e., heat, work, and changes in potential and kinetic energies) are known. Once $\Delta\hat{U}(= \Delta U/m)$ has been determined, $\Delta\hat{H}$ for the same change in state can be calculated as $\Delta\hat{U} + \Delta P\hat{V}$.

A convenient way to tabulate measured changes in $\hat{U}$ or $\hat{H}$ is to choose a temperature, pressure, and state of aggregation as a **reference state,** and to list $\Delta\hat{U}$ or $\Delta\hat{H}$ for changes from this state to a series of other states. Suppose, for example, that the enthalpy changes for carbon monoxide going from a reference state of 0°C and 1 atm to two other states are measured, with the following results:

$$CO(g, 0°C, 1 \text{ atm}) \rightarrow CO(g, 100°C, 1 \text{ atm}): \quad \Delta\hat{H}_1 = 2919 \text{ J/mol}$$

$$CO(g, 0°C, 1 \text{ atm}) \rightarrow CO(g, 500°C, 1 \text{ atm}): \quad \Delta\hat{H}_2 = 15,060 \text{ J/mol}$$

Since $\hat{H}$ cannot be known absolutely, for convenience we may assign a value $\hat{H}_0 = 0$ to the reference state; then $\Delta\hat{H}_1 = \hat{H}_1 - 0 = \hat{H}_1$, $\Delta\hat{H}_2 = \hat{H}_2$, and so on. A table may then be constructed for CO at 1 atm:

$T(°C)$	$\hat{H}(\text{J/mol})$
0	0
100	2919
500	15,060

Note that the value 2919 J/mol for $\hat{H}$ at 100°C does *not* mean that the absolute value of the specific enthalpy of CO at 100°C and 1 atm is 2919 J/mol—we cannot know the absolute value of $\hat{H}$—but rather means that the *change* in $\hat{H}$ when CO goes from the reference state to 100°C and 1 atm is 2919 J/mol. We then say *the specific enthalpy of CO at 100°C and 1 atm relative to CO at 0°C and 1 atm is 2919 J/mol.*

Some enthalpy tables give the reference states on which the listed values of $\hat{H}$ are based and others do not; however, you do not have to know the reference state to calculate $\Delta\hat{H}$ for the transition from one tabulated state to another. If $\hat{H}_1$ is the specific enthalpy at state 1 and $\hat{H}_2$ is that at state 2, then $\Delta\hat{H}$ for the transition from state 1 to state 2 equals $\hat{H}_2 - \hat{H}_1$, regardless of the reference state on which $\hat{H}_1$ and $\hat{H}_2$ are based. (*Caution:* If two different tables are used, be sure $\hat{H}_1$ and $\hat{H}_2$ are based on the same reference state.) For example, $\Delta\hat{H}$ for CO going from 100 to 500°C at 1 atm is $(15{,}060 - 2919)$ J/mol $= 12{,}141$ J/mol. If another reference state had been used to generate the specific enthalpies of CO at 100°C and 500°C, $\hat{H}_1$ and $\hat{H}_2$ would have different values but $\hat{H}_2 - \hat{H}_1$ would still be 12,141 J/mol. (See diagram below.)

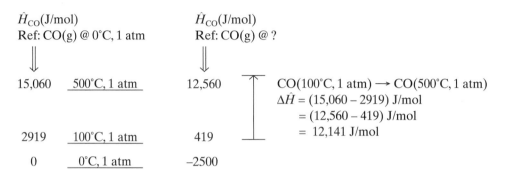

This convenient result is a consequence of the fact that $\hat{H}$, like $\hat{U}$, is a **state property,** or a property of a system component whose value depends only on the state of the system (temperature, pressure, phase, and composition) and not on how the system reached that state.[1] We will have more to say about this concept in Chapter Eight.

EXAMPLE 7.5-1   *Use of Tabulated Enthalpy Data*

The following entries are taken from a data table for saturated methyl chloride:

State	$T(°F)$	P(psia)	$\hat{V}(\mathrm{ft}^3/\mathrm{lb_m})$	$\hat{H}(\mathrm{Btu}/\mathrm{lb_m})$
Liquid	−40	6.878	0.01553	0.000
Vapor	0	18.90	4.969	196.23
Vapor	50	51.99	1.920	202.28

1. What reference state was used to generate the given enthalpies?
2. Calculate $\Delta\hat{H}$ and $\Delta\hat{U}$ for the transition of saturated methyl chloride vapor from 50°F to 0°F.
3. What assumption did you make in solving question 2 regarding the effect of pressure on specific enthalpy?

SOLUTION

1. Liquid at −40°F and 6.878 psia (the state at which $\hat{H} = 0$). You do not need this information to solve part 2.
2. $\Delta\hat{H} = \hat{H}(0°F) - \hat{H}(50°F) = (196.23 - 202.28) = -6.05$ Btu/lb$_m$

$$\Delta\hat{U} = \Delta\hat{H} - \Delta P\hat{V} = \Delta\hat{H} - (P_{final}\hat{V}_{final} - P_{initial}\hat{V}_{initial})$$

$$= -6.05 \text{ Btu/lb}_m$$

$$- \frac{[(18.90)(4.969) - (51.99)(1.920)] \text{ ft}^3\cdot\text{psia/lb}_m}{}\left|\frac{1.987 \text{ Btu}}{10.73 \text{ ft}^3\cdot\text{psia}}\right.$$

$$= \boxed{-4.96 \text{ Btu/lb}_m}$$

---

[1]We will not prove our claim that $\hat{U}$ and $\hat{H}$ satisfy this condition. All references on thermodynamics discuss this point in detail.

The value of the conversion factor Btu/(ft^3·psia) was obtained from the table of gas constants in back of the book. (Verify it!)

3. $\hat{H}$ was assumed independent of P.

Tables of enthalpies and other state properties of many substances may be found in tables B5-B9 of this text and *Perry's Chemical Engineers' Handbook*[2] on pp. 2-206 through 2-316.

TEST YOURSELF

1. What is a state property?

2. The enthalpy of a vapor A relative to liquid A at 0°C and 1 atm is 5000 J/kg at 0°C and 1 atm, and 7500 J/kg at 30°C and 1 atm.

 (a) What is $\hat{H}$ of A(l) at 0°C and 1 atm?

 (b) What is the approximate value of $\hat{H}$ for A(v) at 0°C and 5 atm?

 (c) What is $\Delta\hat{H}$ for the process

$$\text{A(v, 30°C, 1 atm)} \rightarrow \text{A(v, 0°C, 1 atm)}$$

 Does the answer depend on the reference state used to generate the table of enthalpies? Why not?

7.5b Steam Tables

Recall the phase diagram for water (Figure 6.1-1a, p. 239), which has the following appearance:

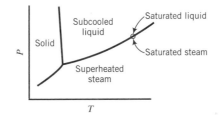

Pure water may coexist as liquid and vapor only at temperature–pressure pairs that fall on the vapor–liquid equilibrium (VLE) curve. At points above the VLE curve (but to the right of the solid–liquid equilibrium curve), water is a **subcooled liquid**. At points on the VLE curve, water may be **saturated liquid** or **saturated steam** (vapor) or a mixture of both. At points below the VLE curve, water is **superheated steam**.

For many years, compilations of physical properties of liquid water, saturated steam, and superheated steam issued in **steam tables** have been standard references for mechanical and chemical engineers involved with steam cycles for electrical power generation. Steam tables are contained in Tables B.5–B.7 of this text. We recommend that you look at these tables as we describe what you can find in them.

Table B.5 lists properties of saturated liquid water and saturated steam at temperatures from 0.01°C (the triple point temperature) to 102°C. The following properties can be determined for each tabulated temperature (and for intermediate temperatures by interpolation):

- *Column 2.* The pressure, P(bar), corresponding to the given temperature on the VLE curve—by definition, the vapor pressure of water at the given temperature. Instead of looking up a given temperature and finding the pressure, you could look up a given pressure in the second column and find the corresponding boiling point temperature in the first column.

[2]R. H. Perry and D. W. Green, Eds., *Perry's Chemical Engineers' Handbook,* 7th Edition, McGraw-Hill, New York, 1997.

- *Columns 3 and 4.* The specific volumes, $\hat{V}(m^3/kg)$, of liquid water and saturated steam at the given temperature. The inverses of these quantities are the densities (kg/m^3) of liquid water and steam.
- *Columns 5 and 6.* The specific internal energies, $\hat{U}(kJ/kg)$, of saturated liquid water and saturated steam at the given temperature *relative to a reference state of liquid water at the triple point.* (Remember, we can never know the absolute value of internal energy or enthalpy, but only how these quantities change when the substance goes from one state to another—in this case, from the reference state to the states listed in the table.)
- *Columns 7–9.* The specific enthalpies, $\hat{H}(kJ/kg)$, of saturated liquid water (Column 7) and saturated steam (Column 9), and the difference between these quantities, known as the *heat of vaporization* (Column 8). The reference point for the tabulated values of $\hat{H}$ is again liquid water at the triple point.

Table B.6 lists the same properties as Table B.5, except that pressure is the first column and temperature the second and the table covers a much broader range of temperatures and pressures. Tables B.5 and B.6 are commonly referred to as the *saturated steam tables.*

Table B.7—which is referred to as the *superheated steam table*—lists $\hat{V}$, $\hat{U}$, and $\hat{H}$ of water (the latter two properties relative to liquid water at the triple point) at *any* temperature and pressure, not just at points on the VLE curve. If you are given a temperature and a pressure, you can locate the properties of water at the intersection of the column corresponding to the given temperature and the row corresponding to the given pressure. If the intersection falls within the closed region on the table bounded by the vertical line to the left of the 50°C column, the horizontal line below the 221.2 bar row, and the zigzag hypotenuse, the water is a liquid; outside this region, it is a superheated vapor.

When you look up a pressure in the first column of Table B.7, you will find just below it in parentheses the boiling point temperature and in Columns 2 and 3 the properties of saturated liquid water and saturated steam at that pressure. If you are at a point in the superheated steam region, you can move all the way to the left to determine the saturation temperature at the same pressure, or the *dew point* of the superheated steam.

The next example illustrates the use of these tables to obtain physical property data for water.

EXAMPLE 7.5-2   *The Steam Tables*

1. Determine the vapor pressure, specific internal energy, and specific enthalpy of saturated steam at 133.5°C.
2. Show that water at 400°C and 10 bar is superheated steam and determine its specific volume, specific internal energy, and specific enthalpy relative to liquid water at the triple point, and its dew point.
3. Show that $\hat{U}$ and $\hat{H}$ for superheated steam depend strongly on temperature and relatively slightly on pressure.

SOLUTION   Verify the results to be given.

1. Table B.5 does not go up to 133.5°C, so we turn to Table B.6. For saturated steam at the given temperature (Column 2),

$$p^* = 3.0 \text{ bar}, \hat{V} = 0.606 \text{ m}^3/\text{kg}, \hat{U} = 2543.0 \text{ kJ/kg}, \hat{H} = 2724.7 \text{ kJ/kg}$$

2. From Table B.7, $[T = 400°C, P = 10 \text{ bar}]$ falls outside the closed region, showing that water is superheated steam at this temperature and pressure. The table also shows that at this condition,

$$\hat{H} = 3264 \text{ kJ/kg}, \hat{U} = 2958 \text{ kJ/kg}, \hat{V} = 0.307 \text{ m}^3/\text{kg}, T_{dp} = 179.9°C$$

3. Look at the properties of water at 400°C and 450°C, both at a pressure of 10 bar. You will see that both $\hat{U}$ and $\hat{H}$ change by about 3% when water goes from the first temperature to the second one (3264 kJ/kg → 3371 kJ/kg for $\hat{H}$, 2958 kJ/kg → 3041 kJ/kg for $\hat{U}$).

Now consider the properties at 10 bar and 20 bar, both at a temperature of 400°C. Even though the pressure has doubled, the values of $\hat{U}$ and $\hat{H}$ change by much less than 1%. Similar results would be obtained for liquid water. The conclusion is that when you need a value of $\hat{U}$ or $\hat{H}$ for water (or for any other species) at a given T and P, you must look it up at the correct temperature—interpolating if necessary—but you don't have to find it at the exact pressure.

The next example illustrates the use of the steam tables to solve energy balance problems.

EXAMPLE 7.5-3   ***Energy Balance on a Steam Turbine***

Steam at 10 bar absolute with 190°C of superheat is fed to a turbine at a rate $\dot{m} = 2000$ kg/h. The turbine operation is adiabatic, and the effluent is saturated steam at 1 bar. Calculate the work output of the turbine in kilowatts, neglecting kinetic and potential energy changes.

SOLUTION   The energy balance for this steady-state open system is

$$\dot{W}_s = -\Delta\dot{H} = -\dot{m}(\hat{H}_{out} - \hat{H}_{in})$$

(Why was the heat term dropped?)

Inlet Steam

Table B.7 indicates that steam at 10 bar is saturated at 180°C (*verify*), so that the inlet steam temperature is 180°C + 190°C = 370°C. Interpolating in the same table,

$$\hat{H}_{in}(10 \text{ bar}, 370°C) = 3201 \text{ kJ/kg}$$

Outlet Steam

From either Table B.6 or B.7, you can find that the enthalpy of saturated steam at 1 bar is

$$\hat{H}_{out}(1 \text{ bar, saturated}) = 2675 \text{ kJ/kg}$$

Energy Balance $\dot{W}_s = -\Delta\dot{H} = -\dfrac{2000 \text{ kg}}{\text{h}} \left| \dfrac{(2675 - 3201) \text{ kJ}}{\text{kg}} \right| \dfrac{1 \text{ h}}{3600 \text{ s}}$

$$= 292 \text{ kJ/s} = \boxed{292 \text{ kW}}$$

The turbine thus delivers 292 kW of work to its surroundings.

The superheated steam table, Table B.7, lists values for both liquid water and steam. If you wish to determine $\hat{H}$ for liquid water at a temperature T and pressure P that cannot easily be found in this table, you may calculate it in the following manner: (1) look up $\hat{U}$ and $\hat{V}$ for *saturated* liquid at the specified temperature in Table B.5; (2) assume these values are independent of pressure, and calculate $\hat{H}(P,T) = \hat{U} + P\hat{V}$. Furthermore, if the pressure is not excessive (say, less than 10 bar) or *if it is unknown,* neglect the $P\hat{V}$ correction and use the saturated liquid enthalpy $\hat{H}(T)$ given in Table B.5.

7.6   ENERGY BALANCE PROCEDURES

A properly drawn and labeled flowchart is essential for the efficient solution of energy balance problems. When labeling the flowchart, be sure to include all of the information you will need to determine the specific enthalpy of each stream component, including known temperatures and pressures. In addition, show states of aggregation of process materials when they are not obvious: do not simply write H_2O, for example, but rather $H_2O(s)$, $H_2O(l)$, or $H_2O(v)$, according to whether water is present as a solid, a liquid, or a vapor.

In the rest of this chapter, we will consider only species (such as water) for which tabulated internal energies or enthalpies are available. In Chapters 8 and 9 we will show how to choose reference states and calculate the required values of $\hat{U}$ and $\hat{H}$ when tabulated values cannot be found.

EXAMPLE 7.6-1   *Energy Balance on a One-Component Process*

Two streams of water are mixed to form the feed to a boiler. Process data are as follows:

Feed stream 1   120 kg/min @ 30°C
Feed stream 2   175 kg/min @ 65°C
Boiler pressure   17 bar (absolute)

The exiting steam emerges from the boiler through a 6-cm ID pipe. Calculate the required heat input to the boiler in kilojoules per minute if the emerging steam is saturated at the boiler pressure. Neglect the kinetic energies of the liquid inlet streams.

SOLUTION

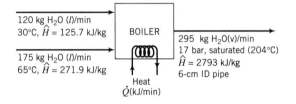

120 kg H_2O (*l*)/min
30°C, $\hat{H}$ = 125.7 kJ/kg

175 kg H_2O (*l*)/min
65°C, $\hat{H}$ = 271.9 kJ/kg

BOILER

Heat
$\dot{Q}$(kJ/min)

295 kg H_2O(v)/min
17 bar, saturated (204°C)
$\hat{H}$ = 2793 kJ/kg
6-cm ID pipe

1. *A first step in solving problems of this sort is to determine (if possible) the flow rates of all stream components using material balances.* In this case, it is a trivial matter to write and solve a mass balance on water to determine that the flow rate of the emerging steam is 295 kg/min.

2. *Next, determine the specific enthalpies of each stream component.* Tables B.5 and B.6 were used to determine $\hat{H}$ for liquid water at 30°C and 65°C and for saturated steam at 17 bar. The latter entry in the table also furnished the temperature of the saturated steam corresponding to this pressure (204°C). Note that the entries for liquid water correspond to pressures that may or may not equal the actual pressures of the inlet streams (which we do not know); we assume, however, that the enthalpy of liquid water is approximately independent of pressure and use the tabulated values.

3. *The final step is to write the appropriate form of the energy balance and solve it for the desired quantity.* For this open process system,

$$\dot{Q} - \dot{W}_s = \Delta\dot{H} + \Delta\dot{E}_k + \Delta\dot{E}_p$$

$\dot{W}_s = 0$ (no moving parts)
$\Delta\dot{E}_p = 0$ (generally assumed unless displacements through large heights are involved)

$$\dot{Q} = \Delta\dot{H} + \Delta\dot{E}_k$$

Evaluate $\Delta\dot{H}$

From Equation 7.4-14a,

$$\Delta\dot{H} = \sum_{\text{outlet}} \dot{m}_i \hat{H}_i - \sum_{\text{inlet}} \dot{m}_i \hat{H}_i$$

$$= \frac{295 \text{ kg}}{\text{min}} \left| \frac{2793 \text{ kJ}}{\text{kg}} - \frac{120 \text{ kg}}{\text{min}} \right| \frac{125.7 \text{ kJ}}{\text{kg}} - \frac{175 \text{ kg}}{\text{min}} \left| \frac{271.9 \text{ kJ}}{\text{kg}} \right.$$

$$= 7.61 \times 10^5 \text{ kJ/min}$$

Evaluate $\Delta \dot{E}_k$

From Table B.6, the specific volume of saturated steam at 17 bar is 0.1166 m³/kg, and the cross-sectional area of the 6-cm ID pipe is

$$A = \pi R^2 = \frac{3.1416}{} \left| \frac{(3.00)^2 \text{ cm}^2}{} \right| \frac{1 \text{ m}^2}{10^4 \text{ cm}^2} = 2.83 \times 10^{-3} \text{ m}^2$$

The steam velocity is

$$u(\text{m/s}) = \dot{V}(\text{m}^3/\text{s})/A(\text{m}^2)$$

$$= \frac{295 \text{ kg}}{\text{min}} \left| \frac{1 \text{ min}}{60 \text{ s}} \right| \frac{0.1166 \text{ m}^3}{\text{kg}} \left| \frac{1}{2.83 \times 10^{-3} \text{ m}^2} \right.$$

$$= 202 \text{ m/s}$$

Then, since the kinetic energies of the inlet streams are assumed negligible,

$$\Delta \dot{E}_k \approx (\dot{E}_k)_{\text{outlet stream}} = \dot{m}u^2/2$$

$$= \frac{295 \text{ kg/min}}{2} \left| \frac{(202)^2 \text{ m}^2}{\text{s}^2} \right| \frac{1 \text{ N}}{1 \text{ kg·m/s}^2} \left| \frac{1 \text{ kJ}}{10^3 \text{ N·m}} \right. = 6.02 \times 10^3 \text{ kJ/min}$$

Finally,

$$\dot{Q} = \Delta \dot{H} + \Delta \dot{E}_k$$

$$= [7.61 \times 10^5 + 6.02 \times 10^3] \text{ kJ/min}$$

$$= \boxed{7.67 \times 10^5 \text{ kJ/min}}$$

Observe that the kinetic energy change is only a small fraction—roughly 0.8%—of the total energy requirement for the process. This is a typical result, and it is not uncommon to neglect kinetic and potential energy changes (at least as a first approximation) relative to enthalpy changes for processes that involve phase changes, chemical reactions, or large temperature changes.

When process streams contain several components, the specific enthalpies of each component must be determined separately and substituted in the energy balance equation when $\Delta \dot{H}$ is evaluated. *For mixtures of near-ideal gases or of liquids with similar molecular structures* (e.g., mixtures of paraffins), *you may assume that $\hat{H}$ for a mixture component is the same as $\hat{H}$ for the pure substance at the same temperature and pressure.* Procedures to follow for solutions of gases or solids in liquids and for mixtures of dissimilar liquids are outlined in Chapter 8.

EXAMPLE 7.6-2   ***Energy Balance on a Two-Component Process***

A liquid stream containing 60.0 wt% ethane and 40.0% *n*-butane is to be heated from 150 K to 200 K at a pressure of 5 bar. Calculate the required heat input per kilogram of the mixture, neglecting potential and kinetic energy changes, using tabulated enthalpy data for C_2H_6 and C_4H_{10} and assuming that mixture component enthalpies are those of the pure species at the same temperature.

SOLUTION   ***Basis: 1 kg/s Mixture***

The enthalpies of *n*-butane at 150 K and 5 bar and at 200 K and 5 bar are given on p 2-223 of *Perry's Chemical Engineers' Handbook* (see footnote 2), and those of ethane at the same conditions are given on p. 2-234 of the *Handbook*. The tabulated enthalpy values are shown in the energy balance.

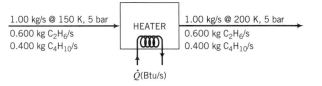

No material balances are necessary since there is only one input stream and one output stream and no chemical reactions, so we may proceed directly to the energy balance:

$$\dot{Q} - \dot{W}_s = \Delta\dot{H} + \Delta\dot{E}_k + \Delta\dot{E}_p$$

$$\left\Vert \begin{array}{l} \dot{W}_s = 0 \quad \text{(no moving parts)} \\ \Delta\dot{E}_k = 0, \Delta\dot{E}_p = 0 \quad \text{(by hypothesis)} \end{array} \right.$$

$$\dot{Q} = \Delta\dot{H}$$

Since the process materials are all gases and we are assuming ideal gas behavior, we may set the enthalpies of each stream equal to the sums of the individual component enthalpies and write

$$\dot{Q} = \Delta\dot{H} = \underbrace{\sum \dot{m}_i \hat{H}_i}_{\substack{\text{outlet} \\ \text{components}}} - \underbrace{\sum \dot{m}_i \hat{H}_i}_{\substack{\text{inlet} \\ \text{components}}}$$

$$= \frac{0.600 \text{ kg C}_2\text{H}_6}{\text{s}} \left| \frac{434.5 \text{ kJ}}{\text{kg}} + \frac{0.400 \text{ kg C}_4\text{H}_{10}}{\text{s}} \right| \frac{130.2 \text{ kJ}}{\text{kg}}$$

$$- [(0.600)(314.3) + (0.400)(30.0)] \text{ kJ/s} = 112 \text{ kJ/s} \implies \frac{112 \text{ kJ/s}}{1.00 \text{ kg/s}} = \boxed{112 \frac{\text{kJ}}{\text{kg}}}$$

In the two previous examples, it was possible to complete all material balance calculations before undertaking the energy balance. In another class of problems one more stream amount or flow rate is unknown than can be determined by material balances alone. To solve problems of this type, you must write and solve material and energy balances simultaneously.

EXAMPLE 7.6-3   *Simultaneous Material and Energy Balances*

Saturated steam at 1 atm is discharged from a turbine at a rate of 1150 kg/h. Superheated steam at 300°C and 1 atm is needed as a feed to a heat exchanger; to produce it, the turbine discharge stream is mixed with superheated steam available from a second source at 400°C and 1 atm. The mixing unit operates adiabatically. Calculate the amount of superheated steam at 300°C produced and the required volumetric flow rate of the 400°C steam.

SOLUTION   Specific enthalpies of the two feed streams and the product stream are obtained from the steam tables and are shown below on the flowchart.

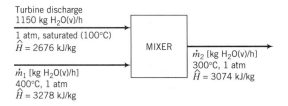

There are two unknown quantities in this process—$\dot{m}_1$ and $\dot{m}_2$—and only one permissible material balance. (Why?) The material and energy balances must therefore be solved simultaneously to determine the two flow rates.

Mass Balance on Water $1150 \text{ kg/h} + \dot{m}_1 = \dot{m}_2$ **(1)**

Energy Balance $\dot{Q} - \dot{W}_s = \Delta\dot{H} + \Delta\dot{E}_k + \Delta\dot{E}_p$

$\dot{Q} = 0$ (process is adiabatic)
$\dot{W}_s = 0$ (no moving parts)
$\Delta\dot{E}_k \approx 0, \Delta\dot{E}_p \approx 0$ (assumption)

$$\Delta\dot{H} = \sum_{\text{outlet}} \dot{m}_i\hat{H}_i - \sum_{\text{inlet}} \dot{m}_i\hat{H}_i = 0$$

$$\frac{1150 \text{ kg}}{\text{h}} \left| \frac{2676 \text{ kJ}}{\text{kg}} \right. + \dot{m}_1(3278 \text{ kJ/kg}) = \dot{m}_2(3074 \text{ kJ/kg}) \qquad \textbf{(2)}$$

Solving Equations 1 and 2 simultaneously yields

$$\dot{m}_1 = 2240 \text{ kg/h}$$

$$\boxed{\dot{m}_2 = 3390 \text{ kg/h}} \quad \text{(product flow rate)}$$

From Table B.7, the specific volume of steam at 400°C and 1 atm ($\approx$1 bar) is 3.11 m³/kg. The volumetric flow rate of this stream is therefore

$$\frac{2240 \text{ kg}}{\text{h}} \left| \frac{3.11 \text{ m}^3}{\text{kg}} \right. = \boxed{6980 \text{ m}^3/\text{h}}$$

If specific-volume data were not available, the ideal gas equation of state could be used as an approximation for the last calculation.

7.7   MECHANICAL ENERGY BALANCES

In chemical process units such as reactors, distillation columns, evaporators, and heat exchangers, shaft work and kinetic and potential energy changes tend to be negligible compared with heat flows and internal energy and enthalpy changes. Energy balances on such units therefore usually omit the former terms and so take the simple form $Q = \Delta U$ (closed system) or $\dot{Q} = \Delta\dot{H}$ (open system).

Another important class of operations is one for which the opposite is true—heat flows and internal energy changes are secondary in importance to kinetic and potential energy changes and shaft work. Most of these operations involve the flow of fluids to, from, and between tanks, reservoirs, wells, and process units. Accounting for energy flows in such processes is most conveniently done with **mechanical energy balances**.

The general form of the mechanical energy balance can be derived starting with the open-system balance and a second equation expressing the law of conservation of momentum, a derivation beyond the scope of this book. This section presents a simplified form for a single incompressible liquid flowing into and out of a process system at steady state.

Consider such a system, letting $\dot{m}$ be the mass flow rate and $\hat{V}$ the specific volume of the liquid. If $\hat{V}$ is replaced by $1/\rho$, where ρ is the liquid density, then the open-system energy balance (Equation 7.4-12) may be written

$$\frac{\Delta P}{\rho} + \frac{\Delta u^2}{2} + g\,\Delta z + \left(\Delta\hat{U} - \frac{\dot{Q}}{\dot{m}}\right) = -\frac{\dot{W}_s}{\dot{m}} \qquad \textbf{(7.7-1)}$$

The shaft work $\dot{W}_s$ is the work done by the fluid on moving elements in the process line.

In many cases only slight amounts of heat are transferred to or from the surroundings, there is little change in temperature from inlet to outlet, and no phase changes or reactions occur. Even under these circumstances, some kinetic or potential energy is always converted to thermal energy as a result of friction due to the movement of the fluid through the system. In consequence, the quantity $(\Delta\hat{U} - \dot{Q}/\dot{m})$ always has a positive component, termed the **friction**

loss and is given the symbol $\hat{F}$. Equation 7.7-1 may therefore be written

$$\frac{\Delta P}{\rho} + \frac{\Delta u^2}{2} + g\,\Delta z + \hat{F} = \frac{-\dot{W}_s}{\dot{m}} \tag{7.7-2}$$

Equation 7.7-2 is referred to as the **mechanical energy balance**. Once again, it is valid for steady-state flow of an incompressible fluid.

Methods of estimating friction losses for flow through straight pipes, orifices, nozzles, elbows, and so on are given in Section 10 of *Perry's Chemical Engineers' Handbook* (see footnote 2) and will not be discussed in this text. In the balance of this book we consider only processes in which friction losses are either specified or neglected.

A simplified form of the mechanical energy balance is obtained for frictionless processes ($\hat{F} \approx 0$) in which no shaft work is performed ($\dot{W}_s = 0$):

$$\frac{\Delta P}{\rho} + \frac{\Delta u^2}{2} + g\,\Delta z = 0 \tag{7.7-3}$$

Equation 7.7-3 is called the **Bernoulli equation**.

EXAMPLE 7.7-1   *The Bernoulli Equation*

Water flows through the system shown here at a rate of 20 L/min. Estimate the pressure required at point ① if friction losses are negligible.

SOLUTION

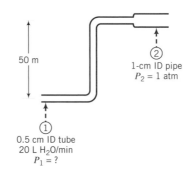

All of the terms of the Bernoulli equation, Equation 7.7-3, are known except ΔP, the variable to be determined, and Δu^2, which must be calculated from the known liquid flow rate and the diameters of the inlet and outlet pipes.

Velocities $\qquad\qquad\qquad\qquad \dot{u}(\text{m/s}) = \dot{V}(\text{m}^3/\text{s}) / A(\text{m}^2)$

The volumetric flow rate must be the same at points ① and ②. (Why?)

$$u_1 = \frac{20\text{ L}}{\text{min}}\left|\frac{1\text{ m}^3}{10^3\text{ L}}\right|\frac{1}{\pi(0.25)^2\text{ cm}^2}\left|\frac{10^4\text{ cm}^2}{\text{m}^2}\right|\frac{1\text{ min}}{60\text{ s}} = 17.0\text{ m/s}$$

$$u_2 = \frac{20\text{ L}}{\text{min}}\left|\frac{1\text{ m}^3}{10^3\text{ L}}\right|\frac{1}{\pi(0.5)^2\text{ cm}^2}\left|\frac{10^4\text{ cm}^2}{1\text{ m}^2}\right|\frac{1\text{ min}}{60\text{ s}} = 4.24\text{ m/s}$$

$$\Downarrow$$

$$\Delta u^2 = (u_2^2 - u_1^2) = (4.24^2 - 17.0^2)\text{ m}^2/\text{s}^2$$
$$= -271.0\text{ m}^2/\text{s}^2$$

Bernoulli Equation   (Equation 7.7-3)

$$\frac{\Delta P(\text{N/m}^2)}{\rho(\text{kg/m}^3)} + \frac{\Delta u^2(\text{m}^2/\text{s}^2)}{2\cdot 1[(\text{kg}\cdot\text{m/s}^2)/\text{N}]} + \frac{g(\text{m/s}^2)\Delta z(\text{m})}{1[(\text{kg}\cdot\text{m/s}^2)/\text{N}]}$$

$$\begin{Vmatrix} \Delta P = P_2 - P_1 \\ \rho = 1000 \text{ kg/m}^3 \\ \Delta u^2 = -271.0 \text{ m}^2/\text{s}^2 \\ g = 9.81 \text{ m/s}^2 \\ \Delta z = z_2 - z_1 \\ \quad = 50 \text{ m} \end{Vmatrix}$$

$$\frac{P_2 - P_1}{1000 \text{ kg/m}^3} - 135.5 \text{ N}\cdot\text{m/kg} + 490 \text{ N}\cdot\text{m/kg} = 0$$

$$\begin{Vmatrix} P_2 = 1 \text{ atm} \\ \quad = 1.01325 \times 10^5 \text{ N/m}^2 \end{Vmatrix}$$

$$P_1 = 4.56 \times 10^5 \text{ N/m}^2$$

$$= 4.56 \times 10^5 \text{ Pa}$$

$$= \boxed{4.56 \text{ bar}}$$

A common type of problem to which the mechanical energy balance is applicable is one that involves drainage or siphoning of a liquid from a container. A suitable choice of the locations of points ① and ② greatly simplifies such problems; it is convenient to choose as point ① a location on the liquid surface in the tank being drained and to take point ② at the discharge stream outlet. If the container is being drained relatively slowly, the kinetic energy at point ① may be neglected. Example 7.7-2 illustrates the computational procedure for such problems.

EXAMPLE 7.7-2   *Siphoning*

Gasoline ($\rho = 50.0$ lb$_\text{m}$/ft^3) is to be siphoned from a tank. The friction loss in the line is $\hat{F} = 0.80$ ft·lb$_\text{f}$/lb$_\text{m}$. Estimate how long it will take to siphon 5.00 gal, neglecting the change in liquid level in the gasoline tank during this process and assuming that both point ① (at the liquid surface in the gas tank) and point ② (in the tube just prior to the exit) are at 1 atm.

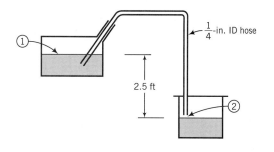

$\frac{1}{4}$-in. ID hose

2.5 ft

SOLUTION

Point ①: $P_1 = 1$ atm, $u_1 \approx 0$ ft/s, $z_1 = 2.5$ ft

Point ②: $P_2 = 1$ atm, $u_2 = ?$, $z_2 = 0$ ft

Mechanical Energy Balance (Equation 7.7-2)

$$\frac{\Delta P}{\rho} + \frac{\Delta u^2}{2} + g\,\Delta z + \hat{F} = \frac{-\dot{W}_s}{\dot{m}}$$

$$
\begin{aligned}
&\Delta P = 0 \\
&\Delta u^2 \approx u_2^2 \\
&g = 32.174 \text{ ft/s}^2 \\
&\Delta z = -2.5 \text{ ft} \\
&\hat{F} = 0.80 \text{ ft·lb}_f/\text{lb}_m \\
&\dot{W}_s = 0
\end{aligned}
$$

$$\frac{u_2^2(\text{ft}^2/\text{s}^2)}{2} \left| \frac{1 \text{ lb}_f}{32.174 \text{ lb}_m \cdot \text{ft/s}^2} \right. + \frac{32.174 \text{ ft/s}^2 \quad \left| \quad -2.5 \text{ ft} \right.}{\quad} \left| \frac{1 \text{ lb}_f}{32.174 \text{ lb}_m \cdot \text{ft/s}^2} \right. + 0.80 \text{ ft·lb}_f/\text{lb}_m$$

$$\Downarrow$$

$$u_2 = 10.5 \text{ ft/s}$$

(Verify that each additive term in the preceding equation has the units ft·lb$_f$/lb$_m$.)
The volumetric flow rate of the fluid in the tube is

$$\dot{V}(\text{ft}^3/\text{s}) = u_2(\text{ft/s}) \cdot A(\text{ft}^2)$$

$$= \frac{10.5 \text{ ft}}{\text{s}} \left| \frac{\pi(0.125)^2 \text{ in.}^2}{} \right| \frac{1 \text{ ft}^2}{144 \text{ in.}^2} = 3.58 \times 10^{-3} \text{ ft}^3/\text{s}$$

$$t(\text{s}) = \frac{\text{volume to be drained (ft}^3)}{\text{volumetric flow rate (ft}^3/\text{s})}$$

$$= \frac{(5.00 \text{ gal})(0.1337 \text{ ft}^3/\text{gal})}{3.58 \times 10^{-3} \text{ ft}^3/\text{s}} = \frac{187 \text{ s}}{60 \text{ s/min}} = \boxed{3.1 \text{ min}}$$

The last illustration we will consider is one in which the potential energy lost by falling water is converted to electrical energy by means of a turbine and generator. The work done by the water in rotating the turbine must be included as shaft work in the energy balance.

EXAMPLE 7.7-3 ***Hydraulic Power Generation***

Water flows from an elevated reservoir through a conduit to a turbine at a lower level and out of the turbine through a similar conduit. At a point 100 m above the turbine the pressure is 207 kPa, and at a point 3 m below the turbine the pressure is 124 kPa. What must the water flow rate be if the turbine output is 1.00 MW?

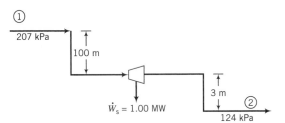

SOLUTION

No data are given for the friction loss so we will set $\hat{F} = 0$, recognizing that doing so introduces an error into the calculation. Since the diameters of the conduits at points ① and ② are the same and water may be considered incompressible, $\Delta u^2 = 0$. Equation 7.7-2 then becomes

$$\frac{\Delta P}{\rho} + g \Delta z = \frac{-\dot{W}_s}{\dot{m}}$$

$$\Downarrow$$

$$\dot{m} = \frac{-\dot{W}_s}{\dfrac{\Delta P}{\rho} + g \Delta z}$$

$$
\begin{aligned}
&\dot{W}_s = 1.00 \text{ MW} = 1.00 \times 10^6 \text{ N·m/s} \quad \text{(convince yourself)} \\
&\Delta P = (124 - 207) \text{ kPa} = -83 \text{ kPa} = -83 \times 10^3 \text{ N/m}^2 \\
&\frac{\Delta P}{\rho} = \frac{-83 \times 10^3 \text{ N/m}^2}{1.00 \times 10^3 \text{ kg/m}^3} = -83 \text{ N·m/kg} \\
&g = 9.81 \text{ m/s}^2 \\
&\Delta z = -103 \text{ m} \\
&g \Delta z = \frac{9.81 \text{ m}}{\text{s}^2} \left| \frac{-103 \text{ m}}{} \right| \frac{1 \text{ N}}{1 \text{ kg·m/s}^2} = -1010 \text{ N·m/kg}
\end{aligned}
$$

$$\dot{m} = \frac{-1.00 \times 10^6 \text{ N·m/s}}{(-83 - 1010) \text{ N·m/kg}} = \boxed{915 \text{ kg/s}}$$

TEST YOURSELF

1. Under what conditions is the mechanical energy balance, Equation 7.7-2, applicable? What is the physical significance of the friction loss term, $\hat{F}$, in this equation?
2. Under what conditions is the Bernoulli equation, Equation 7.7-3, applicable?

7.8 SUMMARY

Considerable amounts of energy are required to run most chemical processes. Engineers use **energy balances** to account for the energy that flows into or out of each unit of a process, to determine the net energy requirement for the process, and to design ways to reduce the energy requirement so as to improve process profitability.

- The total energy of a process system has three components: **kinetic energy**—energy due to motion of the system as a whole; **potential energy**—energy due to the position of the system in a potential field (such as the earth's gravitational field); and **internal energy**—energy due to the translation, rotation, vibration, and electromagnetic interactions of the molecules, atoms, and subatomic particles within the system.

- In a **closed system** (no mass is transferred across the system boundaries while the process is taking place), energy may be transferred between the system and its surroundings in two ways: as **heat**—energy that flows due to a temperature difference between the system and its surroundings, and as **work**—energy that flows in response to any other stimulus such as an applied force, a torque, or a voltage. Heat always flows from a higher temperature to a lower temperature. Heat is always defined to be positive if it flows to a system from the surroundings, and in many engineering references (including this one) work is defined as positive if it flows from the system to the surroundings.

- The **kinetic energy** of a body of mass m moving with velocity u is $E_k = mu^2/2$. The **gravitational potential energy** of the body is $E_p = mgz$, where g is the acceleration of gravity and

z is the height of the object above a reference plane at which E_p is arbitrarily defined to be zero. If a stream at height z flows with mass flow rate $\dot{m}$ and velocity u, $\dot{E}_k = \dot{m}u^2/2$ and $\dot{E}_p = \dot{m}gz$ may be thought of as the rates at which the stream is transporting kinetic energy and gravitational potential energy, respectively.[3]

- The **first law of thermodynamics for a closed system** (which we will generally refer to as the *energy balance*) between two instants of time is

$$\boxed{\Delta U + \Delta E_k + \Delta E_p = Q - W} \tag{7.3-4}$$

where in the context of closed systems, Δ denotes final value minus initial value. This equation states that the total energy transferred to the system in the specified time interval ($Q - W$) equals the gain in the total energy of the system in the same time interval ($\Delta U + \Delta E_k + \Delta E_p$). If energy is transferred *away* from the system, both sides of the equation are negative.

- When writing an energy balance for a closed system, first simplify Equation 7.3-4 by dropping negligible terms, then solve the simplified equation for whichever variable cannot be determined independently from other information in the process description.

    (a) If the system is **isothermal** (constant temperature), no phase changes or chemical reactions take place, and pressure changes are no more than a few atmospheres in magnitude, then $\Delta U \approx 0$.

    (b) If the system is not accelerating, then $\Delta E_k = 0$. If the system is not rising or falling, then $\Delta E_p = 0$. (You will almost always be able to drop these terms when writing balances on closed chemical process systems.)

    (c) If the system and its surroundings are at the same temperature or if the system is perfectly insulated, then $Q = 0$. The system is then termed **adiabatic**.

    (d) If energy is not transmitted across the system boundary by a moving part (such as a piston, an impeller, or a rotor), an electric current, or radiation, then $W = 0$.

- In an open system, work must be done to push input streams into the system $[= \sum_{\text{out}} P_j \dot{V}_j]$ and work is done by the output streams as they leave the system $[= \sum_{\text{in}} P_j \dot{V}_j]$, where P_j is the pressure of input or output stream j and $\dot{V}_j$ is the volumetric flow rate of the stream. The total rate of work done by a system on its surroundings ($\dot{W}$) is customarily divided into **flow work** ($\dot{W}_{fl}$), or work done by the output streams minus work done on the input streams at the system boundary, and **shaft work** ($\dot{W}_s$), all other work transferred across the system boundary by moving parts or as electricity or radiation. Thus

$$\dot{W} = \dot{W}_s + \dot{W}_{fl} = \dot{W}_s + \sum_{\substack{\text{output} \\ \text{streams}}} P_j \dot{V}_j - \sum_{\substack{\text{input} \\ \text{streams}}} P_j \dot{V}_j$$

- The first law of thermodynamics for an open system at steady state resembles the closed-system balance:

$$\Delta \dot{U} + \Delta \dot{E}_k + \Delta \dot{E}_p = \dot{Q} - \dot{W}$$

except that each term now has units of (kJ/s) instead of (kJ) and Δ now signifies (output − input) and not (final − initial). The more commonly used form of the first law is derived by (a) substituting for $\dot{W}$ the previously derived expression in terms of flow work and shaft work; (b) expressing the volumetric flow rate of each input and output stream ($\dot{V}_j$) as $\dot{m}_j \hat{V}_j$, where $\hat{V}_j$ is the **specific volume** (inverse density) of the stream fluid; (c) expressing the rate of transport of internal energy by a stream ($\dot{U}_j$) as $\dot{m}_j \hat{U}_j$, where $\hat{U}_j$ is the **specific internal**

---

[3]The unit conversion factors [(1 N)/(1 kg·m/s^2)] and [(1 kJ)/(1 N·m/s)] must be applied to the right-hand side of each of these equations to express these quantities in kJ/s (kW).

energy of the stream fluid; and (d) defining the **specific enthalpy** ($\hat{H}$) of a substance as $\hat{U} + P\hat{V}$. After some algebraic manipulation (Section 7.4c) the balance equation becomes

$$\boxed{\Delta\dot{H} + \Delta\dot{E}_k + \Delta\dot{E}_p = \dot{Q} - \dot{W}_s} \tag{7.4-15}$$

where

$$\Delta\dot{H} = \sum_{\substack{\text{output} \\ \text{streams}}} \dot{m}_j\hat{H}_j - \sum_{\substack{\text{input} \\ \text{streams}}} \dot{m}_j\hat{H}_j$$

$$\Delta\dot{E}_k = \sum_{\substack{\text{output} \\ \text{streams}}} \dot{m}_j u_j^2/2 - \sum_{\substack{\text{input} \\ \text{streams}}} \dot{m}_j u_j^2/2$$

$$\Delta\dot{E}_p = \sum_{\substack{\text{output} \\ \text{streams}}} \dot{m}_j g z_j - \sum_{\substack{\text{input} \\ \text{streams}}} \dot{m}_j g z_j$$

- When writing an energy balance for an open system at steady state, first simplify Equation 7.4-15 by dropping negligible terms, then solve the simplified equation for whichever variable cannot be determined independently from other information in the process description.

  (a) If no temperature changes, phase changes, or chemical reactions occur in a process and pressure changes from inlet to outlet are no more than a few atmospheres in magnitude, then $\Delta\dot{H} \approx 0$. (Under these circumstances, mechanical energy balances—Section 7.7—tend to be more useful than Equation 7.4-15.)

  (b) If conditions are such that $\Delta\dot{H}$ may not be neglected (i.e., if temperature changes, phase changes, or chemical reactions occur), then $\Delta\dot{E}_k$ and $\Delta\dot{E}_p$ usually may be neglected. In any case, if there are no great vertical distances between the inlets and the outlets of a system, $\Delta\dot{E}_p \approx 0$.

  (c) If the system and its surroundings are at the same temperature or if the system is perfectly insulated, then $\dot{Q} = 0$ and the process is adiabatic.

  (d) If energy is not transmitted across the system boundary by a moving part, an electric current, or radiation, then $\dot{W}_s = 0$.

- The value of $\hat{U}$ for a pure substance in a given state (temperature, pressure, and phase) is the sum of the kinetic and potential energies of the individual molecular, atomic, and subatomic particles in a unit amount of the substance. *It is impossible to determine the true value of $\hat{U}$ for a substance, and hence also impossible to determine the true value of $\hat{H} = \hat{U} + P\hat{V}$.* However, we can measure the *change* in $\hat{U}$ or $\hat{H}$ corresponding to a specified change of state, which is all we ever need to know for energy balance calculations.

- A common practice is to arbitrarily designate a **reference state** for a substance at which $\hat{U}$ or $\hat{H}$ is declared to equal zero, and then tabulate $\hat{U}$ and/or $\hat{H}$ for the substance relative to the reference state. The statement "The specific enthalpy of $CO(g)$ at 100°C and 1 atm relative to $CO(g)$ at 0°C and 1 atm is 2919 J/mol" therefore has the following meaning:

$$CO(g, 0°C, 1\text{ atm}) \rightarrow CO(g, 100°C, 1\text{ atm}) : \Delta\hat{H} = 2919 \text{ J/mol}$$

The statement says nothing about the absolute specific enthalpy of CO at 100°C and 1 atm, which can never be determined.

- Both $\hat{U}$ and $\hat{H}$ are **state properties**, meaning that $\Delta\hat{U}$ and $\Delta\hat{H}$ for a given change in state for a substance are the same regardless of the path the substance follows from the initial state to the final state.

- The **steam tables** (Tables B.5, B.6, and B.7) can be used to estimate $\hat{U}$ and $\hat{H}$ for liquid water and steam (water vapor) at any specified temperature and pressure. The reference state for the tabulated internal energies and enthalpies in the steam tables is liquid water at the triple point—0.01°C and 0.00611 bar.

- At this point, you can perform energy balance calculations only for systems in which ΔU (closed system) or $\Delta \dot{H}$ (open system) can be neglected and for nonreactive systems involving species for which tables of $\hat{U}$ or $\hat{H}$ are available. Energy balance procedures for other types of systems are presented in Chapters 8 and 9.

- **Mechanical energy balances** are useful for open systems in which heat flows and internal energy (and enthalpy) changes are secondary in importance to kinetic and potential energy changes and shaft work. For a liquid of constant density ρ flowing through such a system, the steady-state mechanical energy balance is

$$\boxed{\frac{\Delta P}{\rho} + \frac{\Delta u^2}{2} + g\,\Delta z + \hat{F} = \frac{-\dot{W}_s}{\dot{m}}}$$ **(7.7-2)**

where $\hat{F}$ (N·m/kg) is the **friction loss**—thermal energy generated by friction between adjacent liquid elements moving at different velocities and between liquid elements and the system walls. The friction loss shows up as a heat loss from the system ($\dot{Q} < 0$) and/or a gain in temperature and hence in internal energy from inlet to outlet ($\Delta \dot{U} > 0$). If $\hat{F}$ and $\dot{W}_s$ can be neglected, the resulting form of ¿ Equation 7.7-2 is the **Bernoulli equation**.

- At this point you can solve mechanical energy balances only for systems in which the friction loss ($\hat{F}$) is given or negligible or if it is the only unknown quantity in Equation 7.7-2. Texts on fluid mechanics present methods for estimating $\hat{F}$ from information about fluid flow rates and physical properties and various characteristics of the system through which the fluid is flowing.

PROBLEMS

7.1. A certain gasoline engine has an efficiency of 30%; that is, it converts into useful work 30% of the heat generated by burning a fuel. If the engine consumes 0.80 L/h of a gasoline with a heating value of 3.5×10^4 kJ/L, how much power does it provide? Express the answer both in kW and horse-power.

7.2. Consider an automobile with a mass of 5500 lb_m braking to a stop from a speed of 55 miles/h.
   (a) How much energy (Btu) is dissipated as heat by the friction of the braking process?
   (b) Suppose that throughout the United States, 300,000,000 such braking processes occur in the course of a given day. Calculate the average rate (megawatts) at which energy is being dissipated by the resulting friction.

7.3. A simplified version of the life cycle of grocery sacks is shown below.[4]

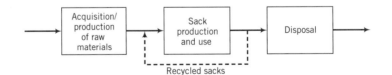

In the late 1970s, supermarkets began to replace paper sacks with polyethylene (plastic) sacks. In the 1980s, a movement to go back to paper arose, primarily inspired by environmental considerations. In the 1990s, a countermovement arose, claiming that paper sacks have a greater negative environmental impact than plastic ones. As this book goes to press, the debate continues.

Following are estimates of the atmospheric emissions and energy consumption associated with the raw materials acquisition and processing (logging, pulping, and papermaking for paper, petroleum production and refining and polymerization for polyethylene) and disposal of sacks, and for the manufacture and use of sacks.

---

[4]Problem adapted from D. T. Allen, N. Bakshani, and K. S. Rosselot, *Pollution Prevention: Homework and Design Problems for Engineering Curricula*, American Institute for Pollution Prevention, New York, 1992. The emissions and energy consumption data are from Franklin Associates, Ltd., *Resource and Environmental Profile Analysis of Polyethylene and Unbleached Paper Grocery Sacks*. Report prepared for the Council for Solid Waste Solutions, Prairie Village, KS, June 1990.

	Emissions (oz/sack)		Energy Consumed (Btu/sack)	
Stages	Paper	Plastic	Paper	Plastic
Raw materials production plus product disposal	0.0510	0.0045	724	185
Sack production and use	0.0516	0.0146	905	464

Assume that atmospheric emissions and energy consumption do not depend on whether the new sacks are made from fresh raw materials or recycled sacks, and that it takes roughly twice as many plastic sacks as paper sacks to hold a given quantity of groceries.

(a) Calculate the air emissions (lb_m) and energy consumption (Btu) per 1000 paper sacks used and per 2000 plastic sacks used, assuming that no sacks are recycled.

(b) Repeat the calculations of part (a) assuming that 60% of the used sacks are recycled. By what percentages are air emissions and energy consumption reduced for each sack material as a consequence of recycling?

(c) Estimate the number of grocery sacks used in a day in the United States (population = 300 million), and calculate the average rate of energy consumption (megawatts, MW) associated with the production, use, and disposal of these sacks, assuming that they are plastic and none are recycled. How many MW would be saved by 60% recycling?

(d) You should have found that both atmospheric emissions and energy consumption are higher when paper is used rather than plastic, although recycling reduces the differences. Nevertheless, deciding to use plastic based entirely on this result could be a serious mistake. List several important factors that were not taken into account in making the decision, including considerations of the potential environmental impact of each type of bag.

7.4. Liquid methanol is pumped from a large storage tank through a 1-in. ID pipe at a rate of 3.00 gal/min.

(a) At what rate in (i) $ft \cdot lb_f/s$ and (ii) hp is kinetic energy being transported by the methanol in the pipe?

(b) The electrical power input to the pump transporting the methanol must be greater than the amount you calculated in part (a). What would you guess becomes of the additional energy? (There are several possible answers.)

7.5. Air at 300°C and 130 kPa flows through a horizontal 7-cm ID pipe at a velocity of 42.0 m/s.

(a) Calculate $\dot{E}_k$(W), assuming ideal gas behavior.

(b) If the air is heated to 400°C at constant pressure, what is $\Delta\dot{E}_k = \dot{E}_k(400°C) - \dot{E}_k(300°C)$?

(c) Why would it be incorrect to say that the rate of transfer of heat to the gas in part (b) must equal the rate of change of kinetic energy?

7.6. Suppose you pour a gallon of water on a yowling cat 10 ft below your bedroom window.

(a) How much potential energy ($ft \cdot lb_f$) does the water lose?

(b) How fast is the water traveling (ft/s) just before impact?

(c) True or false: Energy must be conserved, therefore the kinetic energy of the water before impact must equal the kinetic energy of the cat after impact.

7.7. Methane enters a 3-cm ID pipe at 30°C and 10 bar with an average velocity of 5.00 m/s and emerges at a point 200 m lower than the inlet at 30°C and 9 bar.

(a) Without doing any calculations, predict the signs (+ or −) of $\Delta\dot{E}_k$ and $\Delta\dot{E}_p$, where Δ signifies (outlet − inlet). Briefly explain your reasoning.

(b) Calculate $\Delta\dot{E}_k$ and $\Delta\dot{E}_p$ (W), assuming that the methane behaves as an ideal gas.

7.8. You recently purchased a large plot of land in the Amazon jungle at an extremely low cost. You are quite pleased with yourself until you arrive there and find that the nearest source of electricity is 1500 miles away, a fact that your brother-in-law, the real estate agent, somehow forgot to mention. Since the local hardware store does not carry 1500-mile-long extension cords, you decide to build a small hydroelectric generator under a 75-m high waterfall located nearby. The flow rate of the waterfall is 10^5 m³/h, and you anticipate needing 750 kW·h/wk to run your lights, air conditioner, and television. Calculate the maximum power theoretically available from the waterfall and see if it is sufficient to meet your needs.

7.9. Write and simplify the closed-system energy balance (Equation 7.3-4) for each of the following processes, and state whether nonzero heat and work terms are positive or negative. Begin by defining the system. The solution of part (a) is given as an illustration.

(a) The contents of a closed flask are heated from 25°C to 80°C.

Solution. The system is the flask contents.

$$Q - W = \Delta U + \Delta E_k + \Delta E_p$$

$\quad\quad W = 0$ (no moving parts or generated currents)
$\quad\quad \Delta E_k = 0$ (system is stationary)
$\quad\quad \Delta E_p = 0$ (no height change)

$$\boxed{Q = \Delta U}$$

$\boxed{Q > 0 \quad \text{(heat is transferred to the system)}}$

(b) A tray filled with water at 20°C is put into a freezer. The water turns into ice at −5°C. (*Note:* When a substance expands it does work on its surroundings and when it contracts the surroundings do work on it.)

(c) A chemical reaction takes place in a closed adiabatic (perfectly insulated) rigid container.

(d) Repeat part (c), only suppose that the reactor is isothermal rather than adiabatic and that when the reaction was carried out adiabatically the temperature in the reactor increased.

7.10. A cylinder with a movable piston contains 4.00 liters of a gas at 30°C and 5.00 bar. The piston is slowly moved to compress the gas to 8.00 bar.

(a) Considering the system to be the gas in the cylinder and neglecting ΔE_p, write and simplify the closed-system energy balance. Do not assume that the process is isothermal in this part.

(b) Suppose now that the process is carried out isothermally, and the compression work done on the gas equals 7.65 L·bar. If the gas is ideal so that $\hat{U}$ is a function only of T, how much heat (in joules) is transferred to or from (state which) the surroundings? (Use the gas-constant table in the back of the book to determine the factor needed to convert L·bar to joules.)

(c) Suppose instead that the process is adiabatic and that $\hat{U}$ increases as T increases. Is the final system temperature greater than, equal to, or less than 30°C? (Briefly state your reasoning.)

7.11. A piston-fitted cylinder with a 6-cm inner diameter contains 1.40 g of nitrogen. The mass of the piston is 4.50 kg, and a 20.00-kg weight rests on the piston. The gas temperature is 30°C, and the pressure outside the cylinder is 1.00 atm.

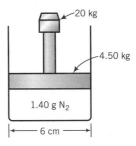

(a) Prove that the absolute pressure of the gas in the cylinder is 1.86×10^5 Pa. Then calculate the volume occupied by the gas, assuming ideal gas behavior.

(b) Suppose the weight is abruptly lifted and the piston rises to a new equilibrium position. Further suppose that the process takes place in two steps: a rapid step in which a negligible amount of heat is exchanged with the surroundings, followed by a slow step in which the gas returns to 30°C. Considering the gas as the system, write the energy balances for step 1, step 2, and the overall process. In all cases, neglect ΔE_k and ΔE_p. If $\hat{U}$ varies proportionally with T, does the gas temperature increase or decrease in step 1? Briefly explain your answer.

(c) The work done by the gas equals the restraining force (the weight of the piston plus the force due to atmospheric pressure) times the distance traveled by the piston. Calculate this quantity and use it to determine the heat transferred to or from (state which) the surroundings during the process.

7.12. Oxygen at 150 K and 41.64 atm has a tabulated specific volume of 4.684 cm³/g and a specific internal energy of 1706 J/mol. Calculate the specific enthalpy of O_2 in this state.

7.13. Values of the specific internal energy of bromine at three conditions are listed here.

State	T(K)	P(bar)	$\hat{V}$(L/mol)	$\hat{U}$(kJ/mol)
Liquid	300	0.310	0.0516	0.000
Vapor	300	0.310	79.94	28.24
Vapor	340	1.33	20.92	29.62

(a) What reference state was used to generate the listed specific internal energies?

(b) Calculate $\Delta\hat{U}$(kJ/mol) for a process in which bromine vapor at 300 K is condensed at constant pressure. Then calculate $\Delta\hat{H}$(kJ/mol) for the same process. (See Example 7.4-1.) Finally, calculate ΔH(kJ) for 5.00 mol of bromine undergoing the process.

(c) Bromine vapor in a 5.00-liter container at 300 K and 0.205 bar is to be heated to 340 K. Calculate the heat (kJ) that must be transferred to the gas to achieve the desired temperature increase, assuming that $\hat{U}$ is independent of pressure.

(d) In reality, more heat than the amount calculated in part (c) would have to be transferred to the container to raise the gas temperature by 40 K, for several reasons. State two of them.

7.14. Prove that for an ideal gas, $\hat{U}$ and $\hat{H}$ are related as $\hat{H} = \hat{U} + RT$, where R is the gas constant. Then:

(a) Taking as given that the specific internal energy of an ideal gas is independent of the gas pressure, justify the claim that $\Delta\hat{H}$ for a process in which an ideal gas goes from (T_1, P_1) to (T_2, P_2) equals $\Delta\hat{H}$ for the same gas going from T_1 to T_2 at a constant pressure of P_1.

(b) Calculate ΔH(cal) for a process in which the temperature of 2.5 mol of an ideal gas is raised by 50°C, resulting in a specific internal energy change $\Delta\hat{U} = 3500$ cal/mol.

7.15. If a system expands in volume by an amount ΔV(m³) against a constant restraining pressure P(N/m²), a quantity $P\Delta V$(J) of energy is transferred as *expansion work* from the system to its surroundings. Suppose that the following four conditions are satisfied for a closed system: (a) the system expands against a constant pressure (so that $\Delta P = 0$); (b) $\Delta E_k = 0$; (c) $\Delta E_p = 0$; and (d) the only work done by or on the system is expansion work. Prove that under these conditions, the energy balance simplifies to $Q = \Delta H$.

7.16. A horizontal cylinder equipped with a frictionless piston contains 785 cm³ of steam at 400 K and 125 kPa. A total of 83.8 joules of heat is transferred to the steam, causing the steam temperature to rise and the cylinder volume to increase. A constant restraining force is maintained on the piston throughout the expansion, so that the pressure exerted by the piston on the steam remains constant at 125 kPa.

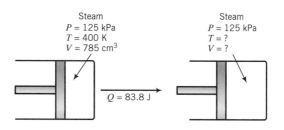

The specific enthalpy of steam at 125 kPa varies with temperature approximately as

$$\hat{H}(\text{J/mol}) = 34{,}980 + 35.5T(\text{K})$$

(a) Taking the steam as the system, convince yourself that $Q = \Delta H$ for this process—that is, the four conditions specified in part (a) of Problem 7.15 are applicable. Then prove that the final steam temperature is 480 K. Finally, calculate (i) the final cylinder volume, (ii) the expansion work done by the steam, and (iii) ΔU(J).

(b) Which of the specified conditions of Problem 7.15 would have been only an approximation if the cylinder were not horizontal?

7.17. You are performing an experiment to measure the specific internal energy of a gas relative to a reference state of 25°C and 1 atm (at which conditions $\hat{U}$ is arbitrarily set equal to 0). The gas is placed in a closed insulated 2.10-liter container at 25°C and 1 atm. A switch is alternately closed and opened, causing a current to flow intermittently through an electrical heating coil in the chamber. The gas temperature, which is monitored with a calibrated thermocouple, increases while the circuit is closed and remains constant while it is open. A wattmeter reads 1.4 W when the circuit is closed; 90% of this power is transferred to the gas as heat. The thermocouple calibration curve is a straight line through the points $(T = 0°C, E = -0.249 \text{ mV})$ and $(T = 100°C, E = 5.27 \text{ mV})$, where E is the thermocouple potentiometer reading.

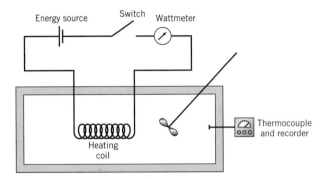

The following data are taken, where t represents the cumulative time during which the circuit was closed:

t(s)	0	30	60	90
E(mV)	1.13	2.23	3.34	4.44

(a) Which given item of information suggests that the chamber may be considered adiabatic? (*Note:* Simply saying the container is insulated does not guarantee that it is adiabatic.)

(b) Write the energy balance for the gas in the chamber and use it to calculate $\hat{U}$(J/mol) at each of the observed temperatures, neglecting the work done on the gas by the stirrer. Express your solution as a table of $\hat{U}$ versus T.

(c) What might the purpose of the stirrer be?

(d) What happens to the 0.14 W of power that does not go to raise the temperature of the gas?

(e) A colleague points out to you that the calculated values of $\hat{U}$ fail to take something into account and so do not precisely correspond to the values at the calculated temperatures and 1 atm. You reply that she is quite correct, but it does not matter. Justify her statement and state the basis of your reply. Suggest several ways to provide quantitative validation of your claim.

7.18. Define a system and simplify the open-system energy balance (Equation 7.4-15) for each of the following cases. State when possible whether nonzero heat and shaft work terms are positive or negative. The solution of part (a) is given as an illustration.

(a) Steam enters a rotary turbine and turns a shaft connected to a generator. The inlet and outlet steam ports are at the same height. Some energy is transferred to the surroundings as heat.

Solution. The system is the steam flowing from the inlet port to the outlet port:

$$\dot{Q} - \dot{W}_s = \Delta\dot{H} + \Delta\dot{E}_k + \Delta\dot{E}_p$$

$$\Big\Downarrow \Delta\dot{E}_p = 0 \quad \text{(no height change)}$$

$$\boxed{\Delta\dot{H} + \Delta\dot{E}_k = \dot{Q} - \dot{W}_s}$$

$$\boxed{\begin{array}{l}\dot{Q} \text{ is negative} \\ \dot{W}_s \text{ is positive}\end{array}}$$

(b) A liquid stream flows through a heat exchanger in which it is heated from 25°C to 80°C. The inlet and outlet pipes have the same diameter, and there is no change in elevation between these points.

(c) Water passes through the sluice gate of a dam and falls on a turbine rotor, which turns a shaft connected to a generator. The fluid velocity on both sides of the dam is negligible, and the water undergoes insignificant pressure and temperature changes between the inlet and outlet. (See Example 7.4-2.)

(d) Crude oil is pumped through a cross-country pipeline. The pipe inlet is 200 m higher than the outlet, the pipe diameter is constant, and the pump is located near the midpoint of the pipeline. Energy dissipated by friction in the line is transferred as heat through the wall.

(e) A chemical reaction takes place in a continuous reactor that contains no moving parts. Kinetic and potential energy changes from inlet to outlet are negligible.

7.19. Air is heated from 25°C to 150°C prior to entering a combustion furnace. The change in specific enthalpy associated with this transition is 3640 J/mol. The flow rate of air at the heater outlet is 1.25 m³/min and the air pressure at this point is 122 kPa absolute.

(a) Calculate the heat requirement in kW, assuming ideal gas behavior and that kinetic and potential energy changes from the heater inlet to the outlet are negligible.

(b) Would the value of $\Delta \dot{E}_k$ [which was neglected in part (a)] be positive or negative, or would you need more information to be able to tell? If the latter, what additional information would be needed?

7.20. A **Thomas flowmeter** is a device in which heat is transferred at a measured rate from an electric coil to a flowing fluid, and the flow rate of the stream is calculated from the measured temperature increase of the fluid. Suppose a device of this sort is inserted in a stream of nitrogen, the current through the heating coil is adjusted until the wattmeter reads 1.25 kW, and the stream temperature goes from 30°C and 110 kPa before the heater to 34°C and 110 kPa after the heater.

(a) If the specific enthalpy of nitrogen is given by the formula

$$\hat{H}(\text{kJ/kg}) = 1.04[T(°\text{C}) - 25]$$

what is the volumetric flow rate of the gas (L/s) upstream of the heater (i.e., at 30°C and 110 kPa)?

(b) List several assumptions made in the calculation of part (a) that could lead to errors in the calculated flow rate.

7.21. The specific enthalpy of liquid *n*-hexane at 1 atm varies linearly with temperature and equals 25.8 kJ/kg at 30°C and 129.8 kJ/kg at 50°C.

(a) Determine the equation that relates $\hat{H}$(kJ/kg) to T(°C) and calculate the reference temperature on which the given enthalpies are based. Then derive an equation for $\hat{U}(T)$(kJ/kg) at 1 atm.

(b) Calculate the average heat transfer rate required to cool 20 kg of liquid *n*-hexane from 80°C to 20°C in 5 min.

7.22. Steam at 260°C and 7.00 bar absolute is expanded through a nozzle to 200°C and 4.00 bar. Negligible heat is transferred from the nozzle to its surroundings. The approach velocity of the steam is negligible. The specific enthalpy of steam is 2974 kJ/kg at 260°C and 7 bar and 2860 kJ/kg at 200°C and 4 bar. Use the open-system energy balance to calculate the exit steam velocity.

7.23. The heart pumps blood at an average rate of 5 L/min. The gauge pressure on the venous (intake) side is 0 mm Hg and that on the arterial (discharge) side is 100 mm Hg. Energy is supplied to the heart as heat released by the absorption of oxygen in the cardiac muscles: 5 mL (STP) O_2/min is absorbed, and 20.2 kJ is released per mL of O_2 absorbed. Part of this absorbed energy is converted to flow work (the work done to pump blood through the circulatory system), and the balance is lost as heat transferred to the tissues surrounding the heart.

(a) Simplify Equation 7.4-12 for this system, assuming (among other things) that there is no change in internal energy from inlet to outlet.

(b) What percentage of the heat input to the heart ($\dot{Q}_{in}$) is converted to flow work? (The answer may be thought of as the efficiency of the heart as a pump.)

7.24. Saturated steam at 100°C is heated to 400°C. Use the steam tables to determine (a) the required heat input (J/s) if a continuous stream flowing at 100 kg/s undergoes the process at constant pressure and (b) the required heat input (J) if 100 kg undergoes the process in a constant-volume container. What is the physical significance of the difference between the numerical values of these two quantities?

7.25. A fuel oil is burned with air in a boiler furnace. The combustion produces 813 kW of thermal energy, of which 65% is transferred as heat to boiler tubes that pass through the furnace. The combustion products pass from the furnace to a stack at 650°C. Water enters the boiler tubes as a liquid at 20°C and leaves the tubes as saturated steam at 20 bar absolute.

    (a) Calculate the rate (kg/h) at which steam is produced.

    (b) Use the steam tables to estimate the volumetric flow rate of the steam produced.

    (c) Repeat the calculation of part (b), only assume ideal gas behavior instead of using the steam tables. Would you have more confidence in the estimate of part (b) or part (c)? Explain.

    (d) What happened to the 35% of the thermal energy released by the combustion that did not go to produce the steam.

7.26. Liquid water is fed to a boiler at 24°C and 10 bar and is converted at constant pressure to saturated steam. Use the steam tables to calculate $\Delta \hat{H}$(kJ/kg) for this process, and then calculate the heat input required to produce 15,000 m³/h of steam at the exiting conditions. Assume that the kinetic energy of the entering liquid is negligible and that the steam is discharged through a 15-cm ID pipe.

7.27. You have been assigned to collect thermodynamic data for a new liquid product your company is about to begin manufacturing and you decide to use a continuous-flow technique to generate a correlation of $\hat{H}$ versus T. You wrap an electrical heating tape around a pipe, cover the tape with a thick layer of insulation, pump the liquid through the pipe at the rate of 228 g/min, and adjust the power input to the heating tape with a variable resistor. For each resistance setting, you record the power input and the temperature of the liquid at the pipe outlet. You multiply the power input by a correction factor of 0.94 to determine the rate of heat input to the liquid. The entering fluid temperature remains at 25°C throughout the experiment.

    The following data are taken:

T_{out}(°C)	Heat Input to the Liquid (W)
25.0	0.0
26.4	17.0
27.8	35.3
29.0	50.9
32.4	94.4

    (a) Generate a table of $\hat{H}$(J/g) versus T(°C), taking 25°C and 1 atm as the reference state.

    (b) Fit a line to the data (either graphically or by the method of least squares) to determine the coefficient b of an expression of the form $\hat{H} = b(T - 25)$.

    (c) Estimate the heat input required to raise 350 kg/min of the liquid from 20°C to 40°C.

    (d) The correction factor of 0.94 accounts for the fact that the rate of energy input to the heating tape is somewhat greater than the rate of energy input to the liquid. Where does the additional energy go? (There are several answers.)

7.28. Saturated steam at a gauge pressure of 2.0 bar is to be used to heat a stream of ethane. The ethane enters a heat exchanger at 16°C and 1.5 bar gauge at a rate of 795 m³/min and is heated at constant pressure to 93°C. The steam condenses and leaves the exchanger as a liquid at 27°C. The specific enthalpy of ethane at the given pressure is 941 kJ/kg at 16°C and 1073 kJ/kg at 93°C.

    (a) How much energy (kW) must be transferred to the ethane to heat it from 16°C to 93°C?

    (b) Assuming that all the energy transferred from the steam goes to heat the ethane, at what rate in m³/s must steam be supplied to the exchanger? If the assumption is incorrect, would the calculated value be too high or too low?

    (c) Should the heat exchanger be set up for cocurrent or countercurrent flow (see schematic diagram below)? Explain. (*Hint:* Remember that heat always flows from a higher temperature to a lower temperature.)

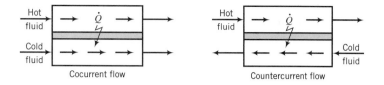

Cocurrent flow   Countercurrent flow

7.29. Superheated steam at 40 bar absolute and 500°C flows at a rate of 250 kg/min to an adiabatic turbine, where it expands to 5 bar. The turbine develops 1500 kW. From the turbine the steam flows to a heater, where it is reheated isobarically to its initial temperature. Neglect kinetic energy changes.

(a) Write an energy balance on the turbine and use it to determine the outlet stream temperature.

(b) Write an energy balance on the heater and use it to determine the required input (kW) to the steam.

(c) Verify that an overall energy balance on the two-unit process is satisfied.

(d) Suppose the turbine inlet and outlet pipes both have diameters of 0.5 meter. Show that it is reasonable to neglect the change in kinetic energy for this unit.

7.30. During a period of relative inactivity, the average rate of transport of enthalpy by the metabolic and digestive waste products leaving the body minus the rate of enthalpy transport by the raw materials ingested and breathed into the body is approximately $\Delta H = -300$ kJ/h. Heat is transferred from the body to its surroundings at a rate given by

$$Q = hA(T_s - T_0)$$

where A is the body surface area (roughly 1.8 m^2 for an adult), T_s is the skin temperature (normally 34.2°C), T_0 is the temperature of the body surroundings, and h is a **heat transfer coefficient**. Typical values of h for the human body are

$$h = 8 \text{ kJ/(m}^2 \cdot \text{h} \cdot °\text{C)} \quad \text{(fully clothed, slight breeze blowing)}$$

$$h = 64 \text{ kJ/(m}^2 \cdot \text{h} \cdot °\text{C)} \quad \text{(nude, immersed in water)}$$

(Data taken from R. C. Seagrave, *Biomedical Applications of Heat and Mass Transfer,* Iowa State University Press, Ames, Iowa, 1971.)

(a) Consider the human body as a continuous system at steady state. Write an energy balance on the body, making all appropriate simplifications and substitutions.

(b) Calculate the surrounding temperature for which the energy balance is satisfied (i.e., at which a person would feel neither hot nor cold) for a clothed person and for a nude person immersed in water.

(c) In terms of the preceding discussion, suggest why you feel colder on a windy day than on a day where the temperature is the same but there is no wind.

7.31. Liquid water at 30.0°C and liquid water at 90.0°C are combined in a ratio (1 kg cold water/2 kg hot water).

(a) Use a *simple* calculation to estimate the final water temperature. For this part, pretend you never heard of energy balances.

(b) Now assume a basis of calculation and write a closed system energy balance for the process, neglecting potential and kinetic energy changes and expansion work and assuming that the mixing is adiabatic. Use the balance to calculate the specific internal energy and hence (from the steam tables) the final temperature of the mixture. What is the percentage difference between your answer and that of part (a)?

7.32. Steam produced in a boiler is frequently "wet"—that is, it is a mist composed of saturated water vapor and entrained liquid droplets. The **quality** of a wet steam is defined as the fraction of the mixture by mass that is vapor.

A wet steam at a pressure of 5.0 bar with a quality of 0.85 is isothermally "dried" by evaporating the entrained liquid. The flow rate of the dried steam is 52.5 m^3/h.

(a) Use the steam tables to determine the temperature at which this operation occurs, the specific enthalpies of the wet and dry steams, and the total mass flow rate of the process stream.

(b) Calculate the heat input (kW) required for the evaporation process.

7.33. Two hundred kg/min of steam enters a steam turbine at 350°C and 40 bar through a 7.5-cm diameter line and exits at 75°C and 5 bar through a 5-cm line. The exiting stream may be vapor, liquid, or "wet steam" (see Problem 7.32).

(a) If the exiting stream were wet steam at 5.0 bar, what would its temperature be?

(b) How much energy is transferred to or from the turbine? (Neglect $\Delta \dot{E}_p$ but not $\Delta \dot{E}_k$.)

7.34. A **steam trap** is a device to purge steam condensate from a system without venting uncondensed steam. In one of the crudest trap types, the condensate collects and raises a float attached to a drain plug. When the float reaches a certain level, it "pulls the plug," opening the drain valve and allowing the liquid to discharge. The float then drops down to its original position and the valve closes, preventing uncondensed steam from escaping.

(a) Suppose saturated steam at 25 bar is used to heat 100 kg/min of an oil from 135°C to 185°C. Heat must be transferred to the oil at a rate of 1.00×10^4 kJ/min to accomplish this task. The steam condenses on the exterior of a bundle of tubes through which the oil is flowing. Condensate collects in the bottom of the exchanger and exits through a steam trap set to discharge when 1200 g of liquid is collected. How often does the trap discharge?

(b) Especially when periodic maintenance checks are not performed, steam traps often fail to close completely and so leak steam continuously. Suppose a process plant contains 1000 leaking traps (not an unrealistic supposition for some plants) operating at the condition of part (a), and that on the average 10% additional steam must be fed to the condensers to compensate for the uncondensed steam venting through the leaks. Further suppose that the cost of generating the additional steam is $1/10^6$ Btu, where the denominator refers to the enthalpy of the leaking steam relative to liquid water at 20°C. Estimate the yearly cost of the leaks based on 24 h/day, 360 day/yr operation.

7.35. A turbine discharges 200 kg/h of saturated steam at 10.0 bar absolute. It is desired to generate steam at 250°C and 10.0 bar by mixing the turbine discharge with a second stream of superheated steam of 300°C and 10.0 bar.

(a) If 300 kg/h of the product steam is to be generated, how much heat must be added to the mixer?

(b) If instead the mixing is carried out adiabatically, at what rate is the product steam generated?

7.36. Liquid water at 60 bar and 250°C passes through an adiabatic expansion valve, emerging at a pressure P_f and temperature T_f. If P_f is low enough, some of the liquid evaporates.

(a) If $P_f = 1.0$ bar, determine the temperature of the final mixture (T_f) and the fraction of the liquid feed that evaporates (y_v) by writing an energy balance about the valve and neglecting $\Delta \dot{E}_k$.

(b) If you took $\Delta \dot{E}_k$ into account in part (a), how would the calculated outlet temperature compare with the value you determined? What about the calculated value of y_v? Explain.

(c) What is the value of P_f above which no evaporation would occur?

(d) Sketch the shapes of plots of T_f versus P_f and y_v versus P_f for 1 bar $\leq P_f \leq 60$ bar. Briefly explain your reasoning.

7.37. A 10.0-m^3 tank contains steam at 275°C and 15.0 bar. The tank and its contents are cooled until the pressure drops to 1.2 bar. Some of the steam condenses in the process.

(a) How much heat was transferred from the tank?

(b) What is the final temperature of the tank contents?

(c) How much steam condensed (kg)?

7.38. Jets of high-speed steam are used in spray cleaning. Steam at 15.0 bar with 150°C of superheat is fed to a well-insulated valve at a rate of 1.00 kg/s. As the steam passes through the valve, its pressure drops to 1.0 bar. The outlet stream may be totally vapor or a mixture of vapor and liquid. Kinetic and potential energy changes may be neglected.

(a) Draw and label a flowchart, assuming that both liquid and vapor emerge from the valve.

(b) Write an energy balance and use it to determine the total rate of flow of enthalpy in the outlet stream ($\dot{H}_{out} = \dot{m}_l \hat{H}_l + \dot{m}_v \hat{H}_v$). Then determine whether the outlet stream is in fact a mixture of liquid and vapor or whether it is pure vapor. Explain your reasoning.

(c) What is the temperature of the outlet stream?

7.39. The following diagram shows a simplified version of how a refrigerator works:

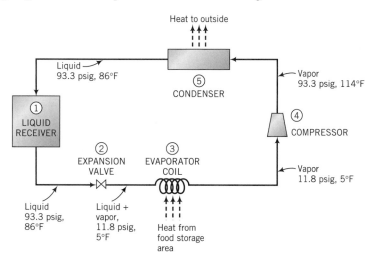

In a **liquid receiver** ①, a liquid refrigerant (any one of a number of halogenated hydrocarbons such as CCl_2F_2) is contained at high pressure and temperature. The liquid passes through an **expansion valve** ②, where it flashes to a low pressure, cooling to its boiling point at this pressure and partially evaporating. The liquid–vapor mixture passes though an **evaporator coil** ③. Air from the food storage area circulates over the coil, and the heat absorbed by the evaporating refrigerant in the coil causes the air to cool. The cold refrigerant vapor emerging from the coil passes to a **compressor** ④, where it is brought back to a high pressure and in the process is raised to a high temperature. The hot vapor then passes through a **condenser** ⑤, where it is cooled and condensed at constant pressure. The air that absorbs the heat given up by the condensing fluid is discharged outside the refrigerator, and the liquefied refrigerant returns to the liquid receiver.

Suppose Refrigerant R-12 (the standard name for CCl_2F_2) undergoes this cycle at a circulation rate of 40 lb_m/min, with the temperatures and pressures at the different points of the cycle being those shown on the flow diagram. Thermodynamic data for Refrigerant R-12 are as follows:

Saturated Fluid: $T = 5°F$, $\hat{H}_{liq} = 9.6$ Btu/lb_m, $\hat{H}_{vap} = 77.8$ Btu/lb_m

$T = 86°F$, $\hat{H}_{liq} = 27.8$ Btu/lb_m, $\hat{H}_{vap} = 85.8$ Btu/lb_m

Superheated Vapor: $T = 114°F$, $P = 93.3$ psig, $\hat{H}_{vap} \approx 90$ Btu/lb_m

(a) Suppose the expansion valve operates adiabatically and $\Delta\dot{E}_k$ is negligible. Use an energy balance about the valve to calculate the fraction of the refrigerant that evaporates in this stage of the process.

(b) Calculate the rate in Btu/min at which heat is transferred to the refrigerant that evaporates in the coil. (This is the useful cooling done in the system.)

(c) If the heat loss in the condenser is 2500 Btu/min, how much horsepower must the compressor deliver to the system? (Use an overall energy balance to solve this problem.)

7.40. Three hundred L/h of a 20 mole% C_3H_8–80% n-C_4H_{10} gas mixture at 0°C and 1.1 atm and 200 L/h of a 40 mole% C_3H_8–60% n-C_4H_{10} mixture at 25°C and 1.1 atm are mixed and heated to 227°C at constant pressure. Calculate the heat requirement in kJ/h. (See Example 7.6-2.) Enthalpies of propane and n-butane are listed below. Assume ideal gas behavior.

$T(°C)$	Propane $\hat{H}$(J/mol)	Butane $\hat{H}$(J/mol)
0	0	0
25	1772	2394
227	20,685	27,442

7.41. Air at 38°C and 97% relative humidity is to be cooled to 18°C and fed into a plant area at a rate of 510 m³/min.

(a) Calculate the rate (kg/min) at which water condenses.

(b) Calculate the cooling requirement in tons (1 ton of cooling = 12,000 Btu/h), assuming that the enthalpy of water vapor is that of saturated steam at the same temperature and the enthalpy of dry air is given by the expression

$$\hat{H}(\text{kJ/mol}) = 0.0291[T(°C) - 25]$$

7.42. A mixture containing 65.0 mole% acetone (Ac) and the balance acetic acid (AA) is separated in a continuous distillation column at 1 atm. A flowchart for the operation is as follows:

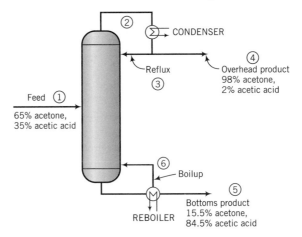

The overhead stream from the column is a vapor that passes though a condenser. The condensed liquid is divided into two equal streams: one is taken off as the overhead product (distillate) and the other (the *reflux*) is returned to the column. The bottom stream from the column is a liquid that is partially vaporized in a reboiler. The liquid stream emerging from the reboiler is taken off as the bottoms product, and the vapor is returned to the column as *boilup*. Negligible heat is lost from the column, so that the only places in the system where external heat transfer takes place are the condenser and the reboiler.

Stream Data

Feed ①	Liquid, 67.5°C, 65 mole% Ac, 35% AA
Overhead ②	Vapor, 63.0°C, 98 mole% Ac, 2% AA
Distillate ③ ⎫	
Reflux ④ ⎬	Liquid, 56.8°C, 98 mole% Ac, 2% AA
Bottoms ⑤	Liquid, 98.7°C, 15.5 mole% Ac, 84.5% AA
Boilup ⑥	Liquid, 98.7°C, 54.4 mole% Ac, 45.6% AA

Thermodynamic Data

$\hat{H}(\text{cal/mol})$

	Acetone		Acetic Acid	
$T(°C)$	$\hat{H}_l$	$\hat{H}_v$	$\hat{H}_l$	$\hat{H}_v$
56.8	0	7205	0	5723
63.0	205	7322	194	6807
67.5	354	7403	335	6884
98.7	1385	7946	1312	7420

(a) Taking 100 mol of feed as a basis, calculate the net heat requirement (cal) for the process. (You may neglect heats of mixing, although doing so for dissimilar liquids like acetone and acetic acid may introduce some error.)

(b) For the same basis, calculate the required heat input to the reboiler and the required heat removal from the condenser.

7.43. Superheated steam at T_1(°C) and 10.0 bar is combined with saturated steam at T_2(°C) and 7.0 bar in a ratio (1.96 kg of steam at 10 bar)/(1.0 kg of steam at 7 bar). The product stream is at 250°C and 7.0 bar. The process operates at steady state.

(a) Calculate T_1 and T_2, assuming that the blender operates adiabatically.

(b) If in fact heat is being lost from the blender to the surroundings, is your estimate of T_1 too high or too low? Briefly explain.

7.44. A 200.0-liter water tank can withstand pressures up to 20.0 bar absolute before rupturing. At a particular time the tank contains 165.0 kg of liquid water, the fill and exit valves are closed, and the absolute pressure in the vapor head space above the liquid (which may be assumed to contain only water vapor) is 3.0 bar. A plant technician turns on the tank heater, intending to raise the water temperature to 155°C, but is called away and forgets to return and shut off the heater. Let t_1 be the instant the heater is turned on and t_2 the moment before the tank ruptures. Use the steam tables for the following calculations.

(a) Determine the water temperature, the liquid and head-space volumes (L), and the mass of water vapor in the head space (kg) at time t_1.

(b) Determine the water temperature, the liquid and head-space volumes (L), and the mass of water vapor (g) that evaporates between t_1 and t_2. (*Hint:* Make use of the fact that the total mass of water in the tank and the total tank volume both remain constant between t_1 and t_2.)

(c) Calculate the amount of heat (kJ) transferred to the tank contents between t_1 and t_2. Give two reasons why the actual heat input to the tank must have been greater than the calculated value.

(d) List three different factors responsible for the increase in pressure resulting from the transfer of heat to the tank. (*Hint:* One has to do with the effect of temperature on the density of liquid water.)

(e) List ways in which this accident could have been avoided.

7.45. A wet steam at 20 bar with a quality of 0.97 (see Problem 7.32) leaks through a defective steam trap and expands to a pressure of 1 atm. The process can be considered to take place in two stages: a rapid adiabatic expansion to 1 atm accompanied by complete evaporation of the liquid droplets in the wet steam, followed by cooling at 1 atm to ambient temperature. $\Delta \dot{E}_k$ may be neglected in both stages.

(a) Estimate the temperature of the superheated steam immediately following the rapid adiabatic expansion.

(b) Someone looking at the steam trap would see a clear space just outside the leak and a white plume forming a short distance away. (The same phenomenon can be observed outside the spout of a kettle in which water is boiling.) Explain this observation. What would the temperature be at the point where the plume begins?

7.46. Eight fluid ounces (1 qt = 32 oz) of a beverage in a glass at 18.0°C is to be cooled by adding ice and stirring. The properties of the beverage may be taken to be those of liquid water. The enthalpy of the ice relative to liquid water at the triple point is -348 kJ/kg. Estimate the mass of ice (g) that must melt to bring the liquid temperature to 4°C, neglecting energy losses to the surroundings. (*Note:* For this isobaric batch process, the energy balance reduces to $Q = \Delta H$.)

7.47. A 25-g block of iron at 175°C is dropped into a liter of water in an insulated flask at 20°C and 1 atm. The specific enthalpy of iron is given by the expression $\hat{H}$(J/g) $= 17.3T$(°C).

(a) What reference temperature was used as the basis for the enthalpy formula?

(b) Calculate the final temperature of the flask contents, assuming that the process is adiabatic, negligible evaporation of water occurs, negligible heat is transferred to the flask wall, and the specific enthalpy of liquid water at 1 atm and a given temperature is that of the saturated liquid at the same temperature. (*Note:* For this isobaric batch process, the energy balance reduces to $Q = \Delta H$.)

7.48. Horatio Meshuggeneh has his own ideas of how to do things. For instance, when given the task of determining an oven temperature, most people would use a thermometer. Being allergic to doing anything most people would do, however, Meshuggeneh instead performs the following experiment. He puts a copper bar with a mass of 5.0 kg in the oven and puts $\Delta \hat{H} = 2919$ an identical bar in a well-insulated 20.0-liter vessel containing 5.00 L of liquid water and the remainder saturated steam at 760 mm Hg absolute. He waits long enough for both bars to reach thermal equilibrium with their surroundings, then quickly takes the first bar out of the oven, removes the second bar from the vessel, drops the first bar in its place, covers the vessel tightly, waits for the contents to come to equilibrium, and notes the reading on a pressure gauge built into the vessel. The value he reads is 50.1 mm Hg. He then uses the facts that copper has a specific gravity of 8.92 and a specific internal energy given by the expression $\hat{U}(kJ/kg) = 0.36T(°C)$ to calculate the oven temperature.

(a) The Meshuggeneh assumption is that the bar can be transferred from the oven to the vessel without any heat being lost. If he makes this assumption, what oven temperature does Meshuggeneh calculate? How many grams of water evaporate in the process? (Neglect the heat transferred to the vessel wall—i.e., assume that the heat lost by the bar is transferred entirely to the water in the vessel. Also, remember that you are dealing with a closed system once the hot bar goes into the vessel.)

(b) In fact, the bar lost 8.3 kJ of heat between the oven and the vessel. What is the true oven temperature?

(c) The experiment just described was actually Meshuggeneh's second attempt. The first time he tried it, the final gauge pressure in the vessel was negative. What had he forgotten to do?

7.49. A perfectly insulated cylinder fitted with a leakproof frictionless piston with a mass of 30.0 kg and a face area of 400.0 cm³ contains 7.0 kg of liquid water and a 3.0-kg bar of aluminum. The aluminum bar has an electrical coil imbedded in it, so that known amounts of heat can be transferred to it. Aluminum has a specific gravity of 2.70 and a specific internal energy given by the formula $\hat{U}(kJ/kg) = 0.94T(°C)$. The internal energy of liquid water at any temperature may be taken to be that of the saturated liquid at that temperature. Negligible heat is transferred to the cylinder wall. Atmospheric pressure is 1.00 atm. The cylinder and its contents are initially at 20°C.

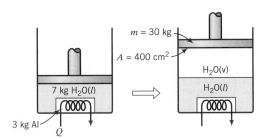

Suppose that 3310 kJ is transferred to the bar from the heating coil and the contents of the cylinder are then allowed to equilibrate.

(a) Calculate the pressure of the cylinder contents throughout the process. Then determine whether the amount of heat transferred to the system is sufficient to vaporize any of the water.

(b) Determine the following quantities: (i) the final system temperature; (ii) the volumes (cm³) of the liquid and vapor phases present at equilibrium; and (iii) the vertical distance traveled by the piston from the beginning to the end of the process. [*Suggestion:* Write an energy balance on the complete process, taking the cylinder contents to be the system. Note that the system is closed and that work is done by the system when it moves the piston through a vertical displacement. The magnitude of this work is $W = P\Delta V$, where P is the constant system pressure and ΔV is the change in system volume from the initial to the final state.]

(c) Calculate an upper limit on the temperature attainable by the aluminum bar during the process, and state the condition that would have to apply for the bar to come close to this temperature.

7.50. A rigid 5.00-liter vessel contains 4.00 L of liquid water in equilibrium with 1.00 L of water vapor at 25°C. Heat is transferred to the water by means of an immersed electrical coil. The volume of the coil is negligible. Use the steam tables to calculate the final temperature and pressure (bar) of the system and the mass of water vaporized (g) if 2915 kJ is added to the water and no heat is transferred from the water to its surroundings. (*Note:* A trial-and-error calculation is required.)

7.51. A liquid mixture of benzene and toluene is to be separated in a continuous single-stage equilibrium flash tank.

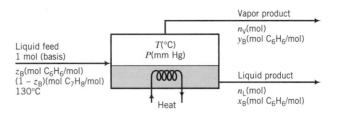

The pressure in the unit may be adjusted to any desired value, and the heat input may similarly be adjusted to vary the temperature at which the separation is conducted. The vapor and liquid product streams both emerge at the temperature T (°C) and pressure P (mm Hg) maintained in the vessel.

Assume that the vapor pressures of benzene and toluene are given by the Antoine equation, Table 6.1-1; that Raoult's law—Equation 6.4-1—applies; and that the enthalpies of benzene and toluene liquid and vapor are linear functions of temperature. Specific enthalpies at two temperatures are given here for each substance in each phase.

$$
\begin{array}{llll}
C_6H_6(l) & (T = 0°C,\ \hat{H} = 0\ \text{kJ/mol}) & (T = 80°C,\ \hat{H} = 10.85\ \text{kJ/mol}) \\
C_6H_6(v) & (T = 80°C,\ \hat{H} = 41.61\ \text{kJ/mol}) & (T = 120°C,\ \hat{H} = 45.79\ \text{kJ/mol}) \\
C_7H_8(l) & (T = 0°C,\ \hat{H} = 0\ \text{kJ/mol}) & (T = 111°C,\ \hat{H} = 18.58\ \text{kJ/mol}) \\
C_7H_8(v) & (T = 89°C,\ \hat{H} = 49.18\ \text{kJ/mol}) & (T = 111°C,\ \hat{H} = 52.05\ \text{kJ/mol})
\end{array}
$$

(a) Suppose the feed is equimolar in benzene and toluene ($z_B = 0.500$). Take a basis of 1 mol of feed and do the degree-of-freedom analysis on the unit to show that if T and P are specified, you can calculate the molar compositions of each phase (x_B and y_B), the moles of the liquid and vapor products (n_L and n_V), and the required heat input (Q). *Don't do any numerical calculations in this part.*

(b) Do the calculations of part (a) for $T = 90°C$ and $P = 652$ mm Hg. (*Suggestion:* First derive an equation for x_B that can be solved by trial and error from known values of T and P.)

(c) For $z_B = 0.5$ and $T = 90°C$, there is a range of feasible operating pressures for the evaporator, $P_{min} < P < P_{max}$. If the evaporator pressure P fell outside this range, no separation of benzene and toluene would be achieved. Why not? What would emerge from the unit if $P < P_{min}$? What would emerge if $P > P_{max}$? [*Hint:* Look at your solution to part (b) and think about how it would change if you lowered P.]

*(d) Set up a spreadsheet to perform the calculation of part (b) and then use it to determine P_{max} and P_{min}. The spreadsheet should appear as follows (some solutions are shown):

Problem 7.51—Flash vaporization of benzene and toluene									
zB	T	P	pB*	pT*	xB	yB	nL	nV	Q
0.500	90.0	652	1021				0.5543		8.144
0.500	90.0	714							−6.093
0.500	90.0								

---

*Computer problem.

Additional columns may be used to store other calculated variables (e.g., specific enthalpies). Briefly explain why Q is positive when $P = 652$ mm Hg and negative when $P = 714$ mm Hg.

(e) In successive rows, repeat the calculation for the same z_B and T at several pressures between p_{min} and p_{max}. Generate a plot (using the spreadsheet program itself, if possible) of n_V versus P. At approximately what pressure is half of the feed stream vaporized?

7.52. An aqueous solution with a specific gravity of 1.12 flows through a channel with a variable cross section. Data taken at two axial positions in the channel are shown here.

	Point 1	Point 2
P_{gauge}	1.5×10^5 Pa	9.77×10^4 Pa
u	5.00 m/s	?

Point 2 is 6.00 meters higher than point 1.

(a) Neglecting friction, calculate the velocity at point 2. (See Example 7.7-1.)

(b) If the pipe diameter at point 2 is 6.00 cm, what is the diameter at point 1?

7.53. A **venturi meter** is a device to measure fluid flow rates, which in its operation resembles the orifice meter (Section 3.2b). It consists of a tapered constriction in a line, with pressure taps leading to a differential manometer at points upstream of the constriction and at the point of maximum constriction (the **throat**). The manometer reading is directly related to the flow rate in the line.

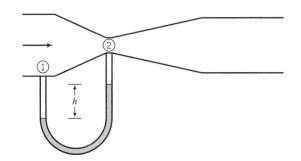

Suppose the flow rate of an incompressible fluid is to be measured in a venturi meter in which the cross-sectional area at point 1 is four times that at point 2.

(a) Derive the relationship between the velocities u_1 and u_2 at points 1 and 2.

(b) Write the Bernoulli equation for the system between points 1 and 2, and use it to prove that to the extent friction is negligible

$$P_1 - P_2 = \frac{15\rho \dot{V}^2}{2A_1^2}$$

where P_1 and P_2 are the pressures at point 1 and 2, ρ and $\dot{V}$ are the density and volumetric flow rate of the flowing fluid, and A_1 is the cross-sectional area of the pipe at point 1.

(c) Suppose this meter is used to measure the flow rate of a stream of water, using mercury as the manometer fluid, and a manometer reading $h = 38$ cm is recorded. What is the volumetric flow rate of the water if the pipe diameter at point 1 is 15 cm? (Recall the differential manometer equation, Equation 3.4-6.)

7.54. Methanol is contained in a large tank under a pressure of 3.1 bar absolute. When a valve on the bottom of the tank is opened, the methanol drains freely though a 1-cm ID tube whose outlet is 7.00 m below the surface of the methanol. The pressure at the outlet of the discharge pipe is 1 atm.

(a) Use the Bernoulli equation to estimate the methanol discharge velocity and flow rate in L/min when the discharge valve is fully opened. Neglect the rate of fall of the methanol level in the tank. (See Example 7.7-2.)

(b) When the discharge valve is partially closed, the flow rate decreases, which means that $\Delta u^2/2$ changes. However, the other two terms in the Bernoulli equation ($\Delta P/\rho$ and $g\,\Delta z$) remain the same. How do you account for this apparently contradictory result? (*Hint:* Examine the assumptions made in the derivation of the Bernoulli equation.)

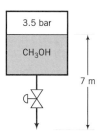

7.55. Water is to be pumped from a lake to a ranger station on the side of a mountain (see figure). The flow rate is to be 95 gal/min, and the flow channel is a standard 1-in. Schedule 40 steel pipe (ID = 1.049 in.). A pump capable of delivering 8 hp ($= -\dot{W}_s$) is available. The friction loss $\hat{F}$ (ft·lb$_f$/lb$_m$) equals $0.041L$, where L(ft) is the length of the pipe. Calculate the maximum elevation, z, of the ranger station above the lake if the pipe rises at an angle of 30°.

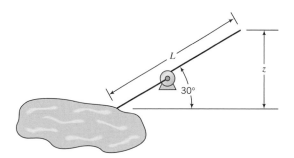

7.56. Water from a reservoir passes over a dam through a turbine and discharges from a 70-cm ID pipe at a point 65 m below the reservoir surface. The turbine delivers 0.80 MW. Calculate the required flow rate of water in m³/min if friction is neglected. (See Example 7.7-3.) If friction were included, would a higher or lower flow rate be required? (*Note:* The equation you will solve in this problem has multiple roots. Find a solution less than 2 m³/s.)

7.57. Cloth bag filters are used to remove particulate matter from the gases passing out through several process and boiler stacks in a large industrial plant. The bags become clogged and must be replaced frequently. Since they are quite expensive, rather than being discarded they are emptied, washed, and reused. In the washing process, a detergent solution with a specific gravity of 0.96 flows from a storage tank to a washing machine. The liquid effluent from the machine is pumped through a filter to remove dirt, and the cleaned detergent is recycled back to the storage tank.

Detergent flows from the storage tank to the washing machine by gravity feed at a rate of 600 L/min. All pipes in the line have 4.0-cm inner diameters. Friction losses are negligible in the line from the tank to the washing machine when the valve is fully opened, and $\hat{F} = 72$ J/kg in the return line, which includes the pump and filter.

(a) Calculate the value of the height H (see figure) required to provide the desired flow rate of detergent into the washing machine when the valve is fully opened.

(b) Suppose the pump has an efficiency of 75%; that is, it delivers 75% of its rated value as shaft work. What must the rated value (kW) of the pump be to return 600 L/min of detergent to the storage tank?

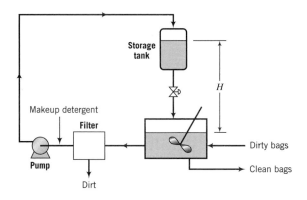

7.58. One thousand liters of a 95 wt% glycerol–5% water solution is to be diluted to 60% glycerol by adding a 35% solution pumped from a large storage tank through a 5-cm ID pipe at a steady rate. The pipe discharges at a point 23 m higher than the liquid surface in the storage tank. The operation is carried out isothermally and takes 13 min to complete. The friction loss ($\hat{F}$ of Equation 7.7-2) is 50 J/kg. Calculate the final solution volume and the shaft work in kW that the pump must deliver, assuming that the surface of the stored solution and the pipe outlet are both at 1 atm.

Data: $\rho_{H2O} = 1.00 \, \text{kg/L}$, $\rho_{gly} = 1.26 \, \text{kg/L}$. (Use to estimate solution densities.)

Chapter 8

Balances on Nonreactive Processes

We have seen that for an open system in which shaft work and kinetic and potential energy changes can be neglected, the energy balance reduces to

$$\dot{Q} = \Delta \dot{H}$$
$$= \sum_{\text{out}} \dot{n}_i \hat{H}_i - \sum_{\text{in}} \dot{n}_i \hat{H}_i$$

where the $\hat{H}_i$'s are the specific enthalpies of the inlet and outlet stream components at their respective process conditions (temperatures, pressures, and states of aggregation) relative to those components at some reference conditions. For an integral balance on a closed constant-volume system ($W = 0$), n (amount) would replace $\dot{n}$ (flow rate) in this equation, U (internal energy) would replace $\dot{H}$ (rate of enthalpy transport) and $\hat{U}$ (specific internal energy) would replace $\hat{H}$ (specific enthalpy).

In Chapter 7, we dealt only with processes involving species for which specific enthalpies and internal energies at specified conditions could be found in tables. Unfortunately, you cannot count on finding such data for every species with which you work. This chapter presents procedures for evaluating $\Delta \dot{H}$ or ΔU when tables of $\hat{H}$ and $\hat{U}$ are not available for all process species. Once these calculations have been performed, the energy balance may be written and solved as before.

8.0 INSTRUCTIONAL OBJECTIVES

After completing this chapter, you should be able to do the following:

- Define both formally (in terms of internal energies and enthalpies) and in words a high school senior could understand the variables $C_v(T)$ (heat capacity at constant volume), $C_p(T)$ (heat capacity at constant pressure), $\Delta \hat{H}_m$ (heat of fusion or heat of melting), $\Delta \hat{H}_v$ (heat of vaporization), standard heats of fusion and vaporization, and $\Delta \hat{H}_s$ (heat of solution or heat of mixing).
- Calculate $\Delta \hat{U}$ and $\Delta \hat{H}$ for the following changes in state of a species, when possible using enthalpies and internal energies, heat capacities, latent heats, and liquid and solid specific gravities tabulated in this text: (a) isothermal changes in pressure, (b) isobaric (constant-pressure) changes in temperature, (c) isothermal isobaric phase changes, and (d) isothermal isobaric mixing of two or more species. State when the formulas you use for these calculations are exact, good approximations, and poor approximations.
- Given a reference state (phase, temperature, and pressure) and a process state for a species, (a) choose a path from the reference state to the process state consisting of a series of

isothermal pressure changes, isobaric temperature changes, and isothermal isobaric phase changes; (b) calculate $\hat{U}$ and $\hat{H}$ for the species at the process state relative to the species at the reference state.

- High school physics and chemistry texts commonly state that the heat (Q) required to raise the temperature of a mass m of a substance by an amount ΔT is $Q = mC_p\Delta T$, where C_p is defined as the heat capacity of the substance. Explain why this formula is only an approximation. List the assumptions required to obtain it from the closed system energy balance ($Q + W = \Delta U + \Delta E_k + \Delta E_p$).

- If your class covers Section 8.3e, evaluate

$$\int_{T_1}^{T_2} C_p(T)\, dT$$

using the trapezoidal rule or Simpson's rule (Appendix A.3) from data for C_p at several temperatures between T_1 and T_2.

- Estimate the heat capacity of a liquid or solid species using Kopp's rule. Estimate the heat of fusion and heat of vaporization of a species using correlations in Section 8.4b.

- Given any nonreactive process for which the required heat transfer Q or heat transfer rate $\dot{Q}$ is to be calculated, (a) draw and label the flowchart, including Q or $\dot{Q}$ in the labeling; (b) carry out a degree-of-freedom analysis; (c) write the material and energy balances and other equations you would use to solve for all requested quantities; (d) perform the calculations; and (e) list the assumptions and approximations built into your calculations.

- Given an adiabatic process or any other nonreactive process for which the value of Q (closed system) or $\dot{Q}$ (open system) is specified, write material and energy balance equations and solve them simultaneously for requested quantities.

- Define the *dry-bulb temperature, wet-bulb temperature,* and *humid volume* of humid air. Given values of any two of the variables plotted on the psychrometric chart (dry-bulb and wet-bulb temperatures, absolute and relative humidity, dew point, humid volume), determine the remaining variable values and the specific enthalpy of the humid air. Use the psychrometric chart to carry out material and energy balance calculations on a heating, cooling, humidification, or dehumidification process involving air and water at 1 atm.

- Explain the meaning of the apparently contradictory term *adiabatic cooling.* Explain how spray cooling and humidification, spray dehumidification, and spray drying work. Explain how it is possible to *de*humidify air by spraying water into it. Use the psychrometric chart to carry out material and energy balance calculations on an adiabatic cooling operation involving air and water at 1 atm.

- Explain to a first-year engineering student why a beaker containing acid gets hot if you add water to it.

- Use the heat of solution data in Table B.10 and solution heat capacity data to (a) calculate the enthalpy of a hydrochloric acid, sulfuric acid, or sodium hydroxide solution of a known composition (solute mole fraction) relative to the pure solute and water at 25°C; (b) calculate the required rate of heat transfer to or from a process in which an aqueous solution of HCl, H_2SO_4, or NaOH is formed, diluted, or combined with another solution of the same species; and (c) calculate the final temperature if an aqueous solution of HCl, H_2SO_4, or NaOH is formed, diluted, or combined with another solution of the same species adiabatically.

- Perform material and energy balance calculations for a process that involves solutions for which enthalpy–concentration charts are available.

8.1 ELEMENTS OF ENERGY BALANCE CALCULATIONS

In this section, we outline a procedure for solving energy balance problems that will be applied to both nonreactive processes (this chapter) and reactive processes (Chapter 9). Section

8.1a reviews the concept of a *reference state* for specific internal energy and enthalpy calculations, and Section 8.1b reviews the fact that $\hat{U}$ and $\hat{H}$ are state properties, so that the values of ΔU or ΔH (closed system) and $\Delta\dot{H}$ (open system) calculated for a process are independent of the reference states chosen for calculations of $\hat{U}_i$ and $\hat{H}_i$. Section 8.1c outlines a procedure for organizing energy balance calculations and presents an extended illustrative example. The remainder of the chapter presents formulas and methods for calculating ΔU, ΔH, and $\Delta\dot{H}$ for processes that involve heating and cooling, compression and decompression, phase changes, mixing of liquids, and dissolving of gases and solids in liquids.

8.1a   Reference States—A Review

Recall that we can never know the absolute values of $\hat{U}$ and $\hat{H}$ for a species at a given state. $\hat{U}$(kJ/mol) is the sum of the energies of motion of all 6.02×10^{23} molecules in one gram-mole of the species plus the intramolecular kinetic and potential energies of all the atoms and subatomic particles, which are quantities we cannot determine. Since $\hat{H} = \hat{U} + P\hat{V}$ and we cannot know the value of $\hat{U}$, we also cannot know the value of $\hat{H}$ at a specified state.

Fortunately, we never need to know the absolute values of $\hat{U}$ or $\hat{H}$ at specified states: we only need to know $\Delta\hat{U}$ and $\Delta\hat{H}$ for specified *changes* of state, and we can determine these quantities experimentally.[1] We may therefore arbitrarily choose a **reference state** for a species and determine $\Delta\hat{U} = \hat{U} - \hat{U}_{\text{ref}}$ for the transition from the reference state to a series of other states. If we set $\hat{U}_{\text{ref}}$ equal to zero, then $\hat{U}(= \Delta\hat{U})$ for a specified state is the *specific internal energy at that state relative to the reference state*. The specific enthalpies at each state can then be calculated from the definition, $\hat{H} = \hat{U} + P\hat{V}$, provided that the specific volume ($\hat{V}$) of the species at the given temperature and pressure is known.

The values of $\hat{U}$ and $\hat{H}$ in the steam tables were generated using this procedure. The reference state was chosen to be liquid water at the triple point [$H_2O(l, 0.01°C, 0.00611 \text{ bar})$], at which point $\hat{U}$ was defined to be zero. According to Table B.7, for water vapor at 400°C and 10.0 bar, $\hat{U} = 2958$ kJ/kg. This does *not* mean that the absolute value of $\hat{U}$ for water in the specified state is 2958 kJ/kg; remember, we cannot know the absolute value of $\hat{U}$. It means that $\hat{U}$ of water vapor at 400°C and 10.0 bar is 2958 kJ/kg *relative to water at the reference state*, or

$$H_2O(l, 0.01°C, 0.00611 \text{ bar}) \longrightarrow H_2O(v, 400°C, 10.0 \text{ bar}), \quad \Delta\hat{U} = 2958 \text{ kJ/kg}$$

Relative to water at the same reference state, the specific enthalpy of water vapor at 400°C and 10.0 bar is

$$\hat{H} = \hat{U} + P\hat{V}$$

$$= 2958 \text{ kJ/kg} + \frac{10 \text{ bar}}{} \left| \frac{0.307 \text{ m}^3}{\text{kg}} \right| \frac{10^3 \text{ L}}{1 \text{ m}^3} \left| \frac{8.314 \times 10^{-3} \text{ kJ/(mol·K)}}{0.08314 \text{ L·bar/(mol·K)}} \right.$$

$$= 3264 \text{ kJ/kg}$$

The quantities 8.314×10^{-3} and 0.08314 are values of the gas constant expressed in different units (see inside back cover).

TEST YOURSELF   Suppose water vapor at 300°C and 5 bar is chosen as a reference state at which $\hat{H}$ is defined to be zero. Relative to this state, what is the specific enthalpy of liquid water at 75°C and 1 bar? What is the specific internal energy of liquid water at 75°C and 1 bar? (Use Table B.7.)

---

[1]The method is to transfer a measured amount of energy, Q, to a known mass of a species, m, in a closed system under conditions such that $W = 0$, $\Delta E_k = 0$, and $\Delta E_p = 0$; measure any changes in temperature, pressure, and phase; and calculate $\Delta\hat{U}$ corresponding to these changes from the energy balance, $Q = m\,\Delta\hat{U}$.

8.1b Hypothetical Process Paths

In Chapter 7, we observed that $\hat{U}$ and $\hat{H}$ are **state properties** of a species: that is, their values depend only on the state of the species—primarily on its temperature and state of aggregation (solid, liquid, or gas) and, to a lesser extent, on its pressure (and for mixtures of some species, on its mole fraction in the mixture). A state property does not depend on how the species reached its state. Consequently, *when a species passes from one state to another, both $\Delta\hat{U}$ and $\Delta\hat{H}$ for the process are independent of the path taken from the first state to the second one.*

In most of this chapter and in Chapter 9, you will learn how to calculate internal energy and enthalpy changes associated with certain processes: specifically,

1. *Changes in P at constant T and state of aggregation* (Section 8.2).
2. *Changes in T at constant P and state of aggregation* (Section 8.3).
3. *Phase changes at constant T and P*—melting, solidifying, vaporizing, condensing, sublimating (Section 8.4).
4. *Mixing of two liquids or dissolving of a gas or a solid in a liquid at constant T and P* (Section 8.5).
5. *Chemical reaction at constant T and P* (Chapter 9).

For example, compressing hydrogen gas from 1 atm to 300 atm at 25°C is a Type 1 process; melting ice at 0°C and then heating the liquid water to 30°C, all at 1 atm, is a Type 3 process followed by a Type 2 process; mixing sulfuric acid and water at a constant temperature of 20°C and a constant pressure of 1 atm is a Type 4 process.

Once we know how to calculate $\Delta\hat{U}$ and $\Delta\hat{H}$ for these five types of processes, we can calculate these quantities for *any* process by taking advantage of the fact that $\hat{U}$ and $\hat{H}$ are state properties. The procedure is to construct a hypothetical **process path** from the initial state to the final state consisting of a series of steps of the given five types. Having done this, we calculate $\Delta\hat{H}$ for each of the steps, and then add the $\Delta\hat{H}$'s for the steps to calculate $\Delta\hat{H}$ for the total process. *Since $\hat{H}$ is a state property, $\Delta\hat{H}$ calculated for the hypothetical process path—which we constructed for convenience—is the same as $\Delta\hat{H}$ for the path actually followed by the process.* The same procedure can be followed to calculate $\Delta\hat{U}$ for any process.

Suppose, for example, that we wish to calculate $\Delta\hat{H}$ for a process in which solid phenol at 25°C and 1 atm is converted to phenol vapor at 300°C and 3 atm. If we had a table of enthalpies for phenol, we could simply subtract $\hat{H}$ at the initial state from $\hat{H}$ at the final state, or

$$\Delta\hat{H} = \hat{H}(\text{vapor}, 300°\text{C}, 3 \text{ atm}) - \hat{H}(\text{solid}, 25°\text{C}, 1 \text{ atm})$$

However, we do not have such a table. Our task is then to construct a hypothetical process path from the solid at 25°C and 1 atm to the vapor at 300°C and 3 atm. To do so, we will look ahead a bit and note that Table B.1 gives enthalpy changes for the melting of phenol at 1 atm and 42.5°C (the normal melting point of phenol) and for the vaporization of phenol at 1 atm and 181.4°C (the normal boiling point of phenol). We therefore choose the following hypothetical process path:

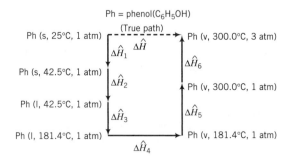

Notice that in this path, the first, third, and fifth steps are Type 2 (change in T at constant P), the second and fourth steps are Type 3 (change in phase at constant T and P), and the sixth step is Type 1 (change in P at constant T). Also notice that the phase changes were made to occur at the conditions for which tabulated enthalpy changes are available.

The next step in the calculation would be to determine the values of $\Delta \hat{H}$ for Steps 1, 3, 5, and 6 using methods to be given in Section 8.2; read the values of $\Delta \hat{H}_2$ and $\Delta \hat{H}_4$ from Table B.1; and then use the fact that enthalpy is a state property to calculate the desired $\Delta \hat{H}$ ($\Delta \hat{H}$ for the upper dashed line in the figure) as

$$\Delta \hat{H} = \Delta \hat{H}_1 + \Delta \hat{H}_2 + \Delta \hat{H}_3 + \Delta \hat{H}_4 + \Delta \hat{H}_5 + \Delta \hat{H}_6$$

TEST YOURSELF

Construct a process path for each of the following processes consisting of sequential steps of the five types listed in the preceding section. (An illustrative solution is given for the first process.)

1. Nitrogen at 20°C and 200 mm Hg is heated and compressed to 140°C and 40 atm. (One of an infinite number of possible solutions: Heat at 200 mm Hg from 20°C to 140°C, then compress isothermally from 200 mm Hg to 40 atm.)
2. Cyclohexane vapor at 180°C and 5 atm is cooled and condensed to liquid cyclohexane at 25°C and 5 atm. The enthalpy change for the condensation of cyclohexane at 80.7°C and 1 atm is known.
3. Water at 30°C and 1 atm and NaOH at 25°C and 1 atm are mixed to form an aqueous NaOH solution at 50°C and 1 atm. The enthalpy change for the dissolution of NaOH in water at 25°C and 1 atm is known.
4. O_2 at 170°C and 1 atm and CH_4 at 25°C and 1 atm are mixed and react completely to form CO_2 and H_2O at 300°C and 1 atm. The enthalpy change for the reaction occurring at 25°C and 1 atm is known.

8.1c   Procedure for Energy Balance Calculations

Most of the problems at the end of this chapter and Chapter 9 look very much like the problems in Chapters 4 to 6: given values of some process variables (feed and product stream temperatures, pressures, phases, amounts or flow rates, and component mole fractions), calculate values of other process variables. Starting in this chapter, you will also be asked to calculate the heat transferred to or from the process system (one additional variable), which will require writing and solving an energy balance (one additional equation).

Here is the procedure to follow for the energy balance calculation.

1. *Perform all required material balance calculations.*
2. *Write the appropriate form of the energy balance (closed or open system) and delete any of the terms that are either zero or negligible for the given process system.* For a stationary closed system, drop ΔE_k and ΔE_p, and neglect W if the system volume is constant, there are no moving parts (such as a mixing impeller in a stirred tank), and no energy is transferred to or from the system by electricity or radiation. For an open system at steady state, drop $\Delta \dot{E}_p$ if there is no appreciable vertical separation between the inlet and outlet ports and drop $\dot{W}_s$ if there are no moving parts (such as a pump or turbine impeller) and no energy is transferred by electricity or radiation. In addition, if temperature changes of more than a few degrees, phase changes, or chemical reactions occur in the process, $\Delta \dot{E}_k$ may usually be neglected.
3. *Choose a reference state—phase, temperature, and pressure—for each species involved in the process.* If $\hat{H}$ or $\hat{U}$ for a species will be looked up in a table (such as the steam tables for water), choose the reference state used to generate the table; otherwise, choose one of the inlet or outlet states as the reference state for the species (so that at least one $\hat{H}$ or $\hat{U}$ may be set equal to zero).

4. *For a closed constant-volume system, construct a table with columns for initial and final amounts of each species (m_i or n_i) and specific internal energies relative to the chosen reference states ($\hat{U}_i$).*[2] *For an open system, construct a table with columns for inlet and outlet stream component flow rates ($\dot{m}_i$ or $\dot{n}_i$) and specific enthalpies relative to the chosen reference states ($\hat{H}_i$).* Insert known values of the amounts or flow rates and the specific internal energies and enthalpies, and insert labels for the entries that must be calculated (e.g., $\hat{H}_1, \hat{H}_2, \ldots$). The next example illustrates the construction of such a table.

5. *Calculate all required values of $\hat{U}_i$ (or $\hat{H}_i$) and insert the values in the appropriate places in the table.* To perform the calculation for a species in a particular state (inlet or outlet), choose any convenient path from the reference state to the process state and determine $\hat{U}_i$ ($\hat{H}_i$) as $\Delta \hat{U}$ ($\Delta \hat{H}$) for that path. Sections 8.2–8.5 outline these calculations for different types of processes.

6. *Calculate*

   Closed System: $\quad \Delta U = \displaystyle\sum_{\text{final}} n_i \hat{U}_i - \sum_{\text{initial}} n_i \hat{U}_i \quad \text{or} \quad \sum_{\text{out}} m_i \hat{U}_i - \sum_{\text{in}} \dot{m}_i \hat{U}_i$

   Open System: $\quad \Delta \dot{H} = \displaystyle\sum_{\text{out}} \dot{n}_i \hat{H}_i - \sum_{\text{in}} \dot{n}_i \hat{H}_i \quad \text{or} \quad \sum_{\text{out}} \dot{m}_i \hat{H}_i - \sum_{\text{in}} \dot{m}_i \hat{H}_i$

7. *Calculate any work, kinetic energy, or potential energy terms that you have not dropped from the energy balance.*

8. *Solve the energy balance for whichever variable is unknown (often Q or $\dot{Q}$).*

   Closed System: $\qquad\qquad Q - W = \Delta U + \Delta E_k + \Delta E_p$

   Open System: $\qquad\qquad \dot{Q} - \dot{W}_s = \Delta \dot{H} + \Delta \dot{E}_k + \Delta \dot{E}_p$

The complete procedure for an open system is illustrated in the next example. We encourage you to take the time to follow each step, even though some parts may not be completely clear until you get to material that comes later in the chapter. The better you understand this example, the easier it will be for you to understand the rest of the chapter and to solve the chapter-end problems.

EXAMPLE 8.1-1   *Energy Balance on a Condenser*

Acetone (denoted as Ac) is partially condensed out of a gas stream containing 66.9 mole% acetone vapor and the balance nitrogen. Process specifications and material balance calculations lead to the flowchart shown below.

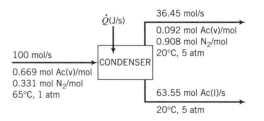

The process operates at steady state. Calculate the required cooling rate.

SOLUTION   We will follow the procedure given preceding this example.

1. *Perform required material balance calculations.* None are required in this example.
2. *Write and simplify the energy balance.*

---

[2]Use $\hat{H}_i$ instead of $\hat{U}_i$ for a closed constant-pressure system, since $Q = \Delta H$ for such systems.

For this open steady-state system, $\dot{Q} - \dot{W}_s = \Delta\dot{H} + \Delta\dot{E}_k + \Delta\dot{E}_p$. There are no moving parts in the system and no energy is transferred by electricity or radiation, so $\dot{W}_s = 0$. No significant vertical distance separates the inlet and outlet ports, so $\Delta\dot{E}_p \approx 0$. Phase changes and nonnegligible temperature changes occur, so $\Delta\dot{E}_k \approx 0$ (relative to $\Delta\dot{H}$). The energy balance reduces to

$$\dot{Q} = \Delta\dot{H} = \sum_{\text{out}} \dot{n}_i \hat{H}_i - \sum_{\text{in}} \dot{n}_i \hat{H}_i$$

3. *Choose reference states for acetone and nitrogen.*

The reference states may be chosen for computational convenience, since the choice has no effect on the calculated value of $\Delta\dot{H}$. You will later learn that Table B.8 lists specific enthalpies of nitrogen relative to N_2(g, 25°C, 1 atm), which makes this state a convenient choice for nitrogen. There are no tabulated enthalpy data for acetone in the text, so we will choose one of the process stream conditions, Ac(l, 20°C, 5 atm), as the reference state for this species, which will enable us to set the corresponding $\hat{H}$ value equal to zero rather than having to calculate it.

4. *Construct an inlet–outlet enthalpy table.*

We first write the chosen reference states, then construct the table shown below:

References: Ac(l, 20°C, 5 atm), N_2(g, 25°C, 1 atm)

Substance	$\dot{n}_{\text{in}}$ (mol/s)	$\hat{H}_{\text{in}}$ (kJ/mol)	$\dot{n}_{\text{out}}$ (mol/s)	$\hat{H}_{\text{out}}$ (kJ/mol)
Ac(v)	66.9	$\hat{H}_1$	3.35	$\hat{H}_3$
Ac(l)	—	—	63.55	0
N_2	33.1	$\hat{H}_2$	33.1	$\hat{H}_4$

Note the following points about the table:

- Nitrogen has only one inlet state (gas, 65°C, 1 atm) and one outlet state (gas, 20°C, 5 atm), so we need only one row in the table for N_2. Acetone has one inlet state (vapor, 65°C, 1 atm) but two outlet states (vapor and liquid, each at 20°C and 5 atm), so we need two rows for this species.
- We mark out (using dashes) the two cells corresponding to $\dot{n}_{\text{in}}$ and $\hat{H}_{\text{in}}$ for liquid acetone, since no liquid acetone enters the system.
- The $\dot{n}$ values are obtained from the flowchart. The flow rate of acetone vapor at the inlet, for example, is determined as (100 mol/s)[0.669 mol Ac(v)/mol] = 66.9 mol Ac(v)/s.
- Since the liquid acetone leaving the system is at the reference state, we set its specific enthalpy equal to zero.
- Four unknown specific enthalpies have been labeled and must be determined in Step 5.

5. *Calculate all unknown specific enthalpies.*

To calculate the four unknown specific enthalpies in the table, we construct hypothetical process paths from the reference states to the states of the species in the process and evaluate $\Delta\hat{H}$ for each path. This is the part of the calculation you have not yet learned to do. We will show you the calculation of $\hat{H}_1$ to illustrate the method, give the results of the other calculations, and go into detail about the required procedures in Sections 8.2–8.5.

$\hat{H}_1$ = specific enthalpy of Ac(v, 65°C, 1 atm) relative to Ac(l, 20°C, 5 atm)

$\quad = \Delta\hat{H}$ for Ac(l, 20°C, 5 atm) $\longrightarrow$ Ac(v, 65°C, 1 atm)

When choosing a process path for the determination of $\Delta\hat{H}$, it helps to know that formulas and data are given in this chapter for enthalpy changes corresponding to certain types of processes:

- Section 8.2 gives the formula $\Delta\hat{H} = \hat{V}\Delta P$ for a change in pressure (ΔP) undergone by a liquid or solid with constant specific volume $\hat{V}$. The value of $\hat{V}$ for liquid acetone may be determined as 0.0734 L/mol from the specific gravity (0.791) given in Table B.1.

- Section 8.3 shows that $\Delta \hat{H} = \int_{T_1}^{T_2} C_p(T)\, dT$ for a change from T_1 to T_2 at constant P. Formulas for $C_p(T)$, the *heat capacity at constant pressure,* are given in Table B.2. The formulas for acetone liquid and vapor are as follows:

$$\text{Ac(l):} \quad C_p\left(\frac{\text{kJ}}{\text{mol·°C}}\right) = 0.123 + 18.6 \times 10^{-5}T$$

$$\text{Ac(v):} \quad C_p\left(\frac{\text{kJ}}{\text{mol·°C}}\right) = 0.07196 + 20.10 \times 10^{-5}T - 12.78 \times 10^{-8}T^2 + 34.76 \times 10^{-12}T^3$$

where T is in °C.

- Section 8.4 defines the *heat of vaporization,* $\Delta \hat{H}_v(T_{bp})$, as $\Delta \hat{H}$ for a change from liquid to vapor at the normal boiling point, T_{bp}. Table B.1 lists T_{bp} for acetone as 56.0°C and $\Delta \hat{H}_v(T_{bp})$ as 30.2 kJ/mol.

The following process path from the reference state [Ac(l), 20°C, 5 atm] to the process state [Ac(v), 56°C, 1 atm)] enables us to use all this information in the determination of $\hat{H}_1$:[3]

$$\text{Ac(l, 20°C, 5 atm)} \xrightarrow{\Delta \hat{H}_{1a}} \text{Ac(l, 20°C, 1 atm)} \xrightarrow{\Delta \hat{H}_{1b}} \text{Ac(l, 56°C, 1 atm)}$$

$$\xrightarrow{\Delta \hat{H}_{1c}} \text{Ac(v, 56°C, 1 atm)} \xrightarrow{\Delta \hat{H}_{1d}} \text{Ac(v, 65°C, 1 atm)}$$

$$\Downarrow$$

$$\hat{H}_1 = \Delta \hat{H}_{path}$$

$$= \Delta \hat{H}_{1a} + \Delta \hat{H}_{1b} + \Delta \hat{H}_{1c} + \Delta \hat{H}_{1d}$$

$$= \hat{V}_{Ac(l)}(1 \text{ atm} - 5 \text{ atm}) + \int_{20°C}^{56°C} (C_p)_{Ac(l)}\, dT + (\Delta \hat{H}_v)_{Ac} + \int_{56°C}^{65°C} (C_p)_{Ac(v)}\, dT$$

When we substitute the values of $\hat{V}_{Ac(l)}$ and $\Delta \hat{H}_v$ and the formulas for $C_p(T)$ into the expression for $\hat{H}_1$ and carry out the necessary unit conversions and integrations, we obtain $\hat{H}_1 = (0.0297 + 4.68 + 30.2 + 0.753)$ kJ/mol = 35.7 kJ/mol.

Proceeding in a similar manner, we obtain the values for $\hat{H}_2$, $\hat{H}_3$, and $\hat{H}_4$ shown in the following revised enthalpy table:

References: Ac(l, 20°C, 5 atm), N$_2$(g, 25°C, 1 atm)

Substance	$\dot{n}_{in}$ (mol/s)	$\hat{H}_{in}$ (kJ/mol)	$\dot{n}_{out}$ (mol/s)	$\hat{H}_{out}$ (kJ/mol)
Ac(v)	66.9	35.7	3.35	32.0
Ac(l)	—	—	63.55	0
N$_2$	33.1	1.16	33.1	−0.10

6. Calculate $\Delta \dot{H}$.

$$\Delta \dot{H} = \sum_{out} \dot{n}_i \hat{H}_i - \sum_{in} \dot{n}_i \hat{H}_i$$

$$= (3.35 \text{ mol/s})(32.0 \text{ kJ/mol}) + [(63.55)(0) + (33.1)(-0.10) - (66.9)(35.7) - (33.1)(1.16)] \text{ kJ/s}$$

$$= -2320 \text{ kJ/s}$$

The factors in the last equation come directly from the inlet–outlet enthalpy table.

7. Calculate nonzero work, kinetic energy, and potential energy terms.

Since there is no shaft work and we are neglecting kinetic and potential energy changes, there is nothing to do in this step.

---

[3]To be completely accurate, we would include a step in which the acetone and nitrogen are mixed since the references are the pure species; however, enthalpy changes when gases are mixed are generally negligible (Section 8.5).

8. *Solve the energy balance for $\dot{Q}$.*

$$\dot{Q} = \Delta\dot{H} = -2320 \text{ kJ/s} = \boxed{-2320 \text{ kW}}$$

Heat must be transferred from the condenser at a rate of 2320 kW to achieve the required cooling and condensation.

Before we leave this section, let us consider what we just did from a different perspective. The process for which we need to calculate $\Delta\dot{H}$ ($= \dot{Q}$) may be depicted as shown below:

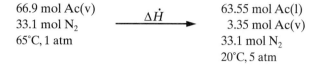

To calculate $\Delta\dot{H}$, in effect we constructed the following process path:

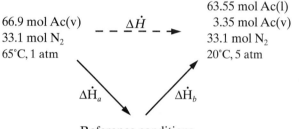

The total enthalpy change for the first step, $\Delta\dot{H}_a$, is the negative of $\Delta\dot{H}$ for the process in which acetone and nitrogen go from the reference conditions to the inlet conditions, or

$$\Delta\dot{H}_a = -\sum_{\text{in}} \dot{n}_i \hat{H}_i$$

Similarly, $\Delta\dot{H}_b$ is the enthalpy change for the process in which acetone and nitrogen go from the reference conditions to the outlet conditions, or

$$\Delta\dot{H}_b = \sum_{\text{out}} \dot{n}_i \hat{H}_i$$

Since enthalpy is a state function, the overall enthalpy change for the process must be

$$\Delta\dot{H} = \Delta\dot{H}_a + \Delta\dot{H}_b = \sum_{\text{out}} \dot{n}_i \hat{H}_i - \sum_{\text{in}} \dot{n}_i \hat{H}_i$$

It remains for us to outline methods for calculating $\Delta\hat{U}$ and $\Delta\hat{H}$ for the different types of processes we have discussed. The methods for the first four (change in P at constant T, change in T at constant P, change in phase at constant T and P, and mixing or dissolving at constant T and P) are outlined in Sections 8.2–8.5 of this chapter, and methods for chemical reactions at constant T and P are given in Chapter 9.

8.2   CHANGES IN PRESSURE AT CONSTANT TEMPERATURE

It has been observed experimentally that internal energy is nearly independent of pressure for solids and liquids at a fixed temperature, as is specific volume. Therefore, *if the pressure of a solid or liquid changes at constant temperature, you may write $\Delta\hat{U} \approx 0$ and $\Delta\hat{H}$ [$= \Delta\hat{U} + \Delta(P\hat{V})$] $\approx \hat{V}\Delta P$.*

Both $\hat{U}$ and $\hat{H}$ are independent of pressure for ideal gases. Consequently, *you may generally assume $\Delta \hat{U} \approx 0$ and $\Delta \hat{H} \approx 0$ for a gas undergoing an isothermal pressure change unless gases at temperatures well below 0°C or well above 1 atm are involved.* [If tables of $\hat{U}(T, P)$ or $\hat{H}(T, P)$ are available for the gas, there is of course no need to make this assumption.] If gases are far from ideal or if they undergo large pressure changes, you must either use tables of thermodynamic properties (such as the steam tables for water) or thermodynamic correlations beyond the scope of this text to determine $\Delta \hat{U}$ or $\Delta \hat{H}$. A good source for such correlations is Chapter 5 of Reid, Prausnitz, and Poling.[4]

TEST YOURSELF

1. Which of the following assumptions appears reasonable for each of the isothermal processes to be described below? (i) $\Delta \hat{U} \approx 0$, $\Delta \hat{H} \approx 0$; (ii) $\Delta \hat{U} \approx 0$, $\Delta \hat{H} \neq 0$; (iii) neither (i) or (ii).
   (a) H_2O (l, 1 atm) $\rightarrow$ H_2O (l, 1200 atm), $T = 25°C$
   (b) N_2 (g, 1 atm) $\rightarrow$ N_2 (g, 1.2 atm), $T = 25°C$
   (c) N_2 (g, 1 atm) $\rightarrow$ N_2 (g, 200 atm), $T = 25°C$
2. Consider the process

$$C_2H_6 \text{ (g, 25°C, 1 atm)} \rightarrow C_2H_6 \text{ (g, 25°C, 30 atm)}$$

How would you use the compressibility charts to determine whether it is reasonable to neglect $\Delta \hat{H}$ for this process?

8.3 CHANGES IN TEMPERATURE

8.3a Sensible Heat and Heat Capacities

The term **sensible heat** signifies heat that must be transferred to raise or lower the temperature of a substance or mixture of substances. The quantity of heat required to produce a temperature change in a system can be determined from the appropriate form of the first law of thermodynamics:

$$Q = \Delta U \quad \text{(closed system)} \qquad \textbf{(8.3-1)}$$

$$\dot{Q} = \Delta \dot{H} \quad \text{(open system)} \qquad \textbf{(8.3-2)}$$

(We have neglected kinetic and potential energy changes and work.) To determine the sensible heat requirement for a heating or cooling process, you must therefore be able to determine ΔU or $\Delta \dot{H}$ for the specified temperature change.

The specific internal energy of a substance depends strongly on temperature. If the temperature is raised or lowered in such a way that the system volume remains constant, the specific internal energy might vary as shown in the following plot:

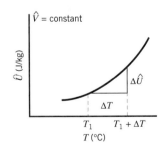

A temperature change ΔT from T_1 leads to a change $\Delta \hat{U}$ in specific internal energy. As $\Delta T \rightarrow 0$, the ratio $\Delta \hat{U}/\Delta T$ approaches a limiting value (i.e., the slope of the curve at T_1), which is by definition the **heat capacity at constant volume** of the substance, denoted by C_v.

[4]R. C. Reid, J. H. Prausnitz, and B. E. Poling, *The Properties of Gases and Liquids,* 4th Edition, McGraw-Hill, New York, 1987.

$$C_v(T) = \left\{ \lim_{\Delta T \to 0} \frac{\Delta \hat{U}}{\Delta T} \right\} = \left(\frac{\partial \hat{U}}{\partial T} \right)_V \tag{8.3-3}$$

Since the plot of $\hat{U}$ versus T is not generally a straight line, C_v (the slope of the curve) is a function of temperature.

The change in $\hat{U}$ for a temperature rise from T to $T + dT$ at constant volume is from Equation 8.3-3

$$d\hat{U} = C_v(T) \, dT \tag{8.3-4}$$

and the change $\Delta \hat{U} = \hat{U}_2 - \hat{U}_1$ associated with a temperature change from T_1 to T_2 at constant volume is, accordingly,

$$\Delta \hat{U} = \int_{T_1}^{T_2} C_v(T) \, dT \tag{8.3-5}$$

Now, suppose both the temperature and the volume of a substance change. To calculate $\Delta \hat{U}$, you may break the process into two steps—a change in $\hat{V}$ at constant T followed by a change in T at constant $\hat{V}$.

$$A(T_1, \hat{V}_1) \xrightarrow{\Delta \hat{U}_1} A(T_1, \hat{V}_2) \xrightarrow{\Delta \hat{U}_2} A(T_2, \Delta \hat{V}_2)$$

$$\downarrow \underline{\hspace{6cm}} \uparrow$$

$$\Delta \hat{U}$$

Since $\hat{U}$ is a state property, $\Delta \hat{U}$ (the desired quantity) equals $\Delta \hat{U}_1 + \Delta \hat{U}_2$. But for ideal gases and (to a good approximation) liquids and solids, $\hat{U}$ depends only on T, so that since T is constant in step 1, $\Delta U_1 \approx 0$ for any substance but a nonideal gas. Moreover, since the second step is a change in temperature at constant volume, $\Delta \hat{U}_2$ is given by Equation 8.3-5. In summary, then, for a temperature change from T_1 to T_2

$$\boxed{\Delta \hat{U} \approx \int_{T_1}^{T_2} C_v(T) \, dT} \quad \begin{array}{l} \text{Ideal gas: exact} \\ \text{Solid or liquid: good approximation} \\ \text{Nonideal gas: valid only if } V \text{ is constant} \end{array} \tag{8.3-6}$$

EXAMPLE 8.3-1   *Evaluation of an Internal Energy Change from Tabulated Heat Capacity*

Calculate the heat required to raise 200 kg of nitrous oxide from 20°C to 150°C in a constant-volume vessel. The constant-volume heat capacity of N_2O in this temperature range is given by the equation

$$C_v \, (\text{kJ/kg·°C}) = 0.855 + 9.42 \times 10^{-4} T$$

where T is in °C.

SOLUTION   From Equation 8.3-6,

$$\Delta \hat{U}(\text{kJ/kg}) = \int_{20°C}^{150°C} (0.855 + 9.42 \times 10^{-4} T) \left(\frac{\text{kJ}}{\text{kg·°C}} \right) dT$$

$$= 0.855 T \big]_{20°C}^{150°C} + \frac{9.42 \times 10^{-4} T^2}{2} \bigg]_{20°C}^{150°C}$$

$$= (111 + 10.4) \, \text{kJ/kg} = 121 \, \text{kJ/kg}$$

The energy balance for this closed system is

$$Q = \Delta U = m(\text{kg}) \, \Delta \hat{U}(\text{kJ/kg}) = (200 \, \text{kg})(121 \, \text{kJ/kg}) = \boxed{24,200 \, \text{kJ}}$$

Next, let us suppose that we heat a substance at constant pressure and consider the resulting change in enthalpy. Like internal energy, enthalpy depends strongly on temperature. If $\Delta \hat{H}$

is the change in specific enthalpy resulting from a temperature increase at constant pressure from T to $T + \Delta T$, then as ΔT approaches zero the ratio $\Delta \hat{H}/\Delta T$ approaches a limiting value, defined as the **heat capacity at constant pressure** and denoted by C_p.

$$C_p(T) = \left\{ \lim_{\Delta T \to 0} \frac{\Delta \hat{H}}{\Delta T} \right\} = \left(\frac{\partial \hat{H}}{\partial T} \right)_P \qquad \textbf{(8.3-7)}$$

Proceeding as before, we observe that the change in $\hat{H}$ for a temperature change at constant pressure from T to $T + dT$ is

$$d\hat{H} = C_p(T)\, dT$$

and so for a change from T_1 to T_2 at constant pressure

$$\Delta \hat{H} = \int_{T_1}^{T_2} C_p(T)\, dT \qquad \textbf{(8.3-8)}$$

For a process $A(T_1, P_1) \rightarrow A(T_2, P_2)$, we may construct a two-step process path

$$A(T_1, P_1) \xrightarrow{\Delta \hat{H}_1} A(T_1, P_2) \xrightarrow{\Delta \hat{H}_2} A(T_2, P_2)$$

$$\underbrace{\qquad\qquad\qquad\qquad\qquad}_{\Delta \hat{H}}$$

The first step is a change in pressure at constant temperature, the type of process described in Section 8.2. We saw in that section that

$$\Delta \hat{H}_1 = 0 \quad \text{(ideal gas)}$$
$$\approx \hat{V}\, \Delta P \quad \text{(solid or liquid)} \qquad \textbf{(8.3-9)}$$

The second step is a change in temperature at constant pressure, so that $\Delta \hat{H}_2$ is given by Equation 8.3-8. Finally, since $\Delta \hat{H} = \Delta \hat{H}_1 + \Delta \hat{H}_2$ (why?), we obtain

$$\boxed{\Delta \hat{H} = \int_{T_1}^{T_2} C_p(T)\, dT} \quad \begin{array}{l} \text{Ideal gas: exact} \\ \text{Nonideal gas: exact only if } P \text{ is constant} \end{array} \qquad \textbf{(8.3-10a)}$$

$$\boxed{\Delta \hat{H} = \hat{V}\, \Delta P + \int_{T_1}^{T_2} C_p(T)\, dT} \quad \text{Solid or liquid} \qquad \textbf{(8.3-10b)}$$

For all but large pressure changes and small temperature changes, the first term of Eq. 8.3-10b is usually negligible relative to the second term.

Evaluating $\Delta \hat{H}$ for a nonideal gas undergoing a temperature and pressure change is best done using tabulated enthalpies. If none are available, a thermodynamic relation for variations of $\hat{H}$ with P must be combined with Equation 8.3-8 to determine the enthalpy change; such relations are given by Reid, Prausnitz, and Poling (see footnote 4).

TEST YOURSELF

1. By definition, what are C_v and C_p?
2. Suppose the formula

$$\Delta \hat{H} = \int_{T_1}^{T_2} C_p(T)\, dT$$

   is used to calculate the specific enthalpy change for a change in temperature *and pressure* undergone by (a) an ideal gas, (b) a highly nonideal gas, and (c) a liquid. For which of these is the formula exact, and for which is it most likely to lead to significant error?
3. If C_p for an ideal gas is 0.5 cal/(g·°C) (i.e., a constant), what is the enthalpy change in calories corresponding to a change from 10°C to 30°C undergone by five grams of the gas?

CREATIVITY EXERCISE

Your task this time is to estimate the heat capacity of an unknown liquid. You have available a laboratory balance, a very well-insulated container, a sensitive thermometer that can measure liquid temperatures, and a thermocouple that can measure temperatures of solids. The container is a very poor conductor of heat, so that virtually all heat transferred to or from its contents goes entirely into changing the temperature of those contents. If you need anything else (within reason), you can get it. Devise as many ways as you can to estimate C_v, which you may assume is independent of temperature. [*Example:* Mix in the insulated flask a known mass, m_1, of your liquid at temperature T_1, and a known mass, m_2, of hot water at temperature T_2, and measure the final temperature T_f. Since you can calculate the heat lost by water, $Q = m_2 C_{vw}(T_2 - T_f)$, and you know Q must also equal the heat gained by the other liquid, $m_1 C_v(T_f - T_1)$, you can solve for C_v.]

8.3b   Heat Capacity Formulas

The heat capacities C_v and C_p are physical properties of materials and are tabulated in standard references such as *Perry's Chemical Engineers' Handbook*.[5] They may be expressed in any units of energy per unit amount per unit temperature interval—for example, J/(mol·K), or Btu/(lb$_m$·°F). The term **specific heat** is also used for this physical property.

Heat capacities are functions of temperature and are frequently expressed in polynomial form ($C_p = a + bT + cT^2 + dT^3$). Values of the coefficients a, b, c, and d are given in Table B.2 of Appendix B for a number of species at 1 atm, and listings for additional substances are given on pp. 2-161 to 2-186 of *Perry's Chemical Engineers' Handbook*.

When reading the coefficients of a heat capacity formula from Table B.2, do not mistake their orders of magnitude: if a value of 72.4 is read from the column labeled $b \cdot 10^5$, then the value of b must be five orders of magnitude *less* than 72.4, or $b = 72.4 \times 10^{-5}$.

Simple relationships exist between C_p and C_v in two cases:

$$\boxed{\textit{Liquids and Solids}: C_p \approx C_v} \qquad \text{(8.3-11)}$$

$$\boxed{\textit{Ideal Gases}: C_p = C_v + R} \qquad \text{(8.3-12)}$$

where R is the gas constant. (Try to prove the second relation.) The relationship between C_p and C_v for nonideal gases is complex and will not be discussed in this text.

EXAMPLE 8.3-2   *Cooling of an Ideal Gas*

Assuming ideal gas behavior, calculate the heat that must be transferred in each of the following cases.

1. A stream of nitrogen flowing at a rate of 100 mol/min is heated from 20°C to 100°C.
2. Nitrogen contained in a 5-liter flask at an initial pressure of 3 bar is cooled from 90°C to 30°C.

SOLUTION

Neglecting kinetic energy changes, the energy balance equation for the open system of part 1 is $Q = \Delta H$, and that for the closed system of part 2 is $Q = \Delta U$. (Prove it.) The problem is therefore to evaluate ΔH and ΔU for the two specified processes.

1. From Table B.2, Appendix B, the heat capacity of N_2 at a constant pressure of 1 atm is

$$C_p[\text{kJ/(mol}\cdot°\text{C})] = 0.02900 + 0.2199 \times 10^{-5}T + 0.5723 \times 10^{-8}T^2 - 2.871 \times 10^{-12}T^3$$

---

[5]R. H. Perry and D. W. Green, Eds., *Perry's Chemical Engineers' Handbook*, 7th Edition, McGraw-Hill, New York, 1997.

where T is in °C. Since we are assuming ideal gas behavior, the enthalpy change for the gas is independent of any pressure change that may occur, and hence, from Equation 8.3-10a,

$$\Delta \hat{H} = \int_{20°C}^{100°C} C_p(T)\, dT$$

$$\Downarrow$$

$$\Delta \hat{H}(\text{kJ/mol}) = 0.02900T \Big|_{20°C}^{100°C} + 0.2199 \times 10^{-5} \frac{T^2}{2} \Big|_{20°C}^{100°C} + 0.5723 \times 10^{-8} \frac{T^3}{3} \Big|_{20°C}^{100°C}$$

$$- 2.871 \times 10^{-12} \frac{T^4}{4} \Big|_{20°C}^{100°C}$$

$$= (2.320 + 0.0106 + 1.9 \times 10^{-3} - 7 \times 10^{-5})\ \text{kJ/mol} = 2.332\ \text{kJ/mol}$$

Finally,

$$\dot{Q} = \Delta \dot{H} = \dot{n}\, \Delta \hat{H}$$

$$= 100\ \frac{\text{mol}}{\text{min}} \left| \frac{2.332\ \text{kJ}}{\text{mol}} \right. = \boxed{233\ \text{kJ/min}}$$

2. To evaluate ΔU, we need the number of moles n, which may be calculated using the ideal gas equation of state, and $\Delta \hat{U}$. To determine the latter quantity we need the constant-volume heat capacity, which from Equation 8.3-12 is

$$C_v = C_p - R$$

$$\Big\Downarrow \quad \begin{array}{l} C_p \text{ given above} \\ R = [8.314\ \text{J}\,(/\text{mol}\cdot\text{K})](1\ \text{K}/1°\text{C})(1\ \text{kJ}/10^3\ \text{J}) \end{array}$$

$$C_v[\text{kg}/(\text{mol}\cdot°\text{C})] = 0.02069 + 0.2199 \times 10^{-5}T + 0.5723 \times 10^{-8}T^2 - 2.871 \times 10^{-12}T^3$$

Calculate $\Delta \hat{U}$

$$\Delta \hat{U} = \int_{90°C}^{30°C} C_v(T)\, dT$$

$$\Downarrow$$

$$\Delta \hat{U} = 0.02069T \Big|_{90°C}^{30°C} + 0.2199 \times 10^{-5} \frac{T^2}{2} \Big|_{90°C}^{30°C} + 0.5723 \times 10^{-8} \frac{T^3}{3} \Big|_{90°C}^{30°C} - 2.871 \times 10^{-12} \frac{T^4}{4} \Big|_{90°C}^{30°C}$$

$$= (-1.241 - 7.92 \times 10^{-3} - 1.34 \times 10^{-3} + 5 \times 10^{-5})\ \text{kJ/mol} = -1.250\ \text{kJ/mol}$$

Calculate n

At the initial condition (the only point at which we know P, V, and T)

$$n = PV/RT$$

$$= \frac{(3.00\ \text{bar})(5.00\ \text{L})}{[0.08314\ \text{L}\cdot\text{bar}/(\text{mol}\cdot\text{K})](363\ \text{K})} = 0.497\ \text{mol}$$

Calculate Q

$$Q = \Delta U = n\, \Delta \hat{U}$$

$$= (0.497\ \text{mol})(-1.250\ \text{kJ/mol}) = \boxed{-0.621\ \text{kJ}}$$

When enthalpies must be calculated frequently for a species, it is convenient to prepare a table of $\hat{H}(T)$ for the species (as was done for water in the steam tables) to avoid having to integrate the formula for $C_p(T)$ again and again. Tables B.8 and B.9 in Appendix B list specific enthalpies of species involved in combustion reactions—air, O_2, N_2, H_2 (a fuel), CO, CO_2, and $H_2O(v)$. The values in these tables were generated by integrating $C_p(T)$ from the specified reference state (25°C for Table B.8, 77°F for Table B.9) to the listed temperatures. The next example illustrates the use of these tables.

EXAMPLE 8.3-3 *Evaluation of $\Delta\dot{H}$ Using Heat Capacities and Tabulated Enthalpies*

Fifteen kmol/min of air is cooled from 430°C to 100°C. Calculate the required heat removal rate using (1) heat capacity formulas from Table B.2 and (2) specific enthalpies from Table B.8.

SOLUTION

$$\text{air(g, 430°C)} \longrightarrow \text{air(g, 100°C)}$$

With $\Delta\dot{E}_k$, $\Delta\dot{E}_p$, and $\dot{W}_s$ deleted, the energy balance is

$$\dot{Q} = \Delta\dot{H} = \dot{n}_{air}\hat{H}_{air,out} - \dot{n}_{air}\hat{H}_{air,in} = \dot{n}_{air}\Delta\hat{H}$$

Assume ideal gas behavior, so that pressure changes (if there are any) do not affect $\Delta\hat{H}$.

1. **The hard way.** Integrate the heat capacity formula in Table B.2.

$$\Delta\hat{H}\left(\frac{kJ}{mol}\right) = \int_{430°C}^{100°C} C_p(T)\,dT$$

$$= \int_{430°C}^{100°C} \left[0.02894 + 0.4147 \times 10^{-5}T + 0.3191 \times 10^{-8}T^2 - 1.965 \times 10^{-12}T^3\right]dT$$

$$= \left[0.02894(100 - 430) + \frac{0.4147 \times 10^{-5}}{2}(100^2 - 430^2)\right.$$

$$\left. + \frac{0.3191 \times 10^{-8}}{3}(100^3 - 430^3) - \frac{1.965 \times 10^{-12}}{4}(100^4 - 430^4)\right] kJ/mol$$

$$= (-9.5502 - 0.3627 - 0.0835 + 0.0167)\,kJ/mol = -9.98\,kJ/mol$$

2. **The easy way.** Use tabulated enthalpies from Table B.8.

$\hat{H}$ for air at 100°C can be read directly from Table B.9 and $\hat{H}$ at 430°C can be estimated by linear interpolation from the values at 400°C (11.24 kJ/mol) and 500°C (14.37 kJ/mol).

$$\hat{H}(100°C) = 2.19\,kJ/mol$$

$$\hat{H}(430°C) = [11.24 + 0.30(14.37 - 11.24)]\,kJ/mol = 12.17\,kJ/mol$$

$$\Downarrow$$

$$\Delta\hat{H} = (2.19 - 12.17)\,kJ/mol = -9.98\,kJ/mol$$

Either way $\Delta\hat{H}$ is determined,

$$\dot{Q} = \Delta\dot{H} = \dot{n}\,\Delta\hat{H} = \frac{15.0\,kmol}{min}\left|\frac{10^3\,mol}{1\,kmol}\right|\frac{-9.98\,kJ}{mol}\left|\frac{1\,min}{60\,s}\right|\frac{1\,kW}{1\,kJ/s} = \boxed{-2500\,kW}$$

Reminder: The enthalpies listed in Tables B.8 and B.9 (and for that matter, the heat capacity formulas of Table B.2) apply strictly to heating and cooling at a constant pressure of 1 atm. The tabulated enthalpies and heat capacities may also be used for nonisobaric heating and cooling of ideal or nearly ideal gases; however, at pressures high enough (or temperatures low enough) for the gases to be far from ideal, enthalpy tables or more accurate heat capacity formulas should be used.

TEST YOURSELF

1. The heat capacity of a species is 28.5 J/(mol·K). Remembering that the temperature unit in the denominator refers to a temperature *interval,* what is the heat capacity of this species in J/(mol·°C)?

2. The gas constant R is approximately equal to 2 cal/(mol·K). If C_p for a vapor is 7 cal/(mol·°C), estimate C_v for the vapor. If C_p for a liquid is 7 cal/(mol·°C), estimate C_v for the liquid.

3. Use Table B.8 or B.9 to calculate the following quantities:
   (a) The specific enthalpy (kJ/mol) of N_2 at 1000°C relative to N_2 at 300°C.
   (b) $\Delta\hat{H}$(kJ/mol) for the process CO_2(g, 800°C, 1 atm) $\longrightarrow$ CO_2(g, 300°C, 1 atm).
   (c) $\Delta\dot{H}$(Btu/h) for 100 lb-mole O_2/h being cooled from 500°F and 1.5 atm to 200°F and 0.75 atm.

8.3c Estimation of Heat Capacities

The polynomial expressions for C_p in Table B.2 are based on experimental data for the listed compounds and provide a basis for accurate calculations of enthalpy changes. Several approximate methods follow for estimating heat capacities in the absence of tabulated formulas.

Kopp's rule is a simple empirical method for estimating the heat capacity of a solid or liquid at or near 20°C. According to this rule, C_p for a molecular compound is the sum of contributions (given in Table B.10) for each element in the compound. For example, the heat capacity of solid calcium hydroxide, $Ca(OH)_2$, would be estimated from Kopp's rule as

$$(C_p)_{Ca(OH)_2} = (C_{pa})_{Ca} + 2(C_{pa})_O + 2(C_{pa})_H$$
$$= [26 + (2 \times 17) + (2 \times 9.6)] \text{ J/(mol·°C)} = 79 \text{ J/(mol·°C)}$$

[The true value is 89.5 J/(mol·°C).]

More accurate heat capacity estimation formulas for various types of gases and liquids are given in Chapter 5 of Reid, Prausnitz, and Poling (see footnote 4), and several correlations are presented by Gold and Ogle.[6]

Suppose you wish to calculate the enthalpy change associated with a change in temperature undergone by a mixture of substances. Enthalpies and heat capacities of certain mixtures are tabulated in standard references. Lacking such data, you may use the following approximation:

Rule 1. *For a mixture of gases or liquids, calculate the total enthalpy change as the sum of the enthalpy changes for the pure mixture components.* You are in effect neglecting enthalpy changes associated with the mixing of the components, which is an excellent approximation for mixtures of gases and for mixtures of similar liquids such as pentane and hexane but a poor one for dissimilar liquids such as nitric acid and water. Enthalpies of mixing for systems of the latter type are discussed in detail in Section 8.5.

Rule 2. *For highly dilute solutions of solids or gases in liquids, neglect the enthalpy change of the solute.* The more dilute the solution, the better this approximation.

The calculation of enthalpy changes for the heating or cooling of a mixture of known composition may often be simplified by calculating a heat capacity for the mixture in the following manner:

$$(C_p)_{\text{mix}}(T) = \sum_{\substack{\text{all} \\ \text{mixture} \\ \text{components}}} y_i C_{pi}(T) \tag{8.3-13}$$

where

$$(C_p)_{\text{mix}} = \text{heat capacity of the mixture}$$
$$y_i = \text{mass or mole fraction of the } i\text{th component}$$
$$C_{pi} = \text{heat capacity of the } i\text{th component}$$

---

[6]P. I. Gold and G. J. Ogle, "Estimating Thermochemical Properties of Liquids, Part 7—Heat Capacity," *Chem. Eng.*, Apr. 7, 1969, p. 130.

If C_{pi} and $(C_p)_{\text{mix}}$ are expressed in molar units, then y_i must be the mole fraction of the ith component, and if the heat capacities are expressed in mass units, then y_i must be the mass fraction of the ith component. Once $(C_p)_{\text{mix}}$ is known, $\Delta \hat{H}$ for a change in temperature from T_1 to T_2 may be calculated as

$$\Delta \hat{H} = \int_{T_1}^{T_2} (C_p)_{\text{mix}}(T)\, dT \tag{8.3-14}$$

Equation 8.3-14 is valid to the extent that enthalpies of mixing may be neglected.

EXAMPLE 8.3-4   *Heat Capacity of a Mixture*

Calculate the heat required to bring 150 mol/h of a stream containing 60% C_2H_6 and 40% C_3H_8 by volume from 0°C to 400°C. Determine a heat capacity for the mixture as part of the problem solution.

SOLUTION

The polynomial heat capacity formulas for ethane and propane given in Table B.2 are substituted into Equation 8.3-13 to yield

$$(C_p)_{\text{mix}}[\text{kJ/(mol}\cdot\text{°C)}] = 0.600(0.04937 + 13.92 \times 10^{-5}T - 5.816 \times 10^{-8}T^2 + 7.280 \times 10^{-12}T^3)$$
$$+ 0.400(0.06803 + 22.59 \times 10^{-5}T - 13.11 \times 10^{-8}T^2 + 31.71 \times 10^{-12}T^3)$$
$$= 0.05683 + 17.39 \times 10^{-5}T - 8.734 \times 10^{-8}T^2 + 17.05 \times 10^{-12}T^3$$

$$\Delta \hat{H} = \int_{0°C}^{400°C} (C_p)_{\text{mix}}\, dT = 34.89 \text{ kJ/mol}$$

where T is in °C. If potential and kinetic energy changes and shaft work are neglected, the energy balance becomes

$$\dot{Q} = \Delta \dot{H} = \dot{n}\, \Delta \hat{H} = \frac{150 \text{ mol}}{\text{h}} \left| \frac{34.89 \text{ kJ}}{\text{mol}} \right. = \boxed{5230 \frac{\text{kJ}}{\text{h}}}$$

As usual, we have assumed that the gases are sufficiently close to ideal for the formulas for C_p at 1 atm to be valid.

TEST YOURSELF

1. Estimate the heat capacity of solid calcium carbonate ($CaCO_3$) using Kopp's rule and Table B.10.
2. Two kilograms of liquid *n*-hexane [$C_p = 2.5$ kJ/(kg·°C)] and 1 kg of liquid cyclohexane [$C_p = 1.8$ kJ/(kg·°C)] are mixed and heated from 20°C to 30°C. Use the rule for liquid mixtures (Rule 1) given in this section to show that $\Delta H \approx 68$ kJ for this process. What is $\Delta \hat{H}$(kJ/kg mixture)?
3. A 0.100 wt% aqueous solution of sodium chloride is heated from 25°C to 50°C. Use the rule for solutions (Rule 2) given in this section to estimate $\Delta \hat{H}$(cal/g) for this process. C_p for water is 1 cal/(g·°C).
4. The heat capacity of liquid water is 1 cal/(g·°C), and that of ethanol is 0.54 cal/(g·°C). Estimate the heat capacity of a mixture containing 50% ethanol and 50% water by mass.

8.3d Energy Balances on Single-Phase Systems

We are now in a position to perform energy balances on any processes that do not involve phase changes, mixing steps for which enthalpy changes cannot be neglected, or chemical reactions.

If a process only involves heating or cooling a single species from T_1 to T_2, the procedure is straightforward:

1. Evaluate $\Delta \hat{U} = \int_{T_1}^{T_2} C_v\, dT$ or $\Delta \hat{H} = \int_{T_1}^{T_2} C_p\, dT$, correcting for pressure changes if necessary.

2. For a closed system at constant volume, calculate $\Delta U = n \Delta \hat{U}$ (where n is the amount of the species being heated or cooled). For a closed system at constant pressure, calculate $\Delta H = n \Delta \hat{H}$. For an open system, calculate $\Delta \dot{H} = \dot{n} \Delta \hat{H}$, where $\dot{n}$ is the species flow rate.

3. Substitute for ΔU, ΔH, or $\Delta \dot{H}$ in the appropriate energy balance equation to determine the required heat transfer, Q, or heat transfer rate, $\dot{Q}$. (See Example 8.3-2.)

If more than one species is involved or if there are several input or output streams instead of just one of each, the procedure given in Section 8.1 should be followed: choose reference states for each species, prepare and fill in a table of amounts and specific internal energies (closed system) or species flow rates and specific enthalpies (open system), and substitute the calculated values into the energy balance equation. The next example illustrates the procedure for a continuous heating process.

EXAMPLE 8.3-5 *Energy Balance on a Gas Preheater*

A stream containing 10% CH_4 and 90% air by volume is to be heated from 20°C to 300°C. Calculate the required rate of heat input in kilowatts if the flow rate of the gas is 2.00×10^3 liters (STP)/min.

SOLUTION *Basis: Given Flow Rate*

Assume ideal gas behavior.

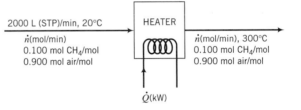

Recall that specifying the flow rate in liters (STP)/min does not imply that the feed gas is at standard temperature and pressure; it is simply an alternative way of giving the molar flow rate.

$$\dot{n} = \frac{2000 \text{ L (STP)}}{\text{min}} \left| \frac{1 \text{ mol}}{22.4 \text{ L (STP)}} \right. = 89.3 \text{ mol/min}$$

The energy balance with kinetic and potential energy changes and shaft work omitted is $\dot{Q} = \Delta \dot{H}$. The task is to evaluate $\Delta \dot{H} = \sum_{\text{out}} \dot{n}_i \hat{H}_i - \sum_{\text{in}} \dot{n}_i \hat{H}_i$. Since each species has only one inlet condition and one outlet condition in the process, two rows are sufficient for the enthalpy table.

References: CH_4(g, 20°C, 1 atm), air(g, 25°C, 1 atm)

Substance	$\dot{n}_{\text{in}}$ (mol/min)	$\hat{H}_{\text{in}}$ (kJ/mol)	$\dot{n}_{\text{out}}$ (mol/min)	$\hat{H}_{\text{out}}$ (kJ/mol)
CH_4	8.93	0	8.93	$\hat{H}_1$
Air	80.4	$\hat{H}_2$	80.4	$\hat{H}_3$

The reference condition for methane was chosen so that $\hat{H}_{\text{in}}$ could be set equal to zero, and that for air was chosen so that $\hat{H}_{\text{in}}$ and $\hat{H}_{\text{out}}$ could be determined directly from Table B.8.

The next step is to evaluate all of the unknown specific enthalpies in the table. $\hat{H}_1$, for example, is the specific enthalpy of methane in the outlet gas mixture at 300°C relative to pure methane at its reference temperature of 20°C. In other words, it is the specific enthalpy change for the process

$$CH_4(g, 20°C, 1 \text{ atm}) \longrightarrow CH_4(g, 300°C, P \text{ in outlet mixture})$$

We neglect the effect of pressure on enthalpy (i.e., we assume ideal gas behavior) and we always neglect heats of mixing of gases, so that the enthalpy change is calculated for the heating of pure methane at 1 atm:

$$\hat{H}_1 = \int_{20°C}^{300°C} (C_p)_{CH_4} \, dT$$

⇓ Substitute for C_p from Table B.2

$$= \int_{20°C}^{300°C} (0.03431 + 5.469 \times 10^{-5}T + 0.3661 \times 10^{-8}T^2 - 11.0 \times 10^{-12}T^3) \, dT$$

$$= 12.09 \text{ kJ/mol}$$

The enthalpies of air at the inlet and outlet conditions relative to air at the reference state ($\hat{H}_2$ and $\hat{H}_3$, respectively) are determined from Table B.8 as

$$\hat{H}_2 = -0.15 \text{ kJ/mol}, \qquad \hat{H}_3 = 8.17 \text{ kJ/mol}$$

The energy balance now yields

$$\dot{Q} = \Delta\dot{H} = \sum_{out} \dot{n}_i \hat{H}_i - \sum_{in} \dot{n}_i \hat{H}_i$$

$$= (8.93 \text{ mol/min})(12.09 \text{ kJ/mol}) + [(80.4)(8.17) - (8.93)(0) - (80.4)(-0.15)] \text{ kJ/min}$$

⇓

$$\dot{Q} = \frac{776 \text{ kJ}}{\text{min}} \left| \frac{1 \text{ min}}{60 \text{ s}} \right| \frac{1 \text{ kW}}{1 \text{ kJ/s}} = \boxed{12.9 \text{ kW}}$$

In the last example the temperatures of all input and output streams were specified, and the only unknown in the energy balance equation was the heat transfer rate required to achieve the specified conditions. You will also encounter problems in which the heat input is known but the temperature of an output stream is not. For these problems, you must evaluate the outlet stream component enthalpies in terms of the unknown T, substitute the resulting expressions in the energy balance equation, and solve for T. Example 8.3-6 illustrates this procedure.

EXAMPLE 8.3-6 | *Energy Balance on a Waste Heat Boiler*

A gas stream containing 8.0 mole% CO and 92.0 mole% CO_2 at 500°C is fed to a **waste heat boiler**, a large metal shell containing a bundle of small-diameter tubes. The hot gas flows over the outside of the tubes. Liquid water at 25°C is fed to the boiler in a ratio 0.200 mol feedwater/mol hot gas and flows inside the tubes. Heat is transferred from the hot gas through the tube walls to the water, causing the gas to cool and the water to heat to its boiling point and evaporate to form saturated steam at 5.0 bar. The steam may be used for heating or power generation in the plant or as the feed to another process unit. The gas leaving the boiler is flared (burned) and discharged to the atmosphere. The boiler operates adiabatically—all the heat transferred from the gas goes into the water, as opposed to some of it leaking through the outside boiler wall. The flowchart for an assumed basis of 1.00 mol feed gas is shown below.

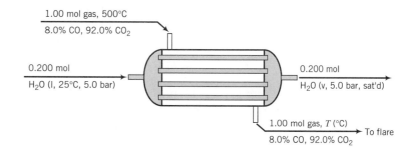

What is the temperature of the exiting gas?

SOLUTION

Since no material balances are required in this problem, we may proceed directly to the energy balance, which for this adiabatic unit reduces to

$$\Delta H = \sum_{out} n_i \hat{H}_i - \sum_{in} n_i \hat{H}_i = 0$$

We do not write $\Delta \dot{H}$ and $\dot{n}_i$ since a quantity (1 mol feed gas) and not a flow rate has been assumed as a basis of calculation.

(*Exercise*: What assumptions have been made in writing the energy balance?)

References: CO(g, 500°C, 1 atm), CO₂(g, 500°C, 1 atm), H₂O(l, triple point)

Substance	n_{in}	$\hat{H}_{in}$	n_{out}	$\hat{H}_{out}$
CO	0.080 mol	0 kJ/mol	0.080 mol	$\hat{H}_1$ (kJ/mol)
CO₂	0.920 mol	0 kJ/mol	0.920 mol	$\hat{H}_2$ (kJ/mol)
H₂O	0.00360 kg	$\hat{H}_3$ (kJ/kg)	0.00360 kg	$\hat{H}_4$ (kJ/kg)

The solution strategy will be to calculate $\hat{H}_1(T)$ and $\hat{H}_2(T)$ by integrating the heat capacity formulas of Table B.2 from the reference temperature (500°C) to the unknown T at the gas outlet, look up $\hat{H}_3$ and $\hat{H}_4$ in the steam tables, substitute for $\hat{H}_1$ through $\hat{H}_4$ in the energy balance, and solve the resulting equation for T using a spreadsheet.

Note the following points about the enthalpy table.

- We chose the reference states for CO and CO₂ as the gas inlet temperature and 1 atm. We assume ideal gas behavior so that deviations of the pressure from 1 atm have no effect on enthalpies, and accordingly set the inlet enthalpies of the gas species equal to zero.
- We will find the enthalpies of the feedwater and product steam in the steam tables. Knowing this, we chose the reference state for the steam tables (liquid water at the triple point) as our reference for water, and knowing that the enthalpies in the steam tables are in kJ/kg, we list the quantity of water in kg ($m = 0.200$ mol H₂O × 0.0180 kg/mol = 0.00360 kg).
- We will integrate the heat capacity formulas of Table B.2 for CO and CO₂ even though enthalpies for those species are listed in Table B.8, since we do not know the temperature at which to look them up.

The specific enthalpies are

$$\hat{H}_1 = \int_{500°C}^{T} (C_p)_{CO}\, dT$$

$$= \int_{500°C}^{T} (0.02895 + 0.4110 \times 10^{-5}T + 0.3548 \times 10^{-8}T^2 - 2.220 \times 10^{-12}T^3)\, dT$$

$$\hat{H}_2 = \int_{500°C}^{T} (C_p)_{CO_2}\, dT$$

$$= \int_{500°C}^{T} (0.03611 + 4.223 \times 10^{-5}T - 2.887 \times 10^{-8}T^2 + 7.464 \times 10^{-12}T^3)\, dT$$

$$\hat{H}_3 = \hat{H}[H_2O(l, 25°C, 5\ bar)] \approx 105\ kJ/kg \quad \text{(Table B.5: neglect effect of pressure on } \hat{H}\text{)}$$

$$\hat{H}_4 = \hat{H}[H_2O(v, 5\ bar, sat'd)] = 2747.5\ kJ/kg \quad \text{(Table B.6)}$$

Integrating the expressions for $\hat{H}_1$ and $\hat{H}_2$ and substituting the resulting expressions and the values of $\hat{H}_3$ and $\hat{H}_4$ into the energy balance ($\Delta H = 0$) yields the following equation:

$$1.672 \times 10^{-12}T^4 - 0.8759 \times 10^{-8}T^3 + 1.959 \times 10^{-5}T^2 + 0.03554T - 12.16 = 0$$

The problem is to find the value of T (°C) that satisfies this equation. The calculations can be done conveniently using a spreadsheet. Let us first estimate the solution by neglecting all terms of second-order and higher in T:

$$0.03554T - 12.16 = 0 \Longrightarrow T \approx 342°C$$

We will use this estimate as the first guess in the spreadsheet solution. In one cell (Cell A1 in this illustration) we insert the initial guess for T (342°C), and in the next cell (B1) we insert the formula for the fourth-order polynomial on the left-hand side of the equation to be solved:

Cell A1: 342

Cell B1: $= 1.672e-12 * A1^4 - 0.8759e-8 * A1^3 + 1.959e-5 * A1^2 + 0.3554 * A1 - 12.16$

(In some spreadsheet programs, the exponentiation function would be ** instead of ^.) The two cells of the spreadsheet would then display the following values:

Our goal is to find the value in Cell A1 that drives the value in Cell B1 to zero. We could perform the trial-and-error search manually, but if the spreadsheet program has a goalseek tool (most programs do), we would select it and use it to perform the search automatically (Set Cell B1 to 0 by varying Cell A1). Either way, at the end of the search the two cells would display values close to those shown below:

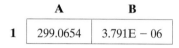

The solution is therefore $\boxed{T = 299°C}$. The heat transferred from the specified quantity of gas as it cools from 500°C to 299°C goes to convert the specified amount of feedwater into steam.

8.3e   Numerical Integration of Tabulated Heat Capacities

You have now seen two ways to evaluate an expression of the type

$$\int_{T_1}^{T_2} C_p(T) \, dT$$

If a functional relation for $C_p(T)$ is available, such as one of the polynomials of Table B.2, the integration can be carried out analytically; and if tabulated specific enthalpies are available for the substance being heated or cooled, a simple subtraction replaces the integration.

Suppose, however, that the only information you have about C_p is its value at a series of temperatures that span the range from T_1 to T_2. The question is how to estimate the value of the integral from these data.

One way would of course be to plot C_p versus T, draw a curve by visual inspection through the points at which C_p is known, and estimate the integral graphically as the area under the curve from T_1 to T_2. This is a cumbersome procedure, however, even if you have access to a *planimeter*, a device that can calculate the area under a traced-out curve.

A better solution is to use one of the many existing **quadrature formulas**—algebraic expressions that provide estimates of the integrals of tabulated data. Several such formulas are presented and illustrated in Appendix A.3; the use of one of them, **Simpson's rule**, is required for the integration of heat capacity data in several problems at the end of this chapter.

8.4   PHASE CHANGE OPERATIONS

Consider liquid water and water vapor, each at 100°C and 1 atm. Which would you expect to be greater, $\hat{U}_{\text{liquid}}$ or $\hat{U}_{\text{vapor}}$? (Recall that $\hat{U}$ is related among other things to the energy of motion of the individual molecules at the specified condition.)

If you said $\hat{U}_{vapor}$, you would be correct. One way to think of it is that the molecules of a vapor, which can move around relatively freely, are much more energetic than the densely packed molecules of a liquid at the same T and P. Also think about the fact that liquid molecules are held in close proximity to each other by attractive forces between the molecules. The energy required to overcome these forces when a liquid is vaporized is reflected in the higher internal energy of the vapor molecules.

Inspection of Table B.5 reveals how dramatic the difference between $\hat{U}_{liquid}$ and $\hat{U}_{vapor}$ can be. For water at 100°C and 1 atm, $\hat{U}_\ell = 419$ kJ/kg and $\hat{U}_v = 2507$ kJ/kg. The difference in specific enthalpy ($= \hat{U} + P\hat{V}$) is even greater, owing to the much greater specific volume of the vapor: at the same temperature and pressure, $\hat{H}_\ell = 419.1$ kJ/kg and $\hat{H}_v = 2676$ kJ/kg.

Phase changes such as melting and evaporation are usually accompanied by large changes in internal energy and enthalpy, as in the example just given. Heat transfer requirements in phase-change operations consequently tend to be substantial, since $Q \approx \Delta U$ (closed constant-volume system) or $\dot{Q} \approx \Delta \dot{H}$ (open system). The paragraphs that follow outline procedures for setting up and solving energy balances on such operations. The discussion will be limited to phase changes between liquid and vapor (evaporation, condensation) and solid and liquid (melting, freezing); however, the methods may be extended directly to other phase changes, such as sublimation (conversion from solid to vapor) and conversion from one solid phase to another.

8.4a   Latent Heats

The specific enthalpy change associated with the transition of a substance from one phase to another at constant temperature and pressure is known as the **latent heat** of the phase change (as distinguished from *sensible heat*, which is associated with temperature changes for a single-phase system). For example, the specific enthalpy change $\Delta \hat{H}$ for the transition of liquid water to steam at 100°C and 1 atm, which equals 40.6 kJ/mol, is by definition the **latent heat of vaporization** (or simply the **heat of vaporization**) of water at this temperature and pressure.

Since condensation is the reverse of vaporization and enthalpy is a state property, the heat of condensation must be the negative of the heat of vaporization. Thus, the heat of condensation of water at 100°C and 1 atm must be -40.6 kJ/mol. Similarly, the heat of solidification is the negative of the heat of fusion at the same temperature and pressure.

Latent heats for the two most commonly encountered phase changes are defined as follows:

1. **Heat of fusion** (or heat of melting). $\Delta \hat{H}_m(T, P)$ is the specific enthalpy difference between the solid and liquid forms of a species at T and P.[7]
2. **Heat of vaporization.** $\Delta \hat{H}_v(T, P)$ is the specific enthalpy difference between the liquid and vapor forms of a species at T and P.

Tabulated values of these two latent heats, such as those in Table B.1 and on pp. 2-151 through 2-160 of *Perry's Chemical Engineers' Handbook,* (see footnote 5) usually apply to a substance at its normal melting or boiling point—that is, at a pressure of 1 atm. These quantities are referred to as *standard* heats of fusion and vaporization.

The latent heat of a phase change may vary considerably with the temperature at which the change occurs but hardly varies at all with the pressure at the transition point. For example, the heat of vaporization of water at 25°C is 2442.5 J/g at $P = 23.78$ mm Hg and 2442.3 J/g at $P = 760$ mm Hg.[8] *When using a tabulated latent heat, you must therefore be sure that the phase*

---

[7] We call it $\Delta \hat{H}_m$ rather than $\Delta \hat{H}_f$ because the latter symbol is used for the *heat of formation*, a quantity defined in Chapter 9.

[8] In a system containing only pure water at 25°C, evaporation can only occur at $P = p_w^*(25°C) = 23.78$ mm Hg, but if the system contains several species evaporation can occur over a range of pressures.

change in question takes place at the temperature for which the tabulated value is reported, but you may ignore moderate variations in pressure.

EXAMPLE 8.4-1   *Heat of Vaporization*

At what rate in kilowatts must heat be transferred to a liquid stream of methanol at its normal boiling point to generate 1500 g/min of saturated methanol vapor?

SOLUTION

From Table B.1, $\Delta \hat{H}_v = 35.3$ kJ/mol at $T_{bp} = 64.7°C$. The energy balance with potential and kinetic energy changes neglected is

$$\dot{Q} = \Delta \dot{H} = \dot{n} \, \Delta \hat{H}_v$$

$$\Downarrow$$

$$\dot{Q} = \frac{1500 \text{ g CH}_3\text{OH}}{\text{min}} \left| \frac{1 \text{ mol}}{32.0 \text{ g CH}_3\text{OH}} \right| \frac{35.3 \text{ kJ}}{\text{mol}} \left| \frac{1 \text{ min}}{60 \text{ s}} \right| \frac{1 \text{ kW}}{1 \text{ kJ/s}} = \boxed{27.6 \text{ kW}}$$

Phase changes often occur at temperatures other than the temperature for which the latent heat is tabulated. When faced with this situation, you must select a hypothetical process path that permits the available data to be used.

Suppose, for example, that a substance is to be vaporized isothermally at 130°C, but the only available value of the heat of vaporization is at 80°C. A process path from the liquid at 130°C to the vapor at the same temperature must then be chosen that includes an isothermal vaporization step at 80°C: specifically, cool the liquid from 130°C to 80°C, vaporize the liquid at 80°C, and then heat the vapor back to 130°C. Summing the changes in enthalpy for each of these steps yields the change in enthalpy for the given process. (By definition, the calculated value is the latent heat of vaporization at 130°C.)

EXAMPLE 8.4-2   *Vaporization and Heating*

One hundred g-moles per hour of liquid *n*-hexane at 25°C and 7 bar is vaporized and heated to 300°C at constant pressure. Neglecting the effect of pressure on enthalpy, estimate the rate at which heat must be supplied.

SOLUTION

An energy balance yields

$$\dot{Q} = \Delta \dot{H} \qquad (\dot{W}_s = \Delta \dot{E}_p = 0, \Delta \dot{E}_k \approx 0)$$

Therefore an evaluation of $\Delta \dot{H}$ will yield the desired value of $\dot{Q}$.

From Figure 6.1-4, the temperature at which the vapor pressure of *n*-hexane is 7 bar (104 psia) is approximately 295°F (146°C), and this is therefore the temperature at which the vaporization actually occurs. However, Table B.1 lists a value of $\Delta \hat{H}_v$ at the normal boiling point of *n*-hexane,

$$\Delta \hat{H}_v = 28.85 \text{ kJ/mol at } 69°C$$

We must therefore find a path that takes hexane from a liquid to a vapor at 69°C, rather than at the true vaporization temperature of 146°C.

As noted previously, the change in enthalpy associated with a process may be determined from any convenient path as long as the initial and final points of the chosen path correspond to those of the process. The diagram shown on the following page illustrates several possible paths from liquid hexane at 25°C to hexane vapor at 300°C.

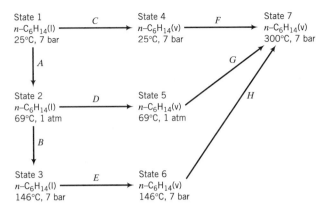

If we knew $\Delta\hat{H}_v$ at 146°C, we would follow path $ABEH$ (the true process path) to evaluate the overall $\Delta\hat{H}$ for the process, or if we knew $\Delta\hat{H}_v$ at 25°C, we would follow path CF, which would require only two calculations. Since we have $\Delta\hat{H}_v$ at 69°C, we must follow path ADG, which includes vaporization at that temperature.

$$n\text{-}C_6H_{14}(l, 25°C, 7 \text{ bar}) \xrightarrow{\Delta\hat{H}_A} n\text{-}C_6H_{14}(l, 69°C, 1 \text{ atm})$$

$$\downarrow \Delta\hat{H}_D$$

$$n\text{-}C_6H_{14}(v, 69°C, 1 \text{ atm}) \xrightarrow{\Delta\hat{H}_G} n\text{-}C_6H_{14}(v, 300°C, 7 \text{ bar})$$

$$\Delta\hat{H}_A = \hat{V}\,\Delta P + \int_{25°C}^{69°C} (C_p)_{C_6H_{14}(l)}\, dT \quad \text{(from Equation 8.3-10b)}$$

Table B.1 $\Longrightarrow$ SG = 0.659 $\Longrightarrow$ ρ = 0.659 kg/L
Table B.2 $\Longrightarrow$ C_p = 0.2163 kJ/(mol·°C)
1 atm = 1.013 bar

$$\Delta\hat{H}_A = \frac{1 \text{ L}}{0.659 \text{ kg}} \left| \frac{(1.013 - 7.0) \text{ bar}}{} \right| \frac{86.17 \text{ kg}}{1000 \text{ mol}} \left| \frac{0.008314 \text{ kJ/(mol·K)}}{0.08314 \text{ L·bar/(mol·K)}} \right.$$

$$+ \frac{0.2163 \text{ kJ}}{\text{mol·°C}} \left| \frac{(69 - 25)°C}{} \right. = (-0.0782 + 9.517) \text{ kJ/mol} = 9.44 \text{ kJ/mol}$$

$$\Delta\hat{H}_D = (\Delta\hat{H}_v)_{C_6H_{14}}(69°C, 1 \text{ atm}) = 28.85 \text{ kJ/kg}$$

$$\Delta\hat{H}_G = \int_{69°C}^{300°C} (C_p)_{C_6H_{14}(v)}\, dT \quad \text{(from Equation 8.3-10a)}$$

$$C_p[\text{kJ/(mol·°C)}] = 0.13744 + 40.85 \times 10^{-5}T - 23.92 \times 10^{-8}T^2 + 57.66 \times 10^{-12}T^3$$

$$\Delta\hat{H}_G = 47.1 \text{ kJ/mol}$$

For the overall process

$$\dot{Q} = \Delta\dot{H} = \dot{n}(\text{mol/h})\Delta\hat{H}(\text{kJ/mol})$$

$$\Delta\hat{H} = \Delta\hat{H}_A + \Delta\hat{H}_D + \Delta\hat{H}_G = 85.5 \text{ kJ/mol}$$

$$\dot{Q} = \frac{100 \text{ mol}}{\text{h}} \left| \frac{85.5 \text{ kJ}}{\text{mol}} \right| \frac{1 \text{ h}}{3600 \text{ s}} \left| \frac{1 \text{ kW}}{1 \text{ kJ/s}} \right. = \boxed{2.38 \text{ kW}}$$

Notice that the pressure change term in the first step ($\hat{V}\,\Delta P = -0.0782$ kJ/mol) accounts for less than 0.1% of the overall process enthalpy change. We will generally neglect the effects of pressure changes on $\Delta\hat{H}$ unless ΔP is on the order of 50 atm or more.

If a phase change takes place in a closed system, you must evaluate $\Delta\hat{U} = \Delta\hat{H} - \Delta(P\hat{V})$ for the phase change to substitute into the energy balance equation. For phase changes such as fusion, which involve only liquids and solids, changes in $P\hat{V}$ are generally negligible compared to changes in $\hat{H}$, so that

$$\Delta\hat{U}_m \approx \Delta\hat{H}_m \qquad\qquad (8.4\text{-}1)$$

For vaporization, $P\hat{V}$ for the vapor (which equals RT if ideal gas behavior may be assumed) is normally orders of magnitude greater than $P\hat{V}$ for the liquid, so that $\Delta(PV) \approx RT$, and

$$\Delta\hat{U}_v \approx \Delta\hat{H}_v - RT \qquad\qquad (8.4\text{-}2)$$

TEST YOURSELF

1. If you are given a value of a heat of vaporization at 100°C and 1 atm, would you be confident about using it to estimate the enthalpy change for a vaporization at 100°C and 2 atm? What about 200°C and 1 atm?
2. The enthalpies of a pure liquid and its vapor at 75°C and 1 atm are 100 J/mol and 1000 J/mol, respectively, both measured relative to the liquid at 0°C.
   (a) What is the enthalpy of the liquid at 0°C?
   (b) What is the heat of vaporization at 75°C?
   (c) Suppose you have heat capacity data for both the liquid and its vapor. What path would you follow to calculate the enthalpy change associated with 100 mol of the vapor at 400°C being cooled and condensed to form a liquid at 25°C?
3. The heat of fusion of zinc chloride at 556 K is $\Delta\hat{H}_m = 5500$ cal/mol, and the heat of vaporization of this substance at 1000 K is $\Delta\hat{H}_v = 28{,}710$ cal/mol. Estimate $\Delta\hat{U}_m(556\ \text{K})$ and $\Delta\hat{U}_v(1000\ \text{K})$ for $ZnCl_2$. [Say $R = 2$ cal/(mol·K).]

8.4b   Estimation and Correlation of Latent Heats

Reid, Prausnitz, and Poling (see footnote 4) review procedures for estimating latent heats of vaporization, fusion, and sublimation. Several of the methods given in this reference are summarized below.

A simple formula for estimating a standard heat of vaporization ($\Delta\hat{H}_v$ at the normal boiling point) is **Trouton's rule**:

$$\Delta\hat{H}_v(\text{kJ/mol}) \quad\begin{cases} \approx 0.088 T_b(\text{K}) & \text{(nonpolar liquids)} \\[4pt] \approx 0.109 T_b(\text{K}) & \text{(water, low molecular weight alcohols)} \end{cases} \qquad (8.4\text{-}3)$$

where T_b is the normal boiling point of the liquid. Trouton's rule provides an estimate of $\Delta\hat{H}_v$ accurate to within 30%. Another formula that provides roughly 2% accuracy is **Chen's equation**:

$$\Delta\hat{H}_v(\text{kJ/mol}) = \frac{T_b[0.0331(T_b/T_c) - 0.0327 + 0.0297\ \log_{10} P_c]}{1.07 - (T_b/T_c)} \qquad (8.4\text{-}4)$$

where T_b and T_c are the normal boiling point and critical temperature in kelvin and P_c is the critical pressure in atmospheres.

A formula for approximating a standard heat of fusion is

$$\Delta\hat{H}_m(\text{kJ/mol}) \quad\begin{cases} \approx 0.0092 T_m(\text{K}) & \text{(metallic elements)} \\[4pt] \approx 0.0025 T_m(\text{K}) & \text{(inorganic compounds)} \\[4pt] \approx 0.050 T_m(\text{K}) & \text{(organic compounds)} \end{cases} \qquad (8.4\text{-}5)$$

Latent heats of vaporization may be estimated from vapor pressure data by using the Clausius–Clapeyron equation, which was discussed in Section 6.1b.

$$\ln p^* = -\frac{\Delta\hat{H}_v}{RT} + B \qquad\qquad (8.4\text{-}6)$$

Provided that $\Delta\hat{H}_v$ is constant over the range of temperatures encompassed by the vapor pressure data, the latent heat of vaporization may be determined from a plot of $\ln p^*$ versus $1/T$. (See Example 6.1-1.)

In many cases the latent heat of vaporization varies considerably with temperature, invalidating Equation 8.4-6. It is then necessary to use the **Clapeyron equation,** from which Equation

8.4-6 was derived. From Equation 6.1-2,

$$\frac{d(\ln p^*)}{d(1/T)} = -\frac{\Delta \hat{H}_v}{R} \tag{8.4-7}$$

The heat of vaporization at a temperature T may be estimated from vapor pressure data by plotting $\ln p^*$ versus $1/T$, determining $[d(\ln p^*)/d(1/T)]$ at the temperature of interest as the slope of the tangent to the curve, and solving Equation 8.4-7 for $\Delta \hat{H}_v$. The slope may be determined graphically or by any of several numerical differentiation techniques outlined in texts on numerical analysis.

A procedure for calculating the latent heat of vaporization at one temperature from a known value at any other temperature was presented in Section 8.4a. The technique outlined is rigorous but time-consuming, and it requires heat capacity data that might not be available for the substance of interest. A useful approximation for estimating $\Delta \hat{H}_v$ at T_2 from a known value at T_1 is **Watson's correlation**:

$$\Delta \hat{H}_v(T_2) = \Delta \hat{H}_v(T_1) \left(\frac{T_c - T_2}{T_c - T_1} \right)^{0.38} \tag{8.4-8}$$

where T_c is the critical temperature of the substance.

EXAMPLE 8.4-3 *Estimation of a Heat of Vaporization*

The normal boiling point of methanol is 337.9 K, and the critical temperature of this substance is 513.2 K. Estimate the heat of vaporization of methanol at 200°C.

SOLUTION We first use Trouton's rule to estimate $\Delta \hat{H}_v$ at the normal boiling point, and then Watson's correlation to estimate $\Delta \hat{H}_v(473 \text{ K})$ from $\Delta \hat{H}_v(337.9 \text{ K})$.

Trouton's Rule $\qquad\qquad \Delta \hat{H}_v(337.9 \text{ K}) = (0.109)(337.9) = 36.8 \text{ kJ/mol}$

(The measured value is 35.3 kJ/mol. Chen's equation yields 37.2 kJ/mol, so in this unusual case Trouton's rule provides the better estimate.)

Watson's Correlation

Using the value of $\Delta \hat{H}_v$ estimated by Trouton's rule

$$\Delta \hat{H}_v(473 \text{ K}) = 36.8 \left(\frac{513.2 - 473}{513.2 - 337.9} \right)^{0.38} = \boxed{21.0 \text{ kJ/mol}}$$

The measured value is 19.8 kJ/mol.

TEST YOURSELF

How would you estimate $\Delta \hat{H}_v$ for a pure hydrocarbon at its normal boiling point under each of the following conditions?

1. You know only the normal boiling point.
2. You know the normal boiling point and the critical constants.
3. You have vapor pressure data over a range that encompasses $p^* = 1$ atm, and a semilog plot of p^* versus $1/T$ is a straight line.
4. Repeat 3, only suppose the plot is curved.
5. You know $\Delta \hat{H}_v$ at a temperature other than T_{bp} and do not know the heat capacities of the substance in its gaseous and liquid forms.
6. Repeat 5, only suppose that you have the heat capacity data.

8.4c Energy Balances on Processes Involving Phase Changes

When writing an energy balance on a process in which a component exists in two phases, you must choose a reference state for that component by specifying both a phase and a temperature and calculate the specific enthalpy of the component in all process streams relative to this state. If the substance is a liquid at its reference state and a vapor in a process stream, $\hat{H}$ may be calculated as outlined in Section 8.4a: that is, bring the liquid from the reference temperature to a point at which $\Delta \hat{H}_v$ is known, vaporize the liquid, bring the vapor to the process stream temperature, and sum the individual enthalpy changes for the three steps.

EXAMPLE 8.4-4   *Partial Vaporization of a Mixture*

An equimolar liquid mixture of benzene (B) and toluene (T) at 10°C is fed continuously to a vessel in which the mixture is heated to 50°C. The liquid product is 40.0 mole% B, and the vapor product is 68.4 mole% B. How much heat must be transferred to the mixture per g-mole of feed?

SOLUTION   *Basis: 1 mol Feed*

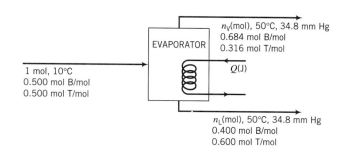

We start with a degree-of-freedom analysis:

$$3 \text{ unknown variables } (n_V, n_L, Q)$$
$$-2 \text{ material balances}$$
$$\underline{-1 \text{ energy balance}}$$
$$= 0 \text{ degrees of freedom}$$

We could count each specific enthalpy to be determined as an unknown variable, but then we would also count the equations for each of them in terms of heat capacities and latent heats, leaving the number of degrees of freedom unchanged.

We next determine n_V and n_L from material balances, and then Q from an energy balance.

Total Mass Balance: $1.00 \text{ mol} = n_V + n_L$ ⎫ $n_V = 0.352 \text{ mol}$

Benzene Balance: $0.500 \text{ mol} = 0.684 n_V + 0.400 n_L$ ⎬ ⟹ $n_L = 0.648 \text{ mol}$

The energy balance for this process has the form $Q = \Delta H$. An enthalpy table for the process appears as follows:

References: B(l, 10°C, 1 atm), T(l, 10°C, 1 atm)

Substance	n_{in} mol	$\hat{H}_{in}$ (kJ/mol)	n_{out} (mol)	$\hat{H}_{out}$ (kJ/mol)
B(l)	0.500	0	0.259	$\hat{H}_1$
T(l)	0.500	0	0.389	$\hat{H}_2$
B(v)	—	—	0.241	$\hat{H}_3$
T(v)	—	—	0.111	$\hat{H}_4$

The values of n_{out} were determined from the known mole fractions of benzene and toluene in the outlet streams and the calculated values of n_V and n_L. We do not know the feed-stream pressure and so we assume that ΔH for the change from 1 atm to P_{feed} is negligible, and since the process is not running at an unusually low temperature or high pressure, we neglect the effects of pressure on enthalpy in the calculations of $\hat{H}_1$ through $\hat{H}_4$. The heat capacity and latent heat data needed to calculate the outlet enthalpies are obtained from Tables B.1 and B.2.

The formulas and values of the unknown specific enthalpies are given below. Convince yourself that the formulas represent $\Delta\hat{H}$ for the transitions from the reference states to the process states.

$$\hat{H}_1 = \int_{10°C}^{50°C} (C_p)_{C_6H_6(l)}\, dT = 5.332 \text{ kJ/mol}$$

$$\hat{H}_2 = \int_{10°C}^{50°C} (C_p)_{C_7H_8(l)}\, dT = 6.340 \text{ kJ/mol}$$

$$\hat{H}_3 = \int_{10°C}^{80.1°C} (C_p)_{C_6H_6(l)}\, dT + (\Delta\hat{H}_v)_{C_6H_6}(80.1°C) + \int_{80.1°C}^{50°C} (C_p)_{C_6H_6(v)}\, dT$$
$$= 37.52 \text{ kJ/mol}$$

$$\hat{H}_4 = \int_{10°C}^{110.62°C} (C_p)_{C_7H_8(l)}\, dT + (\Delta\hat{H}_v)_{C_7H_8}(110.62°C) + \int_{110.62°C}^{50°C} (C_p)_{C_7H_8(v)}\, dT$$
$$= 42.93 \text{ kJ/mol}$$

The energy balance is

$$Q = \Delta H = \sum_{\text{out}} n_i \hat{H}_i - \sum_{\text{in}} n_i \hat{H}_i \implies \boxed{Q = 17.7 \text{ kJ}}$$

CREATIVITY EXERCISE

A gas emerges from a stack at 1200°C. Rather than being released directly to the atmosphere, it can be passed through one or several heat exchangers, and the heat it loses can be put to use in a variety of ways. Think of as many uses of this heat as you can. (*Example:* During the winter, pass the gas through a series of radiators, thereby getting free heating.)

8.4d   Psychrometric Charts

On a **psychrometric chart** (or **humidity chart**) several properties of a gas–vapor mixture are cross-plotted, providing a concise compilation of a large quantity of physical property data. The most common of these charts—that for the air-water system at 1 atm—is used extensively in the analysis of humidification, drying, and air-conditioning processes.

A psychrometric chart in SI units for the air–water system at 1 atm is shown in Figure 8.4-1, and a second chart in American engineering units is shown in Figure 8.4-2. Charts that cover wider temperature ranges are given on pp. 12-4 through 12-7 of *Perry's Chemical Engineers' Handbook* (see footnote 5).

The following paragraphs define and describe the different properties of humid air at 1 atm that appear on the psychrometric chart. Once you know the values of any two of these properties, you can use the chart to determine the values of the others. We will use the abbreviation DA for dry air.

- **Dry-bulb temperature**, T—the abscissa of the chart. This is the air temperature as measured by a thermometer, thermocouple, or other conventional temperature-measuring instrument.
- **Absolute humidity**, h_a [kg $H_2O(v)$/kg DA] (called **moisture content** on Figure 8.4-1)—the ordinate of the chart.

   This ratio can easily be calculated from or converted to the mass fraction of water. If, for example, the absolute humidity is 0.0150 kg H_2O/kg DA, then for every kilogram of dry air there is 0.015 kg of water vapor, for a total of 1.015 kg. The mass fraction of water is (0.0150 kg H_2O)/(1.015 kg humid air) = 0.0148 kg H_2O/kg.

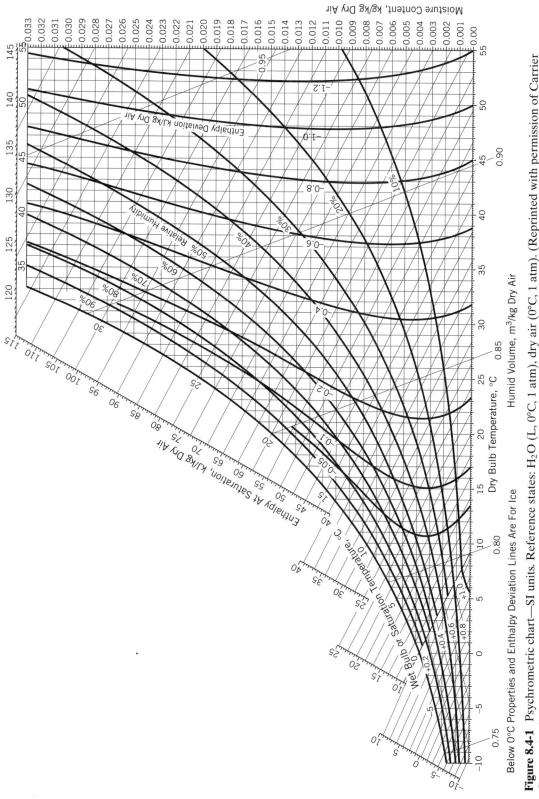

Figure 8.4-1 Psychrometric chart—SI units. Reference states: H₂O (L, 0°C, 1 atm), dry air (0°C, 1 atm). (Reprinted with permission of Carrier Corporation.)

385

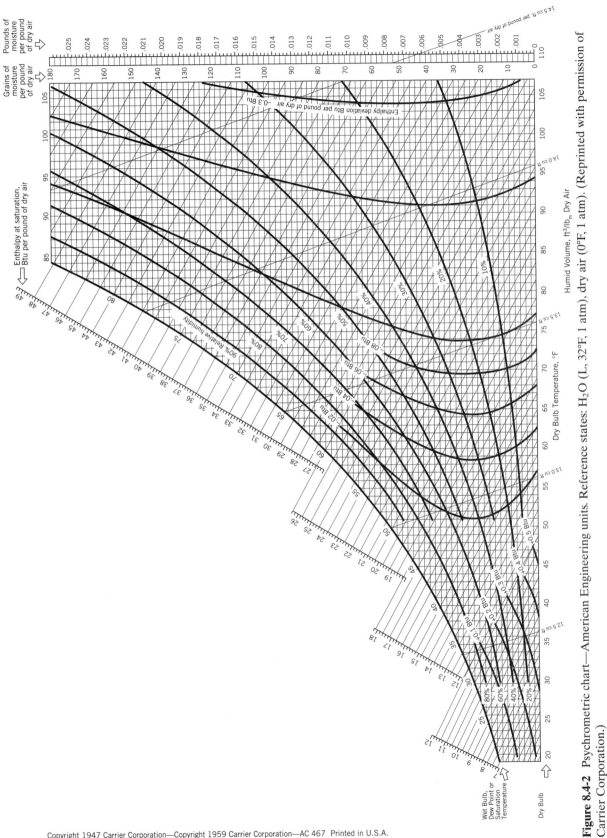

Figure 8.4-2 Psychrometric chart—American Engineering units. Reference states: H_2O (L, 32°F, 1 atm), dry air (0°F, 1 atm). (Reprinted with permission of Carrier Corporation.)

Copyright 1947 Carrier Corporation—Copyright 1959 Carrier Corporation—AC 467 Printed in U.S.A.

- **Relative humidity**, $h_r = [100 \times p_{H_2O}/p^*_{H_2O}(T)]$.

   Curves on the psychrometric chart correspond to specified values of h_r (100%, 90%, 80%, etc.). The curve that forms the left boundary of the chart is the **100% relative humidity curve**, also known as the **saturation curve**.

- **Dew point**, T_{dp}—the temperature at which humid air becomes saturated if it is cooled at constant pressure.

   The dew point of humid air at a given point on the psychrometric chart can easily be determined. For example, locate the point on Figure 8.4-1 corresponding to air at 29°C and 20% relative humidity. Cooling this air at constant pressure ($= 1$ atm) corresponds to moving horizontally (at constant absolute humidity) to the saturation curve. T_{dp} is the temperature at the intersection, or 4°C. (Verify this statement.)

- **Humid volume**, $\hat{V}_H$ (m³/kg DA).

   The humid volume is the volume occupied by 1 kg of dry air plus the water vapor that accompanies it. Lines of constant humid volume on the psychrometric chart are steep and have negative slopes. On Figure 8.4-1, humid volume lines are shown corresponding to 0.75, 0.80, 0.85, and 0.90 m³/kg dry air.

   To determine the volume of a given mass of wet air using the psychrometric chart, you must first determine the corresponding mass of dry air from the absolute humidity, then multiply this mass by $\hat{V}_H$. Suppose, for example, you wish to know the volume occupied by 150 kg of humid air at $T = 30°C$ and $h_r = 30\%$. From Figure 8.4-1, $h_a = 0.0080$ kg H₂O(v)/kg DA and $\hat{V}_H \approx 0.87$ m³/kg DA. The volume may then be calculated as

$$V = \frac{150 \text{ kg humid air}}{} \left| \frac{1.00 \text{ kg DA}}{1.008 \text{ kg humid air}} \right| \frac{0.87 \text{ m}^3}{\text{kg DA}} = 129 \text{ m}^3$$

(In this calculation, we used the fact that if the absolute humidity is 0.008 kg H₂O/kg DA, then 1 kg DA is accompanied by 0.008 kg water for a total of 1.008 kg humid air.)

- **Wet-bulb temperature**, T_{wb}.

   This quantity is best defined in terms of how it is measured. A porous material like cloth or cotton is soaked in water and wrapped around the bulb of a thermometer to form a *wick*, and the thermometer is placed in a stream of flowing air, as in the figure shown below.[9] Evaporation of water from the wick into the flowing air is accompanied by a transfer of heat from the bulb, which in turn causes a drop in the bulb temperature and hence in the thermometer reading.[10] Provided that the wick remains moist, the bulb temperature falls to a certain value and remains there. The final temperature reading is the wet-bulb temperature of the air flowing past the wick.

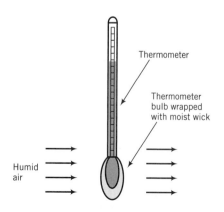

---

[9]Alternatively, the thermometer may be mounted in a *sling psychrometer* and whirled around in stationary air.

[10]Think about what happens when you step out of a shower or swimming pool. Water evaporates, your skin temperature drops, and you feel cold, even if you felt perfectly comfortable when you were dry.

The wet-bulb temperature of humid air depends on both the dry-bulb temperature and the moisture content of the air. If the air is saturated (100% relative humidity), no water evaporates from the wick, and the wet-bulb and dry-bulb temperatures are the same. The lower the humidity, the greater the difference between the two temperatures.

The humid air conditions that correspond to a given wet-bulb temperature fall on a straight line on the psychrometric chart, called a **constant wet-bulb temperature line**. The constant wet-bulb temperature lines for air–water at 1 atm appear on Figures 8.4-1 and 8.4-2 as lines with negative slopes extending beyond the saturation curve that are less steep than the lines of constant humid volume. The value of T_{wb} corresponding to a given line can be read at the intersection of the line with the saturation curve.

For example, suppose you wish to determine the wet-bulb temperature of air at 30°C (dry bulb) with a relative humidity of 30%. Locate the point on Figure 8.4-1 at the intersection of the vertical line corresponding to $T = 30°C$ and the curve corresponding to $h_r = 30\%$. The diagonal line through the point is the constant wet-bulb temperature line for air at the given condition. Follow that line upward to the left until you reach the saturation curve. The temperature value you read on the curve (or vertically down from it on the abscissa) is the wet-bulb temperature of the air. You should get a value of 18°C. This means that if you wrap a wet wick around a thermometer bulb and blow air with $T = 30°C$ and $h_r = 30\%$ past the bulb, the thermometer reading will drop and eventually stabilize at 18°C.

- *Specific enthalpy of saturated air*

   The diagonal scale above the saturation curve on the psychrometric chart shows the enthalpy of a unit mass (1 kg or 1 lb$_m$) of dry air plus the water vapor it contains at saturation. The reference states are liquid water at 1 atm and 0°C (32°F) and dry air at 1 atm and 0°C (Figure 8.4-1) or 0°F (Figure 8.4-2). To determine the enthalpy from the chart, follow the constant wet-bulb temperature line from the saturation curve at the desired temperature to the enthalpy scale.

   For example, saturated air at 25°C and 1 atm—which has an absolute humidity $h_a = 0.0202$ kg H_2O/kg DA—has a specific enthalpy of 76.5 kJ/kg DA. (Verify these values of both h_a and $\hat{H}$ on Figure 8.4-1.) The enthalpy is the sum of the enthalpy changes for 1.00 kg dry air and 0.0202 kg water going from their reference conditions to 25°C. The computation shown below uses heat capacity data from Table B.2 for air and data from the steam tables (Table B.5) for water.

$$1.00 \text{ kg DA}(0°C) \rightarrow 1 \text{ kg DA}(25°C)$$

$$\Downarrow$$

$$\Delta H_{air} = (1.00 \text{ kg DA})\left(\frac{1 \text{ kmol}}{29.0 \text{ kg}}\right)\left[\int_0^{25} C_{p,air}(T)\, dT\right]\left(\frac{\text{kJ}}{\text{kmol}}\right) = 25.1 \text{ kJ}$$

$$0.0202 \text{ kg } H_2O(l, 0°C) \rightarrow 0.0202 \text{ kg } H_2O(v, 25°C)$$

$$\Downarrow$$

$$\Delta H_{water} = (0.0202 \text{ kg})[\hat{H}_{H_2O(v,25°C)} - \hat{H}_{H_2O(l,0°C)}]\left(\frac{\text{kJ}}{\text{kg}}\right) = 51.4 \text{ kJ}$$

$$\hat{H} = \frac{(\Delta H_{air} + \Delta H_{water})(\text{kJ})}{1.00 \text{ kg DA}} = \frac{(25.1 + 51.4) \text{ kJ}}{1.00 \text{ kg DA}} = 76.5 \text{ kJ/kg DA}$$

- *Enthalpy deviation*

   The remaining curves on the psychrometric chart are almost vertical and convex to the left, with labeled values (on Figure 8.4-1) of -0.05, -0.1, -0.2, and so on. (The units of these numbers are kJ/kg DA). These curves are used to determine the enthalpy of humid air that is not saturated. The procedure is as follows: (a) locate the point on the chart corresponding to air at its specified condition; (b) interpolate to estimate the enthalpy deviation at this point;

(c) follow the constant wet-bulb temperature line to the enthalpy scale above the saturation curve, read the value on that scale, and add the enthalpy deviation to it.

For example, air at 35°C and 10% relative humidity has an enthalpy deviation of about -0.52 kJ/kg DA. The specific enthalpy of saturated air at the same wet-bulb temperature is 45.0 kJ/kg DA. (Verify both of these numbers.) The specific enthalpy of the humid air at the given condition is therefore $(45.0 - 0.52)$ kJ/kg DA $= 44.5$ kJ/kg DA.

The basis for the construction of the psychrometric chart is the Gibbs phase rule (Section 6.3a), which states that specifying a certain number of the intensive variables (temperature, pressure, specific volume, specific enthalpy, component mass or mole fractions, etc.) of a system automatically fixes the value of the remaining intensive variables. Humid air contains one phase and two components,[11] so that from Equation 6.2-1 the number of degrees of freedom is

$$F = 2 + 2 - 1 = 3$$

Specifying three intensive variables therefore fixes all other system properties. If the system pressure is fixed at 1 atm, then all other properties may be plotted on a two-dimensional plot, such as those shown in Figures 8.4-1 and 8.4-2.

EXAMPLE 8.4-5   *The Psychrometric Chart*

Use the psychrometric chart to estimate (1) the absolute humidity, wet-bulb temperature, humid volume, dew point, and specific enthalpy of humid air at 41°C and 10% relative humidity, and (2) the amount of water in 150 m^3 of air at these conditions.

SOLUTION

Following is a sketch of the psychrometric chart (Figure 8.4-1) showing the given state of the air:

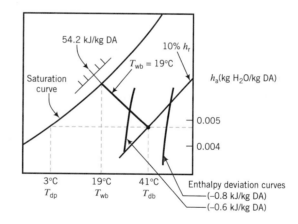

1. Reading from the chart,

$$h_a \approx 0.0048 \text{ kg H}_2\text{O/kg DA}$$
$$T_{wb} = 19°C$$
$$\hat{V}(\text{m}^3/\text{kg DA}) \approx 0.895 \text{ (curve not shown)}$$

The dew point is the temperature at which the air with the given water content would be saturated at the same total pressure (1 atm) and is therefore located at the intersection of the horizontal constant absolute humidity line ($h_a \equiv 0.0048$) and the saturation curve, or

$$T_{dp} = 3°C$$

---

[11] Since the components of dry air do not condense and are present in fixed proportion, dry air may be considered a single species (designated DA) in humidity calculations.

The specific enthalpy of saturated air at $T_{wb} = 19°C$ is 54.2 kJ/kg DA. Since the point corresponding to 41°C and 10% relative humidity falls roughly midway between the enthalpy deviation curves corresponding to −0.6 kJ/kg and −0.8 kJ/kg, we may calculate $\hat{H}$ as

$$\hat{H} = (54.2 - 0.7) \text{ kJ/kg DA}$$

$$\Downarrow$$

$$\boxed{\hat{H} = 53.5 \text{ kJ/kg DA}}$$

2. *Moles of humid air.* From Figure 8.4-1, the humid volume of the air is 0.897 m³/kg DA. We therefore calculate

$$\frac{150 \text{ m}^3}{} \left| \frac{1.00 \text{ kg DA}}{0.897 \text{ m}^3} \right| \frac{0.0048 \text{ kg H}_2\text{O}}{1.00 \text{ kg DA}} = \boxed{0.803 \text{ kg H}_2\text{O}}$$

The psychrometric chart can be used to simplify the solution of material and energy balance problems for constant-pressure air–water systems, at the expense of some precision. Note the following points:

1. Heating or cooling humid air at temperatures above the dew point corresponds to horizontal movement on the psychrometric chart. The ordinate on the chart is the ratio kg H_2O/kg dry air, which does not change as long as no condensation occurs.
2. If superheated humid air is cooled at 1 atm, the system follows a horizontal path to the left on the chart until the saturation curve (dew point) is reached; thereafter, the gas phase follows the saturation curve.
3. Since the psychrometric chart plots the mass ratio kg H_2O/kg dry air rather than the mass fraction of water, it is usually convenient to assume a quantity of dry air in a feed or product stream as a basis of calculation if the chart is to be used in the solution.

EXAMPLE 8.4-6 *Material and Energy Balances on an Air Conditioner*

Air at 80°F and 80% relative humidity is cooled to 51°F at a constant pressure of 1 atm. Use the psychrometric chart to calculate the fraction of the water that condenses and the rate at which heat must be removed to deliver 1000 ft³/min of humid air at the final condition.

SOLUTION

Basis: 1 lb$_m$ Dry Air[12]
A flowchart for the process is shown below. By convention we show heat transfer (Q) into the process unit, but since the air is being cooled we know that Q will be negative.

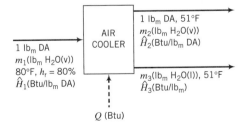

Note: In labeling the outlet gas stream, we have implicitly written a balance on dry air.

---

[12] In assuming this basis, we are temporarily ignoring the specification of the volumetric flow rate at the outlet. After the process is balanced for the assumed basis, we will scale up to an outlet flow rate of 1000 ft³/min.

Degree-of-Freedom Analysis

7 unknowns ($m_1, m_2, m_3, \hat{H}_1, \hat{H}_2, \hat{H}_3, Q$)

-1 material balance (H_2O—dry air is already balanced on the chart)

-2 absolute humidities from psychrometric chart (for inlet and outlet air)

-2 enthalpies from psychrometric chart (for inlet and outlet air)

-1 enthalpy of condensate (from known heat capacity of liquid water)

-1 energy balance

$= 0$ degrees of freedom

Point 1

$$\left. \begin{array}{l} 80°F \\ 80\% \text{ RH} \end{array} \right\} \xrightarrow{\text{Figure 8.4-2}} \begin{array}{l} h_a = 0.018 \text{ lb}_m \text{ H}_2\text{O/lb}_m \text{ DA} \\ \hat{H}_1 = 38.8 \text{ Btu/lb}_m \text{ DA} \end{array}$$

$$m_1 = \frac{1.0 \text{ lb}_m \text{ DA}}{} \frac{0.018 \text{ lb}_m \text{ H}_2\text{O}}{\text{lb}_m \text{ DA}} = 0.018 \text{ lb}_m \text{ H}_2\text{O}$$

Point 2

$$\left. \begin{array}{l} 51°F \\ \text{Saturated} \end{array} \right\} \xrightarrow{\text{Figure 8.4-2}} \begin{array}{l} h_a = 0.0079 \text{ lb}_m \text{ H}_2\text{O/lb}_m \text{ DA} \\ \hat{H}_2 = 20.9 \text{ Btu/lb}_m \text{ DA} \end{array}$$

$$m_2 = \frac{1.0 \text{ lb}_m \text{ DA}}{} \frac{0.0079 \text{ lb}_m \text{ H}_2\text{O}}{\text{lb}_m \text{ DA}} = 0.0079 \text{ lb}_m \text{ H}_2\text{O}$$

Balance on H_2O

$$m_1 = m_2 + m_3$$

$$\left\| \begin{array}{l} m_1 = 0.018 \text{ lb}_m \\ m_2 = 0.0079 \text{ lb}_m \end{array} \right.$$

$$m_3 = 0.010 \text{ lb}_m \text{ H}_2\text{O condensed}$$

Fraction H_2O Condensed

$$\frac{0.010 \text{ lb}_m \text{ condensed}}{0.018 \text{ lb}_m \text{ fed}} = \boxed{0.555}$$

Enthalpy of Condensate

Since the reference condition for water on Figure 8.4-2 is liquid water at 32°F, we must use the same condition to calculate $\hat{H}_3$.

$$\text{H}_2\text{O}(\text{l}, 32°F) \longrightarrow \text{H}_2\text{O}(\text{l}, 51°F)$$

$$\Delta \hat{H} = \hat{H}_3 = 1.0 \frac{\text{Btu}}{\text{lb}_m \cdot °F}(51°F - 32°F) = 19.0 \text{ Btu/lb}_m \text{ H}_2\text{O}$$

Energy Balance

The open-system energy balance with W_s, ΔE_k, and ΔE_p set equal to zero is

$$Q = \Delta H = \sum_{\text{out}} m_i \hat{H}_i - \sum_{\text{in}} m_i \hat{H}_i$$

(There are no dots over the extensive variables in this equation because the basis of calculation is an amount, not a flow rate.) The enthalpy table for the process is shown below. Since (1) the enthalpies ($\hat{H}_i$) of the humid air streams are obtained from the psychrometric chart in Btu/lb$_m$ dry air, and (2) the mass units of m_i and $\hat{H}_i$ must cancel when the two are multiplied in the energy balance, the tabulated values of m_i for these streams must be in lb$_m$ dry air.

References: **Dry air (DA) (g, 0°F, 1 atm), H_2O (l, 32°F, 1 atm)**

Substance	m_{in}	$\hat{H}_{\text{in}}$	m_{out}	$\hat{H}_{\text{out}}$
Humid air	1.0 lb$_m$ DA	38.8 Btu/lb$_m$ DA	1.0 lb$_m$ DA	20.9 Btu/lb$_m$ DA
$H_2O(l)$	—	—	0.010 lb$_m$	19 Btu/lb$_m$

The references were of necessity chosen to be the ones used to generate the psychrometric chart. Substituting the values in the table into the energy balance yields

$$Q = \Delta H = \frac{1.0 \text{ lb}_m \text{ DA}}{} \left| \frac{20.9 \text{ Btu}}{\text{lb}_m \text{ DA}} + \frac{0.010 \text{ lb}_m \text{ H}_2\text{O(l)}}{} \right| \frac{19 \text{ Btu}}{\text{lb}_m \text{ H}_2\text{O}} - \frac{1.0 \text{ lb}_m \text{ DA}}{} \left| \frac{38.8 \text{ Btu}}{\text{lb}_m \text{ DA}} \right.$$

$$= -17.7 \text{ Btu}$$

To calculate the cooling requirement for 1000 ft^3/min of delivered air, we must first determine the volume of delivered air corresponding to our assumed basis and scale the calculated value of Q by the ratio $(1000 \text{ ft}^3/\text{min})/(V_{basis})$. From the psychrometric chart, for humid air saturated at 51°F

$$\hat{V}_H = 13.0 \text{ ft}^3/\text{lb}_m \text{ DA}$$

$$\Downarrow$$

$$V_{basis} = \frac{1.0 \text{ lb}_m \text{ DA}}{} \left| \frac{13.0 \text{ ft}^3}{\text{lb}_m \text{ DA}} \right. = 13.0 \text{ ft}^3$$

$$\Downarrow$$

$$\dot{Q} = \frac{-17.7 \text{ Btu}}{13.0 \text{ ft}^3} \left| \frac{1000 \text{ ft}^3/\text{min}}{} \right. = \boxed{-1360 \text{ Btu/min}}$$

**TEST
YOURSELF**   Air at 25°C and 1 atm has a relative humidity of 20%. Use the psychrometric chart to estimate the absolute humidity, wet-bulb temperature, dew point, humid volume, and specific enthalpy of the air.

8.4e   Adiabatic Cooling

In **adiabatic cooling,** a warm gas is brought into contact with a cold liquid, causing the gas to cool and some of the liquid to evaporate. Heat is transferred from the gas to the liquid but no heat is transferred between the gas–liquid system and its surroundings (hence "adiabatic" cooling). Some common processes of this type are described below.

- *Spray cooling, spray humidification.* Liquid water is sprayed into a relatively dry warm air stream. Some of the water evaporates and the temperature of the air and of the unevaporated liquid both decrease. If the object is to cool the water or the air, the operation is called spray cooling; if the point is to increase the moisture content of the air, the operation is spray humidification.[13]
- *Spray dehumidification.* Warm humid air is *de*humidified by spraying cold water into it. Provided that the liquid temperature is low enough, the air is cooled below its dew point, causing some of the water vapor in it to condense.
- *Drying.* Hot air is blown over wet solids—for example, over a wet cake deposited in a filter or centrifuge. The water evaporates, leaving a dry solid product. Drying is the last step in the production of most crystalline products and powders, including many pharmaceuticals and food products.
- *Spray drying.* A suspension of small solid particles in water is sprayed as a fine mist into a stream of hot air. The water evaporates, larger solid particles settle out of the air and are removed by a conveyor, and fine suspended particles are separated from the air by a bag filter or cyclone separator. Dried milk is produced in this manner.

---

[13]Spraying the water into the air rather than simply blowing air over a water surface provides a large liquid surface-to-volume ratio, greatly increasing the evaporation rate.

Writing material and energy balances on an adiabatic cooling operation is a straightforward but cumbersome procedure. It can be shown, however, that if certain well-justified assumptions are made (we will state them later), *air undergoing adiabatic cooling through contact with liquid water moves along a constant wet-bulb temperature line on the psychrometric chart from its initial condition to the 100% relative humidity curve.* Further cooling of the air below its saturation temperature leads to condensation and hence dehumidification.

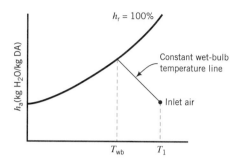

This result (which is far from obvious) allows us to perform adiabatic cooling calculations with relative ease using the psychrometric chart. First locate the initial state of the air on the chart; then locate the final state on the constant wet-bulb temperature line that passes through the initial state (or on the 100% humidity curve if cooling below the adiabatic saturation temperature takes place); and finally perform whatever material and energy balance calculations are required. Example 8.4-7 illustrates such a calculation for an adiabatic humidification operation.

EXAMPLE 8.4-7   *Adiabatic Humidification*

A stream of air at 30°C and 10% relative humidity is humidified in an adiabatic spray tower operating at $P \approx 1$ atm. The emerging air is to have a relative humidity of 40%.

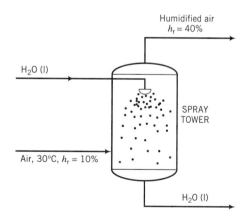

1. Determine the absolute humidity and the adiabatic saturation temperature of the entering air.
2. Use the psychrometric chart to calculate (i) the rate at which water must be added to humidify 1000 kg/h of the entering air, and (ii) the temperature of the exiting air.

SOLUTION   We assume that the heat required to raise the temperature of the liquid in the spray tower is negligible compared with the heat of vaporization of water, so that the air follows an adiabatic saturation curve (constant wet-bulb temperature line) on the psychrometric chart.

1. Air at 30°C, 10% relative humidity

⇓ Figure 8.4-1

$h_a = 0.0026$ kg H_2O/kg DA

$T_{wb} = T_{as} = 13.2°C$

2. The state of the outlet air must lie on the $T_{wb} = 13.2°C$ line. From the intersection of this line with the curve for $h_r = 40\%$, the absolute humidity of the exit gas is determined to be 0.0063 kg H_2O/kg DA. The inlet (and outlet) flow rate of dry air, $\dot{m}_{DA}$, is

$$\dot{m}_{DA} = (1000 \text{ kg air/h})(1 \text{ kg DA}/1.0026 \text{ kg air}) = 997.4 \text{ kg DA/h}$$

The amount of water that must be evaporated, $\dot{m}_{H_2O}$, may be calculated as the difference between the outlet and inlet water flow rates in the air stream.

$$\dot{m}_{H_2O} = (997.4 \text{ kg DA/h})(0.0063 - 0.0026) \frac{\text{kg } H_2O}{\text{kg DA}}$$

$$= \boxed{3.7 \text{ kg } H_2O/\text{h}}$$

From Figure 8.4-1 the temperature of the exiting air is $\boxed{21.2°C}$.

A complete justification of the procedure given above is beyond the scope of this text,[14] but we can at least offer a partial explanation. A flowchart of an adiabatic cooling operation is shown below. A stream of warm air and either a stream of liquid water (spray cooling or spray humidification), a wet solid (drying), or a solid suspension (spray drying) are brought into contact. The air enters at T_1 and leaves at T_3, the water and any solids enter at T_2 and leave at T_4, and entering liquid water evaporates at a rate $\dot{m}_{we}$(kg/s).

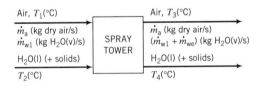

Let us assume:

1. $(C_p)_{air}$, $(C_p)_{H_2O}$, and $(\Delta\hat{H}_v)_{H_2O}$ are independent of temperature at the prevailing process conditions.
2. The enthalpy changes undergone by the unevaporated liquid water and the solid (if there is one) in going from T_2 to T_4 are negligible compared to the changes undergone by the entering wet air and the evaporated water.
3. The heat required to raise liquid water from T_2 to T_3 is negligible compared to the heat of vaporization of water.

If the energy balance equation ($\Delta\dot{H} = 0$) is written for this process and these three assumptions are made, the simplified equation becomes

$$\dot{m}_a(C_p)_{air}(T_3 - T_1) + \dot{m}_{w1}(C_p)_{H_2O(v)}(T_3 - T_1) + \dot{m}_{we}(\Delta\hat{H}_v)_{H_2O} = 0$$

⇓

$$\frac{\dot{m}_{we}}{\dot{m}_a} = \frac{1}{(\Delta\hat{H}_v)_{H_2O}} \left[(C_p)_{air} + \frac{\dot{m}_{w1}}{\dot{m}_a}(C_p)_{H_2O(v)}\right](T_1 - T_3) \qquad \textbf{(8.4-9)}$$

---

[14]One can be found in W. L. McCabe, J. C. Smith, and P. Harriott, *Unit Operations of Chemical Engineering*, 4th Edition, McGraw-Hill, New York, 1985, Chap. 23.

Suppose now that the temperature T_1 and absolute humidity $\dot{m}_{w1}/\dot{m}_a$ of the inlet air are specified, so that the state of the inlet air is fixed on the psychrometric chart. If we specify in addition the outlet air temperature $T_3(< T_1)$, then $\dot{m}_{we}/\dot{m}_a$ may be calculated from Equation 8.4-9, and it may in turn be used to calculate the absolute humidity of the outlet air, $(\dot{m}_{we} + \dot{m}_{w1})/\dot{m}_a$.

The outlet temperature and humidity determined in this manner are represented by a point on the psychrometric chart. If a lower value of T_3 is assumed, a higher outlet humidity would be calculated, yielding another point on the chart. The set of all such points for a specified T_1 and $\dot{m}_{w1}/\dot{m}_a$ defines a curve on the chart, known as the **adiabatic saturation curve**. *If the three stated assumptions are valid, the final state of air undergoing an adiabatic humidification must lie on the adiabatic saturation curve that passes through the inlet state on the psychrometric chart.*

If the outlet temperature T_3 is low enough, the air leaves saturated with water. The temperature corresponding to this condition is called the **adiabatic saturation temperature** and is found at the intersection of the adiabatic saturation curve with the 100% relative humidity curve.

The psychrometric chart for most gas–liquid systems would show a family of adiabatic saturation curves in addition to the families of curves shown on Figures 8.4-1 and 8.4-2. However, *for the air–water system at 1 atm, the adiabatic saturation curve through a given state coincides with the constant wet-bulb temperature line through that state, so that $T_{as} = T_{wb}$.* The simple material and energy balance procedure for adiabatic cooling outlined in this section is possible only because of this coincidence.

TEST YOURSELF

1. **(a)** Under what conditions do the temperature and humidity of a gas undergoing adiabatic cooling follow a unique curve on the psychrometric chart, regardless of the entering liquid temperature?
   (b) Does this curve coincide with a constant wet-bulb temperature line if the gas is air and the liquid is water?
   (c) What if they were nitrogen and acetone?
2. Air at 26°C with a relative humidity of 10% undergoes an adiabatic humidification. Use Figure 8.4-1 to estimate the adiabatic saturation temperature of the air. If the exiting air has a dry-bulb temperature of 14°C, determine its absolute humidity, relative humidity, and specific enthalpy.

8.5 MIXING AND SOLUTION

You may have carried out an experiment in a chemistry laboratory in which you mixed two liquids (such as concentrated sulfuric acid and water) or dissolved a solid in a liquid (such as sodium hydroxide in water) and observed that the mixture or solution became quite hot. The question is, why?

When two different liquids are mixed or when a gas or solid is dissolved in a liquid, bonds are broken between neighboring molecules—and possibly between atoms—of the feed materials, and new bonds are formed between neighboring molecules or ions in the product solution. If less energy is required to break the bonds in the feed materials than is released when the solution bonds form, a net release of energy results. Unless this energy is transferred from the solution to its surroundings as heat, it goes into raising the solution temperature, which is what happened in the experiments described in the first paragraph.

Suppose you mix 1 mol of pure liquid sulfuric acid with water at a specified temperature and pressure and then cool the mixture at constant pressure to bring it back to the initial temperature. The energy balance for this constant-pressure process is

$$Q = \Delta H = H_{H_2SO_4(aq)} - (H_{H_2SO_4(l)} + H_{H_2O})$$

where ΔH—the difference between the enthalpy of the solution at the specified temperature and pressure and the total enthalpy of the pure solute and solvent at the same T and P—is the *heat of solution* at that temperature and pressure. For the sulfuric acid dilution, we know $Q < 0$ (the container must be cooled to keep the solution temperature from rising) and so it follows that ΔH—the heat of solution—is negative for this process.

An **ideal mixture** is one for which the heat of mixing or solution is negligible and so $H_{\text{mixture}} \approx \sum n_i \hat{H}_i$, where n_i is the amount of mixture component i and $\hat{H}_i$ is the specific enthalpy of the pure component at the temperature and pressure of the mixture. Up to now in this text, we have assumed ideal mixture behavior for all mixtures and solutions. This assumption works well for nearly all gas mixtures and for liquid mixtures of similar compounds (such as mixtures of paraffins or of aromatics), but for other mixtures and solutions—such as aqueous solutions of strong acids or bases or certain gases (such as hydrogen chloride) or solids (such as sodium hydroxide)—heats of solution should be included in energy balance calculations. This section outlines the required procedures.

8.5a Heats of Solution and Mixing

The **heat of solution** $\Delta \hat{H}_s(T,r)$ is defined as the change in enthalpy for a process in which 1 mole of a solute (gas or solid) is dissolved in r moles of a liquid solvent at a constant temperature T. As r becomes large, $\Delta \hat{H}_s$ approaches a limiting value known as the **heat of solution at infinite dilution**. The **heat of mixing** has the same meaning as the heat of solution when the process involves mixing two fluids rather than dissolving a gas or solid in a liquid.

Perry's Chemical Engineers' Handbook (see footnote 5) on pp. 2-201 through 2-204 gives heats of solution of various substances in water at either 18°C or "room temperature," which is roughly 25°C. *Caution:* The values given in the *Handbook* are *negatives* of the heats of solution $(-\Delta \hat{H}_s)$, although they are not identified as such explicitly.

As an illustration of how to use these data, suppose you wish to calculate ΔH for a process in which 2 mol of potassium cyanide (KCN) is dissolved in 400 mol of water at 18°C. First, calculate the moles of solvent per mole of solute:

$$r = 400/2 = 200 \text{ mol H}_2\text{O/mol KCN}$$

The value of $-\Delta H_s$ (18°C, $r = 200$) is listed as -3.0 kcal/mol (meaning per mol of KCN dissolved). The total enthalpy change is therefore

$$\Delta H = n\,\Delta \hat{H}_s = \frac{2.0 \text{ mol KCN}}{} \left| \frac{3.0 \text{ kcal}}{\text{mol KCN}} \right. = +6.0 \text{ kcal}$$

Table B.11 lists values of the heats of solution at 25°C of HCl(g) and NaOH(s) in water, and the heat of mixing at 25°C of H_2SO_4(l) and water. Heats of solution such as those given in Table B.11 may be used to determine directly the specific enthalpies of solutions at 25°C relative to the pure solute and solvent at this temperature. Another common choice of reference conditions, however, is the pure solvent and an infinitely dilute solution at 25°C.

Consider, for example, a hydrochloric acid solution for which $r = 10$ moles H_2O/mole HCl. From Table B.11, the specific enthalpy of this solution relative to pure HCl(g) and H_2O(l) at 25°C is $\Delta \hat{H}_s(r = 10) = -69.49$ kJ/mol HCl. Now, the enthalpy of the solution relative to H_2O(l) and a highly dilute solution of HCl (say, $r = 10^6$ moles H_2O/mole HCl) is the enthalpy change for the isothermal process

$$\left\{ \begin{array}{l} 1 \text{ mol HCl} \\ 10^6 \text{ mol H}_2\text{O} \end{array} \right\} \xrightarrow{25°C} \left\{ \begin{array}{l} 1 \text{ mol HCl} \\ 10 \text{ mol H}_2\text{O} \end{array} \right\} + (10^6 - 10) \text{ mol H}_2\text{O(l)}$$

We may evaluate this enthalpy change using any convenient reference state—in particular, taking pure HCl(g) and H_2O(l) at 25°C as references. In view of the latter choice, $\hat{H}$ for the $(10^6 - 10)$ moles of pure water equals zero, and the enthalpy change for the process is therefore

$$\Delta \hat{H} = \Delta \hat{H}_s(r = 10) - \Delta \hat{H}_s(r = \infty)$$
$$= (-69.49 + 75.14) \text{ kJ/mol HCl} = 5.65 \text{ kJ/mol HCl}$$

In general, the enthalpy of a solution containing r moles H_2O/mole solute is for reference states of pure solute and solvent at 25°C and 1 atm

$$\hat{H} = \Delta\hat{H}_s(r) \tag{8.5-1}$$

and for reference states of pure solvent and an infinitely dilute solution at 25°C and 1 atm

$$\hat{H} = \Delta\hat{H}_s(r) - \Delta\hat{H}_s(\infty) \tag{8.5-2}$$

Note again that these enthalpies are expressed per mole of *solute*, not per mole of solution.

TEST YOURSELF

The heat of solution of a solute A in water at 25°C is -40 kJ/mol A for $r = 10$ mol H_2O/mol A and -60 kJ/mol A for infinite dilution.

1. What is the specific enthalpy (kJ/mol A) of an aqueous solution of A for which $r = 10$ mol H_2O/mol A relative to
   (a) pure H_2O and A at 25°C?
   (b) pure H_2O and an infinitely dilute aqueous solution of A?
2. If 5 mol of A is dissolved in 50 mol of H_2O at 25°C, how much heat is evolved or absorbed? (State which, observing that $Q = \Delta H$ for this process.)
3. How much heat is evolved or absorbed if the solution prepared in question 2 is poured into a large tank of water at 25°C?

8.5b Balances on Dissolution and Mixing Processes

When setting up an energy balance on a process that involves forming, concentrating, or diluting a solution for which the heat of solution or mixing cannot be neglected, prepare an inlet–outlet enthalpy table considering the solution as a single substance and the pure components at 25°C as reference states. To calculate the enthalpy of the solution at a temperature $T \neq 25°C$, first calculate its enthalpy at 25°C from tabulated heat of solution data, then add the enthalpy change for the heating or cooling of the solution from 25°C to T. The enthalpy change for the latter step should be calculated from tabulated solution heat capacities if they are available [e.g., if they are listed on pp. 2-184 and 2-185 of *Perry's Chemical Engineers' Handbook* (see footnote 5)]; otherwise, use the average heat capacity determined using Equation 8.3-13 for liquid mixtures or the heat capacity of the pure solvent for dilute solutions.

EXAMPLE 8.5-1 *Production of Hydrochloric Acid*

Hydrochloric acid is produced by absorbing gaseous HCl (hydrogen chloride) in water. Calculate the heat that must be transferred to or from an absorption unit if HCl(g) at 100°C and H_2O(l) at 25°C are fed to produce 1000 kg/h of 20.0 wt% HCl(aq) at 40°C.

SOLUTION

It is advisable to determine the molar amounts or flow rates of the components of all feed and product solutions before drawing and labeling the flowchart. In this case

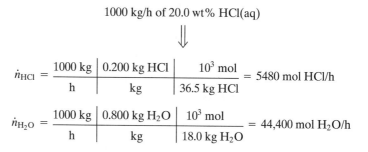

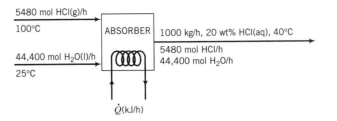

The enthalpy table for the process is shown below. As usual, physical property data valid at $P = 1$ atm are used and the effects on enthalpy of any pressure differences that may occur in the process are neglected. Note that the value of $\dot{n}$ for the product solution is the molar flow rate of the *solute* (HCl) rather than the solution, since the enthalpy will be determined in kJ/mol solute.

References: HCl(g), H$_2$O(l) at 25°C and 1 atm

Substance	$\dot{n}_{\text{in}}$	$\hat{H}_{\text{in}}$	$\dot{n}_{\text{out}}$	$\hat{H}_{\text{out}}$
HCl(g)	5480 mol HCl	$\hat{H}_1$ (kJ/mol HCl)	—	—
H$_2$O(l)	44,400 mol H$_2$O	0	—	—
HCl(aq)	—	—	5480 mol HCl	$\hat{H}_2$(kJ/mol HCl)

Calculate $\hat{H}_1$ and $\hat{H}_2$ $\quad$ HCl(g, 25°C) → HCl(g, 100°C)

$$\hat{H}_1 = \Delta\hat{H} = \int_{25°C}^{100°C} C_p \, dT$$

$$\Downarrow C_p \text{ for HCl(g) from Table B.2}$$

$$\hat{H}_1 = 2.178 \text{ kJ/mol}$$

For the product solution,

$$r = (44{,}400 \text{ mol H}_2\text{O})/(5480 \text{ mol HCl}) = 8.10$$

$$\text{HCl(g, 25°C)} + 8.10 \, \text{H}_2\text{O(l, 25°C)} \xrightarrow{\Delta\hat{H}_a} \text{HCl(aq, 25°C)} \xrightarrow{\Delta\hat{H}_b} \text{HCl(aq, 40°C)}$$

$$\Delta\hat{H}_a = \Delta\hat{H}_s(25°C, r = 8.1) \overset{\text{Table B.11}}{=\!=\!=\!\Longrightarrow} -67.4 \text{ kJ/mol HCl}$$

The heat capacities of aqueous hydrochloric acid solutions are listed on p. 2-184 of *Perry's Chemical Engineers' Handbook* (see footnote 5) as a function of the mole fraction of HCl in the solution, which in our problem is

$$\frac{5480 \text{ mol HCl/h}}{(5480 + 44{,}400) \text{ mol/h}} = 0.110 \text{ mol HCl/mol}$$

$$\Downarrow$$

$$C_p = \frac{0.73 \text{ kcal}}{\text{kg} \cdot °\text{C}} \left| \frac{1000 \text{ kg solution}}{5480 \text{ mol HCl}} \right| \frac{4.184 \text{ kJ}}{\text{kcal}} = 0.557 \frac{\text{kJ}}{\text{mol HCl} \cdot °\text{C}}$$

$$\Delta\hat{H}_b = \int_{25°C}^{40°C} C_p \, dT = 8.36 \text{ kJ/mol HCl}$$

$$\Downarrow$$

$$\hat{H}_2 = \Delta\hat{H}_a + \Delta\hat{H}_b = (-67.4 + 8.36) \text{ kJ/mol HCl} = -59.0 \text{ kJ/mol HCl}$$

Energy Balance

$$\dot{Q} = \Delta\dot{H} = \sum_{\text{out}} \dot{n}_i \hat{H}_i - \sum_{\text{in}} \dot{n}_i \hat{H}_i$$

$$= (5480 \text{ mol HCl/h})(-59.0 \text{ kJ/mol HCl}) - (5480 \text{ mol HCl/h})(2.178 \text{ kJ/mol HCl})$$

$$= \boxed{-3.35 \times 10^5 \text{ kJ/h}}$$

Heat must be transferred out of the absorber at a rate of 335,000 kJ/h to keep the product temperature from rising above 40°C.

8.5c Enthalpy–Concentration Charts—Single Liquid Phase

Energy balance calculations on liquid-phase systems involving mixtures can be cumbersome when heats of mixing are significant. The calculations can be simplified for binary (two-component) systems by using an **enthalpy–concentration chart**, a plot of specific enthalpy versus mole fraction (or mole percent) or mass fraction (or weight percent) of one component. An $\hat{H}$-x chart for aqueous solutions of sulfuric acid at several temperatures is shown in Figure 8.5-1. The reference conditions for the plotted enthalpies are pure liquid H_2SO_4 at 77°F and liquid water at 32°F.

The points on the isotherms of Figure 8.5-1 were determined using the procedure outlined in the last section. Suppose, for example, you wish to calculate the specific enthalpy (Btu/lb_m) of a 40 wt% sulfuric acid solution at 120°F. If you know the heat of mixing of sulfuric acid at 77°F, the process path you would follow would be to bring pure liquid water from its reference temperature of 32°F to 77°F (the sulfuric acid starts at 77°F and so does not require this step), mix the two liquids at 77°F, bring the product solution to 120°F, and calculate and add the enthalpy changes for each of these steps.

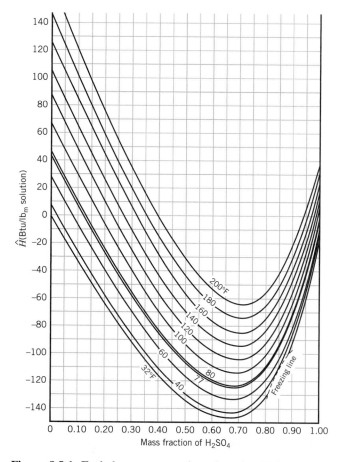

Figure 8.5-1 Enthalpy–concentration chart for H_2SO_4–H_2O. (Redrawn from the data of W. D. Ross, *Chem. Eng. Progr.*, **43**:314, 1952.)

Basis: 1 lb_m solution ($\Longrightarrow$ 0.40 lb_m H_2SO_4 = 4.08 × 10^{-3} lb-mole, 0.60 lb_m H_2O = 3.33 × 10^{-2} lb-mole)

- 0.60 lb_m $H_2O(l, 32°F) \longrightarrow$ 0.60 lb_m H_2O (l, 77°F)

$$\Delta H_1 (\text{Btu}) = (0.60 \ lb_m \ H_2O)\left[\int_{32°F}^{77°F} (C_p)_{H_2O} \ dT\right]\left(\frac{\text{Btu}}{lb_m}\right)$$

The heat capacity of liquid water is approximately 1 Btu/(lb_m·°F).

- 0.40 lb_m $H_2SO_4(77°F)$ + 0.60 lb_m H_2O (77°F) $\longrightarrow$ 1.0 lb_m H_2SO_4 solution (aq, 77°F)

$$\Delta H_2 (\text{Btu}) = (0.40 \ lb_m \ H_2SO_4)\left[\Delta \hat{H}_s\left(77°F, r = 8.2 \ \frac{\text{lb-mole } H_2O}{\text{lb-mole } H_2SO_4}\right)\left(\frac{\text{Btu}}{lb_m \ H_2SO_4}\right)\right]$$

The heat of mixing can be determined from the data in Table B.11 to be −279 Btu/lb_m H_2SO_4.

- 1.0 lb_m H_2SO_4 solution (aq, 77°F) $\longrightarrow$ 1.0 lb_m H_2SO_4 solution (aq, 120°F)

$$\Delta H_3 (\text{Btu}) = (1.0 \ lb_m)\int_{77°F}^{120°F} (C_p)_{40\% \ H_2SO_4(aq)} \ dT$$

The heat capacity of the 40% sulfuric acid solution is roughly 0.67 Btu/(lb_m·°F).[15]

- $\hat{H}$(40% H_2SO_4, 120°F) = $\dfrac{(\Delta H_1 + \Delta H_2 + \Delta H_3)(\text{Btu})}{1.0 \ lb_m \ \text{solution}}$ ≈ $\boxed{-56 \ \text{Btu/}lb_m}$

(Verify that this is the value shown in Figure 8.5-1.)

If any reference temperature but 77°F had been chosen for sulfuric acid, another step would have been included in which H_2SO_4 was brought from T_{ref} to 77°F prior to mixing.

Once someone has gone to the trouble of preparing an enthalpy-concentration chart like that of Figure 8.5-1, energy balance calculations become relatively simple, as shown in Example 8.5-2.

EXAMPLE 8.5-2   ***Concentration of an Aqueous H_2SO_4 Solution***

A 5.0 wt% H_2SO_4 solution at 60°F is to be concentrated to 40.0 wt% by evaporation of water. The concentrated solution and water vapor emerge from the evaporator at 180°F and 1 atm. Calculate the rate at which heat must be transferred to the evaporator to process 1000 lb_m/h of the feed solution.

Basis: Given Feed Rate of 5% Solution

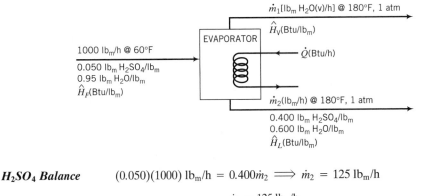

H_2SO_4 Balance   (0.050)(1000) lb_m/h = 0.400$\dot{m}_2$ $\Longrightarrow$ $\dot{m}_2$ = 125 lb_m/h

Total Mass Balance   1000 lb_m/h = $\dot{m}_1$ + $\dot{m}_2$ $\xrightarrow{\dot{m}_2 = 125 \ lb_m/h}$ $\dot{m}_1$ = 875 lb_m/h

---

[15] R. H. Perry and D. W. Green, Eds., *Perry's Chemical Engineers' Handbook*, 7th Edition, McGraw-Hill, New York, 1997, p. 2-184.

Reference States for Energy Balance $H_2O(l, 32°F)$, $H_2SO_4(l, 77°F)$

From Figure 8.5-1:

$$\hat{H}_F = 10 \text{ Btu/lb}_m \quad (5\% \text{ H}_2\text{SO}_4 \text{ at } 60°F)$$
$$\hat{H}_L = -17 \text{ Btu/lb}_m \quad (40\% \text{ H}_2\text{SO}_4 \text{ at } 180°F)$$

The enthalpy of water vapor at 180°F and 1 atm relative to liquid water at 32°F may be obtained from the steam tables in *Perry's Chemical Engineers' Handbook* (see footnote 5) as

$$\hat{H}_V = 1138 \text{ Btu/lb}_m$$

Energy Balance
$$\dot{Q} = \Delta\dot{H} = \dot{m}_1\hat{H}_V + \dot{m}_2\hat{H}_L - (1000\text{lb}_m/h)\hat{H}_F$$
$$= [(875)(1138) + (125)(-17) - (1000)(10)] \text{ Btu/h}$$
$$= \boxed{984{,}000 \text{ Btu/h}}$$

Compare the ease of this computation with that of Example 8.5-1. Having the enthalpy–concentration chart eliminates the need for all of the hypothetical heating, cooling, and isothermal mixing steps that would normally be required to evaluate the total enthalpy change for the process.

Adiabatic mixing processes are particularly simple to analyze when an $\hat{H}$-x chart is available. Suppose x_A is the mass fraction of A in a mixture of two species, A and B, and that a mass m_1 of Solution 1 $(x_{A1}, \hat{H}_1)$ is mixed adiabatically with a mass m_2 of Solution 2 $(x_{A2}, \hat{H}_2)$. We will show that the condition of the product mixture, $(x_{A3}, \hat{H}_3)$, is on a straight line on the $\hat{H}$-x chart between the points corresponding to the feed stream conditions.

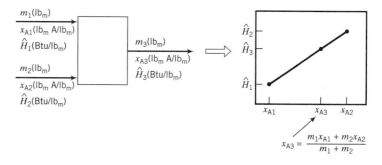

To prove this result, we write a total mass balance, a material balance on species A, and an energy balance ($\Delta H = 0$ for this constant-pressure batch process):

Total Mass Balance: $\qquad\qquad\qquad m_1 + m_2 = m_3 \qquad\qquad\qquad$ **(a)**

A Balance: $\qquad m_1 x_{A1} + m_2 x_{A2} = m_3 x_{A3} \qquad\qquad\qquad$ **(b)**

$$\Downarrow \text{ Substitute for } m_3 \text{ from (a), rearrange}$$

$$m_1(x_{A3} - x_{A1}) = m_2(x_{A2} - x_{A3}) \qquad\qquad\qquad \textbf{(c)}$$

Energy Balance: $\qquad \Delta H = m_3\hat{H}_3 - m_1\hat{H}_1 - m_2\hat{H}_2 = 0$

$$\Downarrow \text{ Substitute for } m_3 \text{ from (a), rearrange}$$

$$m_1(\hat{H}_3 - \hat{H}_1) = m_2(\hat{H}_2 - \hat{H}_3) \qquad\qquad\qquad \textbf{(d)}$$

Dividing (d) by (c) yields

$$\frac{\hat{H}_3 - \hat{H}_1}{x_{A3} - x_{A1}} = \frac{\hat{H}_2 - \hat{H}_3}{x_{A2} - x_{A3}} \qquad\qquad\qquad \textbf{(e)}$$

Since the slope of the line segment from $(x_{A1}, \hat{H}_1)$ to $(x_{A3}, \hat{H}_3)$ (the left-hand side of this equation) equals the slope of the segment from $(x_{A3}, \hat{H}_3)$ to $(x_{A2}, \hat{H}_2)$ (the right-hand side) and the segments have a point in common, the three points must lie on a straight line. The value of x_{A3} can be calculated from Equations a and b:

$$x_{A3} = \frac{m_1 x_{A1} + m_2 x_{A2}}{m_1 + m_2} \tag{8.5-3}$$

It follows that if two feed solutions of known masses and compositions (m_i, x_i, $i = 1, 2$) are mixed adiabatically and you have an $\hat{H}$-x chart, you may (i) calculate x_3 for the product mixture from Equation 8.5-3, (ii) draw a line connecting the points on the chart corresponding to the two feeds, and (iii) read the enthalpy and temperature of the product mixture from the point on the connecting line for which $x = x_3$.

EXAMPLE 8.5-3   *Adiabatic Mixing*

Pure water at 60°F is mixed with 100 g of an aqueous 80 wt% H_2SO_4 solution, also at 60°F. The mixing vessel is insulated well enough to be considered adiabatic.

1. If 250 g H_2O is mixed with the acid, what will the final solution temperature be?
2. What is the maximum attainable solution temperature and how much water must be added to achieve it?

SOLUTION

1. From Equation 8.5-3, the mass fraction of H_2SO_4 in the product solution is

$$x_p = \frac{[(100)(0.80) + (250)(0)] \text{ g } H_2SO_4}{(100 + 250)\text{g}} = 0.23\text{g } H_2SO_4/\text{g}$$

A straight line on Figure 8.5-1 between points at ($x = 0, T = 60°F$) and ($x = 0.80, T = 60°F$) goes through the point ($x = 0.23$, $\boxed{T \approx 100°F}$). (Verify this result.)

2. The line between ($x = 0, T = 60°F$) and ($x = 0.80, T = 60° F$) passes through a temperature maximum at roughly ($x \approx 0.58$, $\boxed{T \approx 150°F}$). (*Verify.*) From Equation 8.5-3,

$$0.58 = \frac{(100)(0.80) \text{ g} + (m_w)(0)}{100 \text{ g} + m_w} \implies \boxed{m_w = 38 \text{ g } H_2O}$$

The graphical construction of these solutions is illustrated below.

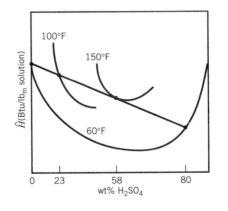

TEST YOURSELF

Use Figure 8.5-1 to answer the following questions.

1. What is the specific enthalpy of 80 wt% H_2SO_4(aq, 110°F) relative to pure H_2SO_4 at 77°F and pure water at 32°F?
2. The 100 wt% intercepts of the isotherms on Figure 8.5-1 are difficult to read. Which isotherm must have an intercept of 0 Btu/lb$_m$? (Your answer should be a temperature.)

3. Pure water at 32°F is used to dilute a 90 wt% H_2SO_4 solution (aq, 32°F). Estimate the maximum temperature the product solution can achieve and the concentration of sulfuric acid (wt%) in this solution.

4. Estimate (a) the specific enthalpy of a 30 wt% H_2SO_4 solution (aq, 77°F) and (b) the specific enthalpy of a 30 wt% solution obtained by mixing pure water at 77°F and pure sulfuric acid at 77°F adiabatically. What is the physical significance of the difference between these two enthalpies?

8.5d   Using Enthalpy–Concentration Charts for Vapor–Liquid Equilibrium Calculations

Enthalpy–concentration charts are particularly useful for two-component systems in which vapor and liquid phases are in equilibrium. The Gibbs phase rule (Equation 6.2-1) specifies that such a system has $(2 + 2 - 2) = 2$ degrees of freedom. If as before we fix the system pressure, then specifying only one more intensive variable—the system temperature, or the mass or mole fraction of either component in either phase—fixes the values of all other intensive variables in both phases. An $\hat{H}$-x diagram for the ammonia–water system at 1 atm is shown in Figure 8.5-2.

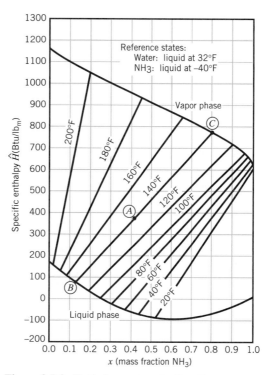

Figure 8.5-2   Enthalpy–concentration diagram for the ammonia–water system at 1 atm. (From G. G. Brown et al., *Unit Operations,* ©1950, Figure 551. Reprinted by permission of John Wiley & Sons.)

The specific enthalpies of aqueous solutions and gaseous mixtures of ammonia and water are shown on the two curves on this figure.

Suppose the mass fraction of ammonia in a liquid solution of NH_3 and H_2O at 1 atm is specified to be 0.25. According to the phase rule, the system temperature and the mass fraction of NH_3 in the vapor phase are uniquely determined by these specifications. (*Verify.*) A **tie line** may therefore be drawn on the enthalpy–concentration chart from $x = 0.25$ on the liquid-phase curve to the corresponding point on the vapor-phase curve, which is at $y = 0.95$; and the tie line may be labeled with the corresponding temperature, 100°F. Several tie lines constructed in this manner are shown in Figure 8.5-2; once drawn, the lines may be used to determine the equilibrium composition and the specific enthalpy of each phase at a specified temperature.

EXAMPLE 8.5-4   *Use of the Enthalpy–Concentration Chart for a Two-Phase System*

An aqueous ammonia solution is in equilibrium with a vapor phase in a closed system at 160°F and 1 atm. The liquid phase accounts for 95% of the total mass of the system contents. Use Figure 8.5-2 to determine the weight percent of NH_3 in each phase and the enthalpy of the system per unit mass of the system contents.

SOLUTION   The mass fractions of ammonia and specific enthalpies of each phase may be read from the intersections of the 160°F tie line with the vapor and liquid equilibrium curves on Figure 8.5-2.

Liquid Phase $\boxed{8\% \ NH_3, \ 92\% \ H_2O}$; $\hat{H}_L = 110 \ \text{Btu/lb}_\text{m}$

Vapor Phase $\boxed{64\% \ NH_3, \ 36\% \ H_2O}$; $\hat{H}_V = 855 \ \text{Btu/lb}_\text{m}$

Basis   1 lb$_\text{m}$ total mass $\Longrightarrow$ 0.95 lb$_\text{m}$ liquid, 0.05 lb$_\text{m}$ vapor

$$\hat{H}(\text{Btu/lb}_\text{m}) = \frac{0.95 \ \text{lb}_\text{m} \ \text{liquid}}{} \left| \frac{110 \ \text{Btu}}{\text{lb}_\text{m}} \right. + \frac{0.05 \ \text{lb}_\text{m} \ \text{vapor}}{} \left| \frac{855 \ \text{Btu}}{\text{lb}_\text{m}} \right.$$

$$= \boxed{147 \ \text{Btu/lb}_\text{m}}$$

If the overall composition of a two-phase two-component system at a given temperature and pressure is known, the fraction of the system that is liquid or vapor may easily be determined from the enthalpy–concentration chart.

Suppose, for example, that a mixture of ammonia and water that is 40% NH_3 by mass is contained in a closed vessel at 140°F and 1 atm. Point *A* on Figure 8.5-2 corresponds to this condition. Since this point lies between the vapor and liquid equilibrium curves, the mixture separates into two phases whose compositions are found at the extremities of the 140°F tie line (points *B* and *C*).

In general, if *F*, *L*, and *V* are the total mass of the mixture, the mass of the liquid phase and the mass of the vapor phase, respectively, and x_F, x_L, and x_V are the corresponding mass fractions of NH_3, then

Total balance: $F = L + V$ (8.5-4)

NH_3 balance: $x_F F = x_L L + x_V V$ (8.5-5)

Substituting the expression of Equation 8.5-4 for *F* into Equation 8.5-5 and rearranging the result yields

$$\frac{L}{V} = \frac{x_V - x_F}{x_F - x_L}$$ (8.5-6)

The tie line in question appears as shown below:

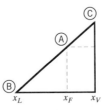

From the properties of similar triangles, the right side of Equation 8.5-6 equals the ratio of distances $\overline{AC}/\overline{AB}$. We have thus proved the following general rule: if A, B, and C are the points on a tie line corresponding to the total mixture, the liquid phase, and the vapor phase, respectively, and if F, L, and V are the corresponding masses, then the liquid-to-vapor mass ratio is

$$\frac{L}{V} = \frac{x_V - x_F}{x_F - x_L} = \frac{\overline{AC}}{\overline{AB}} \tag{8.5-7}$$

This is the *lever rule*. It is also not difficult to prove that the mass fractions of the liquid and vapor phases are

$$\frac{L}{F} = \frac{x_V - x_F}{x_V - x_L} = \frac{\overline{AC}}{\overline{BC}} \tag{8.5-8}$$

$$\frac{V}{F} = \frac{x_F - x_L}{x_V - x_L} = \frac{\overline{AB}}{\overline{BC}} \tag{8.5-9}$$

Once you have located the total mixture on the chart from a specified set of feed conditions, it becomes a simple matter to determine the compositions, enthalpies, and relative proportions of each phase, calculations that would take much more time in the absence of the chart.

EXAMPLE 8.5-5   *Equilibrium Flash Vaporization*

A 30 wt% NH_3 solution at 100 psia is fed at a rate of 100 lb_m/h to a tank in which the pressure is 1 atm. The enthalpy of the feed solution relative to the reference conditions used to construct Figure 8.5-2 is 100 Btu/lb_m. The vapor composition is to be 89 wt% NH_3. Determine the temperature of the stream leaving the tank, the mass fraction of NH_3 in the liquid product, the flow rates of the liquid and vapor product streams, and the rate at which heat must be transferred to the vaporizer.

SOLUTION   ***Basis: 100 lb_m/h Feed***

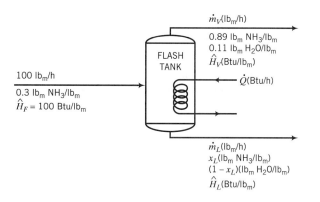

From Figure 8.5-2,

$$x_V = 0.89 \ lb_m \ NH_3/lb_m$$

$$\boxed{T = 120°F}$$

$$\boxed{x_L = 0.185 \ lb_m \ NH_3/lb_m}$$

$$\hat{H}_V = 728 \ \text{Btu/lb}_m$$

$$\hat{H}_L = 45 \ \text{Btu/lb}_m$$

From Equation 8.5-8

$$\frac{\dot{m}_L}{100 \text{ lb}_m/\text{h}} = \frac{x_V - x_F}{x_V - x_L}$$

$$\Downarrow$$

$$\dot{m}_L = (100 \text{ lb}_m/\text{h})\frac{0.89 - 0.30}{0.89 - 0.185} = \boxed{84 \text{ lb}_m/\text{h liquid product}}$$

$$\dot{m}_V = (100 - 84) \text{ lb}_m/\text{h} = \boxed{16 \text{ lb}_m/\text{h vapor product}}$$

Energy Balance $\dot{Q} = \Delta\dot{H} = \dot{m}_V\hat{H}_V + \dot{m}_L\hat{H}_L - 100\hat{H}_F$

$$= [(16)(728) + (84)(45) - (100)(100)] \text{ Btu/h} = \boxed{5400 \text{ Btu/h}}$$

TEST
YOURSELF

Use Figure 8.5-2 for the following calculations.

1. Estimate (a) the temperature at which the mass fraction of NH_3 in the vapor phase of a two-phase system equals 0.85, and (b) the corresponding liquid-phase NH_3 mass fraction.
2. What is the heat of vaporization of NH_3 at its normal boiling point?
3. If an NH_3–H_2O mixture whose overall composition is 50% NH_3–50% H_2O is in equilibrium at 120°F, what fraction of the mixture is a vapor?

8.6   SUMMARY

An integral energy balance (the first law of thermodynamics) for a closed constant-volume system with no kinetic or potential energy changes ($\Delta E_k = 0, \Delta E_p = 0$) and no energy transferred in or out as work ($W = 0$) is

$$Q = \Delta U = \sum_{\text{final}} n_i\hat{U}_i - \sum_{\text{initial}} n_i\hat{U}_i$$

For a closed system expanding or contracting against a constant external pressure, the balance is

$$Q = \Delta H = \sum_{\text{final}} n_i\hat{H}_i - \sum_{\text{initial}} n_i\hat{H}_i$$

For an open system at steady state with negligible kinetic and potential energy changes from inlet to outlet and no energy transfer as shaft work, the balance is

$$\dot{Q} = \Delta\dot{H} = \sum_{\text{outlet}} \dot{n}_i\hat{H}_i - \sum_{\text{inlet}} \dot{n}_i\hat{H}_i$$

In these equations n is the amount (mass or moles) of a species in one of its initial or final states in the process, $\dot{n}$ is the flow rate (mass or molar) of a species in a continuous stream entering or leaving the process, and $\hat{U}$ and $\hat{H}$ are respectively the specific internal energy and specific enthalpy of a species in a process state relative to a specified reference state for the same species.

This chapter presents formulas and methods for evaluating $\hat{U}$ and $\hat{H}$ (and hence ΔU, ΔH, and $\Delta\dot{H}$) when tables of internal energies and enthalpies are not readily available. Here is the general procedure:

1. Choose a reference state (phase, temperature, and pressure) for each species involved in a process.
2. Choose a path from the reference state to each initial and final (or inlet and outlet) process state for each species, and evaluate $\hat{U}_i$ (or $\hat{H}_i$) as $\Delta\hat{U}$ (or $\Delta\hat{H}$) for the transition from the reference state to the process state.

3. Once all of the $\hat{U}_i$ (or all $\hat{H}_i$) values are determined in this manner and all of the n_i (or all $\dot{n}_i$) values are determined from material balances, densities or equations of state, and phase equilibrium relations, calculate ΔU, ΔH, or $\Delta \dot{H}$ and substitute the result in the energy balance to determine whichever variable is unknown (usually the heat, Q, or heat transfer rate, $\dot{Q}$).

Following are points regarding the implementation of this procedure for various types of processes.

- Energy balance calculations for a system (a process unit or combination of units) are conveniently organized through the construction of an *inlet–outlet internal energy table* (or *enthalpy table*). The table lists n (or $\dot{n}$) and $\hat{U}$ (or $\hat{H}$) for each species at each state (phase, temperature, pressure) in which the species is found in process streams. Once all of these variable values have been determined and inserted in the table, the subsequent evaluation of ΔU, ΔH, or $\Delta \dot{H}$ is straightforward.

- The fact that internal energy and enthalpy are *state properties* means that any convenient process path from a reference state to a process state may be chosen, even if the actual process proceeds by a different path. As a rule, you would choose a path that allows you to make use of heat capacities, phase transition temperatures, and latent heats tabulated in an available reference (like this text).

- *Changes in pressure at constant temperature.* For a species undergoing an isothermal pressure change, ΔP,

  $\Delta \hat{U} \approx 0$ for solids, liquids, and nearly ideal gases. For ideal gases, $\Delta \hat{U} = 0$.

  $\Delta \hat{H} \approx \hat{V} \Delta P$ for solids and liquids, where $\hat{V}$ is the (presumably constant) specific volume of the solid or liquid.

  $\Delta \hat{H} \approx 0$ for nearly ideal gases or for moderately small pressure changes (on the order of a few atmospheres). For ideal gases, $\Delta \hat{H} = 0$.

  If gases are at conditions at which they are far from ideal or if they undergo large pressure changes, you must either use tables of thermodynamic properties (such as the steam tables for water) or thermodynamic correlations beyond the scope of this text to determine $\Delta \hat{U}$ or $\Delta \hat{H}$.

- *Changes in temperature.* The specific internal energy of a species increases with increasing temperature. If a species is heated at constant volume and $\hat{U}$ is plotted versus T, the slope of the resulting curve is the *heat capacity at constant volume* of the species, $C_v(T)$, or $C_v = (\partial \hat{U}/\partial T)_{\text{constant } \hat{V}}$. If a species undergoes a change in temperature from T_1 to T_2 without changing phase,

$$\Delta \hat{U} \approx \int_{T_1}^{T_2} C_v(T) \, dT$$

This equation is

  (a) exact for an ideal gas, even if $\hat{V}$ changes during the heating or cooling process. (For an ideal gas, $\hat{U}$ does not depend on $\hat{V}$.)

  (b) a good approximation for a solid or liquid.

  (c) valid for a nonideal gas only if $\hat{V}$ is constant.

- The specific enthalpy of a species ($\hat{H} = \hat{U} + P\hat{V}$) also increases with increasing temperature. If a species is heated at constant pressure and $\hat{H}$ is plotted versus T, the slope of the resulting curve is the *heat capacity at constant pressure* of the species, $C_p(T)$, or $C_p = (\partial \hat{H}/\partial T)_{\text{constant } P}$. It follows that if a *gas* undergoes a change in temperature from T_1 to T_2, with or without a concurrent change in pressure,

$$\Delta \hat{H} \approx \int_{T_1}^{T_2} C_p(T) \, dT$$

This equation is

(a) exact for an ideal gas, even if P changes during the heating or cooling process. (For an ideal gas, $\hat{H}$ does not depend on P.)

(b) valid for a nonideal gas only if P is constant.

If a liquid or solid undergoes a temperature change from T_1 to T_2 and a simultaneous pressure change, ΔP, then

$$\Delta\hat{H} \approx \hat{V}\Delta P + \int_{T_1}^{T_2} C_p(T)\,dT$$

- Table B.2 lists coefficients of polynomial expressions for $C_p(T)[\text{kJ/(mol·°C)}]$ at $P = 1$ atm. The expressions should be accurate for solids, liquids, and ideal gases at any pressure and for nonideal gases only at 1 atm.
- To determine an expression or value for $C_v(T)$ from a known expression or value for $C_p(T)$, use one of the following relationships:

  Liquids and Solids: $C_v \approx C_p$

  Ideal Gases: $C_v = C_p - R$

  where R is the gas constant. Since the degree unit in the denominator of the heat capacity is a temperature interval, R can be subtracted directly from the expressions for C_p in Table B.2.
- The heat capacity of a solid or liquid can be estimated in the absence of tabulated data using *Kopp's rule* (Section 8.3c).
- If only tabulated values of C_p or C_v at discrete temperatures are available, the integrals in the expressions for $\Delta\hat{U}$ and $\Delta\hat{H}$ must be evaluated by *numerical integration,* using formulas such as those given in Appendix A.3.
- *Phase changes at constant temperature and pressure. Latent heats* are changes in specific enthalpy associated with phase changes at constant T and P. For example, the *latent heat of fusion* (more commonly, the *heat of fusion*), $\Delta\hat{H}_m(T, P)$, is the enthalpy change for the process in which a solid at temperature T and pressure P becomes a liquid at the same temperature and pressure, and the *heat of vaporization,* $\Delta\hat{H}_v(T, P)$, is $\Delta\hat{H}$ for the process in which a liquid at T and P becomes a vapor at the same T and P.
- Table B.1 lists *standard heats of fusion and vaporization* for a number of species, or $\Delta\hat{H}_m$ and $\Delta\hat{H}_v$ at the normal melting and boiling point temperatures ($P = 1$ atm), which are also listed in Table B.1. If latent heat data are not available for a species, $\Delta\hat{H}_m$ and $\Delta\hat{H}_v$ may be estimated using formulas given in Section 8.4b.
- You can use the formulas given above to determine the specific enthalpy of any species in one state relative to that species in any other state. For example, to calculate $\hat{H}$ for benzene vapor at temperature 300°C and 15 atm relative to solid benzene at a reference state of -20°C and 1 atm, you would carry out the following steps.

  1. Heat the solid from the reference temperature (-20°C) to its normal melting point T_{mp}, which from Table B.1 is 5.53°C.

  $$\Delta\hat{H}_1 = \int_{0°C}^{5.53°C} (C_p)_{solid}\,dT$$

  $(C_p)_{solid}$ is not listed in Table B.2, so it must either be found elsewhere or estimated using Kopp's rule. The latter provides a crude approximation but a very reasonable one to make in this case, considering how little this step will contribute to the overall enthalpy change.

  2. Melt the solid at T_{mp}. $\Delta\hat{H}_2 = \Delta\hat{H}_m(5.53°C)$, which from Table B.1 is 9.837 kJ/mol.

  3. Heat the liquid from T_{mp} to the normal boiling point, T_{bp}, which from Table B.1 is 80.10°C.

  $$\Delta\hat{H}_3 = \int_{5.53°C}^{80.1°C} (C_p)_{liquid}\,dT$$

A polynomial formula for $(C_p)_{liquid}$ is given in Table B.2. Since it applies to T expressed in kelvin units, the limits of the integral should be changed to their kelvin equivalents.

4. Vaporize the liquid at T_{bp}. $\Delta \hat{H}_4 = \Delta \hat{H}_v(80.1°C)$, which from Table B.1 is 30.765 kJ/mol.
5. Heat the vapor from T_{bp} to 300°C.

$$\Delta \hat{H}_5 = \int_{80.1°C}^{300°C} (C_p)_{vapor} \, dT$$

A formula for $(C_p)_{vapor}$ is given in Table B.2.

6. Bring the vapor from 1 atm to 15 atm at 300°C. $\Delta \hat{H}_6 \approx 0$ as long as the vapor behaves like an ideal gas, which it would at this high temperature.
7. Add the enthalpy changes for each of the preceding steps to calculate the desired specific enthalpy.

- The *psychrometric chart* (or *humidity chart*) contains values of a number of process variables for air–water vapor systems at 1 atm. The values listed on the chart include *dry-bulb temperature* (the temperature measured by common temperature-measurement instruments), *moisture content* or *absolute humidity* (mass ratio of water vapor to dry air), *relative humidity*, *humid volume* (volume per mass of dry air), *wet-bulb temperature* (the temperature reading on a thermometer with a water-saturated wick around the bulb immersed in a flowing stream of humid air), and *enthalpy per mass of dry air.* If you know the values of any two of these variables for humid air at or near 1 atm, you can use the chart to determine the values of the other four, which can greatly simplify material and energy balance calculations.

- In *adiabatic cooling* operations, a stream of warm gas is brought into contact with a stream of cold liquid, causing the gas to cool and some liquid to evaporate. If (a) the gas is dry or humid air, the liquid is water, and the process takes place at about 1 atm, (b) the process is adiabatic, (c) the heat capacities of liquid water, water vapor, and air can be considered constant over the temperature range of the process, and (d) enthalpy changes associated with temperature changes of the liquid may be neglected, then the final state of the air must lie on the same wet-bulb temperature line as the state of the inlet air on the psychrometric chart.

- An enthalpy change known as the *heat of mixing* or *heat of solution* is associated with the mixing of certain liquids (like acids and water) and the dissolving of some gases or solids in a liquid solvent at a given temperature and pressure. An *ideal* solution is one for which the heat of mixing or solution is negligible, so that the enthalpy of the solution is the sum of the enthalpies of the pure solution components at the same temperature and pressure. All gas mixtures are ideal, as are mixtures of structurally similar liquid compounds (like benzene, toluene, and xylene). Table B.11 gives heats of mixing at 25°C and 1 atm for aqueous sulfuric acid solutions and heats of solution at the same temperature and pressure for aqueous solutions of HCl(g) (hydrochloric acid) and NaOH(s) (caustic soda).

- To perform energy balance calculations on processes involving nonideal solutions, take the pure components at 25°C as references. To determine the specific enthalpy of a feed or product solution, look it up on an *enthalpy–concentration chart* if one is available (e.g., Figure 8.5-1 for sulfuric acid solutions or Figure 8.5-2 for aqueous ammonia solutions). Otherwise, form the solution at 25°C [$\Delta \hat{H} = \Delta \hat{H}_s(25°C)$] and heat or cool it to its state in the process ($\Delta \hat{H} = \int_{25°C}^{T} C_p \, dT$). For the latter step, either find heat capacity data for the solution or (for dilute solutions) assume that the heat capacity is that of the pure solvent.

PROBLEMS

8.1. The specific internal energy of formaldehyde (HCHO) vapor at 1 atm and moderate temperatures is given by the formula

$$\hat{U}(J/mol) = 25.96T + 0.02134T^2$$

where T is in °C.

(a) Calculate the specific internal energies of formaldehyde vapor at 0°C and 100°C. What reference temperature was used to generate the given expression for $\hat{U}$?

(b) The value of $\hat{U}$ calculated for 100°C is not the true value of the specific internal energy of formaldehyde vapor at this condition. Why not? (*Hint:* Refer back to Section 7.5a.) Briefly state the physical significance of the calculated quantity.

(c) Use the closed system energy balance equation to calculate the heat (J) required to raise the temperature of 3.0 mol HCHO at constant volume from 0°C to 100°C. List all of your assumptions.

(d) From the definition of heat capacity at constant volume, derive a formula for $C_v(T)$[J/mol·°C]. Then use this formula and Equation 8.3-6 to calculate the heat (J) required to raise the temperature of 3.0 mol of HCHO(v) at constant volume from 0°C to 100°C. [You should get the same result you got in part (c).]

8.2. The heat capacity at constant pressure of hydrogen cyanide is given by the expression

$$C_p[\text{J/(mol·°C)}] = 35.3 + 0.0291T(°C)$$

(a) Write an expression for the heat capacity at constant volume for HCN, assuming ideal gas behavior.

(b) Calculate $\Delta\hat{H}$(J/mol) for the constant-pressure process

$$\text{HCN}(25°C, 1 \text{ atm}) \rightarrow \text{HCN}(100°C, 1 \text{ atm})$$

(c) Calculate $\Delta\hat{U}$(J/mol) for the constant-volume process

$$\text{HCN}(25°C, 1 \text{ m}^3/\text{kmol}) \rightarrow \text{HCN}(100°C, 1 \text{ m}^3/\text{kmol})$$

(d) If the process of part (b) were carried out in such a way that the initial and final pressures were each 1 atm but the pressure varied during the heating, the value of $\Delta\hat{H}$ would still be what you calculated assuming a constant pressure. Why is this so?

8.3. The heat capacity at constant volume of hydrogen sulfide at low pressures is

$$C_v[\text{kJ/(mol·°C)}] = 0.0252 + 1.547 \times 10^{-5}T - 3.012 \times 10^{-9}T^2$$

where T is in °C. A quantity of H_2S is kept in a piston-fitted cylinder with initial temperature, pressure, and volume equal to 25°C, 2.00 atm, and 3.00 liters, respectively.

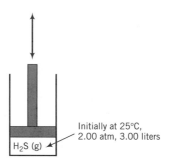

Initially at 25°C,
2.00 atm, 3.00 liters

H_2S (g)

(a) Calculate the heat (kJ) required to raise the gas temperature from 25°C to 1000°C if the heating takes place at constant volume (i.e., if the piston does not move), retaining successively one term, two terms, and all three terms of the heat capacity formula. (See Example 8.3-1.) Determine the percentage errors in Q that result from retaining only one and two terms of the heat capacity formula, assuming that the full expression yields the correct result.

(b) For a closed system at constant pressure with negligible kinetic and potential energy changes, the energy balance equation is $Q = \Delta H$. Use Equation 8.3-12 to determine an expression for the heat capacity at constant pressure (C_p) for H_2S, assuming ideal gas behavior. Then use it to calculate the heat (J) required to raise the gas from 25°C to 1000°C at constant pressure. What would the piston do during this process?

(c) What is the physical significance of the difference between the values of Q calculated in parts (a) and (b)?

8.4. Use data in Table B.2 to calculate the following:

(a) The heat capacity (C_p) of liquid benzene at 40°C.

(b) The heat capacity at constant pressure of benzene vapor at 40°C.

(c) The heat capacity at constant pressure of solid carbon at 40°C.

(d) $\Delta \hat{H}$(kJ/mol) for benzene vapor going from 40°C to 300°C.

(e) $\Delta \hat{H}$(kJ/mol) for solid carbon going from 40°C to 300°C.

8.5. Estimate the specific enthalpy of steam (kJ/kg) at 350°C and 100 bar relative to steam at 100°C and 1 atm using:

(a) The steam tables.

(b) Table B.2 and assuming ideal gas behavior.

What is the physical significance of the difference between the values of $\hat{H}$ calculated by the two methods?

8.6. Calculate $\Delta \hat{H}$ for each of the following processes. In each case, restate your result as a specific enthalpy relative to a reference state. [The solution—which you should verify—and the restatement for part (a) are given as an illustration.] Assume that the process pressures are low enough for $\hat{H}$ to be considered independent of pressure, so that the formulas of Table B.2 (which strictly apply at 1 atm) can be used.

(a) $CH_3COCH_3(l, 15°C) \longrightarrow CH_3COCH_3(l, 55°C)$.

Solution: $\boxed{\Delta \hat{H} = 5.180 \text{ kJ/mol}}$

The specific enthalpy of liquid acetone at 55°C relative to liquid acetone at 15°C is 5.180 kJ/mol.

(b) $n\text{-}C_6H_{14}(l, 25°C) \longrightarrow n\text{-}C_6H_{14}(l, 80°C)$

(c) $n\text{-}C_6H_{14}(v, 500°C) \longrightarrow n\text{-}C_6H_{14}(v, 0°C)$. (Make statements about both the specific enthalpy of hexane vapor at 500°C relative to hexane vapor at 0°C and the specific enthalpy of hexane vapor at 0°C relative to hexane vapor at 500°C.)

8.7. Two formulas for the heat capacity of CO are given here:

$$C_p[\text{cal}/(\text{mol}\cdot°C)] = 6.890 + 0.001436T(°C)$$

$$C_p[\text{Btu}/(\text{lb-mole}\cdot°F)] = 6.864 + 0.0007978T(°F)$$

Starting with the first formula, derive the second. (Recall Section 2.5, and remember that the temperature unit in the denominator of C_p refers to a temperature interval.)

8.8. Table B.2 lists values of the heat capacity of liquid ethanol at two temperatures. Use the tabulated values to derive a linear expression for $C_p(T)$; then use the derived expression and data in Table B.1 to calculate the heat transfer rate (kW) required to bring a stream of liquid ethanol flowing at 55.0 L/s from 20°C to the boiling point at 1 atm.

8.9. Chlorine gas is to be heated from 100°C and 1 atm to 200°C.

(a) Calculate the heat input (kW) required to heat a stream of the gas flowing at 5.0 kmol/s at constant pressure.

(b) Calculate the heat input (kJ) required to raise the temperature of 5.0 kmol of chlorine in a closed rigid vessel from 100°C and 1 atm to 200°C. (Suggestion: Evaluate $\Delta \hat{U}$ directly from the result of the first calculation, so that you do not have to perform another integration.) What is the physical significance of the numerical difference between the values calculated in parts (a) and (b)?

(c) To accomplish the heating of part (b), you would actually have to supply an amount of heat to the vessel greater than the amount calculated. Why?

8.10. The heat required to raise the temperature of m (kg) of a liquid from T_1 to T_2 at constant pressure is

$$Q = \Delta H = m \int_{T_1}^{T_2} C_p(T)\, dT \qquad (1)$$

In high school and in first-year college physics courses, the formula is usually given as

$$Q = mC_p\, \Delta T = mC_p(T_2 - T_1) \qquad (2)$$

(a) What assumption about C_p is required to go from Equation 1 to Equation 2?

(b) The heat capacity (C_p) of liquid n-hexane is measured in a **bomb calorimeter**. A small reaction flask (the bomb) is placed in a well-insulated vessel containing 2.00 L of liquid $n\text{-}C_6H_{14}$ at $T = 300$ K. A combustion reaction known to release 16.73 kJ of heat takes place in the bomb, and the subsequent temperature rise of the system contents is measured and found to be 3.10 K. In a separate experiment, it is found that 6.14 kJ of heat is required to raise the temperature of everything in the system except the hexane by 3.10 K. Use these data to estimate $C_p[kJ/(mol\cdot K)]$ for liquid n-hexane at $T \approx 300$ K, assuming that the condition required for the validity of Equation 2 is satisfied. Compare your result with a tabulated value.

8.11. The heat capacities of a substance have been defined as

$$C_v = \left(\frac{\partial \hat{U}}{\partial T}\right)_V, \quad C_p = \left(\frac{\partial \hat{H}}{\partial T}\right)_P$$

Use the defining relationship between $\hat{H}$ and $\hat{U}$ and the fact that $\hat{H}$ and $\hat{U}$ for ideal gases are functions only of temperature to prove that $C_p = C_v + R$ for an ideal gas.

8.12. Ralph Rackstraw, your next-door neighbor, surprised his wife last January by having a hot tub installed in their back yard while she was away on a business trip. It surprised her, all right, but instead of being pleased she was horrified. "Have you lost your mind, Ralph?" she sputtered. "It will cost a fortune to keep this thing hot." "Don't be silly, Josephine," he replied. "It can't cost more than pennies a day, even in the dead of winter." "No way—and when did you become such an expert, anyway?" "I guarantee it will cost nothing—and I don't see your Ph.D. certificate on the kitchen wall either." They argued for awhile and then, remembering your chemical engineering education, came to ask you to settle it for them. You asked a few questions, made several observations, converted everything to metric units, and arrived at the following data, all corresponding to an average outside air temperature of about 5°C.

- The tub holds 1230 liters of water.
- Rackstraw normally keeps the tub temperature at 29°C, raises it to 40°C when he plans to use it, keeps it at 40°C for about one hour, and drops it back to 29°C when he is finished.
- During heating, it takes about three hours for the water temperature to rise from 29°C to 40°C. When the heat is shut off, it takes eight hours for the water temperature to drop back to 29°C.
- Electricity costs 10 cents per kilowatt-hour.

Taking the heat capacity of the tub contents to be that of pure liquid water and neglecting evaporation, answer the following questions.

(a) What is the average rate of heat loss (kW) from the tub to the outside air? (*Hint:* Consider the period when the tub temperature is dropping from 40°C to 29°C.)

(b) At what average rate (kW) does the tub heater deliver energy to the water when raising the water temperature? What is the total quantity of electricity (kW·h) that the heater must deliver during this period? [Consider the result of part (a) when performing the calculation.]

(c) (These answers should settle the argument.) Consider a day in which the tub is used once. Use the results of parts (a) and (b) to estimate the cost ($) of heating the tub from 29°C to 40°C and the cost ($) of keeping the tub at a constant temperature. (There is no cost for the period in which T is dropping.) What is the total daily cost of running the tub? Assume the rate of heat loss is independent of the tub temperature.

(d) The tub lid, which is an insulator, is removed when the tub is in use. Explain how this fact would probably affect your cost estimates in part (c).

8.13. Use tabulated specific enthalpies from Tables B.8 and B.9 to calculate $\Delta \hat{H}$ for the following processes, all taking place at low pressures:

(a) $N_2(25°C) \rightarrow N_2(700°C)$ **(c)** $CO_2(300°C) \rightarrow CO_2(1250°C)$

(b) $H_2(800°F) \rightarrow H_2(77°F)$ **(d)** $O_2(970°F) \rightarrow O_2(0°F)$

8.14. Calculate the rate of cooling (kW) required to bring 300 kg/min of carbon monoxide from 450°C to 50°C, performing the calculation (a) using Table B.2 and (b) using Table B.8. Considering how much simpler the second calculation is, why would you ever use the polynomial formulas of Table B.2 as opposed to tabulated enthalpies to calculate enthalpy changes?

8.15. A stream of water vapor flowing at a rate of 250 mol/h is brought from 600°C and 10 bar to 100°C and 1 atm.

(a) Estimate the required cooling rate (kW) three ways: (i) from the steam tables, (ii) using heat capacity data in Table B.2, and (iii) using specific enthalpy data in Table B.8.

(b) Which of the answers in part (a) is most accurate, and why?

(c) What is the physical significance of the difference between the values calculated with methods (i) and (ii)?

8.16. A stream of air at 77°F and 1.2 atm absolute flowing at a rate of 200 ft³/h is blown through ducts that pass through the interior of a large industrial motor. The air emerges at 500°F. Using tabulated specific enthalpy data, calculate the rate at which the air is removing heat generated by the motor. What assumption have you made about the pressure dependence of the specific enthalpy of air?

8.17. Calculate the heat required to raise 50 kg of solid sodium carbonate (Na_2CO_3) from 10°C to 50°C at 1 atm using
  (a) the true heat capacity of Na_2CO_3, which is 1.14 kJ/(kg·°C).
  (b) a heat capacity estimated by Kopp's rule. Calculate the percentage error in the latter calculation.

8.18. A liquid mixture of 30 wt% acetone and 70 wt% 2-methyl-1-pentanol ($C_6H_{14}O$) is cooled from 45°C to 20°C. Calculate the associated specific enthalpy change in J/g, using Kopp's rule to estimate any heat capacity for which a tabulated value cannot be found. State all assumptions you make.

8.19. A gas mixture contains one-third methane by volume (recall what that means in terms of mole%) and the balance oxygen at 350°C and 3.0 bar. Calculate the specific enthalpy of this stream in kJ/kg (not per kmol) relative to the pure components at 25°C and 1 atm. *State clearly all assumptions.*

8.20. The radiant energy incident on the earth's surface on a sunny day is roughly 900 W/m². Collecting and focusing sunlight and using the focused beam to heat a fluid is an old idea, and as the environmental costs of fossil fuel combustion mount, solar heating becomes an increasingly attractive alternative.

  Suppose a house is to be designed that will have a circulating forced air central heating unit, and solar energy is contemplated as a heat source (backed up with a conventional furnace to be used on cloudy days). If air is to be fed at a rate of 1000 m³/min at 30°C and 1 atm, and the air is to be heated to 55°C before being discharged into the living space, what is the required area of the solar collector plates? Assume 30% of the radiant energy incident on the plates is used to heat the air.

8.21. Propane is to be burned with 15.0% excess air. Before entering the furnace, the air is preheated from 32°F to 575°F. At what rate (Btu/h) must heat be transferred to the air if the feed rate of propane is 1.35×10^5 SCFH [ft³(STP)/h]?

8.22. A fuel gas containing 95 mole% methane and the balance ethane is burned completely with 25% excess air. The stack gas leaves the furnace at 900°C and is cooled to 450°C in a **waste heat boiler**, a heat exchanger in which heat lost by cooling gases is used to produce steam from liquid water for heating, power generation, or process applications.
  (a) Taking as a basis of calculation 100 mol of the fuel gas fed to the furnace, calculate the amount of heat (kJ) that must be transferred from the gas in the waste heat boiler to accomplish the indicated cooling.
  (b) How much saturated steam at 50 bar can be produced from boiler feedwater at 40°C for the same basis of calculation? (Assume all the heat transferred from the gas goes into the steam production.)
  (c) At what rate (kmol/s) must fuel gas be burned to produce 1250 kg steam per hour (an amount required elsewhere in the plant) in the waste heat boiler? What is the volumetric flow rate (m³/s) of the gas leaving the boiler?
  (d) Briefly explain how the waste heat boiler contributes to the plant profitability. (Think about what would be required in its absence.)

8.23. Twenty liters of liquid *n*-propyl benzoate ($C_6H_5CO_2C_3H_7$, SG = 1.021) and 15 liters of liquid benzene are mixed and heated from 25°C to 75°C. Calculate the required heat input (kJ), using Kopp's rule when necessary. State all assumptions you make.

8.24. Propane gas enters a continuous adiabatic heat exchanger[16] at 40°C and 250 kPa and exits at 240°C. Superheated steam at 300°C and 5.0 bar enters the exchanger flowing countercurrently to the propane and exits as a saturated liquid at the same pressure.
  (a) Taking as a basis 100 mol of propane fed to the exchanger, draw and label a process flowchart. Include in your labeling the volume of propane fed (m³), the mass of steam fed (kg), and the volume of steam fed (m³).
  (b) Calculate values of the labeled specific enthalpies in the following inlet–outlet enthalpy table for this process.

---

[16]An adiabatic heat exchanger is one for which no heat is exchanged with the surroundings. All of the heat lost by the hot stream is transferred to the cold stream.

References: $H_2O(l, 0.01°C)$, $C_3H_8(g, 40°C)$

Species	n_{in}	$\hat{H}_{in}$	n_{out}	$\hat{H}_{out}$
C_3H_8	100 mol	$\hat{H}_a$(kJ/mol)	100 mol	$\hat{H}_c$(kJ/mol)
H_2O	m_w(kg)	$\hat{H}_b$(kJ/kg)	m_w(kg)	$\hat{H}_d$(kJ/kg)

(c) Use an energy balance to calculate the required mass feed rate of the steam. Then calculate the volumetric feed ratio of the two streams (m³ steam fed/m³ propane fed). Assume ideal gas behavior for the propane but not the steam and recall that the exchanger is adiabatic.

(d) Calculate the heat transferred from the water to the propane (kJ/m³ propane fed). (*Hint:* Do an energy balance on either the water or the propane rather than on the entire heat exchanger.)

(e) Over a period of time, scale builds up on the heat transfer surface, resulting in a lower rate of heat transfer between the propane and the steam. What changes in the outlet streams would you expect to see as a result of the decreased heat transfer?

8.25. Saturated steam at 300°C is used to heat a countercurrently flowing stream of methanol vapor from 65°C to 260°C in an adiabatic heat exchanger. The flow rate of the methanol is 5500 standard liters per minute, and the steam condenses and leaves the heat exchanger as liquid water at 90°C.

(a) Calculate the required flow rate of the entering steam in m³/min.

(b) Calculate the rate of heat transfer from the water to the methanol (kW).

8.26. An adiabatic membrane separation unit is used to dry (remove water vapor from) a gas mixture containing 10.0 mole% $H_2O(v)$, 10.0 mole% CO, and the balance CO_2. The gas enters the unit at 30°C and flows past a semipermeable membrane. Water vapor permeates through the membrane into an air stream. The dried gas leaves the separator at 30°C containing 2.0 mole% $H_2O(v)$ and the balance CO and CO_2. Air enters the separator at 50°C with an absolute humidity of 0.002 kg H_2O/kg dry air and leaves at 48°C. Negligible quantities of CO, CO_2, O_2, and N_2 permeate through the membrane. All gas streams are at approximately 1 atm.

(a) Draw and label a flowchart of the process and carry out a degree-of-freedom analysis to verify that you can determine all unknown quantities on the chart.

(b) Calculate (i) the ratio of entering air to entering gas (kg humid air/mol gas) and (ii) the relative humidity of the exiting air.

(c) List several desirable properties of the membrane. (Think about more than just what it allows and does not allow to permeate.)

8.27. A gas containing water vapor has a dry-basis composition of 8.5 mole% CO, 10.5% CO_2, 0.5% O_2, and 80.5% N_2. The gas leaves a catalyst regeneration unit at 620°C and 1 atm with a dew point of 57°C at a flow rate of 28.5 SCMH [m³(STP)/h]. Valuable solid catalyst particles entrained in the gas are to be recovered in an electrostatic precipitator, but the gas must first be cooled to 425°C to prevent damage to the precipitator electrodes. The cooling is accomplished by spraying water at 20°C into the gas.

(a) Use simultaneous material and energy balances on the spray cooler to calculate the required water feed rate (kg/h). Treat the spray cooler as adiabatic and neglect the heat transferred from the entrained solid particles as they cool.

(b) In terms that a high school senior could understand, explain the operation of the spray cooler in this problem. (What happens when the cold water contacts the hot gas?)

8.28. On a cold winter day the temperature is 2°C and the relative humidity is 15%. You inhale air at an average rate of 5500 mL/min and exhale a gas saturated with water at body temperature, roughly 37°C. If the mass flow rates of the inhaled and exhaled air (excluding water) are the same, the heat capacities (C_p) of the water-free gases are each 1.05 J/(g·°C), and water is ingested into the body as a liquid at 22°C, at what rate in J/day do you lose energy by breathing? Treat breathing as a continuous process (inhaled air and liquid water enter, exhaled breath exits) and neglect work done by the lungs.

8.29. Seventy-five liters of liquid ethanol at 70.0°C and 55 L of liquid water at 20.0°C are to be mixed in a well-insulated flask. The energy balance for this constant pressure process is $Q = \Delta H$.

(a) Neglecting evaporation and the heat of mixing, estimate the final mixture temperature. (As part of the calculation, use data in Table B.2 to estimate a linear formula for the heat capacity of liquid ethanol.)

(b) If the experiment were actually performed and the final mixture temperature were measured, it would almost certainly not equal the value estimated in part (a). List as many reasons as you can think of. (There are at least seven of them, most involving approximations made in the estimation.)

8.30. A stream of air at 500°C and 835 torr with a dew point of 30°C flowing at a rate of 1515 L/s is to be cooled in a spray cooler. A fine mist of liquid water at 25°C is sprayed into the hot air at a rate of 110.0 g/s and evaporates completely. The cooled air emerges at 1 atm.

(a) Calculate the final temperature of the emerging air stream, assuming that the process is adiabatic. (*Suggestion*: Derive expressions for the enthalpies of dry air and water at the outlet air temperature, substitute them into the energy balance, and use a spreadsheet to solve the resulting fourth-order polynomial equation.)

(b) At what rate (kW) is heat transferred from the hot air feed stream in the spray cooler? What becomes of this heat?

(c) In a few sentences, explain how this process works in terms that a high school senior could understand. Incorporate the results of parts (a) and (b) in your explanation.

8.31. In the manufacture of nitric acid, ammonia and preheated air are mixed to form a gas containing 10.0 mole% NH_3 at 600°C. The ammonia is then catalytically oxidized to form NO_2, which is absorbed in water to form HNO_3. If ammonia enters the gas blending unit at 25°C at a rate of 520 kg/h and heat is lost from the mixer to its surroundings at a rate of 7.00 kW, determine the temperature to which the air must be preheated. (See Example 8.3-6.)

8.32. A natural gas containing 95 mole% methane and the balance ethane is burned with 20.0% excess air. The stack gas, which contains no unburned hydrocarbons or carbon monoxide, leaves the furnace at 900°C and 1.2 atm and passes through a heat exchanger. The air on its way to the furnace also passes through the heat exchanger, entering it at 20°C and leaving it at 245°C.

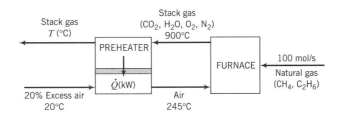

(a) Taking as a basis 100 mol/s of the natural gas fed to the furnace, calculate the required molar flow rate of air, the molar flow rate and composition of the stack gas, the required rate of heat transfer in the preheater, $\dot{Q}$ (write an energy balance on the air), and the temperature at which the stack gas leaves the preheater (write an energy balance on the stack gas). *Note*: The problem statement does not give you the fuel feed temperature. Make a reasonable assumption, and state why your final results should be nearly independent of what you assume.

(b) What would $\dot{Q}$ be if the actual feed rate of the natural gas were 350 SCMH [standard cubic meters per hour, $m^3(STP)/h$]? Scale up the flowchart of part (a) rather than repeating the entire calculation.

8.33. The heat capacity at constant pressure of a gas is determined experimentally at several temperatures, with the following results:

$T(°C)$	0	100	200	300	400	500	600
$C_p[J/(mol·°C)]$	33.5	35.1	36.7	38.4	40.2	42.0	43.9

(a) Calculate the heat (kW) required to raise 150 mol/s of the gas from 0°C to 600°C, using Simpson's rule (Appendix A.3) to integrate the tabulated heat capacities.

(b) Use the method of least squares (Appendix A.1) to derive a linear expression for $C_p(T)$ in the range 0°C to 600°C, and use this expression to estimate once again the heat (kW) required to raise 150 mol/s of the gas from 0°C to 600°C. If the estimates differ, in which one would you have more confidence, and why?

*8.34. As part of a design calculation, you must evaluate an enthalpy change for an obscure organic vapor that is to be cooled from 1800°C to 150°C in a heat exchanger. You search through all the standard references for tabulated enthalpy or heat capacity data for the vapor but have no luck at all, until you finally stumble on an article in the May 1922 *Antarctican Journal of Obscure Organic Vapors* that contains a plot of C_p[cal/(g·°C)] on a logarithmic scale versus $[T(°C)]^{1/2}$ on a linear scale. The plot is a straight line through the points ($C_p = 0.329$, $T^{1/2} = 7.1$) and ($C_p = 0.533$, $T^{1/2} = 17.3$).

    (a) Derive an equation for C_p as a function of T.

    (b) Suppose the relationship of part (a) turns out to be

$$C_p = 0.235 \exp[0.0473T^{1/2}]$$

and that you wish to evaluate

$$\Delta \hat{H}(\text{cal/g}) = \int_{1800°C}^{150°C} C_p \, dT$$

First perform the integration analytically, using a table of integrals if necessary; then write a spreadsheet or computer program to do it using Simpson's rule (Appendix A.3). Have the program evaluate C_p at 11 equally spaced points from 150°C to 1800°C, estimate and print the value of ΔH, and repeat the calculation using 101 points. What can you conclude about the accuracy of the numerical calculation?

8.35. A stream of ethylene glycol vapor at its normal boiling point and 1 atm flowing at a rate of 175 kg/min is to be condensed at constant pressure. The product stream from the condenser is liquid glycol at the condensation temperature.

    (a) Using data in Table B.1, calculate the rate (kW) at which heat must be transferred from the condenser.

    (b) If heat were transferred at a lower rate than that calculated in part (a), what would the state of the product stream be? (Deduce as much as you can about the phase and the temperature of the stream.)

    (c) If heat were transferred at a higher rate than that calculated in part (a), what could you deduce about the state of the product stream? Sketch a phase diagram (see Figure 6.1-1 on p. 241) and use it to explain your answer.

8.36. **(a)** Determine the specific enthalpy (kJ/mol) of *n*-hexane vapor at 200°C and 2.0 atm relative to *n*-hexane liquid at 20°C and 1.0 atm, assuming ideal gas behavior for the vapor. Show clearly the process path you construct for this calculation and give the enthalpy changes for each step. State where you used the ideal gas assumption.

    (b) What is the enthalpy of *n*-hexane liquid at 20°C and 1.0 atm relative to *n*-hexane vapor at 200°C and 2.0 atm? (This part should not take much time to complete.)

    (c) Beginning with the value of $\hat{H}$ calculated in part (a) and still assuming ideal gas behavior, determine the specific internal energy of the vapor at 200°C and 2.0 atm. Once again, state where you used the ideal gas assumption.

8.37. Calculate the heat of vaporization of water (kJ/mol) at 50°C and low pressures from the tabulated heat of vaporization in Table B.1 and data in Table B.2 and Table B.8. Show clearly the process path you construct for the calculation. Compare your answer with the value of $\Delta \hat{H}_v(50°C)$ given in Table B.5 (convert it to kJ/mol for the comparison). What might account for the difference between the two values?

8.38. Benzene vapor at 580°C is cooled and converted to a liquid at 25°C in a continuous condenser. The condensate is drained into 1.75-m³ drums, each of which takes 2.0 minutes to fill. Calculate the rate (kW) at which heat is transferred from the benzene in the condenser.

8.39. In gas adsorption a vapor is transferred from a gas mixture to the surface of a solid. (See Section 6.7.) An approximate but useful way of analyzing adsorption is to treat it simply as condensation of vapor on a solid surface.

    Suppose a nitrogen stream at 35°C and 1 atm containing carbon tetrachloride with a 15% relative saturation is fed at a rate of 10.0 mol/min to a 6-kg bed of activated carbon. The temperature and pressure of the gas do not change appreciably from the inlet to the outlet of the bed, and there

---

*Computer problem.

is no CCl_4 in the gas leaving the adsorber. The carbon can adsorb 40% of its own mass of carbon tetrachloride before becoming saturated. Neglecting the effect of temperature on the heat of vaporization of CCl_4, estimate the rate at which heat must be removed from the adsorber (kJ/min) to keep the process isothermal and the time (min) it will take to saturate the bed.

8.40. If carbon dioxide is cooled at 1 atm, it condenses directly to a solid (**dry ice**) at $-78.4°C$. The heat of sublimation at this temperature is $\Delta \hat{H}_{sub}(-78.4°C) = 6030$ cal/mol.

  (a) Calculate the heat removal rate (kW) required to produce 300 kg/h of dry ice at 1 atm and $-78.4°C$ if $CO_2(v)$ at 20°C is the feed.

  (b) Suppose the process is carried out at 9.9 atm instead of 1 atm with the same initial and final temperatures. Referring to Figure 6.1-1*b* on p. 241, write an expression for the required heat removal rate in terms of heat capacities and latent heats of CO_2 in different phases.

8.41. Molten sodium chloride is to be used as a constant-temperature bath for a high-temperature chemical reactor. Two hundred kilograms of solid NaCl at 300 K is charged into an insulated vessel, and a 3000 kW electrical heater is turned on, raising the salt to its melting point of 1073 K and melting it at a constant pressure of 1 atm.

  (a) The heat capacity (C_p) of solid NaCl is 50.41 J/(mol·K) at $T = 300$ K, and 53.94 J/(mol·K) at $T = 500$ K, and the heat of fusion of NaCl at 1073 K is 30.21 kJ/mol. Use these data to determine a linear expression for $C_p(T)$ and to calculate $\Delta \hat{H}$(kJ/mol) for the transition of NaCl from a solid at 300 K to a liquid at 1073 K.

  (b) Write and solve the energy balance equation for this closed system isobaric process to determine the required heat input in kilojoules.

  (c) If 85% of the full power of 3000 kW goes into heating and melting the salt, how long does the process take?

8.42. Estimate the heat of vaporization of ethyl benzene at its normal boiling point using Trouton's rule and Chen's rule and compare the results with a tabulated value of this quantity. Then estimate $\Delta \hat{H}_v$ at 100°C using Watson's correlation.

8.43. You are writing energy balances for a compound for which you cannot find heat capacity or latent heat data. All you know about the material are its molecular formula ($C_7H_{12}N$) and that it is a liquid at room temperature and has a normal boiling point of 200°C. Use this information to estimate the enthalpy of the vapor of this substance at 200°C relative to the liquid at 25°C. (Recall Section 8.3c.)

8.44. Estimate the heat of vaporization (kJ/mol) of benzene at a pressure of 100 mm Hg, using each of the following correlations and data:

  (a) The heat of vaporization at the normal boiling point given in Table B.1, the boiling point at 100 mm Hg as determined from the Antoine equation, and Watson's correlation.

  (b) The Clausius–Clapeyron equation and the boiling points at 50 mm Hg and 150 mm Hg as determined from the Antoine equation.

  (c) The heat of vaporization at the normal boiling point given in Table B.1, the boiling point at 100 mm Hg as determined from the Antoine equation, and heat capacity data given in Table B. 2.

8.45. A stream of pure cyclopentane vapor flowing at a rate of 1550 L/s at 150°C and 1 atm enters a cooler in which 55% of the feed is condensed at constant pressure.

  (a) What is the temperature at the condenser outlet? Explain how you know (a single sentence should suffice).

  (b) Prepare and fill in an inlet–outlet enthalpy table and calculate the required cooling rate in kW.

8.46. Humid air at 50°C and 1.0 atm with 2°C of superheat is fed to a condenser. Gas and liquid streams leave the condenser in equilibrium at 20°C and 1 atm.

  (a) Assume a basis of calculation of 100 mol inlet air, draw and label a flowchart (including Q in the labeling), and carry out a degree-of-freedom analysis to verify that all labeled variables can be determined.

  (b) Write in order the equations you would solve to calculate the mass of water condensed (kg) per cubic meter of air fed to the condenser. Circle the unknown variable for which you would solve each equation. Do not do any of the calculations.

  (c) Prepare an inlet–outlet enthalpy table, inserting labels for unknown specific enthalpies ($\hat{H}_1$, $\hat{H}_2, \ldots$). Write expressions for the labeled specific enthalpies, substituting values or formulas for heat capacities and latent heats but not calculating the values of the specific enthalpies. Then write an expression for the rate at which heat must be transferred from the unit (kJ) per cubic meter of air fed to the condenser.

 (d) Solve your equations by hand to calculate kg H_2O condensed/m^3 air fed and kJ transferred/m^3 air fed.

 (e) Use an equation-solving program to perform the calculations of part (d).

 (f) What cooling rate (kW) would be required to process 250 m^3 air fed/h?

8.47. An air conditioner cools 226 m^3/min of humid air at 36°C and 98% relative humidity to 10°C.

 (a) Do a degree-of-freedom analysis to prove that enough information is available to determine the required cooling duty (rate of heat transfer).

 (b) Calculate the rate of condensation of water in the unit and the cooling duty in tons (1 ton = 12,000 Btu/h).

8.48. A gas stream containing *n*-hexane in nitrogen with a relative saturation of 90% is fed to a condenser at 75°C and 3.0 atm absolute. The product gas emerges at 0°C and 3.0 atm at a rate of 746.7 m^3/h. Calculate the percentage condensation of hexane (moles condensed/mole fed) and the rate (kW) at which heat must be transferred from the condenser.

8.49. A gas stream containing acetone in air flows from a solvent recovery unit at a rate of 142 L/s at 150°C and 1.3 atm. The stream flows into a condenser which liquefies most of the acetone, and the liquid and gas outlet streams are in equilibrium at −18°C and 5.0 atm. Shaft work is delivered to the system at a rate of 25.2 kW to achieve the compression from 1.3 atm to 5.0 atm. To determine the condenser feed stream composition, a 3.00-liter sample of the gas is taken and cooled to a temperature at which essentially all the acetone in the sample is recovered as a liquid. The liquid is poured into an empty flask with a mass of 4.017 g. The flask containing the liquid acetone is weighed and found to have a mass of 4.973 g.

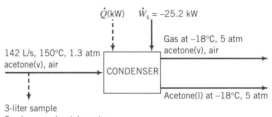

 (a) Carry out a degree-of-freedom analysis to show that enough information is available to determine the compositions of all streams and the required heat transfer rate.

 (b) Write out a complete set of equations for the molar flow rates of all streams, the mole fractions of acetone in the feed and product gas streams, and the rate (kW) at which heat must be removed in the condenser. *Do no calculations.*

 (c) Solve the equations of part (b) by hand.

 (d) Solve the equations of part (b) using an equation-solving program.

8.50. A mixture of *n*-hexane vapor and air leaves a solvent recovery unit and flows through a 70-cm diameter duct at a velocity of 3.00 m/s. At a sampling point in the duct the temperature is 40°C, the pressure is 850 mm Hg, and the dew point of the sampled gas is 25°C. The gas is fed to a condenser in which it is cooled at constant pressure, condensing 60% of the hexane in the feed.

 (a) Perform a degree-of-freedom analysis to show that enough information is available to calculate the required condenser outlet temperature (°C) and cooling rate (kW).

 (b) Perform the calculations.

 (c) If the feed duct diameter were 35 cm for the same molar flow rate of the feed gas, what would the gas velocity be?

8.51. An equimolar liquid mixture of *n*-pentane and *n*-hexane at 80°C and 5.00 atm is fed into a flash evaporator at a rate of 100.0 mol/s. When the feed is exposed to the reduced pressure in the evaporator, a substantial amount is vaporized. The temperature in the tank is maintained at 65°C by adding heat. The vapor and liquid phases, which are in equilibrium with each other, are separated and discharged

---

*Computer problems.

as separate streams. The liquid product stream contains 41.0 mole% pentane. A flowchart and an inlet–outlet enthalpy table for the process are given below.

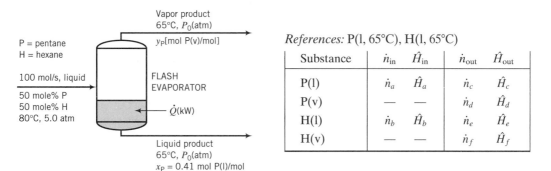

References: P(l, 65°C), H(l, 65°C)

Substance	$\dot{n}_{in}$	$\hat{H}_{in}$	$\dot{n}_{out}$	$\hat{H}_{out}$
P(l)	$\dot{n}_a$	$\hat{H}_a$	$\dot{n}_c$	$\hat{H}_c$
P(v)	—	—	$\dot{n}_d$	$\hat{H}_d$
H(l)	$\dot{n}_b$	$\hat{H}_b$	$\dot{n}_e$	$\hat{H}_e$
H(v)	—	—	$\dot{n}_f$	$\hat{H}_f$

(a) Using Raoult's law for vapor–liquid equilibrium calculations, calculate (i) the system pressure, P_0(atm), (ii) the mole fraction of pentane in the vapor product, y_P, (iii) the volumetric flow rate of the vapor product, $\dot{V}$(L/s), and (iv) the fractional vaporization of pentane, f (mol vaporized/mol fed).

(b) Determine values for all the $\dot{n}$'s and $\hat{H}$'s in the enthalpy table and calculate the required rate of heat addition to the evaporator, $\dot{Q}$(kW).

8.52. A liquid stream containing 50.0 mole% benzene and the balance toluene at 25°C is fed to a continuous single-stage evaporator at a rate of 1320 mol/s. The liquid and vapor streams leaving the evaporator are both at 95.0°C. The liquid contains 42.5 mole% benzene and the vapor contains 73.5 mole% benzene.

(a) Calculate the heating requirement for this process in kW.

(b) Using Raoult's law (Section 6.4b) to describe the equilibrium between the vapor and liquid outlet streams, determine whether or not the given benzene analyses are consistent with each other. If they are, calculate the pressure (torr) at which the evaporator must be operating; if they are not, give several possible explanations for the inconsistency.

8.53. The off-gas from a reactor in a process plant in the heart of an obscure banana republic has been condensing and plugging up the vent line, causing a dangerous pressure buildup in the reactor. Plans have been made to pass the gas directly from the reactor into a cooling condenser in which the gas and liquid condensate will be brought to 25°C.

You have been called in as a consultant to aid in the design of this unit. Unfortunately, the chief (and only) plant engineer has disappeared and nobody else in the plant can tell you what the off-gas is (or what anything else is, for that matter).

However, a job is a job, and you set out to do what you can. You find an elemental analysis in the engineer's notebook indicating that the gas formula is $C_5H_{12}O$. On another page of the notebook, the off-gas flow rate is given as 235 m³/h at 116°C and 1 atm. You take a sample of the gas and cool it to 25°C, where it proves to be a solid. You then heat the solidified sample at 1 atm and note that it melts at 52°C and boils at 113°C. Finally, you make several assumptions and estimate the heat removal rate in kW required to bring the off-gas from 116°C to 25°C. What is your result?

8.54. A sheet of cellulose acetate film containing 5.00 wt% liquid acetone enters an adiabatic dryer where 90% of the acetone evaporates into a stream of dry air flowing over the film. The film enters the dryer at $T_{f1} = 35°C$ and leaves at T_{f2}(°C). The air enters the dryer at T_{a1}(°C) and 1.01 atm and exits the dryer at $T_{a2} = 49°C$ and 1 atm with a relative saturation of 40%. C_p may be taken to be 1.33 kJ/(kg·°C) for dry film and 0.129 kJ/(mol·°C) for liquid acetone. Make a reasonable assumption regarding the heat capacity of dry air. The heat of vaporization of acetone may be considered independent of temperature. Take a basis of 100 kg film fed to the dryer for the requested calculations.

(a) Estimate the feed ratio [liters dry air (STP)/kg dry film].

(b) Derive an expression for T_{a1} in terms of the film temperature change, $(T_{f2} - 35)$, and use it to answer parts (c) and (d).

(c) Calculate the film temperature change if the inlet air temperature is 120°C.

(d) Calculate the required value of T_{a1} if the film temperature falls to 34°C, and the value if it rises to 36°C.

(e) If you solved parts (c) and (d) correctly, you found that even though the air temperature is consistently higher than the film temperature in the dryer, so that heat is always transferred from the air to the film, the film temperature can drop from the inlet to the outlet. How is this possible?

8.55. Saturated propane vapor at 2.00×10^2 psia is fed to a well-insulated heat exchanger at a rate of 3.00×10^3 SCFH (standard cubic feet per hour). The propane leaves the exchanger as a saturated liquid (i.e., a liquid at its boiling point) at the same pressure. Cooling water enters the exchanger at 70°F, flowing cocurrently (in the same direction) with the propane. The temperature difference between the outlet streams (liquid propane and water) is 15°F.

(a) What is the outlet temperature of the water stream? (Use the Cox chart on p. 247.) Is the outlet water temperature less than or greater than the outlet propane temperature? Briefly explain.

(b) Estimate the rate (Btu/h) at which heat must be transferred from the propane to the water in the heat exchanger and the required flow rate (lb_m/h) of the water. (You will need to write two separate energy balances.) Assume the heat capacity of liquid water is constant at 1.00 Btu/($lb_m \cdot °F$) and neglect heat losses to the outside and the effects of pressure on the heat of vaporization of propane.

8.56. An aqueous slurry at 30°C containing 20.0 wt% solids is fed to an evaporator in which enough water is vaporized at 1 atm to produce a product slurry containing 35.0 wt% solids. Heat is supplied to the evaporator by feeding saturated steam at 1.6 bar absolute into a coil immersed in the liquid. The steam condenses in the coil, and the slurry boils at the normal boiling point of pure water. The heat capacity of the solids may be taken to be half that of liquid water.

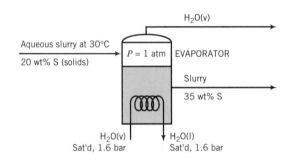

(a) Calculate the required steam feed rate (kg/h) for a slurry feed rate of 1.00×10^3 kg/h.

(b) **Vapor recompression** is often used in the operation of an evaporator. Suppose that the vapor (steam) generated in the evaporator described above is compressed to 1.6 bar and simultaneously heated to the saturation temperature at 1.6 bar, so that no condensation occurs. The compressed steam and additional saturated steam at 1.6 bar are then fed to the evaporator coil, in which isobaric condensation occurs. How much additional steam is required?

(c) What more would you need to know to determine whether or not vapor recompression is economically advantageous in this process?

8.57. A mixture that contains 46 wt% acetone (CH_3COCH_3), 27% acetic acid (CH_3COOH), and 27% acetic anhydride [($CH_3CO)_2O$] is distilled at $P = 1$ atm. The feed enters the distillation column at $T = 348$ K at a rate of 15,000 kg/h. The distillate (overhead product) is essentially pure acetone, and the bottoms product contains 1% of the acetone in the feed.

The vapor effluent from the top of the column enters a condenser at 329 K and emerges as a liquid at 303 K. Half of the condensate is withdrawn as the overhead product, and the remainder is refluxed back to the column. The liquid leaving the bottom of the column goes into a steam-heated reboiler, in which it is partially vaporized. The vapor leaving the reboiler is returned to the column at a temperature of 398 K, and the residual liquid, also at 398 K, constitutes the bottoms product. A flowchart of the process and thermodynamic data for the process materials follow.

(a) Calculate the molar flow rates and compositions of the product streams.

(b) Calculate the condenser cooling requirement $\dot{Q}_c$(kJ/h).

(c) Use an overall energy balance to determine the reboiler heating requirement $\dot{Q}_r$(kJ/h).

(d) If the reboiler heat is provided by the condensation of saturated steam at 10 bar gauge, at what rate must steam be fed?

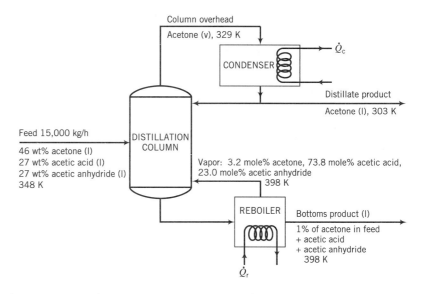

Thermodynamic Data   (All temperatures are in kelvin)

Acetone: $C_{pl} = 2.30$ kJ/(kg·K)

$$C_{pv}[\text{kJ/(kg·K)}] = 0.459 + 3.15 \times 10^{-3}T - 0.790 \times 10^{-6}T^2$$

$$\Delta\hat{H}_v(329 \text{ K}) = 520.6 \text{ kJ/kg}$$

Acetic acid: $C_{pl} = 2.18$ kJ/(kg·K)

$$C_{pv}[\text{kJ/(kg·K)}] = 0.688 + 1.87 \times 10^{-3}T - 0.411 \times 10^{-6}T^2$$

$$\Delta\hat{H}_v(391 \text{ K}) = 406.5 \text{ kJ/kg}$$

Acetic anhydride: $C_{pl}[\text{kJ/(kg·K)}] = ?$ (Estimate it—see Section 8.3c.)

$$C_{pv}[\text{kJ/(kg·K)}] = 0.751 + 1.34 \times 10^{-3}T - 0.046 \times 10^{-6}T^2$$

$$\Delta\hat{H}_v(413 \text{ K}) = ?$$ (Estimate it—see Section 8.4b.)

8.58. A **double-effect evaporator** (two evaporators in series) is used to produce fresh water from seawater containing 3.5 wt% dissolved salts. A flowchart for the process is shown here.

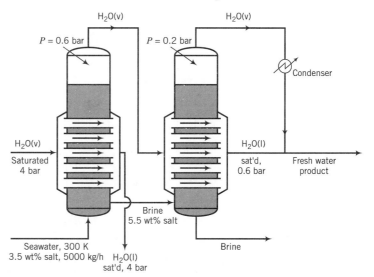

Seawater enters the first effect at 300 K at a rate of 5000 kg/h, and saturated steam at 4.00 bar absolute is fed into a tube bundle in the first effect. The steam condenses at 4.00 bar, and the condensate is withdrawn at the saturation temperature corresponding to this pressure.

The heat given up by the steam condensing in the tubes causes water to evaporate from the brine solution at the pressure of 0.60 bar maintained in the effect. The exiting brine contains 5.5 wt% salt. The steam generated in the first effect is fed to a tube bundle in the second effect. The condensate from the bundle and the steam generated in the second effect at a pressure of 0.20 bar constitute the fresh water produced in the process.

In solving the problems to be given, assume that the brine solutions in both effects have the physical properties of pure water and that the effects operate adiabatically.

(a) Draw and label a flowchart of this process, giving the temperature and specific enthalpy of each stream.

(b) At what rate must steam be fed to the first effect?

(c) What is the production rate of fresh water? What is the salt concentration (weight percent) of the final brine solution?

(d) Why is it necessary that the pressure decrease from one effect to the next?

(e) Suppose a single-effect evaporator were used, operating at $P = 0.20$ bar. Calculate the feed rate of saturated steam at $P = 4.00$ bar that would be required to achieve the same production rate of fresh water. What more would you need to know to determine which process is more economical?

*8.59. Seawater containing 3.5 wt% dissolved salts is to be desalinated in an adiabatic six-effect evaporator. (See Problem 8.58.) Backward feed is to be used: the seawater is fed to the last evaporator, and successively concentrated brine solutions flow countercurrent to the direction of flow of steam from one effect to the next. Saturated steam at $P = 2$ bar is fed to the tube bundle in the first effect. The operating pressures in bars of the six effects are, respectively, 0.9, 0.7, 0.5, 0.3, 0.2, and 0.1. The brine leaving the first effect contains 30 wt% salt. The flowchart shows Effects 1, 5, and 6.

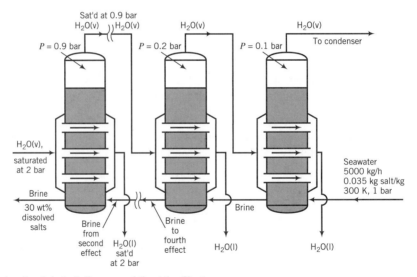

Following is a labeled diagram of the ith effect:

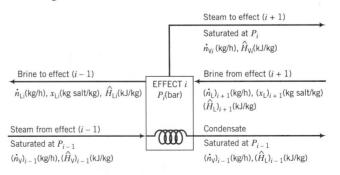

---

*Computer problem.

In terms of the variables defined in this diagram,

$$\dot{n}_{L7} = 5000 \text{ kg/h}$$

$$x_{L7} = 0.035 \text{ kg salt/kg}$$

$$x_{L1} = 0.30 \text{ kg salt/kg}$$

$$\dot{n}_{V0} = \text{feed rate of steam to the first effect}$$

(a) Use a salt balance to calculate $\dot{n}_{L1}$. Then use this result to determine how much fresh water is produced in the process.

(b) Prepare a table as follows:

	P (bar)	T (K)	$\dot{n}_L$ (kg/h)	x_L	$\hat{H}_L$ (kJ/kg)	$\dot{n}_V$ (kg/h)	$\hat{H}_V$ (kJ/kg)
Fresh steam	2.0		—	—	—		
Effect 1	0.9			0.30			
2	0.7						
3	0.5						
4	0.3						
5	0.2						
6	0.1						
(7)	1.0	300	5000	0.035		—	—

Fill in all *known* variable values (do not calculate any yet), including values obtained from the steam tables, *assuming that the physical properties of the brine solution are those of pure water.*

(c) Show that the following equations can be derived from balances:

$$x_{Li} = (\dot{n}_L)_{i+1}(x_L)_{i+1}/\dot{n}_{Li} \tag{1}$$

$$(\dot{n}_V)_{i-1} = \frac{\dot{n}_{Vi}\hat{H}_{Vi} + \dot{n}_{Li}\hat{H}_{Li} - (\dot{n}_L)_{i+1}(\hat{H}_L)_{i+1}}{(\hat{H}_V)_{i-1} - (\hat{H}_L)_{i-1}} \tag{2}$$

$$(\dot{n}_L)_{i-1} = \dot{n}_{Li} - (\dot{n}_V)_{i-1} \tag{3}$$

(d) The missing entries in the table of part (b) may be determined by solving Equations 1 through 3 for all effects simultaneously, but this is a cumbersome task. Instead, a relatively simple trial-and-error method may be used, which consists of assuming a value for $\dot{n}_{V6}$, calculating $\dot{n}_{L6} = \dot{n}_{L7} - \dot{n}_{V6}$, and successively solving Equations 1 through 3 for x_{L6}, $\dot{n}_{V5}$, $\dot{n}_{L5}$, then x_{L5}, $\dot{n}_{V4}$, $\dot{n}_{L4}$, and so on, until finally x_{L1} is calculated. If the calculated value of x_{L1} is less than the known value of 0.3, a higher value of $\dot{n}_{V6}$ is assumed and the process is repeated; if $x_{L1} > 0.3$, a lower value of $\dot{n}_{V6}$ is assumed. A reasonable first guess for $\dot{n}_{V6}$ is 1/6 of the total rate of evaporation [determined in part (a)].

Construct a spreadsheet program to perform these calculations, using the goalseek tool to carry out the trial-and-error search for the correct value of $\dot{n}_{V6}$. Fill in the table of part (b).

(e) Solve the equations of part (c) for all six effects using an equation-solving program. Fill in the table of part (b).

8.60. A liquid is placed in a well-insulated container, which is then sealed. Initially, the container and its contents (the liquid and pure nitrogen) are at 93°C and 1 atm; the liquid volume is 70 cm³, and the gas volume is 3.00 L. The liquid partially evaporates, and the system cools down and eventually comes to thermal equilibrium at 85°C with liquid still present. Physical property data for the liquid and its vapor are:

$$\Delta\hat{U}_v = 20 \text{ kcal/mol at } 90°C$$

$$(C_p)_{\text{liq}} = 20 \text{ cal/(mol·°C)}$$

$$(C_p)_{\text{vap}} = 10 \text{ cal/(mol·°C)}$$

$$(SG)_{\text{liq}} = 0.90$$

$$MW = 42$$

(a) Determine $(C_v)_{\text{liq}}$ and $(C_v)_{\text{vap}}$. (See Equations 8.3-11 and 8.3-12.)
(b) Draw and label a flowchart for this closed system process, and write and simplify the energy balance equation, assuming adiabatic operation.
(c) Use the energy balance to calculate the mass of liquid that evaporates, taking 4.97 cal/(mol·°C) as the heat capacity of nitrogen.
(d) Calculate the vapor pressure of the liquid at 85°C, assuming that the gas volume remains constant at 3.00 L.

8.61. A small pharmaceutical firm plans to manufacture a new drug and has hired you as a consultant to design a condenser to remove the drug from a gas–vapor mixture. The mixture, which contains 20 mole% of the drug and the balance nitrogen, will be fed to the condenser at 510 K and 1 atm at a rate of 3.5 L/s. Of the drug fed to the unit, 90% must be condensed. No physical property data are available for the drug, and part of your job is to acquire the data needed to design the condenser. The company has sent you a large sample of the liquid drug for this purpose.

You acquire an insulated 2.000-liter container with a known heat capacity and a built-in electrical heating coil that can deliver a known heat input to the contents of the container. A calibrated thermocouple is used to measure the temperature in the vessel, and the pressure is measured with a mercury manometer.

You carry out a series of experiments on a day when atmospheric pressure is 763 mm Hg.

Experiment 1. Fill the container with the liquid, then seal and weigh.

$$\text{mass of container} + \text{liquid} = 4.4553 \text{ kg}$$
$$\text{mass of evacuated container} = 3.2551 \text{ kg}$$

Next, starting at each of two temperatures (T_0), add a fixed quantity of heat to the liquid, observe the final temperature (T_f), and subtract the heat absorbed by the container from the total heat input to determine the amount of the heat added to the liquid, Q_a.

$$T_0 = 283.0 \text{ K}, Q_a = 800.0 \text{ J} \Longrightarrow T_f = 285.4 \text{ K}$$
$$T_0 = 330.0 \text{ K}, Q_a = 800.0 \text{ J} \Longrightarrow T_f = 332.4 \text{ K}$$

Assume that the liquid heat capacity may be expressed as a linear function of temperature ($C_v = aT + b$) when analyzing these results.

Experiment 2. Pour a small quantity of the drug into the container, place the container in a liquid nitrogen bath to freeze the drug, evacuate all of the air, and seal the container. Weigh the container after it comes back to room temperature.

$$\text{mass of container} + \text{drug} = 3.2571 \text{ kg}$$

Next heat the sealed container until all of the liquid evaporates, and repeat Experiment 1.

$$T_0 = 363.0 \text{ K}, h_{\text{manometer}} = -500 \text{ mm}, Q_a = 1.30\text{J} \Longrightarrow T_f = 366.9 \text{ K}$$
$$T_0 = 490.0 \text{ K}, h_{\text{manometer}} = -408 \text{ mm}, Q_a = 1.30\text{J} \Longrightarrow T_f = 492.7 \text{ K}$$

Assume that the vapor heat capacity may be expressed as a linear function of temperature when analyzing these results.

Experiment 3. Fill approximately half the container with the drug, freeze, evacuate the air, and seal. Measure the pressure at several temperatures, verifying that liquid is present in the container at each temperature.

$$T = 315.0 \text{ K}, h_{\text{manometer}} = -564 \text{ mm}$$
$$T = 334.0 \text{ K}, h_{\text{manometer}} = -362 \text{ mm}$$
$$T = 354.0 \text{ K}, h_{\text{manometer}} = -2 \text{ mm}$$
$$T = 379.0 \text{ K}, h_{\text{manometer}} = +758 \text{ mm}$$

(a) Using the given data, determine the following physical properties of the drug: (i) liquid specific gravity, (ii) molecular weight, (iii) linear expressions for the heat capacities at constant volume [in J/(mol·K)] for both the liquid and vapor [$C_v = a + bT(\text{K})$], (iv) linear expressions for C_p for both liquid and vapor, (v) a Clausius–Clapeyron expression for $p^*(T)$, (vi) the normal boiling point, and (vii) the heat of vaporization (in J/mol) at the normal boiling point.
(b) Calculate the required condenser temperature, assuming operation at 1 atm.

(c) Calculate the rate at which heat must be removed in the condenser, taking the heat capacity of nitrogen to be constant at 29.0 J/(mol·K).

8.62. Freeze drying is a technique for dehydrating substances at low temperatures, thereby avoiding the degradation that may accompany heating. The material to be dried is cooled to a temperature at which all of the water present turns to ice. The frozen substance is then placed in a vacuum chamber and may also be subjected to radiant or microwave heating; the ice in the food sublimates, and the vapor is carried off by the vacuum pump.

Steaks are to be freeze-dried in a heated chamber at 1 torr (1 mm Hg). The steaks, which contain 72% water by mass, enter the chamber at $-26°C$ at a rate of 50 kg/min. Of the water entering with the steaks, 96% leaves as a vapor at 60°C; the remainder leaves as a liquid with the steaks at 50°C.

(a) Use the heat capacity data given below and additional tabulated data for water to calculate the required heat input in kilowatts.

$$(C_p)_{ice} = 2.17 \text{ J/(g·°C)}$$

$$(C_p)_{dry \ meat} = 1.38 \text{ J/(g·°C)}$$

(b) When large temperature changes are not involved in a phase-change operation, a reasonable estimate of the required heat transfer rate may be obtained by neglecting contributions of temperature changes to the overall process enthalpy change (i.e., by taking only phase changes into account). Moreover, it is often reasonable to use any available values of latent heats, neglecting their dependence on temperature and pressure. In the case of the freeze-drying process, the approximation might be to calculate only the heat needed to melt all the water and vaporize 96% of it, using latent heats at the normal melting and boiling points (Table B.1) and neglecting the heat required to raise the temperature of the meat and water. What percentage error in the calculated value of $\dot{Q}$ would result from this approximation? Take the value determined in part (a) to be exact.

8.63. Freeze concentration is used to produce a fruit-juice concentrate. A stream of fresh juice containing 12 wt% soluble solids in water at 20°C is combined with a recycle stream to form a preconcentrate, which is fed to a crystallizer. The mixture is cooled in the crystallizer to $-7°C$, thereby crystallizing 20,000 kg/h of ice. A slurry leaves the crystallizer containing 10 wt% ice and is fed to a filter. The filtrate, which contains 45 wt% dissolved solids, is removed as the process product. The remaining slurry, which contains all the ice and some concentrate (also containing 45% dissolved solids), is sent to a separator that cleanly removes all of the ice. The residual liquid is the recycle stream that combines with the fresh feed to form the preconcentrate.

(a) Determine the rates (kg/h) at which fresh fruit juice is fed and concentrate is produced, and the mass flow rate (kg/h) and solids concentration of the preconcentrate.

(b) Calculate the cooling requirement (kW) for the freezer, assuming that the temperature of the recycle stream is 0°C and the heat capacity of all solutions is 4.0 kJ/(kg·°C).

8.64. A mixture containing 35.0 mole% n-butane and the balance isobutane at 10°C enters a heat exchanger at a rate of 24.5 kmol/h and a pressure high enough for the mixture to be a liquid. The exchanger has been designed to heat and vaporize the liquid and heat the vapor mixture to 180°C. The heating fluid is a high molecular weight liquid with a constant heat capacity $C_p = 2.62$ kJ/(kg·°C). It enters the exchanger at 215°C and flows countercurrently to the hydrocarbon mixture.

(a) Estimate the minimum pressure (bar) required for the hydrocarbon feed to be a liquid. Use Raoult's law and the Cox chart (Figure 6.1-4) in your calculation.

(b) Assuming that the heat capacities and heats of vaporization of n-butane and isobutane are independent of pressure (so that the values in Tables B.1 and B.2 may be used), calculate the enthalpy change $\Delta\dot{H}$ (kJ/h) undergone by the hydrocarbon mixture in the heat exchanger. Show the process paths you use for n-butane and i-butane in your calculation. (*Hint:* Since you don't have heat capacities for liquid n-butane and i-butane in this text, use process paths that don't require them.)

(c) According to the heat exchanger design calculations, the heating fluid exit temperature should be 45°C. Assuming that all the heat lost by the heating fluid is transferred to the hydrocarbon mixture, what is the required mass flow rate of the heating fluid, $\dot{m}_{hf}$(kg/h)?

(d) When the heat exchanger is run with $\dot{m}_{hf}$ equal to the value calculated in part (b), the exit temperature of the hydrocarbon mixture is measured and found to be only 155°C instead of the design value of 180°C. The process operator observes that the outside of the exchanger is hot to the touch, indicating that some of the heat lost by the heating fluid is escaping into the plant instead of being transferred to the hydrocarbon mixture. After discussing the situation with a production engineer, the operator gradually increases the flow rate of the heating fluid while continuing to monitor the outlet temperature of the hydrocarbon. When the flow rate reaches 2540 kg/h, the outlet fluid temperatures level out at their design values (180°C for the hydrocarbon and 45°C for the heating fluid). At what rate (kJ/h) is heat being transferred from the exchanger to the plant air?

(e) When the heating fluid leaves the exchanger, it passes through a heater, which raises its temperature back to 215°C, and is recycled back to the exchanger. How is the profitability of the process being decreased by the heat loss from the exchanger to the surroundings? (Try to think of two costs that result from the heat loss.)

(f) The engineer proposes adding more insulation to the heat exchanger, which would cut down on the heat loss and reduce the required heating fluid flow rate. What are advantages and disadvantages of the two responses to the heat loss problem (adding insulation versus increasing the heating fluid flow rate)? Which would you guess would be the preferable response in the long run, and why?

8.65. A liquid mixture of benzene and toluene containing 50.0 wt% benzene at 90°C and pressure P_0 is fed at a rate of 32.5 m^3/h into a **flash evaporator**, a heated tank maintained at a pressure $P_{tank} \ll P_0$. When the feed is exposed to the reduced pressure in this unit, a portion of it evaporates. The liquid and vapor product streams are in equilibrium at 75°C and P_{tank}. The liquid product contains 43.9 mole% benzene. When carrying out the requested calculations, assume volume additivity of liquid benzene and toluene, use Raoult's law and the Antoine equation where necessary, and neglect the effect of pressure on enthalpy.

(a) Calculate the molar flow rate (mol/s) and molar composition (component mole fractions) of the feed stream. Then calculate the minimum value of P_0(atm) needed to keep the feed stream in the liquid state until it enters the flash tank.

(b) Calculate P_{tank}(atm), the mole fraction of benzene in the vapor, and the molar flow rates of the liquid and vapor products.

(c) Calculate the required heat input rate in kilowatts.

(d) An hour after the system is started up, a chromatographic analysis of the vapor product is run and the benzene mole fraction is found to be 3% higher than the value calculated in part (b). The system temperature and pressure are rechecked and found to have the correct values. Give several possible explanations of the discrepancy between the calculated and measured values.

(e) Briefly explain why the product temperature is lower than the feed temperature. What would be required to run the unit isothermally?

*8.66. A continuous adiabatic flash tank is to be used to separate a liquid mixture of two substances (A and B). The feed enters at temperature T_F and a high pressure and flashes to a low pressure, P, whereupon its temperature drops to T. For an assumed basis of 1 mol/s of feed, let

$$\dot{n}_L, \dot{n}_V = \text{molar flow rates of liquid product and vapor product}$$

$$x_F, x, y = \text{mole fractions of A in the feed, liquid product, and vapor product}$$

$$p_A^*(T), p_B^*(T) = \text{vapor pressures of A and B}$$

$$T_{RA}, T_{RB} = \text{reference temperatures for enthalpy calculations}$$

$$\left.\begin{array}{l} \hat{H}_{AF}(T_F),\ \hat{H}_{AL}(T),\ \hat{H}_{AV}(T) \\ \hat{H}_{BF}(T_F),\ \hat{H}_{BL}(T),\ \hat{H}_{BV}(T) \end{array}\right\} \begin{array}{l} \text{specific enthalpies of A and B in} \\ \text{the feed, liquid product, and vapor} \\ \text{product, relative to } T_{RA} \text{ and } T_{RB} \end{array}$$

(a) Derive the following relations from Raoult's law and material and energy balances on the flash tank:

$$x = \frac{P - p_B^*(T)}{p_A^*(T) - p_B^*(T)} \tag{1}$$

$$y = x p_A^*(T) / P \tag{2}$$

$$\dot{n}_L = \frac{y - x_F}{y - x} \tag{3}$$

---

*Computer problem.

$$\dot{n}_V = 1 - \dot{n}_L \tag{4}$$

$$\Delta\dot{H} = \dot{n}_L[x\hat{H}_{AL}(T) + (1 - x)\hat{H}_{BL}(T)] + \dot{n}_V[y\hat{H}_{AV}(T) + (1 - y)\hat{H}_{BV}(T)]$$
$$- [x_F\hat{H}_{AF}(T_F) + (1 - x_F)\hat{H}_{BF}(T_F)] = 0 \tag{5}$$

(b) Write a spreadsheet program to perform flash calculations for a feed mixture of *n*-pentane and *n*-hexane. When calculating enthalpies of these species, the following heat capacity formulas should be used for liquid and vapor, respectively:

$$C_{pl} = a_l$$
$$C_{pv} = a_v + b_v T(°C)$$

The spreadsheet should have the following form. Some values are given, others should be looked up in data tables, and the remainder should be calculated from Equations 1 to 5 and appropriate formulas for specific enthalpies.

Chapter 8—Problem 8.66								
Tref = 25 deg. C								
Compound	A	B	C	al	av	bv	Tbp	DHv
n-pentane	6.85221	1064.63	232.000	0.195	0.115	3.41E-4	36.07	25.77
n-hexane				0.216	0.137	4.09E-4		
xF	0.5	0.5	0.5	0.5				
Tf(deg.C)	110	110	150					
P(mm Hg)	760	1000	1000					
HAF (kJ/mol)								
HBF (kJ/mol)								
T(deg.C)	80.0							
pA* (mm Hg)								
pB* (mm Hg)								
x								
y								
nL (mol/s)								
nV (mol/s)								
HAL (kJ/mol)								
HBL (kJ/mol)								
HAV (kJ/mol)								
HBV (kJ/mol)								
DH (kJ/s)	-51.333							

In this table, A, B, and C are Antoine equation constants, al, av, and bv are the coefficients of the given heat capacity formulas; Tbp(°C) and DHv(kJ/mol) ($\Delta\hat{H}_v$) are the normal boiling point and heat of vaporization, xF(mol pentane/mol) is the mole fraction of pentane in the feed, Tf(°C) is the feed temperature, P(mm Hg) is the system pressure, HAF ($\hat{H}_{AF}$) and HBF ($\hat{H}_{BF}$) are the specific enthalpies of pentane and hexane in the feed stream, pA* is the vapor pressure of n-pentane (to be determined using the Antoine equation), x and nL (x and $\dot{n}_L$) are the mole fraction of pentane in the liquid product stream and the molar flow rate of that stream, respectively, y and nV are the corresponding properties of the vapor product stream, HAL is the specific enthalpy of pentane in the liquid product stream, and DH ($\Delta\dot{H}$) is the expression given in Equation 5 for the change in total enthalpy from inlet to outlet.

Enter the appropriate constants and formulas for A, B, C, al, av, bv, Tbp, and DHv for n-pentane and n-hexane, an initial guess for T in Column 2 ($= 80.0$), and the appropriate formulas for the rest of the variables in Column 2. Then vary the value of T until the value of $\Delta\dot{H}$ is suitably close to zero, using the goalseek tool if your spreadsheet program provides it. The value of $\Delta\dot{H}$ (-51.33 kJ/s) corresponding to the initial guess of 80°C is shown in the second column of the table. Your spreadsheet should generate the same value.

After completing the calculations in the second column, copy the formulas into the third and fourth columns and perform the calculations for these two sets of input parameter values. State how increasing the system pressure and feed temperature affect the fraction of the feed vaporized (nV) and the final system temperature (T), and briefly explain why your results make sense.

*(c) Write a computer program to read in and print out values of x_F, T_F(°C), and P(mm Hg), and for each of the feed mixture components the Antoine equation coefficients, a reference temperature (e.g., 25°C), the liquid heat capacity (assume constant), normal boiling point, heat of vaporization, and coefficients of a two-term vapor heat capacity formula $C_p = a + bT$. Then have the program carry out a trial-and-error procedure to calculate and print out the flash tank temperature and the flow rates and compositions of the vapor and liquid product streams.

An inefficient but simple procedure is to assume a value of T (T_F is a reasonable first guess), solve Equations 1 through 4 successively, and then evaluate the left side of Equation 5 (which equals $\Delta\dot{H}$ for the assumed temperature and must be positive when $T = T_F$). The value of T is then decreased by a small fixed amount, such as 0.5°C, and the procedure is repeated until the sign of $\Delta\dot{H}$ changes from positive to negative from one T to the next. The true T (for which $\Delta\dot{H} = 0$) must lie between the two last assumed values and may be estimated by linear interpolation. When calculating specific enthalpies, use the heat capacity formulas for pentane and hexane given in part (b).

A flowchart for a program to implement this procedure is shown in Figure P8.1. Write the program and test it by estimating the flash tank temperature and product stream flow rates (mol/s) and compositions (mole fractions) for the flash vaporization of one mol/s of an equimolar mixture of n-pentane (A) and n-hexane (B), if the feed temperature is 110°C and the tank pressure is 1.0 atm.

8.67. A saturated vapor stream containing 10.9 mole% propane, 75.2% isobutane, and 13.9% n-butane passes from the top of a distillation column to a total condenser. Seventy-five percent of the condensate is returned to the column as reflux, and the remainder is removed as the overhead column product at a rate of 2500 kmol/h.

A decision must be made on whether to use a refrigerant or cooling water in the condenser. If the refrigerant is used, it will be fed to the condenser as a liquid and vaporized by the heat released by the condensing column vapor. The refrigerant pressure will be such that the vaporization takes place at -6°C, at which temperature $\Delta\hat{H}_v = 151$ kJ/kg. The other option calls for cooling water to be taken from a nearby river at its average summer temperature of 25°C. To avoid environmental problems, the temperature of the water returned to the river can be no greater than 34°C. With either system, the temperature of the condensate should be 6°C greater than the outlet temperature of the coolant, so that if the refrigerant is used the saturated condensate should have a temperature of 0°C, and if cooling water is used the saturated condensate should be at 40°C. The condenser pressure will be set to the minimum value needed to condense all of the vapor, which is to say the condensate

---

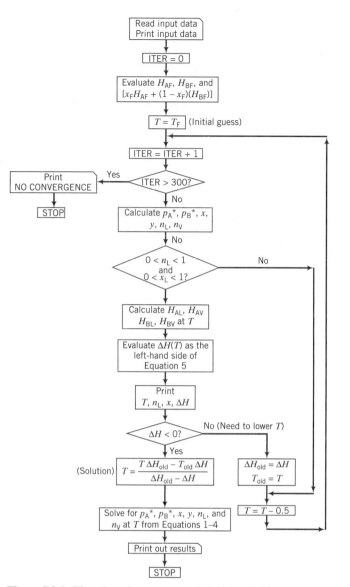

Figure P8.1 Flowchart for program of Problem 8.66.

will be at its bubble-point temperature at the condenser pressure. Raoult's law may be used for all bubble-point and dew-point calculations (see Section 6.4c).

(a) Suppose the refrigerant is used for cooling. Estimate the condenser pressure P(mm Hg); the temperature T_f(°C) of the vapor fed to the condenser, assuming that the vapor is at its dew point at pressure P; and the required coolant flow rate (kg/h).

(b) Repeat part (a) assuming that cooling water is fed to the condenser.

(c) What more would you need to know to be able to choose between the two options?

8.68. Formaldehyde is produced from methanol in a catalytic oxidation reactor. The following reactions take place:

$$CH_3OH \rightarrow HCHO + H_2$$

$$2\,H_2 + O_2 \rightarrow 2\,H_2O$$

A stream of methanol is joined by a recycle stream, also methanol, and the combined stream is fed to the conversion reactor. Also entering the reactor are air (to oxidize some of the hydrogen produced in the methanol conversion reaction) and steam (to control the reactor temperature). The product gas from the reactor contains 19.9 mole% formaldehyde, 8.34% methanol, 30.3% nitrogen, 0.830% oxygen, 5.0% hydrogen, and 35.6% water vapor and is at 600°C and 1 atm.

The following process is used to separate the formaldehyde from the unreacted methanol and noncondensable gases. The gases leaving the reactor are fed to a waste-heat boiler, in which they are cooled to 145°C, in the process generating steam at 3.1 bar from saturated liquid water (i.e., liquid water at its boiling point) at the same pressure. The gases are cooled further to 100°C in a heat exchanger, where they come into thermal contact with cooling water fed at 30°C. To reduce scaling on the heat exchanger tubes, the temperature increase of the cooling water is limited to 15°C. The cooled gases are fed to an absorption column, where methanol and formaldehyde are absorbed in water. Pure water at 20°C is fed to the top of the column. The gas leaving the absorber is saturated with water vapor at 27°C and 1 atm and contains 200 parts of formaldehyde per million parts (by volume) of total gas. The aqueous solution leaving the bottom of the absorber at 88°C is fed to a distillation column that operates at 1 atm. The final product solution, which contains 37 wt% formaldehyde, 1% methanol, and the balance water, is removed from the reboiler at the bottom of the column, while pure methanol vapor emerges as the overhead product and is condensed at 1 atm. A portion of the condensate is refluxed to the top of the column, and the rest is recycled to the methanol conversion reactor. The reflux ratio, or ratio of methanol refluxed to methanol recycled to the reactor, is 2.5:1.

(a) Taking as a basis of calculation 100 mol of gas leaving the conversion reactor, draw and completely label a flowchart of this process. Then calculate the moles of fresh methanol feed, formaldehyde product solution, recycled methanol, and absorber off-gas, the kg of steam generated in the waste-heat boiler, and the kg of cooling water fed to the heat exchanger between the waste-heat boiler and the absorber. Finally, calculate the heat (kJ) that must be removed in the distillation column overhead condenser, assuming that methanol enters as a saturated vapor at 1 atm and leaves as a saturated liquid at the same pressure.

(b) By what factor must all of the calculated quantities be multiplied to scale the flowchart to a production rate of 3.6×10^4 metric tons per year of formaldehyde solution, assuming that the process is on-stream 350 days per calendar year?

8.69. The latest weather report states that the temperature is 24°C and the relative humidity is 50%.

(a) Use the psychrometric chart to estimate the absolute humidity, humid volume, specific enthalpy, wet-bulb temperature, and dew-point temperature of the air.

(b) A thermometer is mounted on the back porch of your house. What temperature would it read?

(c) A sample of outside air is cooled at constant pressure. At what temperature would condensation begin?

(d) You step out of your neighborhood pool and feel quite cold until you dry off. Explain why. Estimate your skin temperature while you were still wet. Explain your answer. What would be different if the relative humidity were 98%?

8.70. An open vessel containing 0.205 lb_m of liquid water is placed in an empty room 5 ft wide, 4 ft deep, and 7 ft high, which initially contains dry air at 90°F. All the water evaporates without changing the room temperature. Use the psychrometric chart to estimate the final relative humidity, wet-bulb temperature, humid volume, dew-point temperature, and specific enthalpy of the room air. Take the molecular weight of dry air to be 29.0, and for simplicity, assume the mass of dry air in the room stays constant at its initial value.

8.71. A **sling psychrometer** is a device to measure the humidity of air. A porous cloth (the **wick**) is wrapped around the bulb of a mercury thermometer, which is then whirled around in the air. As the water in the wick evaporates, the temperature of the thermometer bulb drops, and finally stabilizes at the wet-bulb temperature of the air. The dry-bulb temperature is read from a second thermometer mounted on the sling.

One summer day, the weather bureau reports a temperature of 33°C and a relative humidity of 40%. You wipe the sweat from your forehead and remark to a friend that you'd bet $5 that the report is wrong, and the relative humidity is higher than 80%. He immediately puts a $5 bill on the table, accepting your challenge. You bring out your sling psychrometer, whirl it around, and read a dry-bulb temperature of 35°C and a wet-bulb temperature of 27°C. Who wins the bet?

8.72. Humid air is enclosed in a 2.00-liter flask at 40°C. The flask is slowly cooled. When the temperature reaches 20°C, drops of moisture become visible on the flask wall. Although the pressure in the flask changes when the temperature drops, it remains close enough to 1 atm for the psychrometric chart to provide a close representation of the behavior of the system throughout the process. Use the chart to solve the following problems.

(a) What were the relative humidity, absolute humidity, and wet-bulb temperature of the air at 40°C?

(b) Calculate the mass of the water in the flask. (See Example 8.4-5.)

(c) Calculate the enthalpy change in joules undergone by the air in going from 40°C to 20°C.

(d) Write an energy balance for this closed-system process, taking the wet air in the flask as the system, and use it to calculate the heat in joules that must be transferred from the air to accomplish the cooling. (Assume ideal gas behavior, so that $\hat{H} = \hat{U} + RT$.)

8.73. Wet solids pass through a continuous dryer. Hot dry air enters the dryer at a rate of 400 kg/min and picks up the water that evaporates from the solids. Humid air leaves the dryer at 50°C containing 2.44 wt% water vapor and passes through a condenser in which it is cooled to 10°C. The pressure is constant at 1 atm throughout the system.

(a) At what rate (kg/min) is water evaporating in the dryer?

(b) Use the psychrometric chart to estimate the wet-bulb temperature, relative humidity, dew point, and specific enthalpy of the air leaving the dryer.

(c) Use the psychrometric chart to estimate the absolute humidity and specific enthalpy of the air leaving the condenser.

(d) Use the results of parts (b) and (c) to calculate the rate of condensation of water (kg/min) and the rate at which heat must be transferred from the condenser (kW).

(e) If the dryer operates adiabatically, what can you conclude about the temperature of the entering air? Briefly explain your reasoning. What additional information would you need to calculate this temperature?

8.74. On an uncomfortable summer day, the air is at 87°F and 80% relative humidity. A laboratory air conditioner is to deliver 1.00×10^3 ft³/min of air at 55°F in order to maintain the interior air at an average temperature of 75°F and a relative humidity of 40%.

(a) If the vent switch on the air conditioner is turned to the "open" position, outside air enters the unit as shown below.

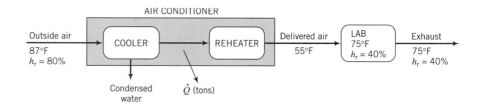

In the air conditioner, the air is cooled to a temperature low enough to condense the necessary amount of water and reheated to 55°F, at which point it has the same absolute humidity as the room air. Use the psychrometric chart to estimate the rate (lb$_m$/min) at which water is condensed, the temperature to which the air must be cooled to condense water at this rate, and the net tons of cooling required ($\dot{Q}$), where 1 ton of cooling $= -12,000$ Btu/h. [*Note:* The humid volume of the delivered air (at 55°F), which is difficult to read from the psychrometric chart, is 13.07 ft³/lb$_m$ dry air, and the heat capacity of liquid water is 1.0 Btu/(lb$_m$·°F).]

(b) If the vent switch is set to the "closed" position (as it normally would be), inside air would be recirculated through the air conditioner as shown in the following diagram.

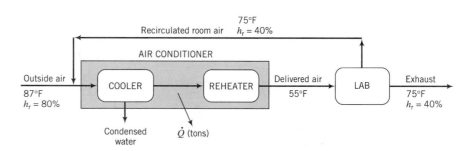

The recycle ratio (ft^3 recirculated/ft^3 exhausted) is 6:1. Calculate the condensation rate and the overall cooling requirement in tons if conditioned air is delivered at the same rate, temperature, and relative humidity as in part (a). What percentage of the cooling load on the air conditioner is saved by recirculating the air? Explain in your own words why the cooling rate is lower when room air is recirculated instead of bringing all the air in from the outside.

(c) An even lower cooling load would be required if *all* of the air passing through the conditioner were recirculated rather than just 6/7 of it, thereby eliminating the need for outside air and exhaust. Why would this be a bad idea? (*Hint:* Think about the people working in the laboratory.)

8.75. Wet wood chips are dried in a continous rotary dryer that operates at atmospheric pressure. The chips enter at 19°C with a water content of 40 wt% and must leave with a moisture content of less than 15%. Hot air is fed to the dryer at a rate of 11.6 m^3(STP)/kg wet chips.

To monitor the performance of the dryer by sampling the exiting chips and determining their moisture content directly would be a cumbersome procedure and almost impossible to automate. Instead, wet- and dry-bulb thermometers are mounted in both the inlet and outlet air lines, and the moisture content of the exiting chips is determined by a material balance.

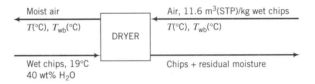

After the unit goes on-stream, the inlet dry-bulb temperature is found to be 100°C, and the wet-bulb temperature is low enough so that the moisture content of the air may be neglected. The dry-bulb temperature of the exiting air is found to be 38°C, and the wet-bulb temperature is 29°C.

(a) Use the psychrometric chart to calculate the absolute humidity (kg H$_2$O/kg dry air) and specific enthalpy (kJ/kg dry air) of the outlet air stream. Then calculate the mass of water in the exiting air per kilogram of wet chips fed, assuming dry air has a molecular weight of 29.0.

(b) Calculate the moisture content of the emerging chips and determine whether the design specification of less than 15% H$_2$O has been achieved.

(c) If the unit is operating adiabatically and the heat capacity of the dry chips is 2.10 kJ/(kg·°C), what is the exit temperature of the chips? (In estimating the specific enthalpy of the entering air, recall that the reference temperature for dry air used in constructing the psychrometric chart of Figure 8.4-1 is 0°C.)

8.76. Air at 45°C (dry bulb) and 10% relative humidity is to be humidified adiabatically to 60% relative humidity.
(a) Use the psychrometric chart to estimate the adiabatic saturation temperature of the air.
(b) Estimate the final temperature of the air and the rate at which water must be added to humidify 15 kg/min of the entering air. (See Example 8.4-7.)

8.77. Air at 50°C with a dew point of 4°C enters a textile dryer at a rate of 11.3 m^3/min and leaves saturated. The dryer operates adiabatically. Use the psychrometric chart to determine the absolute humidity and humid volume of the entering air, and then use the results to determine the flow rate of *dry* air (kg/min) through the dryer, the final temperature of the air, and the rate (kg/min) at which water is evaporated in the dryer. (*Hint:* Refer to Section 8.4e.)

8.78. A solution of sugar in water is to be concentrated from 5 wt% sugar to 20% sugar. The solution is at about 45°C when it is fed continuously to a bubble column. Air at 45°C with a dew point of 4°C is bubbled through the column and emerges saturated. The humidification of the air may be considered adiabatic. (See figure on next page.) Use the psychrometric chart to solve the following problems:
(a) What are the absolute humidities of the entering and exiting air?
(b) How many kilograms of dry air must be fed per kilogram of the entering sugar solution? What is the corresponding volume of the entering wet air? (Use the chart for the latter problem as well.)

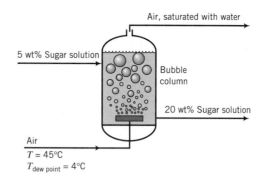

8.79. Cold air at 20°F, 760 mm Hg pressure, and 70% relative humidity is conditioned by being passed through a bank of heating coils, then through a water spray, and finally through a second set of heating coils. In passing through the first coil bank, the air is heated to 75°F. The temperature of the water supplied to the spray chamber is adjusted to the wet-bulb temperature of the air admitted to the chamber, so that the humidifying unit may be assumed to operate adiabatically. It is required that the air emerging from the conditioning unit be at 70°F and 35% relative humidity. Use Figure 8.4-2 to solve the following problems.

(a) Calculate the temperature of the water supplied to the spray chamber and the relative humidity and dry-bulb temperature of the air leaving the spray chamber.

(b) Calculate the mass of water evaporated (lb_m) per cubic foot of air fed to the conditioning unit.

(c) Calculate the required heat transfer rates (Btu/ft^3 entering air) in each of the heating coil banks.

(d) Sketch a psychrometric chart and show the path followed by the air in each of the three steps of this process.

8.80. Spray cooling is a technique for cooling and either humidifying or dehumidifying air by contacting it with a liquid water spray.

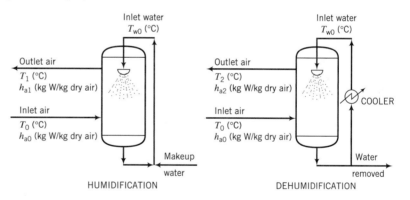

The liquid water leaving the tower is recirculated and, in the case of dehumidification, cooled before reentering the tower.

Two possible paths on the psychrometric chart corresponding to two different entering liquid temperatures are shown below. On the chart, T_0 and T_{dp} are the entering air dry-bulb temperature and dew point, respectively.

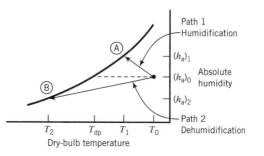

Path (A): The entering liquid temperature (T_{w0}) is above the dew point of the entering air. Liquid water evaporates into the superheated air, causing the absolute humidity of the air to increase (the path rises) and both the evaporation and the contact with the cold liquid cause the air temperature to decrease (the path moves to the left).

Path (B): The entering liquid temperature is below the dew point of the entering air. The temperature of the air contacted by the cold spray drops below the dew point (the path again moves to the left), and water vapor condenses out of the air (the path moves down).

We thus arrive at the interesting conclusion that *you can remove water from air by spraying water into the air,* provided that the entering liquid temperature is below the dew point of the entering air. Use the psychrometric chart to solve the following spray cooling problems.

(a) A spray tower is used to cool and humidify air with dry-bulb and wet-bulb temperatures of 40°C and 18°C, respectively. The air emerges from the tower at 20°C. The tower operation is such that the air follows an adiabatic humidification curve (a constant wet-bulb temperature line on the psychrometric chart). How much water must be added as makeup per kg of dry air treated?

(b) A stream of air at 37°C and 50% relative humidity flowing at a rate of 1250 kg/h is to be cooled to 15°C and dehumidified in a spray tower. The air is saturated as it emerges from the tower. Liquid water leaves the tower at 12°C; some is withdrawn, and the rest is cooled and recirculated. No heat is transferred between the tower and its surroundings. Calculate the rate (kg/h) at which water must be withdrawn from the recirculation loop and the heat duty on the cooler (kW). (*Suggestion:* Use an overall energy balance for the latter calculation.)

8.81. The heat of solution of ammonia in water at 1 atm is

$$\Delta \hat{H}_s(25°C, r = 2 \text{ mol } H_2O/\text{mol } NH_3) = -78.2 \text{ kJ/mol}$$

Calculate the enthalpy change that accompanies the dissolution of 400 mol of NH_3 in 800 mol of water at 25°C and 1 atm.

8.82. Use Table B.11 to determine the specific enthalpy (kJ/mol HCl) of hydrochloric acid containing 1 mol HCl/5 mol H_2O at 25°C relative to:
(a) HCl(g) and H_2O(l) at 25°C.
(b) H_2O (l) and an infinitely dilute HCl solution at 25°C. (Note Equation 8.5-2.)

8.83. Sodium hydroxide is dissolved in enough water to make up a 20.0 mole% solution. If the NaOH and water are initially at 77°F (25°C), how much heat (Btu/lb product solution) must be removed for the solution also to be at 77°F. Assume the process is carried out at constant pressure, so that $Q = \Delta H$, and use Table B.11 to evaluate $\Delta \hat{H}_s$.

8.84. A sulfuric acid solution is labeled 8N (where 1N = 1 g-equivalent /L, and 1 mol of H_2SO_4 contains two g-equivalents). The specific gravity of the solution is 1.230, and its heat capacity is 3.00 J/(g·°C). Calculate the specific enthalpy of this solution (in kJ/mol H_2SO_4) at 60°C relative to pure H_2O and an infinitely dilute solution at 25°C.

8.85. You are about to dilute 2.00 mol of 100% sulfuric acid with enough water to produce a 30 mole% aqueous solution. The acid and water are initially at 25°C.
(a) How much heat would have to be removed to keep the final solution at 25°C?
(b) Suppose the flask has a mass of 150 g, and that the heat capacity of the flask and its contents is 3.30 J/(g·°C). If the flask is sufficiently insulated to be considered adiabatic, what will be the final solution temperature?

8.86. An 8-molar hydrochloric acid solution [SG = 1.12, C_p = 2.76 J/(g·°C)] is produced by absorbing hydrogen chloride [HCl(g)] in water. Liquid water enters the absorber at 25°C and gaseous HCl is fed at 20°C and 790 torr (absolute). Essentially all of the HCl fed to the column is absorbed. Take one liter of product solution as a basis of calculation.
(a) Estimate the volume (liters) of HCl that must be fed to the absorber.
(b) Estimate the heat (kJ) that must be transferred from the absorber if the product solution is to emerge at 40°C.
(c) Estimate the final solution temperature if the absorber operates adiabatically.

8.87. A 0.1 mole% caustic soda (NaOH) solution is to be concentrated in a continuous evaporator. The solution enters the unit at 25°C at a rate of 150 mol/min and is concentrated to 5 mole% at 50°C. Hot dry air at 200°C and 1.1 bar absolute is bubbled through the evaporator and leaves saturated with water at 50°C and 1 atm. Calculate the required volumetric flow rate of the entering air and the rate at which heat must be transferred to or from the unit. Assume that the heat capacity per unit mass of all liquid solutions is that of pure water.

8.88. Water is added to pure sulfuric acid in a well-insulated flask initially at 25°C and 1 atm to produce a 4.00-molar sulfuric acid solution (SG = 1.231). The final temperature of the product solution is to be 25°C, so that the water added must be chilled liquid ($T < 25°C$), or a mixture of liquid water and ice. Take as a basis of calculation one liter of the product solution and assume $Q = \Delta H$ for the process. If you need to know the heat capacity of ice, take it to be half that of liquid water.

(a) If only liquid water is added, what masses (g) of H_2SO_4 and H_2O should be mixed and what should be the initial temperature of the water?

(b) If a mixture of liquid water and ice is added, how many grams of each should be fed?

8.89. Ortho-phosphoric acid (H_3PO_4) is produced as a dilute aqueous solution that must be concentrated before further use. In one facility, 100 tons/day of a 28 wt% P_2O_5 solution [see part (a) of this problem] at 125°F is to be concentrated in a single evaporator to 42 wt% P_2O_5. Heat is supplied to the evaporator by condensing saturated steam at 27.5 psia. The evaporator is to operate at 3.7 psia, and there is a boiling point elevation of 37°F for the 42 wt% P_2O_5 solution in the evaporator (see Section 6.5c). The heat of solution of H_3PO_4 at 77°F may be taken to be -5040 Btu/lb-mole H_3PO_4 relative to H_3PO_4 (l) and H_2O(l). The heat capacity of the 28% solution is 0.705 Btu/(lb$_m$·°F) and that of the 42% solution is 0.583 Btu/(lb$_m$·°F).

(a) It is conventional for the compositions of phosphoric acid solutions to be expressed in terms of wt% P_2O_5. Write the stoichiometric equation for the formation of ortho-phosphoric acid (MW = 98.00) from phosphorus pentoxide (MW = 141.96), and use it to derive the expression

$$\text{wt\% } H_3PO_4 = 1.381(\text{wt\% } P_2O_5)$$

(b) Calculate the ratio (lb$_m$ water evaporated/lb$_m$ feed solution).

(c) Suppose the water evaporated is subsequently condensed at a constant pressure of 3.7 psia. Determine the condensate flow rate in gal/min. How much heat (Btu/min) can be recovered through condensation of this water? At what temperature is this heat available? (To put it another way, if this heat were to be transferred to another stream, what is an upper bound on the temperature of that stream?)

(d) How much steam (lb$_m$/h) must be supplied to the system to evaporate the required amount of water? Recast your answer in terms of lb$_m$ steam per lb$_m$ water evaporated.

8.90. Two hundred kilograms per hour of an aqueous solution containing 20.0 mole% sodium acetate ($NaC_2H_3O_2$) enters an evaporative crystallizer at 60°C. When the solution is exposed to the low pressure in the evaporator, 16.9% of the water evaporates, concentrating the remaining solution and causing crystals of sodium acetate trihydrate ($NaC_2H_3O_2\cdot3\ H_2O$) to form. The product is an equilibrium mixture of crystals and a saturated aqueous solution containing 15.4 mole% $NaC_2H_3O_2$. The effluents (crystals, solution, and water vapor) are all at 50°C.

(a) Calculate the feed rate to the crystallizer in kmol/h.

(b) Calculate the production rate (kg/h) of trihydrate crystals and the mass flow rate (kg/h) of the liquid solution in which the crystals are suspended.

(c) Estimate the rate (kJ/h) at which heat must be transferred to or from the crystallizer (state which), using the following physical property data:

$$(C_p)_{\text{all solutions}} = 3.5 \text{ kJ/(kg·°C)}$$
$$(C_p)_{\text{crystals}} = 1.2 \text{ kJ/(kg·°C)}$$
$$(C_p)_{H_2O(v)} = 32.4 \text{ kJ/(kmol·°C)}$$
$$(\Delta \hat{H}_v)_{H_2O} = 4.39 \times 10^4 \text{ kJ/kmol}$$

Heat of solution of anhydrous sodium acetate:

$$\Delta \hat{H}_s(25°C) = -1.71 \times 10^4 \text{ kJ/kmol } NaC_2H_3O_2$$

Heat of hydration: $NaC_2H_3O_2(s) + 3\ H_2O(l) \rightarrow NaC_2H_3O_2\cdot3\ H_2O(s)$

$$\Delta \hat{H}(25°C) = -3.66 \times 10^4 \text{ kJ/kmol } NaC_2H_3O_2$$

8.91. Fifty milliliters of 100% H_2SO_4 at 25°C and 84.2 mL of liquid water at 15°C are mixed. The heat capacity of the product solution is 2.43 J/(g·°C). Estimate the maximum temperature attainable by the product solution and state the conditions under which this temperature would be attained, using heat of mixing data from Table B.11.

8.92. Suppose m_A(g) of species A {molecular weight M_A, heat capacity C_{pA} [J/(g·°C)]} at temperature T_{A0} (°C) and m_B(g) of species B (M_B, C_{pB}) at temperature T_{B0} are mixed adiabatically. The heat of mixing of A and B at 25°C is $\Delta\hat{H}_m$(r) (J/mol A in solution), where $r = (m_B/M_B)/(m_A/M_A)$. The heat capacity of the product solution is C_{ps}[J/(g·°C)]. All heat capacities may be considered independent of temperature.

(a) Derive an expression for T_{max}, the highest temperature attainable by the product solution, in terms of the other quantities defined. State the conditions that would have to be met for this temperature to be approached.

(b) Use your expression to estimate T_{max} for a process in which 100.0 g of sodium hydroxide at 25°C and 225.0 g of water at 40°C are combined to form a product solution with a heat capacity of 3.35 J/(g·°C).

8.93. One g-mole of pure liquid sulfuric acid at temperature T_0 (°C) is mixed with r g-moles of liquid water, also at temperature T_0(°C), in an adiabatic container. The final solution temperature is T_s(°C). The mass heat capacities of the pure acid, pure water, and the product solution [J/(g·°C)] are C_{pa}, C_{pw}, and C_{ps}, respectively, all of which may be taken to be constant (independent of temperature).

(a) Without doing any calculations, sketch the plot of T_s versus r you would expect to obtain for r varying between 0 and ∞. (*Hint:* Think first about what you would expect T_s to be at the extreme values of r.)

(b) Use an energy balance to derive an expression for T_s in terms of the initial acid and water temperatures, the heat capacities, the water/acid mole ratio (r), and the heat of mixing, $\Delta\hat{H}_m(r, 25\ °C)$ (kJ/mol H_2SO_4).

*(c) A series of 1.00 mol samples of pure liquid sulfuric acid are added to 11 insulated flasks containing varying amounts of water. The quantities of water in the flasks and the mass heat capacities of the product solutions are tabulated below:

r (mol H_2O)	0.5	1.0	1.5	2.0	3.0	4.0	5.0	10.0	25.0	50.0	100.0
C_p [J/(g·°C)]	1.58	1.85	1.89	1.94	2.10	2.27	2.43	3.03	3.56	3.84	4.00

The heat capacities of pure sulfuric acid and pure water may be determined from the molar heat capacities in Table B.2 evaluated at 25°C. All heat capacities should be taken to be independent of temperature.

Unfortunately, the laboratory air conditioner has been out of order for three weeks (Physical Plant promises they will get to it any day now) and the temperature on the July afternoon of the experiment (which also equals the initial acid and water temperatures) is an uncomfortable 40°C. Write a spreadsheet to generate a table and then a plot of T_s, the final temperature in each flask, versus r, the water/acid mole ratio of the solution in the flask. (*Suggestion:* Make the r axis logarithmic.) Assume that mixing is adiabatic.

(d) The actual experimental plot of T_s versus r would lie below the one determined in part (c). Why?

*8.94. A stirred tank with volume V_t(L) is charged with V_l(L) of a liquid, B. The space above the liquid (volume $V_g = V_t - V_l$) is filled with a pure gas, A, at an initial pressure P_0(atm). The initial system temperature is T_0(K). The stirrer in the tank is turned on, and A begins to dissolve in B. The dissolution continues until the liquid is saturated with A at the final system temperature (T) and pressure (P).

The equilibrium solubility of A in B is governed by the following expression, which relates the molar A/B ratio in the liquid to the partial pressure of A in the gas phase (which in turn equals the pressure in the tank, since the gas is pure A):

$$r(\text{mol A/mol B}) = k_s p_A(\text{atm})$$

where

$$k_s[\text{mol A/(mol B·atm)}] = c_0 + c_1 T(\text{K})$$

---

*Computer problems.

When solving the problems to be given, use the following variable definitions:

- M_A, M_B = molecular weights of A and B
- C_{vA}, C_{vB}, C_{vS} [J/(g·K)] =constant-volume heat capacities of A(g), B(l), and solutions of A in B, respectively
- SG_B = specific gravity of B(l)
- $\Delta \hat{U}_s$ (J/mol A dissolved) = internal energy of solution at 298 K (independent of composition over the range of concentrations to be considered)
- n_{A0}, n_{B0} = g-moles of A(g) and B(l) initially charged into the tank
- $n_{A(l)}$, $n_{A(v)}$ = g-moles of A dissolved and remaining in the gas phase at equilibrium, respectively

Make the following assumptions:

- A negligible amount of B evaporates.
- The tank is adiabatic and the work input to the tank from the stirrer is negligible.
- The gas phase behaves ideally.
- The volumes of the liquid and gas phases may be considered constant.
- The heat capacities C_{vA}, C_{vB}, and C_{vs} are constant, independent of temperature and (in the case of C_{vs}) solution composition.

(a) Use material balances, the given equilibrium solubility relation, and the ideal gas equation of state to derive expressions for n_{A0}, n_{B0}, $n_{A(v)}$, $n_{A(l)}$, and P in terms of the final temperature, T, and variables M_A, M_B, SG_B, V_t, V_l, T_0, P_0, c_0, and c_1. Then use an energy balance to derive the following equation:

$$T = 298 + \frac{n_{A(l)}(-\Delta \hat{U}_s) + (n_{A0}M_A C_{vA} + n_{B0}M_B C_{vB})(T_0 - 298)}{n_{A(v)}M_A C_{vA} + (n_{A(l)}M_{A(l)} + n_B M_B)C_{vs}}$$

(b) Write a spreadsheet to calculate T from specified values of $M_A(= 47)$, $M_B(= 26)$, $SG_B(= 1.76)$, $V_t(= 20.0)$, $V_l(= 3.0)$, $c_0(= 1.54 \times 10^{-3})$, $c_1(= -1.60 \times 10^{-6})$, $C_{vA}(= 0.831)$, $C_{vB}(= 3.85)$, $C_{vs}(= 3.80)$, and $\Delta \hat{U}_s(= -1.74 \times 10^5)$, and a number of different values of T_0 and P_0. The spreadsheet should have the structure given below. (Calculated values are shown for one initial temperature and pressure.)

Problem 8.94									
Vt	MA	CvA	MB	CvB	SGB	c0	c1	DUs	Cvs
20.0	47.0	0.831	26.0	3.85	1.76	0.00154	-1.60E-06	-174000	3.80

Vl	T0	P0	Vg	nB0	nA0	T	nA(v)	nA(l)	P	Tcalc
3.0	300	1.0								
3.0	300	5.0								
3.0	300	10.0	17.0	203.1	6.906	320.0	5.222	1.684	8.1	314.2
3.0	300	20.0								
3.0	330	1.0								
3.0	330	5.0								
3.0	330	10.0								
3.0	330	20.0								

The values of V_g, n_{B0}, and n_{A0} should first be calculated from the given values of the other variables. Next, a value of T should be guessed (in the example in the table, the guessed value is 320 K), the values of $n_{A(v)}$, $n_{A(l)}$, and P should be calculated from the equations derived in part (a), and the temperature should be recalculated from the energy balance in the column labeled T_{calc} (it equals 314.2 in the example). The value of T should then be varied until it equals the recalculated value of T_{calc}. (*Suggestion:* Create a new cell as $T - T_{calc}$ and use goalseek to find the value of T that drives $T - T_{calc}$ to zero.)

Enter the formulas in the cells for $V_1 = 3.0$ L, $T_0 = 300$ K, and $P_0 = 10.0$ atm, and verify that your cell values match those shown above. Then find the correct value of T using the procedure just described, copy the formulas into the other rows of the table, and determine T for each set of initial conditions. Summarize the effects of the initial temperature and pressure on the adiabatic temperature rise and briefly explain why your results make sense.

(c) Write a computer program to perform the same calculations done with the spreadsheet in part (b).

- Define values of V_t, M_A, C_{vA}, M_B, C_{vB}, SG_B, c_0, c_1, $\Delta\hat{U}_s$, and C_{vs}. Use the values shown in the fourth row of the spreadsheet.
- Read in a set of values of V_B, T_0, and P_0. Have the program terminate if $V_B \leq 0.0$.
- If a positive value is read in for V_B, calculate V_G, n_B, and n_{A0}.
- Assume a value of T. (Try $1.1T_0$ as a first guess.)
- Calculate $n_{A(v)}$, $n_{A(l)}$, and P from the equations derived in part (a), then recalculate T from the energy balance. Print out the values of T (assumed), P, $n_{A(v)}$, $n_{A(l)}$, and T (recalculated).
- If the assumed and recalculated values of T are within 0.01 K of each other, end the loop and go back to read the next set of input variables. If they are not and more than 15 iterations have been performed, terminate with an error message. Otherwise, repeat the previous step, using the recalculated value of T as the assumed value for this iteration.

Run the program for the eight sets of conditions shown in the spreadsheet table.

8.95. An aqueous solution containing 85.0 wt% H_2SO_4 at 60°F (specific gravity = 1.78) is diluted with pure liquid water at the same temperature. The feed solution volume is 350 mL. The mixing may be considered adiabatic, and the pressure is constant at 1 atm.

(a) The product solution is to contain 30.0 wt% H_2SO_4. Calculate the volume (mL) of water needed for the dilution, ideally using a single-dimensional equation.

(b) Use the enthalpy–concentration chart of Figure 8.5-1 to estimate the specific enthalpies (Btu/lb$_m$) of the feed solution and the water. Then write an energy balance on this closed system constant-pressure process and solve it for the specific enthalpy of the product solution. Finally, use Figure 8.5-1 to verify your calculated value of $\hat{H}_{product}$ and to estimate the product solution temperature. (See Example 8.5-3.)

(c) Use Figure 8.5-1 to estimate the maximum temperature that could be attained by mixing the feed solution with pure water and the concentration (wt% H_2SO_4) of the product solution.

(d) Good laboratory practice calls for adding acid to water when carrying out dilutions rather than vice versa. Use Figure 8.5-1 to justify this rule for the dilution of the feed solution in this problem.

8.96. Aqueous sulfuric acid solutions containing 15.0 wt% H_2SO_4 and 80.0 wt% H_2SO_4 are mixed to form a 60.0 wt% product solution. The 15% solution was in a laboratory in which the temperature was 77°F. The 80% solution had just been taken from a storage cabinet in an air-conditioned stockroom and was at a temperature of 60°F when the mixing occurred.

(a) The mass of the 15% solution is 2.30 lb$_m$. What mass of 60% solution should be weighed out?

(b) Use Figure 8.5-1 to estimate the product solution temperature if the mixing is adiabatic. (See Example 8.5-3.)

(c) The product solution temperature eventually drops to (77°F). How much heat (Btu) is transferred from the solution to the laboratory air in this constant-pressure cooling process?

(d) Which would be safer—adding the 15% solution slowly to the 80% solution or vice versa? Use Figure 8.5-1 to justify your answer.

8.97. You have analyzed an aqueous ammonia solution and find that it contains 30 wt% NH_3.

(a) Use Figure 8.5-2 to determine the mass fraction of NH_3 in the vapor that would be in equilibrium with this solution in a closed flask at 1 atm and the corresponding system temperature.

(b) If the liquid phase in part (a) accounts for 90% of the total system mass, calculate the overall system composition and specific enthalpy using balances. (See Example 8.5-3.)

8.98. An NH_3–H_2O mixture containing 60 wt% NH_3 is brought to equilibrium in a closed container at 140°F. The total mass of the mixture is 250 g. Use Figure 8.5-2 to determine the masses of ammonia and of water in each phase of the system.

8.99. An ammonia solution at a high pressure is flash-vaporized at a rate of 200 lb_m/h. The solution contains 0.70 lb_m NH_3/lb_m, and its enthalpy relative to $H_2O(l, 32°F)$ and $NH_3(l, -40°F)$ is -50 Btu/lb_m. Liquid and gas streams emerge from the unit at 1 atm and 80°F. Use Figure 8.5-2 to determine the mass flow rates and ammonia mass fractions of the vapor and the liquid product streams and the rate (Btu/h) at which heat must be transferred to the vaporizer. (See Example 8.5-4.)

Chapter 9

Balances on Reactive Processes

Consider the familiar reaction in which water is formed from hydrogen and oxygen:

$$2H_2(g) + O_2(g) \longrightarrow 2H_2O(v)$$

On the molecular level, the reaction might be depicted as follows:

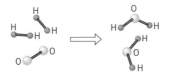

Each time this reaction takes place, three chemical bonds are broken (two between hydrogen atoms and one between oxygen atoms) and four bonds are formed among the atoms of the two water molecules. As it happens, more energy is released when the water molecule bonds form than it takes to break the hydrogen and oxygen molecule bonds. For the reactor temperature to remain constant, the net energy released (about 250 kJ per mol of water formed) must be transferred away from the reactor; otherwise it can raise the reactor temperature by several thousand degrees.

In *any* reaction between stable molecules, energy is required to break the reactant chemical bonds and energy is released when the product bonds form. If the first process absorbs less energy than the second process releases (as in the water formation reaction), the reaction is **exothermic**: the product molecules at a given temperature and pressure have lower internal energies (and hence lower enthalpies) than the reactant molecules at the same temperature and pressure. The net energy released—the **heat of reaction**—must be transferred from the reactor as heat or work, or else the system temperature increases. On the other hand, if less energy is released when the product bonds form than it took to break the reactant bonds, the reaction is **endothermic**: energy must be added to the reactor as heat or work to keep the temperature from decreasing.

The large internal energy and enthalpy changes commonly associated with chemical reactions can play major roles in the design and operation of chemical processes. If a reaction is endothermic, the energy needed to keep the reactor temperature (and hence the reaction rate) from dropping too much may cost enough to turn a profitable process into an unprofitable one. On the other hand, if the reaction is exothermic, heat usually must be transferred away from the reactor to keep the temperature below a value that leads to safety or product quality problems. The heat transferred may be an asset, as when the reactor is a combustion furnace and the heat is used to generate steam in a boiler. It may also be a liability: for example, a momentary failure of the reactor temperature control system can lead to rapid overheating and possibly an explosion.

An energy balance on a reactor tells the process engineer how much heating or cooling the reactor requires in order to operate at the desired conditions. In this chapter we show how enthalpy changes that accompany chemical reactions are determined from tabulated physical properties of the reactants and products and how calculated enthalpies of reaction are incorporated in energy balances on reactive processes.

TEST YOURSELF

1. Explain in your own words the concepts of exothermic and endothermic reactions. The terms "chemical bonds" and "heat of reaction" should appear in your explanation.
2. The following two sentences seem to contain a self-contradiction.

> *In an exothermic reaction, the products are at a* lower *energy level than the reactants. However, if the reactor is not cooled, the products are hotter than the reactants, which means they must be at a* higher *energy level than the reactants.*

Identify the logical error in this paragraph.

CREATIVITY EXERCISE

Suppose an exothermic reaction takes place in a continuous reactor. Think of several ways the heat of reaction might be removed, illustrating your suggestions with sketches. (For example, pass a cold fluid through a hollow metal tube immersed in the reactor, so that heat is transferred from the hot reacting fluid to the coolant.)

9.0 INSTRUCTIONAL OBJECTIVES

After completing this chapter, you should be able to do the following:

- Explain in your own words the concepts of heat of reaction; exothermic and endothermic reactions; heat of formation; combustion; heat of combustion; standard heats of formation, combustion, and reaction; heating value of a fuel; adiabatic flame temperature; ignition temperature; ignition lag; lower and upper flammability limits and flash point of a fuel; a flame; blue and yellow flames; flashback; and detonation.
- Given (a) the amount of any reactant consumed or any product generated in a reaction at a given temperature and pressure and (b) the heat of the reaction at that temperature and pressure, calculate the total enthalpy change.
- Determine a heat of reaction from heats of other reactions using Hess's law. Determine standard enthalpies and internal energies of reaction from known standard heats of formation and heats of combustion.
- Write and solve an energy balance on a chemical reactor using either the heat of reaction method (taking reactant and product species as references for enthalpy calculations) or the heat of formation method (taking elemental species as references), and specify which method is preferable for a given process. Write the process path implicitly adopted when each method is used.
- Solve reactive-system energy balance problems for (a) the heat transfer required for specified inlet and outlet conditions, (b) the outlet temperature corresponding to a specified heat input (e.g., for an adiabatic reactor), and (c) the product composition corresponding to a specified heat input and a specified outlet temperature.
- Solve energy balance problems for processes involving solutions for which heats of solution are significant.
- Convert a higher heating value of a fuel to a lower heating value and vice versa.

9.1 HEATS OF REACTION

Consider the reaction between solid calcium carbide and liquid water to form solid calcium hydroxide and gaseous acetylene:

$$CaC_2(s) + 2H_2O(l) \longrightarrow Ca(OH)_2(s) + C_2H_2(g) \qquad \text{(9.1-1)}$$

The expression *stoichiometric quantities of reactants* means molar amounts of the reactants numerically equal to their stoichiometric coefficients. For the calcium carbide reaction, stoichiometric quantities of the reactants in the SI system of units would be 1 mol of $CaC_2(s)$ and 2 mol of $H_2O(l)$. If stoichiometric quantities of the reactants are fed and the reaction proceeds to completion, both reactants would be completely consumed and stoichiometric quantities of the products would be formed. (Convince yourself.)

The **heat of reaction** (*or* **enthalpy of reaction**), $\Delta \hat{H}_r(T, P)$, *is the enthalpy change for a process in which stoichiometric quantities of reactants at temperature T and pressure P react completely in a single reaction to form products at the same temperature and pressure.* For example, the heat of the calcium carbide reaction at 25°C and 1 atm is

$$\Delta \hat{H}_r(25°C, 1 \text{ atm}) = -125.4 \text{ kJ/mol} \qquad \text{(9.1-2)}$$

Equations 9.1-1 and 9.1-2 together signify that if 1 mol of solid calcium carbide reacts completely with 2 mol of liquid water to form 1 mol of solid calcium hydroxide and 1 mol of gaseous acetylene, and the initial and final temperatures are both 25°C and the initial and final pressures are both 1 atm, then $H_{\text{products}} - H_{\text{reactants}} = -125.4 \text{ kJ}$. If the reaction is run under conditions such that the energy balance reduces to $Q = \Delta H$, then 125.4 kJ of heat must be transferred from the reactor in the course of the reaction. (Recall that a negative Q implies flow of heat out of the system.)

The units of $\Delta \hat{H}_r$ often cause confusion. For example, if the heat of a reaction is reported to be -50 kJ/mol, you might ask "per mol of what?" This difficulty is avoided if you recall that the given $\Delta \hat{H}_r$ applies to stoichiometric quantities of each species. For example,

$$2A + B \rightarrow 3C: \quad \Delta \hat{H}_r(100°C, 1 \text{ atm}) = -50 \text{ kJ/mol}$$

means that the enthalpy change for the given reaction is

$$\frac{-50 \text{ kJ}}{2 \text{ mol A consumed}} = \frac{-50 \text{ kJ}}{1 \text{ mol B consumed}} = \frac{-50 \text{ kJ}}{3 \text{ mol C generated}}$$

If you knew, say, that 150 mol of C/s was generated in the given reaction at 100°C and 1 atm, you could calculate the associated enthalpy change as

$$\Delta \dot{H} = \frac{-50 \text{ kJ}}{3 \text{ mol C generated}} \left| \frac{150 \text{ mol C generated}}{\text{s}} \right. = -2500 \text{ kJ/s}$$

More generally, if ν_A is the stoichiometric coefficient of a reactant or reaction product A (positive if A is a product, negative if it is a reactant) and $n_{A,r}$ moles of A are consumed or generated at $T = T_0$ and $P = P_0$, then the associated enthalpy change is

$$\Delta H = \frac{\Delta \hat{H}_r(T_0, P_0)}{|\nu_A|} n_{A,r}$$

In Chapter 4, we defined the *extent of reaction*, ξ, as a measure of how far a reaction has proceeded. From Equation 4.6-3, this quantity is

$$\xi = \frac{|n_{A,\text{out}} - n_{A,\text{in}}|}{|\nu_A|} = \frac{n_{A,r}}{|\nu_A|} \qquad \text{(9.1-3)}$$

From the preceding two equations, it follows that if a reaction takes place at a temperature T_0 and pressure P_0 and the extent of reaction is ξ, the associated enthalpy change is

$$\boxed{\Delta H = \xi \Delta \hat{H}_r(T_0, P_0)} \qquad \text{(9.1-4)}$$

For a continuous process, $n_{A,r}$(mol) would be replaced by $\dot{n}_{A,r}$(mol/s) in this expression, ξ(mol) would be replaced by $\dot{\xi}$(mol/s), and ΔH(kJ) would be replaced by $\Delta\dot{H}$(kJ/s).

Following are several important terms and observations related to heats of reaction.

1. If $\Delta\hat{H}_r(T, P)$ is negative the reaction is **exothermic** at temperature T and pressure P, and if $\Delta\hat{H}_r(T, P)$ is positive the reaction is **endothermic** at T and P. These definitions of exothermic and endothermic are equivalent to the ones given earlier in terms of chemical bond strengths. (Convince yourself.)

2. *At low and moderate pressures, $\Delta\hat{H}_r(T, P)$ is nearly independent of pressure.* We will presume this independence in the balance of this chapter and write the heat of reaction as $\Delta\hat{H}_r(T)$.

3. *The value of the heat of a reaction depends on how the stoichiometric equation is written.* For example,

$$CH_4(g) + 2\,O_2(g) \longrightarrow CO_2(g) + 2\,H_2O(l): \quad \Delta\hat{H}_{r1}(25°C) = -890.3 \text{ kJ/mol}$$
$$2\,CH_4(g) + 4\,O_2(g) \longrightarrow 2\,CO_2(g) + 4H_2O(l): \quad \Delta\hat{H}_{r2}(25°C) = -1780.6 \text{ kJ/mol}$$

This result should seem reasonable to you if you look back at the definition of $\Delta\hat{H}_r$. The first line states that the combined enthalpy of 1 gram-mole of CO_2 plus 2 gram-moles of liquid water is 890.3 kJ lower than the combined enthalpy of 1 gram-mole of methane plus 2 gram-moles of oxygen at 25°C. Doubling the quantity of reactants at a given condition doubles the total enthalpy of the reactants at that condition, and similarly for the products. The difference between the product and reactant enthalpies in the second reaction (by definition, $\Delta\hat{H}_{r2}$) must therefore be double the enthalpy difference in the first reaction ($\Delta\hat{H}_{r1}$).

4. *The value of a heat of reaction depends on the states of aggregation (gas, liquid, or solid) of the reactants and products.* For example,

$$CH_4(g) + 2\,O_2(g) \longrightarrow CO_2(g) + 2\,H_2O(l): \quad \Delta\hat{H}_r(25°C) = -890.3 \text{ kJ/mol}$$
$$CH_4(g) + 2\,O_2(g) \longrightarrow CO_2(g) + 2\,H_2O(g): \quad \Delta\hat{H}_r(25°C) = -802.3 \text{ kJ/mol}$$

The only difference between the reactions is that the water formed is a liquid in the first one and a vapor in the second. Since enthalpy is a state function, the difference between the two heats of reaction must be the enthalpy change associated with the vaporization of 2 mol of water at 25°C—that is, $2\Delta\hat{H}_v(25°C)$.

5. The **standard heat of reaction**, $\Delta\hat{H}_r^\circ$, is the heat of reaction when both the reactants and products are at a specified reference temperature and pressure, usually (and always in this text) 25°C and 1 atm.

EXAMPLE 9.1-1 | *Calculation of Heats of Reaction*

1. The standard heat of the combustion of *n*-butane vapor is

$$C_4H_{10}(g) + \tfrac{13}{2}O_2(g) \longrightarrow 4\,CO_2(g) + 5\,H_2O(l): \quad \Delta\hat{H}_r^\circ = -2878 \text{ kJ/mol}$$

Calculate the rate of enthalpy change, $\Delta\dot{H}$(kJ/s), if 2400 mol/s of CO_2 is produced in this reaction and the reactants and products are all at 25°C.

2. What is the standard heat of the reaction

$$2\,C_4H_{10}(g) + 13\,O_2(g) \longrightarrow 8\,CO_2(g) + 10\,H_2O(l)$$

Calculate $\Delta\dot{H}$ if 2400 mol/s of CO_2 is produced in *this* reaction and the reactants and products are all at 25°C.

3. The heats of vaporization of *n*-butane and water at 25°C are 19.2 kJ/mol and 44.0 kJ/mol, respectively. What is the standard heat of the reaction

$$C_4H_{10}(l) + \tfrac{13}{2}O_2(g) \longrightarrow 4\,CO_2(g) + 5\,H_2O(v)$$

Calculate $\Delta\dot{H}$ if 2400 mol/s of CO_2 is produced in this reaction and the reactants and products are all at 25°C.

SOLUTION   **1.** From Equation 9.1-3,

$$\dot{\xi} = \frac{(\dot{n}_{CO_2})_r}{|\nu_{CO_2}|} = \frac{2400 \text{ mol/s}}{4} = 600 \text{ mol/s}$$

$$\Big\Downarrow \text{Equation 9.1-4}$$

$$\Delta\dot{H} = \dot{\xi}\,\Delta\hat{H}_r^\circ = \left(600\,\frac{\text{mol}}{\text{s}}\right)\left(-2878\,\frac{\text{kJ}}{\text{mol}}\right) = \boxed{-1.73 \times 10^6 \text{ kJ/s}}$$

2. Since doubling the stoichiometric coefficients of a reaction must double the heat of reaction,

$$\Delta\hat{H}_{r2}^\circ = 2\Delta\hat{H}_{r1}^\circ = 2(-2878 \text{ kJ/mol}) = \boxed{-5756 \text{ kJ/mol}}$$

The enthalpy change associated with the production of 2400 mol/s of CO_2 at 25°C cannot depend on how the stoichiometric equation is written (the same quantities of reactants and products at the same temperatures must have the same enthalpies), and so $\Delta\dot{H}$ must be the value calculated in part (a). Let us do the calculation and prove it, however. From Equation 9.1-3,

$$\dot{\xi} = \frac{(\dot{n}_{CO_2})_{\text{out}}}{|\nu_{CO_2}|} = \frac{2400 \text{ mol/s}}{8} = 300 \text{ mol/s}$$

$$\Big\Downarrow \text{Equation 9.1-4}$$

$$\Delta\dot{H} = \dot{\xi}\,\Delta\hat{H}_r^\circ = \left(300\,\frac{\text{mol}}{\text{s}}\right)\left(-5756\,\frac{\text{kJ}}{\text{mol}}\right) = \boxed{-1.73 \times 10^6 \text{ kJ/s}}$$

3. Compare the two reactions:

$$C_4H_{10}(g) + \tfrac{13}{2}O_2(g) \longrightarrow 4\,CO_2(g) + 5\,H_2O(l): \quad (\Delta\hat{H}_{r1}^\circ) = -2878 \text{ kJ/mol}$$
$$C_4H_{10}(l) + \tfrac{13}{2}O_2(g) \longrightarrow 4\,CO_2(g) + 5\,H_2O(v): \quad (\Delta\hat{H}_{r2}^\circ) = \text{?}$$

The total enthalpy of the products in the second reaction [4 mol CO_2(g) + 5 mol H_2O(g) at 25°C] is greater than that of the products in the first reaction [4 mol CO_2(g) + 5 mol H_2O(l) at 25°C] by five times the heat of vaporization of water. Similarly, the total enthalpy of the reactants in the second reaction is lower than that of the reactants in the first reaction by the heat of vaporization of butane. (Why?) Since $\Delta\hat{H}_r = H_{\text{products}} - H_{\text{reactants}}$, it follows that

$$(\Delta\hat{H}_{r2}^\circ) = (\Delta\hat{H}_{r1}^\circ) + 5(\Delta\hat{H}_v)_{H_2O} + (\Delta\hat{H}_v)_{C_4H_{10}}$$

$$= [-2878 + 5(44.0) + 19.2] \text{ kJ/mol} = \boxed{-2639 \text{ kJ/mol}}$$

$$\Delta\dot{H} = \dot{\xi}\,\Delta\hat{H}_{r2}^\circ = \left(600\,\frac{\text{mol}}{\text{s}}\right)\left(-2639\,\frac{\text{kJ}}{\text{mol}}\right) = \boxed{-1.58 \times 10^6 \text{ kJ/s}}$$

If a reaction takes place in a closed reactor at constant volume, the heat released or absorbed is determined by the change in internal energy between reactants and products. The **internal energy of reaction**, $\Delta\hat{U}_r(T)$, is the difference $U_{\text{products}} - U_{\text{reactants}}$ if stoichiometric quantities of reactants react completely at temperature T.

Suppose a reaction occurs, and ν_i is the stoichiometric coefficient of the ith gaseous reactant or product. If ideal gas behavior can be assumed and specific volumes of liquid and solid reactants and products are negligible compared with those of the gases, the internal energy of reaction is related to the heat of reaction by

$$\Delta\hat{U}_r(T) = \Delta\hat{H}_r(T) - RT\left(\sum_{\substack{\text{gaseous}\\\text{products}}} |\nu_i| - \sum_{\substack{\text{gaseous}\\\text{reactants}}} |\nu_i|\right) \tag{9.1-5}$$

For example, for the reaction

$$C_6H_{14}(l) + \tfrac{19}{2}O_2(g) \to 6\,CO(g) + 7\,H_2O(v)$$

the internal energy of reaction is

$$\Delta \hat{U}_r(T) = \Delta \hat{H}_r(T) - RT(6 + 7 - \tfrac{19}{2})$$
$$= \Delta \hat{H}_r(T) - \tfrac{7}{2}RT$$

If there are no gaseous reactants or products, then to a good approximation $\Delta \hat{U}_r = \Delta \hat{H}_r$.

EXAMPLE 9.1-2   *Evaluation of $\Delta \hat{U}_r$*

The standard heat of the reaction

$$C_2H_4(g) + 2\,Cl_2(g) \rightarrow C_2HCl_3(l) + H_2(g) + HCl(g)$$

is $\Delta \hat{H}_r^\circ = -420.8$ kJ/mol. Calculate $\Delta \hat{U}_r^\circ$ for this reaction.

SOLUTION   From the stoichiometric equation

$$\sum \nu_i \,(\text{product gases}) = 1 + 1 = 2$$
$$\sum \nu_i \,(\text{reactant gases}) = 1 + 2 = 3$$

From Equation 9.1-5

$$\Delta \hat{U}_r = \Delta \hat{H}_r - RT(2 - 3)$$

$$= -420.8 \text{ kJ/mol} - \frac{8.314 \text{ J}}{\text{mol·K}} \left| \frac{298 \text{ K}}{} \right| \frac{-1}{} \left| \frac{1 \text{ kJ}}{10^3 \text{ J}} \right.$$

$$= \boxed{-418.3 \text{ kJ/mol}}$$

TEST YOURSELF
1. What is a heat of reaction? A standard heat of reaction?
2. Suppose $\Delta \hat{H}_r^\circ$ is -40 kJ/mol for the reaction $2A \rightarrow B$.
   (a) What is the value of the ratio (kJ/mol A reacted)?
   (b) Is the reaction exothermic or endothermic at 25°C and 1 atm?
   (c) If the reactants and products are at the same temperature, must heat be added to or withdrawn from the reactor? (Assume that the energy balance reduces to $Q = \Delta H$.)
   (d) If the reactor is adiabatic ($Q = 0$), would the products leave at a higher or a lower temperature than that of the entering reactants?
3. $C_6H_{14}(l) + \tfrac{19}{2}O_2 \rightarrow 6\,CO_2 + 7\,H_2O(l)$: $\Delta \hat{H}_r^\circ = -4163$ kJ/mol
   $C_6H_{14}(g) + \tfrac{19}{2}O_2 \rightarrow 6\,CO_2 + 7\,H_2O(l)$: $\Delta \hat{H}_r^\circ = -4195$ kJ/mol
   The standard state for the heats of reaction is 25° and 1 atm. What is the physical significance of the difference between the two given values of $\Delta \hat{H}_r^\circ$?
4. Write the formula for $\Delta \hat{U}_r(T)$ in terms of $\Delta \hat{H}_r(T)$ for the reaction $A(g) + 2B(g) + C(l) \rightarrow D(g) + 2E(s)$.
5. Derive Equation 9.1-5 from the definition of $\hat{H}$ as $\hat{U} + P\hat{V}$.

9.2   MEASUREMENT AND CALCULATION OF HEATS OF REACTION: HESS'S LAW

A heat of reaction may be measured in a **calorimeter**—a closed reactor immersed in a fluid contained in a well-insulated vessel. The rise or fall of the fluid temperature can be measured and used to determine the energy released or absorbed by the reaction, and the value of $\Delta \hat{H}_r^\circ$ may then be calculated from that energy and known reactant and product heat capacities.

There are serious limitations to this technique, however. Suppose, for example, you wish to determine $\Delta \hat{H}_r^\circ$ for the reaction

$$C(s) + \tfrac{1}{2}O_2(g) \rightarrow CO(g)$$

You could put 1 mole of carbon and 0.5 mole of oxygen together in a reactor, but you would never get 1 mole of carbon monoxide as the final product. If the reactants are at or near 25°C or lower, nothing apparent would occur since the rate at which carbon and oxygen react at this temperature is immeasurably low. If, on the other hand, the mixture were heated to a temperature at which C and O_2 react at a measurable rate, the product would be either pure CO_2 or at best a mixture of CO and CO_2, making it impossible to determine the heat of the CO formation reaction alone.

However, you *can* carry out the reactions

1. $C + O_2 \rightarrow CO_2$: $\Delta \hat{H}_{r1}^{\circ} = -393.51$ kJ/mol
2. $CO + \frac{1}{2}O_2 \rightarrow CO_2$: $\Delta \hat{H}_{r2}^{\circ} = -282.99$ kJ/mol

and determine their heats of reaction experimentally. You may then construct a process path for the reaction

3. $C + \frac{1}{2}O_2 \rightarrow CO$: $\Delta \hat{H}_{r3}^{\circ} = ?$

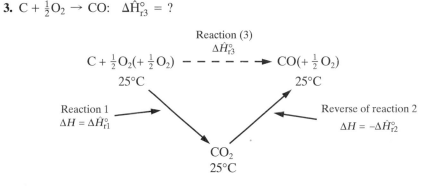

Since $\hat{H}$ is a state function,

$$\Delta \hat{H}_{r3}^{\circ} = \Delta \hat{H}_{r1}^{\circ} + (-\Delta \hat{H}_{r2}^{\circ}) = (-393.51 + 282.99) \text{ kJ/mol} = -110.52 \text{ kJ/mol}$$

You have thus calculated the desired heat of reaction, which could not be measured directly, from two measurable heats of reaction.

This result could have been obtained more concisely by treating the stoichiometric equations for reactions 1 and 2 as algebraic equations. If the equation for reaction 2 is subtracted from that for reaction 1, the result is

$$C + O_2 - CO - \frac{1}{2}O_2 \rightarrow CO_2 - CO_2$$

$$\Downarrow$$

$$C + \frac{1}{2}O_2 \rightarrow CO \text{ (reaction 3)}$$

The standard heat of reaction 3 may be calculated by applying the same operation to the heats of reactions 1 and 2—that is, $\Delta \hat{H}_{r3}^{\circ} = \Delta \hat{H}_{r1}^{\circ} - \Delta \hat{H}_{r2}^{\circ}$—confirming the result previously obtained.

The general statement of the validity of this procedure is called **Hess's law:** *If the stoichiometric equation for reaction 1 can be obtained by algebraic operations (multiplication by constants, addition, and subtraction) on stoichiometric equations for reactions 2, 3, . . . , then the heat of reaction $\Delta \hat{H}_{r1}^{\circ}$ can be obtained by performing the same operations on the heats of reactions $\Delta \hat{H}_{r2}^{\circ}$, $\Delta \hat{H}_{r3}^{\circ}$, ...*

EXAMPLE 9.2-1   *Hess's Law*

The standard heats of the following combustion reactions have been determined experimentally:

1. $C_2H_6 + \frac{7}{2}O_2 \rightarrow 2CO_2 + 3H_2O$: $\Delta \hat{H}_{r1}^{\circ} = -1559.8$ kJ/mol
2. $C + O_2 \rightarrow CO_2$: $\Delta \hat{H}_{r2}^{\circ} = -393.5$ kJ/mol
3. $H_2 + \frac{1}{2}O_2 \rightarrow H_2O$: $\Delta \hat{H}_{r3}^{\circ} = -285.8$ kJ/mol

Use Hess's law and the given heats of reaction to determine the standard heat of the reaction

4. $2\,C + 3\,H_2 \rightarrow C_2H_6: \quad \Delta H^\circ_{r4} = ?$

SOLUTION Since

$$(4) = 2 \times (2) + 3 \times (3) - (1)$$

(*verify*), from Hess's law

$$\Delta\hat{H}^\circ_{r4} = 2\Delta\hat{H}^\circ_{r2} + 3\Delta\hat{H}^\circ_{r3} - \Delta\hat{H}^\circ_{r1} = -84.6 \text{ kJ/mol}$$

This heat of reaction could not have been measured directly, since you cannot react carbon and hydrogen in such a way that ethane is the only reaction product.

TEST YOURSELF

1. What is Hess's law?
2. Suppose heats of reaction at 25°C are measured experimentally for the following set of reactions:

$$2\,A + B \rightarrow 2\,C: \quad \Delta\hat{H}^\circ_{r1} = -1000 \text{ kJ/mol}$$
$$A + D \rightarrow C + 3\,E: \quad \Delta\hat{H}^\circ_{r2} = -2000 \text{ kJ/mol}$$

Use Hess's law to show that for

$$B + 6\,E \rightarrow 2\,D: \quad \Delta\hat{H}^\circ_r = +3000 \text{ kJ/mol}$$

9.3   FORMATION REACTIONS AND HEATS OF FORMATION

A **formation reaction** of a compound is the reaction in which the compound is formed from its elemental constituents as they normally occur in nature (e.g., O_2 rather than O). The enthalpy change associated with the formation of 1 mole of the compound at a reference temperature and pressure (usually 25°C and 1 atm) is the **standard heat of formation** of the compound, $\Delta\hat{H}^\circ_f$.

Standard heats of formation for many compounds are listed in Table B.1 of this text and on pp. 2-187 through 2-198 of *Perry's Chemical Engineers' Handbook*.[1] For example, $\Delta\hat{H}^\circ_f$ for crystalline ammonium nitrate is given in Table B.1 as -365.14 kJ/mol, signifying

$$N_2(g) + 2\,H_2(g) + \tfrac{3}{2}O_2(g) \rightarrow NH_4NO_3(c): \quad \Delta\hat{H}^\circ_r = -365.14 \text{ kJ/mol}$$

Similarly, for liquid benzene $\Delta\hat{H}^\circ_f = 48.66$ kJ/mol, or

$$6\,C(s) + 3\,H_2(g) \rightarrow C_6H_6(l): \quad \Delta\hat{H}^\circ_r = +48.66 \text{ kJ/mol}$$

The standard heat of formation of an elemental species (e.g., O_2) is zero. (Why?)

It may be shown using Hess's law that *if ν_i is the stoichiometric coefficient of the ith species participating in a reaction (+ for products, − for reactants) and $\Delta\hat{H}^\circ_{fi}$ is the standard heat of formation of this species, then the standard heat of the reaction is*

$$\Delta\hat{H}^\circ_r = \sum_i \nu_i \Delta\hat{H}^\circ_{fi} = \sum_{\text{products}} |\nu_i|\Delta\hat{H}^\circ_{fi} - \sum_{\text{reactants}} |\nu_i|\Delta\hat{H}^\circ_{fi} \qquad \textbf{(9.3-1)}$$

The standard heats of formation of all elemental species should be set equal to zero in this formula. The validity of Equation 9.3-1 is illustrated in the next example.

EXAMPLE 9.3-1   *Determination of a Heat of Reaction from Heats of Formation*

Determine the standard heat of reaction for the combustion of liquid *n*-pentane, assuming $H_2O(l)$ is a combustion product.

$$C_5H_{12}(l) + 8\,O_2(g) \rightarrow 5\,CO_2(g) + 6\,H_2O(l)$$

---

[1]R. H. Perry and D. W. Green, eds., *Perry's Chemical Engineer's Handbook*, 7th Edition, McGraw-Hill, New York, 1997.

SOLUTION From Equation 9.3-1

$$\Delta\hat{H}_r^\circ = 5(\Delta\hat{H}_f^\circ)_{CO_2(g)} + 6(\Delta\hat{H}_f^\circ)_{H_2O(l)} - (\Delta\hat{H}_f^\circ)_{C_5H_{12}(l)}$$

⇓ Heats of formation from Table B.1

$$\Delta\hat{H}_r^\circ = [5(-393.5) + 6(-285.84) - (-173.0)] \text{ kJ/mol}$$

$$= \boxed{-3509 \text{ kJ/mol}}$$

To verify the formula for $\Delta\hat{H}_r^\circ$, we may write the stoichiometric equations for the formation reactions of the reactants and products:

1. $5 \text{ C(s)} + 6 \text{ H}_2\text{(g)} \rightarrow \text{C}_5\text{H}_{12}\text{(l)}$: $\quad \Delta\hat{H}_{r1}^\circ = (\Delta\hat{H}_f^\circ)_{C_5H_{12}(l)}$
2. $\text{C(s)} + \text{O}_2\text{(g)} \rightarrow \text{CO}_2\text{(g)}$: $\quad \Delta\hat{H}_{r2}^\circ = (\Delta\hat{H}_f^\circ)_{CO_2(g)}$
3. $\text{H}_2\text{(g)} + \frac{1}{2}\text{O}_2\text{(g)} \rightarrow \text{H}_2\text{O(l)}$: $\quad \Delta\hat{H}_{r3}^\circ = (\Delta\hat{H}_f^\circ)_{H_2O(l)}$

The desired reaction,

4. $\text{C}_5\text{H}_{12}\text{(l)} + 8 \text{ O}_2\text{(g)} \rightarrow 5 \text{ CO}_2\text{(g)} + 6 \text{ H}_2\text{O(l)}$: $\quad \Delta\hat{H}_r^\circ = ?$

may be obtained as $5 \times (2) + 6 \times (3) - (1)$ (*verify*), and the given formula for $\Delta\hat{H}_r^\circ$ then follows from Hess's law.

Techniques for estimating heats of formation of compounds with known molecular structures are given by Reid, Prausnitz, and Poling.[2]

TEST YOURSELF

1. The standard heat of the reaction

$$2 \text{ CO} \rightarrow 2 \text{ C} + \text{O}_2$$

is $\Delta\hat{H}_r^\circ = +221.0$ kJ/mol. Use this result to calculate the standard heat of formation of CO and check your result with a tabulated value.

2. $\Delta\hat{H}_f^\circ$ is -28.64 kcal/mol for $\text{C}_3\text{H}_8\text{(l)}$ and -24.82 kcal/mol for $\text{C}_3\text{H}_8\text{(g)}$. What is the physical significance of the difference between these values?

3. Consider the reaction

$$\text{CH}_4 + 2 \text{ O}_2 \rightarrow \text{CO}_2 + 2 \text{ H}_2\text{O(v)}$$

Write the formula for $\Delta\hat{H}_r^\circ$ in terms of the standard heats of formation of the reactants and products.

9.4 HEATS OF COMBUSTION

The **standard heat of combustion** of a substance, $\Delta\hat{H}_c^\circ$, is the heat of the combustion of that substance with oxygen to yield specified products [e.g., $\text{CO}_2\text{(g)}$ and $\text{H}_2\text{O(l)}$], with both reactants and products at 25°C and 1 atm (the arbitrary but conventional reference state).

Table B.1 lists standard heats of combustion for a number of substances. The given values are based on the following assumptions: (a) all carbon in the fuel forms $\text{CO}_2\text{(g)}$, (b) all hydrogen forms $\text{H}_2\text{O(l)}$, (c) all sulfur forms $\text{SO}_2\text{(g)}$, and (d) all nitrogen forms $\text{N}_2\text{(g)}$. The standard heat of combustion of liquid ethanol, for example, is given in Table B.1 as $\Delta\hat{H}_c^\circ = -1366.9$ kJ/mol, which signifies

$$\text{C}_2\text{H}_5\text{OH(l)} + 3 \text{ O}_2\text{(g)} \rightarrow 2 \text{ CO}_2\text{(g)} + 3 \text{ H}_2\text{O(l)}: \quad \Delta\hat{H}_r(25°C, 1 \text{ atm}) = -1366.9 \text{ kJ/mol}$$

Additional heats of combustion are given on pp. 2-195 through 2-199 of *Perry's Chemical Engineers' Handbook* (see footnote 1).

---

[2]R. C. Reid, J. M. Prausnitz, and B. E. Poling, *The Properties of Gases and Liquids,* 4th Edition, McGraw-Hill, New York, 1987.

Standard heats of reactions that involve only combustible substances and combustion products can be calculated from tabulated standard heats of combustion, in another application of Hess's law. A hypothetical reaction path may be constructed in which (a) all combustible reactants are burned with O_2 at 25°C and (b) CO_2 and H_2O combine to form the reaction products plus O_2. Step (b) involves the reverse of the combustion reactions of the reaction products. Since both steps involve only combustion reactions, the total enthalpy change—which equals the desired heat of reaction—can be determined entirely from heats of combustion as

$$\Delta \hat{H}_r^\circ = -\sum_i \nu_i (\Delta \hat{H}_c^\circ)_i = \sum_{\text{reactants}} |\nu_i|(\Delta \hat{H}_c^\circ)_i - \sum_{\text{products}} |\nu_i|(\Delta \hat{H}_c^\circ)_i \qquad \textbf{(9.4-1)}$$

If any of the reactants or products are themselves combustion products $[CO_2, H_2O(l), SO_2, \ldots]$, their $\Delta \hat{H}_c^\circ$ terms in Equation 9.4-1 should be set equal to 0.

Note that this formula is similar to that used to determine $\Delta \hat{H}_r^\circ$ from heats of formation, except that in this case the negative of the sum is taken. The validity of this formula is illustrated in the next example.

EXAMPLE 9.4-1   *Calculation of a Heat of Reaction from Heats of Combustion*

Calculate the standard heat of reaction for the dehydrogenation of ethane:

$$C_2H_6 \rightarrow C_2H_4 + H_2$$

SOLUTION   From Table B.1,

$$(\Delta \hat{H}_c^\circ)_{C_2H_6} = -1559.9 \text{ kJ/mol}$$
$$(\Delta \hat{H}_c^\circ)_{C_2H_4} = -1411.0 \text{ kJ/mol}$$
$$(\Delta \hat{H}_c^\circ)_{H_2} = -285.84 \text{ kJ/mol}$$

From Equation 9.4-1, therefore,

$$\Delta \hat{H}_r^\circ = (\Delta \hat{H}_c^\circ)_{C_2H_6} - (\Delta \hat{H}_c^\circ)_{C_2H_4} - (\Delta \hat{H}_c^\circ)_{H_2} = \boxed{136.9 \text{ kJ/mol}}$$

As an illustration, let us demonstrate the validity of this formula using Hess's law. The combustion reactions are

1. $C_2H_6 + \frac{7}{2}O_2 \rightarrow 2\, CO_2 + 3\, H_2O$
2. $C_2H_4 + 3\, O_2 \rightarrow 2\, CO_2 + 2\, H_2O$
3. $H_2 + \frac{1}{2}O_2 \rightarrow H_2O$

It is easy to show that

4. $C_2H_6 \rightarrow C_2H_4 + H_2$

is obtained as $(1) - (2) - (3)$. (Show it.) The desired result follows from Hess's law.

One of the principal applications of Equation 9.4-1 is to determine heats of formation for combustible substances whose formation reactions do not occur naturally. For example, the formation reaction of pentane

$$5\, C(s) + 6\, H_2(g) \rightarrow C_5H_{12}(l): \quad \Delta \hat{H}_f^\circ = ?$$

cannot be carried out in a laboratory, but carbon, hydrogen, and pentane can all be burned and their standard heats of combustion determined experimentally. The heat of formation of pentane may then be calculated from Equation 9.4-1 as

$$(\Delta \hat{H}_f^\circ)_{C_5H_{12}(l)} = 5(\Delta \hat{H}_c^\circ)_{C(s)} + 6(\Delta \hat{H}_c^\circ)_{H_2(g)} - (\Delta \hat{H}_c^\circ)_{C_5H_{12}(l)}$$

CREATIVITY EXERCISE

When an exothermic reaction takes place, the energy released raises the temperature of the reactor contents unless the reactor is cooled. Suppose such a reaction occurs in a batch reactor.

1. Think of as many reasons as you can why you might not want the reactor temperature to increase. (*Example:* The product might degrade or decompose at higher temperatures.)
2. Think of as many ways as you can to keep the reactor temperature from increasing as the reaction proceeds. (*Example:* Throw in some ice.)

9.5   ENERGY BALANCES ON REACTIVE PROCESSES

9.5a   General Procedures

To perform energy balance calculations on a reactive system, proceed much as you did for nonreactive systems: (a) draw and label a flowchart; (b) use material balances and phase equilibrium relationships such as Raoult's law to determine as many stream component amounts or flow rates as possible; (c) choose reference states for specific enthalpy (or internal energy) calculations and prepare and fill in an inlet–outlet enthalpy (or internal energy) table; and (d) calculate $\Delta \dot{H}$ (or $\Delta \dot{U}$ or ΔH), substitute the calculated value in the appropriate form of the energy balance equation, and complete the required calculation.

Two methods are commonly used to choose reference states for enthalpy calculations and to calculate specific enthalpies and $\Delta \dot{H}$.[3] We outline the two approaches below, using a propane combustion process to illustrate them. For simplicity, the material balance calculations for the illustrative process have been performed and the results incorporated into the flowchart.

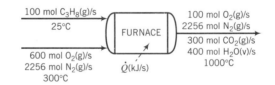

$$C_3H_8(g) + 5\,O_2(g) \longrightarrow 3\,CO_2(g) + 4\,H_2O(l): \quad \Delta \hat{H}_r^\circ = -2220 \text{ kJ/mol}$$

Heat of Reaction Method.   This method is generally preferable when there is a single reaction for which $\Delta \hat{H}_r^\circ$ is known.

1. *Complete the material balance calculations on the reactor to the greatest extent possible.*
2. *Choose reference states for specific enthalpy calculations.* The best choices are generally reactant and product species at 25°C and 1 atm in the states for which the heat of reaction is known [$C_3H_8(g)$, $O_2(g)$, $CO_2(g)$, and $H_2O(l)$ in the example process], and nonreacting species at any convenient temperature, such as the reactor inlet or outlet temperature or the reference condition used for the species in an available enthalpy table [$N_2(g)$ at 25°C and 1 atm, the reference state for Table B.8].
3. *For a single reaction in a continuous process, calculate the extent of reaction, $\dot{\xi}$, from Equation 9.1-3.*[4] When writing the equation, choose as species A any reactant or product for

---

[3] In what follows, we presume that the value of $\Delta \dot{H}$ is needed for the energy balance. If ΔU or ΔH is required, replace each $\dot{H}$ that appears with U or H.

[4] If multiple reactions occur you would calculate the extents of each independent reaction, $\dot{\xi}_1, \dot{\xi}_2, \ldots$ (Equation 4.6-6 on p. 123), but for such processes you are generally better off using the heat of formation method to be described.

which the feed and product flow rates are known. In the example, we may choose any reactant or product since we know all inlet and outlet species flow rates and calculate the rate of consumption or generation of A ($\dot{n}_{A,r}$ in Equation 9.1-3) as $|(\dot{n}_A)_{out} - (\dot{n}_A)_{in}|$. If A is propane,

$$\dot{\xi} = \frac{|(\dot{n}_{C_3H_8})_{out} - (\dot{n}_{C_3H_8})_{in}|}{|\nu_{C_3H_8}|} = \frac{|0 - 100| \text{ mol/s}}{1} = 100 \text{ mol/s}$$

As an exercise, let A be O_2, CO_2, or H_2O and verify that the value of $\dot{\xi}$ is independent of the species chosen.

4. *Prepare the inlet–outlet enthalpy table, inserting known molar amounts (n_i) or flow rates ($\dot{n}_i$) for all inlet and outlet stream components.* If any of the components is at its reference state, insert 0 for the corresponding $\hat{H}_i$. For the example process, the table would appear as follows:

References: $C_3H_8(g)$, $O_2(g)$, $N_2(g)$, $CO_2(g)$, $H_2O(l)$ at 25°C and 1 atm

Substance	$\dot{n}_{in}$ (mol/s)	$\hat{H}_{in}$ (kJ/mol)	$\dot{n}_{out}$ (mol/s)	$\hat{H}_{out}$ (kJ/mol)
C_3H_8	100	0	—	—
O_2	600	$\hat{H}_2$	100	$\hat{H}_4$
N_2	2256	$\hat{H}_3$	2256	$\hat{H}_5$
CO_2	—	—	300	$\hat{H}_6$
H_2O	—	—	400	$\hat{H}_7$

5. *Calculate each unknown stream component enthalpy, $\hat{H}_i$, as $\Delta\hat{H}$ for the species going from its reference state to the process state, and insert the enthalpies in the table.* In the example,

$$\hat{H}_2 = \Delta\hat{H} \text{ for } O_2(25°C) \rightarrow O_2(300°C) = 8.47 \text{ kJ/mol (from Table B.8)}$$

We proceed in the same manner to calculate $\hat{H}_3 = 8.12$ kJ/mol, $\hat{H}_4 = 32.47$ kJ/mol, $\hat{H}_5 = 30.56$ kJ/mol, $\hat{H}_6 = 48.60$ kJ/mol, and $\hat{H}_7 = 81.71$ kJ/mol.

Consider the last result. By definition

$$\hat{H}_7 = \Delta\hat{H} \text{ for } H_2O(l, 25°C) \rightarrow H_2O(g, 1000°C)$$

We could either use steam tables to determine $\Delta\hat{H}$ in one step or heat the liquid water from 25°C to 100°C, vaporize it, heat the vapor from 100°C to 1000°C, and calculate $\hat{H}_7 = \int_{25°C}^{100°C} C_{pl} \, dT + \Delta\hat{H}_v(100°C) + \int_{100°C}^{1000°C} C_{pv} \, dT$.

6. *Calculate $\Delta\dot{H}$ for the reactor.* Use one of the following formulas:

$$\Delta\dot{H} = \dot{\xi} \Delta\hat{H}_r° + \sum \dot{n}_{out}\hat{H}_{out} - \sum \dot{n}_{in}\hat{H}_{in} \quad \text{(single reaction)} \tag{9.5-1a}$$

$$\Delta\dot{H} = \sum_{\text{reactions}} \dot{\xi}_j \Delta\hat{H}_{rj}° + \sum \dot{n}_{out}\hat{H}_{out} - \sum \dot{n}_{in}\hat{H}_{in} \quad \text{(multiple reactions)} \tag{9.5-1b}$$

A derivation of these equations is outlined following the presentation of the heat of formation method. Substitution of the previously calculated values into Equation 9.5-1a yields $\Delta\dot{H} = -1.26 \times 10^5$ kJ/s.

7. Substitute the calculated value of $\Delta\dot{H}$ in the energy balance ($\dot{Q} - \dot{W}_s = \Delta\dot{H} + \Delta\dot{E}_k + \Delta\dot{E}_p$ for an open system) and complete the required calculations.

Heat of Formation Method: This method is generally preferable for multiple reactions and single reactions for which $\Delta\hat{H}_r$ is not readily available.

1. *Complete the material balance calculations on the reactor to the greatest extent possible.*
2. *Choose reference states for enthalpy calculations.* (This is the step that distinguishes the preceding method from this one.) The choices should be the elemental species that con-

stitute the reactants and products in the states in which the elements are found at 25°C and 1 atm [C(s), H$_2$(g), etc.] and nonreacting species at any convenient temperature. In the example, the reference states would be C(s), H$_2$(g), and O$_2$(g) at 25°C (the elemental species constituting the reactants and products), and N$_2$ at 25°C (the reference temperature of Table B.8).

3. *Prepare the inlet–outlet enthalpy table, inserting known molar amounts (n_i) or flow rates ($\dot{n}_i$) for all inlet and outlet stream components.* For the example process, the table would appear as follows:

References: C(s), H$_2$(g), O$_2$(g), N$_2$(g) at 25°C and 1 atm

Substance	$\dot{n}_{in}$ (mol/s)	$\hat{H}_{in}$ (kJ/mol)	$\dot{n}_{out}$ (mol/s)	$\hat{H}_{out}$ (kJ/mol)
C$_3$H$_8$	100	$\hat{H}_1$	—	—
O$_2$	600	$\hat{H}_2$	100	$\hat{H}_4$
N$_2$	2256	$\hat{H}_3$	2256	$\hat{H}_5$
CO$_2$	—	—	300	$\hat{H}_6$
H$_2$O	—	—	400	$\hat{H}_7$

4. *Calculate each unknown specific enthalpy.* For a reactant or product, start with the elemental species at 25°C and 1 atm (the references) and form 1 mol of the process species at 25°C and 1 atm ($\Delta\hat{H} = \Delta\hat{H}_f^\circ$ from Table B.1). Then bring the species from 25°C and 1 atm to its process state, calculating $\Delta\hat{H}$ using the appropriate heat capacities from Table B.2, specific enthalpies from Table B.8 or B.9, and latent heats from Table B.1. The specific enthalpy that goes in the inlet–outlet table is the sum of the enthalpy changes for each step in the process path.

In the example, we would first calculate the specific enthalpy of the entering propane ($\hat{H}_1$) as follows:

$$3\,C(s)(25°C, 1\text{ atm}) + 4\,H_2(g)(25°C, 1\text{ atm}) \rightarrow C_3H_8(g)(25°C, 1\text{ atm})$$

$$\Downarrow$$

$$\hat{H}_1 = (\Delta\hat{H}_f^\circ)_{C_3H_8(g)} = -103.8 \text{ kJ/mol (from Table B.1)}$$

This is the enthalpy of propane at 25°C (the process state) relative to C(s) and H$_2$(g) at 25°C (the reference states). If the propane had entered at a temperature T_0 other than 25°C, a term of the form $\int_{25°C}^{T_0} C_p\, dT$ would be added to the heat of formation of propane.

Next, we calculate the specific enthalpy of O$_2$ at 300°C (the process state) relative to O$_2$ at 25°C (the reference state) as $\hat{H}_2 = 8.47$ kJ/mol (from Table B.8). There is no heat of formation term, since O$_2$ is an elemental species. We proceed in the same manner to calculate $\hat{H}_3 = 8.12$ kJ/mol, $\hat{H}_4 = 32.47$ kJ/mol, $\hat{H}_5 = 30.56$ kJ/mol, $\hat{H}_6 = -344.9$ kJ/mol, and $\hat{H}_7 = -204.1$ kJ/mol. To calculate $\hat{H}_6$ and $\hat{H}_7$, we form the corresponding species [CO$_2$(g) and H$_2$O(v)] at 25°C from their elements ($\Delta\hat{H} = \Delta\hat{H}_f^\circ$), then heat them from 25°C to 1000°C ($\Delta\hat{H} = \hat{H}_{1000°C}$ from Table B.8), and add the formation and heating terms.

5. *Calculate $\Delta\dot{H}$ for the reactor.* For both single and multiple reactions, the formula is

$$\Delta\dot{H} = \sum \dot{n}_{out}\hat{H}_{out} - \sum \dot{n}_{in}\hat{H}_{in} \qquad \textbf{(9.5-2)}$$

Note that heat of reaction terms are not required if the elements are chosen as references. The heats of reaction are implicitly included when the heats of formation of the reactants (included in the $\hat{H}_{in}$ terms) are subtracted from those of the products (in the $\hat{H}_{out}$ terms) in the expression for $\Delta\dot{H}$. Substituting the calculated $\dot{n}$ and $\hat{H}$ values into Equation 9.5-2 yields $\Delta\dot{H} = -1.26 \times 10^5$ kJ/s.

6. *Substitute the calculated value of $\Delta\dot{H}$ in the energy balance equation and complete the required calculations.*

The process paths that correspond to the heat of reaction and heat of formation methods are shown below.

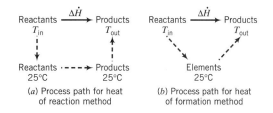

(a) Process path for heat
of reaction method

(b) Process path for heat
of formation method

The heat of reaction method amounts to bringing the reactants from their inlet conditions to their reference states at 25°C ($\Delta\dot{H} = -\sum \dot{n}_{in}\hat{H}_{in}$), carrying out the reaction at 25°C (from Equation 9.1-3, $\Delta\dot{H} = \dot{\xi}\,\Delta\hat{H}_r^\circ$ or the summation of such terms for multiple reactions), bringing the products from their reference states at 25°C to their outlet states ($\Delta\dot{H} = \sum \dot{n}_{out}\hat{H}_{out}$), and summing the enthalpy changes for these three steps to calculate $\Delta\dot{H}$ for the overall process. The heat of formation method amounts to bringing the reactants from their inlet conditions to their constituent elements at 25°C ($\Delta\dot{H} = -\sum \dot{n}_{in}\hat{H}_{in}$), taking the elements to the products at their outlet states ($\Delta\dot{H} = \sum \dot{n}_{out}\hat{H}_{out}$), and summing the enthalpy changes for these two steps to calculate $\Delta\dot{H}$ for the overall process.

EXAMPLE 9.5-1   *Energy Balance About an Ammonia Oxidizer*

The standard heat of reaction for the oxidation of ammonia is given below:

$$4\,NH_3(g) + 5\,O_2(g) \longrightarrow 4\,NO(g) + 6\,H_2O(v): \quad \Delta\hat{H}_r^\circ = -904.7\ kJ/mol$$

One hundred mol NH_3/s and 200 mol O_2/s at 25°C are fed into a reactor in which the ammonia is completely consumed. The product gas emerges at 300°C. Calculate the rate at which heat must be transferred to or from the reactor, assuming operation at approximately 1 atm.

SOLUTION   *Basis: Given Feed Rates*

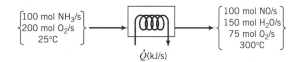

(Verify the product flow rates.) Since only one reaction takes place and $\Delta\hat{H}_r^\circ$ is known, we will use the heat of reaction method for the energy balance, choosing as references the reactant and product species in the states for which the heat of reaction is given. The enthalpy table appears as follows:

References: $NH_3(g)$, $O_2(g)$, $NO(g)$, $H_2O(v)$ at 25°C and 1 atm

Substance	$\dot{n}_{in}$ (mol/s)	$\hat{H}_{in}$ (kJ/mol)	$\dot{n}_{out}$ (mol/s)	$\hat{H}_{out}$ (kJ/mol)
NH_3	100	0	—	—
O_2	200	0	75	$\hat{H}_1$
NO	—	—	100	$\hat{H}_2$
H_2O	—	—	150	$\hat{H}_3$

Calculate Unknown Enthalpies

$O_2(g, 300°C)$: From Table B.8, $\hat{H}_1 = 8.470$ kJ/mol (Insert value in enthalpy table)

$NO(g, 300°C)$: $\hat{H}_2 = \int_{25°C}^{300°C} (C_p)_{NO}\, dT \xrightarrow{\text{Table B.2}} \hat{H}_2 = 8.453$ kJ/mol (Insert in table)

$H_2O(v, 300°C)$: From Table B.8, $\hat{H}_3 = 9.570$ kJ/mol (Insert in table)

Calculate $\dot{\xi}$ and $\Delta\dot{H}$

Since 100 mol NH_3/s is consumed in the process ($A = NH_3$, $\dot{n}_{A,r} = 100$ mol NH_3 consumed/s), Equation 9.1-3 becomes

$$\dot{\xi} = \frac{\dot{n}_{NH_3,r}}{|\nu_{NH_3}|} = \frac{100 \text{ mol/s}}{4} = 25 \text{ mol/s}$$

$$\Big\Downarrow \text{Equation 9.5-1a}$$

$$\Delta\dot{H} = \dot{\xi}\Delta\hat{H}_r^\circ + \underbrace{\sum \dot{n}_{out}\hat{H}_{out} - \sum \dot{n}_{in}\hat{H}_{in}}_{\text{from table}}$$

$$= (25 \text{ mol/s})(-904.7 \text{ kJ/mol}) + [(75)(8.470) + (100)(8.453)$$
$$+ (150)(9.570) - (100)(0) - (200)(0)] \text{ kJ/s} = -19{,}700 \text{ kJ/s}$$

Energy Balance

For this open system,

$$\dot{Q} - \dot{W}_s = \Delta\dot{H} + \Delta\dot{E}_k + \Delta\dot{E}_p$$

$$\Big\Downarrow \begin{array}{l} \dot{W}_s = 0 \quad \text{(no moving parts)} \\ \Delta\dot{E}_p = 0 \quad \text{(horizontal unit)} \\ \Delta\dot{E}_k \approx 0 \quad \text{(neglect kinetic energy changes)} \end{array}$$

$$\dot{Q} \approx \Delta\dot{H} = -19{,}700 \text{ kJ/s} = \boxed{-19{,}700 \text{ kW}}$$

Thus, 19,700 kW of heat must be transferred from the reactor to maintain the product temperature at 300°C. If less heat were transferred, more of the heat of reaction would go into the reaction mixture and the outlet temperature would increase.

The heat of formation method, which involves taking elemental constituents of the reactants and products in their naturally occurring states as references for enthalpy calculations, is usually convenient for processes that involve several simultaneous reactions. The next example illustrates this approach.

EXAMPLE 9.5-2 **Energy Balance on a Methane Oxidation Reactor**

Methane is oxidized with air to produce formaldehyde in a continuous reactor. A competing reaction is the combustion of methane to form CO_2.

1. $CH_4(g) + O_2 \rightarrow HCHO(g) + H_2O(v)$
2. $CH_4(g) + 2O_2 \rightarrow CO_2 + 2H_2O(v)$

A flowchart of the process for an assumed basis of 100 mol of methane fed to the reactor is shown here.

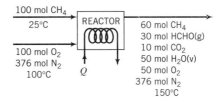

SOLUTION

Basis: 100 mol CH₄ Fed

Since the component amounts of all streams are known, we may proceed directly to the energy balance. We choose as references the elemental species that form the reactants and products at 25°C and 1 atm (the state for which heats of formation are known) and the nonreactive species—$N_2(g)$—also at 25°C and 1 atm (the reference state for Table B.8). The inlet–outlet enthalpy table is shown below.

References: C(s), O_2(g), H_2(g), N_2(g) at 25°C and 1 atm

Substance	n_{in} (mol)	$\hat{H}_{in}$ (kJ/mol)	n_{out} (mol)	$\hat{H}_{out}$ (kJ/mol)
CH₄	100	$\hat{H}_1$	60	$\hat{H}_4$
O₂	100	$\hat{H}_2$	50	$\hat{H}_5$
N₂	376	$\hat{H}_3$	376	$\hat{H}_6$
HCHO	—	—	30	$\hat{H}_7$
CO₂	—	—	10	$\hat{H}_8$
H₂O	—	—	50	$\hat{H}_9$

Calculate Unknown Enthalpies

In the following calculations, values of $\Delta\hat{H}_f^\circ$ come from Table B.1, formulas for $C_p(T)$ come from Table B.2, and values of $\hat{H}(T)$ for O_2 and N_2 are specific enthalpies relative to the gaseous species at 25°C taken from Table B.8. Effects of any pressure changes on enthalpies are neglected, and the details of the calculations are not shown.

$$\text{CH}_4(25°\text{C}):\quad \hat{H}_1 = (\Delta\hat{H}_f^\circ)_{\text{CH}_4} = -74.85\ \text{kJ/mol}$$

$$\text{O}_2(100°\text{C}):\quad \hat{H}_2 = \hat{H}_{\text{O}_2}(100°\text{C}) = 2.235\ \text{kJ/mol}$$

$$\text{N}_2(100°\text{C}):\quad \hat{H}_3 = \hat{H}_{\text{N}_2}(100°\text{C}) = 2.187\ \text{kJ/mol}$$

$$\text{CH}_4(150°\text{C}):\quad \hat{H}_4 = (\Delta\hat{H}_f^\circ)_{\text{CH}_4} + \int_{25°\text{C}}^{150°\text{C}} (C_p)_{\text{CH}_4}\, dT$$

$$= (-74.85 + 4.90)\ \text{kJ/mol} = -69.95\ \text{kJ/mol}$$

$$\text{O}_2(150°\text{C}):\quad \hat{H}_5 = \hat{H}_{\text{O}_2}(150°\text{C}) = 3.758\ \text{kJ/mol}$$

$$\text{N}_2(150°\text{C}):\quad \hat{H}_6 = \hat{H}_{\text{N}_2}(150°\text{C}) = 3.655\ \text{kJ/mol}$$

$$\text{HCHO}(150°\text{C}):\quad \hat{H}_7 = (\Delta\hat{H}_f^\circ)_{\text{HCHO}} + \int_{25°\text{C}}^{150°\text{C}} (C_p)_{\text{HCHO}}\, dT$$

$$= (-115.90 + 4.75)\ \text{kJ/mol} = -111.15\ \text{kJ/mol}$$

$$\text{CO}_2(150°\text{C}):\quad \hat{H}_8 = (\Delta\hat{H}_f^\circ)_{\text{CO}_2} + \hat{H}_{\text{CO}_2}(150°\text{C})$$

$$= (-393.5 + 4.75)\ \text{kJ/mol} = -388.6\ \text{kJ/mol}$$

$$\text{H}_2\text{O(v, 150°C)}:\quad \hat{H}_9 = (\Delta\hat{H}_f^\circ)_{\text{H}_2\text{O(v)}} + \hat{H}_{\text{H}_2\text{O(v)}}(150°\text{C})$$

$$= (-241.83 + 4.27)\ \text{kJ/mol} = -237.56\ \text{kJ/mol}$$

As each of these values is calculated, it should be substituted in the inlet–outlet enthalpy table. The table finally appears as follows:

References: C(s), O_2(g), H_2(g), N_2(g) at 25°C and 1 atm

Substance	n_{in} (mol)	$\hat{H}_{in}$ (kJ/mol)	n_{out} (mol)	$\hat{H}_{out}$ (kJ/mol)
CH_4	100	–74.85	60	–69.95
O_2	100	2.235	50	3.758
N_2	376	2.187	376	3.655
HCHO	—	—	30	–111.15
CO_2	—	—	10	–388.6
H_2O	—	—	50	–237.56

Evaluate ΔH

From Equation 9.5-2,

$$\Delta H = \sum n_{out}\hat{H}_{out} - \sum n_{in}\hat{H}_{in} = -15,300 \text{ kJ}$$

If molecular species had been chosen as references for enthalpy calculations, the extents of each reaction (ξ_1 and ξ_2) would have had to be calculated and Equation 9.5-1b used to determine ΔH. When more than one reaction occurs in a process, you are advised to choose elemental species as references and avoid these complications.

Energy Balance

Remember that we are dealing with a continuous process and hence an open system. [The reason we use n(mol) and not $\dot{n}$(mol/s) is that we took 100 mol CH_4 as a basis of calculation.] With ΔE_k, ΔE_p, and W_s neglected, the open system energy balance yields

$$\boxed{Q = \Delta H = -15,300 \text{ kJ}}$$

9.5b Processes with Unknown Outlet Conditions: Adiabatic Reactors

In the reactive systems we have looked at so far, the inlet and outlet conditions were specified and the required heat input could be determined from an energy balance.

In another important class of problems, the input conditions, heat input, and product composition are specified, and the outlet temperature is to be calculated. To solve such problems, you must evaluate the enthalpies of the products relative to the chosen reference states in terms of the unknown final temperature, and then substitute the resulting expressions into the energy balance ($\dot{Q} = \Delta\dot{H}$, or $\Delta\dot{H} = 0$ for an adiabatic reactor) to calculate T_{out}.

EXAMPLE 9.5-3 *Energy Balance on an Adiabatic Reactor*

The dehydrogenation of ethanol to form acetaldehyde

$$C_2H_5OH(v) \rightarrow CH_3CHO(v) + H_2(g)$$

is carried out in a continuous adiabatic reactor. Ethanol vapor is fed to the reactor at 400°C, and a conversion of 30% is obtained. Calculate the product temperature.

SOLUTION

Basis: 100 mol Feed

Material balances lead to the information on the flowchart shown here.

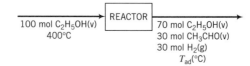

100 mol C_2H_5OH(v)
400°C

REACTOR

70 mol C_2H_5OH(v)
30 mol CH_3CHO(v)
30 mol H_2(g)
T_{ad}(°C)

Since only one reaction occurs, we could equally well choose the reactants and products [$C_2H_5OH(v)$, $CH_3CHO(v)$, $H_2(g)$] or their elemental constituents [$C(s)$, $H_2(g)$, $O_2(g)$] as references for enthalpy calculations. Let us choose the molecular species.

References: $C_2H_5OH(v)$, $CH_3CHO(v)$, $H_2(g)$ at 25°C and 1 atm

Substance	n_{in} (mol)	$\hat{H}_{in}$ (kJ/mol)	n_{out} (mol)	$\hat{H}_{out}$ (kJ/mol)
C_2H_5OH	100.0	$\hat{H}_1$	70.0	$\hat{H}_2$
CH_3CHO	—	—	30.0	$\hat{H}_3$
H_2	—	—	30.0	$\hat{H}_4$

The open system energy balance neglecting kinetic and potential energy changes and shaft work and setting $Q = 0$ for this adiabatic reactor is

$$\Delta H = \xi \, \Delta \hat{H}_r^\circ + \sum n_{out} \hat{H}_{out} - \sum n_{in} \hat{H}_{in} = 0$$

The expression for ΔH is that of Equation 9.5-1a.

Calculate Extent of Reaction

We could use any reactant or product as a basis for calculating ξ. Let us use acetaldehyde. From Equation 9.1-3,

$$\xi = \frac{|(n_{CH_3CHO})_{out} - (n_{CH_3CHO})_{in}|}{|\nu_{CH_3CHO}|} = \frac{|30.0 \text{ mol} - 0 \text{ mol}|}{1} = 30.0 \text{ mol}$$

Calculate Standard Heat of Reaction

From Equation 9.3-1 and Table B.1 (heats of formation),

$$\Delta \hat{H}_r^\circ = \sum \nu_i \, \Delta \hat{H}_f^\circ = (-1)(\Delta \hat{H}_f^\circ)_{C_2H_5OH(v)} + (1)(\Delta \hat{H}_f^\circ)_{CH_3CHO(v)} + (1)(\Delta \hat{H}_f^\circ)_{H_2(g)}$$

$$= [(-1)(-235.31) + (1)(-166.2) + (1)(0)] \text{ kJ/mol} = 69.11 \text{ kJ/mol}$$

Calculate Inlet Enthalpy

$$\hat{H}_1 = \int_{25°C}^{400°C} (C_p)_{C_2H_5OH} \xrightarrow{C_p \text{ from Table B.2}} \hat{H}_1 = 33.79 \text{ kJ/mol}$$

Calculate Outlet Enthalpies

The heat capacities of ethanol vapor and hydrogen are given in Table B.2. For acetaldehyde vapor, the heat capacity is given by Reid, Prausnitz, and Poling:[5]

$$(C_p)_{CH_3CHO}\left(\frac{kJ}{mol \cdot °C}\right) = 0.05048 + 1.326 \times 10^{-4} T - 8.050 \times 10^{-8} T^2 + 2.380 \times 10^{-11} T^3$$

where T is in °C. For the three species in the product stream,

$$\hat{H}_i = \int_{25°C}^{T_{ad}} C_{pi}(T) \, dT, \quad i = 1, 2, 3$$

If the heat capacity formulas for the three species are substituted in this expression and the integrals are evaluated, the results are three fourth-order polynomial expressions for $\hat{H}_2(T_{ad})$, $\hat{H}_3(T_{ad})$, and $\hat{H}_4(T_{ad})$:

C_2H_5OH:

$$\hat{H}_2(kJ/mol) = 4.958 \times 10^{-12} T_{ad}^4 - 2.916 \times 10^{-8} T_{ad}^3$$

$$+ 7.860 \times 10^{-5} T_{ad}^2 + 0.06134 T_{ad} - 1.582$$

---

[5]R. C. Reid, J. M. Prausnitz, and B. E. Poling, *The Properties of Gases and Liquids,* 4th Edition, McGraw-Hill, New York, 1987. The formula given has been derived from the one shown in this reference, which is for the heat capacity in J/(mol·K) with the temperature expressed in kelvin.

CH₃CHO: $\hat{H}_3(kJ/mol) = 5.950 \times 10^{-12}T_{ad}^4 - 2.683 \times 10^{-8}T_{ad}^3$
$$+ 6.630 \times 10^{-5}T_{ad}^2 + 0.05048T_{ad} - 1.303$$

H₂: $\hat{H}_4(kJ/mol) = -0.2175 \times 10^{-12}T_{ad}^4 + 0.1096 \times 10^{-8}T_{ad}^3$
$$+ 0.003825 \times 10^{-5}T_{ad}^2 + 0.02884T_{ad} - 0.7210$$

Solve the Energy Balance for T_{ad}

$$\Delta H = \xi \, \Delta\hat{H}_r^{\circ} + (70.0 \text{ mol})\hat{H}_2 + (30.0 \text{ mol})\hat{H}_3 + (30.0 \text{ mol})\hat{H}_4 - (100.0 \text{ mol})\hat{H}_1 = 0$$

$\Downarrow$ Substitute for $\xi \, (= 30.0$ mol$)$, $\Delta\hat{H}_r^{\circ} \, (= 69.11$ kJ/mol$)$, $\hat{H}_1 \, (= 33.79$ kJ/mol$)$, and $\hat{H}_2$ through $\hat{H}_4$

$$\Delta H = 5.190 \times 10^{-10}T_{ad}^4 + 2.879 \times 10^{-6}T_{ad}^3 + 7.492 \times 10^{-3}T_{ad}^2 + 6.673T_{ad} - 1477 = 0$$

This equation can be solved using an equation-solving program or a spreadsheet.[6] The solution is

$$\boxed{T_{ad} = 185°C}$$

Another class of problems involves processes for which the heat input and outlet temperature are specified but the extent of reaction and product composition are not. Solving such problems requires the simultaneous solution of material and energy balance equations, as the next example illustrates.

EXAMPLE 9.5-4   ***Simultaneous Material and Energy Balances***

The ethanol dehydrogenation reaction of Example 9.5-3 is carried out with the feed entering at 300°C. The feed contains 90.0 mole% ethanol and the balance acetaldehyde and enters the reactor at a rate of 150 mol/s. To keep the temperature from dropping too much and thereby decreasing the reaction rate to an unacceptably low level, heat is transferred to the reactor. When the heat addition rate is 2440 kW, the outlet temperature is 253°C. Calculate the fractional conversion of ethanol achieved in the reactor.

SOLUTION

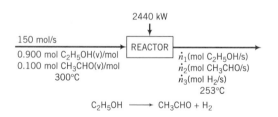

A degree-of-freedom analysis based on atomic species balances (see Section 4.7) is as follows:

$$3 \text{ unknown labeled variables } (\dot{n}_1, \dot{n}_2, \dot{n}_3)$$
$$- 2 \text{ independent atomic species balances (C and H)}$$
$$\underline{- 1 \text{ energy balance}}$$
$$= 0 \text{ degrees of freedom}$$

(Convince yourself that there are only two independent atomic balances by writing the C and O balances and observing that they yield the same equation.)

---

[6] To obtain the solution using a spreadsheet, put a guessed value of T_{ad} in one cell and the expression for ΔH in an adjacent cell, and use the goalseek tool to determine the value of T_{ad} for which the expression for ΔH equals zero. A first guess might be the value of T_{ad} obtained by dropping all higher-order terms in the expression, leaving $6.673T_{ad} - 1134 = 0 \Longrightarrow T_{ad} \approx 170°C$.

Balance on C

$$\frac{150 \text{ mol}}{\text{s}} \bigg| \frac{0.900 \text{ mol C}_2\text{H}_5\text{OH}}{\text{mol}} \bigg| \frac{2 \text{ mol C}}{1 \text{ mol C}_2\text{H}_5\text{OH}} + \frac{150 \text{ mol}}{\text{s}} \bigg| \frac{0.100 \text{ mol CH}_3\text{CHO}}{\text{mol}} \bigg| \frac{2 \text{ mol C}}{1 \text{ mol CH}_3\text{CHO}}$$

$$= \frac{\dot{n}_1(\text{mol C}_2\text{H}_5\text{OH})}{\text{s}} \bigg| \frac{2 \text{ mol C}}{1 \text{ mol C}_2\text{H}_5\text{OH}} + \frac{\dot{n}_2(\text{mol CH}_3\text{CHO})}{\text{s}} \bigg| \frac{2 \text{ mol C}}{1 \text{ mol CH}_3\text{CHO}}$$

$$\Downarrow$$

$$\dot{n}_1 + \dot{n}_2 = 150 \text{ mol/s} \qquad \textbf{(1)}$$

Balance on H

$$[(150)(0.900)(6) + (150)(0.100)(4)] \text{ mol H/s} = 6\dot{n}_1 + 4\dot{n}_2 + 2\dot{n}_3 \quad \text{(Convince yourself)}$$

$$\Downarrow$$

$$3\dot{n}_1 + 2\dot{n}_2 + \dot{n}_3 = 435 \text{ mol H/s} \qquad \textbf{(2)}$$

Energy Balance

In the last example we used molecular species as references for specific enthalpy calculations. This time we will use elemental species [C(s), H_2(g), O_2(g)] at 25°C and 1 atm. (For a single reaction both choices require about the same computational effort.) The energy balance neglecting shaft work and kinetic and potential energy changes becomes

$$\dot{Q} = \Delta\dot{H} = \sum \dot{n}_{\text{out}}\hat{H}_{\text{out}} - \sum \dot{n}_{\text{in}}\hat{H}_{\text{in}}$$

The value of $\dot{Q}$ is 2440 kJ/s and the expression for $\Delta\dot{H}$ is that of Equation 9.5-2. The specific enthalpies of the species at the inlet and outlet of the process relative to their elemental constituents are calculated as

$$\hat{H}_i = \Delta\hat{H}_{fi}^\circ + \int_{25°C}^{T} C_{pi}(T)\,dT$$

where T is 300°C at the inlet and 253°C at the outlet. Taking standard heats of formation from Table B.1 and formulas for C_p from Table B.2 and (for acetaldehyde vapor) Example 9.5-3, we calculate the values of $\hat{H}_i$ shown in the inlet–outlet enthalpy table.

References: C(s), H_2(g), O_2(g) at 25°C and 1 atm

Substance	$\dot{n}_{\text{in}}$ (mol/s)	$\hat{H}_{\text{in}}$ (kJ/mol)	$\dot{n}_{\text{out}}$ (mol/s)	$\hat{H}_{\text{out}}$ (kJ/mol)
C_2H_5OH	135	−212.19	$\dot{n}_1$	−216.81
CH_3CHO	15	−147.07	$\dot{n}_2$	−150.90
H_2	—	—	$\dot{n}_3$	6.595

The energy balance ($\dot{Q} = \sum \dot{n}_{\text{out}}\hat{H}_{\text{out}} - \sum \dot{n}_{\text{in}}\hat{H}_{\text{in}}$) becomes

$$2440 \text{ kJ/s} = [-216.81\dot{n}_1 - 150.90\dot{n}_2 + 6.595\dot{n}_3 - (135)(-212.19) - (15)(-147.07)] \text{ kJ/s}$$

$$\Downarrow$$

$$216.81\dot{n}_1 + 150.90\dot{n}_2 - 6.595\dot{n}_3 = 28{,}412 \text{ kJ/s} \qquad \textbf{(3)}$$

Solving Equations 1 through 3 simultaneously yields

$$\dot{n}_1 = 92.0 \text{ mol C}_2\text{H}_5\text{OH/s}$$
$$\dot{n}_2 = 58.0 \text{ mol CH}_3\text{CHO/s}$$
$$\dot{n}_3 = 43.0 \text{ mol H}_2\text{/s}$$

The fractional conversion of ethanol is

$$x = \frac{(\dot{n}_{C_2H_5OH})_{in} - (\dot{n}_{C_2H_5OH})_{out}}{(\dot{n}_{C_2H_5OH})_{in}} = \frac{(135 - 92.0)\ mol/s}{135\ mol/s} = \boxed{0.319}$$

9.5c   Thermochemistry of Solutions[7]

The enthalpy change associated with the formation of a solution from the solute elements and the solvent at 25°C is called the **standard heat of formation of the solution**. If a solution contains n moles of solvent per mole of solute, then

$$\boxed{(\Delta\hat{H}_f^\circ)_{solution} = (\Delta\hat{H}_f^\circ)_{solute} + \Delta\hat{H}_s^\circ(n)} \tag{9.5-3}$$

where $\Delta\hat{H}_s^\circ$ is the heat of solution at 25°C (Section 8.5). From the definitions of $\Delta\hat{H}_f^\circ$ and $\Delta\hat{H}_s^\circ$, the dimensions of the heat of formation of the solution are (energy)/(mole of solute).

The standard heat of a reaction involving solutions may be calculated from heats of formation of the solutions. For example, for the reaction

$$2\ HNO_3(aq, r = 25) + Ca(OH)_2(aq, r = \infty) \rightarrow Ca(NO_3)_2(aq, r = \infty) + 2\ H_2O(l)$$

the standard heat of reaction is

$$\Delta\hat{H}_r^\circ = (\Delta\hat{H}_f^\circ)_{Ca(NO_3)_2(aq)} + 2(\Delta\hat{H}_f^\circ)_{H_2O(l)} - 2(\Delta\hat{H}_f^\circ)_{HNO_3(aq,\ r=25)} - (\Delta\hat{H}_f^\circ)_{Ca(OH)_2(aq,\ r=\infty)}$$
$$= -114.2\ kJ/mol$$

The last equation signifies that if a solution containing 2 mol of HNO_3 in 50 mol of $H_2O(r = 25)$ is neutralized at 25°C with 1 mol of $Ca(OH)_2$ dissolved in enough water so that the addition of more water would not cause a measurable enthalpy change ($r = \infty$), the enthalpy change is -114.2 kJ.

If a standard heat of formation is tabulated for a solution involved in a reaction, the tabulated value may be substituted directly into the expression for $\Delta\hat{H}_r$; otherwise, $(\Delta\hat{H}_f^\circ)_{soln}$ must first be calculated by adding the standard heat of formation of the pure solute to the standard heat of solution.

EXAMPLE 9.5-5   *Standard Heat of a Neutralization Reaction*

1. Calculate $\Delta\hat{H}_r^\circ$ for the reaction

$$H_3PO_4(aq, r = \infty) + 3\ NaOH(aq, r = 50) \rightarrow Na_3PO_4(aq, r = \infty) + 3\ H_2O(l)$$

2. If 5.00 mol of NaOH dissolved in 250 mol of water is neutralized completely at 25°C with dilute phosphoric acid, what is the attendant enthalpy change?

SOLUTION

1. $H_3PO_4(aq)$: $\Delta\hat{H}_f^\circ = -309.3$ kcal/mol $= -1294$ kJ/mol [from p. 2-189 of *Perry's Chemical Engineers' Handbook* (see footnote 1)].

NaOH(aq, $r = 50$): $(\Delta\hat{H}_f^\circ)_{NaOH(aq)} = (\Delta\hat{H}_f^\circ)_{NaOH(s)} + \Delta\hat{H}_s^\circ(r = 50)$

$\left\Vert\begin{array}{l} \text{Table B.1 } (\Delta\hat{H}_f^\circ) \\ \text{Table B.11 } (\Delta\hat{H}_s^\circ) \end{array}\right.$

$= (-426.6 - 42.51)$ kJ/mol $= -469.1$ kJ/mol

$Na_3PO_4(aq)$: $(\Delta\hat{H}_f^\circ)_{NaOH(aq)} = -471.9$ kcal/mol $= -1974$ kJ/mol (from p. 2-193 of *Perry's Chemical Engineers' Handbook*).

---

[7]Reviewing Sections 8.5a and 8.5b before reading this section might prove helpful.

$H_2O(l)$: $\Delta \hat{H}_f^\circ = -285.8 \text{ kJ/mol}$ (from Table B.1)

$$\Delta \hat{H}_r^\circ = (\Delta \hat{H}_f^\circ)_{\text{Na}_3\text{PO}_4(\text{aq})} + 3(\Delta \hat{H}_f^\circ)_{\text{H}_2\text{O}(l)} - (\Delta \hat{H}_f^\circ)_{\text{H}_3\text{PO}_4(\text{aq})} - 3(\Delta \hat{H}_f^\circ)_{\text{NaOH}(\text{aq},\, r\,=\,50)}$$

$$= \boxed{-130.1 \text{ kJ/mol}}$$

2. If 5 mol of dissolved NaOH is neutralized, then

$$\Delta H(25^\circ\text{C}) = \frac{-130.1 \text{ kJ}}{3.00 \text{ mol NaOH}} \left| \frac{5.00 \text{ mol NaOH}}{} \right. = \boxed{-217 \text{ kJ}}$$

When you calculate ΔH for a reactive process as

$$\sum_{\text{products}} n_i \hat{H}_i - \sum_{\text{reactants}} n_i \hat{H}_i$$

and one of the reactant or product species is a solution, its specific enthalpy usually has the dimensions (energy)/(mole of solute), so that the corresponding value of n_i must be moles or molar flow rate of the *solute*, and not of the total solution. A complicating factor may be that while the heat of formation of a solution is always obtained in the desired units, solution heat capacities are normally based on a unit mass of the total solution rather than of the solute. To calculate the specific enthalpy of a solution at a temperature T in (energy)/(mole of solute), you must first calculate m, the mass of solution corresponding to 1 mole of dissolved solute, and then add

$$m \int_{25^\circ\text{C}}^{T} (C_p)_{\text{solution}} \, dT$$

to the standard heat of formation of the solution. The next example illustrates this calculation.

EXAMPLE 9.5-6   *Energy Balance on a Neutralization Process*

A 10.0 wt% aqueous solution of H_2SO_4 at 40°C is to be neutralized with a 20.0 wt% aqueous solution of NaOH at 25°C in a continuous reactor. At what rate in kJ/kg H_2SO_4 solution must heat be removed from the reactor if the product solution emerges at 35°C?

SOLUTION   *Basis: 1 kg H_2SO_4 Solution*

$$H_2SO_4(\text{aq}, 10\%) + 2\,\text{NaOH}(\text{aq}, 20\%) \rightarrow \text{Na}_2\text{SO}_4(\text{aq}) + 2\,H_2O(l)$$

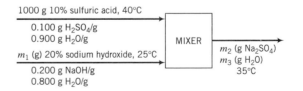

This is a straightforward problem, but the number of intermediate calculations required to solve it might make it appear more difficult than it is. Let us summarize what must be done.

1. Solve for m_1, m_2, and m_3 by material balances.
2. Calculate the solvent-to-solute mole ratios of all solutions. (The quantities are needed to determine the solution enthalpies from tabulated heats of solution.)
3. Calculate the enthalpies of the solutions. (This will require additional composition calculations to allow the use of tabulated solution heat capacities.)
4. Write the energy balance equation and solve it for the heat removal rate.

Observe that nothing here is really new, and, as we work our way through to the final result, recognize that most of the calculations are simply conversions of solution compositions from mass fractions to mole ratios back to mass ratios—conversions required by the nature of the available data for properties of solutions.

1. *Solve for m_1, m_2, and m_3 by material balances, and calculate the amount of water formed.*

$$\textit{S balance:}\quad \frac{100\text{ g H}_2\text{SO}_4}{}\left|\frac{32.0\text{ g S}}{98.1\text{ g H}_2\text{SO}_4}\right. = \frac{m_2(\text{g Na}_2\text{SO}_4)}{}\left|\frac{32.0\text{ g S}}{142\text{ g Na}_2\text{SO}_4}\right.$$

$$\Downarrow$$

$$m_2 = 145\text{ g Na}_2\text{SO}_4$$

$$\textit{Na balance:}\quad \frac{0.200\,m_1(\text{g NaOH})}{}\left|\frac{23.0\text{ g Na}}{40.0\text{ g NaOH}}\right. = \frac{145\text{ g Na}_2\text{SO}_4}{}\left|\frac{46.0\text{ g Na}}{142\text{ g Na}_2\text{SO}_4}\right.$$

$$\Downarrow$$

$$m_1 = 408\text{ g NaOH(aq)}$$

Total mass balance: $\quad 1000\text{ g} + 408\text{ g} = 145\text{ g} + m_3 \implies m_3 = 1263\text{ g H}_2\text{O(l)}$

Mass of product solution: $\quad m = m_2 + m_3 = 1408\text{ g}$

$$\text{Water formed by reaction} = \frac{145\text{ g Na}_2\text{SO}_4\text{ formed}}{}\left|\frac{1\text{ mol}}{142\text{ g}}\right|\frac{2\text{ mol H}_2\text{O}}{1\text{ mol Na}_2\text{SO}_4}$$

$$= 2.04\text{ mol H}_2\text{O}$$

2. *Calculate solvent/solute mole ratios* (needed to determine heats of solution).

H_2SO_4 *(aq):* $\qquad (900\text{ g H}_2\text{O})/(18.0\text{ g/mol}) = 50.0\text{ mol H}_2\text{O}$

$\qquad\qquad\qquad (100\text{ g H}_2\text{SO}_4)/(98.1\text{ g/mol}) = 1.02\text{ mol H}_2\text{SO}_4$

$$\Downarrow$$

$$r = 50.0\text{ mol H}_2\text{O}/1.02\text{ mol H}_2\text{SO}_4 = 49.0\text{ mol H}_2\text{O/mol H}_2\text{SO}_4$$

NaOH (aq): $\qquad [(0.800 \times 408)\text{g H}_2\text{O}]/(18.0\text{ g/mol}) = 18.1\text{ mol H}_2\text{O}$

$\qquad\qquad\qquad [(0.200 \times 408)\text{g NaOH}]/(40.0\text{ g/mol}) = 2.04\text{ mol NaOH}$

$$\Downarrow$$

$$r = 18.1\text{ mol H}_2\text{O}/2.04\text{ mol NaOH} = 8.90\text{ mol H}_2\text{O/mol NaOH}$$

Na_2SO_4 *(aq):* $\qquad (1263\text{ g H}_2\text{O})/(18.0\text{ g/mol}) = 70.2\text{ mol H}_2\text{O}$

$\qquad\qquad\qquad (145\text{ g Na}_2\text{SO}_4)/(142\text{ g/mol}) = 1.02\text{ mol Na}_2\text{SO}_4$

$$\Downarrow$$

$$r = 70.2\text{ mol H}_2\text{O}/1.02\text{ mol Na}_2\text{SO}_4 = 68.8\text{ mol H}_2\text{O/mol Na}_2\text{SO}_4$$

3. *Calculate extent of reaction.* To calculate ξ, we note that 1.02 mol H_2SO_4 reacted. From Equation 9.1-3,

$$\xi = \frac{(n_{\text{H}_2\text{SO}_4})_{\text{reacted}}}{|\nu_{\text{H}_2\text{SO}_4}|} = \frac{1.02\text{ mol}}{1} = 1.02\text{ mol}$$

4. *Calculate ΔH.* This problem is made tricky by the fact that water is not just the solvent in the solutions involved but is also formed as a reaction product. We will take as references the reactant and product solutions at 25°C and evaluate ΔH using Equation 9.5-1a:

$$\Delta H = \xi\,\Delta\hat{H}_r^\circ + \sum n_{\text{out}}\hat{H}_{\text{out}} - \sum n_{\text{in}}\hat{H}_{\text{in}}$$

It is convenient in solution chemistry calculations to tabulate the products $n\hat{H}$ rather than n and $\hat{H}$ separately. The completed enthalpy table is shown below, followed by the calculations that led to the entries.

References: $H_2SO_4(aq, r = 49)$,
$NaOH(aq, r = 8.9)$, $Na_2SO_4(aq, r = 69)$ at 25°C

Substance	$n_{in}\hat{H}_{in}$	$n_{out}\hat{H}_{out}$	
$H_2SO_4(aq)$	57.8	—	
$NaOH(aq)$	0	—	$n\hat{H}$ in kJ
$Na_2SO_4(aq)$	—	58.9	

$H_2SO_4(aq, r = 49, 40\,°C)$: From Table 2.217, p. 2-185 of *Perry's Chemical Engineers' Handbook* (see footnote 1), the heat capacity of a sulfuric acid solution with the given composition is 3.85 J/(g·°C).

$$n\hat{H} = m \int_{25°C}^{40°C} C_p\, dT$$

$$= \frac{1000\ \text{g}}{} \left| \frac{3.85\ \text{J}}{\text{g·°C}} \right| \frac{(40-25)°C}{} \left| \frac{1\ \text{kJ}}{1000\ \text{J}} \right| = 58.9\ \text{kJ}$$

$NaOH(aq, r = 8.9, 25\,°C)$: $n\hat{H} = 0$

$Na_2SO_4(aq, r = 69, 35\,°C)$: In the absence of better information, we will assume that the heat capacity of the solution is that of pure water, 4.184 J/(g·°C).

$$n\hat{H} = m \int_{25°C}^{35°C} C_p\, dT$$

$$= \frac{1408\ \text{g}}{} \left| \frac{4.184\ \text{J}}{\text{g·°C}} \right| \frac{(35-25)°C}{} \left| \frac{1\ \text{kJ}}{1000\ \text{J}} \right| = 58.9\ \text{kJ}$$

The heats of formation of $H_2SO_4(l)$ and $NaOH(c)$ are given in Table B.1, and the heats of solution of these species are given in Table B.11. *Perry's Chemical Engineers' Handbook* (see footnote 1) on p. 2-193 gives the standard heat of formation of $Na_2SO_4(aq, r = 1100)$ as -330.82 kcal/mol $Na_2SO_4 = -1384$ kJ/mol Na_2SO_4. In the absence of heat of solution data, we will assume that this value also applies to the solution for which $r = 69$ moles of water per mole of solute. The standard heats of formation of the species involved in the reaction

$$H_2SO_4(aq,\ r = 49) + 2\,NaOH(aq,\ r = 8.9) \rightarrow Na_2SO_4(aq) + 2\,H_2O(l)$$

are obtained from Equation 9.5-4 (heat of formation of the solution equals heat of formation of the solute plus heat of solution) as

$H_2SO_4(aq)$: $\Delta\hat{H}_f^\circ = [(-811.3) + (-73.3)]$ kJ/mol $H_2SO_4 = -884.6$ kJ/mol H_2SO_4

$NaOH(aq)$: $\Delta\hat{H}_f^\circ = [(-426.6) + (-41.5)]$ kJ/mol $NaOH = -468.1$ kJ/mol $NaOH$

$Na_2SO_4(aq)$: $\Delta\hat{H}_f^\circ = -1384$ kJ/mol Na_2SO_4

$H_2O(l)$: $\Delta\hat{H}_f^\circ = -285.84$ kJ/mol H_2O

and the standard heat of reaction is therefore

$$\Delta\hat{H}_r^\circ = [(-1384)(1) + (-285.84)(2) - (-884.6)(1) - (-468.1)(2)]\ \text{kJ/mol}$$
$$= -1349\ \text{kJ/mol}$$

5. ***Energy balance.***

$$Q = \Delta H = \xi\,\Delta\hat{H}_r^\circ + \sum n_{out}\hat{H}_{out} - \sum n_{in}\hat{H}_{in}$$
$$= (1.02\ \text{mol})(-1349\ \text{kJ/mol}) + (58.9 - 57.8)\ \text{kJ} = \boxed{-1375\ \text{kJ}}$$

When a strong acid or base is dissolved in water, it dissociates into ionic species; for example, dissolved NaOH exists as Na^+ and OH^- in a dilute solution. **Heats of formation of ions** may be determined from heats of solutions of such substances and may be used to calculate heats of formation of dilute solutions of highly dissociated materials. A good discussion of this topic and a table of heats of formation of ions is given by Hougen, Watson, and Ragatz.[8]

TEST YOURSELF

The heat of formation of A(s) is $(\Delta \hat{H}_f^\circ)_A = -100$ kJ/mol; heats of solution of A in a solvent B are $\Delta \hat{H}_s^\circ(r = 50$ mol B/mol A$) = -10$ kJ/mol and $\Delta \hat{H}_s^\circ(r = \infty) = -15$ kJ/mol.

1. **(a)** What is the standard heat of formation of A(soln, $r = 50$) relative to B and the elements of A(s)?
   (b) What is $\Delta \hat{H}_f^\circ$ for A(soln, $r = \infty$) relative to the same references?
2. **(a)** What is the enthalpy (kJ/mol A) of a solution of A in B at 25°C for which $r = 50$, relative to B and the elements of A at 25°C?
   (b) What is the enthalpy (kJ) of a solution containing 5 mol of A in 250 mol of B at 25°C, relative to A(s) and B(l) at 25°C? What is it relative to B(l) and the elements of A at 25°C?

9.6 FUELS AND COMBUSTION

The use of heat generated by a combustion reaction to produce steam, which drives turbines to produce electricity, may be the single most important commercial application of chemical reactions. (See Chapter 14.)

The analysis of fuels and combustion reactions and reactors has always been an important activity for chemical engineers. In this section, we review the properties of the fuels most often used for power generation and outline techniques for energy balances on combustion reactors.

9.6a Fuels and Their Properties

Fuels burned in power-plant furnaces may be solids, liquids, or gases. Some of the more common fuels are:

Solid fuels: Principally coal (a mixture of carbon, water, noncombustible ash, hydrocarbons, and sulfur), coke (primarily carbon—the solid residue left after coal or petroleum is heated, driving off volatile substances and decomposing hydrocarbons), and to a small extent wood and solid waste (garbage).

Liquid fuels: Principally hydrocarbons obtained by distilling crude oil (petroleum); also coal tars and shale oil. There is also a strong worldwide interest in the use of alcohols obtained by fermenting grains.

Gaseous fuels: Principally natural gas (80% to 95% methane, the balance ethane, propane, and small quantities of other gases); also light hydrocarbons obtained from petroleum or coal treatment, acetylene, and hydrogen (the latter two are relatively expensive to produce).

QUESTIONS FOR DISCUSSION

Coal contains primarily carbon and combustible hydrocarbons, but also contains substantial amounts of noncombustible ash and as much as 5% sulfur by weight.

1. What becomes of the sulfur when coal is burned? What about the ash? (Suggest two possibilities for the ash.)

---

[8]O. A. Hougen, K. M. Watson, and R. A. Ragatz, *Chemical Process Principles*, Part I, Wiley, New York, 1954, pp. 315–317.

2. In view of the answer to the preceding question, why might coal be less desirable as a fuel than natural gas?
3. What might prompt a power company to use coal as its primary fuel despite its drawbacks relative to liquid or gaseous fuels?

The **heating value** of a combustible material is the negative of the standard heat of combustion. The **higher heating value** (or **total heating value** or **gross heating value**) is $-\Delta \hat{H}_c^\circ$ with $H_2O(l)$ as a combustion product, and the **lower heating value** (or **net heating value**) is the value based on $H_2O(v)$ as a product. Since $\Delta \hat{H}_c^\circ$ is always negative, the heating value is positive.

To calculate a lower heating value of a fuel from a higher heating value or vice versa, you must determine the moles of water produced when one mole of the fuel is burned. If this quantity is designated n, then

$$HHV = LHV + n\,\Delta \hat{H}_v(H_2O,\ 25°C) \tag{9.6-1}$$

(Try to prove this relationship from the definitions of LHV and HHV and Hess's law.) The heat of vaporization of water at 25°C is

$$\Delta \hat{H}_v(H_2O,\ 25°C) = 44.013\ \text{kJ/mol} \tag{9.6-2a}$$
$$= 18{,}934\ \text{Btu/lb-mole} \tag{9.6-2b}$$

If a fuel contains a mixture of combustible substances, its heating value (lower or higher) is

$$HV = \sum x_i (HV)_i \tag{9.6-3}$$

where $(HV)_i$ is the heating value of the ith combustible substance. If the heating values are expressed in units of (energy)/(mass), then the x_is are the mass fractions of the fuel components, while if the dimensions of the heating values are (energy)/(mole) then the x_is are mole fractions.

EXAMPLE 9.6-1 | *Calculation of a Heating Value*

A natural gas contains 85% methane and 15% ethane by volume. The heats of combustion of methane and ethane at 25°C and 1 atm with water *vapor* as the assumed product are given below:

$$CH_4(g) + 2\,O_2(g) \rightarrow CO_2(g) + 2\,H_2O(v): \quad \Delta \hat{H}_c^\circ = -802\ \text{kJ/mol}$$
$$C_2H_6(g) + \tfrac{7}{2}O_2(g) \rightarrow 2\,CO_2(v) + 3\,H_2O(v): \quad \Delta \hat{H}_c^\circ = -1428\ \text{kJ/mol}$$

Calculate the higher heating value (kJ/g) of the natural gas.

SOLUTION

Since the heating value per unit mass of the fuel is desired, we will first calculate the composition on a mass basis:

$$1\ \text{mol fuel} \Longrightarrow \begin{array}{l} 0.85\ \text{mol CH}_4 \Longrightarrow 13.6\ \text{g CH}_4 \\ 0.15\ \text{mol C}_2\text{H}_6 \Longrightarrow \underline{\ \ 4.5\ \text{g C}_2\text{H}_6} \\ \hphantom{0.15\ \text{mol C}_2\text{H}_6 \Longrightarrow } 18.1\ \text{g total} \end{array}$$

Thus
$$x_{CH_4} = 13.6\ \text{g CH}_4 / 18.1\ \text{g} = 0.751\ \text{g CH}_4/\text{g fuel}$$
$$x_{C_2H_6} = 1 - x_{CH_4} = 0.249\ \text{g C}_2\text{H}_6/\text{g fuel}$$

The higher heating values of the components are calculated from the given heats of combustion (which are the negatives of the lower heating values) as follows:

$$(HHV)_{CH_4} = (LHV)_{CH_4} + n_{H_2O}(\Delta \hat{H}_v)_{H_2O}$$

$$= \left[802\ \frac{\text{kJ}}{\text{mol CH}_4} + \frac{2\ \text{mol H}_2\text{O}}{\text{mol CH}_4}\left(44.013\ \frac{\text{kJ}}{\text{mol H}_2\text{O}} \right) \right] \frac{1\ \text{mol}}{16.0\ \text{g CH}_4}$$

$$= 55.6\ \text{kJ/g}$$

$$(HHV)_{C_2H_6} = \left[1428\ \frac{\text{kJ}}{\text{mol C}_2\text{H}_6} + \frac{3\ \text{mol H}_2\text{O}}{\text{mol C}_2\text{H}_6}\left(44.013\ \frac{\text{kJ}}{\text{mol H}_2\text{O}} \right) \right] \frac{1\ \text{mol}}{30.0\ \text{g C}_2\text{H}_6}$$

$$= 52.0\ \text{kJ/g}$$

The higher heating value of the mixture is from Equation 9.6-3:

$$HHV = x_{CH_4}(HHV)_{CH_4} + x_{C_2H_6}(HHV)_{C_2H_6}$$

$$= [(0.751)(55.6) + (0.249)(52.0)] \text{ kJ/g} = \boxed{54.7 \text{ kJ/g}}$$

Higher heating values for common solid, liquid, and gaseous fuels are tabulated in Section 27 of *Perry's Chemical Engineers' Handbook* (see footnote 1). Representative values are given in Table 9.6-1. From the standpoint of heating value per unit mass, hydrogen is clearly the best fuel; however, it does not occur naturally in appreciable quantities and the current cost of producing it makes it less economical than the other fuels in Table 9.6-1.

Table 9.6-1 Typical Heating Values of Common Fuels

	Higher Heating Value	
Fuel	kJ/g	Btu/lb$_m$
Wood	17	7700
Soft coal	23	10,000
Hard coal	35	15,000
Fuel oil, gasoline	44	19,000
Natural gas	54	23,000
Hydrogen	143	61,000

TEST YOURSELF

1. The standard heat of the reaction

$$n\text{-}C_4H_{10}(v) + \tfrac{13}{2} O_2 \longrightarrow 4 CO_2 + 5 H_2O(v)$$

   is -2658 kJ/mol. What is the lower heating value per mol of *n*-butane vapor? The higher heating value?
2. A gas mixture contains 40.0 wt% H_2 ($HHV = 143$ kJ/g) and 60.0 wt% CH_4 ($HHV = 55$ kJ/g). Calculate the higher heating value of this mixture in kJ/g.
3. In 1998, bituminous coal for residential heating use cost roughly $150 per ton. What would natural gas have to cost ($/ton) to be as economical as coal on a $/Btu basis? (Use Table 9.6-1.)

9.6b Adiabatic Flame Temperature

When a fuel is burned, a considerable amount of energy is released. Some of this energy is transferred as heat through the reactor walls, and the remainder raises the temperature of the reaction products; the less heat transferred, the higher the product temperature. The highest achievable temperature is reached if the reactor is adiabatic and all of the energy released by the combustion goes to raise the temperature of the combustion products. This temperature is called the **adiabatic flame temperature**, T_{ad}.

The calculation of an adiabatic flame temperature follows the general procedure outlined in Section 9.5b. Unknown stream flow rates are first determined by material balances. Reference conditions are chosen, specific enthalpies of feed components are calculated, and specific enthalpies of product components are expressed in terms of the product temperature, T_{ad}. Finally, $\Delta\dot{H}(T_{ad})$ for the process is evaluated and substituted into the energy balance equation ($\Delta\dot{H} = 0$), which is solved for T_{ad}.

Suppose $\dot{n}_f$(mol/s) of a fuel species with heat of combustion $\Delta\hat{H}_c^\circ$ is burned completely with pure oxygen or air in a continuous adiabatic reactor. If the reference states of the molecular feed and product species are those used to determine $\Delta\hat{H}_c^\circ$, the enthalpy change from inlet to outlet is determined from Equation 9.5-2a to be[9]

$$\Delta \dot{H} = \dot{n}_f \Delta \hat{H}_c^\circ + \sum_{out} \dot{n}_i \hat{H}_i(T_{ad}) - \sum_{in} \dot{n}_i \hat{H}_i(T_{feed})$$

Since the reactor is adiabatic, $\dot{Q} = 0$ in the energy balance. If shaft work and kinetic and potential energy changes ($\dot{W}_s$, $\Delta \dot{E}_k$, $\Delta \dot{E}_p$) are negligible compared to each of the first two terms in the expression for $\Delta \dot{H}$, the energy balance simplifies to $\Delta \dot{H} = 0$, which in turn leads to

$$\sum_{out} \dot{n}_i \hat{H}_i(T_{ad}) = -\dot{n}_f \Delta \hat{H}_c^\circ + \sum_{in} \dot{n}_i \hat{H}_i(T_{feed}) \tag{9.6-4}$$

Once again, the reference states for determination of the specific enthalpies in this equation must be those used to determine the value of $\Delta \hat{H}_c^\circ$. If the heats of combustion in Table B.1 are used, the reference states would be the fuel, combustion products (including liquid water), and inert species at 25°C and 1 atm. The fuel would be in whichever state (solid, liquid, or gas) Table B.1 specifies.

If the third-order polynomial heat capacity formulas in Table B.2 are used to determine $\hat{H}_i(T_{ad})$ for each product species, Equation 9.6-4 becomes a fourth-order polynomial equation. Solving this equation for T_{ad} is easily accomplished with a spreadsheet or equation-solving program. The next example illustrates the procedure.

EXAMPLE 9.6-2   *Calculation of an Adiabatic Flame Temperature*

Liquid methanol is to be burned with 100% excess air. The engineer designing the furnace must calculate the highest temperature that the furnace walls will have to withstand so that an appropriate material of construction can be chosen. Perform this calculation, assuming that the methanol is fed at 25°C and the air enters at 100°C.

SOLUTION

Basis: 1 mol CH₃OH Burned

Assume complete combustion. From Table B.1,

$$CH_3OH(l) + \tfrac{3}{2} O_2 \rightarrow CO_2 + 2 H_2O(l): \quad \Delta \hat{H}_c^\circ = -726.6 \text{ kJ/mol}$$

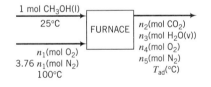

Calculate Component Amounts

$$(n_{O_2})_{theoretical} = 1.50 \text{ mol}$$

$$n_1 = (2)(1.50 \text{ mol}) = 3.00 \text{ mol O}_2 \text{ fed}$$

$$\Downarrow$$

$$(3.76 \text{ mol N}_2/\text{mol O}_2)(3.00 \text{ mol O}_2) = 11.28 \text{ mol N}_2 \text{ fed}$$

Material balances yield

$$n_2 = 1.00 \text{ mol CO}_2$$

$$n_3 = 2.00 \text{ mol H}_2O$$

$$n_4 = 1.50 \text{ mol O}_2$$

$$n_5 = 11.28 \text{ mol N}_2$$

---

[9]In the equation that follows, we make use of the fact that the tabulated values of $\Delta \hat{H}_c^\circ$ presume a stoichiometric coefficient of 1 for the fuel species, so that $\dot{n}_f$ may be substituted for $\dot{\xi} = \dot{n}_f/|\nu_f|$ in Equation 9.5-2a.

Calculate Feed Component Enthalpies

References: $CH_3OH(l)$, O_2, N_2 at 25°C
$CH_3OH(l, 25°C)$: $\hat{H} = 0$
Air (100°C): $\hat{H} = 2.191$ kJ/mol   (from Table B.8)

Evaluate the Right Side of Equation 9.6-4

$$-n_f\Delta\hat{H}_c^\circ + \sum_{\text{in}} n_i\hat{H}_i = -(1.00 \text{ mol } CH_3OH)\left(-726.6 \frac{\text{kJ}}{\text{mol}}\right)$$

$$+ (1.00 \text{ mol } CH_3OH)\left(0 \frac{\text{kJ}}{\text{mol}}\right) + (14.28 \text{ mol air})\left(2.191 \frac{\text{kJ}}{\text{mol}}\right)$$

$$= 757.9 \text{ kJ}$$

Calculate Enthalpy of Product Stream

References: $CO_2(g)$, $O_2(g)$, $N_2(g)$, and $H_2O(l)$ at 25°C. (Liquid water is assumed since this is the state for which $\Delta\hat{H}_c$ is known.)

From Table B.2, the heat capacities of the product gases in kJ/(mol·°C) in terms of T(°C) are

$$(C_p)_{CO_2} = 0.03611 + 4.233 \times 10^{-5}T - 2.887 \times 10^{-8}T^2 + 7.464 \times 10^{-12}T^3$$

$$(C_p)_{H_2O(g)} = 0.03346 + 0.688 \times 10^{-5}T + 0.7604 \times 10^{-8}T^2 - 3.593 \times 10^{-12}T^3$$

$$(C_p)_{O_2} = 0.02910 + 1.158 \times 10^{-5}T - 0.6076 \times 10^{-8}T^2 + 1.311 \times 10^{-12}T^3$$

$$(C_p)_{N_2} = 0.02900 + 0.2199 \times 10^{-5}T + 0.5723 \times 10^{-8}T^2 - 2.871 \times 10^{-12}T^3$$

We could integrate each of these formulas from the reference temperature of 25°C to the unknown T_{ad} to obtain expressions for each $(\hat{H}_i)_{out}$, and then substitute in $\sum_{out} n_i\hat{H}_i$; however, we can save some calculation time by summing before integrating. Substituting the values of n_i calculated previously and the tabulated heat capacities, and recalling that to calculate $(\hat{H})_{H_2O(g)}$ we must first vaporize the liquid at 25°C (using Equation 9.6-2a for $\Delta\hat{H}_v$), we obtain

$$\sum n_iC_{pi} = 0.4378 + 9.826 \times 10^{-5}T + 4.178 \times 10^{-8}T^2 - 30.14 \times 10^{-12}T^3$$

$$\Downarrow$$

$$\sum_{\text{out}} n_i\hat{H}_i = n_3(\Delta\hat{H}_v)_{H_2O} + \int_{25°C}^{T_{ad}}\left(\sum n_iC_{pi}\right)dT$$

$$= 88.026 + 0.4378T_{ad} + 4.913 \times 10^{-5}T_{ad}^2 + 1.393 \times 10^{-8}T_{ad}^3$$

$$- 7.535 \times 10^{-12}T_{ad}^4 - 11.845$$

$$\Downarrow$$

$$\sum_{\text{out}} n_i\hat{H}_i = 76.18 + 0.4738T_{ad} + 4.913 \times 10^{-5}T_{ad}^2 + 1.393 \times 10^{-8}T_{ad}^3$$

$$- 7.535 \times 10^{-12}T_{ad}^4$$

Energy Balance

From Equation 9.6-4

$$\sum_{\text{out}} n_i\hat{H}_i = -n_f \Delta\hat{H}_c^\circ + \sum_{\text{in}} n_i\hat{H}_i = 757.9 \text{ kJ}$$

$$\Downarrow$$

$$7.535 \times 10^{-12}T_{ad}^4 - 1.393 \times 10^{-8}T_{ad}^3 - 4.913 \times 10^{-5}T_{ad}^2 - 0.4738T_{ad} + 681.7 = 0$$

This quartic equation may be solved using a spreadsheet, an equation-solving program, or a calculator programmed to solve polynomial equations. The solution is

$$\boxed{T_{ad} = 1256°C}$$

The furnace walls will therefore never be exposed to a temperature greater than 1256°C as long as the feed and air properties remain the same.

The adiabatic flame temperature is much greater when pure oxygen rather than air is fed to the reactor and is greatest when the fuel and oxygen are fed in stoichiometric proportion.

TEST YOURSELF

1. What is the adiabatic flame temperature of a fuel?
2. Suppose T_{ad} is the adiabatic flame temperature calculated for a given fuel + air feed to a furnace. Give two reasons why the actual furnace temperature might be less than T_{ad}.
3. Why should the adiabatic flame temperature be much higher for a pure oxygen feed than for an air feed?

9.6c   Flammability and Ignition

In this section and the one that follows, we discuss qualitatively what happens during the rapid chemical reaction between a fuel and oxygen. Along the way, we provide answers to the following questions:

1. What is a flame? Why are some flames blue and some yellow?
2. If you light a match in a mixture of methane and air that contains 10% CH_4 by volume, the mixture will burn explosively, but if the mixture contains 20% CH_4 nothing will happen. Why?
3. What is an explosion? What is the loud noise you hear when something explodes?
4. Hydrogen and oxygen react explosively to form water, yet if you mix these two gases in a flask, nothing happens. Why not?

We have so far in this text only considered the initial and final conditions in a chemical reactor, and not how long it may have taken to get from one to the other. When you study **chemical reaction kinetics**, you will learn that the rate of a reaction depends strongly on the reaction temperature; for many reactions, a temperature rise of only 10°C is enough to double the rate.

Suppose a mixture of methane and air containing 10 mole% CH_4 is heated by a central heat source (e.g., an electrical coil) at atmospheric pressure, beginning at room temperature. Although methane reacts with oxygen

$$CH_4 + 2\,O_2 \rightarrow CO_2 + 2\,H_2O$$

the reaction proceeds at an immeasurably low rate at room temperature, and to an observer nothing would seem to be happening in the reactor.

As the temperature increases, the rate of the oxidation reaction also increases, and measurable amounts of CO_2 and H_2O appear. However, if the heat source is turned off, the reactor temperature drops again—the rate at which heat is generated by the reaction alone is not enough to compensate for the rate at which heat is lost from the reaction zone.

However, if the temperature at any point in the reactor reaches about 640°C or higher, the rate of heat generation by the reaction exceeds the rate of heat loss from the reaction zone. The gas adjacent to this zone is then heated above 640°C, causing the zone of rapid reaction to spread. The temperature of the gas rapidly rises by several hundred or even a thousand degrees in a fraction of a second; even if the heating source is turned off, the rate of heat generation by the now rapidly occurring reaction is enough to maintain the system at its high temperature until the reactants are exhausted.

Combustion is defined as a rapid, high-temperature oxidation reaction. What happens in the reactor just described after the reaction rate accelerates dramatically is combustion, whereas the initial slow oxidation reaction between methane and oxygen to form CO_2 and H_2O and other reactions between these species, such as the formation reaction of formaldehyde

$$CH_4 + O_2 \rightarrow HCHO + H_2O$$

are not classified as combustion reactions.

The rapid increase in the rate of an oxidation reaction when the reaction mixture exceeds a certain temperature is called **ignition**; the temperature at which this phenomenon occurs is called the **ignition temperature**, and the time between the instant when the mixture reaches the ignition temperature and the moment of ignition is the **ignition lag**. The ignition temperature and lag are shown here on a representative plot of the temperature of a fuel mixture that is being heated.

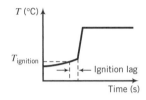

The value of the ignition temperature depends on a number of things for a given fuel, including the fuel-to-air ratio, the total pressure in the reactor, and even the reactor geometry. For any given fuel, there is a lower limit to this quantity called the **autoignition temperature**. Representative values of this quantity for stoichiometric fuel–air mixtures at 1 atm are 400°C for H_2, 540°C for CH_4, and 515°C for C_2H_6. Ignition lags are typically 0.1–10 s in duration and decrease with increasing temperature above the autoignition temperature.

We saw in Section 9.6b that the highest attainable temperature in a combustion reaction—the adiabatic flame temperature—depends on the fuel-to-air ratio, and we stated but did not prove that this upper temperature limit is a maximum when the fuel and oxygen are present in stoichiometric proportion. If the mixture is either **rich** (fuel in excess) or **lean** (O_2 in excess), the adiabatic flame temperature decreases.

There exist two values of the mole percent of fuel in a reaction mixture—the **lower** or **lean flammability limit** and the **upper** or **rich flammability limit**—for which the adiabatic flame temperature equals the ignition temperature of the mixture. *A fuel–air mixture whose composition falls outside these limits is incapable of igniting or exploding, even if exposed to a spark or flame.* The composition range between the two flammability limits is called the **explosive range** of the mixture.

For example, the stoichiometric percentage of methane in a methane–air mixture is 9.5 mole%. (Prove it.) Experimentally, it is found that the lower flammability limit of CH_4–air mixtures at 1 atm is approximately 5% CH_4 and the upper flammability limit is approximately 15% CH_4. Thus, a CH_4–air mixture containing between 5% CH_4 and 15% CH_4 must be considered a fire or explosion hazard, while a mixture containing 3% CH_4 may be considered safe, and a mixture containing 18% CH_4 may also be considered safe as long as it is not brought into contact with additional oxygen.

Flammability limits of a number of hydrocarbon–air mixtures are listed in tables on pp. 26-53 and 26-54 of *Perry's Chemical Engineers' Handbook* (see footnote 1). The given values apply to an initial temperature of roughly 25°C and a pressure of 1 atm.

EXAMPLE 9.6-3 *Ignition Temperature and Flammability Limits*

Propane gas and air are to be mixed and fed to a combustion reactor. The combustion is to be initiated with a gas torch. Determine the minimum and maximum percentages of propane in the feed to the reactor and the minimum required temperature of the torch flame.

SOLUTION   From Table 26-10 of *Perry's Chemical Engineer's Handbook* (see footnote 1),

Minimum mole% C_3H_8 for combustion $= 2.1\%$

Maximum mole% C_3H_8 for combustion $= 9.5\%$

The torch flame temperature must be at least as high as the autoignition temperature of a propane–air mixture, which from Table 26-10 is $\boxed{450°C}$.

If a liquid (or a volatile solid) is exposed to air, the vapor given off could form a combustible mixture with the air adjacent to it, and a spark or match lit in the vicinity of the liquid could cause the mixture to ignite or explode. The **flash point** of a liquid is the temperature at which the liquid gives off enough vapor to form an ignitable mixture with the air above the liquid surface. The flash point of gasoline, for example, is roughly $-42°C$, and that of ethanol is $13°C$, so that these liquids constitute fire hazards at room temperature, while the flash points of fuel oils vary from $38°C$ to $55°C$, making the hazards associated with these materials considerably less.

TEST YOURSELF

1. Briefly define the following terms: (a) ignition, (b) autoignition temperature, (c) ignition lag, (d) flammability limits of a fuel–air mixture, and (e) flash point of a liquid.
2. The flammability limits of methane–air mixtures are 5% CH_4 and 15% CH_4 at 1 atm, and the autoignition temperature is 540°C.
   (a) What would happen if a spark were struck in a methane–air mixture containing 10% CH_4? What about a mixture containing 20% CH_4?
   (b) If a methane–air mixture containing 20% CH_4 were heated to 700°C, would the combustion reaction take place? What would happen if the heat source were turned off?
   (c) Pure methane is clearly not within the explosive range of methane–air mixtures, yet if pure methane is blown out of a cylinder into a room and a match is lit nearby, a flame is observed that persists after the match is withdrawn. How is this possible?

9.6d   Flames and Detonations

Suppose a combustible gas–air mixture is contained in an open-ended tube, and a match or another ignition source is applied to one end of the tube. The gas mixture at this end is heated and eventually ignites. The intense heat generated by the combustion reaction raises the chemical species formed during the reaction to high energy states. When these species return to lower energy states, some of the energy they lose is given off in the form of light. The result is a visible **flame** accompanying the combustion.

Initially the flame is located at the end of the tube that was ignited. However, the heat of combustion quickly raises the adjacent unburned gas to its ignition point, causing the flame to "travel" toward the other end of the tube. At some point, the tube appears as follows.

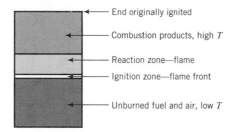

The flame front moves in the direction of the unburned gases at a velocity called the **flame velocity**, which typically has a value of 0.3 to 1 m/s. The exact value of the flame velocity depends on a number of things, including the type of fuel, fuel-to-air ratio, initial temperature and pressure of the unburned gases, and the geometry of the combustion chamber.

Suppose now that instead of being stationary in the tube, the combustion mixture is fed continuously into the bottom (as in a Bunsen burner), and the top is ignited. If the velocity with which the gases leave the tube equals the velocity with which the flame would travel down in a stationary gas, a **stationary flame** is achieved at the top. The wall of the tube lowers the flame velocity, so that the flame burns at the end but does not penetrate into the tube.

If the gas flow rate is increased, the flame size and rate of heat generation both increase, since a larger quantity of gas is being burned. However, once the flow rate reaches a critical value, the flame can no longer travel back as fast as the combustion region is transported away from the burner. The gases in the combustion region become increasingly diluted with air, until the region finally falls outside the flammability limits and the fire is literally blown out.

On the other hand, if the gas flow rate to the burner tube is decreased, the gas velocity in the tube may become lower than the flame propagation velocity in the tube. The result is **flashback**—the flame travels back through the tube toward the fuel source. Flashback is extremely dangerous, and any flow system involving combustible gases must be designed to guarantee that the flow rate stays above the flame propagation velocity.

When combustion of a well-mixed fuel–air mixture occurs, the fuel rapidly reacts with oxygen to form a number of unstable intermediate species (such as oxygen and hydrogen atoms, and OH and H_2O radicals), which then proceed through a complicated chain mechanism to form CO_2 and H_2O. Some of these species undergo transitions that cause them to emit radiation whose wavelength falls within the blue region of the visible spectrum. The result is that the flame appears blue.

On the other hand, when the fuel and air are not well mixed (such as when a pure hydrocarbon gas is burned as it emerges from a stack and mixes with atmospheric air), the combustion proceeds relatively slowly, and some of the hydrocarbon fuel decomposes to form elementary carbon and hydrogen before oxidation takes place. The heat of reaction is sufficient to raise the temperature to a point where the carbon particles glow incandescently. A yellow flame is the result.

Finally, suppose ignition of a gas takes place in a confined or partially confined space. The large temperature rise in the combustion region causes a rapid buildup of pressure in this region. If the combustion is fast enough and the heat of reaction is high enough, a **detonation** may result, wherein a sharply defined high-pressure front, or **shock wave**, travels through the gas at a velocity well in excess of the flame propagation velocity in the gas. The shock wave rapidly compresses and ignites the gas as it passes through, giving the appearance of an instantaneous combustion.

Even after the combustion reaction that gave rise to the detonation has consumed all the available fuel, the shock wave can persist for large distances, carrying with it considerable energy. The energy of even a small shock wave is sufficient to vibrate the eardrums of anyone near the site of the detonation, producing the bang that always accompanies an explosion. The energy of a large shock wave may be sufficient to demolish a city.

TEST YOURSELF

You are the guest lecturer at a junior high school science fair. The following questions are put to you. How would you answer them in terms an intelligent 14-year-old might understand?

1. What is a flame?
2. What is a yellow flame? A blue flame?
3. What is detonation? What is the loud noise you hear when something explodes? What is it that knocks buildings over when dynamite is set off? How does a gun work?
4. (This one from one of the science teachers who thought they should have asked *him* to give the lecture.) You say that hydrogen and oxygen react explosively to form water. Why is it that I can mix hydrogen and oxygen in a flask and nothing will happen? (*Hint:* Reread the beginning of Section 9.6c.)

9.7 SUMMARY

Large internal energy and enthalpy changes are often associated with chemical reactions, leading to substantial heat transfer (heating or cooling) requirements for chemical reactors. This chapter outlines methods of calculating $\Delta \dot{H}$ for open reactive systems and ΔU and ΔH for closed systems. Once the appropriate quantity has been determined, it may be substituted into the energy balance to determine the required heat transfer.

- The **heat of reaction** (or **enthalpy of reaction**), $\Delta \hat{H}_r (T, P)$, is the enthalpy change when stoichiometric quantities of reactants at temperature T and pressure P are consumed completely to form products at the same temperature and pressure. The **standard heat of reaction**, $\Delta \hat{H}_r^\circ$, is the heat of reaction at a specified reference temperature and pressure, in this text 25°C and 1 atm. At low to moderate pressures, the heat of reaction is nearly independent of P.

    For example, the standard heat of reaction for the complete combustion of methane is

$$CH_4(g) + 2\,O_2(g) \longrightarrow CO_2(g) + 2\,H_2O(l): \quad \Delta \hat{H}_r^\circ = -890.3 \text{ kJ/mol}$$

    which signifies that if 1 g-mole of gaseous methane and 2 g-moles of gaseous oxygen at 25°C and 1 atm react completely to form 1 g-mole of gaseous carbon dioxide and 2 g-moles of liquid water and the products are brought back to 25°C and 1 atm, the net enthalpy change would be $\Delta H = -890.3$ kJ. If the energy balance reduces to $Q = \Delta H$, 890.3 kJ of heat would have to be transferred away from the reactor to keep the products at 25°C.

- If $\Delta \hat{H}_r (T, P) < 0$, the reaction is **exothermic** at T and P: less energy is required to break the bonds holding the reactant molecules together than is released when the product bonds form, resulting in a net release of energy as the reaction proceeds. This energy may be transferred from the reactor as heat or it may serve to raise the temperature of the reaction mixture.

- Similarly, if $\Delta \hat{H}_r (T, P) > 0$, the reaction is **endothermic**: more energy is required to break the reactant bonds than is released when the product bonds form, leading to a net absorption of energy as the reaction proceeds. Unless this energy is supplied to the reactor as heat, the mixture temperature decreases.

- Provided that gaseous reactants and products behave ideally and the specific volumes of liquid and solid reactants and products are negligible compared with the specific volumes of the gases, the **internal energy of reaction** may be calculated from Equation 9.1-5. (This quantity is required for energy balances on constant-volume batch reactors.)

- According to **Hess's law**, if a stoichiometric equation for a reaction can be obtained as a linear combination of the equations for other reactions (i.e., by adding and subtracting those equations), the heat of the first reaction can be calculated as the same linear combination of the heats of the other reactions.

- The **standard heat of formation** of a species, $\Delta \hat{H}_f^\circ$, is the heat of the reaction in which one mole of the species is formed from its constituent elemental species in their naturally occurring states at 25°C and 1 atm. Standard heats of formation of many species are listed in Table B.1.
- A consequence of Hess's law is that the standard heat of any reaction may be calculated as

$$\Delta \hat{H}_r^\circ = \sum \nu_i \, \Delta \hat{H}_{fi}^\circ$$

where ν_i is the stoichiometric coefficient of reactant or product species i (positive for products, negative for reactants), and $\Delta \hat{H}_{fi}^\circ$ is the standard heat of formation of that species.

- The **standard heat of combustion** of a species, $\Delta \hat{H}_c^\circ$, is the heat of the reaction in which one mole of the species undergoes complete combustion to form products in specified states. Standard heats of combustion of many species are listed in Table B.1, with the presumed combustion products being CO_2, $H_2O(l)$, SO_2 for species containing sulfur, and N_2 for species containing nitrogen. A consequence of Hess's law is that the standard heat of any reaction involving only oxygen and combustible species may be calculated as

$$\Delta \hat{H}_r^\circ = -\sum \nu_i \, \Delta \hat{H}_{ci}^\circ$$

As before, ν_i is the stoichiometric coefficient of species i.

- When performing energy balances on a reactive chemical process, two procedures may be followed in the calculation of $\Delta \dot{H}$ (or ΔH or ΔU) that differ in the choice of reference states for enthalpy or internal energy calculations. In the **heat of reaction method**, the references are the reactant and product species at 25°C and 1 atm in the phases (solid, liquid, or gas) for which the heat of reaction is known. In the **heat of formation method**, the references are the elemental species that constitute the reactant and product species [e.g., C(s), O_2(g), H_2(g), etc.] at 25°C and 1 atm. In both methods, reference states for nonreactive species may be chosen for convenience, as was done for the nonreactive processes of Chapters 7 and 8.
- The heat of reaction method may be slightly easier when only one reaction occurs and the heat of reaction is known. When this method is used, the specific enthalpy of each species in every feed or product stream is calculated by choosing a process path from the reference state to the process state, calculating $\Delta \hat{H}$ for each heating and cooling step and each phase change in the path, and summing the enthalpies for the steps. When specific enthalpies have been calculated for all species in all of their inlet and outlet states, $\Delta \dot{H}$ for a continuous process is calculated as

$$\Delta \dot{H} = \dot{\xi} \, \Delta \hat{H}_r^\circ + \sum_{\text{out}} \dot{n}_i \hat{H}_i - \sum_{\text{in}} \dot{n}_i \hat{H}_i$$

In this equation, $\dot{\xi}$ is the extent of reaction (determined from Equation 9.1-3); $\dot{n}_i$ and $\hat{H}_i$ are respectively the molar flow rate and specific enthalpy of a process species in an inlet or outlet stream; and the summations are taken over all species in all of their inlet and outlet states. Once calculated, $\Delta \dot{H}$ is substituted in the open-system energy balance, which is solved for $\dot{Q}$ or whichever other variable is unknown.[10]

- The heat of formation method is generally easier when there are multiple reactions. When this method is used, the specific enthalpy of a species in a feed or product stream is calculated by choosing a process path from the reference state (the elements at 25°C) to the process state, beginning with the formation of the species from the elements ($\Delta \hat{H} = \Delta \hat{H}_f^\circ$); evaluating $\Delta \hat{H}$ for each subsequent heating and cooling step and each phase change in the path; and summing the enthalpies for the steps (including the formation step). When specific enthalpies have been calculated for all species in all of their inlet and outlet states, $\Delta \dot{H}$ is calculated for

---

[10]If the system in question is a closed system at constant pressure, the dots over variables would be deleted in the above formulas, and if the system is a closed system at constant volume, U would replace H.

an open system as

$$\Delta \dot{H} = \sum_{\text{out}} \dot{n}_i \hat{H}_i - \sum_{\text{in}} \dot{n}_i \hat{H}_i$$

As before, the summations are taken over all species in all of their inlet and outlet states. Once calculated, $\Delta \dot{H}$ is substituted in the open-system energy balance, which is solved for $\dot{Q}$ or whichever other variable is unknown (see footnote 10).

- Sometimes the feed conditions and heat input to a reactor are specified (as in an adiabatic reactor) and the outlet temperature, T_{out}, is to be determined. The procedure is to derive expressions for the specific enthalpies of the reactor outlet species in terms of T_{out}; substitute these expressions into the summation $\sum_{\text{out}} \dot{n}_i \hat{H}_i$ in the expression for $\Delta \dot{H}$; substitute in turn for $\Delta \dot{H}(T_{\text{out}})$ in the energy balance, and solve the resulting equation for T_{out}.

- The **standard heat of formation of a liquid solution** is the sum of the standard heat of formation of the solute and the standard heat of solution calculated using the methods of Section 8.5. The standard heat of a reaction involving solutions may be determined as the weighted sum of the heats of formation of the reactants and products (including the solutions), with the weighting factors being the stoichiometric coefficients (positive for products, negative for reactants). An energy balance for a reactor in which solutions react or form may be written by taking the feed and product solutions at 25°C and 1 atm as references and using the heat of reaction method.

- **Combustion** is a rapid high-temperature reaction between a fuel and oxygen. The **higher heating value** of a fuel is the negative of the standard heat of combustion of the fuel($-\Delta \hat{H}_c^\circ$) with $H_2O(l)$ as a combustion product, and the **lower heating value** is the negative of the standard heat of combustion based on $H_2O(v)$ as a product. The relationship between the two heating values is given by Equation 9.6-1.

- The **adiabatic flame temperature** of a fuel is the temperature that would be attained if the fuel were burned in an adiabatic combustion chamber and all of the energy released went into raising the temperature of the reaction products (as opposed to being absorbed by or transferred through the reactor wall).

- When the temperature of a combustible mixture exceeds a certain value, after a short time lag the rate of the reaction and the reaction temperature increase extremely rapidly. This phenomenon is called **ignition**, and the time interval after T_{ignition} is reached and before the temperature increases sharply is the **ignition lag**. The lowest temperature at which ignition may occur for a fuel is the **autoignition temperature** of the fuel.

- If the mole percentage of a fuel in a fuel-air mixture falls below a certain value (the **lower flammability limit**) or above another value (the **upper flammability limit**), the mixture will not ignite or explode, even if exposed to a flame or spark. The composition range between the flammability limits is called the **explosive range** of the mixture.

PROBLEMS   **9.1.** The standard heat of the reaction

$$4\,NH_3(g) + 5\,O_2(g) \longrightarrow 4\,NO(g) + 6\,H_2O(g)$$

is

$$\Delta \hat{H}_r^\circ = -904.7 \text{ kJ/mol}$$

(a) Briefly explain what that means. Your explanation may take the form "When _____ (specify quantities of reactant species and their physical states) react to form _____ (quantities of product species and their physical state), the change in enthalpy is _____."

(b) Is the reaction exothermic or endothermic at 25°C? Would you have to heat or cool the reactor to keep the temperature constant? What would the temperature do if the reactor ran adiabatically? What can you infer about the energy required to break the molecular bonds of the reactants and that released when the product bonds form?

(c) What is $\Delta \hat{H}_r^\circ$ for

$$2\,NH_3(g) + \tfrac{5}{2}\,O_2 \longrightarrow 2\,NO(g) + 3\,H_2O(g)$$

(d) What is $\Delta \hat{H}_r^\circ$ for

$$NO(g) + \tfrac{3}{2}\,H_2O(g) \longrightarrow NH_3(g) + \tfrac{5}{4}\,O_2$$

(e) Estimate the enthalpy change associated with the consumption of 340 g NH_3/s if the reactants and products are all at 25°C. (See Example 9.1-1.) What have you assumed about the reactor pressure? (You don't have to assume that it equals 1 atm.)

(f) The values of $\Delta \hat{H}_r^\circ$ given in this problem apply to water vapor at 25°C and 1 atm, and yet the normal boiling point of water is 100°C. Can water exist as a vapor at 25°C and a total pressure of 1 atm? Explain your answer.

9.2. The standard heat of reaction for the combustion of liquid n-nonane to form CO_2 and liquid water at 25°C and 1 atm is $\Delta \hat{H}_r^\circ = -6124$ kJ/mol.

(a) Briefly explain what that means. Your explanation may take the form "When _____ (specify quantities of reactant species and their physical states) react to form _____ (quantities of product species and their physical state), the change in enthalpy is _____."

(b) Is the reaction exothermic or endothermic at 25°C? Would you have to heat or cool the reactor to keep the temperature constant? What would the temperature do if the reactor ran adiabatically? What can you infer about the energy required to break the molecular bonds of the reactants and that released when the product bonds form?

(c) If 25.0 mol/s of liquid nonane is consumed and the reactants and products are all at 25°C, estimate the required rate of heat input or output (state which) in kilowatts, assuming that $\dot{Q} = \Delta \dot{H}$ for the process. What have you also assumed about the reactor pressure in your calculation? (You don't have to assume that it equals 1 atm.)

(d) The standard heat of combustion of n-nonane *vapor* is $\Delta \hat{H}_r^\circ = -6171$ kJ/mol. What is the physical significance of the 47 kJ/mol difference between this heat of combustion and the one given previously?

(e) The value of $\Delta \hat{H}_r^\circ$ given in part (d) applies to n-nonane vapor at 25°C and 1 atm, and yet the normal boiling point of n-nonane is 150.6°C. Can n-nonane exist as a vapor at 25°C and a total pressure of 1 atm? Explain your answer.

9.3. The standard heat of the combustion reaction of liquid n-hexane to form $CO_2(g)$ and $H_2O(l)$, with all reactants and products at 77°F and 1 atm, is $\Delta \hat{H}_r^\circ = -1.791 \times 10^6$ Btu/lb-mole. The heat of vaporization of hexane at 77°F is 13,550 Btu/lb-mole and that of water is 18,934 Btu/lb-mole.

(a) Is the reaction exothermic or endothermic at 77°F? Would you have to heat or cool the reactor to keep the temperature constant? What would the temperature do if the reactor ran adiabatically? What can you infer about the energy required to break the molecular bonds of the reactants and that released when the product bonds form?

(b) Use the given data to calculate $\Delta \hat{H}_r^\circ$(Btu/lb-mole) for the combustion of n-hexane vapor to form $CO_2(g)$ and $H_2O(g)$.

(c) If $\dot{Q} = \Delta \dot{H}$, at what rate in Btu/s is heat absorbed or released (state which) if 120 lb$_m$/s of O_2 is consumed in the combustion of hexane vapor, water vapor is the product, and the reactants and products are all at 77°F?

9.4. The standard heat of the reaction

$$CaC_2(s) + 5\,H_2O(l) \longrightarrow CaO(s) + 2\,CO_2(g) + 5\,H_2(g)$$

is $\Delta \hat{H}_r^\circ = +69.36$ kJ/mol.

(a) Is the reaction exothermic or endothermic at 25°C? Would you have to heat or cool the reactor to keep the temperature constant? What would the temperature do if the reactor ran adiabatically? What can you infer about the energy required to break the molecular bonds of the reactants and that released when the product bonds form?

(b) Calculate $\Delta \hat{U}_r^\circ$ for this reaction. (See Example 9.1-2.) Briefly explain the physical significance of your calculated value.

(c) Suppose you charge 150.0 g of CaC_2 and liquid water into a rigid container at 25°C, heat the container until the calcium carbide reacts completely, and cool the products back down to 25°C, condensing essentially all the unconsumed water. Write and simplify the energy balance equation for this closed constant-volume system and use it to determine the net amount of heat (kJ) that must be transferred to or from the reactor (state which).

9.5. Use Hess's law to calculate the standard heat of the water-gas shift reaction

$$CO(g) + H_2O(v) \longrightarrow CO_2(g) + H_2(g)$$

from each of the two sets of data given here.

(a) $CO(g) + H_2O(l) \longrightarrow CO_2(g) + H_2(g)$: $\Delta \hat{H}_r^\circ = +1226$ Btu/lb-mole
   $H_2O(l) \longrightarrow H_2O(v)$: $\Delta \hat{H}_v = +18{,}935$ Btu/lb-mole

(b) $CO(g) + \frac{1}{2} O_2(g) \longrightarrow CO_2(g)$: $\Delta \hat{H}_r^\circ = -121{,}740$ Btu/lb-mole
   $H_2(g) + \frac{1}{2} O_2(g) \longrightarrow H_2O(v)$: $\Delta \hat{H}_r^\circ = -104{,}040$ Btu/lb-mole

9.6. Formaldehyde may be produced in the reaction between methanol and oxygen:

$$2\,CH_3OH(l) + O_2(g) \longrightarrow 2\,HCHO(g) + 2\,H_2O(l): \quad \Delta \hat{H}_r^\circ = -326.2 \text{ kJ/mol}$$

The standard heat of combustion of hydrogen is

$$H_2(g) + \frac{1}{2} O_2(g) \longrightarrow H_2O(l): \quad \Delta \hat{H}_r^\circ = -285.8 \text{ kJ/mol}$$

(a) Use these heats of reaction and Hess's law to determine the standard heat of the direct decomposition of methanol to form formaldehyde:

$$CH_3OH(l) \longrightarrow HCHO(g) + H_2(g)$$

(b) Explain why you would probably use the method of part (a) to determine the heat of the methanol decomposition reaction experimentally rather than carrying out the decomposition reaction and measuring $\Delta \hat{H}_r^\circ$ directly.

9.7. Use tabulated heats of formation (Table B.1) to determine the standard heats of the following reactions in kJ/mol, letting the stoichiometric coefficient of the first reactant in each reaction equal one.

(a) Nitrogen + oxygen react to form nitric oxide (NO).

(b) Gaseous *n*-pentane + oxygen react to form carbon monoxide + liquid water.

(c) Liquid *n*-hexane + oxygen react to form carbon dioxide + water vapor. After doing the calculation, write the stoichiometric equations for the formation of the reactant and product species, then use Hess's law to derive the formula you used to calculate $\Delta \hat{H}_r^\circ$.

(d) Liquid sodium sulfate + carbon monoxide react to form liquid sodium sulfide + carbon dioxide. (Note that Table B.1 only lists the heats of formation of the solid sodium salts. To estimate the required heat of reaction, you will also need to use tabulated heats of fusion.)

9.8. Trichloroethylene, a widely used degreasing solvent for machine parts, is produced in a two-step reaction sequence. Ethylene is first chlorinated to yield tetrachloroethane, which is dehydrochlorinated to form trichloroethylene.

$$C_2H_4(g) + 2\,Cl_2(g) \rightarrow C_2H_2Cl_4(l) + H_2(g): \quad \Delta \hat{H}_r^\circ = -385.76 \text{ kJ/mol}$$
$$C_2H_2Cl_4(l) \rightarrow C_2HCl_3(l) + HCl(g)$$

The standard heat of formation of liquid trichloroethylene is -276.2 kJ/mol.

(a) Use the given data and tabulated standard heats of formation of ethylene and hydrogen chloride to calculate the standard heat of formation of tetrachloroethane and the standard heat of the second reaction.

(b) Use Hess's law to calculate the standard heat of the reaction

$$C_2H_4(g) + 2\,Cl_2(g) \rightarrow C_2HCl_3(l) + H_2(g) + HCl(g)$$

(c) If 300 mol/h of $C_2HCl_3(l)$ is produced in the reaction of part (b) and the reactants and products are all at 25°C and 1 atm, how much heat is evolved or absorbed in the process? (Assume $\dot{Q} = \Delta \dot{H}$.)

9.9. The standard heat of combustion of gaseous acetylene is listed in Table B.1 as -1299.6 kJ/mol.

(a) In your own words, briefly explain what that means. (Your explanation should mention the reference states used to define the tabulated heats of combustion.)

(b) Use tabulated heats of formation to verify the given value of $\Delta \hat{H}_c^\circ$.

(c) Calculate the standard heat of the acetylene hydrogenation reaction

$$C_2H_2(g) + 2H_2(g) \longrightarrow C_2H_6(g)$$

using (i) tabulated heats of formation and (ii) tabulated heats of combustion (Equation 9.4-1).

(d) Write the stoichiometric equations for the combustion reactions of acetylene, hydrogen, and ethane, and use Hess's law to derive the formula you used in part (c-ii).

9.10. The standard heat of combustion $(\Delta \hat{H}_c^\circ)$ of liquid 2,3,3-trimethylpentane $[C_8H_{18}]$ is reported in a table of physical properties to be -4850 kJ/mol. A footnote indicates that the reference temperature for the reported value is 25°C and the presumed combustion products are $CO_2(g)$ and $H_2O(g)$.

(a) In your own words, briefly explain what all that means.

(b) There is some question about the accuracy of the reported value, and you have been asked to determine the heat of combustion experimentally. You burn 2.010 grams of the hydrocarbon with pure oxygen in a constant-volume calorimeter and find that the net heat released when the reactants and products $[CO_2(g)$ and $H_2O(g)]$ are all at 25°C is sufficient to raise the temperature of 1.00 kg of liquid water by 21.34°C. Write an energy balance to show that the heat released in the calorimeter equals $n_{C_8H_{18}} \Delta \hat{U}_c^\circ$, and calculate $\Delta \hat{U}_c^\circ$ (kJ/mol). Then calculate $\Delta \hat{H}_c^\circ$. (See Example 9.1-2.) By what percentage of the measured value does the tabulated value differ from the measured one?

(c) Use the result of part (b) to estimate $\Delta \hat{H}_f^\circ$ for 2,3,3-trimethylpentane. Why would the heat of formation of 2,3,3-trimethylpentane probably be determined this way rather than directly from the formation reaction?

9.11. n-Butane is converted to isobutane in a continuous isomerization reactor that operates isothermally at 149°C. The feed to the reactor contains 93 mole% n-butane, 5% isobutane, and 2% HCl at 149°C, and a 40% conversion of n-butane is achieved.

(a) Taking a basis of 1 mol of feed gas, calculate the moles of each component of the feed and product mixtures and the extent of reaction, ξ(mol).

(b) Calculate the standard heat of the isomerization reaction (kJ/mol). Then, taking the feed and product species at 25°C as references, prepare an inlet–outlet enthalpy table and calculate and fill in the component amounts (mol) and specific enthalpies (kJ/mol). (See Example 9.5-1.)

(c) Calculate the required rate of heat transfer (kJ) to or from the reactor (state which it is). Then determine the required heat transfer rate (kW) for a reactor feed of 325 mol/h.

(d) Use your calculated results to determine the heat of the isomerization reaction at 149°C, $\Delta \hat{H}_r(149°C)$(kJ/mol).

9.12. In the production of many microelectronic devices, continuous **chemical vapor deposition** (CVD) processes are used to deposit thin and exceptionally uniform silicon dioxide films on silicon wafers. One CVD process involves the reaction between silane and oxygen at a very low pressure.

$$SiH_4(g) + O_2(g) \longrightarrow SiO_2(s) + 2 H_2(g)$$

The feed gas, which contains oxygen and silane in a ratio 8.00 mol O_2/mol SiH_4, enters the reactor at 298 K and 3.00 torr absolute. The reaction products emerge at 1375 K and 3.00 torr absolute. Essentially all of the silane in the feed is consumed.

(a) Taking a basis of 1 m^3 of feed gas, calculate the moles of each component of the feed and product mixtures and the extent of reaction, ξ(mol).

(b) Calculate the standard heat of the silane oxidation reaction (kJ/mol). Then, taking the feed and product species at 298 K (25°C) as references, prepare an inlet–outlet enthalpy table and calculate and fill in the component amounts (mol) and specific enthalpies (kJ/mol). (See Example 9.5-1.)

Data

$$(\Delta \hat{H}_f^\circ)_{SiH_4(g)} = -61.9 \text{ kJ/mol}, \quad (\Delta \hat{H}_f^\circ)_{SiO_2(s)} = -851 \text{ kJ/mol}$$

$$(C_p)_{SiH_4(g)}[\text{kJ/(mol·K)}] = 0.01118 + 12.2 \times 10^{-5}T - 5.548 \times 10^{-8}T^2 + 6.84 \times 10^{-12}T^3$$

$$(C_p)_{SiO_2(s)}[\text{kJ/(mol·K)}] = 0.04548 + 3.646 \times 10^{-5}T - 1.009 \times 10^3/T^2$$

The temperatures in the formulas for C_p are in kelvin.

(c) Calculate the heat (kJ) that must be transferred to or from the reactor (state which it is). Then determine the required heat transfer rate (kW) required for a reactor feed of 27.5 m^3/h.

9.13. The production of most of the steel manufactured in the United States begins with the reduction of hematite ore (mostly ferric oxide) with coke (carbon) in a blast furnace to obtain pig iron. The basic reaction is

$$Fe_2O_3(s) + 3\,C(s) \longrightarrow 2\,Fe(s) + 3\,CO(g): \quad \Delta\hat{H}_r(77°F) = 2.111 \times 10^5 \text{ Btu/lb-mole}$$

Suppose that stoichiometric amounts of ferric oxide and carbon are fed at 77°F, the reaction is complete, the iron emerges as a liquid at 2800°F, and CO emerges at 570°F. Perform the following calculations for a basis of 1 ton of iron produced.

(a) Draw and label a flowchart and perform all the material balance calculations needed to determine the amounts (lb-mole) of each feed and product stream component.

(b) Taking the reactant and product species in their normal states at 77°F as references, prepare an inlet–outlet enthalpy table and calculate and fill in all unknown component specific enthalpies (Btu/lb-mole). Use the following physical property data for iron:

$$Fe(s): \quad C_p[\text{Btu/(lb-mole·°F)}] = 5.90 + 1.50 \times 10^{-3}\, T(°F)$$
$$T_m = 2794°F, \quad \Delta\hat{H}_m(T_m) = 6496 \text{ Btu/lb-mole}$$
$$Fe(l): \quad C_p[\text{Btu/(lb-mole·°F)}] = 8.15$$

(c) Estimate the furnace heat requirement (Btu/ton Fe produced).

(d) List the assumptions that make the value calculated in part (b) only an approximate estimate of the furnace heat requirement. (One of the assumptions has something to do with the reactor pressure.)

9.14. Normal heptane is dehydrocyclicized to toluene in a continuous vapor-phase reaction:

$$C_7H_{16} \longrightarrow C_6H_5CH_3 + 4\,H_2$$

Pure heptane at 400°C is fed to the reactor. The reactor operates isothermally at 400°C and the reaction goes to completion.

(a) Taking a basis of 1 mol of heptane fed, draw and label a flowchart.

(b) Taking elemental species [C(s), H_2(g)] at 25°C as references, prepare and fill in an inlet–outlet enthalpy table. (See Example 9.5-2.)

(c) Calculate the required heat transfer to or from the reactor (state which it is) in kJ.

(d) What is the heat of the heptane dehydrocyclization reaction ($\Delta\hat{H}_r$) at 400°C and 1 atm?

9.15. The thermal decomposition of dimethyl ether

$$(CH_3)_2O(g) \longrightarrow CH_4(g) + H_2(g) + CO(g)$$

is to be carried out in an isothermal 2.00-liter laboratory reactor at 600°C. The reactor is charged with pure dimethyl ether at a pressure of 350 torr. After about two hours, the reactor pressure is 875 torr.

(a) Has the reaction proceeded to completion at the end of the two-hour period? If not, what percentage of the dimethyl ether has decomposed?

(b) Taking elemental species [C(s), H_2(g), O_2(g)] at 25°C as references, prepare and fill in an inlet–outlet enthalpy table. (See Example 9.5-2.) Use tabulated data for methane, hydrogen, and carbon monoxide, and the following data for dimethyl ether:

$$\Delta\hat{H}_f^\circ = -180.16 \text{ kJ/mol}$$
$$C_p[\text{J/(mol·K)}] = 26.86 + 0.1659T - 4.179 \times 10^{-5}T^2 \quad (T \text{ in kelvin})$$

(c) Calculate $\Delta\hat{H}_r(600°C)$ and $\Delta\hat{U}_r(600°C)$ for the dimethyl ether decomposition reaction.

(d) How much heat (kJ) was transferred to or from the reactor (state which it is) during the two-hour period of the reaction?

9.16. Sulfur dioxide is oxidized to sulfur trioxide in a small pilot-plant reactor. SO_2 and 100% excess air are fed to the reactor at 450°C. The reaction proceeds to a 65% SO_2 conversion, and the products emerge from the reactor at 550°C. The production rate of SO_3 is 1.00×10^2 kg/min. The reactor is surrounded by a water jacket into which water at 25°C is fed.

(a) Calculate the feed rates (standard cubic meters per second) of the SO_2 and air feed streams and the extent of reaction, $\dot{\xi}$(kmol/s).

(b) Calculate the standard heat of the SO_2 oxidation reaction, $\Delta \hat{H}_r^\circ$ (kJ/mol). Then, taking molecular species at 25°C as references, prepare and fill in an inlet–outlet enthalpy table and write an energy balance to calculate the heat (kW) that must be transferred from the reactor to the cooling water.

(c) Calculate the minimum flow rate of the cooling water if its temperature rise is to be kept below 15°C.

(d) Briefly state what would have been different in your calculations and results if you had taken elemental species as references in part (b).

9.17. Carbon monoxide at 25°C and steam at 150°C are fed to a continuous water-gas shift reactor. The product gas, which contains 40.0 mole% H_2, 40.0% CO_2, and the balance $H_2O(v)$, emerges at 500°C at a rate of 2.50 SCMH (standard cubic meters per hour) and goes to a condenser. The gas and liquid streams leaving the condenser are in equilibrium at 15°C and 1 atm. The liquid may be taken to be pure water (no dissolved gases).

(a) Calculate the percent excess steam fed to the reactor and the rate of condensation of the water (kg/h).

(b) Calculate the rate (kW) at which heat must be transferred from the condenser.

(c) Taking atomic species at 25°C as references, prepare and fill in an inlet–outlet enthalpy table and calculate the required rate of heat transfer (kW) to or from the reactor.

(d) It has been suggested that the carbon monoxide feed stream could pass through a heat exchanger prior to entering the reactor and the reactor product gas could pass through the same heat exchanger before entering the condenser. Sketch and label a flowchart of the heat exchanger and state economic benefits that might result from its use. (*Hint:* Heating and cooling operations are both expensive.)

*9.18. Metallic iron is produced in the reaction between ferrous oxide and carbon monoxide:

$$FeO(s) + CO(g) \longrightarrow Fe(s) + CO_2(g), \qquad \Delta \hat{H}_r^\circ = -16.480 \text{ kJ/mol}$$

The flowchart shown below depicts this process for a basis of 1 mol FeO fed.

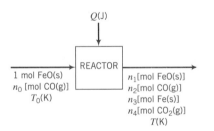

(a) We wish to explore the effects of the variables n_0 (the molar feed ratio of CO to FeO), T_0 (the feed temperature of the carbon monoxide), X (the fractional conversion of FeO), and T (the product temperature) on Q (the heat duty on the reactor). *Without doing any calculations,* sketch the shapes of the curves you would expect to obtain for the following plots:

(i) Let $n_0 = 1$ mol CO fed/mol FeO fed, $T_0 = 400$ K, and $X = 1$. Vary T from 298 K to 1000 K, calculate Q for each T, and plot Q versus T.

(ii) Let $n_0 = 1$ mol CO fed/mol FeO fed, $T = 700$ K, and $X = 1$. Vary T_0 from 298 K to 1000 K, calculate Q for each T_0, and plot Q versus T_0.

(iii) Let $n_0 = 1$ mol CO fed/mol FeO fed, $T_0 = 400$ K, and $T = 500$ K. Vary X from 0 to 1, calculate Q for each X, and plot Q versus X.

(iv) Let $X = 0.5$, $T_0 = 400$ K, and $T = 400$ K. Vary n_0 from 0.5 to 2 mol CO fed/mol FeO fed, calculate Q for each n_0, and plot Q versus n_0.

---

*Computer problem.

(b) Following is an inlet-outlet enthalpy table for the process:

References: FeO(s), CO(g), Fe(s), CO$_2$(g) at 25°C

Substance	n_{in} (mol)	$\hat{H}_{in}$ (kJ/mol)	n_{out} (mol)	$\hat{H}_{out}$ (kJ/mol)
FeO	1.00	0	n_1	$\hat{H}_1$
CO	n_0	$\hat{H}_0$	n_2	$\hat{H}_2$
Fe	—	—	n_3	$\hat{H}_3$
CO$_2$	—	—	n_4	$\hat{H}_4$

Write an expression for the heat duty on the reactor, Q(kJ), in terms of the n's and $\hat{H}$'s in the table, the standard heat of the given reaction, and the extent of reaction, ξ. Then derive expressions for the quantities ξ, n_1, n_2, n_3, and n_4 in terms of the variables n_0 and X. Finally, derive expressions for $\hat{H}_0$ as a function of T_0 and for $\hat{H}_1$, $\hat{H}_2$, $\hat{H}_3$, and $\hat{H}_4$ as functions of T. In the latter derivations, use the following formulas for C_p[kJ/(mol·K)] in terms of T(K) adapted from Table 2-194 of *Perry's Chemical Engineers' Handbook* (see footnote 1):

$$\text{FeO(s):} \quad C_p = 0.05280 + 6.243 \times 10^{-6} T - 3.188 \times 10^2\, T^{-2}$$
$$\text{Fe(s):} \quad C_p = 0.01728 + 2.67 \times 10^{-5}\, T$$
$$\text{CO(g):} \quad C_p = 0.02761 + 5.02 \times 10^{-6}\, T$$
$$\text{CO}_2\text{(g):} \quad C_p = 0.04326 + 1.146 \times 10^{-5} T - 8.180 \times 10^2 T^{-2}$$

(c) Calculate the heat duty, Q(kJ), for $n_0 = 2.0$ mol CO, $T_0 = 350$ K, $T = 550$ K, and $X = 0.700$ mol FeO reacted/mol FeO fed.

(d) Prepare a spreadsheet that has the following format (a partial solution is given for one set of process variables):

Ferrous oxide reduction problem													
	FeO + CO → Fe + CO2							DHr=	−16480	J/mol			
n0	T0	X	T	Xi	n1	n2	n3	n4	H0	H1	H2	H3	H4	Q
(mol)	(K)		(K)	(mol)	(mol)	(mol)	(mol)	(mol)	(kJ/mol)	(kJ/mol)	(kJ/mol)	(kJ/mol)	(kJ/mol)	(kJ)
2	350	0.7	550	0.7	0.3	1.3	0.7	0.7	1520	13482	...	...	...	11863

where DHr ($= \Delta\hat{H}_r^\circ$) denotes the standard heat of the FeO reduction reaction and Xi (ξ) is the extent of reaction. Use the spreadsheet to generate the four plots described in part (a). If the shapes of the plots do not match your predictions, explain why.

9.19. Ethyl alcohol (ethanol) can be produced by the **fermentation** of sugars derived from grains and other agricultural products. Some countries without large petroleum and natural gas reserves—such as Brazil—have found it profitable to convert a portion of their abundant grains to ethanol for blending with gasoline as an octane enhancer or for use as a feedstock in the synthesis of other chemicals.

In one such process, a portion of the starch in corn is converted to ethanol in two consecutive reactions. In a *saccharification* reaction, starch decomposes in the presence of certain enzymes (biological catalysts) to form an aqueous *mash* containing maltose ($C_{12}H_{22}O_{11}$, a sugar) and several other decomposition products. The mash is cooled and combined with additional water and a yeast culture in a batch fermentation tank (fermentor). In the fermentation reaction (actually a complex series of reactions), the yeast culture grows and in the process converts maltose to ethanol and carbon dioxide:

$$C_{12}H_{22}O_{11} + H_2O \longrightarrow 4\,C_2H_5OH + 4\,CO_2$$

The fermentor is a 550,000 gallon tank filled to 90% of its capacity with a suspension of mash and yeast in water. The mass of the yeast is negligible compared to the total mass of the tank contents. Thermal energy is released by the exothermic conversion of maltose to ethanol. In an adiabatic operating stage, the temperature of the tank contents increases from an initial value of 85°F to 95°F, and in a second stage the temperature is kept at 95°F by a reactor cooling system. The final reaction mixture contains carbon dioxide dissolved in a slurry containing 7.1 wt% ethanol, 6.9 wt% soluble and suspended solids, and the balance water. The mixture is pumped to a flash evaporator in which CO_2 is vaporized, and the ethanol product is then separated from the remaining mixture components in a series of distillation and stripping operations.

Data

- One bushel (56 lb$_m$) of corn yields 25 gallons of mash fed to the fermentor, which in turn yields 2.6 gallons of ethanol. Roughly 101 bushels of corn is harvested from an acre of land.
- A batch fermentation cycle (charging the fermentation tank, running the reaction, discharging the tank, and preparing the tank to receive the next load) takes eight hours. The process operates 24 hours per day, 330 days per year.
- The specific gravity of the fermentation reaction mixture is approximately constant at 1.05. The average heat capacity of the mixture is 0.95 Btu/(lb$_m$·°F).
- The standard heat of combustion of maltose to form $CO_2(g)$ and $H_2O(l)$ is $\Delta\hat{H}_c^\circ = -5649.1$ kJ/mol.

(a) Calculate (i) the quantity of ethanol (lb$_m$) produced per batch, (ii) the quantity of water (gal) that must be added to the mash and yeast in the fermentation tank, and (iii) the acres of land that must be harvested per year to keep the process running.

(b) Calculate the standard heat of the maltose conversion reaction, $\Delta\hat{H}_r^\circ$(Btu/lb-mole).

(c) Estimate the total amount of heat (Btu) that must be transferred from the fermentor during the reaction period. Take only the maltose conversion into account in this calculation (i.e., neglect the yeast growth reaction and any other reactions that may occur in the fermentor), assume that the heat of reaction is independent of temperature in the range from 77°F (= 25°C) to 95°F, and neglect the heat of solution of carbon dioxide in water.

(d) Although Brazil and Venezuela are neighboring countries, producing ethanol from grain for use as a fuel is an important process in Brazil and an almost nonexistent one in Venezuela. What difference between the two countries probably accounts for this observation?

9.20. Ammonia is oxidized with air to form nitric oxide in the first step of the production of nitric acid. Two principal reactions occur:

$$4\,NH_3 + 5\,O_2 \longrightarrow 4\,NO + 6\,H_2O$$
$$2\,NH_3 + \tfrac{3}{2}\,O_2 \longrightarrow N_2 + 3\,H_2O$$

A flowchart of the reactor follows.

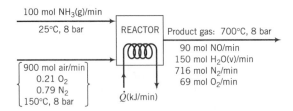

(a) Taking elemental species [$N_2(g)$, $H_2(g)$, $O_2(g)$] at 25°C as references, prepare and fill in an inlet–outlet enthalpy table.

(b) Calculate the required rate of heat transfer to or from the reactor in kW.

(c) What would have been different in your calculations and results in parts (a) and (b) if you had taken molecular species as references in part (a)?

9.21. Ethanol is produced commercially by the hydration of ethylene:

$$C_2H_4(g) + H_2O(v) \rightleftharpoons C_2H_5OH(v)$$

Some of the product is converted to diethyl ether in the undesired side reaction

$$2\,C_2H_5OH(v) \rightleftharpoons (C_2H_5)_2O(v) + H_2O(v)$$

The combined feed to the reactor contains 53.7 mole% C_2H_4, 36.7% H_2O and the balance nitrogen which enters the reactor at 310°C. The reactor operates isothermally at 310°C. An ethylene conversion of 5% is achieved, and the yield of ethanol (moles ethanol produced/mole ethylene consumed) is 0.900.

Data for Diethyl Ether

$$\Delta\hat{H}_f^\circ = -272.8 \text{ kJ/mol } \textit{for the liquid}$$

$$\Delta\hat{H}_v = 26.05 \text{ kJ/mol} \quad (\text{assume independent of } T)$$

$$C_p[\text{kJ/(mol·°C)}] = 0.08945 + 40.33 \times 10^{-5}T(\text{°C}) - 2.244 \times 10^{-7}T^2$$

(a) Calculate the reactor heating or cooling requirement in kJ/mol feed.

(b) Why would the reactor be designed to yield such a low conversion of ethylene? What processing step (or steps) would probably follow the reactor in a commercial implementation of this process?

9.22. Benzaldehyde is produced from toluene in the catalytic reaction

$$C_6H_5CH_3 + O_2 \rightarrow C_6H_5CHO + H_2O$$

Dry air and toluene vapor are mixed and fed to the reactor at 350°F and 1 atm. Air is supplied in 100% excess. Of the toluene fed to the reactor, 13% reacts to form benzaldehyde and 0.5% reacts with oxygen to form CO_2 and H_2O. The product gases leave the reactor at 379°F and 1 atm. Water is circulated through a jacket surrounding the reactor, entering at 80°F and leaving at 105°F. During a four-hour test period, 29.3 lb_m of water is condensed from the product gases. (Total condensation may be assumed.) The standard heat of formation of benzaldehyde vapor is $-17{,}200$ Btu/lb-mole; the heat capacities of both toluene and benzaldehyde vapors are approximately 31 Btu/(lb-mole·°F); and that of liquid benzaldehyde is 46 Btu/(lb-mole·°F).

(a) Calculate the volumetric flow rates (ft³/h) of the combined feed stream to the reactor and the product gas.

(b) Calculate the required rate of heat transfer from the reactor (Btu/h) and the flow rate of the cooling water (gal/min).

9.23. Lime (calcium oxide) is widely used in the production of cement, steel, medicines, insecticides, plant and animal food, soap, rubber, and many other familiar materials. It is usually produced by heating and decomposing limestone ($CaCO_3$), a cheap and abundant mineral, in a *calcination* process:

$$CaCO_3(s) \xrightarrow{\text{heat}} CaO(s) + CO_2(g)$$

(a) Limestone at 25°C is fed to a continuous calcination reactor. The calcination is complete, and the products leave at 900°C. Taking 1 metric ton (1000 kg) of limestone as a basis and elemental species [Ca(s), C(s), $O_2(g)$] at 25°C as references for enthalpy calculations, prepare and fill in an inlet–outlet enthalpy table and prove that the required heat transfer to the reactor is 2.7×10^6 kJ.

(b) In a common variation of this process, hot combustion gases containing oxygen and carbon monoxide (among other components) are fed into the calcination reactor along with the limestone. The carbon monoxide is oxidized in the reaction

$$CO(g) + \tfrac{1}{2} O_2(g) \longrightarrow CO_2(g)$$

Suppose

- the combustion gas fed to a calcination reactor contains 75 mole% N_2, 2.0% O_2, 9.0% CO, and 14% CO_2;
- the gas enters the reactor at 900°C in a feed ratio of 20 kmol gas/kmol limestone;
- the calcination is complete;
- all of the oxygen in the gas feed is consumed in the CO oxidation reaction;
- the reactor effluents are at 900°C.

Again taking a basis of 1 metric ton of limestone calcined, prepare and fill in an inlet–outlet enthalpy table for this process [don't recalculate enthalpies already calculated in part (a)] and calculate the required heat transfer to the reactor.

(c) You should have found that the heat that must be transferred to the reactor is significantly lower with the combustion gas in the feed than it is without the gas. By what percentage is the heat requirement reduced? Give two reasons for the reduction.

*9.24. A pair of gas-phase reactions with the following stoichiometric equations take place in a continuous reactor:

$$A + B \longrightarrow C$$
$$2\,C \longrightarrow D + B$$

The reactions of ethylene and hydrogen to form ethanol and of ethanol to form diethyl ether and hydrogen constitute such a reaction system. (See Problem 9.21.)

(a) Suppose the reactor feed contains A, B, and inerts (I), with mole fractions x_{A0}, x_{B0}, and x_{I0}, respectively. Letting f_A denote the fractional conversion of A (mol A consumed/mol A fed) and Y_C the yield of C based on consumption of A (mol C generated/mol A consumed), prove that for a basis of 1 mol of feed, the number of moles of each species at the outlet are as follows:

$$n_A = x_{A0}(1 - f_A)$$
$$n_C = x_{A0} f_A Y_C$$
$$n_D = \tfrac{1}{2}(x_{A0} f_A - n_C)$$
$$n_B = x_{B0} - x_{A0} f_A + n_D$$
$$n_I = x_{I0}$$

(b) Write a spreadsheet to perform material and energy balance calculations for a basis of 1.00 mol feed. The program should take as inputs
  (i) the standard heats of formation (kJ/mol) of A(g), B(g), C(g), and D(g);
  (ii) the coefficients (a, b, c, d) of the formulas $C_p = a + bT + cT^2 + dT^3$ for gaseous A, B, C, D, and I, where C_p has units of kJ/(mol·°C);
  (iii) the feed and product temperatures, $T_f(°C)$ and $T_p(°C)$;
  (iv) x_{A0}, x_{B0}, f_A, and Y_C.
It should generate an inlet–outlet enthalpy table based on elemental species at 25°C as references and then calculate the required heat transfer to or from the reactor, $Q(kJ)$. The spreadsheet should be tested using the species and reactions of Problem 9.21 and should appear as shown below. (Some of the input data and calculated results are shown.)

---

*Computer problem.

Chapter 9—Problem 9.24						
Species	Formula	DHf	a	b	c	d
A	C2H4(v)	+52.28	0.04075	11.47e-5	-6.891e-8	17.66e-12
B	H2O(v)					
C	C2H5OH(v)					
D	(C4H10)O(v)	-246.8	0.08945	40.33e-5	-2.244e-7	
I	N2(g)					
Tf	Tp	xA0	xB0	xI0	fA	YC
310	310	0.537	0.367	0.096	0.05	0.90
	n(in)	H(in)	n(out)	H(out)		
Species	(mol)	(kJ/mol)	(mol)	(kJ/mol)		
A						
B						
C						
D						
I						
Q(kJ)=	-1.31					

where DHf $[= \Delta \hat{H}_f^\circ \text{(kJ/mol)}]$ denotes the standard heat of formation.

(c) Use the program to calculate Q at the reactor conditions shown in the spreadsheet, then for a feed temperature of 125°C and all other input parameters the same. (The enthalpy table and the value of Q should automatically correct themselves as soon as you type in the new value of T_f.) Print out and turn in your program output for the second feed temperature.

(d) Run the program for several different values of T_p, f_A, and Y_C. Summarize the effects of each of these parameters on Q and briefly explain why your results make sense.

9.25. A gas mixture containing 85 mole% methane and the balance oxygen is to be charged into an evacuated well-insulated 10-liter reaction vessel at 25°C and 200 kPa. An electrical coil in the reactor, which delivers heat at a rate of 100 watts, will be turned on for 85 seconds and then turned off. Formaldehyde will be produced in the reaction

$$CH_4 + O_2 \longrightarrow HCHO + H_2O$$

The reaction products will be cooled and discharged from the reactor.

(a) Calculate the maximum pressure that the reactor is likely to have to withstand, assuming that there are no side reactions. If you were ordering the reactor, why would you specify an even greater pressure in your order? (Give several reasons.)

(b) Why would heat be added to the feed mixture rather than running the reactor adiabatically?

(c) Suppose the reaction is run as planned, the reaction products are analyzed chromatographically, and some CO_2 is found. Where did it come from? If you had taken it into account, would your calculated pressure in part (a) have been larger, smaller, or can't you tell without doing the detailed calculations?

9.26. Ethylene oxide is produced by the catalytic oxidation of ethylene:

$$C_2H_4(g) + \tfrac{1}{2}O_2(g) \longrightarrow C_2H_4O(g)$$

An undesired competing reaction is the combustion of ethylene to CO_2.

The feed to a reactor contains 2 mol C_2H_4/mol O_2. The conversion and yield in the reactor are respectively 25% and 0.70 mol C_2H_4O produced/mol C_2H_4 consumed. A multiple-unit process separates the reactor outlet stream components: C_2H_4 and O_2 are recycled to the reactor, C_2H_4O is sold, and CO_2 and H_2O are discarded. The reactor inlet and outlet streams are each at 450°C, and the fresh feed and all species leaving the separation process are at 25°C. The combined fresh feed-recycle stream is preheated to 450°C.

(a) Taking a basis of 2 mol of ethylene entering the reactor, draw and label a flowchart of the complete process (show the separation process as a single unit) and calculate the molar amounts and compositions of all process streams.

(b) Calculate the heat requirement (kJ) for the entire process and that for the reactor alone.

Data for C_2H_4O (g)

$$\Delta\hat{H}_f^\circ = -51.00 \text{ kJ/mol}$$

$$C_p[\text{J/(mol·K)}] = -4.69 + 0.2061T - 9.995 \times 10^{-5}T^2$$

where T is in kelvin.

(c) Calculate the flow rate (kg/h) and composition of the fresh feed and the overall and reactor heat requirements (kW) for a production rate of 1500 kg C_2H_4O/day.

9.27. Cumene ($C_6H_5C_3H_7$) is produced by reacting benzene with propylene $[\Delta\hat{H}_r^\circ(77°F) = -39,520$ Btu/lb-mole].

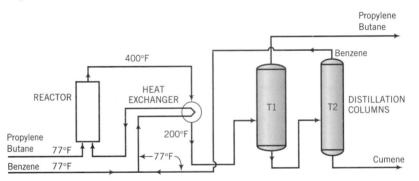

A liquid feed containing 75 mole% propylene and 25% *n*-butane and a second liquid stream containing essentially pure benzene are fed to the reactor. Fresh benzene and recycled benzene, both at 77°F, are mixed in a 1:3 ratio (1 mole fresh feed/3 moles recycle) and passed through a heat exchanger, where they are heated by the reactor effluent before being fed to the reactor. The reactor effluent enters the exchanger at 400°F and leaves at 200°F. The pressure in the reactor is sufficient to maintain the effluent stream as a liquid.

After being cooled in the heat exchanger, the reactor effluent is fed to a distillation column (T1). All of the butane and unreacted propylene are removed as overhead product from the column, and the cumene and unreacted benzene are removed as bottoms product and fed to a second distillation column (T2) where they are separated. The benzene leaving the top of the second column is the recycle that is mixed with the fresh benzene feed. Of the propylene fed to the process, 20% does not react and leaves in the overhead product from the first distillation column. The production rate of cumene is 1200 lb$_m$/h.

(a) Calculate the mass flow rates of the streams fed to the reactor, the molar flow rate and composition of the reactor effluent, and the molar flow rate and composition of the overhead product from the first distillation column, T1.

(b) Calculate the temperature of the benzene stream fed to the reactor and the required rate of heat addition to or removal from the reactor. Use the following approximate heat capacities in your calculations: C_p [Btu/(lb$_m$·°F)] = 0.57 for propylene, 0.55 for butane, 0.45 for benzene, and 0.40 for cumene.

9.28. Ethylbenzene is converted to styrene in the catalytic dehydrogenation reaction

$$C_8H_{10}(g) \rightarrow C_8H_8(g) + H_2: \quad \Delta\hat{H}_r^\circ(600°C) = +124.5 \text{ kJ/mol}$$

A flowchart of a simplified version of the commercial process is shown here.

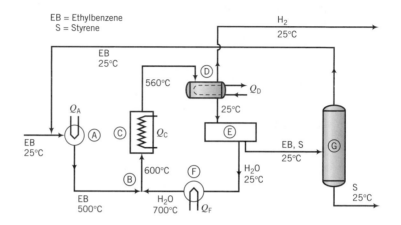

Fresh and recycled liquid ethylbenzene combine and are heated from 25°C to 500°C Ⓐ, and the heated ethylbenzene is mixed adiabatically with steam at 700°C Ⓑ to produce the feed to the reactor at 600°C. (The steam suppresses undesired side reactions and removes carbon deposited on the catalyst surface.) A once-through conversion of 35% is achieved in the reactor Ⓒ, and the products emerge at 560°C. The product stream is cooled to 25°C Ⓓ, condensing essentially all of the water, ethylbenzene, and styrene and allowing hydrogen to pass out as a recoverable by-product of the process.

The water and hydrocarbon liquids are immiscible and are separated in a settling tank decanter Ⓔ. The water is vaporized and heated Ⓕ to produce the steam that mixes with the ethylbenzene feed to the reactor. The hydrocarbon stream leaving the decanter is fed to a distillation tower Ⓖ (actually, a series of towers), which separates the mixture into essentially pure styrene and ethylbenzene, each at 25°C after cooling and condensation steps have been carried out. The ethylbenzene is recycled to the reactor preheater, and the styrene is taken off as a product.

(a) Calculate on the basis of 100 kg/h styrene produced the required fresh ethylbenzene feed rate, the flow rate of recycled ethylbenzene, and the circulation rate of water, all in mol/h. (Assume $P = 1$ atm.)

(b) Calculate the required rates of heat input or withdrawal in kJ/h for the ethylbenzene preheater Ⓐ, steam generator Ⓕ, and reactor Ⓒ.

(c) Suggest possible ways to improve the energy economy of this process.

Physical Property Data

$$\text{Ethylbenzene:} \quad (C_p)_{liquid} = 182 \text{ J/(mol·°C)}$$
$$(\Delta\hat{H}_v) = 36.0 \text{ kJ/mol at } 136°C$$
$$(C_p)_{vapor}[\text{J/(mol·°C)}] = 118 + 0.30T(°C)$$

$$\text{Styrene:} \quad (C_p)_{liquid} = 209 \text{ J/(mol·°C)}$$
$$(\Delta\hat{H}_v) = 37.1 \text{ kJ/mol at } 145°C$$
$$(C_p)_{vapor}[\text{J/(mol·°C)}] = 115 + 0.27T(°C)$$

9.29. Formaldehyde is produced by decomposing methanol over a silver catalyst:

$$CH_3OH \rightarrow HCHO + H_2$$

To provide heat for this endothermic reaction, some oxygen is included in the feed to the reactor, leading to the partial combustion of the hydrogen produced in the methanol decomposition.

The feed to an adiabatic formaldehyde production reactor is obtained by bubbling a stream of air at 1 atm through liquid methanol. The air leaves the vaporizer saturated with methanol and contains 42% methanol by volume. The stream then passes through a heater in which its temperature is raised to 145°C. To avoid deactivating the catalyst, the maximum temperature attained in the reactor must be limited to 600°C. For this purpose, saturated steam at 145°C is metered into the air–methanol stream, and the combined stream enters the reactor. A fractional methanol conversion of 70.0% is achieved in the reactor, and the product gas contains 5.00 mole% hydrogen. The product gas is cooled to 145°C in a waste heat boiler in which saturated steam at 3.1 bar is generated from liquid water at 30°C. Several absorption and distillation units follow the waste heat boiler, and formaldehyde is ultimately recovered in an aqueous solution containing 37.0 wt% HCHO. The plant is designed to produce 36 metric kilotons of this solution per year, operating 350 days/yr.

(a) Draw the process flowchart and label it completely. Show the absorption/distillation train as a single unit with the reactor product gas and additional water entering and the formaldehyde solution and a gas stream containing methanol, oxygen, nitrogen, and hydrogen leaving.

(b) Calculate the operating temperature of the methanol vaporizer.

(c) Calculate the required feed rate of steam to the reactor (kg/h) and the molar flow rate and composition of the product gas.

(d) Calculate the rate (kg/h) at which steam is generated in the waste heat boiler.

9.30. The synthesis of ethyl chloride is accomplished by reacting ethylene with hydrogen chloride in the presence of an aluminum chloride catalyst:

$$C_2H_4(g) + HCl(g) \xrightarrow{\text{catalyst}} C_2H_5Cl(g): \quad \Delta\hat{H}_r(0°C) = -64.5 \text{ kJ/mol}$$

Process data and a simplified schematic flowchart are given here.

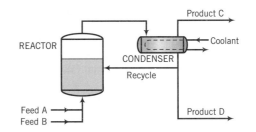

Data

Reactor: adiabatic, outlet temperature = 50°C
Feed A: 100% HCl(g), 0°C
Feed B: 93 mole% C_2H_4, 7% C_2H_6, 0°C
Product C: Consists of 1.5% of the HCl, 1.5% of the C_2H_4, and all of the C_2H_6 that enter the reactor
Product D: 1600 kg $C_2H_5Cl(l)/h$, 0°C
Recycle to reactor: $C_2H_5Cl(l)$, 0°C
C_2H_5Cl: $\Delta\hat{H}_v = 24.7$ kJ/mol (assume independent of T)
$(C_p)_{C_2H_5Cl(v)}[\text{kJ/(mol·°C)}] = 0.052 + 8.7 \times 10^{-5}T(°C)$

The reaction is exothermic, and if the heat of reaction is not removed in some way the reactor temperature could increase to an undesirably high level. To avoid this occurrence, the reaction is carried out with the catalyst suspended in liquid ethyl chloride. As the reaction proceeds, most of the heat liberated goes to vaporize the liquid, making it possible to keep the reaction temperature at or below 50°C.

The stream leaving the reactor contains ethyl chloride formed by reaction and that vaporized in the reactor. This stream passes through a heat exchanger where it is cooled to 0°C, condensing essentially all of the ethyl chloride and leaving only unreacted C_2H_4, HCl, and C_2H_6 in the gas phase. A portion of the liquid condensate is recycled to the reactor at a rate equal to the rate at which ethyl chloride is vaporized, and the rest is taken off as product. At the process conditions, heats of mixing and the influence of pressure on enthalpy may be neglected.

(a) At what rates (kmol/h) do the two feed streams enter the process?

(b) Calculate the composition (component mole fractions) and molar flow rate of product stream C.

(c) Write an energy balance around the reactor and use it to determine the rate at which ethyl chloride must be recycled.

(d) A number of simplifying assumptions were made in the process description and the analysis of this process system, so the results obtained using a more realistic simulation would differ considerably from those you should have obtained in parts (a)–(c). List as many of these assumptions as you can think of.

9.31. Ammonia is oxidized in a well-insulated continuous reactor:

$$4 \, NH_3(g) + 5 \, O_2(g) \longrightarrow 4 \, NO(g) + 6 \, H_2O(v): \quad \Delta\hat{H}_r^\circ = -904.7 \; kJ/mol$$

The feed stream enters at 200°C and the products leave at temperature T_{out}(°C). The inlet–outlet enthalpy table for the reactor appears as follows:

References: $NH_3(g)$, $O_2(g)$, $NO(g)$, $H_2O(v)$ at 25°C, 1 atm

Substance	$\dot{n}_{in}$ (mol/s)	$\hat{H}_{in}$ (kJ/mol)	$\dot{n}_{out}$ (mol/s)	$\hat{H}_{out}$ (kJ/mol)
$NH_3(g)$	4.00	$\hat{H}_1$	—	—
$O_2(g)$	6.00	$\hat{H}_2$	$\dot{n}_3$	$\hat{H}_3$
$NO(g)$	—	—	$\dot{n}_4$	$\hat{H}_4$
$H_2O(v)$	—	—	$\dot{n}_5$	$\hat{H}_5$

(a) Draw and label a process flowchart and calculate the molar amounts of the product stream components and the extent of reaction, $\dot{\xi}$(mol/s). Fill in the values of $\dot{n}_3$, $\dot{n}_4$, and $\dot{n}_5$ on the enthalpy table.

(b) The energy balance for this reactor reduces to $\Delta\dot{H} \approx 0$. Summarize the assumptions that must be made to obtain this result.

(c) Calculate the values of $\hat{H}_1$ and $\hat{H}_2$ and write expressions for $\hat{H}_3$, $\hat{H}_4$, and $\hat{H}_5$ in terms of the outlet temperature, T_{out}. Then calculate T_{out} from the energy balance, using either a spreadsheet or a programmable calculator. (See Example 9.5-3.)

(d) A design engineer obtained a preliminary estimate of the reactor outlet temperature using only the first terms of the heat capacity formulas in Table B.2. [For example, $(C_p)_{NH_3} \approx 0.03515$ kJ/(mol·°C).] What value did she calculate? Taking the result of part (c) to be correct, determine the percentage error in T_{out} that results from using the one-term heat capacity formulas.

(e) The preliminary estimate of T_{out} was mistakenly used as the basis of the design and construction of the reactor. Was this a dangerous error from the standpoint of reactor safety or did it in fact lower the hazard potential? Explain.

9.32. Coke can be converted into CO—a fuel gas—in the reaction

$$CO_2(g) + C(s) \rightarrow 2 \, CO(g)$$

A coke that contains 84% carbon by mass and the balance noncombustible ash is fed to a reactor with a stoichiometric amount of CO_2. The coke is fed at 77°F, and the CO_2 enters at 400°F. Heat is transferred to the reactor in the amount of 5859 Btu/lb$_m$ coke fed. The gaseous products and the solid reactor effluent (the ash and unburned carbon) leave the reactor at 1830°F. The heat capacity of the solid is 0.24 Btu/(lb$_m$·°F).

(a) Calculate the percentage conversion of the carbon in the coke.

(b) The carbon monoxide produced in this manner can be used as a fuel for residential home heating, as can the coke. Speculate on the advantages and disadvantages of using the gas. (There are several of each.)

9.33. The synthesis of methanol from carbon monoxide and hydrogen is carried out in a continuous vapor-phase reactor at 5.00 atm absolute. The feed contains CO and H_2 in stoichiometric proportion and enters the reactor at 25°C and 5.00 atm at a rate of 17.1 m³/h. The product stream emerges from the reactor at 127°C. The rate of heat transfer from the reactor is 17.05 kW. Calculate the fractional conversion achieved and the volumetric flow rate (m³/h) of the product stream. (See Example 9.5-4.)

9.34. Carbon disulfide, a key component in the manufacture of rayon fibers, is produced in the reaction between methane and sulfur vapor over a metal oxide catalyst:

$$CH_4(g) + 4\,S(v) \rightarrow CS_2(g) + 2\,H_2S(g)$$

$$\Delta\hat{H}_r(700°C) = -274 \text{ kJ/mol}$$

Methane and molten sulfur, each at 150°C, are fed to a heat exchanger in stoichiometric proportion. Heat is exchanged between the reactor feed and product streams, and the sulfur in the feed is vaporized. The gaseous methane and sulfur leave the exchanger and pass through a second preheater in which they are heated to 700°C, the temperature at which they enter the reactor. Heat is transferred from the reactor at a rate of 41.0 kJ/mol of feed. The reaction products emerge from the reactor at 800°C, pass through the heat exchanger, and emerge at 200°C with sulfur as a liquid. Use the heat capacity data given below to perform the following calculations.

Heat capacities: C_p [J/(mol·°C)] ≈ 29.4 for S(l), 36.4 for S(v), 71.4 for $CH_4(g)$, 31.8 for CS_2, and 44.8 for $H_2S(g)$.

(a) Estimate the fractional conversion achieved in the reactor.

(b) Estimate the quantity of heat (kJ/mol feed) that must be transferred in the second preheater to bring the feed up to 700°C.

(c) Suggest a method to improve the energy economy of the process.

*9.35. The equilibrium constant for the ethane dehydrogenation reaction,

$$C_2H_6(g) \rightleftharpoons C_2H_4(g) + H_2(g)$$

is defined as

$$K_p(\text{atm}) = \frac{x_{C_2H_4}\,x_{H_2}}{x_{C_2H_6}} P$$

where $P(\text{atm})$ is the total pressure and x_i is the mole fraction of the ith substance in an equilibrium mixture. The equilibrium constant has been found experimentally to vary with temperature according to the formula

$$K_p(T) = 7.28 \times 10^6 \exp[-17{,}000/T(\text{K})] \tag{1}$$

The heat of reaction at 1273K is +145.6 kJ/mol, and the heat capacities of the reactive species may be approximated by the formulas

$$\left. \begin{aligned} (C_p)_{C_2H_4} &= 9.419 + 0.1147T(\text{K}) \\ (C_p)_{H_2} &= 26.90 + 4.167 \times 10^{-3}T(\text{K}) \\ (C_p)_{C_2H_6} &= 11.35 + 0.1392T(\text{K}) \end{aligned} \right\} \text{[J/(mol·K)]}$$

Suppose pure ethane is fed to a continuous constant-pressure adiabatic reactor at 1273 K and pressure $P(\text{atm})$, the products emerge at $T_f(\text{K})$ and $P(\text{atm})$, and the residence time of the reaction mixture in the reactor is large enough for the outlet stream to be considered an equilibrium mixture of ethane, ethylene, and hydrogen.

(a) Prove that the fractional conversion of ethane in the reactor is

$$f = \left(\frac{K_p}{P + K_p}\right)^{1/2} \tag{2}$$

(b) Write an energy balance on the reactor, and use it to prove that

$$f = \frac{1}{1 + \phi(T_f)} \tag{3}$$

where

$$\phi(T_f) = \frac{\Delta\hat{H}_r(1273 \text{ K}) - \displaystyle\int_{T_f}^{1273\text{K}} [(C_p)_{C_2H_4} + (C_p)_{H_2}]\,dT}{\displaystyle\int_{T_f}^{1273\text{K}} (C_p)_{C_2H_6}\,dT} \tag{4}$$

---

*Computer problem.

Finally, substitute for $\Delta\hat{H}_r$ and the heat capacities in Equation 4 to derive an explicit expression for $\phi(T_f)$.

(c) We now have two expressions for the fractional conversion f: Equation 2 and Equation 3. If these expressions are equated, K_p is replaced by the expression of Equation 1, and $\phi(T_f)$ is replaced by the expression derived in part (b), the result is one equation in one unknown, T_f. Derive this equation, and transpose the right side to obtain an expression of the form

$$\psi(T_f) = 0 \tag{5}$$

(d) Write a spreadsheet program to take P as input, solve Equation 5 for T_f (use goalseek), and determine the final fractional conversion, f. (*Suggestion:* Set up columns for P, T_f, f, K_p, ϕ, and ψ.) Run the program for $P(\text{atm}) = 0.01, 0.05, 0.10, 0.50, 1.0, 5.0$, and 10.0. Plot T_f versus P and f versus P, using a logarithmic coordinate scale for P.

(e) Use a symbolic mathematics program or write a program in Fortran or some other language to perform the calculations of part (d).

9.36. You are checking the performance of a reactor in which acetylene is produced from methane in the reaction

$$2\ CH_4(g)\ \rightarrow\ C_2H_2(g) + 3\ H_2(g)$$

An undesired side reaction is the decomposition of acetylene:

$$C_2H_2(g)\ \rightarrow\ 2\ C(s) + H_2(g)$$

Methane is fed to the reactor at 1500°C at a rate of 10.0 mol CH₄/s. Heat is transferred to the reactor at a rate of 975 kW. The product temperature is 1500°C and the fractional conversion of methane is 0.600. A flowchart of the process and an enthalpy table are shown below.

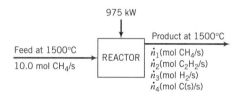

References: C(s), H₂(g), at 25°C, 1 atm

Substance	$\dot{n}_{in}$ (mol/s)	$\hat{H}_{in}$ (kJ/mol)	$\dot{n}_{out}$ (mol/s)	$\hat{H}_{out}$ (kJ/mol)
CH₄	10.0	41.65	$\dot{n}_1$	$\hat{H}_1$
C₂H₂	—	—	$\dot{n}_2$	$\hat{H}_2$
H₂	—	—	$\dot{n}_3$	$\hat{H}_3$
C	—	—	$\dot{n}_4$	$\hat{H}_4$

(a) Using the heat capacities given below for enthalpy calculations, write and solve material balances and an energy balance to determine the product component flow rates and the yield of acetylene (mol C₂H₂ produced/mol CH₄ consumed).

$$CH_4(g):\quad C_p \approx 0.079\ \text{kJ/(mol·°C)}$$
$$C_2H_2(g):\quad C_p \approx 0.052\ \text{kJ/(mol·°C)}$$
$$H_2(g):\quad C_p \approx 0.031\ \text{kJ/(mol·°C)}$$
$$C(s):\quad C_p \approx 0.022\ \text{kJ/(mol·°C)}$$

For example, the specific enthalpy of methane at 1500°C relative to methane at 25°C is [0.079 kJ/(mol·°C)](1500°C −25°C) = 116.5 kJ/mol.

(b) The reactor efficiency may be defined as the ratio (actual acetylene yield/acetylene yield with no side reaction). What is the reactor efficiency for this process?

9.37. Hydrogen is produced in the steam reforming of propane:

$$C_3H_8(g) + 3\,H_2O(v) \rightarrow 3\,CO(g) + 7\,H_2(g)$$

The water–gas shift reaction also takes place in the reactor, leading to the formation of additional hydrogen:

$$CO(g) + H_2O(v) \rightarrow CO_2(g) + H_2(g)$$

The reaction is carried out over a nickel catalyst in the tubes of a shell-and-tube reactor. The feed to the reactor contains steam and propane in a 6:1 molar ratio at 125°C, and the products emerge at 800°C. The excess steam in the feed assures essentially complete consumption of the propane. Heat is added to the reaction mixture by passing a hot gas over the outside of the tubes that contain the catalyst. The gas is fed at 4.94 m³/mol C_3H_8, entering the unit at 1400°C and 1 atm and leaving at 900°C. The unit may be considered adiabatic.

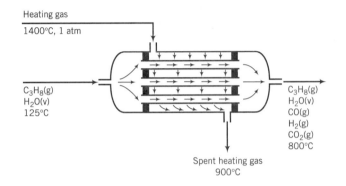

Calculate the molar composition of the product gas, assuming that the heat capacity of the heating gas is 0.040 kJ/(mol·°C).

9.38. In a **coal gasification** process, carbon (the primary constituent of coal) reacts with steam to produce carbon monoxide and hydrogen (*synthesis gas*). The gas may either be burned or subjected to further processing to produce any of a variety of chemicals.

A coal contains 10.5 wt% moisture (water) and 22.6 wt% noncombustible ash. The remaining fraction of the coal contains 81.2 wt% C, 13.4 wt% O, and 5.4% H. A coal slurry containing 2.00 kg coal/kg water is fed at 25°C to an adiabatic gasification reactor along with a stream of pure oxygen at the same temperature. The following reactions take place in the reactor:

$$C(s) + H_2O(v) \longrightarrow CO(g) + H_2(g): \quad \Delta\hat{H}_r^\circ = +131.3 \text{ kJ/mol} \tag{1}$$

$$C(s) + O_2(g) \longrightarrow CO_2(g): \quad \Delta\hat{H}_r^\circ = -393.5 \text{ kJ/mol} \tag{2}$$

$$2H(\text{in coal}) + \tfrac{1}{2}O_2(g) \longrightarrow H_2O(v): \quad \Delta\hat{H}_r^\circ \approx -242 \text{ kJ/mol} \tag{3}$$

Gas and slag (molten ash) leave the reactor at 2500°C. The gas contains CO, H_2, CO_2, and H_2O.[11]

(a) Feeding oxygen to the reactor lowers the yield of synthesis gas, but no gasifier ever operates without supplementary oxygen. Why does the oxygen lower the yield? Why it is nevertheless always supplied. (*Hint:* All the necessary information is contained in the first two stoichiometric equations and associated heats of reaction shown above.)

(b) Suppose the oxygen gas fed to the reactor and the oxygen in the coal combine with all the hydrogen in the coal (Reaction 3) and with some of the carbon (Reaction 2), and the remainder of the carbon is consumed in Reaction 1. Taking a basis of 1.00 kg coal fed to the reactor and letting n_0 equal the moles of O_2 fed, draw and label a flowchart. Then derive expressions for the molar flow rates of the four outlet gas species in terms of n_0. (Partial solution: $n_{H_2} = 51.5 - n_0$.)

---

[11] In a real coal gasification reactor, sulfur in the coal would form hydrogen sulfide in the product gas, nitrogen in the coal would form N_2, some of the carbon monoxide formed in the first reaction would react with steam to form carbon dioxide and more hydrogen, and some of the carbon in the coal would react with hydrogen to form methane. For simplicity, we are ignoring these reactions.

(c) The standard heat of combustion of the coal has been determined to be $-21,400$ kJ/kg, taking $CO_2(g)$ and $H_2O(l)$ to be the combustion products. Use this value and the given elemental composition of the coal to prove that the standard heat of formation of the coal is -1510 kJ/kg. Then use an energy balance to calculate n_0, using the following approximate heat capacities in your calculation:

Species	O_2	CO	H_2	CO_2	$H_2O(v)$	Slag(l)
C_p[kJ/(mol·°C)]	0.0336	0.0332	0.0300	0.0508	0.0395	—
C_p[kJ/(kg·°C)]	—	—	—	—	—	1.4

Take the heat of fusion of ash (the heat required to convert ash to slag) to be 710 kJ/kg.

9.39. Three cubic meters of a 1.00-molar aqueous sulfuric acid solution (SG = 1.064) is stored at 25°C. Use data in Tables B.1 and B.11 to calculate the standard heat of formation of the solution in kJ/mol H_2SO_4 relative to the solute elements and water, and the total enthalpy of the solution relative to the same reference conditions.

9.40. Calculate the standard heat (kJ/mol) of the neutralization reaction between dilute hydrochloric acid and dilute sodium hydroxide solutions, taking the heat of solution of NaCl to be $+4.87$ kJ/mol. Then calculate the standard heat of the reaction between hydrogen chloride gas and solid sodium hydroxide to form solid sodium chloride and liquid water. What is the physical significance of the difference between the two calculated values?

9.41. A 10.0 mole% aqueous sulfuric acid solution (SG = 1.27) is to be titrated to neutrality at 25°C with a 3.00-molar caustic soda (sodium hydroxide) solution (SG = 1.13):

$$H_2SO_4(aq) + 2\,NaOH(aq) \longrightarrow Na_2SO_4(aq) + 2\,H_2O(l)$$

(a) Calculate the required volume ratio (cm^3 caustic solution/cm^3 acid solution).

(b) Calculate the standard heats of formation (kJ/mol solute) of each of the three solutions involved in this process, and the heat released (kJ/cm^3 acid solution neutralized) if the reactants and products are all at 25°C. The heat of solution of sodium sulfate is -1.17 kJ/mol. Assume that $Q = \Delta H$ for the process. (See Example 9.5-5.)

9.42. Most of the world's chlorine and sodium hydroxide are obtained through the electrolysis of brine:

$$NaCl(aq) + H_2O(l) \rightarrow \tfrac{1}{2}\,H_2(g) + \tfrac{1}{2}\,Cl_2(g) + NaOH(aq)$$

(a) Determine the standard heats of formation of NaCl(aq) and NaOH(aq) and then the standard heat of the brine electrolysis reaction. The standard heat of solution of NaCl is $\Delta \hat{H}_s(r = \infty) = +4.87$ kJ/mol NaCl.

(b) Several years ago, the annual production of chlorine in the United States was roughly 8500 metric kilotons. Calculate the energy requirement in MW·h/yr corresponding to this production rate, assuming that all of the chlorine is produced by electrolysis at 25°C and the required energy input equals ΔH for the process.

9.43. You have been given the task of determining the standard heat of the reaction in which calcium chloride hexahydrate is formed from anhydrous calcium chloride:

$$CaCl_2(s) + 6\,H_2O(l) \longrightarrow CaCl_2 \cdot 6H_2O(s): \quad \Delta \hat{H}_r^\circ \text{ (kJ/mol)} = ?$$

By definition, the desired quantity is the *heat of hydration* of calcium chloride hexahydrate. You cannot carry out the hydration reaction directly, so you resort to an indirect method. You first dissolve 1.00 mol of anhydrous $CaCl_2$ in 10.0 mol of water in a calorimeter and determine that 64.85 kJ of heat must be transferred away from the calorimeter to keep the solution temperature at 25°C. You next dissolve 1.00 mol of the hexahydrate salt in 4.00 mol of water and find that 32.1 kJ of heat must be transferred *to* the calorimeter to keep the temperature at 25°C.

(a) Use these results to calculate the desired heat of hydration. (*Suggestion:* Begin by writing out the stoichiometric equations for the two dissolution processes.)

(b) Calculate the standard heat of formation of $CaCl_2(aq, r = 10)$ in kJ/mol $CaCl_2$ relative to Ca(s), $Cl_2(g)$, and $H_2O(l)$ at 25°C.

9.44. A dilute aqueous solution of sulfuric acid at 25°C is used to absorb ammonia in a continuous reactor, thereby producing ammonium sulfate, a fertilizer:

$$2 NH_3(g) + H_2SO_4(aq) \rightarrow (NH_4)_2SO_4(aq)$$

    (a) If the ammonia enters the absorber at 75°C, the sulfuric acid enters at 25°C, and the product solution emerges at 25°C, how much heat must be withdrawn from the unit per mol of $(NH_4)_sO_4$ produced? (All needed physical property data may be found in Appendix B.)

    (b) Estimate the final temperature if the reactor of part (a) is adiabatic and the product of the solution contains 1.00 mole % ammonium sulfate. Take the heat capacity of the solution to be that of pure liquid water [4.184 kJ/(kg·°C)].

    (c) In a real (imperfectly insulated) reactor, would the final solution temperature be less than, equal to, or greater than the value calculated in part (b), or is there no way to tell without more information? Briefly explain your answer.

9.45. A 2.00 mole% sulfuric acid solution is neutralized with a 5.00 mole% sodium hydroxide solution in a continuous reactor. All reactants enter at 25°C. The standard heat of solution of sodium sulfate is –1.2 kJ/mol Na_2SO_4, and the heat capacities of all solutions may be taken to be that of pure liquid water [4.184 kJ/(kg·°C)].

    (a) How much heat (kJ/kg acid solution fed) must be transferred to or from the reactor contents (state which it is) if the product solution emerges at 40°C?

    (b) Estimate the product solution temperature if the reactor is adiabatic, neglecting heat transferred between the reactor contents and the reactor wall.

9.46. A 12.0-molar solution of sodium hydroxide (SG = 1.37) is neutralized with 75.0 mL of a 4.0 molar solution of sulfuric acid (SG = 1.23) in a well-insulated container.

    (a) Estimate the volume of the sodium hydroxide solution and the final solution temperature if both feed solutions are at 25°C. The heat capacity of the product solution may be taken to be that of pure liquid water, and the energy balance reduces to $Q = \Delta H$ for this constant-pressure batch process.

    (b) List several additional assumptions you made to arrive at your estimated volume and temperature.

9.47. Ammonia scrubbing is one of many processes for removing sulfur dioxide from flue gases. The gases are bubbled through an aqueous solution of ammonium sulfite, and the SO_2 reacts to form ammonium bisulfite:

$$(NH_4)_2SO_3(aq) + SO_2(g) + H_2O(l) \rightarrow 2NH_4HSO_3(aq)$$

Subsequent process steps yield concentrated SO_2 and regenerate ammonium sulfite, which is recycled to the scrubber. The sulfur dioxide is either oxidized and absorbed in water to form sulfuric acid or reduced to elemental sulfur.

    Flue gas from a power plant boiler containing 0.30% SO_2 by volume enters a scrubber at a rate of 50,000 mol/h at 50°C. The gas is bubbled through an aqueous solution containing 10.0 mole% ammonium sulfite that enters the scrubber at 25°C. The gas and liquid effluents from the scrubber both emerge at 35°C. The scrubber removes 90% of the SO_2 entering with the flue gas. The effluent liquid is analyzed and is found to contain 1.5 moles $(NH_4)_2SO_3$ per mole of NH_4HSO_3. The heat of formation of $(NH_4)_2SO_3(aq)$ at 25°C is -890.0 kJ/mol, and that of $NH_4HSO_3(aq)$ is -760 kJ/mol. The heat capacities of all liquid solutions may be taken to be 4.0 J/(g·°C) and that of the flue gas may be taken to be that of nitrogen. Evaporation of water may be neglected. Calculate the required rate of heat transfer to or from the scrubber (kW).

9.48. A natural gas is analyzed and found to consist of 87.5% v/v (volume percent) methane, 7.0% ethane, 2.0% propane, and 3.5% N_2 (noncombustible).

    (a) Calculate the higher heating value and the lower heating value of this fuel in kJ/mol, using the heats of combustion in Table B.1.

    (b) Calculate the lower heating value of the fuel in kJ/kg.

    (c) State in your own words the meaning of the quantity calculated in part (b).

9.49. An **ultimate analysis** of a coal is a series of operations that yields the percentages by mass of carbon, hydrogen, nitrogen, oxygen, and sulfur in the coal. The heating value of a coal is best determined in a calorimeter, but it may be estimated with reasonable accuracy from the ultimate analysis using the **Dulong formula:**

$$HHV(kJ/kg) = 33,801(C) + 144,158[(H) - 0.125(O)] + 9413(S)$$

where (C), (H), (O), and (S) are the mass fractions of the corresponding elements. The 0.125(O) term accounts for the hydrogen bound in the water contained in the coal.

(a) Derive an expression for the higher heating value (*HHV*) of a coal in terms of C, H, O, and S, and compare your result with the Dulong formula. Suggest a reason for the difference.

(b) A coal with an ultimate analysis of 75.8 wt% C, 5.1% H, 8.2% O, 1.5% N, 1.6% S, and 7.8% ash (noncombustible) is burned in a power plant boiler furnace. All of the sulfur in the coal forms SO_2. The gas leaving the furnace is fed through a tall stack and discharged to the atmosphere. The ratio ϕ(kg SO_2 in the stack gas/kJ heating value of the fuel) must be below a specified value for the power plant to be in compliance with Environmental Protection Agency regulations regarding sulfur emissions. Estimate ϕ, using the Dulong formula for the heating value of the coal.

(c) An earlier version of the EPA regulation specified that the mole fraction of SO_2 in the stack gas must be less than a specified amount to avoid a costly fine and the required installation of an expensive stack gas scrubbing unit. When this regulation was in force, a few unethical plant operators blew clear air into the base of the stack while the furnace was operating. Briefly explain why they did so and why they stopped this practice when the new regulation was introduced.

9.50. A fuel gas is known to contain methane, ethane, and carbon monoxide. A sample of the gas is charged into an initially evacuated 2.000-liter vessel at 25°C and 2323 mm Hg absolute. The vessel is weighed before and after being charged, and the mass difference is found to be 4.929 g. Next, the higher heating value of the gas is determined in a calorimeter to be 841.9 kJ/mol. Calculate the molar composition of the gas.

9.51. A fuel gas containing 85.0 mole% methane and the balance ethane is burned completely with pure oxygen at 25°C, and the products are brought back down to 25°C.

(a) Suppose the reactor is continuous. Take a basis of calculation of 1 mol/s of the fuel gas, assume some value for the percent excess oxygen fed to the reactor (the value you choose will not affect the results), and calculate $-\dot{Q}(kW)$, the rate at which heat must be transferred from the reactor.

(b) Now suppose the combustion takes place in a constant-volume batch reactor. Take a basis of calculation of 1 mol of the fuel gas charged into the reactor, assume any percent excess oxygen, and calculate $-Q(kJ)$. (*Hint:* Recall Equation 9.1-5.)

(c) Briefly explain why the results in parts (a) and (b) do not depend on the percent excess O_2 and why they would not change if air rather than pure oxygen were fed to the reactor.

9.52. A mixture of air and a fine spray of gasoline at ambient (outside air) temperature is fed to a set of piston-fitted cylinders in an automobile engine. Sparks ignite the combustible mixtures in one cylinder after another, and the consequent rapid increase in temperature in the cylinders causes the combustion products to expand and drive the pistons. The back-and-forth motion of the pistons is converted to rotary motion of a crank shaft, motion that in turn is transmitted through a system of shafts and gears to propel the car.

Consider a car driving on a day when the ambient temperature is 298 K and suppose that the rate of heat loss from the engine to the outside air is given by the formula

$$-\dot{Q}_l \left(\frac{kJ}{h}\right) \approx \frac{15 \times 10^6}{T_a(K)}$$

where T_a is the ambient temperature.

(a) Take gasoline to be a liquid with a specific gravity of 0.70 and a higher heating value of 49.0 kJ/g, assume complete combustion and that the combustion products leaving the engine are at 298 K, and calculate the minimum feed rate of gasoline (gal/h) required to produce 100 hp of shaft work.

(b) If the exhaust gases are well above 298 K (which they are), is the work delivered by the pistons more or less than the value determined in part (a)? Explain.

(c) If the ambient temperature is much lower than 298 K, the work delivered by the pistons would decrease. Give two reasons.

9.53. The heating value of a fuel oil is to be measured in a constant-volume bomb calorimeter. The bomb is charged with oxygen and 0.00215 lb$_m$ of the fuel and is then sealed and immersed in an insulated container of water. The initial temperature of the system is 77.00°F. The fuel–oxygen mixture is ignited, and the fuel is completely consumed. The combustion products are CO_2(g) and H_2O(v). The final calorimeter temperature is 87.06°F. The mass of the calorimeter, including the bomb and its contents, is 4.62 lb$_m$, and the average heat capacity of the system (C_v) is 0.900 Btu/(lb$_m$·°F).

   (a) Calculate $\Delta \hat{U}_c^{\circ}$ (Btu/lb$_m$ oil) for the combustion of the fuel oil at 77°F. Briefly explain your calculation.

   (b) What more would you need to know to determine the higher heating value of the oil?

9.54. Methanol vapor is burned with excess air in a catalytic combustion chamber. Liquid methanol initially at 25°C is vaporized at 1.1 atm and heated to 100°C; the vapor is mixed with air that has been preheated to 100°C, and the combined stream is fed to the reactor at 100°C and 1 atm. The reactor effluent emerges at 300°C and 1 atm. Analysis of the product gas yields a dry-basis composition of 4.8% CO_2, 14.3% O_2, and 80.9% N_2.

   (a) Calculate the percentage excess air supplied and the dew point of the product gas.

   (b) Taking a basis of 1 g-mole of methanol burned, calculate the heat (kJ) needed to vaporize and heat the methanol feed, and the heat (kJ) that must be transferred from the reactor.

9.55. Methane at 25°C is burned in a boiler furnace with 10.0% excess air preheated to 100°C. Ninety percent of the methane fed is consumed, the product gas contains 10.0 mol CO_2/mol CO, and the combustion products leave the furnace at 400°C.

   (a) Calculate the heat transferred from the furnace, $-\dot{Q}$(kW), for a basis of 1000 mol CH$_4$ fed/s. (The greater the value of $-\dot{Q}$, the more steam is produced in the boiler.)

   (b) Would the following changes increase or decrease the rate of steam production? (Assume the fuel feed rate and fractional conversion of methane remain constant.) Briefly explain your answers. (i) Increasing the temperature of the inlet air; (ii) increasing the percent excess air for a given stack gas temperature; (iii) increasing the selectivity of CO_2 to CO formation in the furnace; and (iv) increasing the stack gas temperature.

9.56. A gaseous fuel containing methane and ethane is burned with excess air. The fuel enters the furnace at 25°C and 1 atm, and the air enters at 200°C and 1 atm. The stack gas leaves the furnace at 800°C and 1 atm and contains 5.32 mole% CO_2, 1.60% CO, 7.32% O_2, 12.24% H_2O, and the balance N_2.

   (a) Calculate the molar percentages of methane and ethane in the fuel gas and the percentage excess air fed to the reactor.

   (b) Calculate the heat (kJ) transferred from the reactor per cubic meter of fuel gas fed.

9.57. A coal contains 73.0 wt% C, 4.7% H (not including the hydrogen in the coal moisture), 3.7% S, 6.8% H_2O, and 11.8% ash. The coal is burned at a rate of 50,000 lb$_m$/h in a power plant boiler with air 50% in excess of that needed to oxidize all the carbon in the coal to CO_2. The air and coal are both fed at 77°F and 1 atm. The solid residue from the furnace is analyzed and is found to contain 28.7 wt% C, 1.6% S, and the balance ash. The coal sulfur oxidized in the furnace is converted to SO_2(g). Of the ash in the coal, 30% emerges in the solid residue and the balance is emitted with the stack gases as fly ash. The stack gas and solid residue emerge from the furnace at 600°F. The higher heating value of the coal is 18,000 Btu/lb$_m$.

   (a) Calculate the mass flow rates of all components in the stack gas and the volumetric flow rate of this gas. (Ignore the contribution of the fly ash in the latter calculation, and assume that the stack gas contains a negligible amount of CO.)

   (b) Assume that the heat capacity of the solid furnace residue is 0.22 Btu/(lb$_m$·°F), that of the stack gas is the heat capacity per unit mass of nitrogen, and 35% of the heat generated in the furnace is used to produce electricity. At what rate in MW is electricity produced?

   (c) Calculate the ratio (heat transferred from the furnace)/(heating value of the fuel). Why is this ratio less than one?

   (d) Suppose the air fed to the furnace were preheated rather than being fed at ambient temperature, but that everything else (feed rates, outlet temperatures, and fractional coal conversion) were the same. What effect would this change have on the ratio calculated in part (c)? Explain. Suggest an economical way in which this preheating might be accomplished.

9.58. A mixture of methane, ethane, and argon at 25°C is burned with excess air in a power plant boiler. The hydrocarbons in the fuel are completely consumed. The following variable definitions will be used throughout this problem:

x_m = mole fraction of methane in the fuel
x_a = mole fraction of argon in the fuel
$P_{xs}(\%)$ = percent excess air fed to the furnace
$T_a(°C)$ = temperature of the entering air
$T_s(°C)$ = stack gas temperature
r = ratio of CO_2 to CO in the stack gas (mol CO_2/mol CO)
$\dot{Q}(kW)$ = rate of heat transfer from the furnace to the boiler tubes

(a) Without doing any calculations, sketch the shapes of the plots you would expect to obtain for plots of $\dot{Q}$ versus (i) x_m, (ii) x_a, (iii) P_{xs}, (iv) T_a, (v) T_s, and (vi) r, assuming in each case that the other variables are held constant. Briefly state your reasoning for each plot.

(b) Take a basis of 1.00 mol/s of fuel gas, draw and label a flowchart, and derive expressions for the molar flow rates of the stack gas components in terms of x_m, x_a, P_{xs}, and r. Then take as references the elements at 25°C, prepare and fill in an inlet–outlet enthalpy table for the furnace, and derive expressions for the specific molar enthalpies of the feed and stack gas components in terms of T_a and T_s.

(c) Calculate $\dot{Q}(kW)$ for x_m = 0.85 mol CH_4/mol, x_a = 0.05 mol Ar/mol, P_{xs} = 5%, r = 10.0 mol CO_2/mol CO, T_a = 150°C, and T_s = 700°C. (*Solution: $\dot{Q}$ = −655 kW.*)

*(d) Prepare a spreadsheet that has columns for x_m, x_a, P_{xs}, T_a, r, T_s, and $\dot{Q}$, plus columns for any other variables you might need for the calculation of $\dot{Q}$ from given values of the preceding six variables (e.g., component molar flow rates and specific enthalpies). Use the spreadsheet to generate plots of $\dot{Q}$ versus each of the following variables over the specified ranges:

x_m = 0.00–0.85 mol CH_4/mol
x_a = 0.01–0.05 mol Ar/mol
P_{xs} = 0%–100%
T_a = 25°C–250°C
r = 1–100 mol CO_2/mol CO (make the r axis logarithmic)
T_s = 500°C–1000°C

When generating each plot, use the variable values given in part (c) as base values. (For example, generate a plot of $\dot{Q}$ versus x_m for x_a = 0.05, P_{xs} = 5%, and so on, with x_m varying from 0.00 to 0.85 on the horizontal axis.) If possible, include the plots on the same spreadsheet as the data.

9.59. A gas stream consisting of *n*-hexane in methane is fed to a condenser at 60°C and 1.2 atm. The dew point of the gas (considering hexane as the only condensable component) is 55°C. The gas is cooled to 5°C in the condenser, recovering pure hexane as a liquid. The effluent gas leaves the condenser saturated with hexane at 5°C and 1.1 atm and is fed to a boiler furnace at a rate of 207.4 L/s, where it is burned with 100% excess air that enters the furnace at 200°C. The stack gas emerges at 400°C and 1 atm and contains no carbon monoxide or unburned hydrocarbons. The heat transferred from the furnace is used to generate saturated steam at 10 bar from liquid water at 25°C.

(a) Calculate the mole fractions of hexane in the condenser feed and product gas streams and the rate of hexane condensation (liters condensate/s).

(b) Calculate the rate at which heat must be transferred from the condenser (kW) and the rate of generation of steam in the boiler (kg/s).

9.60. In the preliminary design of a furnace for industrial boiler, methane at 25°C is burned completely with 20% excess air, also at 25°C. The feed rate of methane is 450 kmol/h. The hot combustion gases leave the furnace at 300°C and are discharged to the atmosphere. The heat transferred from the furnace ($\dot{Q}$) is used to convert boiler feedwater at 25°C into superheated steam at 17 bar and 250°C.

Use the following approximate heat capacities [kJ/(mol·°C)] in your solution of this problem: $CH_4(g)$ = 0.0431, $CO_2(g)$ = 0.0423, $H_2O(g)$ = 0.0343, $O_2(g)$ = 0.0312, $N_2(g)$ = 0.0297.

(a) Draw and label a flowchart of this process [the chart should look like the one shown in part (b) without the preheater] and calculate the composition of the gas leaving the furnace. Then, using the given constant heat capacities, calculate the average molar heat capacity of the gas. (See Equation 8.3-13.) Finally, calculate $\dot{Q}(kJ/h)$ and the rate of steam production in the boiler (kg/h).

---

*Computer problem.

(b) In the actual boiler design, the air feed at 25°C and the combustion gas leaving the furnace at 300°C pass through a heat exchanger (the *air preheater*). The combustion (flue) gas is cooled to 150°C in the preheater and is then discharged to the atmosphere, and the heated air is fed to the furnace.

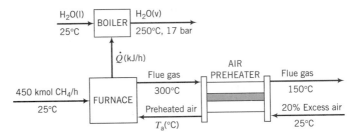

Calculate the temperature of the air entering the furnace (a trial-and-error solution is required) and the rate of steam production (kg/h).

(c) Explain why preheating the air increases the rate of steam production. (*Suggestion:* Use the energy balance on the furnace in your explanation.) Why does it make sense economically to use the combustion gas as the heating medium?

9.61. A bituminous coal is burned with air in a boiler furnace. The coal is fed at a rate of 40,000 kg/h and has an ultimate analysis of 76 wt% C, 5% H, 8% O, negligible amounts of N and S, and 11% noncombustible ash (see Problem 9.49), and a higher heating value of 25,700 kJ/kg. Air enters a preheater at 30°C and 1 atm with a relative humidity of 30%, exchanges heat with the hot flue gas leaving the furnace, and enters the furnace at temperature T_a(°C). The flue gas contains 7.71 mole% CO_2 and 1.29 mole% CO *on a dry basis,* and the balance a mixture of O_2, N_2, and H_2O. It emerges from the furnace at 260°C and is cooled to 150°C in the preheater. Noncombustible residue (*slag*) leaves the furnace at 450°C and has a heat capacity of 0.97 kJ/(kg·°C).

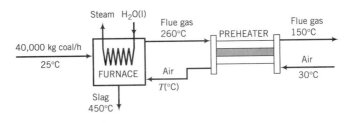

(a) Prove that the air-to-fuel ratio is 16.1 standard cubic meters/kg coal and that the flue gas contains 4.6% H_2O by volume.

(b) Calculate the rate of cooling required to cool the flue gas from 260°C to 150°C and the temperature to which the air is preheated. (*Note:* A trial-and-error calculation is required.)

(c) If 60% of the heat transferred from the furnace $(-\dot{Q})$ goes into producing saturated steam at 30 bar from liquid boiler feedwater at 50°C, at what rate (kg/h) is steam generated?

9.62. Carbon monoxide is burned with excess air at 1 atm in an adiabatic reactor. The reactants are fed at 25°C and the final temperature (i.e., the adiabatic flame temperature) is 1400°C.

(a) Calculate the percentage excess air fed to the reactor.

(b) If the percentage excess air were increased, how would the adiabatic flame temperature change and why would it change in this manner?

9.63. A natural gas containing 82.0 mole% CH_4 and the balance C_2H_6 is burned with 20% excess air in a boiler furnace. The fuel gas enters the furnace at 298 K, and the air is preheated to 423 K. The heat capacities of the stack gas components may be assumed to have the following constant values:

$$CO_2: \quad C_p = 50.0 \text{ J/(mol·K)}$$
$$H_2O(v): \quad C_p = 38.5 \text{ J/(mol·K)}$$
$$O_2: \quad C_p = 33.1 \text{ J/(mol·K)}$$
$$N_2: \quad C_p = 31.3 \text{ J/(mol·K)}$$

(a) Assuming complete combustion of the fuel, calculate the adiabatic flame temperature.

(b) How would the flame temperature change if the percent excess air were increased? How would it change if the percentage of methane in the fuel increased? Briefly explain both of your answers.

9.64. In a surface coating operation, a polymer (plastic) dissolved in liquid acetone is sprayed on a solid surface and a stream of hot air is then blown over the surface, vaporizing the acetone and leaving a residual polymer film of uniform thickness. Because environmental standards do not allow discharging acetone into the atmosphere, a proposal to incinerate the stream is to be evaluated.

The proposed process uses two parallel columns containing beds of solid particles. The air–acetone stream, which contains acetone and oxygen in stoichiometric proportion, enters one of the beds at 1500 mm Hg absolute at a rate of 1410 standard cubic meters per minute. The particles in the bed have been preheated and transfer heat to the gas. The mixture ignites when its temperature reaches 562°C, and combustion takes place rapidly and adiabatically. The combustion products then pass through and heat the particles in the second bed, cooling down to 350°C in the process. Periodically the flow is switched so that the heated outlet bed becomes the feed gas preheater/combustion reactor and vice versa.

Use the following average values for C_p[kJ/(mol·°C)] in solving the problems to be given: 0.126 for C_3H_6O, 0.033 for O_2, 0.032 for N_2, 0.052 for CO_2, and 0.040 for $H_2O(v)$.

(a) If the relative saturation of acetone in the feed stream is 12.2%, what is the stream temperature?

(b) Determine the composition of the gas after combustion, assuming that all of the acetone is converted to CO_2 and H_2O, and estimate the temperature of this stream.

(c) Estimate the rates (kW) at which heat is transferred from the inlet bed particles to the feed gas prior to combustion and from the combustion gases to the outlet bed particles. Suggest an alternative to the two-bed feed switching arrangement that would achieve the same purpose.

9.65. Liquid *n*-pentane at 25°C is burned with 30% excess oxygen (not air) fed at 75°C. The adiabatic flame temperature is T_{ad}(°C).

(a) Take as a basis of calculation 1.00 mol C_5H_{12}(l) burned and use an energy balance on the adiabatic reactor to derive an equation of the form $f(T_{ad}) = 0$, where $f(T_{ad})$ is a fourth-order polynomial $[f(T_{ad}) = c_0 + c_1 T_{ad} + c_2 T_{ad}^2 + c_3 T_{ad}^3 + c_4 T_{ad}^4]$. If your derivation is correct, the ratio c_0/c_4 should equal -6.892×10^{14}. Then solve the equation to determine T_{ad}. (This solution is easily obtained using the goalseek tool on a spreadsheet.)

(b) Repeat the calculation of part (a) using successively the first two terms, the first three terms, and the first four terms of the fourth-order polynomial equation. If the solution of part (a) is taken to be exact, what percentage errors are associated with the linear (two-term), quadratic (three-term), and cubic (four-term) approximations?

(c) Determine the fourth-order solution using Newton's method (Appendix A.2), taking the linear approximation as the first guess and stopping when $|f(T_{ad})| < 0.01$.

(d) Why is the fourth-order solution at best an approximation and quite possibly a poor one? (*Hint:* Examine the conditions of applicability of the heat capacity formulas you looked up in Table B.2.)

9.66. Methane is burned with 25% excess air in a continuous adiabatic reactor. The methane enters the reactor at 25°C and 1.10 atm at a rate of 5.50 L/s, and the entering air is at 150°C and 1.1 atm. Combustion in the reactor is complete, and the reactor effluent gas emerges at 1.05 atm. Calculate (a) the temperature and (b) the degrees of superheat of the reactor effluent. (Consider water to be the only condensable species in the effluent.)

[12]**9.67.** Methane and 30% excess air are to be fed to a combustion reactor. An inexperienced technician mistakes his instructions and charges the gases together in the required proportion into an evacuated closed tank. (The gases were supposed to be fed directly into the reactor.) The contents of the charged tank are at 25°C and 4.00 atm absolute.

(a) Calculate the standard internal energy of combustion of the methane combustion reaction, $\Delta\hat{U}_c^\circ$(kJ/mol), taking CO_2(g) and H_2O(v) as the presumed products. Then prove that if the constant-pressure heat capacity of an ideal gas species is independent of temperature, the specific internal energy of that species at temperature T(°C) relative to the same species at 25°C is

---

[12]Modified version of a problem in D. A. Crowl, D. W. Hubbard, and R. M. Felder, *Problem Set: Stoichiometry,* AIChE Center for Chemical Process Safety, New York, 1993.

given by the expression

$$\hat{U} = (C_p - R_g)(T - 25°C)$$

where R_g is the gas constant. Use this formula in the next part of the problem.

(b) You wish to calculate the maximum temperature, $T_{max}(°C)$, and corresponding pressure, $P_{max}(atm)$, that the tank would have to withstand if the mixture it contains were to be accidentally ignited. Taking molecular species at 25°C as references and treating all species as ideal gases, prepare an inlet–outlet internal energy table for the closed system combustion process. In deriving expressions for each $\hat{U}_i$ at the final reactor condition (T_{max}, P_{max}), use the following approximate values for $C_{pi}[kJ/(mol·°C)]$: 0.033 for O_2, 0.032 for N_2, 0.052 for CO_2, and 0.040 for $H_2O(v)$. Then use an energy balance and the ideal gas equation of state to perform the required calculations.

(c) Why would the actual temperature and pressure attained in a real tank be less than the values calculated in part (a)? (State several reasons.)

(d) Think of ways that the tank contents might be accidentally ignited. The list should suggest why accepted plant safety regulations prohibit the storage of combustible vapor mixtures.

*9.68. Natural gas that contains methane, ethane, and propane is to be burned with humid air. The adiabatic flame temperature is to be calculated from specified values of the following quantities:

$$y_{CH_4}, y_{C_2H_6}, y_{C_3H_8} = \text{mole fractions of fuel components}$$

$$T_f, T_a = \text{inlet temperatures of fuel and air, °C}$$

$$P_{xs} = \text{percent excess air}$$

$$y_{w0} = \text{mole fraction of water in the inlet air}$$

(a) Without doing any calculations, predict the direction of change (increase, decrease, no change) in the adiabatic flame temperature you would expect for an increase in (i) y_{CH_4} with $y_{C_3H_8}$ held constant, (ii) T_f, (iii) T_a, (iv) P_{xs}, and (v) y_{w0}. Briefly state your reasoning for each variable.

(b) For a basis of 1 g-mole of natural gas, calculate the gram-moles of each molecular species in the feed and product streams, assuming complete combustion and negligible CO formation. The answer should be expressed in terms of the variables given above.

(c) Given here are expressions for the specific enthalpies of the feed and product components, relative to their elements at 25°C.

$$\hat{H}_i(kJ/mol) = a_i + b_iT + c_iT^2 + d_iT^3 + e_iT^4, \quad T \text{ in °C}$$

Substance (i)	a	$b \times 10^2$	$c \times 10^5$	$d \times 10^8$	$e \times 10^{12}$
(1) CH_4	-75.72	3.431	2.734	0.122	-2.75
(2) C_2H_6	-85.95	4.937	6.96	-1.939	1.82
(3) C_3H_8	-105.6	6.803	11.30	-4.37	7.928
(4) N_2	-0.7276	2.900	0.110	0.191	-0.7178
(5) O_2	-0.7311	2.910	0.579	-0.2025	0.3278
(6) $H_2O(v)$	-242.7	3.346	0.344	0.2535	-0.8982
(7) CO_2	-394.4	3.611	2.117	-0.9623	1.866

Derive the given expression for the specific enthalpy of methane from the heat capacity data in Table B.2. Then show that ΔH for the reactor is given by an expression of the form

$$\Delta H = \alpha_0 + \alpha_1 T + \alpha_2 T^2 + \alpha_3 T^3 + \alpha_4 T^4$$

---

*Computer problem.

where T is the product temperature, and

$$\alpha_0 = \sum_{i=4}^{7} (n_i)_{\text{out}} a_i - \sum_{i=1}^{3} (n_i)_{\text{in}} \hat{H}_i(T_f) - \sum_{i=4}^{7} (n_i)_{\text{in}} \hat{H}_i(T_a)$$

$$\alpha_1 = \sum_{i=4}^{7} (n_i)_{\text{out}} b_i \qquad \alpha_3 = \sum_{i=4}^{7} (n_i)_{\text{out}} d_i$$

$$\alpha_2 = \sum_{i=4}^{7} (n_i)_{\text{out}} c_i \qquad \alpha_4 = \sum_{i=4}^{7} (n_i)_{\text{out}} e_i$$

(d) Write a spreadsheet program to take as input values of y_{CH_4}, $y_{C_3H_8}$, T_f, T_a, P_{xs}, and y_{w0}, and to solve the energy balance equation [$\Delta H(T) = 0$] to determine the adiabatic flame temperature. Run the program for the following sets of input variable values:

Variable	Run 1	Run 2	Run 3	Run 4	Run 5	Run 6
y_{CH_4}	0.75	0.86	0.75	0.75	0.75	0.75
$y_{C_3H_8}$	0.04	0.04	0.04	0.04	0.04	0.04
$T_f(°C)$	40	40	150	40	40	40
$T_a(°C)$	150	150	150	250	150	150
P_{xs}	25%	25%	25%	25%	100%	25%
y_{w0}	0.0306	0.0306	0.0306	0.0306	0.0306	0.10

Suggestion: Near the top of the spreadsheet, enter the values of a, b, c, d, and e for each species. Starting several rows below the last of these entries, list in Column A labels for the input variables and all calculated variables (component molar flow rates, specific enthalpies, $T_{ad}, \alpha_0, \alpha_1, \ldots, \alpha_4, \Delta H$), and enter in adjacent columns the corresponding values or formulas for these variables in successive runs. (*Solution for Run 1: $T_{ad} = 1743.1°C$*.)

(e) Write a computer program to carry out the above calculations using Newton's rule (Appendix A.2), with an initial guess of 1000°C for each run. Build in a limit on the number of iterations in case the program fails to converge.

9.69. Acetylene is produced by pyrolyzing—decomposing at high temperature—natural gas (predominantly methane):

$$2\,CH_4(g) \rightarrow C_2H_2(g) + 3\,H_2$$

The heat required to sustain this endothermic reaction is provided by feeding oxygen to the reactor and burning a portion of the methane to form primarily CO and some CO_2.

A simplified version of the process is as follows. A stream of natural gas, which for the purposes of this problem may be considered pure methane, and a stream containing 96.0 mole% oxygen and the balance nitrogen are each preheated from 25°C to 650°C. The streams are combined and fed into an adiabatic converter, in which most of the methane and all of the oxygen are consumed, and the product gas is rapidly quenched to 38°C as soon as it emerges from the converter. The residence time in the converter is less than 0.01 s, low enough to prevent most but not all of the methane from decomposing to form hydrogen and solid carbon particles (soot). Of the carbon in the feed, 5.67% emerges as soot.

The cooled effluent passes through a carbon filter in which the soot is removed. The clean gas is then compressed and fed to an absorption column, where it is contacted with a recycled liquid solvent, dimethylformamide, or DMF (MW = 73.09). The off-gas leaving the absorber contains all of the hydrogen and nitrogen, 98.8% of the CO, and 95% of the methane in the gas fed to the column. The "lean" solvent fed to the absorber is essentially pure DMF; the "rich" solvent leaving the column contains all of the water and CO_2 and 99.4% of the acetylene in the gas feed. This solvent is analyzed and found to contain 1.55 mole% C_2H_2, 0.68% CO_2, 0.055% CO, 0.055% CH_4, 5.96% H_2O, and 91.7% DMF.

The rich solvent goes to a multiple-unit separation process from which three streams emerge. One—the *product gas*—contains 99.1 mole% C_2H_2, 0.059% H_2O, and the balance CO_2; the second—the *stripper off-gas*—contains methane, carbon monoxide, carbon dioxide, and water; and the third—the *regenerated solvent*—is the liquid DMF fed to the absorber.

A plant is designed to produce 5 metric tons/day of product gas. Calculate the following quantities.

(a) The required flow rates (SCMH) of the methane and oxygen feed streams.

(b) The molar flow rates (kmol/h) and compositions of the gas feed to the absorber, the absorber off-gas, and the stripper off-gas.

(c) The DMF circulation rate (kmol/h).

(d) The overall product yield (mol C_2H_2 in product gas/mol CH_4 in the feed to the reactor), and the fraction that this quantity represents of its theoretical maximum value.

(e) The total heating requirements (kW) for the methane and oxygen feed preheaters.

(f) The temperature attained in the converter.

[13]**9.70.** The wastewater treatment plant at the Ossabaw Paper Company paper mill generates about 24 tonnes of sludge per day (1 tonne = 1 metric ton = 1000 kg). The *consistency* of the sludge is 35%, meaning that the sludge contains 35 wt% solids and the balance liquids. The mill currently spends $40/tonne to dispose of the sludge in a landfill. The plant environmental engineer has determined that if the sludge consistency could be increased to 75%, the sludge could be incinerated (burned) to generate useful energy and to eliminate the environmental problems associated with landfill disposal.

A flowchart for a preliminary design of the proposed sludge treatment process follows. For simplicity, we will assume that the liquid in the sludge is just water.

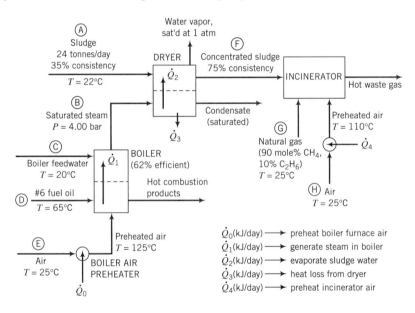

Process description: The sludge from the wastewater treatment plant (Stream (A)) passes through a dryer where a portion of the water in the sludge is vaporized. The heat required for the vaporization comes from condensing saturated steam at 4.00 bar (Stream (B)). The steam fed to the dryer is produced in the plant's oil-fired boiler from feedwater at 20°C (Stream (C)). The heat required to produce the steam is transferred from the boiler furnace, where fuel oil (Stream (D)) is burned with 25% excess air (Stream (E)). The concentrated sludge coming from the dryer (Stream (F)), which has a consistency of 75%, is fed to an incinerator. The heating value of the sludge is insufficient to keep the incinerator temperature high enough for complete combustion, so natural gas (Stream (G)) is used as a supplementary fuel. A stream of outside air at 25°C (Stream (H)) is heated to 110°C and fed to the incinerator along with the concentrated sludge and natural gas. The waste gas from the incinerator is discharged to the atmosphere.

---

[13]Problem based on material contributed by Joseph Lemanski of the Kimberly-Clark Corporation and Morton Barlaz of North Carolina State University.

Fuel oil: The oil is a low-sulfur No. 6 fuel oil. Its ultimate (elemental) analysis on a weight basis is 87% C, 10% H, 0.84% S, and the balance oxygen, nitrogen, and nonvolatile ash. The higher heating value of the oil is 3.75×10^4 kJ/kg and the heat capacity is $C_p = 1.8$ kJ/(kg·°C).

Boiler: The boiler has an efficiency of 62%, meaning that 62% of the heating value of the fuel oil burned is used to produce saturated steam at 4.00 bar from boiler feedwater at 20°C. Fuel oil at 65°C and dry air at 125°C are fed to the boiler furnace. The air feed rate is 25% in excess of the amount theoretically required for complete consumption of the fuel.

Sludge: The sludge from the wastewater treatment plant contains 35% w/w solids (S) and the balance liquids (which for the purposes of this problem may be treated as only water) and enters the dryer at 22°C. The sludge includes a number of volatile organic species, some of which may be toxic, and has a terrible odor. The heat capacity of the solids is approximately constant at 2.5 kJ/(kg·°C).

Dryer: The dryer has an efficiency of 55%, meaning that the heat transferred to the sludge, $\dot{Q}_2$, is 55% of the total heat lost by the condensing steam, and the remainder, $\dot{Q}_3$, is lost to the surroundings. The dryer operates at 1 atm, and the water vapor and concentrated sludge emerge at the corresponding saturation temperature. The steam condensate leaves the dryer as a liquid saturated at 4.00 bar.

Incinerator: The concentrated sludge has a heating value of 19,000 kJ/kg dry solids. For a feed sludge of 75% consistency, the incinerator requires 195 SCM natural gas/tonne wet sludge [1 SCM = 1 m³(STP)]. The theoretical air requirement for the sludge is 2.5 SCM air/10,000 kJ of heating value. Air is fed in 100% excess of the amount theoretically required to burn the sludge and the natural gas.

(a) Use material and energy balances to calculate the mass flow rates (tonnes/day) of Streams Ⓑ, Ⓒ, Ⓓ, Ⓔ, Ⓕ, Ⓖ, and Ⓗ, and heat flows $\dot{Q}_0, \dot{Q}_1, \ldots, \dot{Q}_4$(kJ/day). Take the molecular weight of air to be 29.0. (*Caution:* Before you start doing lengthy and unnecessary energy balance calculations on the boiler furnace, remember the given furnace efficiency.)

(b) The money saved by implementing this process will be the current cost of disposing of the wastewater plant sludge in a landfill. Two major costs of implementing the process are the installed costs of the new dryer and incinerator. What other costs must be taken into account when determining the economic feasibility of the process? Why might management decide to go ahead with the project even if it proves to be unprofitable?

(c) What opportunities exist for improving the energy economy of the process? (*Hint:* Think about the need to preheat the fuel oil and the boiler and incinerator air streams and consider heat exchange possibilities.)

(d) The driving force for the introduction of this process is to eliminate the environmental cost of sludge disposal. What is that cost—that is, what environmental penalties and risks are associated with using landfills for hazardous waste disposal? What environmental problems might incineration introduce?

Chapter 10

Computer-Aided Balance Calculations

If you have worked your way through the preceding nine chapters, you are aware—maybe painfully so—that solving material and energy balance problems manually for even relatively simple systems can be a cumbersome and time-consuming task. Imagine what it must be like to do it for, say, an entire petroleum refinery! An alternative approach is to devise an *algorithm* (a systematic procedure) to perform the calculations and then to program a computer to implement it. This now common technique, referred to as *flowsheeting,* is the principal topic of this chapter.

10.0 INSTRUCTIONAL OBJECTIVES

After completing this chapter, you should be able to do the following:

- Clearly define in your own words the terms *design variables* and *state variables, sequential modular flowsheet simulation, equation-based flowsheet simulation, tear stream, convergence block,* and *design specification.*
- Given a description of a single-unit chemical process, determine the number of degrees of freedom and identify a set of design variables that leads to an efficient calculation procedure (one that minimizes the number of equations that must be solved simultaneously).
- Given a description of a multiple-unit process, determine the number of degrees of freedom, identify a set of feasible design variables, and if there are cycles in the flowchart, identify reasonable tear stream variables and outline the solution procedure. Draw a sequential modular block diagram for the process, inserting necessary convergence blocks.
- Set up a spreadsheet to perform material and energy balance calculations for single-unit and multiple-unit processes.

10.1 DEGREE-OF-FREEDOM ANALYSIS REVISITED

As defined in Section 4.3d, the *degrees of freedom* of a system, n_{df}, is the number of system variables whose values must be specified before the remaining variables can be calculated. *If a system is described by n_e independent equations in n_v variables, then $n_{df} = n_v - n_e$.*

- If $n_{df} = 0$ (e.g., if there are three independent equations in three unknowns), then the unknown system variables can in principle be calculated.
- If $n_{df} > 0$—for example, three equations in five unknowns, so that $n_{df} = 2$—then n_{df} variable values must be specified as part of the system definition, and the remaining unknown

variables may then be calculated from the system equations. Variables that are externally specified are called **design variables,** and those that are then calculated from the system equations are called **state variables.**

- If $n_{df} < 0$, there are more equations than variables, and the process is *overdefined:* either you have written more equations than you are entitled to write (e.g., three material balances on a two-component system), or you have miscounted the variables, probably by not fully labeling the flowchart.

The remainder of this section illustrates degree-of-freedom analyses for increasingly complex systems: first a single nonreactive process unit, then a single reactive unit, and finally a system of several interconnected units.

10.1a   Single Process Unit

The number of unknown variables for a single unit is the sum of the unknown component amounts or flow rates for all inlet and outlet streams, plus all unknown stream temperatures and pressures, plus the rates of energy transfer as heat and work. The equations available to determine these unknowns include material balances for each independent species, an energy balance, phase and chemical equilibrium relations, and additional specified relationships among the process variables.

Let us first consider a heated mixer in which a stream of oxygen and nitrogen at 25°C is blended with a stream of pure oxygen at 40°C, and the combined stream is heated to 50°C.

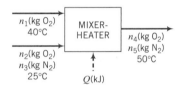

A degree-of-freedom analysis on this simple process would proceed as follows:

$$6 \text{ variables } (n_1, \ldots, n_5, \text{ and } Q)$$
$$\underline{-3 \text{ relations (2 material balances and 1 energy balance)}}$$
$$= 3 \text{ degrees of freedom}$$

Thus, three specifications would have to be provided in the problem statement, following which the system equations could be solved for all remaining unknowns. Specifying values of three of the variables would suffice, as would giving values for two of the variables and a relationship between the masses of the two feed streams. (Convince yourself—choose values for any three variables and mentally go through the calculation of the remaining three from the system equations.)

The following example illustrates a degree-of-freedom analysis for a process that involves vapor–liquid equilibria between effluent process streams.

EXAMPLE 10.1-1   *Degree-of-Freedom Analysis of a Flash Vaporizer*

A liquid mixture of *n*-hexane (HX) and *n*-heptane (HP) at a high pressure is abruptly exposed to a lower pressure. A portion of the mixture evaporates, yielding a vapor mixture relatively rich in hexane (the more volatile of the two feed components) and a residual liquid mixture relatively rich in heptane. The two product streams are in equilibrium at temperature T and pressure P; their compositions are related by Raoult's law (Section 6.4b).

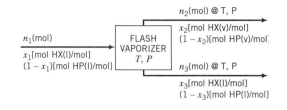

The relations among the system variables are as follows:

$$n_1 = n_2 + n_3 \qquad \text{(total mole balance)} \qquad \textbf{(1)}$$

$$n_1 x_1 = n_2 x_2 + n_3 x_3 \qquad \text{(hexane balance)} \qquad \textbf{(2)}$$

$$\left.\begin{array}{l} x_2 P = x_3 p_{HX}^*(T) \\[4pt] (1 - x_2)P = (1 - x_3)p_{HP}^*(T) \end{array}\right\} \qquad \text{(Raoult's law)} \qquad \begin{array}{l} \textbf{(3)} \\[4pt] \textbf{(4)} \end{array}$$

$$\left.\begin{array}{l} \log[p_{HX}^*(T)] = A_{HX} - B_{HX}/(T + C_{HX}) \\[4pt] \log[p_{HP}^*(T)] = A_{HP} - B_{HP}/(T + C_{HP}) \end{array}\right\} \qquad \text{(Antoine equation)} \qquad \begin{array}{l} \textbf{(5)} \\[4pt] \textbf{(6)} \end{array}$$

Perform a degree-of-freedom analysis on this system, and outline a solution procedure for the following sets of design variables:

1. n_1, x_1, T, and x_3.
2. n_2, x_2, n_3, and x_3.
3. n_1, n_2, n_3, and T.

SOLUTION

Ten variables—$n_1, n_2, n_3, x_1, x_2, x_3, T, P, p_{HX}^*$, and p_{HP}^*—appear in these six equations. The system consequently has $10 - 6 = 4$ degrees of freedom. Four of these variables must be specified in a problem statement, following which the remaining six variables can be calculated from the given six equations. Notice how the difficulty of the calculation depends dramatically on which four variables are specified (i.e., which ones are the design variables).

1. **Design variables n_1, x_1, T, and x_3.** If values of these four variables are specified, the following relatively straightforward solution procedure could be used to solve for the remaining (state) variables.

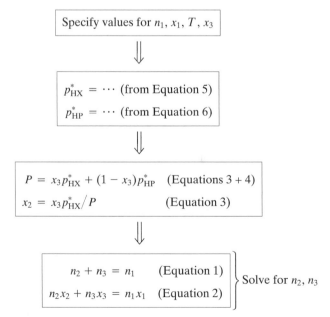

The first four equations each contain one unknown and can be solved directly; the final two equations in two unknowns are linear, so they also can be solved without difficulty.

2. **Design variables n_2, x_2, n_3, and x_3.** For this choice of design variables, a relatively lengthy trial-and-error procedure is required to solve for the remaining variables. One possible procedure is shown here.

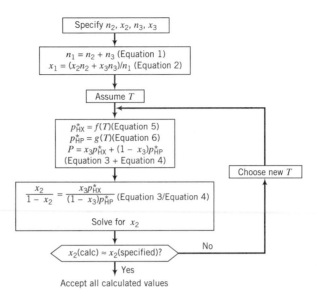

Several methods can be used for the trial-and-error calculation, such as defining $F(T) = x_2(\text{calc}) - x_2(\text{specified})$ and using the goalseek tool on a spreadsheet to find the value of T for which $F(T) \approx 0$.

3. **Design variables n_1, n_2, n_3, and T.** Some choices of design variables lead to contradictions of the system equations and so are not allowed. For example, choosing n_1, n_2, and n_3 as design variables is forbidden: once any two of these variables are specified, the third may not be assigned a value independently, but must be calculated from Equation 1.

TEST YOURSELF

1. Consider the following set of equations:

$$x_1 + 2x_2 - x_3^2 = 0$$
$$5x_1 - x_2^3 + 4 = 0$$

   (a) How many degrees of freedom does this system possess?
   (b) Which choice of a design variable would lead to an easier procedure to solve for the state variables, x_1 or x_3?

2. A liquid stream containing species A, B, C, and D enters a vaporizer (flow rates $n_1, \ldots, n_4$ kg/s). Essentially all of the A, 25% of the B, and none of the D is vaporized. The C is distributed between the liquid and vapor products according to a known equilibrium relationship. How many degrees of freedom does this process possess?

3. Consider the equations (i) $5x - 3y = 7$, (ii) $10x - 3y - 6z = 14$, and (iii) $y = 2z$.
   (a) Try to solve this system of equations for x, y, and z, and explain why you can't do it.
   (b) Prove that the set of equations has one degree of freedom, choose a design variable, specify a value for it, and determine the corresponding state variable values.

10.1b   Multiple-Unit Processes

A process that contains two or more interconnected units has fewer degrees of freedom than the sum of the degrees of freedom of the individual unconnected units. Suppose, for example, that an output stream from the first of two units is the feed stream to the second unit and that no reactions occur in either unit.

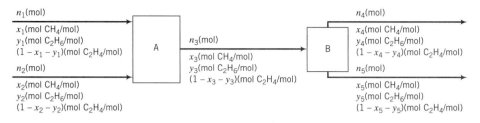

Let us first perform a degree-of-freedom analysis for unit A. We count nine variables (three for each stream entering or leaving the unit) and three equations (mass balances on methane, ethane, and ethylene) to obtain six degrees of freedom. The same result is obtained for unit B. Individually, then, there are 12 *local* degrees of freedom—local signifying that we are considering each unit as isolated, unconnected to the others.

In fact, the units *are* linked so that in the preceding analysis we have overcounted variables: specifically, the three variables associated with the product stream from unit A (n_3, x_3, and y_3) are identical to the three associated with the feed to unit B. The analysis proceeds as follows:

$$\text{system variables} = 18 \text{ local} - 3 \text{ overcounted} = 15$$
$$\underline{\text{system equations} = 3 \text{ balances for each unit} = 6}$$
$$\text{degrees of freedom} = 9$$

Thus, nine process stream variable values must be specified for the given system, at which point balances can be used to determine the remaining six variables.

In general, the number of degrees of freedom of a system of linked units equals the number of degrees of freedom of the individual units—that is, the total local degrees of freedom—minus the number of **ties** (variables common to two or more units, and hence overcounted), minus the number of any additional relations that were not counted in the individual unit analyses.

The procedure for performing a degree-of-freedom analysis on a multi-unit process may be summarized as follows:

1. ***Draw and completely label the flowchart.***
2. ***Determine the local degrees of freedom for each process unit, mixing point, and stream-splitting point in the process.*** The local degrees of freedom equal the number of unknown variables involved in the unit equations minus the number of independent relations among these variables. The relations may include material balances, an energy balance, equilibrium relations, and specifications of conversions, yields, selectivities, and mass or mole ratios. If a relation involves any variables belonging to units other than the one in question, however, count it in step 3 rather than here.
3. ***Determine the degrees of freedom of the total process.*** This quantity equals the sum of the local degrees of freedom, minus the number of independent relations among process variables that were not counted in step 2 (such as overall reactant conversions or percentage recoveries), minus the number of ties (overcounted variables) between units.

If the number of degrees of freedom determined in this manner equals zero, all unknown variables can in principle be determined. If n_{df} is greater than zero, then n_{df} variable values must be specified as design variables, and the remaining state variables may then be calculated.

The following example illustrates this procedure.

EXAMPLE 10.1-2   ***Degree-of-Freedom Analysis of a Multiple-Unit Process***

Methanol may be produced from carbon monoxide and hydrogen in the reaction

$$CO + 2H_2 \rightarrow CH_3OH$$

The fresh feed, which contains CO and H_2 in stoichiometric proportion, enters the process at a rate of 2.2 m³/s at 25°C and 6.0 MPa and combines adiabatically with a recycle stream. The combined stream is heated to 250°C and fed to the reactor. The reactor effluent emerges at the same temperature and is cooled to 0°C at $P = 6.0$ MPa, partially condensing the methanol product. The gas leaving the condenser is saturated with methanol: 1% is taken off for process monitoring purposes and the remainder is recycled. An overall CO conversion of 98% is achieved. The ratio of H_2 to CO is 2 mol H_2/1 mol CO everywhere in the process system. Ideal gas behavior may be assumed.

Carry out a degree-of-freedom analysis for this process and show that the given information is sufficient to allow the calculation of the component flow rates for all streams, the required heat duties for all process units, and the temperature of the combined feed stream to the reactor pre-heater.

SOLUTION

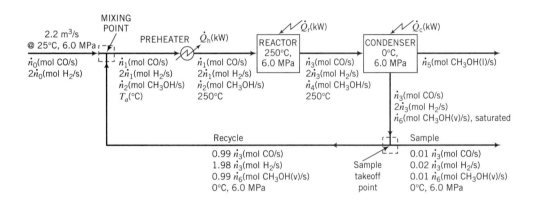

In labeling the flowchart, we have taken advantage of the fact that CO and H_2 remain in stoichiometric proportion throughout the process, which cuts down on the number of unknown variables that must be defined: if the flow rate of CO is $\dot{n}_i$, the flow rate of H_2 must be $2\dot{n}_i$. However, this also means that we cannot write independent balances on these two species. We must therefore allow for no more than *two* independent material balances on each process unit when we do the degree-of-freedom analysis, rather than the three that would normally be allowed when three species are involved in a process.

Mixing Point	6 variables ($\dot{n}_0, \dot{n}_1, \dot{n}_2, \dot{n}_3, \dot{n}_6, T_a$)
	− 4 relations (2 material balances, 1 energy balance, and the ideal gas equation of state for the fresh feed)
	= 2 local degrees of freedom
Preheater	4 variables ($\dot{n}_1, \dot{n}_2, T_a, \dot{Q}_h$)
	− 1 relation (The energy balance. We cannot count material balances, since in labeling the outlet flow rates the same as the inlet flow rates, we have implicitly used up the two allowed material balances.)
	= 3 local degrees of freedom
Reactor	5 variables ($\dot{n}_1, \dot{n}_2, \dot{n}_3, \dot{n}_4, \dot{Q}_r$)
	− 3 relations (2 molecular species balances, 1 energy balance)
	+ 1 chemical reaction
	= 3 local degrees of freedom
Condenser	5 variables ($\dot{n}_3, \dot{n}_4, \dot{n}_5, \dot{n}_6, \dot{Q}_c$)
	− 3 relations [1 material balance (Can you see why?), 1 energy balance, saturation condition at outlet]
	= 2 local degrees of freedom

Purge Point	2 variables ($\dot{n}_3$, $\dot{n}_6$)
	− 0 relations (In labeling the chart, we have built in all allowed balances. If this is not clear to you, try writing a balance on, say, hydrogen, and observe the result.)
	= 2 local degrees of freedom
Process	12 local degrees of freedom ($2 + 3 + 3 + 2 + 2$)
	− 11 ties ($\dot{n}_1$, $\dot{n}_2$, and $\dot{n}_6$ were each counted three times, $\dot{n}_3$ was counted four times, and $\dot{n}_4$ and T_a were each counted twice)
	− 1 additional relation (98% overall conversion)
	= 0 degrees of freedom

The process is therefore completely specified.

The solution of the system equations for all unknown variables is straightforward. The ideal gas equation of state applied to the fresh feed stream yields $\dot{n}_0$. The specified overall CO conversion yields $\dot{n}_3$ from the equation $0.01\dot{n}_3 = (1 − 0.98)\dot{n}_0$. Raoult's law at the condenser outlet combined with the calculated value of $\dot{n}_3$ yields $\dot{n}_6$, and an overall carbon balance yields $\dot{n}_5$. Balances on CO and CH_3OH at the mixing point yield $\dot{n}_1$ and $\dot{n}_2$, and an energy balance for the same subsystem yields T_a. An energy balance on the preheater then yields $\dot{Q}_h$. A methanol balance on the condenser yields $\dot{n}_4$, and then energy balances on the reactor and the condenser yield $\dot{Q}_r$ and $\dot{Q}_c$, respectively.

TEST YOURSELF

Consider the flowchart shown here for a nonreactive mixing–separation process.

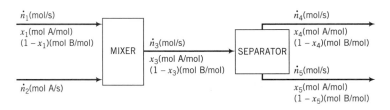

1. How many local degrees of freedom does each unit possess?
2. How many total degrees of freedom does the entire process system possess?

10.1c Solving Algebraic Equations

Once you have specified values for your design variables, you are faced with the task of solving the system equations for the state variables. For relatively simple processes, the manual techniques presented in previous chapters are applicable and should be used.

Unfortunately, chemical processes are seldom relatively simple. For example, most problems you have encountered in this text could eventually be reduced to the solution of linear algebraic equations.

$$0.21 n_{\text{air}} = 2015 \qquad \text{Solve for } n_{\text{air}}$$

$$\left.\begin{array}{r} n_1 + n_2 = 100 \\ 0.96 n_1 + 0.05 n_2 = 50 \end{array}\right\} \qquad \text{Solve for } n_1 \text{ and } n_2$$

However, some could not. You encountered nonlinear equations, for example, when learning about equations of state:

$$\frac{P\hat{V}}{RT} = 1 + \frac{B}{\hat{V}} + \frac{C}{\hat{V}^2} + \frac{D}{\hat{V}^4} + \frac{E}{\hat{V}^5} \qquad \text{Given } P \text{ and } T, \text{ solve for } \hat{V}$$

and when studying vapor–liquid equilibrium

$$\log_{10} p^* = A + B/(C + T) \qquad \text{Given } T, \text{ solve for } p^*$$

and when applying the first law of thermodynamics

$$\Delta\dot{H} = a(T − 25) + b(T^2 − 25^2) + c(T^3 − 25^3) \qquad \text{Given } \Delta\dot{H}, \text{ solve for } T$$

Solving sets of linear equations is tedious but straightforward. Solving nonlinear equations, on the other hand, may or may not be straightforward. Doing so in all but the simplest cases involves trial-and-error, and there is usually no guarantee that you will be able to find a solution, or that a solution you find is the only possible solution, or even that a solution exists.

We suggest the following guidelines for solving n algebraic equations in n unknowns.

- For linear equations with $n \leq 3$ or a single simple nonlinear equation, (e.g., $y^2 - 2y - 3 = 0$, $e^{-x} = 0.2$, or $\sin \theta = 0.5$), solve manually using high school algebra methods.
- For a single complex nonlinear equation of the form $f(x) = 0$, use a spreadsheet or an equation-solving program. If you use a spreadsheet, put an estimated value of x in one cell and the formula for $f(x)$ in a second cell, then use the goalseek tool to set the value in the second cell equal to zero by varying the value in the first cell. The final value in the first cell is the desired solution.
- For linear equations with $n > 3$ or nonlinear equations with $n \geq 2$, use an equation-solving program.

10.2   SEQUENTIAL MODULAR SIMULATION

As we noted at the beginning of this chapter, there are two broad approaches to the automated solution of the balance equations for a process system: the **sequential modular approach** and the **equation-based approach**. This section outlines the first of these methods. The balance equations (and any other equations that may arise from physical considerations or process specifications) for each unit are written and solved. If there are no recycle streams, the calculation moves from one unit to another, until all units have been covered. If there is a **cycle** (the conventional term for a recycle loop in a process flowchart), a trial-and-error procedure is required: values of one or more stream variables in the cycle are assumed; the balance equations for units in the cycle are solved, one unit at a time, until the values of the assumed variables are recalculated; new variable values are assumed; and the procedure is repeated until the assumed and calculated values agree.

The object of the presentation that follows is not to enable you to go out and simulate complex processes—the treatment is too brief and general for that—but simply to give you a sense of the structure of process simulation programs and calculations and to provide an idea of what such programs can do.

10.2a   Elements of a Modular Simulation

The first step in setting up a process for the sequential modular approach is to reconstruct the process flowchart in terms of **blocks** or **modules** (process units or operations) and streams connecting them. Several types of blocks and names that might be given to them are as follows:

MIX	Mix several inlet streams adiabatically to form one product stream.
SPLIT	Split a single inlet stream into two or more product streams with the same composition and temperature.
COMPRESS	Raise the pressure of a gas by a specified amount.
PUMP	Raise the pressure of a liquid by a specified amount.
FLASH	Convert a liquid stream at one pressure to liquid and vapor streams in equilibrium at a lower pressure.
DISTILL EXTRACT CRYSTAL ABSORB	Simulate the separation processes of distillation, extraction, crystallization, and absorption.
REACT	Simulate a chemical reactor.

An additional block type—the convergence block—has a function to be described later.

The simulation program contains a built-in subprogram corresponding to each block type. To simulate a process, you would use the simulation program to build a flowchart and then enter known block and stream variable values in forms provided by the program. When you subsequently run the simulation, a series of calls to the block subroutines would lead to the solution of the process material and energy balance equations.

For example, suppose two streams labeled S1 and S2 are mixed adiabatically to form a third stream S3. A MIX block might be used to simulate this operation. (What the subprogram for this block might look like is the subject of the next example.)

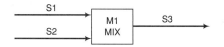

The step in the program that would perform the material and energy balances for this unit would be initiated by a statement equivalent to the program statement.

CALL MIX(M1,S1,S2,S3)

M1 is a label, identifying the unit that has streams S1 and S2 as inputs and stream S3 as an output. In the subprogram, S1, S2, and S3 would each be arrays, the elements of which contain enough information to specify the flow rate, composition, and temperature of each stream. Prior to this statement in the program, the values of the elements of arrays S1 and S2 would have been either specified or determined as the outputs of other blocks. The subprogram would calculate the component flow rates of stream S3 by solving material balance equations and (if required) the temperature of the stream by solving the energy balance equation ($\Delta \dot{H} = 0$).

EXAMPLE 10.2-1   *Structure of a Mixing Block Routine*

Two streams are mixed adiabatically. Each stream may contain any of five components (A, B, C, D, and E). No phase changes take place. The heat capacities of all components may be approximated as constants, and the heat of mixing may be neglected. A computer program is to be written to calculate the component molar flow rates and temperature of the product stream from specified values of these quantities for the feed streams.

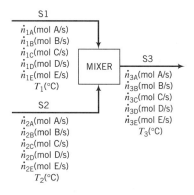

1. Write equations for the product stream component flow rates and temperature.
2. Create a spreadsheet that would determine the product stream variables from given values of the feed stream variables.
3. Write a Fortran subroutine that would perform the same calculations.

SOLUTION

1. A degree-of-freedom analysis of this unit reveals 18 variables (six for each stream) and six equations (five material balances and an energy balance) for a net 12 degrees of freedom. The design variables are the 12 feed stream variables, leaving the product stream component flow rates and temperature as the state variables.

The material balance equations are trivial:

$$\dot{n}_{3A} = \dot{n}_{1A} + \dot{n}_{2A} \tag{1}$$

$$\dot{n}_{3B} = \dot{n}_{1B} + \dot{n}_{2B} \tag{2}$$

$$\dot{n}_{3C} = \dot{n}_{1C} + \dot{n}_{2C} \tag{3}$$

$$\dot{n}_{3D} = \dot{n}_{1D} + \dot{n}_{2D} \tag{4}$$

$$\dot{n}_{3E} = \dot{n}_{1E} + \dot{n}_{2E} \tag{5}$$

Let us choose as the reference state for each component the system phase (liquid or gas), temperature (T_1), and pressure (1 atm). The specific enthalpy of, say, component A in the product stream would be $\hat{H}_{3A} = C_{pA}(T_3 - T_1)$. (Remember that there are no phase changes and we are assuming constant heat capacities.) The energy balance for this open adiabatic system is $\Delta\dot{H} = 0$, or

$$\Delta\dot{H} = \sum \dot{n}_{\text{out}}\hat{H}_{\text{out}} - \sum \dot{n}_{\text{in}}\hat{H}_{\text{in}}$$
$$= [\dot{n}_{A3}C_{pA} + \dot{n}_{B3}C_{pB} + \cdots + \dot{n}_{E3}C_{pE}](T_3 - T_1)$$
$$- [\dot{n}_{A2}C_{pA} + \dot{n}_{B2}C_{pB} + \cdots + \dot{n}_{E2}C_{pE}](T_2 - T_1) = 0$$

(All of the enthalpies of the stream 1 components are zero). Solving this equation for T_3 yields

$$T_3 = T_1 + \frac{\dot{n}_{A2}C_{pA} + \dot{n}_{B2}C_{pB} + \dot{n}_{C2}C_{pC} + \dot{n}_{D2}C_{pD} + \dot{n}_{E2}C_{pE}}{\dot{n}_{A3}C_{pA} + \dot{n}_{B3}C_{pB} + \dot{n}_{C3}C_{pC} + \dot{n}_{D3}C_{pD} + \dot{n}_{E3}C_{pE}}(T_2 - T_1) \tag{6}$$

Equations 1 to 6 may be solved sequentially for the product stream variables.

2. A spreadsheet program that solves the six system equations is easy to construct. With a little more effort, a flowchart can be imbedded into the spreadsheet in such a way that the product stream variables are automatically updated if any of the input stream variables are changed. In the example that follows, the following heat capacities are used:

Species	A	B	C	D	E
$C_p[\text{J}/(\text{mol}\cdot°\text{C})]$	77.3	135.0	159.1	173.2	188.7

The spreadsheet might appear as follows.[1]

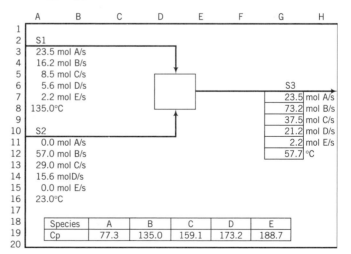

The flowchart shows a set of specified input variables for streams S1 and S2 and the calculated product stream variables determined from Equations 1 to 6. The formula of Equation 1 for $\dot{n}_{A3}(= \dot{n}_{A1} + \dot{n}_{A2})$ would be entered in Cell G7 as

$$= \text{A3 + A11}$$

---

[1]The spreadsheet was created with Microsoft Excel®. The values and formulas were first entered, and then the line and rectangle drawing tools were used to create the flowchart. For clarity, most grid lines are not shown.

indicating that the value in Cell G7 would be obtained as the sum of the values in Cells A3 and A11. If the value in either of these cells is changed on the spreadsheet, the value in G7 would change accordingly. Similarly, the formula of Equation 6 for T_3 would be entered in Cell G12 as

```
= A8 + (A11*C19 + A12*D19 + A13*E19 + A14*F19 + A15*G19)*(A16-A8)/
   (G7*C19 + G8*D19 + G9*E19 + G10*F19 + G11*G19)
```

(*Verify.*) Changing either T_1 (in Cell A8) or T_2 (in Cell A16) on the spreadsheet would cause the new value of T_3 to appear in Cell G12.

3. A Fortran subroutine that performs the same calculations follows.

```
              SUBROUTINE MIX(M,S1,S2,S3)
      C****   M IS A UNIT LABEL
      C****   ARRAYS S1 AND S2 ARE INPUT, AND S3 IS OUTPUT
              DIMENSION S1(6),S2(6),S3(6)
              COMMON CP(5)
      C****   SOLVE MATERIAL BALANCES
              DO 100J = 1,5
      100     S3(J) = S1(J) + S2(J)
      C****   IF T1 = T2, SET T3 = T1 AND RETURN
              T1 = S1(6)
              T2 = S2(6)
              IF (T1.EQ.T2) GO TO 200
      C****   EVALUATE SUMS FOR ENERGY BALANCES
              SUM2 = 0.0
              SUM3 = 0.0
              DO 150J = 1,5
              SUM2 = SUM2 + S2(J)*CP(J)
      150     SUM3 = SUM3 + S3(J)*CP(J)
      C****   SOLVE ENERGY BALANCE FOR T3
              T3 = T1 + SUM2*(T2-T1)/SUM3
              S3(6) = T3
              RETURN
      200     S3(6) = T1
              RETURN
              END
```

The next example illustrates the structuring of a sequential modular process simulation using blocks of the types just described.

EXAMPLE 10.2-2   *Simulation of an Acyclic Process*

The flowchart shown here depicts a multi-unit separation process. Three liquid streams are mixed adiabatically; the product stream is pumped through a heater to a distillation column, and the overhead product from the column is partially condensed to yield liquid and vapor products. Using the blocks MIX (mix two streams to form a third), PUMP, HEAT, DISTILL, and CNDS, construct a block diagram for the simulation of this process.

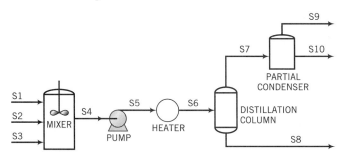

SOLUTION   Since three streams are mixed in the process but the MIX routine can only handle two feed streams, we need to use this block twice. In other respects, the block diagram looks like the process flowchart.

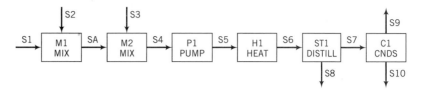

In the main simulation program, a series of statements resembling the following would be included (the first argument of each calling statement is the unit identifier):

```
CALL MIX(M1,S1,S2,SA)
CALL MIX(M2,S3,SA,S4)
CALL PUMP(P1,S4,S5,parameters)
CALL HEAT(H1,S5,S6,parameters)
CALL DISTILL(ST1,S6,S7,S8,parameters)
CALL CNDS(C1,S7,S9,S10,parameters)
```

The PUMP calculation would cause the pressure of the liquid stream to be raised by a specified amount; the routine might also calculate the required shaft work for this operation and could go as far as to specify the type and size of pump needed for the job. The HEAT routine would calculate the heat input required to achieve the required temperature change and might also perform heat exchanger design calculations.

The DISTILL subroutine would solve material and energy balance equations to determine the flow rates and compositions of both product streams and the heat requirement for the distillation column. Depending on the level of sophistication of the simulation program, the subroutine might then go on to determine details of the column design for the given separation.

The CNDS subroutine would proceed in a similar manner for the condenser calculations—first solving material and energy balances to determine unknown flow rates and the heat duty, and then possibly performing design calculations.

10.2b   Cyclic Systems and the Convergence Block

Suppose the block diagram of a process has the following appearance:

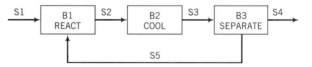

Further suppose that you know the values of the variables associated with stream S1 and you wish to calculate the remaining stream variables.

The procedure outlined in the previous section will not work, since there is no way to get the calculation started. To solve the equations for block B1 you must know the variable values for both S1 and S5, but you cannot know S5 until you solve B3, for which you need S3 and hence the solution of B2, which in turn requires a knowledge of S2 and hence the solution of B1, and you are back where you began.

If the calculations were to be done by hand, overall system and subsystem balances would eventually yield *n* equations in *n* unknowns, and the equations could then in principle be solved for all the desired process variables. It would be difficult to write a sequential modular program to implement this method for an arbitrary process, however. Instead, the following iterative approach is used.

1. Assume variable values for a stream within the cycle (this is known as *tearing the cycle*).
2. Solve the system balance equations, working around the cycle from unit to unit until the tear stream variables are recalculated.

3. If the assumed and calculated tear stream variables agree within a specified tolerance, the solution is complete; if they do not, use the new values or some combination of the new and old values to initiate another swing around the cycle. Iterate in this manner until convergence is achieved.

Suppose, for example, that in the process described above we choose S5 as the tear stream. The flowchart might then appear as shown here:

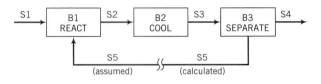

The calculation would begin with an assumption of values of the variables associated with stream S5. The equations for blocks B1, B2, and B3 could then be solved sequentially. The solution for B3 would yield calculated variable values for stream S5. New assumed variable variables would be determined by one of several methods to be outlined, and the procedure would be repeated until the assumed and calculated values of the S5 variables agree to within a specified tolerance.

The application of this procedure to a single tear stream variable is tantamount to solving an equation of the form $x = f(x)$, where $f(x)$ is the "function" that generates a new value of the tear stream variable x by working around the cycle. Techniques described in Appendix A.2—successive substitution and Wegstein's algorithm—can be used to perform this calculation.

Most simulation programs have a block, called a **convergence block**, that performs such calculations using the Wegstein algorithm. The output stream from this block contains the assumed set of tear stream variables, and the input stream contains the values calculated by working around the cycle. The block diagram for the simulated process would appear as follows:

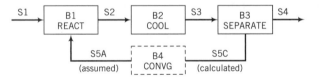

It is conventional to show the convergence block using dashed lines, since it does not correspond to a physical operation in the process.

EXAMPLE 10.2-3   *Simulation of a Cyclic Process*

The gas-phase dehydrogenation of isobutane to isobutene

$$C_4H_{10} \rightarrow C_4H_8 + H_2$$

is carried out in a continuous reactor. A stream of pure isobutane (the fresh feed to the process) is mixed adiabatically with a recycle stream containing 90 mole% isobutane and the balance isobutene, and the combined stream goes to a catalytic reactor. The effluent from this process goes through a multistage separation process: one product stream containing all of the hydrogen and 1% of the isobutane leaving the reactor as well as some isobutene is sent to another part of the plant for additional processing, and the other product stream is the recycle to the reactor. The single-pass isobutane conversion in the reactor is 35%.

The process flowchart for a fresh feed of 100 mol isobutane is shown below. For simplicity, the symbol A will be used to denote isobutane, B denotes isobutene, and C denotes hydrogen.

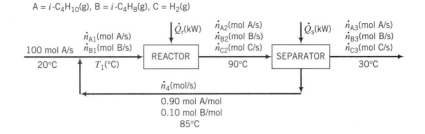

All streams shown are gases. On the flowchart, $\dot{Q}_r$ is the required rate of heat transfer to the reactor and $\dot{Q}_s$ is the net rate of heat transfer to the separation process (which involves several heat addition and removal steps).

1. Perform a degree-of-freedom analysis on the process.
2. Choose a tear stream variable and convert the flowchart into a block diagram for a sequential modular simulation, using blocks MIX, REACT, SEP, and a convergence block CONVG.
3. Write a spreadsheet program to determine the labeled process variables on the chart.

SOLUTION

1. Three "units" comprise this process: the reactor, the separation process, and the mixing point, which in the process may be a stirred tank or simply a tee in the feed line. Let us first calculate the local degrees of freedom associated with each unit, and then the net degrees of freedom for the process.

   Mixing Point   4 unknowns ($\dot{n}_{A1}, \dot{n}_{B1}, \dot{n}_4, T_1$)

   $-$ 3 balances (2 material balances, 1 energy balance)

   1 local degree of freedom

   Reactor   (Count material balances on molecular species.)

   7 unknowns ($\dot{n}_{A1}, \dot{n}_{B1}, \dot{n}_{A2}, \dot{n}_{B2}, \dot{n}_{C2}, T_1, \dot{Q}_r$)

   $-$ 4 balances (3 molecular species balances, 1 energy balance)

   $-$ 1 additional relation (35% single-pass conversion)

   $+$ 1 chemical reaction

   3 local degrees of freedom

   Separator   8 unknowns ($\dot{n}_{A2}, \dot{n}_{B2}, \dot{n}_{C2}, \dot{n}_{A3}, \dot{n}_{B3}, \dot{n}_{C3}, \dot{n}_4, \dot{Q}_s$)

   $-$ 4 balances (3 material balances, 1 energy balance)

   $-$ 1 additional relation (isobutane split)

   3 local degrees of freedom

   (We used the information that the recycle contains no hydrogen when we did not label a variable n_{C4} on the flowchart, so we should not count this fact as a second additional relation.)

   Overall Process   7 local degrees of freedom (1 + 3 + 3)

   $-$ 7 ties ($\dot{n}_{A1}, \dot{n}_{B1}, \dot{n}_{A2}, \dot{n}_{B2}, \dot{n}_{C2}, \dot{n}_4,$ and T_1 were each counted twice)

   0 net degrees of freedom

   The problem can in principle be solved for all labeled variables.

2. Before planning the solution strategy, let us observe the flowchart and confirm that we cannot proceed in a unit-to-unit manner without trial and error.

   The mixing unit has one local degree of freedom and three unknowns. We cannot solve for $\dot{n}_{A1}$ and $\dot{n}_{B1}$ until we know $\dot{n}_4$. To determine this variable, we must solve the separation process balance equations, but we cannot do so unless we know $\dot{n}_{A2}, \dot{n}_{B2},$ and $\dot{n}_{C2}$. To calculate *these* variables, we must solve the reactor equations. Unfortunately, we cannot solve them either, since we must first know $\dot{n}_{A1}$ and $\dot{n}_{A2}$, for which we must solve the mixer equations, and we are back where we started.

The solution is to tear the cycle. We can tear it in any of three places: between the mixing point and the reactor, between the reactor and the separation process, or between the separation process and the mixing point. The first choice involves the trial-and-error determination of two variables, the second one involves three variables, and the third involves only one ($\dot{n}_4$). The fewer variables you have to determine by trial and error, the more likely you are to succeed. Let us therefore choose the recycle stream as the tear stream.

The block diagram for the simulation now appears as follows:

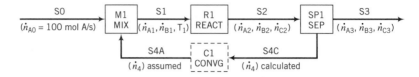

The program would take as input a guessed value for the tear stream variable $\dot{n}_4$ (stream S4A) and might contain the following sequence of statements:

```
CALL MIX(M1,S0,S4A,S1)
CALL REACT(R1,S1,S2,QR)
CALL SEP(SP1,S2,S3,S4R,QS)
CALL CONVG(C1,S4R,S4A)
```

(The first unit called is always the one following the tear stream.) The call of SEP would result in the recalculation of $\dot{n}_4$ from the separation process balances (stream S4R). CONVG would compare this value with the one initially assumed. If the two values agree to within a specified tolerance, the calculation would be terminated; otherwise, the new value would be used to begin another journey around the cycle. The nature of the iterative procedure performed by CONVG is discussed in greater detail in Appendix A.2.

3. Equations could be derived and solved for all of the unknown process variables, making trial-and-error solution unnecessary. However, for illustrative purposes we will set up the spreadsheet to parallel the sequential modular solution procedure of part 2.

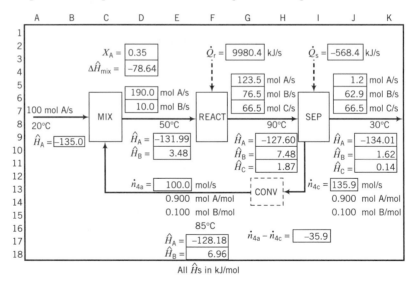

The spreadsheet shows an initially assumed value of the recycle flow rate ($\dot{n}_{4a}$) of 100.0 mol/s (Cell E13) and an assumed value of the mixing point outlet temperature of 50°C (Cell D8). The value of $\dot{n}_{4a}$ will be varied until the calculated recycle flow rate ($\dot{n}_{4c}$) in Cell J13 equals the assumed value, which it now does not. (The actual calculation will be done by finding the value of $\dot{n}_{4a}$ that drives the value of $\dot{n}_{4a} - \dot{n}_{4c}$ in Cell I17 to zero.) Once the flow rates are correct, the mixing point temperature will be varied to determine the value that drives $\Delta \dot{H} = \sum \dot{n}_{out}\hat{H}_{out} - \sum \dot{n}_{in}\hat{H}_{in}$ (in Cell D4) to zero for the adiabatic mixer.

The entries in selected spreadsheet cells are shown below, in the order in which they might be entered. The specific enthalpies are for the gaseous species at the stream temperatures relative to the elements at 25°C, with heats of formation taken from Table B.1 and heat capacity formulas taken from Table B.2. Their numerical values are in kJ/mol. The other cells contain labels to provide clarity (such as $[\hat{H}_A =]$ in Cell A9). A spreadsheet drawing tool is used to draw the lines (denoting streams) and boxes (denoting process steps).

D3 = 0.35   (given fractional conversion of A)
A7 = 100   (basis of calculation)
A8 = 20   (given fresh feed temperature)
G8 = 90   (given reactor effluent temperature)
J8 = 30   (given product stream temperature)
E13 = 100   (initial guess for recycle stream flow rate)
E14 = 0.900, J14 = 0.900   (given mole fraction of A in recycle stream)
E15 = 0.100, J15 = 0.100   (calculated mole fraction of B in recycle stream)
E16 = 85   (given temperature of recycle stream)
D6 = A7 + E13*E14   (A balance on mixing point)
D7 = E13*E15   (B balance on mixing point)
D8 = 50   (initial guess for combined stream temperature. Later vary to make
$\Delta\dot{H}_{\mathrm{mixer}} = 0.$)
G5 = D6*(1-D3)   (unreacted A)
G6 = D7 + D6*D3   (B balance on reactor: output = input + generation)
G7 = D6*D3   (C balance on reactor: output = generation)
J5 = 0.01*G5   (1% of A in reactor effluent)
J13 = (G5-J5)/J14   (from A balance on separator)
J6 = G6 - H13*H15   (B balance on separator)
J7 = G7   (C balance on separator)
B9 = -134.5 + 0.08946*(A8-25) + 30.13e-5*(A8^2-25^2)/2-18.91e-8*
   (A8^3-25^3)/3 + 49.87e-12*(A8^4 - 25^4)/4 $[= (\Delta\hat{H}_f^\circ)_A + \int C_{pA}dT,$
   specific enthalpy of A at 20°C relative to elements at 25°C.]
F17, F18, E9, E10, H9, H10, H11, K9, K10, K11 =   specific enthalpies of A, B,
   and C at process temperatures (see formula for B9)
D4 = D6*E9 + D7*E10 - A7*B9 - E13*E14*F17 - E13*E15*F18 $(\Delta\dot{H}_{\mathrm{mixing\ point}})$
G3 = G5*H9 + G6*H10 + G7*H11 - D6*E9 - D7*E10 $(\Delta\dot{H}_{\mathrm{reactor}})$
J3 = J5*K9 + J6*K10 + J7*K11 + J13*J14*F17 + J13*J15*F18 - G5*H9
   - G6*H10 - G7*H11 $(\Delta\dot{H}_{\mathrm{separator}})$
I17 = E13 - J13   (assumed recycle rate - recalculated recycle rate)

As noted previously, once the given spreadsheet has been set up, the value in Cell E13 ($=\dot{n}_{4a}$) is varied until the value in Cell I17 ($=\dot{n}_{4a} - \dot{n}_{4c}$) equals zero. This calculation is easily done using the goalseek tool in most spreadsheets. Then, the value in Cell D8 (the combined stream temperature) is varied until the value in Cell D4 ($\Delta\dot{H}_{\mathrm{mixing\ point}}$) equals zero, reflecting the energy balance on the adiabatic mixing point. The resulting spreadsheet is the correct flowchart for the process, and appears as follows:

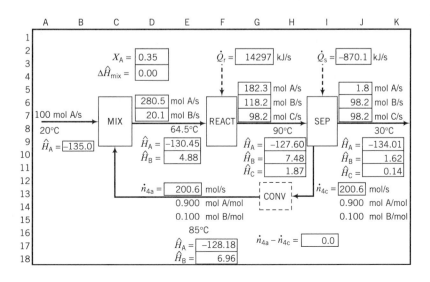

The recycle rate is 200.6 mol/s, and the adiabatic mixing temperature is 64.5°C. Any of the input variable values could now be changed and the flowchart easily recalculated.

Networks of recycle loops are commonly encountered in large processes, and a suitable choice of a tear stream may minimize the number of iterations required to solve the balance equations of such systems. For example, consider the block diagram shown below. There are two cycles in this process: S2-S3-S4 and S3-S5-S7. To solve the system equations you could, for example, tear both S4 and S7, which would require the inclusion of two convergence blocks and hence the simultaneous solution of two iterative loops; however, you can instead tear one stream common to both cycles (S3), probably decreasing the computation time required to achieve the solution.

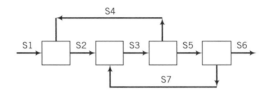

On the other hand, you will commonly encounter systems where you have no choice but to tear two or more streams, as in the next process.

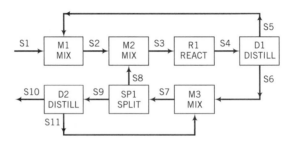

There are three cycles: S2-S3-S4-S5, S7-S9-S11, and S3-S4-S6-S7-S8, and no single stream that if torn would permit the solution of all the system equations. For instance, if you tear S3 you could work your way around the first cycle to unit M2 but you would be stuck there for lack of knowledge of S8, and you would be similarly stuck at unit M3 in the third cycle since you would not know S11.

One of many possible ways to analyze this system is to tear S4 and S7 (so that at least one stream in each cycle is torn). You must call the block programs in an order for which they can be solved. Remember that S1, S4B, and S7B are presumably known, and convince yourself that the program following the block diagram would in principle work.

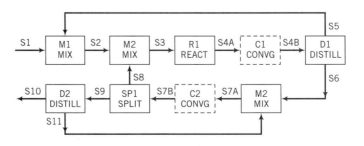

```
CALL DISTILL (D1,S4B,S5,S6)     (Calculates S5,S6)
CALL MIX (M1,S1,S5,S2)          (Calculates S2)
CALL SPLIT (SP1,S7B,S8,S9)      (Calculates S8,S9)
CALL MIX (M2,S2,S8,S3)          (Calculates S3)
CALL REACT (R1,S3,S4A)          (Calculates S4A)
CALL CONVG (C1,S4A,D1,S4B)      (Forces convergence of S4)
CALL DISTILL (D2,S9,S10,S11)    (Calculates S10,S11)
CALL MIX (M3,S6,S11,S7A)        (Calculates S7A)
CALL CONVG (C2,S7A,D1,S7B)      (Forces convergence of S7)
```

Other choices of tear streams could be made, and some simulation programs would permit you to use a single convergence block to force convergence of both tear streams simultaneously.

Techniques exist to determine systematically how many streams must be torn when multiple cycles occur and the order in which to carry out the subsequent calculations. We will not discuss these matters here.

10.2c   Design Specifications

In the process simulations we have discussed so far, the direction of flow of information corresponded to the direction of flow of the process streams—from feeds to products and around cycles. This mode of information flow is appropriate if the object is to calculate the output of a process for a given input and set of process parameters (temperatures, pressures, etc.); however, it often happens that a desired output is specified and input or process unit parameters required to achieve this output are to be calculated. A feature of process simulation programs called a *design specification* is used for calculations of this type.

Consider, for example, a single adiabatic flash evaporator for which the flow rate, composition, temperature, and pressure of the feed stream are known.

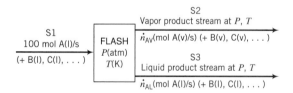

If the pressure in the evaporator is specified, a single call to the FLASH subprogram (which contains applicable vapor–liquid equilibrium correlations) yields the component flow rates in both product streams as well as the evaporator temperature. Suppose, however, that one of the component flow rates in one of the product streams is specified (e.g., $\dot{n}_{AV} = 65.0$ mol/s) and you wish to determine the value of P required to achieve the desired separation. The FLASH program by itself is incapable of performing this determination: it is set up to perform the forward calculation only. The approach is to set up an artificial cycle, varying the value of P (the *manipulated variable*) until the calculated output variable (the *sampled variable*) is within a specified tolerance of the desired value. For example, you might specify

$$\left| \dot{n}_{AV} - 65.0 \text{ mol A/s} \right| \leq 0.001 \text{ mol A/s (i.e., 64.999 mol A/s} \leq \dot{n}_{AV} \leq 65.001 \text{ mol A/s)}$$

This inequality is called the *design specification* for the process.

A block diagram for the simulation with this design specification incorporated would be as follows:

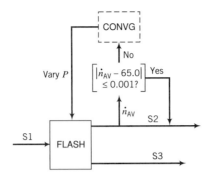

A value of the manipulated variable (P) is initially guessed and the flash block equations are solved to generate the variables associated with streams S2 (including $\dot{n}_{AV}$) and S3. The value of the sampled variable ($\dot{n}_{AV}$) is substituted into the design specification ($|\dot{n}_{AV} - 65.0 \text{ mol A/s}| \leq 0.001$). If the specification is not satisfied, the convergence block generates a new value of P and the calculations are repeated. When the sampled variable value satisfies the design specification, the value of the manipulated variable is accepted and the simulation proceeds to downstream blocks. The sampled variables in a design specification (there may be several of them) may be associated with the product streams of the unit for which the manipulated variable is being adjusted (as in this example), or they may be associated with streams far removed from this unit.

Any number of design specifications may be built into a simulation, providing considerable flexibility for process design. You should remember, however (and if you do much simulation you will certainly be reminded), that whenever convergence blocks are part of a simulation there is no guarantee that convergence will be achieved, and the more convergence blocks you have the less likely you are to achieve it. In general, the better your initial guesses of the manipulated variable values, the more likely you are to achieve convergence and to achieve it rapidly.

An example of a sequential modular simulation of a relatively large process is given in Example 10.3-3, following a discussion of the second broad approach to process simulation.

10.3 EQUATION-BASED SIMULATION

The sequential modular approach to process simulation solves system equations in blocks corresponding to the unit operations that make up the process. The block diagram for the process looks very much like the traditional process flowchart. Since engineers are accustomed to viewing chemical processes as sequences of unit operations, they tend to feel comfortable with this approach.

In the **equation-based approach**, the equations for all units are collected and solved simultaneously. The natural decomposition of the system into its constituent unit operations is therefore lost. Moreover, the simultaneous solution of large numbers of equations, some of which may be nonlinear, can be a cumbersome and time-consuming problem, even for a powerful computer. For all these reasons, most commercial simulation programs were still based on the sequential modular approach when this text was written.

However, there are disadvantages associated with the sequential modular method. The block subroutines for simulating specific unit processes are designed to operate in a forward calculation mode only: that is, they calculate product stream variables from given values of feed stream variables and process conditions. Frequently, the problems the process engineer wants to solve fall into either of two other categories: (a) given process conditions and product stream variables, calculate feed stream variables; (b) given feed and product stream variables, calculate process conditions. In both cases, iterative calculations using design specifications and convergence blocks are required to solve the problems using a sequential modular program.

These difficulties vanish if the system equations are simply collected and solved for all unknown variables. Several powerful equation-solving algorithms are available in commercial programs like Maple®, Mathematica®, Matlab®, Mathcad®, and E-Z Solve® that make the equation-based approach competitive with the sequential modular approach. Many researchers in the field believe that as this trend continues, the former approach will replace the latter one as the standard method for flowsheet simulation. (Engineers are also working on "simultaneous modular" methods, which combine features of both sequential modular and equation-based approaches. We will not deal with these refinements here, however.)

The following example illustrates the equation-based approach.

EXAMPLE 10.3-1 **Simulation and Design of a Two-Column Separation Process**

A stream containing 35.0 wt% benzene (B), 50.0% toluene (T), and the balance xylene (X) is fed to a distillation column. The overhead product from the column contains 67.3 wt% benzene and 30.6% toluene. The bottoms product is fed to a second column. The overhead product from the second column contains 5.9 wt% benzene and 92.6% toluene. Of the toluene fed to the process, 10.0% is recovered in the bottoms product from the second column, and 90.0% of the xylene fed to the process is recovered in the same stream.

1. Perform a degree-of-freedom analysis on the process and outline a manual calculation procedure to determine the compositions of all streams.
2. Construct a set of n equations in the n unknown stream variables identified in part 1.

SOLUTION **Basis: 100 kg Feed**

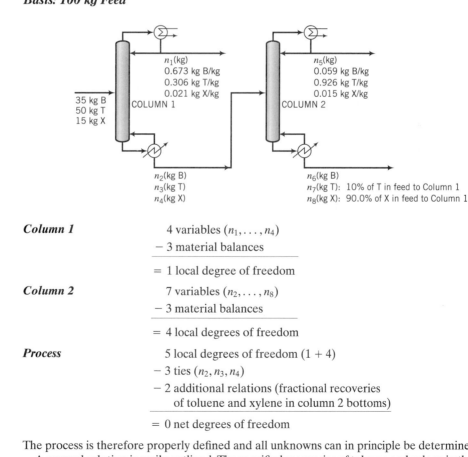

Column 1	4 variables ($n_1, \ldots, n_4$)
	$-$ 3 material balances
	$=$ 1 local degree of freedom
Column 2	7 variables ($n_2, \ldots, n_8$)
	$-$ 3 material balances
	$=$ 4 local degrees of freedom
Process	5 local degrees of freedom $(1 + 4)$
	$-$ 3 ties (n_2, n_3, n_4)
	$-$ 2 additional relations (fractional recoveries of toluene and xylene in column 2 bottoms)
	$=$ 0 net degrees of freedom

The process is therefore properly defined and all unknowns can in principle be determined.

A manual solution is easily outlined. The specified recoveries of toluene and xylene in the bottoms from the second column allow the immediate determination of n_7 and n_8, leaving three unknowns—

n_1, n_5, and n_6—in the overall process system. Overall toluene and xylene balances provide two equations in two unknowns—n_1 and n_5—and an overall mass balance then yields n_6. Finally, benzene, toluene, and xylene balances on either Column 1 or Column 2 yield solutions for n_2, n_3, and n_4.

The system equations are given below. We first write the allowed balances for column 1, then those for column 2, and finally the additional process specifications.

Column 1 Balances B: $35.0 = 0.673n_1 + n_2$ **(1)**

T: $50.0 = 0.306n_1 + n_3$ **(2)**

X: $15.0 = 0.021n_1 + n_4$ **(3)**

Column 2 Balances B: $n_2 = 0.059n_5 + n_6$ **(4)**

T: $n_3 = 0.926n_5 + n_7$ **(5)**

X: $n_4 = 0.015n_5 + n_8$ **(6)**

10% T recovery: $n_7 = 0.100(50.0) = 5.00$ **(7)**

93.3% X recovery: $n_8 = 0.933(15.0) = 14.0$ **(8)**

We thus have eight equations in eight unknowns. The equations may easily be entered into an equation-solving program and solved to obtain $n_1 = 48.1$, $n_2 = 2.65$, $n_3 = 35.3$, $n_4 = 14.0$, $n_5 = 32.7$, $n_6 = 0.72$, $n_7 = 5.0$, and $n_8 = 14.0$.

Although a computer is not needed to solve this particular set of equations, the same method can be used to derive sets of equations for much more complex processes—processes that do not easily yield to manual solution techniques.

The next example illustrates the equation-based approach for a system that involves nonlinear relations and recycle.

EXAMPLE 10.3-2 **Simulation of an Equilibrium Reaction/Separation Process**

Ethane is dehydrogenated to ethylene and acetylene in the following pair of catalytic reactions:

$$C_2H_6 \rightleftharpoons C_2H_4 + H_2 \qquad [A \rightleftharpoons B + D]$$
$$C_2H_6 \rightleftharpoons C_2H_2 + 2H_2 \qquad [A \rightleftharpoons C + 2D]$$

The reactions take place at 977°C and 1 atm and proceed to a point such that the product gas composition satisfies the following equilibrium conditions:

$$\frac{y_B y_D}{y_A} = 3.75 \qquad \frac{y_C y_D^2}{y_A} = 0.135$$

where y denotes mole fraction. The product gas goes to a complex separation process that separates 95% of the unreacted ethane from the ethlyene, acetylene, and hydrogen and recycles the separated ethane to the reactor.

Perform a degree-of-freedom analysis on this process, set up the equations for all unknown stream variables, and outline a method of solution.

SOLUTION **Basis: 100 mol Ethane Fed to the Reactor**

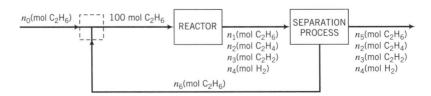

The degree-of-freedom analysis proceeds in the customary manner.

Mixing Point   2 variables (n_0, n_6)

$-$ 1 balance (ethane)

$=$ 1 local degree of freedom

Reactor   (Count balances on molecular species)

4 variables $(n_1, \ldots, n_4)$

$-$ 4 molecular species balances

$-$ 2 equilibrium relations

$+$ 2 chemical reactions

$=$ 0 local degrees of freedom

Separation Process   6 variables $(n_1, \ldots, n_6)$

$-$ 1 balance (Ethane. We implicitly used up the balances on ethylene, acetylene, and hydrogen when we labeled the outlet flow rates of these species n_2, n_3, and n_4.)

$-$ 1 additional relation (95% of the ethane fed to the separator leaves with the product.)

$=$ 4 local degrees of freedom

Process   5 local degrees of freedom $(1 + 0 + 4)$

$-$ 5 ties $(n_1, \ldots, n_4, n_6)$

$=$ 0 net degrees of freedom

The system equations are as follows.

Mixing Point $n_0 + n_6 = 100$ **(1)**

Reactor

We will use the extent of reaction method for the reactor analysis (see Section 4.7e). From the two given stoichiometric equations,

$$n_1 = 100 \text{ mol} - \xi_1 - \xi_2 \tag{2}$$

$$n_2 = \xi_1 \tag{3}$$

$$n_3 = \xi_2 \tag{4}$$

$$n_4 = \xi_1 + 2\xi_2 \tag{5}$$

$$n_{\text{tot}} = n_1 + n_2 + n_3 + n_4 = 100 \text{ mol} + \xi_1 + 2\xi_2 \tag{6}$$

The mole fraction of the ith component in the reactor effluent is n_i/n_{tot}. The mole fractions of all four species can be expressed in terms of ξ_1 and ξ_2 using Equations 2 through 6, and the resulting expressions can be substituted into the given equilibrium relations to yield the following equations:

$$\frac{\xi_1(\xi_1 + 2\xi_2)}{(100 - \xi_1 - \xi_2)(100 \text{ mol} + \xi_1 + 2\xi_2)} = 3.75 \tag{7}$$

$$\frac{\xi_2(\xi_1 + 2\xi_2)^2}{(100 - \xi_1 - \xi_2)(100 \text{ mol} + \xi_1 + 2\xi_2)^2} = 0.135 \tag{8}$$

Separation Process   Ethane split: $n_5 = 0.05 n_1$ **(9)**

Ethane balance: $n_1 = n_5 + n_6$ **(10)**

We now have 10 equations in 10 unknowns $(n_0, \ldots, n_6, n_{\text{tot}}, \xi_1, \text{ and } \xi_2)$. Once the equations are solved, any desired quantities may be determined, such as the component mole fractions in the reactor effluent, the overall and single-pass conversions, yields and selectivities, and (if stream phases and temperatures are specified) required heat duties on the reactor and separation process.

Examining the 10 equations, we see that two are nonlinear (Equations 7 and 8) and the remainder are linear. The full set of 10 simultaneous equations could be entered into an equation-solving program, or the following procedure involving no more than two simultaneous equations could be used:

Solve equations 7 and 8 simultaneously for ξ_1 and ξ_2 (trial and error—see below)

$$\begin{aligned}
n_1 &= 100 - \xi_1 - \xi_2 \quad &\textbf{(2)}\\
n_2 &= \xi_1 \quad &\textbf{(3)}\\
n_3 &= \xi_2 \quad &\textbf{(4)}\\
n_4 &= \xi_1 + 2\xi_2 \quad &\textbf{(5)}\\
n_{tot} &= n_1 + n_2 + n_3 + n_4 \quad &\textbf{(6)}
\end{aligned}$$

$$\begin{aligned}
n_5 &= 0.05 n_1 \quad &\textbf{(9)}\\
n_6 &= n_1 - n_5 \quad &\textbf{(10)}
\end{aligned}$$

$$n_0 = 100 - n_6 \quad \textbf{(1)}$$

The only problem that remains is the simultaneous solution of Equations 7 and 8. We first rewrite each of the two equations in the form $f(\xi_1, \xi_2) = 0$ by multiplying through by the denominators of the left sides and bringing all terms to the left of the equal sign. Equation 7 becomes

$$f_1(\xi_1, \xi_2) = \xi_1(\xi_1 + 2\xi_2) - 3.75(100 - \xi_1 - \xi_2)(100 + \xi_1 + 2\xi_2) = 0$$

and from Equation 8

$$f_2(\xi_1, \xi_2) = \xi_2(\xi_1 + 2\xi_2)^2 - 0.135(100 - \xi_1 - \xi_2)(100 + \xi_1 + 2\xi_2)^2 = 0$$

These two equations may be solved simultaneously using an equation-solving program or a numerical procedure like the Newton–Raphson algorithm (Section A.2h). The solution is $\xi_1 = 83.06$ mol, $\xi_2 = 6.127$ mol. Equations 2–5, 9–10, and 1 then yield $n_1 = 10.81$ mol, $n_2 = 83.06$ mol, $n_3 = 6.127$ mol, $n_4 = 95.32$ mol, $n_5 = 0.5405$ mol, $n_6 = 10.27$ mol, $n_0 = 89.73$ mol.

The final example illustrates the analysis of a multiple-unit process with several internal cycles, using both the sequential modular and equation-based approaches.

EXAMPLE 10.3-3 **Simulation of an Ammonium Nitrate Plant[2]**

Ammonium nitrate, a constituent of many fertilizers, is manufactured in the reaction of ammonia and aqueous nitric acid:

$$NH_3(g) + HNO_3(aq) \rightarrow NH_4NO_3(aq)$$

A labeled flowchart is shown in Figure 10.3-1. The principal features of the process are summarized in the paragraphs that follow.

The nitric acid fed to the process is a 59.5 wt% solution that enters a charge tank at a rate of 10,970 kg/h. Also fed to the tank are two recycle streams coming from different parts of the process. Both of these streams contain ammonium nitrate, and one also contains dissolved ammonia.

---

[2]This example is based on a process description in L. B. Andersen and L. A. Wenzel, *Introduction to Chemical Engineering,* McGraw-Hill, New York, 1961.

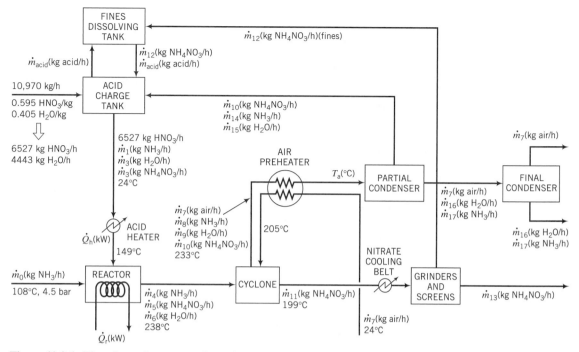

Figure 10.3-1 Flowchart of an ammonium nitrate production process.

A liquid stream from the charge tank is heated from 24°C to 149°C and fed to the reactor, along with a stream of ammonia vapor at 108°C and 4.5 bar. The total ammonia fed to the reactor is 5% in excess of the amount needed to react completely with the nitric acid in the feed. At the reactor operating conditions, the ammonium nitrate is formed as liquid droplets and most of the water in the acid is vaporized. The reaction goes to completion.

The reactor effluent—a gas–liquid mixture—leaves the reactor at a temperature of 238°C and flows into a cyclone separator. The tangential flow of the mixture generates a centrifugal force on the liquid droplets that impels them to the wall. The droplets adhere to and flow down the wall to the base of the unit where they are contacted with hot air, which vaporizes essentially all of the water and a small amount of the ammonium nitrate, leaving the remaining ammonium nitrate in a molten condition. The mass ratio of air to reactor effluent entering the cyclone is 0.045:1. Before entering the cyclone, the air is heated from 24°C to 205°C by exchange of heat with the gases leaving the cyclone.

The molten ammonium nitrate stream leaves the separator at 199°C and is air-cooled on a slowly moving belt to a temperature at which it all solidifies. The solid ammonium nitrate is then ground and screened. All but the smallest particles (the "fines") go through a coating process and then to bagging and shipping. The fines, which account for 16.4% of the nitrate fed to the grinding mill, are recycled to a tank where they are redissolved in acid and recycled to the charge tank.

The gas stream leaving the cyclone contains hot air, the excess ammonia, water evaporated from the nitric acid solution in the reactor and from the collected liquid in the cyclone, and 3% of the ammonium nitrate in the reactor effluent. The stream leaves the separator at 233°C, passes through the air preheater, and enters a partial condenser where some of the water and ammonia and essentially all of the nitrate are condensed. The equilibrium relationship between the compositions of the vapor and liquid streams leaving this unit may be expressed in the form

$$y_{H_2O} = f_1(x_{H_2O}, x_{NH_3})$$

$$y_{NH_3} = f_2(x_{H_2O}, x_{NH_3})$$

where y and x denote mass fractions in the gas and liquid phases, respectively. The gas stream leaving the partial condenser passes to a second condenser, which removes the remaining water and ammonia, leaving essentially pure air to be discharged to the atmosphere.

It is desired to calculate the mass flow rates $\dot{m}_0, \ldots, \dot{m}_{17}$ (see Figure 10.3-1), the temperature of the gas entering the partial condenser (T_a), and the heat duties on the acid heater ($\dot{Q}_h$) and the reactor ($\dot{Q}_r$).

1. Carry out a degree-of-freedom analysis on the process.
2. Set up a sequential modular simulation of the process, using the following blocks:

MIX   Mix two or more streams of known composition adiabatically to form a single outlet stream. Calculate the component flow rates and temperature of the outlet stream.

REACT   Take as input a stream of reactants at a specified temperature and carry the reaction to completion at a specified outlet temperature.

SEPR   Separate a feed stream with specified component flow rates into two outlet streams. The fractions of each component that go into the first outlet stream are specified.

CONDNS Carry out a partial condensation on a multicomponent feed stream.

EXCH   Exchange heat between two streams. The flow rates of both streams are specified, as are the inlet and outlet temperatures of one stream and the inlet temperature of the second.

HEAT   Raise or lower the temperature of a stream by a specified amount.

CONVG   Perform a Wegstein search to determine the value of one or more tear stream variables.

CALC   Perform a simple calculation of a specified stream variable from input stream variable values (a user-written subroutine).

3. Write a set of equations for all unknown variables; that is, set up the problem in a form suitable for solution using an equation-solving algorithm.

SOLUTION

1. *Degree-of-freedom analysis.* We will present the analysis with a minimum of comment. Try to justify to yourself the given number of relations for each process unit.

Reactor (Count balances on molecular species.)

> 8 variables ($\dot{m}_0, \ldots, \dot{m}_6, \dot{Q}_r$)
> − 4 balances (ammonia, water, nitrate, energy) (We do not include a nitric acid balance, since we did not label an outlet flow rate for this species.)
> − 2 additional relations (% excess ammonia, complete conversion)
> + 1 chemical reaction
> _____
> = 3 local degrees of freedom

Cyclone

> 8 variables ($\dot{m}_4, \ldots, \dot{m}_{11}$)
> − 3 balances (ammonia, water, ammonium nitrate)(Why not air?)
> − 2 additional relations (air/reactor effluent mass ratio, ammonium nitrate split)
> _____
> = 3 local degrees of freedom

Grinders

> 3 variables ($\dot{m}_{11}, \ldots, \dot{m}_{13}$)
> − 1 balance (ammonium nitrate)
> − 1 additional relation(% fines)
> _____
> = 1 local degree of freedom

Preheater

> 5 variables ($\dot{m}_7, \ldots, \dot{m}_{10}, T_a$)
> − 1 balance (energy—Why no material balances?)
> _____
> = 4 local degrees of freedom

Partial Condenser

9 variables ($\dot{m}_7, \ldots, \dot{m}_{10}, \dot{m}_{14}, \ldots, \dot{m}_{17}, T_a$)
$-$ 2 balances (ammonia, water)
$-$ 2 additional relations (equilibrium relations for ammonia and water)

$=$ 5 local degrees of freedom

Final Condenser

3 variables ($\dot{m}_7, \dot{m}_{16}, \dot{m}_{17}$)
$-$ 0 balances (Why?)

$=$ 3 local degrees of freedom

Fines Dissolving Tank

1 variable ($\dot{m}_{12}$) (We are ignoring $\dot{m}_{acid}$,
   for reasons to be discussed below)
$-$ 0 balances

$=$ 1 local degree of freedom

Acid Charge Tank

7 variables ($\dot{m}_1, \dot{m}_2, \dot{m}_3, \dot{m}_{10}, \dot{m}_{12}, \dot{m}_{14}, \dot{m}_{15}$)
$-$ 3 balances (ammonia, water, ammonium nitrate)

$=$ 4 local degrees of freedom

Acid Heater

4 variables ($\dot{m}_1, \dot{m}_2, \dot{m}_3, \dot{Q}_h$)
$-$ 1 balance (energy)

$=$ 3 local degrees of freedom

Process

27 local degrees of freedom
$-$ 27 ties (Referring to Figure 10.3-1, count them.)

$=$ 0 net degrees of freedom

The process is therefore well defined.

In our analysis of the fines dissolving tank, we paid no attention to $\dot{m}_{acid}$. The reason is that from the standpoint of the requested process analysis, the value of this flow rate is both indeterminate and immaterial. Whatever the quantity of acid that flows to the fines dissolving tank from the acid charge tank, the same quantity flows back. Since the value of $\dot{m}_{acid}$ has no bearing on any of the other material flows and we have no way of calculating this value from the given information about the process, we may as well ignore it when setting up the simulation.

2. **Sequential modular simulation.** Examination of the flowchart of Figure 10.3-1 shows that there are two cycles in the process. The first includes the reactor, cyclone, preheater, partial condenser, acid charge tank, and acid heater, and the second includes the reactor, cyclone, ammonium nitrate cooler, grinders, fines dissolving tank, acid charge tank, and acid heater. We can tear both cycles at once between the charge tank and the acid heater, between the heater and the reactor, or between the reactor and the cyclone. The stream between the reactor and the cyclone involves three variables ($\dot{m}_4$, $\dot{m}_5$, and $\dot{m}_6$), one of which may immediately be eliminated by virtue of the specified percentage excess of ammonia in the reactor feed. Since no other potential tear stream involves fewer variables, we will choose this location as the tear point.

Figure 10.3-2 shows a block diagram of the simulation. Note the following features of the diagram.

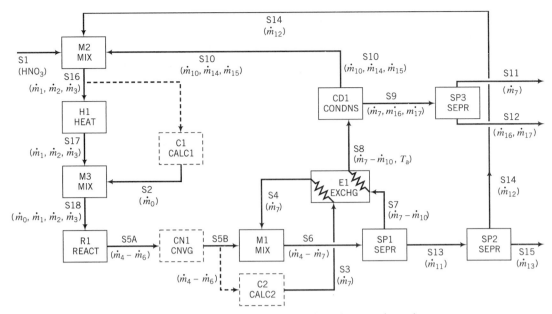

Figure 10.3-2 Block diagram for sequential modular simulation of ammonium nitrate process.

1. Blocks are not included for the fines dissolver and the ammonium nitrate cooler, since both of these units serve simply to pass along a stream of material from one unit to another. If energy balances on these units were required, we would then include blocks for them.

2. Two calculation blocks are included. In the first (C1), the value of the ammonia fresh feed rate ($\dot{m}_0$) will be calculated from the known value of the nitric acid feed rate, the calculated quantity of recycled ammonia ($\dot{m}_1$), and the specified percentage excess ammonia in the reactor feed (5%). In the second block (C2), the flow rate of air ($\dot{m}_7$) will be calculated from the known mass flow rate of the reactor effluent ($\dot{m}_4 + \dot{m}_5 + \dot{m}_6$) and the specified mass ratio of air to reactor effluent (0.045).

3. The SEPR block must be used to simulate the cyclone. However, since this block only accepts a single input stream, the reactor effluent and air feed must first be combined in a MIX operation, even though there is no physical mixing stage in the process.

The simulation would begin with a guess at the tear stream variables (S5B). The subroutine calling sequence for the simulation would then be as follows.

CALL CALC2(C1,S5B,S3)

Calculates $\dot{m}_7$, hence streams S3, S4, and S11

CALL MIX(M1,S5B,S4,S6)
CALL SEPR(SP1,S6,S7,S13)

Takes as input the flow rates of air ($\dot{m}_7$), ammonia ($\dot{m}_4$), ammonium nitrate ($\dot{m}_5$), and water ($\dot{m}_6$) entering the cyclone and the known splits of air, ammonia, water, and ammonium nitrate, and calculates the component flow rates of streams S7 and S13 ($\dot{m}_8, \ldots, \dot{m}_{11}$).

CALL SEPR(SP2,S13,S14,S15)

Calculates $\dot{m}_{12}$ and $\dot{m}_{13}$ from the specified fraction of fines in the grinder product.

CALL EXCH(E1,S3,S4,S7,S8)

Calculates T_a from the known stream flow rates and the three other stream temperatures.

CALL CONDS(CD1,S8,S9,S10)

Calculates $\dot{m}_{14}, \ldots, \dot{m}_{17}$ from the known flow rates of stream S8 and the equilibrium relations for ammonia and water. If the temperatures of S9 and S10 were specified, the subroutine could also determine the required heat duty on the condenser.

At this point in the calculation, we could call SEPR for unit SP3. However, since this calculation has no bearing on the convergence of the iterative cyclic calculations, we may as well hold off on it and do it only after the iterative procedure has converged.

CALL MIX(M2,S1,S10,S14,S16)

Calculates $\dot{m}_1$, $\dot{m}_2$, and $\dot{m}_3$. Notice that in previous flowsheet simulation examples, our MIX block could only handle two inlet streams. If that were the case here, we would need two block calls—one to mix S1 and S14, and the next to mix the product of the first stage with S10.

CALL CALC1(C1,S16,S2)

Calculates $\dot{m}_0$ from the known value of the nitric acid feed rate, the current value of the ammonia recycle rate ($\dot{m}_1$), and the specified value of 5% excess ammonia in the reactor feed.

We will bypass the call of HEAT for the acid heater at this point, for the same reason we did not call SEPR for the final condenser. This calculation is not needed to complete the cycle, and so we need only do it after the cyclic calculation has converged.

CALL MIX(M3,S2,S17,S18)

Calculates the combined feed to the reactor.

CALL REACT(R1,S18,S5A)

Calculates the reactor effluent (S5A) flow rates $\dot{m}_4, \dot{m}_5$, and $\dot{m}_6$ and the required heat duty on the reactor, $\dot{Q}_r$.

CALL CONVG(CN1,S5A,S5B)

Compares the estimated (S5B) and recalculated (S5A) values of the tear stream variables. If they do not agree within a specified tolerance, the program generates a new set of estimated values using the Wegstein algorithm and goes back to the first step [CALL MIX(M1,...)] to begin a new cycle. If convergence is achieved, the program proceeds to the remaining calculations.

CALL SEPR(SP3,S9,S11,S12)

Calculates the flow rates of two of the process product streams. If temperatures of S9, S11, and S12 were specified, an energy balance could be performed on this unit as well.

CALL HEAT(H1,S16,S17)

Calculates the required heat duty on the acid heater, $\dot{Q}_h$.

3. *Equation-based simulation.* The system equations are presented below. Verify that each set of equations is consistent with the degree-of-freedom analysis for that unit.

Reactor

Nitric acid fed (and reacted) = (6527 kg/h)/(63.0 kg/kmol) = 103.6 kmol/h

Excess ammonia: $\dfrac{\dot{m}_0 + \dot{m}_1}{17.0 \text{ kg/kmol}} = 1.05 \times 103.6 \text{ kmol/h}$

$$\Downarrow$$

$$\dot{m}_0 + \dot{m}_1 = 1849 \text{ kg/h} \tag{1}$$

Water balance: $\dot{m}_2 = \dot{m}_6$ \hfill **(2)**

Ammonium nitrate balance: $\dot{m}_5 = 103.6$ kmol/h formed (80.0 kg/kmol)

$$\Downarrow$$

$$\dot{m}_5 = 8288 \text{ kg/h}$$

Ammonia balance: $m_4 = 1849$ kg/h fed $- (103.6$ kmol/h react$)(17.0$ kg/kmol$)$

$$\Downarrow$$

$$\dot{m}_4 = 88 \text{ kg/h}$$

Energy balance: $\quad \dot{Q}_r = \sum_{\text{out}} \dot{m}_i \hat{H}_i - \sum_{\text{in}} \dot{m}_i \hat{H}_i + [103.6 \text{ kmol HNO}_3/\text{h}]\Delta\hat{H}_r^\circ$ $\qquad$ **(3)**

The specified enthalpies of the feed and product components would be calculated in kJ/kg, relative to the components at 25°C (liquid nitric acid, gaseous ammonia, and solid ammonium nitrate), and the standard heat of reaction would be determined in kJ/kmol.

Centrifuge

Air feed: $\quad \dot{m}_7 = 0.045(\dot{m}_4 + \dot{m}_5 + \dot{m}_6)$

$$\Downarrow \dot{m}_4 = 88 \text{ kg/h}, \dot{m}_5 = 8288 \text{ kg/h}$$

$\qquad\qquad \dot{m}_7 = 0.045\dot{m}_6 + 377 \text{ kg/h}$ $\qquad$ **(4)**

Ammonia balance: $\quad \dot{m}_8 = \dot{m}_4 = 88 \text{ kg/h}$

Water balance: $\quad \dot{m}_9 = \dot{m}_6$ $\qquad$ **(5)**

Ammonium nitrate vaporization: $\quad \dot{m}_{10} = 0.03\dot{m}_5 = 249 \text{ kg/h}$

Ammonium nitrate balance: $\quad \dot{m}_{11} = \dot{m}_5 - \dot{m}_{10} = (8288 - 249) \text{ kg/h} = 8039 \text{ kg/h}$

Grinders

Fines: $\quad \dot{m}_{12} = 0.164\dot{m}_{11} = 1318 \text{ kg/h}$

Ammonium nitrate balance: $\quad \dot{m}_{13} = \dot{m}_{11} - \dot{m}_{12} = 6721 \text{ kg/h}$

Air Preheater

Energy balance: $\quad \left[\sum_i \dot{m}_i C_{pi}\right](233°\text{C} - T_a) = \dot{m}_7 C_{p7}(205°\text{C} - 24°\text{C})$ $\qquad$ **(6)**

Partial Condenser

$y_{\text{H}_2\text{O}} = \dot{m}_{16}/(\dot{m}_7 + \dot{m}_{16} + \dot{m}_{17})$ $\qquad$ **(7)**

$y_{\text{NH}_3} = \dot{m}_{17}/(\dot{m}_7 + \dot{m}_{16} + \dot{m}_{17})$ $\qquad$ **(8)**

$x_{\text{H}_2\text{O}} = \dot{m}_{15}/(\dot{m}_{10} + \dot{m}_{14} + \dot{m}_{15})$ $\qquad$ **(9)**

$x_{\text{NH}_3} = \dot{m}_{14}/(\dot{m}_{10} + \dot{m}_{14} + \dot{m}_{15})$ $\qquad$ **(10)**

$y_{\text{H}_2\text{O}} = f_1(x_{\text{H}_2\text{O}}, x_{\text{NH}_3})$ $\qquad$ **(11)**

$y_{\text{NH}_3} = f_2(x_{\text{H}_2\text{O}}, x_{\text{NH}_3})$ $\qquad$ **(12)**

NH_3 *balance:* $\quad \dot{m}_8 = \dot{m}_{14} + \dot{m}_{17}$ $\qquad$ **(13)**

H_2O *balance:* $\quad \dot{m}_9 = \dot{m}_{15} + \dot{m}_{16}$ $\qquad$ **(14)**

Acid Charge Tank

NH_3 *balance:* $\quad \dot{m}_{14} = \dot{m}_1$ $\qquad$ **(15)**

H_2O *balance:* $\quad 4443 \text{ kg/h} + \dot{m}_{15} = \dot{m}_2$ $\qquad$ **(16)**

Ammonium nitrate balance: $\quad \dot{m}_{10} + \dot{m}_{12} = \dot{m}_3$

$$\Downarrow \begin{array}{l} \dot{m}_{10} = 249 \text{ kg/h} \\ \dot{m}_{12} = 1318 \text{ kg/h} \end{array}$$

$\qquad\qquad \dot{m}_3 = 1567 \text{ kg/h}$

Acid heater

Energy balance: $\quad \dot{Q}_h = \left[\sum_{\text{all species}} \dot{m}_i C_{pi}\right](149°\text{C} - 24°\text{C})$ $\qquad$ **(17)**

At this point we have 17 equations in 17 unknowns: $\dot{m}_0, \dot{m}_1, \dot{m}_2, \dot{m}_6, \dot{m}_7, \dot{m}_9, \dot{m}_{14}$ to $\dot{m}_{17}, T_a, \dot{Q}_r, \dot{Q}_h$, and the four mass fractions defined by Equations 7 through 10. The system of equations can be solved by an equation-based algorithm.

10.4   COMMERCIAL PROCESS SIMULATION PACKAGES

A number of programs have been written to simulate steady-state processes using a sequential modular approach (some with imbedded simultaneous equation solution capability). At the time this text was written, the most frequently used general purpose programs and the companies that distributed them were ASPEN PLUS® (Aspen Tech), CHEMCAD® (Chemstations), HYSYS® (Hyprotech), DESIGN II® (WinSim), and PROVISION® (Simulation Sciences). Other programs have been written specifically for particular process industries, such as the petroleum processing and pulp and paper manufacturing industries.

Besides having provisions for simulating most of the principal types of process equipment, the more extensive of these programs have routines that generate physical properties of hundreds of chemical species, either from built-in data tables or by using estimation formulas. Included in the stored data are boiling and melting points, liquid densities, critical constants, heat capacities, latent heats, solubilities, and vapor pressures.

The simulation program is written so that a block subroutine that requires a physical property automatically retrieves the property from the data bank. In addition, it is possible for the user to have direct access to the data, and it is not uncommon for a simulation program to be used just to generate physical properties of materials without actually carrying out a simulation.

The design of a process unit to achieve a desired output from a specified input normally includes a calculation of the unit size, which is the principal determinant of the installation and operating costs of the unit. A number of general estimation formulas have been proposed to predict the cost of a specific type of unit from a given size-dependent parameter (e.g., volume, heat transfer area, or power rating of a pump). These formulas have been built into some simulation programs. The normal procedure for using them is to solve all system material and energy balances in the manner we have described and then to call the subroutines that determine the sizes and costs of the units.

Equation-based flowsheet simulators have not reached the level of commercialization of their sequential modular counterparts as of the time of this writing, and the development of efficient equation-solving algorithms is the subject of much continuing research.

10.5   FINAL CONSIDERATIONS

This chapter has introduced computer process simulation—using computers to carry out material and energy balance calculations on processes at steady state. When thinking about the chapter material, remember the following points:

- Whether you do process analysis manually or with a spreadsheet or simulation program, you can only determine all unknown process variables associated with a process if the process has zero degrees of freedom. It is therefore always a good idea to perform a degree-of-freedom analysis before attempting to solve the system equations.
- If a process has positive degrees of freedom, you must select as many design variables as there are degrees of freedom, and then solve the system state equations for the remaining (state) variables. Which variables you select as design variables can have a dramatic effect on the difficulty of the subsequent calculations.
- For single-unit processes, it is usually desirable to pick design variables that minimize the number of state equations that must be solved simultaneously. For multiple-unit processes, it is desirable to pick design variables to minimize the number of cycles on the flowchart. When doing calculations for cycles on flowcharts, it is desirable to tear as many cycles as possible with the minimum number of tear stream variables. These rules of thumb may not always lead to the most efficient calculations and they may contradict one another, however, so that finding the best combination of design and tear stream variables may require considerable trial-and-error for complex processes.
- When you first construct a simulation, don't believe any of the initial results it gives you until you have checked as many of them as possible with manual calculations. Once the first set

of results has been verified, it is probably safe to believe results for additional cases. (There are never any absolute guarantees, however.)

You are advised at this point to go back over the instructional objectives at the beginning of the chapter and review the chapter material on any objectives you feel you have not yet mastered.

PROBLEMS 10.1. Draw and label a flowchart and determine the number of degrees of freedom for each of the given systems. Give a feasible set of design variables and, if possible, an infeasible set. The solution to part (a) is given as an example.

(a) An aqueous solution of sulfuric acid at temperature T_0 is diluted with pure water at the same temperature in an adiabatic mixer. Calculate the final mixture temperature.

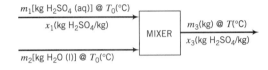

Solution: 7 variables ($m_1, x_1, m_2, m_3, x_3, T_0, T$)

$-$ 3 relations (2 material balances and 1 energy balance)

$=$ 4 degrees of freedom

One feasible set of design variables (there are others) is

$$\{m_1,\ x_1,\ m_2,\ T_0\}$$

If you are given values of these variables you can calculate m_3 (total mass balance), x_3 (H_2SO_4 balance), and T (energy balance). An infeasible set is

$$\{m_1,\ m_2,\ m_3,\ T_0\}$$

Once m_1 and m_2 have been specified, m_3 is fixed by a total material balance and may not be independently assigned a value.

(b) A natural gas containing methane, ethane, and propane at temperature T_1 is mixed with pre-heated air at temperature T_2, and the mixture is heated to 200°C. Calculate the required heat input.

(c) A stream containing hexane vapor in nitrogen at temperature T_1 is cooled at constant pressure, condensing 95% of the hexane. Calculate the product temperature.

10.2. Benzene (B) and chlorobenzene (C) are being separated in a distillation column. Vapor and liquid streams, each containing both species, are fed to one of the trays of the column, and liquid and vapor streams are taken off the tray. The tray functions as an **ideal stage** (see Problem 6.63): the effluent streams are in equilibrium at temperature T and pressure P, with compositions related by Raoult's law, Equation 6.4-1.

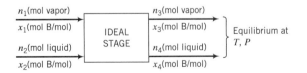

Calculate the number of degrees of freedom. Then specify sets of design variables for which the solution for the remaining state variables would be (a) straightforward, (b) iterative, and (c) impossible.

10.3. Barium sulfate is reduced with carbon as part of the manufacture of lithopone, a white pigment:

$$2\,BaSO_4(s) + 4\,C(s) \rightarrow 2\,BaS(s) + 4\,CO_2(g)$$

A mixture of crushed barite (an ore containing x_B kg $BaSO_4$/kg barite) and coal (containing x_C kg C/kg coal) at temperature $T_0(K)$ is fed to a furnace. Essentially all of the barium sulfate in the feed is converted. As part of the furnace design, you must specify for a basis of 100 kg barite ore the masses of coal fed and of carbon, barium sulfide, carbon dioxide, and other solids emerging from the reactor, as well as the product temperature and required heat input to the furnace.

(a) Draw and label a flowchart, and calculate the degrees of freedom of the process.

(b) Outline the solution procedure (i.e., list in order the equations you would write and the variables you would determine) if you are given values of x_B, x_C, T_0, the percent excess coal fed, and the product temperature. If a solution involves iteration or is impossible, say so.

(c) Repeat part (b), assuming that the specified variables are x_B, x_C, T_0, the mass of BaS produced, and the heat input to the furnace.

(d) Repeat part (b), assuming that the specified variables are x_B, x_C, T_0, the percent excess coal fed, and the heat input to the furnace.

10.4. Ethyl alcohol can be bacterially oxidized to acetic acid in the following two-step fermentation sequence:

$$2\,C_2H_5OH + O_2 \rightarrow 2\,CH_3CHO + 2\,H_2O$$

$$2\,CH_3CHO + O_2 \rightarrow 2\,CH_3COOH$$

If the alcohol-containing feedstock is wine, cider, or a malt solution, the resulting solution is *vinegar*.

An aqueous solution containing ethyl alcohol in water is fermented to produce dilute acetic acid. The feed mixture (the ethanol solution and the bacteria that make the fermentation occur) and air are fed at a temperature T_0. The product solution contains ethanol, acetaldehyde (CH_3CHO), acetic acid, and water. All liquid and gaseous effluents are at temperature T. The variables involved in the process are n_f (mol feed solution), x_{ef} (mol ethanol/mol feed solution), n_{air} (mol air fed), P_{xs} (percent excess air), n_e, n_{ah}, n_{aa}, n_w (gram-moles of ethanol, acetaldehyde, acetic acid, and water, respectively, in the product mixture), n_{ox}, n_n (gram-moles of oxygen and nitrogen, respectively, emerging from the reactor), T_0, T, and Q (kJ heat transferred).

(a) Calculate the number of degrees of freedom of the process. How would the answer differ if the reactor were adiabatic?

(b) Outline the solution procedure to determine all unknown variables from given values of n_f, x_{ef}, P_{xs}, n_e, n_{ah}, T_0, and T.

(c) Specify a set of design variables that would require an iterative solution for the remaining state variables.

(d) Specify an infeasible set of design variables.

10.5. Propylene is converted to butyraldehyde and n-butanol in the following reaction sequence:

$$C_3H_6 + CO + H_2 \rightarrow C_3H_7CHO \quad \text{(butyraldehyde)}$$

$$C_3H_7CHO + H_2 \rightarrow C_4H_9OH \quad \text{(n-butanol)}$$

Liquid propylene, gaseous carbon monoxide and hydrogen, and a soluble cobalt catalyst are fed to a high-pressure catalytic reactor. The reactor effluent goes to a flash tank, where all of the solution constituents are vaporized except the catalyst, which is recycled to the reactor. The reaction products are separated from unconsumed reactants in a multiple-unit process, and the product stream, which contains both butyraldehyde and n-butanol, is subjected to additional hydrogenation with excess hydrogen, converting all of the butyraldehyde to butanol.

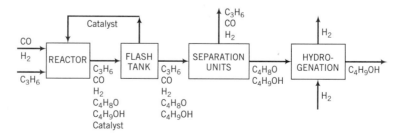

(A more detailed flowchart is shown on p. 769 of G. T. Austin, *Shreve's Chemical Process Industries*, 5th Edition, McGraw-Hill, New York, 1984.)

(a) Redraw and label the flowchart, including in the labeling the molar flow rates of all stream components, the temperatures of each stream, and the heat duties ($\dot{Q}$) for each unit. If all of a species entering a process unit leaves in a single product stream, use the same variable label at the inlet and outlet. Calculate the number of degrees of freedom for each subprocess (the reactor, flash tank, separation process, and hydrogenator) and then the number for the entire process. (See Example 10.3-1.)

(b) Why must the catalyst circulation rate be a design variable?

10.6. Normal butane is catalytically isomerized to isobutane. A fresh feed stream containing pure n-butane at temperature T_1(°C) is mixed adiabatically with a recycle stream containing n-butane and isobutane, and the combined stream is fed to the reactor, where some but not all of the n-butane in the feed is converted. The reactor effluent is pumped to a distillation column. The overhead distillation product contains primarily isobutane and a small amount of n-butane. The bottoms product, which also contains both species, is the stream recycled to the reactor. The streams leaving the distillation column are at different temperatures.

(a) Determine the number of degrees of freedom associated with the feed mixer, the reactor, the distillation column, and the entire process. Include all unknown process stream temperatures and process unit heat duties in your analysis.

(b) Suppose a single-pass conversion of 35.0% is attained in the reactor, the overhead distillation column product contains 88.5 mole% isobutane, and the recycle stream contains 85.0 mole% n-butane. Calculate the overall conversion, the recycle ratio (mol recycle/mol fresh feed), and the gram-moles of overhead product produced per 100 mol fresh feed.

(c) Solve the problem of part (b) again, only this time do it as it might be done by a sequential modular simulation of the process. Take a basis of calculation of 100 mol fresh feed and choose n_r (mol recycle) as the tear stream variable. Begin by assuming $n_r = 100$ mol and go through three successive substitution iterations (Appendix A.2). What is the percentage error in the third calculated value of n_r?

(d) Repeat part (c), only perform a Wegstein acceleration step (Appendix A.2) after the first two successive substitution iterations. What is the percentage error in the last calculated value of n_r?

(e) Set up a spreadsheet to do the calculations of parts (c) and (d), using Example 10.2-3 as a model. Proceed to convergence.

10.7. Simulate a module called SPLIT, which takes a feed stream (SF) and splits it into two streams (S1 and S2), each having the composition and temperature of the feed stream. A fraction X1 of the feed stream becomes stream S1, and a fraction (1−X1) becomes S2. Up to four components may be present in the feed stream, so that each stream has associated with it five attributes: four component molar flow rates and the stream temperature. Test your simulation for a feed stream flowing at a rate of 150 mol/h at $T = 315$ K containing 35 mole% n-octane (A), 57% iso-octane (B), and the balance a chemically inert species (C), which is to be split into two streams in a 60:40 ratio.

(a) Write a spreadsheet to perform the simulation. Draw a flowchart on the spreadsheet and label each stream with the component flow rates and temperature (see Example 10.2-1). Include the split ratio (X1) in a separate cell and vary it to show its effect on the component flow rates in streams S1 and S2.

(b) Write a computer subprogram to simulate the module (see Example 10.2-1). The arguments of the subprogram should be SF, S1, and S2 (each five-membered arrays), X1, and N, the number of components in the feed stream (which may be any number from one to four). X1, N, and the elements of SF are input variables, and the elements of S1 and S2 are output variables. Write and run a calling program that assigns the test case values to the input variables, calls the module subprogram, and prints out the attributes of S1 and S2.

10.8. You have been assigned to simulate a flash evaporator that separates a liquid feed stream containing benzene and toluene at temperature T_F(°C) into liquid and vapor product streams in equilibrium at temperature T(°C) and pressure P(mm Hg). The compositions of the product streams are related by Raoult's law (Equation 6.4-1), and the component vapor pressures are expressed by the Antoine equation (Table B.4).

A spreadsheet that performs the required material and energy balances and vapor–liquid equilibrium calculations on this process unit is shown on the next page. In the test case, a 40 mole% benzene–60 mole% toluene mixture is fed to the evaporator at $T_F = 120$°C and a pressure high enough to assure that the feed stream remains in the liquid state. The unit operates at $T = 100$°C and $P = 800$ mm Hg.

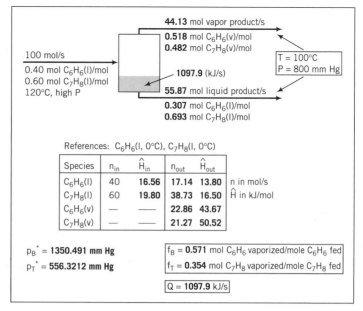

The heat capacities of liquid benzene and toluene have been taken to be 0.138 kJ/(mol·°C) and 0.165 kJ/(mol·°C), respectively, and the vapor heat capacities and heats of vaporization of both species are those given in Appendix B.

(a) Derive expressions for the quantities in boldface on the spreadsheet—that is, the flow rates and compositions of the liquid and vapor product streams, the vapor pressures of benzene and toluene at the evaporator temperature, the fractional vaporizations of benzene and toluene, the specific enthalpies of benzene and toluene liquid at T_F and T and of benzene and toluene vapor at T (all relative to the liquid species at 0°C), and the required rate of heat transfer to the evaporator.

(b) Create a spreadsheet that replicates the one shown above, entering the formulas derived in part (a) and, if possible, imbedding the graphics for the streams and the process unit. In the formulas, enter the cell addresses for variables appearing elsewhere on the spreadsheet: for example, if a formula involves the mole fraction of benzene in the feed, do not enter "0.40" but the cell address of this variable on the spreadsheet. In this way, variables like the benzene mole fraction in the feed and the temperature and pressure of the evaporator may be changed and the new values of the calculated variables will be instantly determined. When the spreadsheet is complete, use it to determine the bubble-point and dew-point temperatures of a 40 mole% benzene–60 mole% toluene mixture at $P = 800$ mm Hg. Print out the spreadsheets for $T = 100°C$ and for the bubble-point and dew-point temperatures.

(c) Write the code for a subprogram called FLASH to simulate the evaporator. The input variables should be the attributes of the feed stream (benzene and toluene flow rates and T_F), T, P, and the physical properties of the feed stream species (Antoine constants, heat capacity formula coefficients for liquid and vapor, and heats of vaporization). The output variables are the attributes of the vapor and liquid product streams and Q. Write and run a calling program that defines the values of the feed stream variables and other input variables (use the test case values), calls the subprogram, and prints out the output variables. The physical properties may be passed to the subprogram either as arguments or through a COMMON or GLOBAL block.

10.9. You are to simulate a flash evaporator that converts a liquid feed stream (SF) containing N species at a high pressure to liquid and vapor product streams (SL, SV) in equilibrium at temperature $T(°C)$ and pressure P(mm Hg). The compositions of the liquid and vapor product streams are related by Raoult's law (Equation 6.4-1), and the component vapor pressures are expressed by the Antoine equation, Table B.4.

(a) Write the system mass and energy balances and equilibrium relations in terms of the following variables:

NF,NL,NV   The molar flow rates (mol/s) of feed, liquid product, and vapor product.

XF(I)   The mole fractions of the *i*th component in the feed,
XL(I)   liquid product, and vapor product, where I runs from
XV(I)   1 to $(N − 1)$.

TF,T,P   Feed temperature, vaporizer temperature, and pressure, respectively.

PV(I)   Vapor pressures of the N species at temperature T, where I runs from 1 to N.

A(I)
B(I)   Antoine equation constants for the N species involved in the process (transmitted to the subroutine via a COMMON or GLOBAL
C(I)   statement).

CP(I)   Liquid-phase heat capacities [kJ/(mol·°C)] of the N species (transmitted via COMMON or GLOBAL). Assume independent of temperature.

HV(I)   Heats of vaporization (kJ/mol) of the N species (transmitted via COMMON or GLOBAL). Assume independent of temperature.

Q   The required heat input (kW) to the vaporizer.

Show that the system has $(N + 3)$ degrees of freedom, counting as the system variables three stream flow rates, $3(N − 1)$ mole fractions, N vapor pressures, TF, T, P, and Q. Then work out a trial-and-error procedure for determining the product stream flow rates and mole fractions and Q from specified values of TF, T, P, and the feed stream molar flow rate and component mole fractions.

(b) Write a module subroutine FLASHN to implement the procedure outlined in part (a) for a feed stream containing up to seven components. The arguments of the subroutine should be N, SF, SV, SL, P, and Q, where SF, SV, and SL are eight-membered arrays. The first N elements of each array are the component molar flow rates and the $(N + 1)$st element is the stream temperature. The input variables are SL(N + 1) and SV (N + 1) (both of which equal the vaporizer temperature), P, and the attributes of SF (N molar flow rates and the feed temperature), and the outlet variables are Q and the remaining attributes of SV and SL.

(c) Test your code for a feed stream containing 34.8 mole% *n*-pentane, 30.0 mole% *n*-hexane, and the balance *n*-heptane flowing at a rate of 1.00 mol/s which is to be flashed from 363 K and a high pressure to 338 K and 611 mm Hg. The heat capacities of liquid pentane, hexane, and heptane [in kJ/(mol·K)] may be taken to be 0.188, 0.216, and 0.213, respectively, and the heats of vaporization may be taken to have their values at the normal boiling points of these substances. Write and run a calling program that defines the attributes of SF and other input parameters (including the heat capacities and heats of vaporization), calls the module subroutine, and prints out the attributes of the product streams and the required heat input.

(d) Use a process simulator to perform the same calculations.

(e) Use an equation-solving program to perform the same calculations.

10.10. You are to simulate a partial condenser module that converts a vapor feed stream (SF) containing a single condensable species at temperature TF(K) to liquid and vapor product streams (SL, SV) in equilibrium at a temperature T(K). The process takes place at a constant pressure P(atm). The compositions of the liquid and vapor product streams are related by Raoult's law (Equation 6.4-1), and the component vapor pressures are correlated with temperature by the Antoine equation, Table B.4.

The system variables are as follows:

NF,NL,NV   mol/s of feed, liquid product, and vapor product.

XF   Mole fractions of the condensable substance in the feed and the vapor
XV   product.

TF,T,P   Feed temperature (K), condenser temperature (K), and condenser pressure (atm), respectively.

PV   Vapor pressure (mm Hg) of the condensable substance at temperature T.

A,B,C   Antoine equation constants for the condensable substance.

CPL   Liquid-phase heat capacity [kJ/(mol·K)] of the condensable substance. Assume independent of temperature.

HV   Heat of vaporization (kJ/mol) of the condensable substance. Assume independent of temperature.

CPV ⌠ Heat capacities [kJ/(mol·K)] of the condensable vapor and the noncondensable

CPG ⌡ gas. Assume independent of temperature.

Q   Required heat input (kW) to the condenser. (Q will be negative.)

(a) Show that the system has five degrees of freedom, counting as the system variables three stream flow rates, two mole fractions, one vapor pressure, TF, T, P, and Q.

(b) The input variables to the module are to be the molar flow rates of the condensable and non-condensable feed stream constituents [SF(1) and SF(2)], the feed temperature [SF(3) = TF], the operating pressure of the condenser (P), and the fraction YC of the condensable species in the feed that is to be condensed. (The heat capacities and heat of vaporization must also be supplied.) The output variables are the molar flow rate of the liquid product stream [SL(1)], the molar flow rates of the vapor product stream constituents [SV(1) and SV(2)], the operating temperature of the condenser [SL(2) = SV(3) = T], and the required heat transfer rate (Q). Outline the required calculations.

(c) Write a spreadsheet to perform the calculations of part (b) for a feed stream flowing at 1.00 mol/s containing methyl alcohol in air at 60°C and 1 atm with a relative saturation of 85%, from which 90% of the methanol is to be removed by partial condensation at constant pressure. The heat capacities in kJ/(mol·K) of liquid methanol, methanol vapor, and air may be taken to be 0.078, 0.050, and 0.030, respectively. After you have generated the solution and recorded the output variable values, use the spreadsheet to generate a plot of condenser temperature (T) versus fractional methanol removal (YC), and briefly explain why the plot looks the way it does.

(d) Use an equation-solving program to perform the calculations outlined in part (c).

(e) Write a computer subprogram CNDNS to implement the procedure of part (b) for the test case of part (c). The subroutine arguments should be SF, SV, SL, P, YC, and Q. Input variables are YC, P, and the attributes of SF, and output variables are Q and the attributes of SV and SL. The values of physical property parameters, A, B, C, CPL, CPV, CPG, and HV should be transmitted to the subprogram either as additional arguments or through a COMMON or GLOBAL statement. Then write and run a calling program that defines the attributes of SF and other input and physical property parameters, calls the subprogram, and prints out the input and output variable values for the test case of part (c).

10.11. You are to simulate a single-phase reactor in which a single reaction takes place. The reaction has the general form.

$$\nu_1 A_1 + \nu_2 A_2 + \nu_3 A_3 + \cdots + \nu_m A_m = 0$$

In this equation A_i is the ith reactant or product and ν_i is the stoichiometric coefficient of this species (negative for reactants and positive for products). It is also convenient to define ν_i for each inert species in the feed to the reactor, assigning it a value of 0. The inputs to the module are the feed stream flow rate, composition, and temperature, the fractional conversion of one of the reactants, and the product stream temperature. The module is to calculate the product stream component flow rates and the required heat transfer to the reactor.

The module equations should be written in terms of the following variables:

N	The total number of reactive and inert species present in the feed and product streams.
SF(I),I = 1,...,N+1	The molar flow rates of the feed stream components (mol/s) and the feed stream temperature (K).
SP(I),I = 1,...,N+1	The molar flow rates of the product stream components (mol/s) and the product stream temperature (K).
NU(I),I = 1,...,N	The stoichiometric coefficients of all species (− for reactants, + for products, 0 for inerts).
X	The fractional conversion of one of the reactants.
IX	The number of the reactant for which the fractional conversion is specified.
CP(I),I = 1,...,N	The heat capacities of all species, kJ/(mol·K) (assume constant).

HF(I),I = 1,...,N   The standard heats of formation of all species, kJ/mol.

Q   The heat input to the reactor, kW. (If Q is negative, heat must be transferred from the reactor.)

(a) Write the equations you would use to calculate Q and the first N components of SP from specified values of all of the other variables defined above.

(b) Write a spreadsheet to perform the calculations of part (a) for a reactor in which propane flowing at a rate of 270 standard cubic meters per hour is burned with 20% excess air. The combined stream enters the reactor at 423 K and the stack gas leaves at 1050 K. Ninety percent of the propane fed is converted, and no CO is contained in the stack gas. Take the heat capacity of each species to be its value at 700 K as calculated from Table B.2 [so that, for example, CP(1) = 0.1431 kJ/(mol·K), where (1) refers to propane]. After you have performed the calculations and recorded the output variable values, use the spreadsheet to generate a plot of Q versus stack gas temperature and briefly explain why the plot looks the way it does.

(c) Use an equation-solving program to perform the calculations outlined in part (b).

(d) Write a computer subprogram REACTS to implement the procedure of part (a). The subprogram arguments should be SF, SP, NU, N, X, IX, and Q. The arrays CP and HF should either be transmitted as additional arguments or via a COMMON or GLOBAL statement. Write and run a calling program that defines the input variables, calls the subprogram, and prints out the required output variables for the test case of part (b). Number the species involved in the process as follows: 1—C_3H_8, 2—O_2, 3—N_2, 4—CO_2, 5—H_2O. For example, NU(1) = −1, NU(2) = −5, SF(1) = 3.348, SF(3) = 75.54, SF(6) = 423, and SP(6) = 1050. (Verify these values as part of your problem solution.)

10.12. You are to simulate a single-phase adiabatic reactor in which a single gas-phase reaction takes place. The reaction has the general form

$$\nu_1 A_1 + \nu_2 A_2 + \nu_3 A_3 + \cdots + \nu_m A_m = 0$$

In this equation A_i is the ith reactant or product and ν_i is the stoichiometric coefficient of this species; ν_i is negative for reactants and positive for products. It is also convenient to define ν_i for each inert species in the feed to the reactor, assigning it a value of 0. The inputs to the program are the stoichiometric coefficients, feed stream flow rate, composition, and temperature, and the fractional conversion of one of the reactants. The program is to calculate the product stream component flow rates and temperature.

The program equations should be written in terms of the following variables:

N   The total number of reactive and inert species present in the feed and product streams.

SF(I),1 = I,...,N+1   The molar flow rates of the feed stream components (mol/s) and the feed stream temperature (K).

SP(I),1 = I,...,N+1   The molar flow rates of the product stream components (mol/s) and the product stream temperature (K).

NU(I),1 = I,...,N   The stoichiometric coefficients of all species (− for reactants, + for products, 0 for inerts).

X   The fractional conversion of one of the reactants.

IX   The number of the reactant for which the fractional conversion is specified.

ACP(I),1 = I,...,N
BCP(I),1 = I,...,N   Coefficients of the heat capacity formula $C_p[kJ/(mol\cdot°C)] = a + bT + cT^2 + dT^3$.
CCP(I),1 = I,...,N
DCP(I),1 = I,...,N

DHF(I),1 = I,...,N   The standard heats of formation of all species in the gaseous state, kJ/mol.

(a) Write the equations you would use to calculate the attributes of SP from specified values of all the other listed variables. The last equation you derive should be a fourth-order equation for the reactor temperature,

$$(\Delta H =) \alpha T^4 + \beta T^3 + \gamma T^2 + \delta T + \varepsilon = 0$$

where $\alpha, \beta, \gamma, \delta$, and ε involve most of the system variables.

(b) Write a spreadsheet to perform the calculations of part (a) for a reactor in which carbon monoxide is oxidized with 25% excess air at 1 atm to form carbon dioxide. The combined feed stream enters the reactor at 650°C at a rate of 23.0 kmol/h, and a CO conversion of 45% is obtained. Use the goalseek tool to solve the fourth-order energy balance equation. After you have performed the calculations and recorded the output variable values, use the spreadsheet to generate a plot of product gas temperature versus percentage CO conversion and briefly explain why the plot looks the way it does.

(c) Use an equation-solving program to perform the calculations outlined in part (b).

(d) Write a computer subprogram REACTAD to implement the procedure of part (a). The subprogram arguments should be SF, SP, NU, N, X, and IX. The arrays ACP, BCP, CCP, DCP, and DHF should either be transmitted as additional arguments or via COMMON or GLOBAL. All arguments but the attributes of SP should be considered input variables. Use Newton's rule (Appendix A.2) to solve the fourth-order energy balance equation. Write and run a calling program that defines the input variables, calls the subprogram, and prints out the required output variables for the test case of part (b). Number the species involved in the process as follows: 1—CO, 2—O_2, 3—N_2, 4—CO_2. For example, NU(1) = −1, NU(2) = −1/2, SF(1) = 1.607, and SF(3) = 3.777. (Verify these values as part of your problem solution.)

10.13. The following two reactions occur in an ethylene oxide production process:

$$2C_2H_4 + O_2 \rightarrow 2C_2H_4O$$

$$C_2H_4 + 3O_2 \rightarrow 2CO_2 + 2H_2O$$

A stream containing equimolar amounts of ethylene and oxygen is joined by a recycle stream containing pure ethylene, and the combined stream is fed to the reactor. The single-pass conversion of ethylene in the reactor and the single-pass ethylene oxide yield based on ethylene consumption are defined as X_{sp} (mol C_2H_4 react/mol C_2H_4 fed) and Y_{sp} (mol C_2H_4O formed/mol C_2H_4 react). The reactor effluent goes through a multiple-unit separation process that has three outlet streams. The first stream, which is sold, is pure ethylene oxide; the second, which is discarded, contains all the carbon dioxide, oxygen, water, and 5% of the unreacted ethylene leaving the reactor; and the third stream, which is the recycle stream, contains the remaining unreacted ethylene.

(a) Taking 100 mol/s fresh feed as a basis of calculation and letting X_{sp} = 0.20 and Y_{sp} = 0.90, construct a spreadsheet to simulate this process, treating the separation process as a single unit. Use the recycle stream flow rate as a tear stream variable: that is, assume a recycle flow rate entering the recycle–fresh feed mixing point, solve the system balance equations to recalculate the recycle flow rate at the outlet of the separation process and find the assumed value for which the assumed and recalculated values are the same. Determine the overall conversion and yield,

$$X_0 \text{ (mol } C_2H_4 \text{ consumed in process/mol } C_2H_4 \text{ in fresh feed)}$$

$$Y_0 \text{ (mol } C_2H_4O \text{ produced in process/mol } C_2H_4 \text{ in fresh feed)}$$

and the molar flow rate of the recycle stream.

(b) For the given fresh feed and single-pass conversion of ethylene, use the spreadsheet to find the range of single-pass yields that are physically possible. (A value is not possible if it leads to a negative species flow rate at any point in the process.) Note the values of Y_0 and the recycle flow rate at the extremes of this range. Then repeat the calculation for a 30% single-pass conversion of ethylene. Show your results in a table with the following form:

X_{sp}	Y_{sp}	Y_0	n_{rec}
0.2	(Y_{min})		
0.2	(Y_{max})		
0.3	(Y_{min})		
0.3	(Y_{max})		

Explain why a species flow rate becomes negative when Y falls below a certain value.

(c) Summarize the equations that describe the behavior of this system and use an equation-solving program to solve them for the variable values of part (a).

10.14. You are to write the code for a convergence module that can deal with one to three tear stream variables using the Wegstein algorithm, as outlined in Appendix A.2. The object is to determine the values of one, two, or three of the variables x_1, x_2, and x_3 that satisfy the relations

$$x_1 = f_1(x_1, x_2, x_3)$$
$$x_2 = f_2(x_1, x_2, x_3)$$
$$x_3 = f_3(x_1, x_2, x_3)$$

where f_1, f_2, and f_3 may be simple algebraic functions, or they may represent the composite set of calculations involved in proceeding around a cycle of unit operations to recalculate tear stream variables. The calculation consists of assuming values for the tear stream variables $[x_{ia}, i = 1, \ldots, 3]$; computing $[x_{ic} = f_i(x_{1a}, x_{2a}, x_{3a}), i = 1, \ldots, 3]$; determining whether x_{ia} and x_{ic} are within a specified tolerance of each other for each i; and if they are not, using the Wegstein algorithm to estimate the values of x_{1a}, x_{2a}, and x_{3a} to be used in the next iteration.

The module code should be written in terms of the following variables:

N	Number of tear stream variables (1, 2, or 3).
XA(I), I = 1, ..., N	Assumed values of x_i (initially guessed, subsequently calculated by the Wegstein algorithm).
XC(I), I = 1, ..., N	Recalculated values of x_i obtained from the functions f_i (or by proceeding around a cycle).
KMAX	Maximum number of iterations to be allowed.
EPS	Relative convergence tolerance (see Equation A.2-9). The difference between XA(I) and XC(I) divided by the value of XC(I) must be less than EPS for convergence to be declared.
IPR	A printout parameter. The code should be written such that if IPR = 1 the values of x_i are printed after each iteration, while if IPR = 0 the printing is suppressed.

(a) Write the code for a module subprogram called CONVG that has as its arguments XA, XC, N, KMAX, EPS, and IPR. The values of N, KMAX, EPS, IPR, and the elements of XA are input variables, and the final estimated values of x_i are to be returned in array XC. The code should execute the procedure outline in Section A.2g separately for each of the N variables, up to a maximum of KMAX iterations. If this limit is reached and convergence has not been attained, an error message should be printed and the program terminated.

In a real flowchart simulator, the code would be written so that the module could be inserted as a block in the flowchart (see, for example, the flowchart block diagram on p. xxx), and the tear stream variables XC(I) would then be recalculated from assumed values of XA(I) by proceeding around the cycle. To simplify the coding in this problem, however, have the subprogram call a function-generating subroutine of the form

CALL FUNCGEN(N,XA,XC)

You will need to write Subroutine FUNCGEN to evaluate

$$XC(1) = f_1[XA(1), XA(2), XA(3)]$$

and similarly for XC(2) and XC(3).

(b) Apply your code to the problem of Example A.2-2. Build in an upper limit of 20 iterations (KMAX = 20).

10.15. Use a commercial flowchart simulation program such as HYSYS or ASPEN to simulate the ammonium nitrate manufacturing process described in Example 10.3-3.

Chapter 11

Balances on Transient Processes

A system is said to be in a *transient* (or *unsteady-state*) condition if the value of any system variable changes with time. Batch and semibatch process systems are always transient: in a batch system, if nothing is changing with time then nothing is happening, and in a semibatch system (which has an input stream but no output stream or vice versa) at least the mass of the system contents must vary with time. Continuous systems are always transient when they are started up and shut down, and they become transient at other times due to planned or unexpected changes in operating conditions.[1]

The procedures for deriving balances on transient systems are essentially those developed in Chapters 4 (material balances) and 7 (energy balances). The main difference is that transient balances have nonzero accumulation terms that are derivatives, so that instead of algebraic equations the balances are differential equations.

11.0 INSTRUCTIONAL OBJECTIVES

After completing this chapter, you should be able to do the following:

- Derive material balance equations and provide initial conditions for well-mixed transient single-unit processes, and derive energy balance equations and provide initial conditions for well-mixed transient single-unit nonreactive processes.
- Predict transient system behavior by inspecting balance equations. For example, given an equation that has the form $[dC_A/dt = 4 - 2C_A, C_A(0) = 0]$, sketch the expected plot of C_A versus t without integrating the equation.
- Obtain analytical solutions to problems that involve single separable first-order differential balance equations.
- Derive balance equations for systems that involve several dependent variables [e.g., $y_1 = C_A(t)$, $y_2 = C_B(t)$, $y_3 = T(t)$] and express the equations in a form suitable for solution using equation-solving software [$dy_i/dt = f_i(y_1, y_2, \ldots, y_n, t), i = 1, 2, \ldots, n$].

---

[1] In reality, the concept of a true steady state is fiction, since there are *always* process variable fluctuations in real systems. When you assume steady-state operation, you are assuming that these fluctuations are small enough to be neglected without causing serious errors in calculated values.

11.1   THE GENERAL BALANCE EQUATION...AGAIN

In Section 4.2, the general balance equation (4.2-1) was given as

$$\text{accumulation} = \text{input} + \text{generation} - \text{output} - \text{consumption}$$

Two forms of this equation were discussed: differential balances, which relate instantaneous rates of change at a moment in time, and integral balances, which relate changes that occur over a finite time period. We examine in this section the nature of the relationship between these two types of balances; in doing so, we belatedly show why they are called differential and integral.

11.1a   Differential Balances

Suppose a species A is involved in a process. Let $\dot{m}_{in}$(kg/s) and $\dot{m}_{out}$(kg/s) be the rates at which A enters and leaves the process by crossing the boundaries, and let $\dot{r}_{gen}$(kg/s) and $\dot{r}_{cons}$(kg/s) be the rates of generation and consumption of A within the system by chemical reaction. Any or all of the variables $\dot{m}_{in}$, $\dot{m}_{out}$, $\dot{r}_{gen}$, and $\dot{r}_{cons}$ may vary with time.

Let us now write a balance on A for a period of time from t to $t + \Delta t$, supposing that Δt is small enough for the quantities $\dot{m}_{in}$, $\dot{m}_{out}$, $\dot{r}_{gen}$, and $\dot{r}_{cons}$ to be considered constant. (Since we will eventually let Δt approach 0, this assumption is not restrictive.) The terms of a balance on A are easily calculated.

$$\text{input(kg)} = \dot{m}_{in}\text{(kg/s)}\,\Delta t(s)$$

$$\text{output} = \dot{m}_{out}\,\Delta t$$

$$\text{generation} = \dot{r}_{gen}\,\Delta t$$

$$\text{consumption} = \dot{r}_{cons}\,\Delta t$$

We also suppose that the mass of A in the system changes by an amount ΔM(kg) during this small time interval. By definition, ΔM is the accumulation of A in the system. From the balance equation, (4.2-1),

$$\Delta M = (\dot{m}_{in} + \dot{r}_{gen} - \dot{m}_{out} - \dot{r}_{cons})\,\Delta t \tag{11.1-1}$$

If now we divide by Δt and then let Δt approach 0, the ratio $\Delta M / \Delta t$ becomes the derivative of M with respect to t (dM/dt), and the balance equation becomes

$$\boxed{\frac{dM}{dt} = \dot{m}_{in} + \dot{r}_{gen} - \dot{m}_{out} - \dot{r}_{cons}} \tag{11.1-2}$$

This is the general differential balance equation: M is the amount of the balanced quantity in the system, and the four terms on the right side are rates that may vary with time.

If Equation 11.1-2 is applied to a continuous system at steady state, the quantity M must be a constant, its time derivative therefore equals zero, and the equation reduces to the familiar equation introduced in Chapter 4:

$$\text{input} + \text{generation} = \text{output} + \text{consumption}$$

However, as long as any term varies with time, the derivative on the left side of Equation 11.1-2 remains part of the equation. We thus conclude that *the balance equation for an unsteady-state system at an instant of time is a differential equation* (hence the term differential balance).

Equation 11.1-2 is an ordinary first-order differential equation. Before it can be solved to yield an expression for $M(t)$, a **boundary condition** must be provided—a specified value of the dependent variable (M) at some value of the independent variable (t). Frequently, the value of M at time $t = 0$ (an "initial condition") is specified. The complete balance equation would be

Equation 11.1-2 followed by

$$t = 0, \quad M = \ldots$$

or simply

$$M(0) = \ldots$$

When you analyze a transient system, your analysis is not complete unless every differential equation you derive is accompanied by a boundary condition similar to one of those just given.

EXAMPLE 11.1-1 **_Differential Balances on a Chemical Reactor_**

A continuous stirred-tank reactor is used to produce a compound R in the liquid-phase reaction A → R. Feed enters the reactor at a rate of $\dot{v}_0$(L/s); the concentration of the reactant in the feed is C_{A0}(mol A/L). The volume of the tank contents is V (L). The vessel may be considered perfectly mixed, so that the concentration of A in the product stream equals that in the tank. For this process the rate of consumption of A equals kC_A[mol/(s·L of reaction volume)]. All fluids (the feed, the tank contents, and the product) may be taken to have the same density, ρ(g/L).

Write differential balances on total mass and on moles of A, expressing the balances in terms of the variables shown on the following diagram:

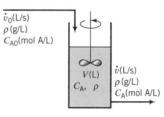

SOLUTION

Basis: Given Quantities

Total Mass Balance (generation = 0, consumption = 0)

$$\text{accumulation} = \text{input} - \text{output}$$

Mass in reactor: $\quad M\text{(g)} = V\text{(L)}\rho\text{(g/L)}$

$$\Downarrow$$

$$\text{accumulation (g/s)} = \frac{dM}{dt} = \frac{d(V\rho)}{dt} = \rho\frac{dV}{dt} \quad \text{(since } \rho \text{ is constant)}$$

$$\text{input (g/s)} = \dot{v}_0\text{(L/s)}\rho\text{(g/L)}$$

$$\text{output (g/s)} = \dot{v}\text{(L/s)}\rho\text{(g/L)}$$

$$\Downarrow$$

$$\rho\frac{dV}{dt} = \dot{v}_0\rho - \dot{v}\rho$$

$$\Downarrow \text{cancel } \rho$$

$$\boxed{\begin{array}{l} dV/dt = \dot{v}_0 - \dot{v} \\ \quad t = 0, \quad V = V_0 \end{array}}$$

where V_0 is the initial volume of the tank contents.

Question: If $\dot{v}_0 = \dot{v}$, what does the mass balance tell you?

Balance on A

$$\text{accumulation} = \text{input} - \text{output} - \text{consumption}$$

$$\text{Moles of A in the reactor} = V(\text{L})C_A(\text{mol/L})$$

$$\Downarrow$$

$$\text{accumulation (mol A/s)} = \frac{d(VC_A)}{dt}$$

$$\text{input (mol A/s)} = \dot{v}_0(\text{L/s})C_{A0}(\text{mol A/L})$$

$$\text{output (mol A/s)} = \dot{v}\,(\text{L/s})C_A(\text{mol A/L})$$

$$\text{consumption (mol A/s)} = kC_A[\text{mol A}/(\text{s} \cdot \text{L})]V(\text{L})$$

$$\Downarrow$$

$$\boxed{\begin{aligned} \frac{d(VC_A)}{dt} &= \dot{v}_0 C_{A0} - \dot{v}C_A - kC_A V \\ t = 0, \quad C_A &= C_A(0) \end{aligned}}$$

where $C_A(0)$ is the concentration of A in the initial tank contents. How you would proceed to solve this equation for the output concentration $C_A(t)$ depends on how the quantities $\dot{v}_0$, $\dot{v}$, and C_{A0} vary with time.

11.1b Integral Balances

Let us reconsider the form of the differential balance given in Equation 11.1-3.

$$\frac{dM}{dt} = \dot{m}_{\text{in}} + \dot{r}_{\text{gen}} - \dot{m}_{\text{out}} - \dot{r}_{\text{cons}} \tag{11.1-3}$$

The equation may be rewritten as

$$dM = \dot{m}_{\text{in}}\,dt + \dot{r}_{\text{gen}}\,dt - \dot{m}_{\text{out}}\,dt - \dot{r}_{\text{cons}}\,dt$$

and integrated from an initial time t_0 to a later time t_f, to obtain

$$\int_{t_0}^{t_f} dM = M(t_f) - M(t_0) = \int_{t_0}^{t_f} \dot{m}_{\text{in}}\,dt + \int_{t_0}^{t_f} \dot{r}_{\text{gen}}\,dt - \int_{t_0}^{t_f} \dot{m}_{\text{out}}\,dt - \int_{t_0}^{t_f} \dot{r}_{\text{cons}}\,dt \tag{11.1-4}$$

This is the integral balance equation. The left side is the accumulation of the balanced quantity in the system between t_0 and t_f. The term ($\dot{m}_{\text{in}}\,dt$) is the amount of the balanced quantity that enters the system in the infinitesimal interval from t to $t + dt$, so that the integral

$$\int_{t_0}^{t_f} \dot{m}_{\text{in}}\,dt$$

is the total amount that enters between t_0 and t_f. Similar reasoning can be applied to the other terms, leading to the conclusion that Equation 11.1-4 is simply another statement of the general balance equation

$$\text{accumulation} = \text{input} + \text{generation} - \text{output} - \text{consumption} \tag{11.1-5}$$

only now each term represents an amount of the balanced quantity rather than a rate. For a closed (batch) system, if the balanced quantity is mass (as opposed to energy), $\dot{m}_{\text{in}} = \dot{m}_{\text{out}} = 0$, and the equation may be written

$$M_{\text{initial}} + \int_{t_0}^{t_f} \dot{r}_{\text{generation}}\,dt = M_{\text{final}} + \int_{t_0}^{t_f} \dot{r}_{\text{consumption}}\,dt$$

or

$$\text{initial input} + \text{generation} = \text{final output} + \text{consumption}$$

This is the form of the integral balance equation given in Chapter 4 for a closed system.

The procedure to follow when writing a balance on a transient system is to derive the differential balance equation, integrate it between initial and final times, and solve for the unknown quantity in the resulting equation.

EXAMPLE 11.1-2   *Water Balance on a City Reservoir*

The water level in a municipal reservoir has been decreasing steadily during a dry spell, and there is concern that the drought could continue for another 60 days. The local water company estimates that the consumption rate in the city is approximately 10^7 L/day. The State Conservation Service estimates that rainfall and stream drainage into the reservoir coupled with evaporation from the reservoir should yield a net water input rate of $10^6 \exp(-t/100)$ L/day, where t is the time in days from the beginning of the drought, at which time the reservoir contained an estimated 10^9 liters of water.

1. Write a differential balance on the water in the reservoir.
2. Integrate the balance to calculate the reservoir volume at the end of the 60 days of continued drought.

SOLUTION

1. We will write a balance on the mass M(kg) of water in the reservoir, but will express the equation in terms of volumes to make use of the given data, using the relationship $M(\text{kg}) = \rho(\text{kg/L})V(\text{L})$. The differential balance equation is as follows:

$$\frac{dM}{dt} = \dot{m}_{\text{in}} + \dot{r}_{\text{gen}} - \dot{m}_{\text{out}} - r_{\text{cons}} \qquad \text{(each term in kg/day)}$$

$$\left\| \begin{array}{l} \dfrac{dM}{dt} = \dfrac{d}{dt}(\rho V) = \rho(\text{kg/L}) \dfrac{dV}{dt}(\text{L/day}) \qquad (\text{since } \rho \text{ is constant}) \\[2mm] \dot{m}_{\text{in}} = \rho(\text{kg/L})[10^6 e^{-t/100}(\text{L/day})] \\[2mm] \dot{m}_{\text{out}} = \rho(\text{kg/L})(10^7 \text{ L/day}) \\[2mm] \dot{r}_{\text{gen}} = \dot{r}_{\text{cons}} = 0 \qquad (\text{water is not produced or consumed in the reservoir}) \\[2mm] \text{Cancel } \rho \end{array} \right.$$

$$\boxed{\begin{array}{l} \dfrac{dV(t)}{dt} = 10^6 \exp(-t/100) - 10^7 \\[3mm] t = 0, \quad V = 10^9 \text{ L} \end{array}}$$

2. We now separate variables and integrate the differential balance equation from $t = 0$ to $t = 60$ days.

$$\int_{V(0)}^{V(60)} dV = \int_0^{60 \text{ d}} [10^6 \exp(-t/100) - 10^7] \, dt$$

$$\Downarrow$$

$$V(60 \text{ days}) - V(0) = \int_0^{60 \text{ d}} 10^6 e^{-t/100} \, dt - \int_0^{60 \text{ d}} 10^7 \, dt$$

$$\Downarrow V(0) = 10^9 \text{ liters}$$

$$V(60 \text{ days}) = 10^9 - 10^6(10^2)e^{-t/100}\Big|_0^{60 \text{ d}} - 10^7 t\Big|_0^{60 \text{ d}}$$

$$= \boxed{4.45 \times 10^8 \text{ L}} \quad (\textit{verify})$$

TEST YOURSELF

A liquid, A, is poured at a rate of 10 kg/h into a tank containing a second liquid, B. The two substances react and A is consumed, but at a lower rate than that at which it is added. Liquid is withdrawn from the reaction vessel at a rate of 10 kg/h.

1. Which terms of the general balance equation

$$\text{accumulation} = \text{input} + \text{generation} - \text{output} - \text{consumption}$$

   do *not* equal zero in each of the following balances on the reaction vessel?
   (a) Total mass.   **(b)** Moles of A.   **(c)** Moles of B.
2. Write a differential mass balance on the system, letting $m(t)$ be the total mass of the system contents.

11.2   MATERIAL BALANCES

11.2a   Total Mass Balances

A total mass balance necessarily has the form [accumulation = input − output], since mass can neither be generated nor consumed.[2] The accumulation term is always dM/dt, where $M(t)$ is the mass of the system contents. Once you have determined $M(t)$ by solving the differential balance equation, you may have to verify that the mathematical solution remains within the bounds of physical reality—that it does not become negative, for example, or that it does not exceed the total capacity of the system.

EXAMPLE 11.2-1   *Mass Balance on a Water Storage Tank*

A 12.5-m³ tank is being filled with water at a rate of 0.050 m³/s. At a moment when the tank contains 1.20 m³ of water, a bottom leak develops and gets progressively worse with time. The rate of leakage can be approximated as $0.0025t$ (m³/s), where t(s) is the time from the moment the leak begins.

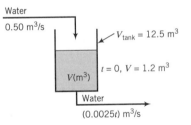

1. Write a mass balance on the tank and use it to obtain an expression for dV/dt, where V is the volume of water in the tank at any time. Provide an initial condition for the differential equation.
2. Solve the balance equation to obtain an expression for $V(t)$ and draw a plot of V versus t.

SOLUTION

1. The total mass of the tank contents is $M(\text{kg}) = \rho(\text{kg/m}^3)V(\text{m}^3)$, where $\rho = 0.00100 \text{ kg/m}^3$ is the density of liquid water. Then

$$\text{accumulation (kg/s)} = \frac{d(\rho V)}{dt} = \rho \frac{dV}{dt}$$

   (The second step follows from the fact that the density of the liquid water in the tank is independent of time and so may be taken out of the derivative.)

$$\text{input (kg/s)} = \rho(\text{kg/m}^3)(0.05 \text{ m}^3/\text{s}) = 0.05\rho$$

$$\text{output (kg/s)} = \rho(\text{kg/m}^3)[0.0025t \text{ (m}^3/\text{s})] = 0.0025\rho t$$

---

[2]We are excluding nuclear reactions from consideration.

Substituting these terms into the water balance equation (accumulation = input − output) and canceling ρ yields the differential equation

$$\frac{dV}{dt} = 0.050 \text{ m}^3/\text{s} - 0.0025t$$

$$t = 0, \quad V = 1.2 \text{ m}^3$$

Verify that each term in the equation (including dV/dt) has units of m³/s.

2. To solve the equation, we separate variables (bring dt to the right-hand side) and integrate from the initial condition ($t = 0$, $V = 1.2$ m³) to an arbitrary time, t, and corresponding volume, V.

$$dV(\text{m}^3) = (0.050 - 0.0025t)\,dt \Longrightarrow \int_{1.2 \text{ m}^3}^{V} dV = \int_{0}^{t} (0.050 - 0.0025t)\,dt$$

$$\Longrightarrow V\big]_{1.2 \text{ m}^3}^{V} = \left(0.050t - 0.0025\frac{t^2}{2}\right)\Big]_{0}^{t}$$

$$\Longrightarrow V(\text{m}^3) = 1.2 + 0.050t - 0.00125t^2$$

Check 1: When $t = 0$, $V = 1.2$ m³ (confirming the given initial condition).
Check 2: $dV/dt = 0.050 - 0.0025t$ [differentiating $V(t)$ yields the original equation for dV/dt].

A plot of the derived expression for $V(t)$ is as follows:

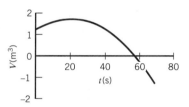

Initially, the filling causes the volume of the tank contents to increase, but as the leak gets larger the tank begins to drain. The maximum volume is 1.7 m³, well below the tank capacity of 12.5 m³. At about $t = 57$ s the contents drain completely. The mathematical formula for V predicts negative volumes after this time, but physically the volume must remain at zero (the liquid discharges as fast as it is poured in). The actual solution of the balance equation is therefore

$$V(\text{m}^3) = 1.2 + 0.050t - 0.00125t^2 \qquad 0 \le t \le 57 \text{ s}$$
$$= 0 \qquad\qquad\qquad\qquad\qquad t > 57 \text{ s}$$

The plot shown above should be changed in the range $t > 57$ s to a line coincident with the t axis.

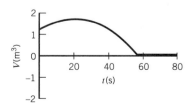

11.2b A Brief Look Back at Calculus

As you will see, balances on transient systems often lead to differential equations that look like this:

$$\frac{d(VC_A)}{dt} = 1.50 \text{ mol/s} - (\text{s}^{-1})0.200VC_A \tag{11.2-1}$$

$$C_A(0) = 2.00 \text{ mol/L}$$

In this equation, 1.50 mol/s would be the sum of the input and generation terms in the balance on species A and $0.200C_A$ would be the sum of the output and consumption terms. The goal would be to solve the differential balance to determine the concentration $C_A(\text{mol/L})$ as a function of time $t(\text{s})$, either as an analytical function or in the form of a table or plot.

This section reviews some calculus rules and procedures for solving differential equations like Equation 11.2-1. In what follows, x is an independent variable, $y(x)$ is a dependent variable, and a is a constant.

Rule 1: Derivative of a constant times a function

$$\frac{d(ay)}{dx} = a\frac{dy}{dx} \tag{11.2-2}$$

If the system volume in Equation 11.2-1 is constant, the equation would become

$$V\frac{dC_A}{dt} = 1.50 \text{ mol/s} - 0.200VC_A$$

We will shortly review how to solve this equation.

Rule 2: Product rule for differentiation

$$\frac{d(y_1 y_2)}{dx} = y_1\frac{dy_2}{dx} + y_2\frac{dy_1}{dx} \tag{11.2-3}$$

If the system volume in Equation 11.2-1 changes with time (e.g., because the tank is being filled or discharged, or the reactor is a cylinder with a moving piston), the product rule yields

$$\frac{d(VC_A)}{dt} = V\frac{dC_A}{dt} + C_A\frac{dV}{dt} = 1.50 \text{ mol/s} - 0.200C_A$$

or

$$\frac{dC_A}{dt} = \frac{1}{V}[1.50 \text{ mol/s} - 0.200C_A] - \frac{C_A}{V}\left(\frac{dV}{dt}\right)$$

If V is constant, this equation reduces to the one given following Rule 1 (convince yourself). To solve for $C_A(t)$, you would have to obtain an independent expression for dV/dt and solve both equations simultaneously. We will discuss problems of this type in Section 11.5.

Rule 3: Solution of separable first-order differential equations

The general form of a first-order differential equation is

$$\frac{dy}{dx} = f(x, y)$$

Consider a specific example:

$$\frac{dy}{dx} = 3xy$$

Novice calculus students are often tempted to solve this equation by doing something like

$$y = \int (3xy)\,dx$$

which is correct but useless, since you cannot evaluate that integral without first substituting for $y(x)$, the function you are trying to determine.

A *separable* first-order differential equation is one that can be written in the form

$$\frac{dy}{dx} = f_1(x)f_2(y)$$

$$x = 0, \quad y = y(0)$$

The procedure for solving a separable equation is to bring all terms involving y (including dy) to one side of the equation and all terms involving x (including dx) to the other side, and then to integrate each side over its respective variable from the initial value [0 for x, $y(0)$ for y] to an arbitrary value:

$$\frac{dy}{dx} = f_1(x)f_2(y) \xrightarrow{\text{separate}} \frac{dy}{f_2(y)} = f_1(x)\,dx \xrightarrow{\text{integrate}} \int_{y(0)}^{y} \frac{dy}{f_2(y)} = \int_{0}^{x} f_1(x)\,dx \quad \textbf{(11.2-4)}$$

$$x = 0, \quad y = y(0)$$

Each integral involves a function of only the variable of integration (y on the left, x on the right), and so both integrals can be evaluated to obtain an expression relating x and y.

Reconsider Equation 11.2-1 once more, letting the system volume V equal 1.00 liter.

$$\left\{ \begin{array}{l} \dfrac{dC_A}{dt} = 1.50 \text{ mol/s} - (0.200 \text{ L/s})C_A \\[2mm] t = 0, \quad C_A = 2.00 \text{ mol/L} \end{array} \right\}$$

$$\xrightarrow{\text{separate}} \frac{dC_A}{1.50 - 0.200C_A} = dt \xrightarrow{\text{integrate}} \int_{2.00}^{C_A} \frac{dC_A}{1.50 - 0.200C_A} = \int_{0}^{t} dt$$

$$\implies -\frac{1}{0.200} \ln(1.50 - 0.200C_A)\Big|_{2.00}^{C_A} = t \implies \ln\left(\frac{1.50 - 0.200C_A}{1.50 - 0.400}\right) = -0.200t$$

$$\implies \frac{1.50 - 0.200C_A}{1.10} = e^{-0.200t} \implies \boxed{C_A(\text{mol/L}) = \frac{1}{0.200}(1.50 - 1.10e^{-0.200t(\text{s})})}$$

Try to follow each step of this procedure. It will be used to solve almost every balance equation in the remainder of this chapter.

TEST YOURSELF

Separate variables for each of the following equations to obtain integrals of the form of Equation 11.2-4. Then proceed as far as you can to obtain expressions for $y(t)$.

1. $(dy/dt) = 2 - t$, $y(0) = 1$.
2. $(dy/dt) = 2 - y$, $y(0) = 1$.
3. $(dy/dt) = (2 - t)(2 - y)$, $y(0) = 1$.

11.2c Balances on Single Well-Mixed Process Units

Here is the general procedure for writing and solving a transient material balance equation:

1. *Eliminate terms in the general balance equation that equal zero* (input and output for batch systems, generation and consumption for balances on total mass and nonreactive species).
2. *Write an expression for the total amount of the balanced species in the system* [$V(\text{m}^3)\rho(\text{kg/m}^3)$ for total mass, $V(\text{m}^3)C_A(\text{mol A/m}^3)$ or $n_{\text{total}}(\text{mol})x_A(\text{mol A/mol})$ for species A]. *Differentiate the expression with respect to time to obtain the accumulation term in the balance equation.*
3. *Substitute system variables into the remaining terms* (input, generation, output, consumption) *in the balance equation.* Make sure that all terms have the same units (kg/s, lb-mole/h, etc.).

4. *If $y(t)$ is the dependent variable to be determined* (e.g., the mass of the system contents, the concentration of species A, the mole fraction of methane), *rewrite the equation to obtain an explicit expression for dy/dt. Supply a boundary condition*—the value of the dependent variable at a specified time (usually $t = 0$). This condition may be expressed as $[t = 0, y = y_0]$ or simply $[y(0) = y_0]$, where y_0 is a number.

5. *Solve the equation—analytically if possible, otherwise numerically.* In the processes to be analyzed in this chapter, you will generally be able to use separation of variables to obtain the solution analytically.

6. *Check the solution.* You can do so using any (and preferably all) of these methods:
   (a) Substitute $t = 0$ and verify that the known initial condition $[y(0) = y_0]$ is obtained.
   (b) Find the long-time asymptotic (steady-state) value of the dependent variable by setting dy/dt equal to 0 in the original balance equation and solving the resulting algebraic equation for y_{ss}, and then verify that if you let $t \to \infty$ in your solution, $y \to y_{ss}$. (The equation may not have a steady-state solution, in which case this method will not work.)
   (c) Differentiate your solution to obtain an expression for dy/dt, substitute for y and dy/dt in the original differential equation, and verify that the equation is satisfied.

7. *Use your solution to generate a plot or table of y versus t.*

The next example illustrates this procedure.

EXAMPLE 11.2-2   **Transient Behavior of a Stirred-Tank Reactor**

A liquid-phase reaction with stoichiometry A→B takes place in a continuous well-mixed 10.0-liter stirred-tank reactor. A schematic diagram of the process is shown below.

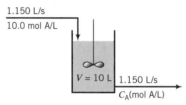

A ⟶ B, rate = $0.0050C_A$[mol A react/(L·s)]

The reactor may be considered perfectly mixed, so that the contents are uniform and the concentration of A in the product stream equals that inside the tank. The tank is initially filled with a solution that contains 2.00 mol A/L, and the inlet and outlet flows then begin.

1. Write a balance on species A in the tank and provide an initial condition.
2. Calculate C_{AS}, the steady-state concentration of A in the tank (the value approached as $t \to \infty$).
3. Sketch the shape expected for a plot of C_A versus t.
4. Solve the balance equation for $C_A(t)$, check the solution, and draw the actual plot of C_A versus t.

SOLUTION   Compare what follows with the general procedure given prior to this example.

1. The total moles of A in the reactor at any time equals $(10.0\text{ L})[C_A(\text{mol/L})] = 10.0C_A$ (mol A). Therefore,

Accumulation: $\dfrac{d(10.0C_A)}{dt} = 10.0\dfrac{dC_A}{dt}\left(\dfrac{\text{mol A}}{\text{s}}\right)$

Input: $(0.150\text{ L/s})(10.0\text{ mol A/L}) = 1.50\text{ mol A/s}$

Output: $(0.150\text{ L/s})[C_A(\text{mol A/L})] = 0.150C_A(\text{mol A/s})$

Generation: 0 mol A/s   (A is not a reaction product)

Consumption: $(10.0\text{ L})[0.0050C_A(\text{mol A/(L·s)})] = 0.050C_A$ (mol A/s)

These terms are substituted into the balance equation on A (accumulation = input − output − consumption), which is then divided through by 10.0 to obtain an expression for dC_A/dt. The result along with the initial condition for the equation [$C_A(0)$ = 2.00 mol A/L] is

$$\frac{dC_A}{dt} = 0.150 \text{ mol A/s} - (0.0200 \text{ L/s})C_A$$

$$t = 0, \quad C_A = 2.00 \text{ mol A/L}$$

2. At steady state nothing varies with time, so that the derivative of C_A (and of every other system variable) with respect to time must equal zero. Setting dC_A/dt = 0 in the balance equation and letting $C_A = C_{AS}$ (steady state) in the resulting equation yields

$$0 = 0.150 \text{ mol/s} - 0.0200C_{AS} \implies \boxed{C_{AS} = 7.50 \text{ mol A/L}}$$

If C_{AS} had not had an asymptotic (steady-state) limit, the equation would not have had a finite solution.

3. We can now deduce quite a lot about the plot of C_A versus t, even though we have not yet solved the differential balance equation. We know a point at t = 0 (the initial condition) and the asymptotic value as $t \to \infty$ (the steady-state solution), and we also have an expression for the slope of the plot at any time (dC_A/dt) as a function of the concentration (slope = $0.150 - 0.0200C_A$). Let us summarize what we can deduce.

- The plot originates at (t = 0, C_A = 2.00 mol/L).
- At t = 0, the slope of the plot is $[0.150-0.0200](2.00 \text{ mol/s}) = 0.110 \text{ mol/s}$. Since it is positive, C_A must increase as t increases.
- As t continues to increase and C_A also increases, the slope of the curve ($0.150 - 0.0200C_A$) gets progressively less positive. The curve must therefore be concave down.

- At long times, the plot asymptotes to C_A = 7.50 mol/L.

Combining all these observations leads to the following sketch:

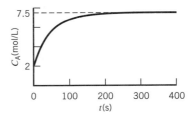

4. We may now solve the differential balance to determine the values of C_A at specific values of t or vice versa. Reconsider the equation

$$\frac{dC_A}{dt} = 0.150 - 0.0200C_A$$

$$t = 0, \quad C_A = 2.00$$

Separating variables and integrating as in Equation 11.2-4 yields

$$\frac{dC_A}{0.150 - 0.0200C_A} = dt \implies \int_{2.00}^{C_A} \frac{dC_A}{0.150 - 0.0200C_A} = \int_0^t dt = t$$

$$\implies -\frac{1}{0.0200} \ln(0.150 - 0.0200C_A)\Big|_{2.00}^{C_A} = t \implies \ln \frac{0.150 - 0.0200C_A}{0.150 - 0.0200(2.00)} = -0.0200t$$

$$\implies \frac{0.150 - 0.0200C_A}{0.110} = e^{-0.0200t} \implies \boxed{C_A(\text{mol/L}) = 7.50 - 5.50e^{-0.0200t}}$$

(Verify each step.)

Whenever you obtain the solution of a differential equation, you should check it as many ways as you can. In this case, we have three ways:

Check 1: Substitute $t = 0$ into the solution $\implies C_A(0) = 2.00$ mol/L (the correct initial value).

Check 2: Substitute $t \to \infty$ into the solution to obtain $C_A(\infty) = 7.50$ mol/L (the previously determined steady-state value).

Check 3: Differentiate the equation to obtain an expression for dC_A/dt, and then substitute for both dC_A/dt and $C_A(t)$ in the original equation $[dC_A/dt = 1.50 - 0.200C_A]$ to show that the solution satisfies the equation.

Verify that the derived solution satisfies each of these conditions.

A plot of the solution matches the form of the plot previously sketched.

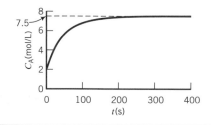

11.3 ENERGY BALANCES ON SINGLE-PHASE NONREACTIVE PROCESSES

The general energy balance has the form

$$\text{accumulation} = \text{input} - \text{output} \qquad (11.3\text{-}1)$$

since energy can neither be generated nor consumed.

Suppose $E_{\text{sys}}(t)$ is the total energy (internal + kinetic + potential) of a system, and $\dot{m}_{\text{in}}$ and $\dot{m}_{\text{out}}$ are the mass flow rates of the system input and output streams. (If the system is closed, these quantities each equal zero.) Proceeding as in the development of the transient mass balance equation, we apply the general energy balance equation (11.3-1) to the system in a small time interval from t to $t + \Delta t$, during which time the properties of the input and output streams remain approximately constant. The terms of the equation are as follows (see Section 7.4):

$$\text{accumulation} = \Delta E_{\text{sys}} = \Delta U_{\text{sys}} + \Delta E_{k,\text{sys}} + \Delta E_{p,\text{sys}}$$

$$\text{input} = \dot{m}_{\text{in}}\left(\hat{H}_{\text{in}} + \frac{u_{\text{in}}^2}{2} + gz_{\text{in}}\right)\Delta t + \dot{Q}\,\Delta t$$

$$\text{output} = \dot{m}_{\text{out}}\left(\hat{H}_{\text{out}} + \frac{u_{\text{out}}^2}{2} + gz_{\text{out}}\right)\Delta t + \dot{W}_s\,\Delta t$$

where the quantities in parentheses are the specific enthalpies, kinetic energies, and potential energies of the input and output streams, and $\dot{Q}$ and $\dot{W}_s$ are the rates of transfer of heat and shaft work. (See Section 7.4c.) Any or all of the variables $\dot{m}$, $\hat{H}$, u, z, $\dot{Q}$, and $\dot{W}_s$ may vary with time.

If we now substitute the given expressions for accumulation, input, and output into Equation 11.3-1, divide by Δt, and let Δt approach zero, we obtain the general differential energy balance:

$$\frac{dU_{sys}}{dt} + \frac{dE_{k,sys}}{dt} + \frac{dE_{p,sys}}{dt} = \dot{m}_{in}\left(\hat{H}_{in} + \frac{u_{in}^2}{2} + gz_{in}\right)$$

$$- \dot{m}_{out}\left(\hat{H}_{out} + \frac{u_{out}^2}{2} + gz_{out}\right) + \dot{Q} - \dot{W}_s \qquad \textbf{(11.3-2)}$$

If there are several input and output streams, a term of the form

$$\dot{m}\left(\hat{H} + \frac{u^2}{2} + gz_{in}\right)$$

must be included in Equation 11.3-2 for each stream.

It is quite difficult to solve Equation 11.3-2 unless a number of simplifications are made. We will restrict our consideration to systems that satisfy the following conditions:

1. The system has at most a single input stream and a single output stream, each with the same mass flow rate.

$$\dot{m}_{in} = \dot{m}_{out} = \dot{m} \qquad \textbf{(11.3-3)}$$

A consequence of this assumption is that the mass of the system contents does not change with time.

2. Kinetic and potential energy changes in the system and between the inlet and outlet streams are negligible.

$$\frac{dE_{k,sys}}{dt} \approx \frac{dE_{p,sys}}{dt} \approx 0 \qquad \textbf{(11.3-4)}$$

$$\dot{m}\left(\frac{u_{in}^2}{2} - \frac{u_{out}^2}{2}\right) \approx 0 \qquad \textbf{(11.3-5)}$$

$$\dot{m}(gz_{in} - gz_{out}) \approx 0 \qquad \textbf{(11.3-6)}$$

Under these conditions, Equation 11.3-2 simplifies to

$$\frac{dU_{sys}}{dt} = \dot{m}(\hat{H}_{in} - \hat{H}_{out}) + \dot{Q} - \dot{W}_s \qquad \textbf{(11.3-7)}$$

If the equation is applied to a closed system, $\dot{m} = 0$, and the rate of transfer of energy as shaft work, $\dot{W}_s$, must be replaced by the total rate of energy transfer as work, $\dot{W}$.

Equation 11.3-7 is simple in appearance, but its solution is still generally difficult to obtain. If, for example, the composition or temperature of the system contents varies with position in the system, it is difficult to express the total internal energy U_{sys} in terms of measurable quantities, and a similar problem occurs if phase changes or chemical reactions take place in the course of the process. To illustrate the solution of energy balance problems without becoming too involved in the thermodynamic complexities, we will impose the additional restrictions that follow.

3. The temperature and composition of the system contents do not vary with position within the system (i.e., the system is perfectly mixed). In consequence, the outlet stream and the system contents must be at the same temperature, or

$$T_{out} = T_{sys} = T \qquad \textbf{(11.3-8)}$$

4. No phase changes or chemical reactions take place within the system; $\hat{U}$ and $\hat{H}$ are independent of pressure; and the mean heat capacities C_v and C_p of the system contents (and of the inlet and outlet streams) are independent of composition and temperature, and hence unchanging with time. Then if T_r is a reference temperature at which $\hat{H}$ is defined to be zero and M is the mass (or number of moles) of the system contents,

$$U_{\text{sys}} = M\hat{U}_{\text{sys}} = M[\hat{U}(T_r) + C_v(T - T_r)]$$

$\Big\Downarrow M, \hat{U}(T_r),$ and C_v are constant

$$\frac{dU_{\text{sys}}}{dt} = MC_v \frac{dT}{dt} \qquad \textbf{(11.3-9)}$$

$$\hat{H}_{\text{in}} = C_p(T_{\text{in}} - T_r) \qquad \textbf{(11.3-10)}$$

$$\hat{H}_{\text{out}} = C_p(T_{\text{out}} - T_r)$$

$\Big\Downarrow$ Equation 11.3-8

$$\hat{H}_{\text{out}} = C_p(T - T_r) \qquad \textbf{(11.3-11)}$$

Finally, we may substitute the expressions of Equations 11.3-3 through 11.3-11 into the general energy balance (Equation 11.3-2) to obtain for an open system

Open System:
$$\boxed{MC_v \frac{dT}{dt} = \dot{m}C_p(T_{\text{in}} - T) + \dot{Q} - \dot{W}_s} \qquad \textbf{(11.3-12)}$$

(Verify this result for yourself.) For a closed system, the equation is

Closed System:
$$\boxed{MC_v \frac{dT}{dt} = \dot{Q} - \dot{W}} \qquad \textbf{(11.3-13)}$$

To summarize, the conditions under which Equations 11.3-12 and 11.3-13 are valid are: (a) negligible kinetic and potential energy changes, (b) no accumulation of mass in the system, (c) pressure independence of $\hat{U}$ and $\hat{H}$, (d) no phase changes or chemical reactions, and (e) a spatially uniform system temperature. Any or all of the variables T, T_{in}, $\dot{Q}$ and $\dot{W}_s$ (or $\dot{W}$) may vary with time, but the system mass, M, the mass throughput rate, $\dot{m}$, and the heat capacities, C_v and C_p, must be constants.

The following example illustrates the derivation and solution of an energy balance on a closed system that satisfies these restrictions.

EXAMPLE 11.3-1 **Startup of a Batch Reactor**

A well-stirred batch reactor wrapped in an electrical heating mantle is charged with a liquid reaction mixture. The reactants must be heated from an initial temperature of 25°C to 250°C before the reaction can take place at a measurable rate. Use the data given below to determine the time required for this heating to take place.

Reactants: Mass = 1.50 kg
              $C_v = 0.900$ cal/(g·°C)

Reactor: Mass = 3.00 kg
              $C_v = 0.120$ cal/(g·°C)

Heating rate: $\dot{Q} = 500.0$ W

Negligible reaction and no phase changes during heating.
Negligible energy added to the system by the stirrer.

SOLUTION

We first note that the conditions of validity of the simplified closed system energy balance equation, Equation 11.3-13, are all satisfied (*verify*); moreover, since the system has constant volume and the energy input due to the stirrer is presumed negligible, $\dot{W} \approx 0$. The equation therefore becomes

$$MC_v \frac{dT}{dt} = \dot{Q}$$

$$t = 0, \quad T_{sys} = 25°C$$

The task is now to integrate this equation from the initial state of the system ($t = 0$, $T = 25°C$) to the final state ($t = t_f$, $T = 250°C$), and to solve the integrated equation for the heating time t_f. Rearranging the equation,

$$MC_v \, dT = \dot{Q} \, dt$$

$$\Big\Downarrow \text{Integrate}$$

$$\int_{25°C}^{250°C} MC_v \, dT = \int_0^{t_f} \dot{Q} \, dt$$

$$\Big\Downarrow \dot{Q}, M, \text{and } C_v \text{ are constant}$$

$$MC_v(250°C - 25°C) = \dot{Q} t_f$$

$$\Big\Downarrow$$

$$t_f = \frac{225 M C_v}{\dot{Q}}$$

The heat capacity of the system is obtained from Equation 8.3-13 as

$$C_v = \frac{M_{\text{reactants}}}{M}(C_v)_{\text{reactants}} + \frac{M_{\text{reactor}}}{M}(C_v)_{\text{reactor}}$$

$$\Big\Downarrow$$

$$MC_v = (1500 \text{ g})\left(0.900 \, \frac{\text{cal}}{\text{g·°C}}\right) + (3000 \text{ g})\left(0.120 \, \frac{\text{cal}}{\text{g·°C}}\right)$$

$$= (1710 \text{ cal/°C})(4.184 \text{ J/cal})$$

$$= 7150 \text{ J/°C}$$

The final result is

$$t_f = \frac{225 M C_v}{\dot{Q}}$$

$$\Big\Downarrow \begin{array}{l} MC_v = 7150 \text{ J/°C} \\ \dot{Q} = 500 \text{ W} = 500 \text{ J/s} \end{array}$$

$$t_f = \frac{7150 (\text{J/°C})}{500 (\text{J/s})}(225°C)$$

$$= 3220 \text{ s} \Longrightarrow \boxed{53.7 \text{ min}}$$

Question: Which restriction on the energy balance (Equation 11.3-13) would probably be violated if the reactants were not stirred?

The final example illustrates a transient energy balance on a continuous system.

EXAMPLE 11.3-2 **Transient Behavior of an Air-Cooling System**

An air-cooled engine generates heat at a constant rate $\dot{Q}_{\text{gen}} = 8530$ Btu/min.

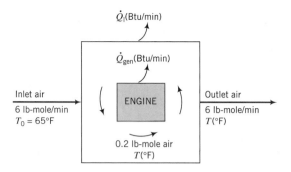

The air in the engine housing is circulated rapidly enough for its temperature to be considered uniform and equal to the temperature of the outlet air. Air passes through the engine housing at a rate of 6.00 lb-mole/min, entering at a temperature of 65°F, and an average of 0.200 lb-mole of air is contained within the engine housing. (We will neglect the variation of this quantity with the changing temperature of the gas.) Heat is lost from the housing to its surroundings at a rate

$$\dot{Q}_{lost}(\text{Btu/min}) = [33.0\ \text{Btu/(°F·min)}](T - 65°F)$$

Suppose the engine is started with the inside air temperature equal to 65°F.

1. Calculate the steady-state air temperature if the engine runs continuously for an indefinite period of time, assuming

$$C_v = 5.00\ \text{Btu/(lb-mole·°F)}$$

2. Derive a differential equation for the variation of the outlet temperature with the time from startup and solve it.

SOLUTION   System = the air within the engine housing.

1. The steady-state energy balance equation may be obtained by setting dT/dt equal to zero in Equation 11.3-12.

$$0 = \dot{m}C_p(T_{in} - T) + \dot{Q} - \dot{W}_s$$

$$\left\|\begin{array}{l} T_{in} = 65°F \ \text{(given)} \\ T = T_s(°F) \ \ \text{(steady-state outlet temperature)} \\ \dot{W}_s = 0 \ \ \text{(no moving parts)} \\ \dot{Q} = \dot{Q}_{gen} - \dot{Q}_{lost} \end{array}\right.$$

$$\dot{m}C_p(T_s - 65.0°F) = 8530\ \text{Btu/min} - 33.0(T_{ss} - 65.0°F)$$

Assuming ideal gas behavior

$$C_p = C_v + R = (5.00 + 1.99)\text{Btu/(lb-mole·°F)} \quad \text{(Equation 8.3-12)}$$

$$\dot{m}C_p = \frac{6.00\ \text{lb-mole}}{\text{min}}\left|\frac{6.99\ \text{Btu}}{\text{lb-mole·°F}}\right. = 41.9\ \frac{\text{Btu}}{\text{min·°F}}$$

The energy balance equation then becomes

$$41.9(T_s - 65.0°F) = 8530\ \text{Btu/min} - 33.0(T_s - 65.0°F)$$

$$\Downarrow$$

$$\boxed{T_s = 179°F}$$

2. The unsteady-state balance equation (11.3-12) is, for our system,

$$MC_v \frac{dT}{dt} = \dot{m}C_p(65°F - T) + \dot{Q}_{gen} - \dot{Q}_{lost}$$

$$
\begin{array}{l}
M = 0.2 \text{ lb-mole} \\
C_v = 5.00 \text{ Btu/(lb-mole·°F)} \\
\dot{m}C_p = 41.9 \text{ Btu/(min·°F)} \quad \text{[from part 1]} \\
\dot{Q}_{gen} = 8530 \text{ Btu/min} \\
\dot{Q}_{lost} = 33.0(T - 65)\text{(Btu/min)}
\end{array}
$$

$$
\boxed{
\begin{array}{c}
\dfrac{dT}{dt} = -74.9T + 13,400°C/\text{min} \\[2mm]
t = 0, \quad T = 65°F
\end{array}
}
$$

You can check this result by setting $dT/dt = 0$ and solving the resulting equation for the steady-state value of T. The result is $(13,400/74.9)°F = 179°F$, in agreement with the result of part 1.

The solution of the equation is obtained by separating variables and integrating:

$$\int_{65°F}^{T} \frac{dT}{13,400 - 74.9T} = \int_{0}^{t} dt$$

$$\Downarrow$$

$$-\frac{1}{74.9}\ln(13,400 - 74.9T)\Big]_{65°F}^{T} = t$$

$$\Downarrow$$

$$\ln(13,400 - 74.9T) = -74.9t + \ln(8530)$$

$$\Downarrow$$

$$13,400 - 74.9T = 8530\exp(-74.9t)$$

$$\Downarrow$$

$$\boxed{T(°F) = 179 - 114\exp(-74.9t)}$$

A check on the solution may be obtained by substituting $t = 0$ and verifying that T equals the specified initial value, 65°F.

If you evaluate T from this equation for a number of values of t (5), a plot would appear as follows:

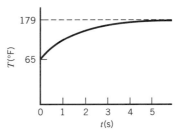

The plot starts from the initial condition of 65°F and asymptotically approaches the steady-state value of 179°F. It takes 3.8 seconds. for the temperature to reach 178°F, one degree short of its final value.

TEST YOURSELF

1. Under what conditions is the simplified form of the energy balance (Equation 11.3-12) valid?
2. Suppose the heat capacity C_v is a function of T, which varies with time. Where would the derivation of Equation 11.3-12 break down?

11.4 SIMULTANEOUS TRANSIENT BALANCES

Throughout this book, we have seen that when more than one species is involved in a process or when energy balances are required, several balance equations must be derived and solved simultaneously. For steady-state systems the equations are algebraic, but when the systems are transient, simultaneous differential equations must be solved. For the simplest systems, analytical solutions may be obtained by hand, but more commonly numerical solutions are required. Software packages that solve general systems of ordinary differential equations—such as Mathematica®, Maple®, Matlab®, TK-Solver®, Polymath®, and EZ-Solve®—are readily obtained for most computers. Other software packages have been designed specifically to simulate transient chemical processes. Some of these dynamic process simulators run in conjunction with the steady-state flowsheet simulators mentioned in Chapter 10 (e.g., SPEEDUP®, which runs with Aspen Plus, and a dynamic component of HYSYS®) and so have access to physical property databases and thermodynamic correlations.

Suppose $y_1(t), y_2(t), \ldots, y_n(t)$ are dependent variables in a process system (such as species flow rates or concentrations or mole fractions or temperature), and that at time $t = t_0$ (usually but not always 0) these variables have the values $y_{1i}, y_{2i}, \ldots, y_{ni}$. The goal is to derive a set of n differential equations that have the following form:

$$\frac{dy_1}{dt} = f_1(y_1, y_2, \ldots, y_n, t) \tag{11.4-1}$$
$$y_1(t_0) = y_{1i}$$

$$\frac{dy_2}{dt} = f_2(y_1, y_2, \ldots, y_n, t) \tag{11.4-2}$$
$$y_2(t_0) = y_{2i}$$

$$\vdots \qquad\qquad\qquad\qquad\qquad \vdots$$

$$\frac{dy_n}{dt} = f_n(y_1, y_2, \ldots, y_n, t) \tag{11.4-n}$$
$$y_n(t_0) = y_{ni}$$

The functions on the right-hand sides of these equations are derived from the input, output, generation, and consumption terms in the balance equations. The solutions of the equations may be expressed as a table of $y_1, y_2, \ldots, y_n$ for increasing values of t or as plots of y_1 versus t, y_2 versus $t, \ldots, y_n$ versus t. The next example provides an illustration.

EXAMPLE 11.4-1 **Transient Balances on a Semibatch Reactor**

An aqueous solution containing 0.015 mol/L of species A is fed into a holding tank that initially contains 75 liters of pure water. The reactant decomposes at a rate

$$r[\text{mol A/(L·s)}] = 0.0375C_A$$

where $C_A(\text{mol A/L})$ is the concentration of A in the tank. The volumetric feed rate of the solution, $\dot{v}(t)$, increases linearly over a 10-second period from 0 to 25 L/s and stays constant at that rate thereafter until the tank is filled to the desired level. The density of the feed stream is constant.

1. Write transient balances for the total mass of the tank contents and the mass of A in the tank. Convert the equations to differential equations for $V(t)$ (the volume of the tank contents) and $C_A(t)$ (the concentration of A in the tank) that have the form of Equations 11.4-1 and 11.4-2, and provide initial conditions.

2. Sketch the shapes of the plots you would expect for the volume of the tank contents, V (L), and the concentration of A in the tank, C_A (mol/L), versus time.
3. Outline how the equations would be solved to derive an expression for $C_A(t)$ for the period from $t = 0$ to $t = 60$ s.

SOLUTION   A flowchart of the process is as follows:

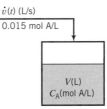

The concentration of A in the tank changes with time because both the moles of A in the tank and the volume of the tank contents are changing.

1. ***Total mass balance***: accumulation = input (kg/s). The total mass of the tank contents (kg) at any time is ρ(kg/L)V(L), and the mass flow rate of the feed stream (kg/s) is ρ(kg/L)$\dot{v}$(L/s). The mass balance thus becomes $d(\rho V)/dt = \rho \dot{v}$, or, bringing ρ (which we are told is constant) out of the derivative and canceling it,

$$\frac{dV}{dt} = \dot{v}$$

$$V(0) = 75.0 \text{ L}$$

(1)

Balance on A: accumulation (mol A/s) = input − consumption. The number of moles of A in the tank at any time equals V(L)C_A(mol A/L). The balance equation therefore becomes

$$\frac{d}{dt}(VC_A) = \dot{v}\left(\frac{L}{s}\right) \cdot 0.015 \frac{\text{mol A}}{L} - (0.0375 C_A)\left(\frac{\text{mol A}}{L \cdot s}\right)V(L)$$

$$\Big\Downarrow \text{product rule}$$

$$V\frac{dC_A}{dt} + C_A\frac{dV}{dt} = 0.015\dot{v} - 0.0375 VC_A$$

$$\Big\Downarrow \text{Substitute for } dV/dt \text{ from Equation 1, solve for } dC_A/dt$$

$$\frac{dC_A}{dt} = \frac{\dot{v}}{V}(0.015 \text{ mol A/L} - C_A) - 0.0375 C_A$$

$$C_A(0) = 0 \text{ mol A/L}$$

(2)

The initial condition in Equation 2 follows from the statement that the tank initially contains pure water. In Equations 1 and 2,

$$\dot{v}(t) = 2.5t \qquad 0 \le t \le 10\text{s}$$

(3a)

$$= 25 \text{ L/s} \qquad t > 10 \text{ s}$$

(3b)

(Verify Equation 3a.)

Equations 1 and 2 are two differential equations in two dependent variables that have the form of Equations 11.4-1 and 11.4-2, where V and C_A correspond to y_1 and y_2, respectively. The equations may therefore be solved with any of the computer programs mentioned at the beginning of this section.[3]

---

[3]It would be easier in this particular problem to solve Equation 1 analytically and substitute for $V(t)$ in Equation 2. The methods we are illustrating now would work even if an analytical solution to Equation 1 could not be found.

2. To predict the shape of the curve on a plot of V versus t, we need only remember that the slope of the curve is dV/dt, which in turn equals $\dot{v}(t)$ (from Equation 1). Try to follow this chain of reasoning:

- A point on the plot of V versus t is the initial condition ($t = 0$, $V = 75$ L).
- During the first 10 seconds, $dV/dt = 2.5t$ (from Equations 1 and 3a). The slope of the curve therefore equals zero at $t = 0$ (so that the curve is horizontal at the V axis) and increases over the first 10 seconds (so that the curve is concave up).
- At $t = 10$ seconds, dV/dt reaches a value of 25 L/s and thereafter remains constant at that value. A curve with a constant slope is a straight line. The plot of V versus t for $t \geq 10$ s must therefore be a straight line with a slope of 25 L/s.
- Putting the preceding observations together, we conclude that the plot of V versus t starts horizontally at ($t = 0$, $V = 75$ L), curves up for 10 seconds, and then becomes a straight line with a slope of 25 L/s. It should have the following appearance:

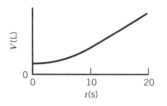

- The plot of C_A versus t must begin at ($t = 0$, $C_A = 0$), since the tank initially contains pure water.
- At $t = 0$, the expression of Equation 2 for dC_A/dt equals zero since both t and C_A are zero at this point. (*Verify.*) The plot of C_A versus t is therefore horizontal at the C_A axis. Since we are adding A to the tank, its concentration must increase and so the curve must be concave up.
- As time proceeds, more and more of the tank volume is occupied by fluid in which the A has had a long time to react. We could anticipate that at a *very* long time, the tank would contain a huge volume with very little A in it, and the A being added would be diluted down to a concentration approaching zero. C_A should therefore increase near $t = 0$, rise to a maximum, start decreasing, and approach zero at long times.
- Furthermore, the concentration in the tank can never be greater than that in the feed stream (0.015 mol/L) and, in fact, must always be less than this amount since (a) the feed is diluted by the water initially in the tank and (b) some of the A in the feed reacts once it is in the tank. The maximum value of C_A must therefore be less than 0.015 mol A/L.
- All of these observations combine to predict a plot with the following shape:

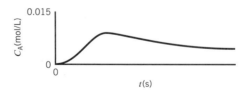

3. The system of equations must be solved in two stages—the first from $t = 0$ to $t = 10$ s (when $v = 2.5t$) and the second for $t > 10$ s, when $\dot{v} = 25$ L/s. The procedure is as follows:

- Substitute $2.5t$ for $\dot{v}(t)$ in Equations 1 and 2.

$$\frac{dV}{dt} = 2.5t \tag{1a}$$

$$V(0) = 75.0 \text{ L}$$

$$\frac{dC_A}{dt} = \frac{2.5t}{V}(0.015 - C_A) - 0.0375C_A \tag{2a}$$

$$C_A(0) = 0$$

When this pair of equations is solved for $V(t)$ and $C_A(t)$ (we will omit details of the solution procedure), we determine that $V(10\text{ s}) = 200$ L and $C_A(10\text{ s}) = 0.00831$ mol A/L.

- Substitute $\dot{v}(t) = 25$ L/s in Equations 1 and 2 and substitute the dependent variable values at $t = 10$ s for the initial conditions:

$$\frac{dV}{dt} = 25 \text{ L/s} \tag{1b}$$

$$V(10) = 200 \text{ L}$$

$$\frac{dC_A}{dt} = \frac{25}{V}(0.015 - C_A) - 0.0375C_A \tag{2b}$$

$$C_A(10) = 0.0831 \text{ mol A/L}$$

These equations may be solved for $V(t)$ and $C_A(t)$ for $t > 10$ s. These solutions along with the previous solutions for $t \leq 10$ s are shown in the following plots:

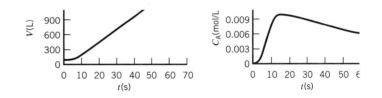

11.5 SUMMARY

All batch and semibatch processes are transient, as are continuous processes being started up, shut down, or in transition from one operating state to another. The accumulation terms in balance equations for transient systems are not zero (as they are for steady-state systems) but rather are derivatives of system variables with respect to time, and balance equations are consequently differential rather than algebraic.

The procedure for writing and solving transient balances is as follows:

- Write an expression for the amount of the balanced quantity in the system (mass, moles of a particular species, energy) and set the accumulation term in the balance equation equal to the derivative of that amount with respect to time.
- Substitute for the input, output, generation, and consumption terms in the balance equation, and convert the resulting equation into one with the form

$$\frac{dy}{dt} = f(y, t)$$

$$y(0) = y_0$$

where $y(t)$ is the dependent system variable to be determined (total mass or volume of the system contents, concentration or mole fraction of a species, temperature) and y_0 is the specified initial value of y.

- Sketch the anticipated plot of y versus t, using the initial condition to locate the starting point and using what you know about the slope (which equals dy/dt) to predict the shape of the curve.
- If the balance equation can be solved analytically (e.g., by separation of variables and integration), do so; otherwise, solve it using differential equation-solving software.
- If the system involves more than one dependent variable (such as a semibatch unit in which both the volume and composition of the system contents vary or a reactor in which several

reactions occur simultaneously), write balance equations for all the dependent variables and convert them to the form

$$\frac{dy_1}{dt} = f_1(y_1, y_2, \ldots, y_n, t)$$

$$y_1(t_0) = y_{1i}$$

$$\vdots$$

$$\frac{dy_n}{dt} = f_n(y_1, y_2, \ldots, y_n, t)$$

$$y_n(t_0) = y_{ni}$$

Differential equation-solving software may then be used to generate tables and/or plots of the dependent variables versus time.

PROBLEMS   Most of the following problems ask you to write one or more transient balances, provide initial conditions for each derived equation, and integrate the equations. Whether or not such a problem explicitly requests it, you should always check your solution by verifying that (a) the initial condition is satisfied, (b) differentiating your solution yields the original balance equation, and (c) the previously determined steady-state solution (if one has been determined) is approached as $t \to \infty$ in the transient solution.

11.1. A solution containing hydrogen peroxide with a mass fraction x_{p0} (kg H_2O_2/kg solution) is added to a storage tank at a steady rate $\dot{m}_0$(kg/h). During this process, the liquid level reaches a corroded spot in the tank wall and a leak develops. As the filling continues, the leak rate $\dot{m}_1$(kg/h) becomes progressively worse. Moreover, once it is in the tank the peroxide begins to decompose at a rate

$$r_d(\text{kg/h}) = kM_p$$

where M_p(kg) is the mass of peroxide in the tank. The tank contents are well mixed, so that the peroxide concentration is the same at all positions. At a time $t = 0$ the liquid level reaches the corroded spot. Let M_0 and M_{p0} be the total liquid mass and mass of peroxide, respectively, in the tank at that moment, and let $M(t)$ be the total mass of liquid in the tank at any time thereafter.
   (a) Show that the leakage rate of hydrogen peroxide at any time is $\dot{m}_1 M_p/M$.
   (b) Write differential balances on the total tank contents and on the peroxide in the tank, and provide initial conditions. Your solution should involve only the quantities $\dot{m}_0$, $\dot{m}_1$, x_{p0}, k, M, M_0, M_p, M_{p0}, and t.

11.2. One hundred fifty kmol of an aqueous phosphoric acid solution contains 5.00 mole% H_3PO_4. The solution is concentrated by adding pure phosphoric acid at a rate of 20.0 L/min.
   (a) Write a differential mole balance on phosphoric acid and provide an initial condition. [Start by defining n_p(kmol) to be the total quantity of phosphoric acid in the tank at any time.]
   (b) Solve the balance to obtain an expression for $n_p(t)$. Use the result to derive an expression for $x_p(t)$, the mole fraction of phosphoric acid in the solution.
   (c) How long will it take to concentrate the solution to 15% H_3PO_4?

11.3. Methanol is added to a storage tank at a rate of 1200 kg/h and is simultaneously withdrawn at a rate $\dot{m}_w(t)$ (kg/h) that increases linearly with time. At $t = 0$ the tank contains 750 kg of the liquid and $\dot{m}_w = 750$ kg/h. Five hours later $\dot{m}_w$ equals 1000 kg/h.
   (a) Calculate an expression for $\dot{m}_w(t)$, letting $t = 0$ signify the time at which $\dot{m}_w = 750$ kg/h, and incorporate it into a differential methanol balance, letting M (kg) be the mass of methanol in the tank at any time.
   (b) Integrate the balance equation to obtain an expression for $M(t)$ and check the solution two ways. (See Example 11.2-1.) For now, assume that the tank has an infinite capacity.
   (c) Calculate how long it will take for the mass of methanol in the tank to reach its maximum value, and calculate that value. Then calculate the time it will take to empty the tank.
   (d) Now suppose the tank volume is 3.40 m^3. Draw a plot of M versus t, covering the period from $t = 0$ to an hour after the tank is empty. Write expressions for $M(t)$ in each time range when the function changes.

11.4. A 10.0-ft^3 compressed-air tank is being filled. Before the filling begins, the tank is open to the atmosphere. The reading on a Bourdon gauge mounted on the tank increases linearly from an initial value of 0.0 to 100 psi after 15 seconds. The temperature is constant at 72°F, and atmospheric pressure is 1 atm.

   (a) Calculate the rate $\dot{n}$(lb-mole/s) at which air is being added to the tank, assuming ideal gas behavior. (*Suggestion:* Start by calculating how much is in the tank at $t = 0$.)

   (b) Let $N(t)$ equal the number of lb-moles of air in the tank at any time. Write a differential balance on the air in the tank in terms of N and provide an initial condition.

   (c) Integrate the balance to obtain an expression for $N(t)$. Check your solution two ways.

   (d) Estimate the number of lb-moles of *oxygen* in the tank after two minutes.

11.5. A gas storage tank with a floating roof receives a steady input of 540 m^3/h of a natural gas. The rate of withdrawal of gas from the tank, $\dot{v}_w$, varies more or less randomly during the day and is recorded at 10-min intervals. At 8:00 a.m. one morning the volume of stored gas is 3.00×10^3 m^3. The withdrawal rate data for the next 4 hours are as follows:

Hour Beginning at	$\dot{v}_w$(m^3/min)
8:00	11.4, 11.9, 12.1, 11.8, 11.5, 11.3
9:00	11.4, 11.1, 10.6, 10.8, 10.4, 10.2
10:00	10.2, 9.8, 9.4, 9.5, 9.3, 9.4
11:00	9.5, 9.3, 9.6, 9.6, 9.4, 9.9
12:00	9.8

The temperature and pressure of the inlet, stored, and outlet gases are equal and nearly constant throughout the given time period.

   (a) Write a differential balance on the moles of gas in the tank, and prove that when integrated it yields the following equation for the gas volume:

$$V(t) = 3.00 \times 10^3 + 9.00t - \int_0^t \dot{v}_w \, dt$$

   where t(min) is the time elapsed since 8:00 a.m.

   (b) Calculate the stored gas volume at noon, using Simpson's rule (Appendix A.3) to evaluate the integral.

   (c) Although a running estimate of the tank volume is important to have, in practice it would probably not be obtained in the manner indicated. How would it be obtained? What might you infer if the value estimated in part (b) is greater than that obtained by the more accurate method?

   ***(d)** Create a spreadsheet or a computer program to read in the data for $\dot{v}_w(t)$ and to estimate and print out the gas volume at each time for which $\dot{v}_w$ is recorded, using the trapezoidal rule (Appendix A.3) to evaluate the integral over each successive 10-min interval. What is the percentage difference between the volumes at noon estimated by Simpson's rule and the trapezoidal rule? Which estimate should be more accurate?

11.6. Water is added at varying rates to a 300-liter holding tank. When a valve in a discharge line is opened, water flows out at a rate proportional to the height and hence to the volume V of water in the tank. The flow of water into the tank is slowly increased and the level rises in consequence, until at a steady input rate of 60.0 L/min the level just reaches the top but does not spill over. The input rate is then abruptly decreased to 20.0 L/min.

   (a) Write the equation that relates the discharge rate, $\dot{v}_{out}$(L/min), to the volume of water in the tank, V(L), and use it to calculate the steady-state volume when the input rate is 20 L/min.

   (b) Write a differential balance on the water in the tank for the period from the moment the input rate is decreased ($t = 0$) to the attainment of steady state ($t \rightarrow \infty$), expressing it in the form $dV/dt = \cdots$. Provide an initial condition.

   (c) Without integrating the equation, use it to confirm the steady-state value of V calculated in part (a) and then to predict the shape you would anticipate for a plot of V versus t. Explain your reasoning.

---

*Computer problem.

(d) Separate variables and integrate the balance equation to derive an expression for $V(t)$. Calculate the time in minutes required for the volume to decrease to within 1% of its steady-state value.

11.7. The production supervisor of a small pharmaceutical firm has observed a decreasing demand for potassium regurgitol (PRG) over a two-month period, and since the plant manager has been throwing up the low sales of this product at the weekly staff meetings, the supervisor decides to discontinue its production immediately. On the day of this decision, the inventory of PRG is 18,000 kg. Based on the orders on hand, the manager projects the following weekly demand for the next six weeks:

Week	1	2	3	4	5	6
Demand $\dot{D}$(kg/wk)	2385	1890	1506	1196	950	755

(a) Use a semilog plot of the projected demand figures to derive an equation for $\dot{D}$ as a function of t(weeks) from the present time.

(b) Write a differential balance on the inventory I(kg) of PRG, and integrate it to determine I as a function of t.

(c) If the demand continues to follow the projected trend of the next six weeks, how much PRG will eventually have to be discarded?

11.8. A ventilation system has been designed for a large laboratory with a volume of 1100 m³. The volumetric flow rate of ventilation air is 700 m³/min at 22°C and 1 atm. (The latter two values may also be taken as the temperature and pressure of the room air.) A reactor in the laboratory is capable of emitting as much as 1.50 mol of sulfur dioxide into the room if a seal ruptures. An SO_2 mole fraction in the room air greater than 1.0×10^{-6} (1 ppm) constitutes a health hazard.

(a) Suppose the reactor seal ruptures at a time $t = 0$ and the maximum amount of SO_2 is emitted and spreads uniformly throughout the room almost instantaneously. Assuming that the air flow is sufficient to make the room air composition spatially uniform, write a differential SO_2 balance, letting N be the total moles of gas in the room (assume constant) and $x(t)$ the mole fraction of SO_2 in the laboratory air. Convert the balance into an equation for dx/dt and provide an initial condition. (Assume that all of the SO_2 emitted is in the room at $t = 0$.)

(b) Predict the shape of a plot of x versus t. Explain your reasoning, using the equation of part (a) in your explanation.

(c) Separate variables and integrate the balance to obtain an expression for $x(t)$. Check your solution.

(d) Convert the expression for $x(t)$ into an expression for the concentration of SO_2 in the room, C_{SO_2} (mol SO_2/L). Calculate (i) the concentration of SO_2 in the room two minutes after the rupture occurs, and (ii) the time required for the SO_2 concentration to reach the "safe" level.

(e) Why would it probably not yet be safe to enter the room after the time calculated in part (d)? (*Hint:* One of the assumptions made in the problem is probably not a good one.)

11.9. A gas leak has led to the presence of 1.00 mole% carbon monoxide in a 350-m³ laboratory.[4] The leak was discovered and sealed, and the laboratory is to be purged with clean air to a point at which the air contains less than the OSHA (Occupational Safety and Health Administration)-specified Permissible Exposure Level (PEL) of 35 ppm (molar basis). Assume that the clean air and the air in the laboratory are at the same temperature and pressure and that the laboratory air is perfectly mixed throughout the purging process.

(a) Let t_r(h) be the time required for the specified reduction in the carbon monoxide concentration. Write a differential CO mole balance, letting N equal the total moles of gas in the room (assume constant), x the mole fraction of CO in the room air, and $\dot{v}_p$(m³/h) the flow rate of purge air entering the room (and also the flow rate of laboratory air leaving the room). Convert the balance into an equation for dx/dt and provide an initial condition.

(b) Integrate the balance to derive an equation for t_r in terms of $\dot{v}_p$.

(c) If the volumetric flow rate is 700 m³/h (representing a turnover of two room volumes per hour), how long will the purge take? What would the volumetric flow rate have to be to cut the purge time in half?

---

[4]D. A. Crowl, D. W. Hubbard, and R. M. Felder, *Problem Set: Stoichiometry,* AIChE/CCPS, New York, 1993.

(d) Give several reasons why it might not be safe to resume work in the laboratory after the calculated purge time has elapsed? What precautionary steps would you advise taking at this point?

11.10. Ninety kilograms of sodium nitrate is dissolved in 110 kg of water. When the dissolution is complete (at time $t = 0$), pure water is fed to the tank at a constant rate $\dot{m}$(kg/min), and solution is withdrawn from the tank at the same rate. The tank may be considered perfectly mixed.

(a) Write a total mass balance on the tank and use it to prove that the total mass of liquid in the tank remains constant at its initial value.

(b) Write a balance on sodium nitrate, letting $x(t, \dot{m})$ equal the mass fraction of $NaNO_3$ in the tank and outlet stream. Convert the balance into an equation for dx/dt and provide an initial condition.

(c) On a single graph of x versus t, sketch the shapes of the plots you would expect to obtain for $\dot{m} = 50$ kg/min, 100 kg/min, and 200 kg/min. (Don't do any calculations.) Explain your reasoning, using the equation of part (b) in your explanation.

(d) Separate variables and integrate the balance to obtain an expression for $x(t, \dot{m})$. Check your solution. Then generate the plots of x versus t for $\dot{m} = 50$ kg/min, 100 kg/min, and 200 kg/min and show them on a single graph. (A spreadsheet is a convenient tool for carrying out this step.)

(e) If $\dot{m} = 100$ kg/min, how long will it take to flush out 90% of the sodium nitrate originally in the tank? How long to flush out 99%? 99.9%?

11.11. A *tracer* is used to characterize the degree of mixing in a continuous stirred tank. Water enters and leaves the mixer at a rate of $\dot{v}$(m³/min). Scale has built up on the inside walls of the tank, so that the effective volume V(m³) of the tank is unknown. At time $t = 0$, a mass m_0(kg) of the tracer is injected into the tank and the tracer concentration in the outlet stream, C(kg/m³), is monitored.

(a) Write a differential balance on the tracer in the tank in terms of V, C, and $\dot{v}$, assuming that the tank contents are perfectly mixed, and convert the balance into an equation for dC/dt. Provide an initial condition, assuming that the injection is rapid enough so that all of the tracer may be considered to be in the tank at $t = 0$.

(b) Integrate the balance to prove that

$$C(t) = (m_0/V) \exp(-\dot{v}t/V)$$

(c) Suppose the flow rate through the mixer is $\dot{v} = 30.0$ m³/min and that the following data are taken:

Time from injection, t(min)	1	2	3	4
$C \times 10^3$ (kg/m³)	0.223	0.050	0.011	0.0025

(For example, at $t = 1$ min, $C = 0.223 \times 10^{-3}$ kg/m³.) Verify graphically that the tank is functioning as a perfect mixer—that is, that the expression of part (b) fits the data—and determine the effective volume V(m³) from the slope of your plot.

11.12. A 40.0-ft³ oxygen tent initially contains air at 68°F and 14.7 psia. At a time $t = 0$ an enriched air mixture containing 35.0% v/v O_2 and the balance N_2 is fed to the tent at 68°F and 1.3 psig at a rate of 60.0 ft³/min, and gas is withdrawn from the tent at 68°F and 14.7 psia at a molar flow rate equal to that of the feed gas.

(a) Calculate the total lb-moles of gas ($O_2 + N_2$) in the tent at any time.

(b) Let $x(t)$ equal the mole fraction of oxygen in the outlet stream. Write a differential mole balance on oxygen, assuming that the tent contents are perfectly mixed (so that the temperature, pressure, and composition of the contents are the same as those properties of the exit stream). Convert the balance into an equation for dx/dt and provide an initial condition.

(c) Integrate the equation to obtain an expression for $x(t)$. How long will it take for the mole fraction of oxygen in the tent to reach 0.27?

11.13. A radioactive isotope decays at a rate proportional to its concentration. If the concentration of an isotope is C(mg/L), then its rate of decay may be expressed as

$$r_d[\text{mg/(L·s)}] = kC$$

where k is a constant.

(a) A volume V(L) of a solution of a radioisotope whose concentration is C_0(mg/L) is placed in a closed vessel. Write a balance on the isotope in the vessel and integrate it to prove that the **half-life** $t_{1/2}$ of the isotope—by definition, the time required for the isotope concentration to decrease to half of its initial value—equals $(\ln 2)/k$.

(b) The half-life of ^{56}Mn is 2.6 h. A batch of this isotope that was used in a radiotracing experiment has been collected in a holding tank. The radiation safety officer declares that the activity (which is proportional to the isotope concentration) must decay to 1% of its present value before the solution can be discarded. How long will this take?

11.14. A chemical reaction with stoichiometry A $\rightarrow$ products is said to follow an nth-order rate law if A is consumed at a rate proportional to the nth power of its concentration in the reaction mixture. If r_A is the rate of consumption of A per unit reactor volume, then

$$r_A[\text{mol/(L·s)}] = kC_A^n$$

where C_A(mol/L) is the reactant concentration, and the constant of proportionality k is the reaction *rate constant*. A reaction that follows this law is referred to as an nth order reaction. The rate constant is a strong function of temperature but is independent of the reactant concentration.

(a) Suppose a first-order reaction ($n = 1$) is carried out in an isothermal batch reactor of constant volume V. Write a material balance on A and integrate it to derive the expression

$$C_A = C_{A0} \exp(-kt)$$

where C_{A0} is the concentration of A in the reactor at $t = 0$.

(b) The gas-phase decomposition of sulfuryl chloride

$$SO_2Cl_2 \rightarrow SO_2 + Cl_2$$

is thought to follow a first-order rate law. The reaction is carried out in a constant-volume isothermal batch reactor and the concentration of SO_2Cl_2 is measured at several reaction times, with the following results:

t(min)	4.0	21.3	39.5	63.4	120.0	175.6
C_A(mol/L)	0.0279	0.0262	0.0246	0.0226	0.0185	0.0152

Verify the proposed rate law graphically [i.e., demonstrate that the expression given in part (a) fits the data for $C_A(t)$] and determine the rate constant k, giving both its value and its units.

11.15. A gas-phase decomposition reaction with stoichiometry 2A $\rightarrow$ 2B + C follows a second-order rate law (see Problem 11.14):

$$r_d[\text{mol/(m}^3\text{·s)}] = kC_A^2$$

where C_A is the reactant concentration in mol/m^3. The rate constant k varies with the reaction temperature according to the **Arrhenius law**

$$k[\text{m}^3\text{/(mol·s)}] = k_0 \exp(-E/RT)$$

where

$$k_0[\text{m}^3\text{/(mol·s)}] = \text{the \textit{preexponential factor}}$$
$$E(\text{J/mol}) = \text{the reaction \textit{activation energy}}$$
$$R = \text{the gas constant}$$
$$T(\text{K}) = \text{the reaction temperature}$$

(a) Suppose the reaction is carried out in a batch reactor of constant volume V(m^3) at a constant temperature T(K), beginning with pure A at a concentration C_{A0}. Write a differential balance on A and integrate it to obtain an expression for $C_A(t)$ in terms of C_{A0} and k.

(b) Let P_0(atm) be the initial reactor pressure. Prove that $t_{1/2}$, the time required to achieve a 50% conversion of A in the reactor, equals RT/kP_0, and derive an expression for $P_{1/2}$, the reactor pressure at this point, in terms of P_0. Assume ideal gas behavior.

(c) The decomposition of nitrous oxide (N_2O) to nitrogen and oxygen is carried out in a 5.00-liter batch reactor at a constant temperature of 1015 K, beginning with pure N_2O at several initial pressures. The reactor pressure $P(t)$ is monitored, and the times ($t_{1/2}$) required to achieve 50% conversion of N_2O are noted.

P_0(atm)	0.135	0.286	0.416	0.683
$t_{1/2}$(s)	1060	500	344	209

Use these results to verify that the N_2O decomposition reaction is second-order and determine the value of k at $T = 1015$ K.

(d) The same experiment is performed at several other temperatures at a single initial pressure of 1.00 atm, with the following results:

T(K)	900	950	1000	1050
$t_{1/2}$(s)	5464	1004	219	55

Use a graphical method to determine the Arrhenius law parameters (k_0 and E) for the reaction.

(e) Suppose the reaction is carried out in a batch reactor at $T = 980$ K, beginning with a mixture at 1.20 atm containing 70 mole% N_2O and the balance a chemically inert gas. How long (minutes) will it take to achieve a 90% conversion of N_2O?

11.16. In an enzyme-catalyzed reaction with stoichiometry A → B, A is consumed at a rate given by an expression of the **Michaelis–Menten** form:

$$r_A[\text{mol/(L·s)}] = \frac{k_1 C_A}{1 + k_2 C_A}$$

where C_A(mol/L) is the reactant concentration, and k_1 and k_2 depend only on temperature.

(a) The reaction is carried out in an isothermal batch reactor with constant reaction mixture volume V (liters), beginning with pure A at a concentration C_{A0}. Write a balance on A and integrate it to obtain an expression for the time required to achieve a specified concentration C_A.

(b) Use the expression of part (a) to devise a graphical method of determining k_1 and k_2 from data for C_A versus t. Your plot should involve fitting a straight line and determining the two parameters from the slope and intercept of the line. (There are several possible solutions.) Then apply your method to determine k_1 and k_2 for the following data taken in a 2.00-liter reactor, beginning with A at a concentration $C_{A0} = 5.00$ mol/L.

t(s)	60.0	120.0	180.0	240.0	480.0
C_A(mol/L)	4.484	4.005	3.561	3.154	1.866

11.17. Phosgene ($COCl_2$) is formed by CO and Cl_2 reacting in the presence of activated charcoal:

$$CO + Cl_2 \rightarrow COCl_2$$

At $T = 303.8$ K the rate of formation of phosgene in the presence of 1 gram of charcoal is

$$R_f(\text{mol/min}) = \frac{8.75 C_{CO} C_{Cl_2}}{(1 + 58.6 C_{Cl_2} + 34.3 C_{COCl_2})^2}$$

where C denotes concentration in mol/L.

(a) Suppose the charge to a 3.00-liter batch reactor is 1.00 g of charcoal and a gas containing 60 mole% CO and 40 mole% Cl_2, and that the initial reactor conditions are 303.8 K and 1 atm. Calculate the initial concentrations (mol/L) of both reactants, neglecting the volume occupied by the charcoal. Then, letting $C_P(t)$ be the concentration of phosgene at an arbitrary time t, derive relations for C_{CO} and C_{Cl_2} in terms of C_P.

(b) Write a differential balance on phosgene and show that it simplifies to

$$\frac{dC_P}{dt} = \frac{2.92(0.02407 - C_P)(0.01605 - C_P)}{(1.941 - 24.3 C_P)^2}$$

Provide an initial condition for this equation.

(c) Starting with the equation of part (b), derive an expression for the time required to achieve a 75% conversion of the limiting reactant. *Your solution should have the form t = a definite integral.*

*(d) The integral you derived in part (c) can be evaluated analytically; however, more complex rate laws than the one given for the phosgene formation reaction would yield an integral that must be evaluated numerically. One procedure is to evaluate the integrand at a number of points between the limits of integration and to use a quadrature formula such as the trapezoidal rule or Simpson's rule (Appendix A.3) to estimate the value of the integral.

Use a spreadsheet or write a computer program to evaluate the integrand of the integral of part (c) at n_p equally spaced points between and including the limits of integration, where n_p is an odd number, and then to evaluate the integral using Simpson's rule. Perform the calculation for $n_p = 5, 21,$ and 51, and compare the results with the exact value of the integral.

11.18. A gas that contains CO_2 is contacted with liquid water in an agitated batch absorber. The equilibrium solubility of CO_2 in water is given by Henry's law (Section 6.4b)

$$C_A = p_A/H_A$$

where
$$C_A(\text{mol/cm}^3) = \text{concentration of } CO_2 \text{ in solution}$$
$$p_A(\text{atm}) = \text{partial pressure of } CO_2 \text{ in the gas phase}$$
$$H_A[\text{atm/(mol/cm}^3)] = \text{Henry's law constant}$$

The rate of absorption of CO_2 (i.e., the rate of transfer of CO_2 from the gas to the liquid per unit area of gas–liquid interface) is given by the expression

$$r_A[\text{mol/(cm}^2\cdot\text{s})] = k(C_A^* - C_A)$$

where
$$C_A = \text{actual concentration of } CO_2 \text{ in the liquid}$$
$$C_A^* = \text{concentration of } CO_2 \text{ that would be in equilibrium with the } CO_2 \text{ in the gas phase } (C_A^* = p_A/H_A)$$
$$k(\text{cm/s}) = \text{a } mass\ transfer\ coefficient$$

The gas phase is at a total pressure $P(\text{atm})$ and contains $y_A(\text{mol } CO_2/\text{mol gas})$, and the liquid phase initially consists of $V(\text{cm}^3)$ of pure water. The agitation of the liquid phase is sufficient for its composition to be considered spatially uniform, and the amount of CO_2 absorbed is low enough for P, V, and y_A to be considered constant throughout the process.

(a) Write a differential balance on CO_2 in the liquid phase and integrate it to derive the relation

$$C_A(t) = C_A^*[1 - \exp(-kSt/V)]$$

where $S(\text{cm}^2)$ is the effective contact area between the gas and liquid phases.

(b) Suppose the system pressure is 20.0 atm, the liquid volume is 5.00 liters, the tank diameter is 10.0 cm, the gas contains 30.0 mole% CO_2, the Henry's law constant is 9230 atm/(mole/cm³), and the mass transfer coefficient is 0.020 cm/s. Calculate the time required for C_A to reach 0.620 mol/L if the gas-phase properties remain essentially constant.

11.19. A liquid-phase chemical reaction with stoichiometry $A \rightarrow B$ takes place in a semibatch reactor. The rate of consumption of A per unit volume of the reactor contents is given by the first-order rate expression (see Problem 11.14)

$$r_A[\text{mol/(L}\cdot\text{s})] = kC_A$$

where C_A (mol A/L) is the reactant concentration. The tank is initially empty. Beginning at a time $t = 0$, a solution containing A at a concentration C_{A0}(mol A/L) is fed to the tank at a steady rate $\dot{v}(\text{L/s})$.

(a) Write a differential balance on the total mass of the reactor contents. Assuming that the density of the contents always equals that of the feed stream, convert the balance into an equation for dV/dt, where V is the total volume of the contents, and provide an initial condition. Then write a differential mole balance on the reactant, A, letting $N_A(t)$ equal the total moles of A in the vessel, and provide an initial condition. Your equations should contain only the variables N_A, V, and t and the constants $\dot{v}$ and C_{A0}. (You should be able to eliminate C_A as a variable.)

—————

*Computer problem.

(b) Without attempting to integrate the equations, derive a formula for the steady-state value of N_A.

(c) Integrate the two equations to derive expressions for $V(t)$ and $N_A(t)$, and then derive an expression for $C_A(t)$. Determine the asymptotic value of N_A as $t \to \infty$ and verify that the steady-state value obtained in part (b) is correct. Briefly explain how it is possible for N_A to reach a steady value when you keep adding A to the reactor and then give two reasons why this value would never be reached in a real reactor.

(d) Determine the limiting value of C_A as $t \to \infty$ from your expressions for $N_A(t)$ and $V(t)$. Then explain why your result makes sense in light of the results of part (c).

11.20. A kettle containing 3.00 liters of water at a temperature of 18°C is placed on an electric stove and begins to boil in four minutes.

(a) Write an energy balance on the water, neglecting evaporation that may occur during the heating.

(b) What is the average rate (W) at which heat is added to the water during this period? (See Example 11.3-1.)

(c) The rate of heat output from the stove element differs significantly from the heating rate calculated in part (b). In which direction, and why?

11.21. An electrical coil is used to heat 20.0 kg of water in a closed well-insulated vessel. The water is initially at 25°C and 1 atm. The coil delivers a steady 2.50 kW of power to the vessel and its contents.

(a) Write a differential energy balance on the water, assuming that 97% of the energy delivered by the coil goes into heating the water. What happens to the other 3%?

(b) Integrate the equation of part (a) to derive an expression for the water temperature as a function of time.

(c) How long will it take for the water to reach the normal boiling point? Will it boil at this temperature? Why not?

11.22. An iron bar 2.00 cm × 3.00 cm × 10.0 cm at a temperature of 95°C is dropped into a barrel of water at 25°C. The barrel is large enough so that the water temperature rises negligibly as the bar cools. The rate at which heat is transferred from the bar to the water is given by the expression

$$\dot{Q}(\text{J/min}) = UA(T_b - T_w)$$

where $U[= 0.050 \text{ J/(min·cm}^2\text{·°C)}]$ is a *heat transfer coefficient*, $A(\text{cm}^2)$ is the exposed surface area of the bar, and $T_b(\text{°C})$ and $T_w(\text{°C})$ are the surface temperature of the bar and the water temperature, respectively.

The heat capacity of the bar is 0.460 J/(g·°C). Heat conduction in iron is rapid enough for the temperature $T_b(t)$ to be considered uniform throughout the bar.

(a) Write an energy balance on the bar, assuming that all six sides are exposed. Your result should be an expression for dT_b/dt and an initial condition.

(b) Without integrating the equation, sketch the expected plot of T_b versus t, labeling the values of T_b at $t = 0$ and $t \to \infty$.

(c) Derive an expression for $T_b(t)$ and check it three ways. How long will it take for the bar to cool to 30°C?

11.23. A steam coil is immersed in a stirred heating tank. Saturated steam at 7.50 bar condenses within the coil, and the condensate emerges at its saturation temperature. A solvent with a heat capacity of 2.30 kJ/(kg·°C) is fed to the tank at a steady rate of 12.0 kg/min and a temperature of 25°C, and the heated solvent is discharged at the same flow rate. The tank is initially filled with 760 kg of solvent at 25°C, at which point the flows of both steam and solvent are commenced. The rate at which heat is transferred from the steam coil to the solvent is given by the expression

$$\dot{Q} = UA(T_{\text{steam}} - T)$$

where UA (the product of a heat transfer coefficient and the coil surface area through which the heat is transferred) equals 11.5 kJ/(min·°C). The tank is well stirred, so that the temperature of the contents is spatially uniform and equals the outlet temperature.

(a) Prove that an energy balance on the tank contents reduces to the equation given below and supply an initial condition.

$$\frac{dT}{dt} = 1.50°\text{C/min} - 0.0224T$$

(b) Without integrating the equation, calculate the steady-state value of T and sketch the expected plot of T versus t, labeling the values of T_b at $t = 0$ and $t \rightarrow \infty$.

(c) Integrate the balance equation to obtain an expression for $T(t)$ and calculate the solvent temperature after 40 minutes.

(d) A new stirrer assembly has been installed in the heating tank. The process described above is run again with the same steam conditions, solvent flow rate, and mass of solvent charged to the tank, and the temperature after 40 minutes is 55°C instead of the value calculated in part (c). One of the system variables listed in the problem statement must have changed as a result of the change in the stirrer. Which variable would you guess it to be, and by what percentage of its initial value did it change?

11.24. At 9:30 one morning, a graduate student measures 350 grams of liquid benzene at 20°C into a glass flask dirty enough that its contents cannot be seen, puts the open flask on a bunsen burner, turns the burner on, and goes for a coffee break. The conversation at the break is lively and he doesn't get back until 10:10 a.m. He looks down into the flask, sees the liquid is boiling, turns the burner off, feels a little irritation in his eye and rubs the eye with his hand, picks up the flask, says "Ouch" (or something roughly equivalent), puts the flask down on his laboratory partner's thermodynamics homework, and starts to prepare the next step of the experiment.

(a) Suppose the heat input rate to the flask contents is 40.2 W. Calculate the time at which the benzene temperature reached 40°C. Neglect evaporation of benzene during the heating and take the heat capacity of liquid benzene to be constant at 1.77 J/(g·°C).

(b) Calculate the quantity of benzene left in the flask at 10:10 a.m., assuming that once the benzene starts boiling, the rate of heat input to the flask (40.2 W) equals the rate of vaporization (g/s) times the heat of vaporization (J/g).

(c) The graduate student was lucky. First, neither his supervisor nor the university safety officer came into the laboratory during this episode. More importantly, he was still alive and well at the end of the day. Identify as many of his safety violations as you can, explaining the danger and suggesting for each violation what he should have done instead.

11.25. A steam radiator is used to heat a 60-m^3 room. Saturated steam at 3.0 bar condenses in the radiator and emerges as a liquid at the saturation temperature. Heat is lost from the room to the outside at a rate

$$\dot{Q}(kJ/h) = 30.0(T - T_0)$$

where $T(°C)$ is the room temperature and $T_0 = 0°C$ is the outside temperature. At the moment the radiator is turned on, the temperature in the room is 10°C.

(a) Let $\dot{m}_s(kg/h)$ denote the rate at which steam condenses in the radiator and $n(kmol)$ the quantity of air in the room. Write a differential energy balance on the room air, assuming that n remains constant at its initial value, and evaluate all numerical coefficients. Take the heat capacity of air (C_v) to be constant at 20.8 J/(mol·°C).

(b) Write the steady-state energy balance on the room air and use it to calculate the steam condensation rate required to maintain a constant room temperature of 24°C.

(c) Integrate the transient balance to calculate the time required to achieve a temperature of 23°C, assuming that the steam rate is that calculated in part (b).

11.26. An immersed electrical heater is used to raise the temperature of a liquid from 20°C to 60°C in 20.0 min. The combined mass of the liquid and the container is 250 kg, and the mean heat capacity of the system is 4.00 kJ/(kg·°C). The liquid decomposes explosively at 85°C.

At 10.00 a.m. a batch of liquid is poured into the vessel and the operator turns on the heater and leaves to make a phone call. Ten minutes later, his supervisor walks by and looks at the strip chart record of the power input. This what she sees.

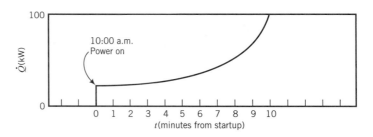

The supervisor immediately shuts off the heater and charges off to pass on to the operator several brief observations that come to her mind.

(a) Calculate the required constant power input $\dot{Q}(kW)$, neglecting energy losses from the container.

(b) Write and integrate using Simpson's rule (Appendix A.3) an energy balance on the system to estimate the system temperature at the moment the heater is shut off. Use the following data from the recorder chart:

$t(s)$	0	30	60	90	120	150	180	210	240	270	300
$\dot{Q}(kW)$	33	33	34	35	37	39	41	44	47	50	54

$t(s)$	330	360	390	420	450	480	510	540	570	600
$\dot{Q}(kW)$	58	62	66	70	75	80	85	90	95	100

(c) Suppose that if the heat had not been shut off, $\dot{Q}$ would have continued to increase linearly at a rate of 10 kW/min. At what time would everyone in the plant realize that something was wrong?

11.27. A 2000-liter tank initially contains 400 liters of pure water. Beginning at $t = 0$, an aqueous solution containing 1.00 g/L of potassium chloride flows into the tank at a rate of 8.00 L/s and an outlet stream simultaneously starts flowing at a rate of 4.00 L/s. The contents of the tank are perfectly mixed, and the density of the feed stream and of the tank solution, $\rho(g/L)$, may be considered constant. Let $V(t)(L)$ denote the volume of the tank contents and $C(t)(g/L)$ the concentration of potassium chloride in the tank contents and outlet stream.

(a) Write a total mass balance on the tank contents, convert it to an equation for dV/dt, and provide an initial condition. Then write a potassium chloride balance, convert it to an equation of the form $dC/dt = f(C, V)$, and provide an initial condition. (See Example 11.4-1.)

(b) Without solving either equation, sketch the plots you would expect to obtain for V versus t and C versus t. Briefly explain your reasoning.

(c) Solve the mass balance equation to obtain an expression for $V(t)$. Then substitute for V in the potassium chloride balance and solve for $C(t)$. Calculate the KCl concentration in the tank at the moment the tank overflows.

11.28. The diagram below shows three continuous stirred tanks connected in series.

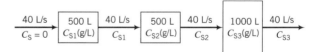

The flow and mixing patterns in this system are studied by dissolving 1500 g of a salt (S) in the first tank, filling each of the other two tanks with pure solvent, and then starting the 40 L/s flow through the system. Each tank outlet stream is monitored with an on-line thermal conductivity detector calibrated to provide instantaneous readings of salt concentration. The data are plotted versus time and the results are compared with the plots that would be expected if the tanks are all perfectly mixed. Your job is to generate the latter plots.

(a) Assuming that pure solvent is fed to the first tank and that each tank is perfectly mixed (so that the salt concentration in a tank is uniform and equal to the concentration in the outlet stream from that tank), write salt balances on each of the three tanks, convert them to expressions for dC_{S1}/dt, dC_{S2}/dt, and dC_{S3}/dt, and provide appropriate initial conditions.

(b) Without doing any calculations, on a single graph sketch the forms of the plots of C_{S1} versus t, C_{S2} versus t, and C_{S3} versus t you would expect to obtain. Briefly explain your reasoning.

*(c) Use a differential equation-solving program to solve the three equations, proceeding to a time at which C_{S3} has fallen below 0.01 g/L, and plot the results.

---

*Computer problems.

11.29. The following chemical reactions take place in a liquid-phase batch reactor of constant volume V.

$$A \rightarrow 2B \qquad r_1[\text{mol A consumed}/(\text{L}\cdot\text{s})] = 0.100C_A$$

$$B \rightarrow C \qquad r_2[\text{mol C generated}/(\text{L}\cdot\text{s})] = 0.200C_B^2$$

where the concentrations C_A and C_B are in mol/L. The reactor is initially charged with pure A at a concentration of 1.00 mol/L.

(a) Write expressions for (i) the rate of generation of B in the first reaction and (ii) the rate of consumption of B in the second reaction. (If this takes you more than about 10 seconds, you're missing the point.)

(b) Write mole balances on A, B, and C, convert them into expressions for dC_A/dt, dC_B/dt, and dC_C/dt, and provide boundary conditions.

(c) Without doing any calculations, sketch on a single graph the plots you would expect to obtain of C_A versus t, C_B versus t, and C_C versus t. Clearly show the function values at $t = 0$ and $t \rightarrow \infty$ and the curvature (concave up, concave down, or linear) in the vicinity of $t = 0$. Briefly explain your reasoning.

*(d) Solve the equations derived in part (b) using a differential equation-solving program. On a single graph, show plots of C_A versus t, C_B versus t, and C_C versus t from $t = 0$ to $t = 50$ s. Verify that your predictions in part (c) were correct. If they were not, change them and revise your explanation.

11.30. A liquid mixture containing 70.0 mol of n-pentane and 30.0 mol of n-hexane initially at 46°C is partially vaporized at $P = 1$ atm in a single-stage distillation apparatus (**Rayleigh still**).

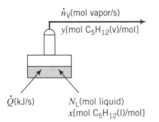

The heat added to the system, $\dot{Q}$, vaporizes liquid at the rate $\dot{n}_V$(mol/s). The vapor product and remaining liquid at a given moment are always in equilibrium with each other. The relationship between the mole fraction of pentane in the liquid (x) and that in the vapor (y) is of the form

$$y = \frac{ax}{x + b}$$

so that the system involves four time-dependent variables—N_L, $\dot{n}_V$, x, and y, where N_L is the total moles of liquid in the still at any time. (We will suppose that the rate of heat transfer to the evaporator, $\dot{Q}$, is constant and known.) Four equations relating the unknowns will be required to determine these variables. The equations are two material balances, an energy balance, and the vapor–liquid equilibrium relationship just given.

(a) When $x = 1$, what must y equal? (Think of the definitions of these quantities.) Use your answer and the vapor–liquid equilibrium expression to derive an equation relating the parameters a and b.

(b) Use Raoult's law (Equation 6.4-1) and the Antoine equation to calculate the mole fraction of pentane in the vapor phase in equilibrium with the 70% pentane–30% hexane feed mixture at the initial system temperature of 46°C and a pressure of 1 atm. Then use this result and that of part (a) to estimate a and b. (We will assume that these values remain the same over the range of compositions and temperatures to be undergone by the system.)

(c) Taking the residual liquid in the still as your system, write a differential balance on total moles to obtain an expression for dN_L/dt. Then write a balance on pentane, recognizing that both N_L and x are functions of time. (*Hint:* Remember the product rule for differentiation.) Prove that the pentane balance can be converted into the following equation:

$$\frac{dx}{dt} = \frac{\dot{n}_V}{N_L}\left(\frac{ax}{x+b} - x\right)$$

Supply initial conditions for your two differential equations.

(d) In part (c), you derived two equations in three unknown dependent variables—$\dot{n}_V(t)$, $N_L(t)$, and $x(t)$. To determine these variables, we need a third relationship. An energy balance provides it.

A rigorous energy balance would take into account the changing composition of the liquid, the slightly different heats of vaporization of pentane and hexane, and the enthalpy changes associated with temperature changes, and would make the problem relatively hard to solve. A reasonable approximation is to assume that (i) the liquid has a constant heat of vaporization of 27.0 kJ/mol, independent of composition and temperature; and (ii) all heat supplied to the still $[\dot{Q}(\text{kJ/s})]$ goes to vaporize liquid (i.e., we neglect energy that goes into raising the temperature of the liquid or the vapor). Make these assumptions, consider $\dot{Q}$ to be constant and known, and derive a simple expression for $\dot{n}_V$ that can be used to eliminate this variable in the differential equations of part (c). From there, derive the following expression:

$$\frac{dx}{dt} = -\frac{\dot{Q}/27.0}{100.0 \text{ mol} - \dot{Q}t/27.0}\left(\frac{ax}{x+b} - x\right)$$

*(e) Use a differential equation-solving program to calculate x, y, N_L, and $\dot{n}_V$ from $t = 0$ until the time at which the liquid completely evaporates. Do the calculation for (i) $\dot{Q} = 1.5$ kJ/s and (ii) $\dot{Q} = 3.0$ kJ/s. On a single graph, plot x and y versus t, showing curves for both values of $\dot{Q}$.

(f) In a short paragraph, describe what happens to the compositions of the vapor product and residual liquid over the course of a run. Include a statement of what the initial and final vapor compositions are and how the heating rate affects the system behavior.

---

*Computer problem.

Part Four

Case Studies

Chapter 12

Production of Chlorinated Polyvinyl Chloride[1]

Certain types of molecules can react to form **polymers**—long molecular chains with thousands of links. For example, vinyl chloride (C_2H_3Cl) may react in the presence of an *initiator* (I) to form polyvinyl chloride, or PVC.

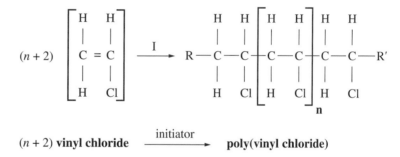

($n + 2$) **vinyl chloride** $\xrightarrow{\text{initiator}}$ **poly(vinyl chloride)**

Several terms associated with polymer production processes may be defined in terms of this reaction. The **monomer** is vinyl chloride, the molecular species that reacts to form the high molecular weight polymer, polyvinyl chloride. The **initiator**, I, may be a reactive chemical species that decomposes into fragments that initiate the chain growth, or it may be a type of radiation (such as ultraviolet light) that has the same effect. Once a chain has been initiated, it steadily grows as monomer molecules continue to add to it. Chain growth **termination** occurs when two growing chains react with each other, or when another chemical species in the reaction mixture attaches to the active end of a growing chain. The symbols R and R′ denote **end groups**, molecular fragments at the two ends of the polymer molecule that were involved in the chain growth initiation or termination steps. The repeating units in the chain (C_2H_3Cl) are called **mers**. (Monomer = single mer, polymer = multiple mers.) The symbol n denotes the number of mers in the polymer molecule, not counting the two at the end of the chain. The **polymer molecular weight** is approximately equal to n times the molecular weight of the monomer, since if n is in the thousands or higher, the contribution of the end groups to the molecular weight is negligible.

[1]This case study was prepared by Gary S. Huvard, HRC, Inc., Richmond, VA, and Richard M. Felder.

All rubbers, glasses, and plastics are polymers. You are probably familiar with natural polymers like cellulose (the building block of plant fibers) and synthetic polymers like polyethylene (plastic milk cartons), polyisoprene (automobile tires), polyethylene terephthalate (soft drink bottles), polymethyl methacrylate (Plexiglas®), polyvinylidene chloride (transparent plastic wrap), polytetrafluoroethylene (Teflon®), and various polyesters (fabrics). Polyvinyl chloride, the polymer shown above, is used to make rigid pipes, house siding, and protective coverings for automobile seats and dashboards, among many other applications.

PVC can be made in several different ways, depending on the intended end use. One method, **suspension polymerization**, produces a dry white powder called a **resin**. In this process, droplets of liquid vinyl chloride are suspended in water (in which vinyl chloride is insoluble) in a stirred-tank reactor. Polymerization occurs in each droplet, and the water surrounding the droplets is used to convey away the heat generated by the exothermic polymerization reaction. Under a microscope, the suspension PVC resin looks like a tiny bundle of grapes. Each *macroparticle* in the bundle is about 100–150 microns (micrometers, μm) in diameter, a bit larger than a grain of sand, and is itself composed of hundreds of *microparticles*, each about 1–3 microns in diameter, tightly stuck together and covered by a very thin external membrane.

At room temperature, PVC is rigid. When raised to sufficiently high temperatures it becomes molten and can be **extruded** into a variety of different shapes, including pipe and pipe fittings. PVC pipe has become the material of choice for residential cold water plumbing; it is less expensive than copper pipe and easier to install, since pipe fittings can be glued onto pipe sections instead of having to be soldered like copper. One problem with PVC pipe, however, is that it begins to soften and become rubbery at about 80°C, and once it softens it loses strength and can no longer be trusted to hold pressure. Hot water piping requires a material with a higher softening temperature. A candidate for this application is **chlorinated polyvinyl chloride**, or **CPVC**.

PVC CHLORINATION REACTION CHEMISTRY

CPVC is produced by reacting suspension PVC resin with chlorine. The reaction is initiated by decomposing diatomic chlorine dissolved in water into highly unstable and reactive chlorine atoms,

$$Cl_2 \rightarrow 2\,Cl\cdot$$

with either high heat or ultraviolet radiation. Once this **initiation step** has taken place, a chlorine atom attacks a PVC chain at a random location and replaces a hydrogen atom at the attack position.

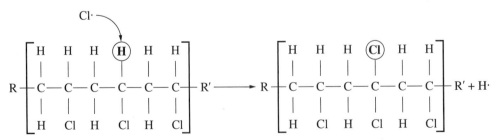

The liberated hydrogen atom reacts with a chlorine molecule to form a molecule of hydrogen chloride and another chlorine atom:

$$H\cdot + Cl_2 \rightarrow HCl + Cl\cdot$$

The new chlorine atom attacks a polymer molecule and replaces another hydrogen atom, and this **chain propagation sequence** continues as long as both chlorine gas and hydrogen atoms on polymer chains remain in the system. If enough chlorine is added to the reactor, a structure similar to the following one might be obtained (the added chlorines are circled):

"67% Cl CPVC":

$$R-\left[\begin{array}{ccccccccc} H & H & \textcircled{Cl} & H & H & H & H & \textcircled{Cl} & H \\ | & | & | & | & | & | & | & | & | \\ C-C-C-C-C-C-C-C-C \\ | & | & | & | & | & | & | & | & | \\ H & Cl & H & Cl & H & Cl & \textcircled{Cl} & Cl & H \end{array}\right]-R'$$

When the polymer has this structure (which requires replacing roughly one out of every five hydrogen atoms on the PVC chains with chlorine atoms) it contains between 67 and 68 wt% chlorine and its softening temperature is about 110°C, high enough for the material to be used for hot water piping. If the chlorine content rises above 70 wt% Cl, the polymer becomes too difficult to process. In this case study, we analyze a process for producing CPVC from PVC.

PROCESS DESCRIPTION

Step 1. Solid PVC resin is combined with water in a stirred tank to make a **slurry** (a suspension of solid particles in a liquid) containing about 10 wt% PVC. The slurry is pumped to a continuous **chlorine absorber**, a vessel equipped with an impeller that keeps the contents agitated. A stream of chlorine vapor at 25°C also enters the absorber. The absorber operates at 25°C.

> The chlorine absorber is enclosed by a jacket through which a hot or cold fluid can be pumped to maintain the absorber contents at any desired temperature. A **pressure relief pipe** emerging from a point near the top of the absorber is equipped with a **rupture disk**, a thin metal membrane designed to break if the absorber pressure rises above a certain value. If this happens, gas flows from the absorber through the relief pipe to a **surge tank**, lowering the pressure in the absorber while containing the released gas.

Step 2. The chlorine that will eventually be fed to the absorber is stored in a tank as a liquid under pressure at 22°C. Liquid is pumped from this tank through a **chlorine flow control valve** to a **chlorine vaporizer**, a tall tank about half of which is filled with chlorine liquid and the other half with chlorine vapor. The liquid and vapor are in equilibrium at 5°C. Saturated steam at 2 bar absolute is fed through a **steam flow control valve** into a coil submerged in the liquid in the vaporizer, and the heat transferred through the coil wall vaporizes chlorine. Chlorine vapor flows from the top of the vaporizer.

> The pressure and liquid level in the chlorine vaporizer are controlled automatically. As vapor flows out the top, the pressure in the tank begins to decrease. A pressure sensor detects the drop and sends a signal to the steam flow control valve. The valve opens and feeds more steam to the coil, causing more liquid chlorine to evaporate and raising the pressure back to the desired value. As the liquid evaporates, the liquid level in the tank begins to drop. A liquid level sensor detects the drop and sends a signal to the liquid chlorine flow control valve. The valve opens, and more liquid enters the tank to replace the liquid that evaporated, raising the liquid level back to the set-point (desired) value.

Step 3. The chlorine vapor leaving the vaporizer passes through an adiabatic **expansion valve**, then through a **chlorine vapor heat exchanger**. In the exchanger, the vapor comes in thermal contact with a hot flue gas from a furnace to be described in Step 15. The heat transferred from the flue gas is used to raise the temperature of the chlorine vapor to 25°C. The heated vapor is the chlorine feed to the absorber mentioned in Step 1.

Step 4. The chlorine enters the absorber through a pipe in the bottom that ends in a **sparger** (something like a shower head nozzle) located close to the impeller. The chlorine comes out of the sparger in thousands of tiny bubbles, which are then dispersed by the impeller throughout the slurry in the absorber. Chlorine is soluble in water and even more soluble in the PVC resin, and so dissolves in both the liquid and solid phases of the slurry.

The feed rate of chlorine to the absorber is automatically controlled to maintain the absorber pressure at a specified value. If the pressure begins to fall, a pressure sensor detects the drop and sends a signal to the expansion valve described in Step 3. The valve opens, causing the flow rate of chlorine to increase and raising the absorber pressure back to the set-point value. If the pressure rises above the set point, the pressure sensor sends a signal that adjusts the control valve to decrease the flow rate of chlorine and the pressure comes back down to the set-point value.

Step 5. The PVC resin slurry that leaves the chlorine absorber at 25°C is heated to 50°C in the **reactor preheater**. A very small amount of a chemical *initiator* is added to the heated slurry leaving the preheater. The amount of initiator added is small enough to be neglected in material balance calculations on the reactor.

Step 6. The heated slurry flows to a **chlorination reactor**, which is a long jacketed pipe. In the reactor, the initiator—which is unstable at 50°C and higher—decomposes and begins the PVC chlorination chain reaction described in the introductory section. As the slurry moves through the reactor, dissolved chlorine in the resin particles reacts with the polymer, producing hydrogen chloride as a by-product. The HCl—which is essentially insoluble in PVC and CPVC—diffuses out of the particles and into the liquid phase, forming hydrochloric acid (by definition an aqueous solution of hydrogen chloride), and the chlorine that had been dissolved in the liquid phase diffuses into the particles and reacts. All of the chlorine fed to the reactor is eventually consumed. The exit stream from the reactor is a slurry of CPVC particles containing 67–68 wt% chlorine suspended in aqueous hydrochloric acid.

The chlorination reaction is highly exothermic. Some of the heat of reaction raises the slurry temperature to its exit value of 65°C, and cooling water flowing through the reactor jacket removes the additional heat generated. (If the reactor temperature were to rise above 65°C, the rate of the chlorination reaction would become extremely high, chlorine dissolved in the liquid phase would all react near the outer surface of the particles rather than diffusing through the pores, and the quality of the final product would suffer.)

Step 7. The CPVC slurry flows to the **reactor centrifuge**, a spinning horizontal drum. The solids are forced against the circular drum wall and are compressed there. Liquid hydrochloric acid **(waste liquor)** collects in the drum and overflows through an opening at one end, and the **wetcake** retained on the wall, which contains 90 wt% CPVC resin and 10% hydrochloric acid, is scraped out by a large interior screw conveyor.

Step 8. The waste liquor from the reactor centrifuge flows to a **waste liquor neutralizer**, where the HCl is neutralized with a caustic soda (sodium hydroxide) solution:

$$HCl(aq) + NaOH(aq) \rightarrow NaCl(aq) + H_2O(l)$$

A stream of CO_2 produced in the resin neutralizer (Step 10) also enters the neutralizer liquid through a sparger and reacts with sodium hydroxide:

$$CO_2(g) + 2\,NaOH(aq) \rightarrow Na_2CO_3(aq) + H_2O(l)$$

The caustic solution fed to the neutralizer, which comes from the NaOH feed drum (Step 9), contains 10 wt% NaOH and also some dissolved NaCl and Na_2CO_3.

Step 9. The solution of sodium chloride and sodium carbonate leaving the waste liquor neutralizer is combined with the waste liquor from the product centrifuge (Step 12). The combined solution is split into two streams. One goes to a waste treatment facility, and the other goes to the **NaOH makeup tank**, an agitated jacketed batch vessel. Solid pellets of sodium hydroxide are added to this tank to bring the NaOH concentration in the solution up to 10.0 wt%. Cooling water is circulated through the makeup tank jacket to remove the considerable heat generated by the dissolution of NaOH. Batches of solution are pumped from the NaOH makeup tank to the **NaOH feed drum** (a holding tank), and solution is pumped from the feed drum to the waste liquor neutralizer (Step 8). The feed drum always contains enough inventory to provide a continuous feed to the neutralizer.

Step 10. The wetcake produced in the reactor centrifuge is conveyed to a **resin neutralizer**, another large, jacketed, agitated vessel, where the HCl in the resin is neutralized with sodium bicarbonate in the reaction

$$HCl(aq) + NaHCO_3(aq) \rightarrow NaCl(aq) + CO_2(g) + H_2O(l)$$

(Sodium bicarbonate, a weak base, is used because a strong base like sodium hydroxide can damage the resin.) A 10.0 wt% aqueous $NaHCO_3$ solution at 21°C pumped from the $NaHCO_3$ feed drum (Step 11) is augmented by a clean water stream at 60°C, and the combined stream is fed to the tank. Enough bicarbonate solution is fed to neutralize all the HCl in the wetcake, and enough additional water is added to make a 15.0 wt% CPVC slurry in the vessel.

A pH controller monitors the neutralizer vessel contents and sends a signal to a flow control valve in the bicarbonate solution feed line if more or less solution is needed. The resin remains in the neutralizer long enough for even the HCl trapped in the resin pores to be neutralized. The neutralizer exit stream is a slurry containing 15.0 wt% CPVC in an aqueous sodium chloride solution at 90°C and pH = 7.

Steam fed to the vessel jacket maintains the contents at 90°C. At this temperature, the reaction proceeds to completion and essentially all of the carbon dioxide formed in the reaction between HCl and $NaHCO_3$ is driven out of the aqueous phase. The CO_2 leaving the resin neutralizer is recycled to the waste liquor neutralizer (Step 8).

Step 11. The sodium bicarbonate solution fed to the resin neutralizer (Step 10) is produced by dissolving $NaHCO_3$ powder from 25-kg bags in clean water in the **$NaHCO_3$ makeup tank**, an agitated batch vessel. This makeup tank is not jacketed like the NaOH makeup tank, as the heat of solution of sodium bicarbonate is low enough for the heat generated in the tank to be neglected. Batches of solution are pumped from the makeup tank to the **$NaHCO_3$ feed drum**, and solution is pumped continuously from the feed drum to the resin neutralizer.

Step 12. The slurry leaving the resin neutralizer goes to the **product centrifuge**. The dilute NaCl solution wrung from the resin in the centrifuge is cooled to 25°C in a heat exchanger. The cooled solution combines with the effluent stream from the waste liquor neutralizer (Step 8).

Step 13. The wetcake from the product centrifuge, which contains 90 wt% CPVC, is conveyed to a **rotary dryer**. This unit functions like a clothes dryer. The resin is tumbled down an inclined, rotating drum with hot air flowing through the drum in the opposite direction. Heat transferred from the air vaporizes most of the remaining water in the resin. The CPVC product resin emerges from the dryer at 120°C containing roughly 0.1 wt% moisture.

Step 14. The humid air leaving the rotary dryer at 110°C, 1.3 atm, and 20% relative humidity is split into two streams. Ten percent of the air is vented to the atmosphere, and the other 90% is combined with **makeup dryer air**, heated in the **indirect contact furnace** described in the next step, and fed back to the rotary dryer (Step 13).

Step 15. In the **indirect contact furnace**, natural gas containing 92.5 mole% methane, 4.8 mole% ethane, and the remainder propane is burned completely with excess air (**combustion air**). The dryer air feed stream (the combination of makeup dryer air and humid air recycled from the dryer outlet) passes through metal channels within the furnace combustion zone and then out of the furnace to the rotary dryer air inlet. The hot gases in the furnace combustion zone heat the metal channel walls, and the walls in turn heat the dryer air.

Step 16. The combustion products leave the furnace at 250°C and flow through the chlorine vapor heat exchanger (this stream is the flue gas referred to in Step 3) and then to a **flare**, where any carbon monoxide and residual hydrocarbons in the gas are burned and the products released to the atmosphere. The flare is a safety precaution: if the furnace operates as intended, the CO and hydrocarbon content of the flue gas should be negligible.

PROBLEMS* **12.1.** On an oversized page, draw a flowchart for this process. Label each stream with an identifying symbol (e.g., S1, S2, S3, . . .) and known information about what the stream is and/or what it contains [e.g., slurry, 10 wt% PVC in water; wetcake, 90 wt% CPVC, 10 wt %HCl(aq); 10 wt% NaHCO$_3$(aq)].

12.2. On a second page (which you may find convenient to generate with a spreadsheet program), prepare a table to accompany the flowchart. The column headings should be the stream identifiers from the flowchart (S1, S2, S3, or whatever system you used). The row labels should be as follows: T(°C), P(atm), PVC(kg/h), CPVC(kg/h), H$_2$O(kg/h), Cl$_2$(kg/h), HCl(kg/h), NaOH(kg/h), NaHCO$_3$(kg/h), NaCl(kg/h), Na$_2$CO$_3$(kg/h), CH$_4$(kg/h), C$_2$H$_6$(kg/h), C$_3$H$_8$(kg/h), O$_2$(kg/h), N$_2$(kg/h), CO$_2$(kg/h), TOTAL(kg/h).

Begin by filling in all known quantities in the appropriate location in the table, and put dashes (—) in every cell of a stream column corresponding to species not present in that stream. Thereafter, whenever you determine the temperature, pressure, flow rate, and/or composition of a process stream, calculate and enter the corresponding values in the table. When you have worked through the entire case study, every cell should contain either dashes or a number. *When doing mass/mole conversions, use 35.45 as the atomic weight of chlorine.*

12.3. In a page or less, summarize the overall process. In your summary, answer the following questions:
   (a) What is the reaction that this process is designed to carry out?
   (b) What are the feeds to the process?
   (c) What is the condition of the final product?
   (d) What other streams leave the process, what do they contain, and what happens to them?
   (e) Why is the process carried out? (What is the point of chlorinating PVC?)

12.4. Neglecting the end groups, calculate the mass fractions of carbon, hydrogen, and chlorine in PVC and in the "67% Cl" produced in this process. (The molecular structure of this product is shown in the section on PVC chlorination reaction chemistry.)

12.5. The plant is to produce 3.5×10^6 kg/year of "67% Cl." Assuming that the plant operates 300 days per year, 24 hours per day, calculate the required hourly feed rates (kg/h) of the 10 wt% PVC slurry and of Cl$_2$(g) to the absorber and the production rate of hydrogen chloride (kg/h) in the chlorination reactor. (*Suggestion:* Write balances around the system composed of the chlorine absorber and chlorination reactor.)

12.6. Vapor pressure data for chlorine are given below.

T (°C)	−20.0	−10.0	0.0	10.0	20.0	30.0	40.0	60.0	80.0	100.0
p^*(atm)	1.78	2.59	3.64	4.99	6.69	8.78	11.32	17.95	27.04	39.06

   (a) Use these data and the Clausius–Clapeyron equation (Equation 6.1-3) to estimate the heat of vaporization of chlorine (kJ/mol) and to obtain an expression for $p^*_{Cl_2}(T)$.
   (b) What is the operating pressure in the chlorine vaporizer (torr)?
   (c) At what rate (kW) must heat be added to the chlorine in the vaporizer?

---

*Problems 12.1–12.21 involve material from Chapters 2–6 of the text. Energy balance calculations (Chapters 7–9) are required in Problems 12.22–12.29.

(d) What is the percentage difference between the calculated value of $\Delta \hat{H}_v$ and the one given in Table B.1? Why might the two values differ? (Think of several possible reasons.)

12.7. Speculate on why chlorine is stored as a liquid rather than being purchased and stored as a gas at the absorber temperature (25°C), when storing it as a gas would eliminate the need for the chlorine vaporizer and possibly for the heat exchanger in the chlorine feed line.

Problems 12.8–12.12 deal with the chlorine absorber, depicted in the flowchart below.

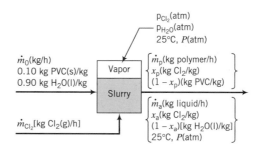

The chlorine fed to the tank dissolves in both the liquid phase and the polymer phase. The slurry inside the tank is well mixed, so that its composition is the same as the composition of the exiting stream. The **head space** above the slurry contains chlorine vapor in equilibrium with the dissolved chlorine in both condensed phases and water vapor in equilibrium with the liquid water in the aqueous phase. The condition inside the absorber is shown schematically in the following diagram:

Head space $p_{Cl_2}(atm)$, $p_{H_2O}(atm)$ 25°C, $P(atm)$	
Polymer phase(solid) x_p (kg Cl_2/kg polymer) $(1 - x_p)$(kg PVC/kg polymer) 25°C	**Aqueous phase(liquid)** x_a (kg Cl_2/kg liquid) $(1 - x_a)$[kg $H_2O(l)$/kg liquid] 25°C

The following physical property data pertain to this system:

- The solubilities of Cl_2 in PVC and in water are correlated by a form of Henry's law:

$$x_p(\text{kg } Cl_2/\text{kg PVC phase}) = \alpha_p p_{Cl_2}, \qquad x_a(\text{kg } Cl_2/\text{kg aqueous phase}) = \alpha_a p_{Cl_2}$$

  where p_{Cl_2} is the partial pressure of chlorine in the head space. The coefficients α_p and α_a depend on the system temperature.
- The solubility data shown below have been obtained for chlorine in PVC at 25°C:

$p_{Cl_2}(atm)$	0.75	1.18	1.23	2.95	3.03	3.87	4.82
wt% Cl_2	2.9%	4.6%	5.1%	11.9%	12.3%	15.4%	19.8%

- At 25°C, chlorine gas is 2.68 times more soluble in PVC than in water for any chlorine partial pressure.
- Raoult's law provides a good correlation of the equilibrium between liquid water and water vapor in the system.

12.8. Use the given data to calculate the modified Henry's law coefficients, α_p and α_a, for chlorine in PVC and in water at 25°C. When determining α_p, plot the solubility data to make sure that the Henry's law correlation is valid, and use the method of least squares to force a fit through the origin to obtain the slope of the plot. (See Appendix A.1.)

12.9. The flowchart for the absorber involves seven unknown variables ($\dot{m}_0$ and $\dot{m}_{Cl_2}$ having been determined in Problem 12.5). Write the seven equations you would use to calculate those variables, noting that the amounts of chlorine and water in the head space are constant and so do not enter into steady-state material balances. Then solve the equations. (*Note:* Trial-and-error will be required as part of the solution. An equation-solving program or spreadsheet is a convenient tool for performing the required calculations.)

12.10. Starting with the mole fraction of dissolved chlorine in the aqueous phase calculated in Problem 12.9 (x_a), use Raoult's law to estimate the partial pressure of chlorine in the gas phase. What percentage error results from this estimation, presuming that the value of p_{Cl_2} from Problem 12.9 is correct. Why is such a large error not surprising?

12.11. The slurry occupies 70% of the total tank volume and the head space occupies the remaining 30%. The *mean residence time* of the slurry in the tank (which equals the volume of the slurry in the tank divided by the volumetric flow rate of the stream leaving the tank) is 12 minutes. Use this information, the quantities calculated in Problem 12.9, and the specific volume data given below to estimate the volume (m^3) of slurry in the tank, the head space volume, and the total tank volume.

> *Specific volume of PVC* $= 0.709$ L/kg
>
> *Specific volume of liquid water* $= 1.00$ L/kg
>
> *Specific volume of dissolved chlorine (in either PVC or water)* $= 0.690$ L/(kg dissolved)

12.12. Calculate the total mass of chlorine (kg) in the chlorine absorber tank, assuming ideal gas behavior in the head space. (*Suggestion:* Take as a basis of calculation a specified mass of slurry, calculate the volume your basis amount would occupy and the mass of chlorine it would contain, and scale the results to the actual volume of the slurry in the tank calculated in Problem 12.11. Then calculate the mass of chlorine in the head space.)

12.13. List the reaction and mass transfer processes occurring in the chlorination reactor. (What reacts and what is formed? Which species diffuse into and out of the resin particles?)

12.14. The length of the reactor is an important design parameter. What would be the drawbacks of making the reactor (i) too short, and (ii) too long?

12.15. Hydrogen chloride is highly soluble in water and essentially insoluble in CPVC. Estimate the pH of the aqueous phase (hydrochloric acid) leaving the reactor, taking the density of this phase to be 1.005 g/cm^3 and assuming that the HCl dissociates completely into H$^+$ and Cl$^-$.

12.16. Recalling that the reactor centrifuge separates the slurry leaving the reactor into a wetcake that contains 90 wt% CPVC and a waste liquor stream of hydrochloric acid, calculate the mass flow rate of the wetcake and the mass flow rate and composition (component mass fractions) of the waste liquor.

12.17. The wetcake leaving the reactor centrifuge is conveyed to the resin neutralizer, a jacketed reactor in which the hydrochloric acid is neutralized with a weak base, sodium bicarbonate:

$$HCl(aq) + NaHCO_3(aq) \rightarrow NaCl(aq) + CO_2(g) + H_2O(l)$$

The wetcake enters the neutralizer at 60°C. A stream consisting of a 10.0 wt% aqueous NaHCO$_3$ solution at 21°C coming from the NaHCO$_3$ feed drum is diluted with water at 60°C, and the combined stream is fed to the resin neutralizer. The product slurry leaving the neutralizer contains 15.0 wt% CPVC and has a pH of 7.0. Steam circulated through the neutralizer jacket keeps the contents at 90°C. Carbon dioxide is essentially insoluble in aqueous sodium chloride at this temperature, so that all of the CO$_2$ formed in the reaction may be assumed to be released from the neutralizer as a gas. Calculate the mass flow rates of the 10 wt% bicarbonate solution and the dilution water fed to the neutralizer, the mass flow rates of the CPVC slurry and the carbon dioxide stream leaving the neutralizer, and the mass fraction of NaCl in the product slurry liquid. (*Reminder:* When writing a balance on water, don't forget the water generated in the reaction.)

12.18. CPVC is degraded by both strong acids and strong bases at elevated temperatures; acids corrode process units unless very expensive corrosion-resistant construction materials are used; and aqueous slurries with high solids content are difficult to pump. Use these facts to explain why the wetcake from the centrifuge is not sent directly to the dryer instead of going through the resin neutralizer, why sodium bicarbonate is used to achieve the neutralization, and why additional water is fed to the neutralizer.

12.19. The 10% sodium bicarbonate solution is made in the plant by adding two 50-lb$_m$ bags of NaHCO$_3$ powder to a batch dissolving tank filled to roughly 80% of its capacity. From the dissolving tank, the solution is sent to a feed drum from which solution is pumped continuously to the resin neutralizer. It takes an operator four hours to make a batch of solution, pump the solution to the feed drum, and rinse the dissolving tank to prepare it for the next batch. The feed drum can hold a 2.5-day supply of solution. The addition of each new batch of solution brings the feed drum contents up to a 2-day (48-hour) supply, so that the plant can continue to run for at least 40 hours if the bicarbonate powder inventory runs low or the dissolver is shut down for maintenance. (*Note: The specific gravity of a 10 wt% sodium bicarbonate solution is 1.08.*)

   (a) Calculate the feed drum volume (liters), the volume of solution made in each batch (liters), and the probable volume of the dissolving tank (liters).

   (b) What is the minimum dissolving tank volume needed to meet the requirements of the process, assuming that roughly four hours would be required to prepare a batch of any size and the tank is always filled to 80% of its capacity? What are the benefits of using a larger tank?

   (c) Suppose the dissolver agitator breaks down just after an operator finishes transferring a batch of solution to the feed drum and it takes 32 hours to repair the agitator. How quickly can the operators replenish the 48-hour supply of solution in the feed drum once the agitator has been repaired? What would have happened if the minimum tank volume calculated in part (b) had been used?

   (d) The feed drum is outdoors. Neglecting dissociation of NaHCO$_3$ in solution, estimate the outside temperature (°C) at which freezing of the tank contents would become a concern. (Use Equation 6.5-11.) The calculated temperature would be a conservative estimate, since dissociation of the salt would lower the freezing point of water even more.

12.20. The slurry exiting the resin neutralizer, which contains 15.0 wt% CPVC in an aqueous NaCl solution, is concentrated in the product centrifuge. The wetcake leaving the centrifuge contains 90.0 wt% CPVC and 10.0% solution and the effluent liquor contains the remainder of the entering solution. Calculate the mass flow rates (kg/h) of both streams leaving the centrifuge.

12.21. In the waste liquor neutralizer, both the hydrochloric acid separated from the wetcake in the reactor centrifuge and the CO$_2$ generated in the resin neutralizer are neutralized with an aqueous solution containing 10 wt% NaOH and also some dissolved NaCl and Na$_2$CO$_3$.

$$HCl(aq) + NaOH(aq) \rightarrow NaCl(aq) + H_2O$$

$$CO_2 + 2\,NaOH \rightarrow Na_2CO_3 + H_2O$$

The exit stream from the neutralizer joins with the effluent liquor from the product centrifuge. Most of the combined stream is fed to a waste treatment facility, but a portion of it is fed to the NaOH makeup tank (a batch stirred tank) to prepare more caustic feed solution for the neutralizer. Enough NaOH(s) is added to the makeup tank to bring the solution concentration back up to 10 wt% NaOH. Each batch prepared in this manner contains enough solution to supply the waste liquor neutralizer for eight hours of operation (one shift). When the NaOH pellets have completely dissolved and a batch is ready, a valve in the tank exit line is opened and the batch is fed by gravity to the NaOH feed drum, from which solution is pumped continuously to the waste liquor neutralizer.

   (a) Draw and label a flowchart for this part of the process.

   (b) Speculate on the reason that the HCl must be neutralized.

   (c) Calculate the required mass flow rate (kg/h) and composition (component mass fractions) of the caustic solution fed to the neutralizer, the mass flow rate and composition of the solution leaving the neutralizer, and the fraction of the combined salt solution sent to the makeup tank.

   (d) Calculate (i) the mass of solid NaOH (kg) required per batch of solution prepared in the tank, and (ii) the required tank size if the specific gravity of the exiting solution is 1.11 and a batch occupies 60% of the total tank volume.

   (e) Determine the metric tons of NaCl and Na$_2$CO$_3$ that must be processed each year in the waste treatment facility. (Recall that the plant operates 300 days per year.)

12.22. Liquid chlorine at 22°C is pumped continuously to the chlorine vaporizer. The vaporization process causes the temperature of the chlorine to drop. Take the heat of vaporization of chlorine to be 290 kJ/kg and the heat capacity of chlorine vapor to be 0.48 kJ/(kg·°C).

   (a) The heat required to keep the exiting chlorine temperature at 5°C is supplied by saturated steam at 2 bar (absolute) condensing in a coil submerged in the liquid chlorine in the tank. The condensate emerges at the saturation temperature of the steam. Calculate the required rate of steam condensation.

(b) If the vaporization process were adiabatic, what would be the temperature of the exiting chlorine vapor? (*Suggestion:* Take a basis of 1 kg of chlorine liquid at 22°C, calculate the enthalpy gain when chlorine vaporizes at that temperature, and equate it to the enthalpy loss when the vapor cools from 22°C to the adiabatic vaporization temperature.) Is the result physically possible?

12.23. The chlorine vapor leaving the vaporizer passes through an adiabatic expansion valve in which its pressure drops by 640 mm Hg and its temperature drops to 3°C. It then goes through a heat exchanger in which its temperature is raised to 25°C, and from there it proceeds to the chlorine absorber. Calculate the rate (kJ/h) at which heat must be transferred to the chlorine vapor in the heat exchanger, using the approximate heat capacity given in Problem 12.22.

12.24. The stream leaving the chlorine absorber must be heated from 25°C to 50°C before it enters the chlorination reactor. The heat capacity of PVC is 1.2 kJ/(kg·°C) and the heat capacity of dissolved chlorine (in either PVC or water) is 0.96 kJ/(kg·°C). At what rate (kJ/h) must heat be transferred to the slurry in the reactor preheater?

12.25. In the chlorination reactor, the feed slurry enters at 50°C and the products leave at 65°C. Most of the heat generated by the reaction is transferred to cooling water flowing through a jacket surrounding the reactor. The cooling water enters at 15°C and leaves at 45°C.

(a) What becomes of the heat generated by the reaction that is not transferred to the cooling water?

(b) Using data given below and neglecting heat of solution effects, estimate the heat transferred to the cooling water (kJ/h) and the required flow rate (kg/h) of the cooling water.

Data

Heat of reaction at 50°C = −1770 kJ/(kg Cl_2 consumed)
Heat capacity of CPVC = 1.9 kJ/(kg·°C)
Average heat capacity of hydrochloric acid = 4.0 kJ/(kg·°C)

(c) Estimate the temperature to which the product stream could rise if the cooling water flow were interrupted (i.e., if the reactor operation were to become adiabatic). List several serious problems that could result if this situation arises.

12.26. Batches of caustic solution for the resin neutralizer are prepared by filling the NaOH makeup tank with the required amount of mixed salt solution at 25°C, then adding the required mass of NaOH pellets, also at 25°C (see Problem 12.21), and immediately starting an agitator to aid the dissolution process. When solid NaOH dissolves in water, a great deal of thermal energy is released. To keep the batch from getting too hot, cooling water initially at 15°C is pumped through the jacket surrounding the makeup tank. The dissolution is complete after four hours. The flow of coolant is continued after that point until the tank contents are brought back down to 25°C, whereupon the cooling water flow and the tank agitator are turned off and the batch is dumped to the NaOH feed drum.

An integral energy balance on the tank contents from $t = 0$ to an arbitrary time t has the form

$$\dot{Q}t - \dot{W}_s t = N_{NaOH}\, \Delta\hat{H}^\circ_s + M_{soln} C_p(T - 25)$$

where

$\dot{Q}(kJ/h)$ = rate of heat transfer from the tank jacket to the tank
$-\dot{W}_s(kJ/h)$ = shaft work, rate of addition of energy to the tank contents by the agitator
$N_{NaOH}(\text{mol NaOH})$ = g-mole of sodium hydroxide dissolved in the batch at time t
$\Delta\hat{H}^\circ_s(kJ/\text{mol NaOH})$ = heat of solution of sodium hydroxide at 25°C, obtainable from Table B.11.
$M_{soln}(kg)$ = total mass of the tank contents at time t
$C_p[kJ/(kg·°C)]$ = heat capacity of the solution. Assume independent of temperature.
$T(°C)$ = solution temperature at time t.

(a) Heat is transferred from the tank to the coolant at an average rate of 2.5×10^5 kJ/h during the production of a batch of solution. The agitator has a maximum power rating of 20 kW. At the agitation rate used in the mixing process, about 65% of the rated power is actually drawn, of which 80% is transferred as shaft power to the tank contents and the remaining 20% is lost as wasted heat to the surroundings. Use this information to calculate the values of $\dot{Q}(kJ/h)$ and $\dot{W}_s(kJ/h)$ in the given energy balance equation. (Make sure the values have the correct signs.)

(b) Calculate the temperature in the reactor at the moment the NaOH has completely dissolved ($t = 4$ h). Use heat of solution data from Table B.11 in your calculation and take the average heat capacity of the tank contents to be 3.8 kJ/(kg·°C).

(c) How long must the cooling be continued after dissolution is complete to bring the solution temperature down to 25°C? (Remember that the agitator is turned off during this phase of the operation.) What is the total time required to produce a batch from the moment the sodium hydroxide pellets are added?

(d) Starting with the first law of thermodynamics for a closed system operating at constant pressure $[Q(kJ) - W_s(kJ) = \Delta H(kJ)$, where $-W_s$ is the total shaft work added to the system], derive the given energy balance equation by defining a suitable process path for the determination of ΔH.

12.27. The CPVC wetcake leaves the product centrifuge at 80°C and 1 atm and is fed to the elevated end of a rotary dryer (a rotating cylinder inclined about 35° from horizontal). A fresh air stream at 27°C, 1 atm, and 60% relative humidity is mixed adiabatically with an air stream recycled from the dryer exhaust, and the combined stream is heated in an air furnace and fed to the lower end of the dryer. Water evaporates from the wetcake moving downward in the rotating drum into the hot air moving in the opposite direction. The exhaust air leaves the dryer at 110°C, 1.3 atm, and 20% relative humidity. *The dryer operates adiabatically.* Ninety percent of the exhaust air is recycled, and the rest is vented to the atmosphere. The dried polymer resin leaves the dryer at 120°C containing 0.1 wt% water.

(a) In your own words, explain the purpose of this portion of the process. Your explanation should include statements about why the dryer must be tilted, why the air temperature drops as the air moves through the dryer, and why the preheater (air furnace) is necessary.

(b) Draw and fully label a flowchart of this portion of the process, substituting known variable values from previous calculations and including the required rate of heat transfer to the air in the preheater in the labeling. Use molar flow rates and mole fractions when labeling gas streams.

(c) Carry out the process degree-of-freedom analysis to verify that enough information is available to determine all unknown stream flow rates and component mole fractions (air) and mass fractions (wet resin), the temperature of the air entering the dryer, and the heat transferred to the air in the furnace. Write in order the equations you would solve to determine the unknown variables, circling in each equation the variable for which you would solve. Use as references for enthalpy calculations solid CPVC at 80°C, liquid water at 0°C, and dry air at 110°C (all at 1 atm), neglect the contribution of sodium chloride to the energy balance, use the steam tables to determine vapor pressures and the heat of vaporization of water, and use the following heat capacities: CPVC(s), 1.88 kJ/(kg·°C); dry air, 29.2 kJ/(kmol·°C); $H_2O(l)$, 4.2 kJ/(kg·°C); $H_2O(v)$, 34.0 kJ/(kmol·°C). Do not perform the calculations in this part.

(d) Solve the equations of part (c) for the unknown stream flow rates, component fractions and temperature of the air entering the dryer, and rate of heat transfer to the air in the furnace (kW).

12.28. In the dryer air furnace, a natural gas containing 92.5 mole% methane, 4.8% ethane, and the remainder propane is burned with 20% excess air. The fuel enters the furnace at 25°C and the air enters at 27°C, 1 atm, and 60% relative humidity. Essentially all of the fuel is consumed, and the flue gas contains a negligible amount of carbon monoxide. Heat is transferred from the hot combustion products through the furnace wall to the dryer air. The combustion products leave the furnace at 250°C. Calculate the fuel feed rate (kg/h) required to achieve the heating rate calculated in Problem 12.27(d). (*Suggestion:* Assume a fuel feed of 100 mol/s, calculate the energy transfer corresponding to this basis of calculation, and then scale the results to determine the actual fuel feed rate.)

12.29. After several years of research, the company has developed a new grade of CPVC that contains 70.0 wt% chlorine. The high softening temperature and nonflammability of the new material make it suitable for computer cabinets and other potentially profitable products. Plant engineering has determined that the plant can make both the old and the new grades of CPVC, converting the same total amount of PVC per year. The problem is to determine whether or not the production of the 70% Cl grade can be justified economically, and if it can, the fraction of the operating year that should be dedicated to producing each grade.

PVC currently costs $0.70/kg and chlorine costs $0.15/kg. The 67% Cl CPVC sells for $2.50/kg, and 70% Cl CPVC can be sold for $2.90/kg. The manufacturing costs of the two products (which include costs for equipment maintenance and replacement, personnel, utilities, and marketing but do not include the costs of PVC and chlorine) are given by the following functions:

$$C_{67}(\$/\text{kg } 67\% \text{ Cl CPVC}) = 1.80 - (6.0 \times 10^{-8})M_{67}$$

$$C_{70}(\$/\text{kg } 70\% \text{ Cl CPVC}) = 1.85 - (5.0 \times 10^{-8})M_{70} + (4.0 \times 10^{-14})(M_{70})^2$$

where M_{67}(kg) and M_{70}(kg) are the masses of each product produced. The second term of each function accounts for the fact that some manufacturing costs (like salaries) are relatively independent of how much product is produced, and so the cost per kilogram of product (C) decreases as the amount of product produced (M) increases. The third term of the second function is a penalty for producing too much of the new product, reflecting costs of additional advertising, sales staff retraining, and additional technical support. (These added costs are expected to disappear after about three years.)

(a) PVC contains 38.4 wt% carbon and 56.8 wt% chlorine; 67% Cl CPVC contains 29.4 wt% carbon and 67.6 wt% chlorine; and 70.0 wt% Cl CPVC contains 27.4 wt% carbon and 70.0 wt% chlorine. Calculate the total chlorine consumption (kg/year) required to produce each grade of CPVC.

(b) Let X equal the fraction of the 300-day year in which 67% Cl CPVC is produced and $(1 - X)$ the fraction in which 70% Cl CPVC is produced. Derive an expression for the gross annual profit $[P(X) = \text{sales revenue} - \text{raw material cost} - \text{manufacturing cost}]$, assuming that all CPVC produced can be sold. Then determine the optimal value of X (the value that maximizes P) and the corresponding percentage increase in P over the profit attained if only 67% Cl CPVC is produced.

(c) There is another complication. If the plant produces the new material for more than 30% of the year, there will be insufficient 67% Cl polymer to sell to existing customers. Marketing could put all customers on reduced allocation (so that all customers get a fixed percentage of the quantity they have received in the past), but the company could risk losing the business of its biggest customers. Alternatively, the company could continue to supply its biggest customers and stop selling the 67% material to some small customers. Unfortunately, most of the small customers have optimized their processes to use this polymer, and the cost of adapting their processes to a different polymer may be high enough to drive them out of business. Discuss these and other options available to the company, make a recommendation, and explain your reasoning.

Chapter 13

Steam Reforming of Natural Gas and Subsequent Synthesis of Methanol[1]

The industrial capacity for worldwide production of methanol in 1994 was 2.42×10^7 metric tons per year,[2] about 85% of which was used as a starting material in the production of other chemicals or as a solvent.[3] Methanol is used as a raw material in the manufacture of formaldehyde, acetic acid, methyl *tert*-butyl ether (MTBE), dimethyl terephthalate, methyl chloride, methyl amines, and many other chemicals. It can also be used as a clean-burning fuel.

Processes for manufacturing methanol have evolved over time and the history provides an informative perspective on the link between improving technology and the economics of the chemical business.[4] During the 1800s and early 1900s, methanol was produced primarily by the dry distillation of wood, and it is still sometimes referred to as "wood alcohol." In 1913, BASF successfully synthesized methanol by reacting carbon monoxide and hydrogen over a zinc chromite catalyst at high temperatures and pressures (320°C to 450°C; 25 to 35 MPa). At that point, the industry shifted abruptly from wood-based to catalyst-based technologies using **synthesis gas** (i.e., a mixture of CO, CO_2, and H_2) as feedstock. This revolution in process technology dramatically decreased the cost of producing methanol, and production volumes greatly increased.

In 1966, ICI commercialized a methanol process using a much more active copper–zinc oxide catalyst. Although the activity of this catalyst is more sensitive to impurities (poisoning), the copper–zinc oxide-based technology uses more moderate temperatures and pressures (200° to 300°C; 5 to 10 MPa) than the zinc chromite technology, thereby significantly reducing manufacturing costs. Many producers now offer the copper–zinc oxide catalyst and that technology currently dominates the worldwide methanol industry.

---

[1]Prepared by Ronald W. Rousseau and Brian Keyes, Celanese Ltd., Corpus Christi, TX.

[2]*Chemical Marketing Reporter,* p. 3, Apr. 4, 1994.

[3]E. Fiedler, G. Grossman, B. Kersebohm, G. Weiss, and C. Witte, Methanol, in *Ullmann's Encyclopedia of Industrial Chemistry,* 5th Edition, Vol. A 16, Wiley-VCH, New York, 1990, p. 465.

[4]R. J. Hawkins, R. J. Kane, W. E. Slinkard, and J. L. Trumbley, Methanol, in *Encyclopedia of Chemical Processing and Design,* J. J. McKetta and W. A. Cunningham, Eds., Vol. 29, Marcel Dekker, New York, 1988, p. 418.

There are different types of adiabatic and nonadiabatic methanol synthesis reactors used in industry. Some are shell-and-tube reactors that are designed to remove heat from the reaction side (in the tubes) by boiling water on the shell side. Others use the shell side for reaction and the tube side for preheating the feeds. In the quench-style reactor, which is the focus of this chapter, the synthesis gas reacts in successive adiabatic catalyst beds. In each catalyst bed, the exothermic chemical reactions increase the temperature of the reactor gases by 40°C to 90°C. Between beds, fresh feed at a lower temperature is injected and blended with the hotter reactor gases. This form of direct-contact cooling is used to control the temperature in the reactor beds, thereby protecting the catalyst and limiting the formation of by-products. Regardless of the reactor style, single-pass conversions of CO and CO_2 are moderate (e.g., 30% to 70%).

Commonly, methanol plants are very large. The rationale for this is based on a number of factors, including the decrease in capital costs per unit of product with an increase in the capacity of the plant (See, for example, R. H. Perry and D. Green, *Perry's Chemical Engineers' Handbook,* 7th Edition, McGraw-Hill, New York, 1997, pp. 9–63ff.). A modern methanol plant can produce 3×10^5 to 8×10^5 metric tons per year of specification-grade methanol (>99.85 wt% methanol), and the synthesis gas from which methanol is generated can be produced from a variety of raw materials, including natural gas, petroleum residues, coal, and naphtha. Natural gas feedstock accounts for 70% of worldwide methanol capacity,[5] and it is the manufacture of methanol from natural gas feedstock that will be described here.

PROCESS DESCRIPTION

Incompletely labeled flowcharts for the overall process and simplified versions of the reformer, heat-recovery and compression, and converter loop units are given in Figures 13.1 through 13.4. Below we provide a process description that includes details that may be added to the process and/or unit flowcharts.

Reformer

A mixture of carbon monoxide, hydrogen, and carbon dioxide is produced by **steam reforming**, a process in which natural gas and steam are mixed and reacted in a **reformer** operated at 1.6 MPa. Natural gas may be assumed to consist entirely of methane (CH_4), although other compounds may be present in small concentrations. In the present process, steam and natural gas are fed to the reformer in a ratio of 3.0 moles of steam per mole of methane. The reformer consists of an arrangement of vertical tubes filled with nickel-impregnated ceramic catalyst. Rows of these tubes are located inside an insulated firebox, where they are heated by the combustion of natural gas.

The natural gas and steam that are blended to become the reformer feed enter the process at 30°C and 210°C, respectively. The mixture is preheated to 450°C by exhaust gas from the firebox, and it is introduced to the reformer through a header that distributes the mixture evenly among the parallel reformer tubes. Two key reactions occur: the **steam-reforming reaction** itself,

$$CH_4 + H_2O(g) \rightleftharpoons CO + 3\,H_2 \tag{13.1}$$

and the **water–gas shift reaction**,

$$CO + H_2O \rightleftharpoons CO_2 + H_2 \tag{13.2}$$

The product gas leaves the reformer at 855°C and 1.6 MPa.

Energy efficiency in steam reforming is improved by recovering heat from the burner exhaust gas, which leaves the firebox at 960°C. The exhaust gas is cooled in a series of

---

[5]L. W. Wade, R. B. Gengelback, J. L. Trumbley, and W. L. Hallbauer, Methanol, in *Kirk-Othmer Encyclopedia of Chemical Technology,* 3rd Edition, Vol. 15, Wiley, New York, 1981, p. 398.

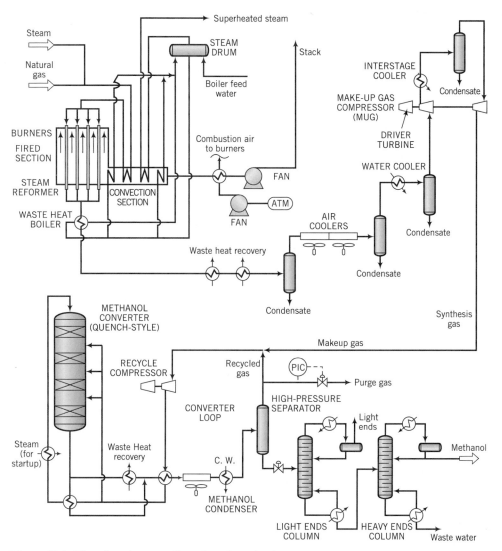

Figure 13.1 Flowchart for overall methanol synthesis process.

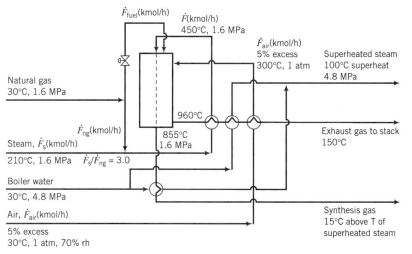

Figure 13.2 Detailed schematic diagram of reformer.

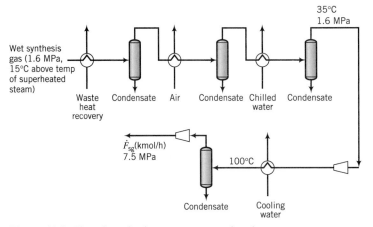

Figure 13.3 Flowchart for heat recovery and makeup gas compressor.

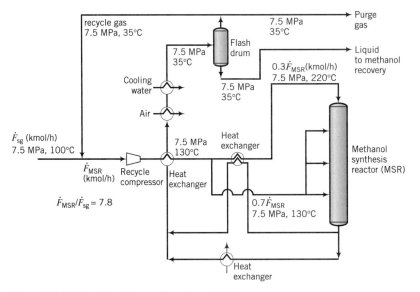

Figure 13.4 Flowchart for methanol synthesis reactor.

heat-exchange operations that preheat the reformer feed streams to 450°C, produce super-heated steam at 4.8 MPa and 100°C superheat from boiler feedwater at 30°C, and preheat the combustion air to 300°C. The superheated steam is used to drive turbines elsewhere in the process or it can be exported, for example to generate electricity. The burner exhaust gas leaves the heat-recovery units and enters a stack at 150°C for release to the atmosphere.

Heat Recovery and Compression

The product gas leaving the reformer contains water that should be removed to reduce the amount of gas that must be compressed and to minimize the impact on subsequent conversion of CO to methanol. Heat is removed from the gas by generating superheated steam (at 4.8 MPa, 100°C superheat), cooling the gas to 15°C above the temperature of the generated steam. Then, three steps occur in recovering heat and, concomitantly, reducing the water content: first, heat recovery for use elsewhere in the process; second, cooling by ambient air in an air cooler; and third, use of cooling water to reduce the temperature of the synthesis gas to 35°C. Condensed water is separated from the gas in each of these steps and collected in a condensate drum. With much of the water now removed, the product can properly be referred to as **synthesis gas**.

The makeup gas (MUG) compressor increases the pressure of the synthesis gas from 1.6 MPa to 7.5 MPa in two stages so that it can be injected into the **converter loop**. Between compressor stages, cooling water is used to reduce the temperature of the gas to 100°C, and any condensate formed is removed. The compressed synthesis gas is introduced into the converter loop, where it is combined with recycle gas.

Converter Loop

The converter loop consists of a recycle compressor, whose primary purpose is to provide the pressure head required for the gas to flow through the system, the methanol synthesis reactor (MSR), heat exchangers, a methanol condenser, and a gas–liquid separator (flash drum). The mixture that is to become the feed to the MSR consists of recycle gas and fresh synthesis gas. After the recycle gas and fresh synthesis gas are blended, the mixture flows through the recycle compressor and then is heated to 130°C by a partially cooled product stream leaving the MSR. (The partial cooling will be described later.) The recycle compressor is sized to circulate the recycle stream at a rate that is 7.8 times the rate at which fresh synthesis gas is fed to the converter loop. The blended recycle–fresh feed mixture leaving the heat exchanger following the compressor is split into two streams: one, containing 30% of the mixture, is sent to another heat exchanger where its temperature is raised to 220°C by a fraction of the product stream from the MSR and injected into the first stage of the MSR; the remaining 70%, which is still at 130°C, is injected at various locations along the MSR.

The key reactions occurring in the MSR are

$$CO + 2\,H_2 \rightleftharpoons CH_3OH(g) \tag{13.3}$$

$$CO_2 + 3\,H_2 \rightleftharpoons CH_3OH(g) + H_2O(g) \tag{13.4}$$

$$CO_2 + H_2 \rightleftharpoons CO + H_2O(g) \tag{13.5}$$

Note that the reaction in Equation 13.5 is the reverse of the water–gas shift reaction in Equation 13.2.

The product gas leaving the MSR is partially cooled by being split into two streams, each of which passes through a heat exchanger before being recombined; one is used to heat the feed stream to the first stage of the MSR to 220°C, and the other passes through a waste-heat recovery unit. The recombined product stream is cooled further in an air-cooled exchanger before being brought to 35°C by cooling water. At 35°C, a liquid consisting of the condensed methanol and dissolved gases is separated from the gas stream in a flash drum and sent to a methanol purification unit. The uncondensed gases are split, with a portion being purged from the system and the remainder forming the recycle gas that is blended with fresh synthesis gas to form the feed to the recycle compressor.

After the condensed crude methanol is recovered in the high-pressure separator, it is sent to a methanol purification column. Typically, methanol purification requires two columns, one to remove the light ends (mainly by-products generated in the methanol synthesis reactor such as dimethyl ether and dissolved gases) and another to separate methanol and water and any other by-products with a lower volatility than methanol. Specification-grade methanol (greater than 99.85 wt% methanol) is recovered as the overhead product of the heavy ends column and sent to storage.

PROBLEMS

Your company is considering the purchase of a plant that uses the previously described technology. The company from which the purchase may be made has indicated that the plant will produce 4.3 $\times 10^5$ metric tons per year of specification-grade methanol. The plant may be assumed to operate 350 days per year.

You have been asked to perform an analysis that will be a key part of determining the price that will be offered for the plant. The most important objectives in completing the assignment are:

- To quantify flow rates and compositions of streams in the process.
- To determine energy and raw materials requirements.

The following problems have been formulated in strategy sessions by the team negotiating the purchase and should be helpful in completing your task. The ordering of the problems follows the progression of the process (i.e., from reformer through to methanol separation and purification). As you structure your work efforts, you may find it possible to solve or at least set up solution procedures for problems late in the sequence while simultaneously developing skills required to solve earlier problems.

13.1. The process objective can be described most simply as converting methane and water into methanol and hydrogen, and then purifying the methanol so that it meets specifications. The overall process stoichiometry is given by the following relationship:

$$CH_4 + H_2O \rightarrow CH_3OH + H_2$$

From this statement, estimate the feed rates of the natural gas (kmol/h, SCMM) and steam (kmol/h, kg/h) fed as reactants (as opposed to fuel) to the reformer. (***Note:*** *The requested estimate neglects formation of by-products and the carbon monoxide and carbon dioxide lost in the purge stream.*)

13.2. Five percent excess air is used in burning the reformer fuel; it is drawn into the system at 30°C and 70% relative humidity. Estimate the average molecular weight of the air. Why does it differ from the value of 29 determined in Example 3.3-4 even though the ratio of nitrogen to oxygen is the same? Determine the flow rate of this stream (kmol, m³) per kmol of natural gas burned.

13.3. What are the compositions (mole and mass fractions) and volumetric flow rates (m³/kmol CH₄ fed to burners) of (a) the effluent gas from the reformer burners and (b) the gas entering the stack? What is the specific gravity, relative to ambient air (30°C, 1 atm, 70% rh), of the stack gas as it enters the stack? Why is this quantity of importance in designing the stack? Why might there be a lower limit on the temperature to which the gas can be cooled prior to introducing it to the stack?

Use a methane feed rate to the reformer of 1600 kmol/h as a basis for subsequent calculations. When all calculations have been completed, scale the results based on the required production rate of specification-grade methanol.

13.4. The primary purpose of the reformer is to convert methane and water to carbon monoxide and hydrogen (Equation 13.1). The extent of this reaction is limited by chemical equilibrium.[6]

$$K_{P_{13.1}} = \frac{y_{CO} y_{H_2}^3}{y_{CH_4} y_{H_2O}} P^2 \tag{13.6}$$

where

$$\log_{10} K_{P_{13.1}} = -\frac{11{,}769}{T(K)} + 13.1927 \tag{13.7}$$

Subscript 13.1 refers to the steam-reforming reaction (Equation 13.1), y_i is the mole fraction of species i, P is the system pressure (atm), and T is the temperature (K).

(a) If Equation 13.1 were the only reaction occurring in the reformer, estimate the composition of the product gas that would be leaving the reformer and the conversion of CH₄, assuming the product stream has achieved chemical equilibrium at 855°C and 1.6 MPa. What would be the total flow rate of this stream (kmol/h, kg/h)?

(b) It is specified that the molar ratio of steam to methane fed to the reformer is 3.0, whereas the stoichiometric ratio for the reforming reaction (Equation 13.1) is 1 mole of water per mole of methane. Estimate the conversion of methane for steam-to-methane feed ratios of 1:1 and 2:1, and compare these to the conversion in part a. Based on your results, explain in your own words why you think the ratio of 3 moles of steam per mole of methane was chosen for the process.

---

[6] P. L. Morse, *Hydrocarbon Processing,* **52**(1): 113, 1973.

13.5. As pointed out in the Process Description, the water–gas shift reaction (Equation 13.2) occurs in the reformer along with the reforming reaction (Equation 13.1). It too is controlled by chemical equilibrium.

$$K_{p_{13.2}} = \frac{y_{CO_2} y_{H_2}}{y_{CO} y_{H_2O}} \tag{13.8}$$

$$\log_{10} K_{p_{13.2}} = \frac{1197.8}{T(K)} - 1.6485 \tag{13.9}$$

where the nomenclature is analogous to that in the preceding problem.

(a) Taking into account the occurrence of reactions given by both Equations 13.1 and 13.2, estimate the composition of the product gas leaving the reformer and the conversion of CH_4, assuming the product stream leaving the reformer has achieved chemical equilibrium at 855°C and 1.6 MPa. What is the total flow rate of this stream in both kmol/h and kg/h? What effect does the water–gas shift reaction have on the production of CO at the reformer conditions?

(b) The ratio of CO to H_2 can be an important variable in efficient use of raw materials. In this case study a 3:1 steam-to-methane molar ratio of feed streams was specified. Determine how this feed ratio affects the ratio of CO to H_2 in the product from the reformer assuming the reaction products are in chemical equilibrium at 855°C and 1.6 MPa.

13.6. Quantitatively demonstrate that high temperatures and low pressures favor the formation of CO and H_2 in the reformer. Do this by calculating and then plotting the production rates (kmol/kmol of CH_4 fed) of CO and H_2 in the reformer product stream over the temperature range 750°C to 950°C at 1.2, 1.6, and 2.0 MPa. Furthermore, construct plots showing the effect of temperature and pressure on selectivity (defined as kmol CO formed per kmol CO_2 formed) over the same range of conditions. Assuming that your results support the hypothesis that high temperatures and low pressures favor the formation of CO and H_2, speculate as to why the temperature and pressure are at the values specified in the Process Description (855°C and 1.6 MPa) rather than at a higher temperature and lower pressure.

13.7. The reformer product gas leaves the reformer at 855°C.

(a) Using the flow rate of the product gas determined in Problem 13.5, calculate the rate (kJ/h) at which heat must be transferred from the combustion gases to the gases flowing through the inside of the reformer tubes.

(b) What is the required flow rate of natural gas (kmol/h and SCMM) to the reformer burners? Assume that the natural gas is burned to completion in the reformer firebox and that the combustion gases leave the firebox at 960°C.

(c) The **thermal efficiency** of the firebox may be defined as the percentage of the lower heating value of the fuel transferred to the reformer gases. Estimate the lower heating value of methane and, assuming the combustion gases leave the firebox at 960°C, the corresponding thermal efficiency of the firebox.

13.8. The heated tube length in the reformer is 10 m and the external diameter of the tubes is 10.5 cm. If the rate of heat transfer ($\dot{Q}$) from the combustion gases in the firebox to the reformer gases were accomplished entirely by convection,[7] the following equation would apply:

$$\dot{Q} = U_o A_o \Delta T_{lm}$$

where U_o is an overall **heat transfer coefficient** based on the external surface area of the reformer tubes in the firebox, A_o is the total external surface area of the tubes, and ΔT_{lm} is an average difference between temperatures of the heat source (combustion gases) and the heat sink (reformer reaction gases):

$$\Delta T_{lm} = \frac{(\Delta T_1 - \Delta T_2)}{\ln(\Delta T_1 / \Delta T_2)}$$

where ΔT_1 and ΔT_2 are differences in temperature between the reformer gas and combustion gas at the inlet and at the outlet of the firebox. If the combustion gases are assumed to have a constant

---

[7]**Convection** is a heat transfer mechanism involving bulk movement of fluid at an elevated temperature to a region of lower temperature.

temperature in the firebox of 960°C (i.e., they are perfectly mixed), and $U_o \approx 50$ Btu/(°F·ft²·h),[8] what is the required number of tubes in the firebox? In fact, a large fraction of the heat transferred to the tubes is accomplished by a mechanism other than convection. What is that mechanism? (*Hint:* Think of your experience in being close to a flame associated with combustion of natural gas.) What will consideration of this additional mechanism mean in terms of the number of tubes required in the firebox?

13.9. Operating data from the plant are available for a past period of operation. For that period, the plant operators determined the composition of the reformer product gas by sampling the product gas stream and analyzing it with a gas chromatograph. The overall flow rates of the methane and steam feed streams and of the product stream were determined from calibrated flow meters. Collected data show that the stream leaving the reformer was at 900°C and 1.6 MPa.

  (a) Component flow rates shown in the table below were calculated from the measured compositions and flow rates. Assuming the temperature, pressure, flow rate, and composition measurements are accurate, has the product gas reached equilibrium?

  (b) If the calculations in part (a) are inconsistent with the product gas being at equilibrium, give at least two possible reasons and suggest steps by which your hypotheses can be tested.

Component	Inlet (kmol/h)	Outlet (kmol/h)
Methane (CH_4)	1600.0	203.6
Carbon dioxide (CO_2)	0.0	500.0
Carbon monoxide (CO)	0.0	896.4
Hydrogen (H_2)	0.0	4689.2
Water (H_2O)	4800.0	2903.6

13.10. The generation of steam from the heat produced in the reformer is essential for economic viability of this energy-intensive operation.

  (a) At what rate (kg/h) is superheated steam at 4.8 MPa and 100°C superheat generated by recovering heat from the reformer combustion gases? The superheated steam is formed from boiler feed water at 30°C and 4.8 MPa.

  (b) How much additional superheated steam (at 100°C superheat and 4.8 MPa) is produced by cooling the hot synthesis gas leaving the reformer?

  (c) The process specification states that the hot synthesis gas is to be cooled in the steam generator to 15°C above the temperature of the superheated steam in the waste-heat boiler. Provide both an advantage and a disadvantage to changing this specification so that the synthesis gas is cooled to within 5°C of the generated steam.

13.11. Assume that CO, CO_2, H_2, and CH_4 are insoluble in liquid water.

  (a) Determine the dew-point temperature of the product gas from the reformer at 1.6 MPa.

  (b) What is the composition of the gas stream after the reformer gas has been cooled and equilibrated at 1.6MPa and 35°C? At what rate (kg/h) has water been removed from the stream?

  (c) Suppose that one-third of the total water removal determined in part (b) occurs in each of the units (waste-heat recovery, air cooler, water cooler) preceding the makeup gas compressor. Estimate the temperatures of the gas and liquid streams leaving each condensate recovery drum in this part of the process. At what rate is heat removed (kJ/h) in the waste-heat recovery, air-cooler, and water-cooler units? Speculate on why heat removal is done in stages; in other words, why not use either air or cooling water alone to reduce the temperature?

13.12. Each compressor stage in the MUG compressor unit operates adiabatically. If ideal gas behavior is assumed, the temperature of the gas leaving each compressor stage (T_{out}) is given by the expression

$$T_{out} = T_{in}\left(\frac{P_{out}}{P_{in}}\right)^{(k-1)/k}$$

---

[8]We recognize that American engineering units have been introduced here, but many common references still present process parameters in such units.

where T_{out} and T_{in} are absolute temperatures, P_{out} and P_{in} are absolute pressures, and k is $C_p/C_v \approx$ 1.4. The ratio of outlet to inlet pressures is the same for each compressor stage. Determine if water condenses as the synthesis gas is cooled between stages of the makeup gas compressor. At what rate is heat removed between compressor stages? What is the composition (mole fractions) and flow rate (kmol/h and m³/min) of the synthesis gas as it enters the converter loop?

13.13. For the purpose of analyzing the effects of recycle on process economics, consider a variation from the process flowchart that takes the synthesis gas leaving the MUG compressor directly to the MSR. Suppose that the synthesis gas fed to the MSR is 5 mole % methane, 25% CO, 5% CO_2, and the remainder hydrogen, and that the product stream leaving the MSR is at 250°C and 7.5 MPa. Also, for the time being, ignore all reactions in the MSR except for that given by Equation 13.3, which is governed by the equilibrium relationship:

$$K_{a_{13.3}} = \frac{y_{CH_3OH}}{y_{CO}\, y_{H_2}^2} \frac{K_{\phi_{13.3}}}{P^2} \tag{13.10}$$

where

$$K_{a_{13.3}} = \exp\left(21.225 + \frac{9143.6}{T} - 7.492 \ln T + 4.076 \times 10^{-3} T - 7.161 \times 10^{-8} T^2\right) \tag{13.11}$$

T is temperature in K, P is pressure in atm, and $K_{\phi_{13.3}}$ is a term that accounts for deviations from ideal gas behavior. Assume $K_{\phi_{13.3}} = 1.0$ and determine the composition of the product stream from the MSR and the conversions (%) of CO and H_2. Based on these results, comment on the rationale for the recycle loop discussed in the process description.

13.14. Again consider the modified system configuration described in the previous problem and estimate the effects of temperature and pressure on conversion. Do this by calculating CO and H conversions and methanol production for the following conditions:

$T(°C)$	250	250	200	300
$P(MPa)$	5	10	7.5	7.5

Describe qualitatively the effects of pressure and temperature on conversion, and suggest reasons for using the intermediate values of the process description.

13.15. As covered in the Process Description, there are three primary reactions that occur in the MSR. These are given by Equations 13.3, 13.4, and 13.5. However, determination of chemical equilibrium among the species H_2, CO, CO_2, H_2O, and CH_3OH involves only two of the three reactions because each reaction is a linear combination of the other two. Cherednichenko[9] gives an approximation for the equilibrium relationships in Equation 13.3 (see Problem 13.13) and in Equation 13.5:

$$K_{a_{13.5}} = \exp\left(13.148 - \frac{5639.5}{T} - 1.077 \ln T - 5.44 \times 10^{-4} T + 1.125 \times 10^{-7} T^2 + \frac{49170}{T^2}\right) \tag{13.12}$$

The equilibrium constant $K_{a_{13.5}}$ is defined by the relationship

$$K_{a_{13.5}} = \frac{y_{CO}\, y_{H_2O}}{y_{CO_2}\, y_{H_2}} K_{\phi_{13.5}} \tag{13.13}$$

where T is in kelvin and $K_{\phi_{13.5}}$ accounts for nonideal behavior of the gas phase.

(a) As in Problems 13.13 and 13.14, suppose the synthesis gas leaving the MUG compressor is fed directly to the MSR and that the composition of this gas is 5 mole% methane, 25% CO, 5% CO_2, and the remainder hydrogen. The product stream leaving the MSR is at 250°C and 7.5 MPa and you may assume $K_{\phi_{13.3}} = K_{\phi_{13.5}} = 1.0$. Again determine the conversions of CO and H_2 with this process configuration, and use these results to justify utilization of the recycle loop.

---

[9]V. M. Cherednichenko, Dissertation, Karpova, Physico Chemical Institute, Moscow, U.S.S.R., 1953.

(b) Show that removal of water from the reformer product gas minimizes the impact on conversion of CO to methanol by determining the effect on CO and H_2 conversions and selectivity as defined by moles CH_3OH formed per mole of CO reacted if the given composition of MSR feed gas in part (a) is on a dry basis and the stream itself contained 5 mole% water.

13.16. A situation often encountered with thermodynamic data is that different sources may present different correlations for a given quantity. In this case study, for example, two different expressions are given for the equilibrium constant for the water–gas shift reaction, Equations 13.9 and 13.12. By what percentages do the conversions of CO and H_2 and production of methanol differ if, instead of using Equation 13.12 in Problem 13.15(a), Equation 13.9 is used. Provide at least two reasons for the variations in the two equations for the water–gas shift equilibrium constant.

13.17. Values of $K_{\phi 13.3}$ and $K_{\phi 13.5}$ can deviate significantly from unity; typical values for $K_{\phi 13.3}$ range from 0.7 to 0.9 at a pressure of 7.5 MPa and typical MSR temperatures from 200°C to 320°C, while values for $K_{\phi 13.5}$ range from 0.8 to 0.95. Estimates of $K_{\phi i}$ may be obtained from an equation of state, but such calculations are beyond the scope of the present discussion.[10] Furthermore, computational alternatives often require a choice between simplified and more realistic, but complex, models. An instance that illustrates this situation is estimation of $K_{\phi 13.3}$ and $K_{\phi 13.5}$ for use in determining reaction equilibrium constants.
(a) What physical model is invoked if it is assumed that $K_{\phi 13.3} = K_{\phi 13.5} = 1.0$?
(b) In the calculations of Problem 13.15, $K_{\phi 13.3}$ and $K_{\phi 13.5}$ were taken to be 1.0. Determine the significance of the ideal gas assumptions by reestimating the conversions of CO and H_2 first with $K_{\phi 13.3} = 0.7$ and $K_{\phi 13.5} = 1.0$, then with $K_{\phi 13.3} = 1.0$ and $K_{\phi 13.5} = 0.8$, and finally with $K_{\phi 13.3} = 0.7$ and $K_{\phi 13.5} = 0.8$.
(c) From the preceding parts of this problem, what would be your decision (and why?) regarding inclusion of methods to estimate $K_{\phi 13.3}$ and $K_{\phi 13.5}$ in the process analysis?

13.18. You need to examine various ways of controlling the temperature in the MSR because the catalyst is sensitive to high temperatures and it has been reported that elevated temperatures lead to the production of unacceptable by-products. In the analysis, suppose that the feed gas to the MSR is 5 mole% methane, 25% CO, 5% CO_2, and the remainder hydrogen, and that the MSR operates at 7.5 MPa. Take $K_{\phi 13.3} = K_{\phi 13.5} = 1.0$.
(a) Determine the temperature of the stream leaving the MSR if no heat is removed from the reactor (i.e., it is operated adiabatically), the CO conversion is 45%, the CO_2 conversion is 30%, and all of the feed enters the top of the MSR at 210°C and 7.5 MPa.
(b) How much heat must be removed from the MSR in part (a) to ensure that the temperature does not exceed 250°C? Give your answer in kJ/kmol of feed.
(c) Repeat the calculations in part (a) if 70% of the total feed to the MSR enters as quench streams (see the Process Description) at 130°C and 7.5 MPa, while the remainder enters the top of the converter at 210°C. Assume the exit pressure is 7.5 MPa.
(d) Suppose that 70% of the feed to the MSR enters as quench streams at 130°C and 7.5 MPa, while the remainder of the feed enters at the top of the converter at 210°C. Determine the equilibrium conversions of CO and CO_2 if the MSR is to operate adiabatically.

13.19. In terms of Le Châtelier's Principle, explain why steam reforming is done at low pressures while methanol synthesis is performed at moderate to high pressures. Explain why reforming is done at high temperature and methanol synthesis is performed at low to moderate temperatures.

13.20. Perform an analysis of the converter loop by determining the composition and flow rate of the purge stream using a basis of 100 kmol of feed from the MUG compressor.
(a) Assume that the feed has a composition that is 3 mole% methane, 8% CO_2, 15% CO, and the remainder hydrogen.
(b) Revise the feed composition to that calculated in Problem 13.12. As given in the Process Description, the molar flow rate of material to the MSR is 7.8 times the flow rate of fresh feed.
    To simplify calculations, assume that the liquid leaving the flash drum contains no methane, CO_2, CO, or hydrogen and that the gas contains no water or methanol. The single-pass conversions

---

[10]T. Chang, R. W. Rousseau, and P. K. Kilpatrick, *Industrial and Engineering Chemistry Process Design and Development,* **25**, 477 (1986).

of CO and CO_2 are 15% and 10%, respectively. You may be assisted in your calculations by assuming flow rates of components in the recycle stream mixed with fresh synthesis gas entering the converter loop. In this approach the recycle stream is known as a **tear stream**, and an iterative solution will be required to determine the requested values. Such calculations are easily performed using a simulation program of the type described in Chapter 10, or you may write your own program or spreadsheet to obtain the desired results. If you develop a spreadsheet to perform the calculations, direct substitution of calculated values of component flow rates in the tear stream for new estimates may suffice. (This is the method of successive substitution described in Appendix A.2.)

13.21. The purge stream may be used to replace a portion of the methane burned in the reformer. Assuming that the purge gas is at 35°C and that it is fed directly to the reformer burners with 5% excess air at 300°C, how much heat would be released per 100 kmol of purge gas? Estimate the savings ($ per kmol of purge gas) if the purge stream is used in place of natural gas for fuel. (Take $3 per million Btu as the approximate cost of natural gas.)

13.22. How much heat must be removed to cool the reaction products from the MSR to 35°C? Express your answer in kJ/kmol of reaction products from the MSR.

13.23. The liquid leaving the flash tank in the converter loop is fed to a distillation column in which specification-grade methanol is produced. The column is designed to recover 95% of the methanol in an overhead stream that is 99.85 wt% methanol. The remaining methanol and water leave the distillation column in a bottom stream. Estimate the flow rates of the overhead and bottom streams (kmol/h and kg/h) and the composition (mole fractions) of the bottom stream.

13.24. Scale the results of your calculations in Problems 13.5(a); 13.7(a),(b); 13.10(a),(b); 13.20(b); 13.22; and 13.23 so that the quantities requested in each correspond to the specified production rate of methanol.

Additional Problems and Issues

13.25. Relax the assumptions regarding the split of components at the flash tank by incorporating the information below into your calculations. Using all other conditions given in Problem 13.20, reestimate the composition and flow rate of the purge stream. We have the following K values at the high-pressure separator (y_i/x_i):

Component	CH_4	CO_2	CO	H_2	H_2O	CH_3OH
K	29.8	5.12	39.9	86.3	2.85×10^{-3}	6.80×10^{-3}

13.26. The converter loop is a net generator of energy (why?) and proper utilization of that energy improves process economics. Assuming that the makeup gas is at 100°C and that the purge gas and crude methanol are at 35°C, what is the net rate of energy released from the converter loop in Problem 13.20? As the recycle compressor only serves to move the recycle gas, you may neglect the work input from this unit. Assume ideal gas behavior.

Chapter 14

The Use of Limestone Slurry Scrubbing to Remove Sulfur Dioxide from Power Plant Flue Gases[1]

Protection of the environment through the development of processes that have minimal ecological impact is one of the most important responsibilities facing chemical engineers. Moreover, it is often necessary to use chemical engineering principles to remedy existing environmental problems. In this chapter, we examine a situation in which a coal-burning power plant is designed so that the impact on its surroundings is minimized.

Coal is an abundant and practical source of energy, and it can be a source of raw materials used in chemical production. Coal occurs in various forms,[2] and the form used in a particular application often depends on its availability in the region of the application. Coal is predominantly carbon. However, it also contains numerous other elements, including metals, nitrogen compounds, and sulfur, as well as hydrocarbon volatile matter that can be burned to produce heat, water, and carbon dioxide (CO_2).

Coal is burned in many of the world's power plants to generate electricity. Significant air pollution is produced if the stack gases from coal-fired furnaces are not treated before being released to the atmosphere. Untreated gases contain soot (fine unburned carbon particles), nitrogen oxides (NO_x), ash, and sulfur dioxide (SO_2).

Sulfur dioxide emissions from coal-fired furnaces are regulated in the United States by the Environmental Protection Agency (EPA) and appropriate state and local agencies. Current EPA standards require that gases released to the atmosphere from new and some existing power plants (some existing plants are subject to a less rigorous standard) contain no more than 520 nanograms SO_2 per joule of higher heating value of the fuel fed to the furnace (1.2 lb_m SO_2/million Btu). In addition, these plants must remove at least 90% of the SO_2 released

---

[1]Prepared by Ronald W. Rousseau and Jack Winnick, School of Chemical Engineering, Georgia Institute of Technology, Atlanta, GA; and Norman Kaplan, National Risk Management Research Laboratory, U.S. Environmental Protection Agency, Research Triangle Park, NC.

[2]R. H. Perry and D. W. Green, Eds., *Perry's Chemical Engineers' Handbook,* 7th Edition, McGraw-Hill, New York, 1997, pp. 27-4 to 27-6.

when burning the coal.[3] The emissions standard for some existing plants may be satisfied by burning low-sulfur coal, by removing sulfur from the coal prior to combustion, or by removing SO_2 from the product gases before they are released to the atmosphere. The emission standard for new plants is more stringent and may only be achieved in practice by removing SO_2 from stack gases. The technology for removing SO_2 from stack gases is more advanced than that for separating high percentages of sulfur directly from coal, and a large number of stack-gas desulfurization processes are available commercially. We examine one of these processes in this case study.

Commercial processes for SO_2 removal are classified as regenerative or throwaway, depending on whether or not the agent used to remove SO_2 is reusable. Some throwaway processes can be modified to produce a saleable product (gypsum) without regeneration of the absorbent material. Regenerative processes have two major steps: the removal of dilute SO_2 from stack gases by a separating agent followed by removal of SO_2 in concentrated form from the separating agent, which is then recycled to the first step. An example of such a procedure is the Wellman–Lord process, which first absorbs the SO_2 in an aqueous solution of sodium sulfite (Na_2SO_3). In the solution, the following reaction occurs:

$$Na_2SO_3 + SO_2 + H_2O \rightarrow 2\,NaHSO_3 \qquad \textbf{(14.1)}$$

Subsequently, the solution is heated and, at the resulting elevated temperature, the process is reversed:

$$2\,NaHSO_3 \rightarrow Na_2SO_3 + SO_2 + H_2O \qquad \textbf{(14.2)}$$

The regenerated Na_2SO_3 solution is recycled to the absorber, and the concentrated SO_2 produced is captured and processed further to recover elemental sulfur.

Throwaway processes utilize a separating agent to remove SO_2 from the stack gases followed by disposal of both the SO_2 in an innocuous form ($CaSO_3 \cdot \frac{1}{2} H_2O$) and the slurried separating agent, calcium carbonate ($CaCO_3$). Although there are many processes that involve injection of either wet or dry limestone into the power plant furnace, we have chosen to use limestone-slurry scrubbing in the present case study.

PROCESS DESCRIPTION

Five hundred megawatts of electrical power (500 MWe)[4] is to be generated in the present facility. Coal with properties given in Table 14.1 is fed at 25°C to a furnace where it is burned with 15% excess air. During combustion of the coal, sulfur reacts to form SO_2 and a negligible amount of sulfur trioxide (SO_3), while carbon and hydrogen are oxidized completely to CO_2 and H_2O. Essentially all of the nitrogen in the coal leaves the furnace as N_2.[5] The ash in the coal leaves the furnace in two streams: 80% leaves as fly ash in the furnace flue gas, and the remainder leaves the furnace as bottom ash at 900°C.

Combustion air is brought into the process at 25°C and 50% relative humidity and sent to a heat exchanger, where its temperature is increased to 315°C by exchanging heat with the furnace flue gas. It is then fed to the boiler, where it reacts with coal. The flue gas leaves the furnace at 330°C, goes to an electrostatic precipitator where 99.9% of the particulate material

---

[3]Federal New Source Performance Standards for control of SO_2 from power plants are applicable to electric utility steam-generating units that are capable of firing more than 73 MW (250 million Btu/h) heat input of fossil fuel and for which construction or modification was commenced after September 18, 1978. This regulation was proposed in the *Federal Register,* Vol. 43, No. 182, Tuesday, Sept. 19, 1978. The 520 ng/J limit applies to solid fuels.

[4]The notation MWe should be read as megawatts of electricity.

[5]A very small fraction of the nitrogen in the coal and/or air is converted to gaseous NO_x during combustion. These are pollutants that must also be removed from the flue gas or controlled by combustion modification, but that technology is not within the scope of this case study.

Table 14.1 Average Properties of Coal

Composition (Ultimate Analysis)	
Component	Dry Wt %
Carbon	75.2
Hydrogen	5.0
Nitrogen	1.6
Sulfur	3.5
Oxygen	7.5
Ash	7.2

Moisture: 4.80 kg/100 kg dry coal
HHV: 30,780 kJ/kg dry coal (see Section 9.6a)
Dry coal: $C_p = 1.046$ kJ/(kg·°C)
Ash: $C_p = 0.921$ kJ/(kg·°C)

is removed, and then to the air preheater where it exchanges heat with the combustion air. The flue gas leaves the air preheater and is split into two equal streams, with each being the feed stream to one of two identical scrubber trains. Each of the trains is sized to process 60% of the flue gas, giving flexibility and partial redundancy should one of the trains require downtime.

In each of the scrubber trains, the divided off-gas stream is fed to a scrubber, where it contacts an aqueous slurry of limestone and undergoes adiabatic cooling to 53°C. Sulfur dioxide is absorbed in the slurry and reacts with the limestone:

$$CaCO_3 + SO_2 + \tfrac{1}{2} H_2O \rightarrow CaSO_3 \cdot \tfrac{1}{2} H_2O + CO_2 \qquad (14.3)$$

The solid–liquid limestone slurry enters the scrubber at 50°C; the liquid portion of the slurry flows at a rate of 15.2 kg liquid/kg inlet gas and the solid-to-liquid ratio in the slurry is 1:9 by weight. The liquid is saturated with $CaCO_3$ and $CaSO_3$. The cleaned flue gas meets the EPA standard on SO_2 emissions; it leaves the scrubber saturated with water at 53°C, containing the CO_2 generated in the scrubbing but none of the entering fly ash. Sulfur dioxide is the only constituent of the flue gas that is absorbed in the scrubber. The cleaned flue gas is then reheated to 80°C, blended with the cleaned flue-gas stream from the other train, and sent to the stack where it is released to the atmosphere.

The solids in the spent aqueous slurry leaving the scrubber contain unreacted $CaCO_3$, fly ash removed from the flue gas, inert materials that enter the process with fresh limestone, and $CaSO_3$ formed by the reaction in Equation 14.3. The liquid portion of the slurry is saturated with $CaCO_3$ and $CaSO_3$ and has a specific gravity of 0.988. The spent slurry is split into two streams. One of these streams is sent to a blending tank where it is mixed with fresh finely ground limestone, makeup water, and a recycle stream whose source will be described shortly. A fresh slurry stream from the blending tank is fed to the top of the scrubber. The other spent slurry stream is sent to a filter where wet solids containing fly ash, inert materials, $CaSO_3$, and $CaCO_3$ are separated from a filtrate, which is saturated with the latter two components and is the recycle stream fed to the blending tank. The wet solids contain 50.2 wt% liquid that has the same composition as the filtrate.

The fresh ground limestone is fed to the blending tank at a rate that is 5.2% in excess of that required to react with the SO_2 absorbed from the flue gas. The limestone material fed consists of 92.1% $CaCO_3$ and the remainder is inert insoluble material.

The generation of steam and its utilization in the production of electricity in this facility are typical of many power cycles.[6] The boiler used in the present situation generates steam at super-

---

[6] *Steam: Its Generation and Use,* 39th Edition, Babcock & Wilcox, New York, 1978.

critical conditions: 540°C and 24.1 MPa absolute. Mechanical work is derived by expanding the steam through a power-generating system of turbines; appropriate reheaters and moisture separators are used to facilitate that operation. The low-pressure steam extracted from the power system contains 27.5% liquid water at 6.55 kPa absolute. Heat is removed from the wet low-pressure steam in a condenser by cooling water that enters the condenser at 25°C and leaves at 28°C. Saturated condensate at 38°C is produced by the condenser and pumped back to the boiler.

PROBLEMS

Problems 14.2 through 14.10 should be solved using a basis of 100 kg dry coal/min fed to the furnace.

14.1. Construct a flowchart of the process and completely label the streams. Show the details of only one train in the SO_2 scrubber operation.

14.2. Estimate the molar flow rate (kmol/min) of each element in the coal (other than those in the ash).

14.3. Determine the feed rate (kmol/min) of O_2 required for complete combustion of the coal.

14.4. If 15% excess O_2 is fed to the combustion furnace, estimate the following:
   (a) The oxygen and nitrogen feed rates (kmol/min).
   (b) The mole fraction of water in the wet air, the average molecular weight, the dew point and degrees of superheat of the wet air, and the molar flow rate (kmol/min) of water in the air stream.
   (c) The air feed rate (kmol/min, standard cubic meters/min, cubic meters/min).

14.5. Estimate the flow rate (kg/min and kmol/min) of each component and the composition (mole fractions) of the furnace flue gas. (Ignore the fly ash in calculating mole fractions.) At what rate (kg/min) is fly ash removed from the flue gas by the electrostatic precipitator?

14.6. The system may be assumed to meet the standard of 90% removal of the SO_2 released upon combustion.
   (a) Determine the flow rate (kg/min and kmol/min) of each component in the cleaned flue gas leaving the scrubber.
   (b) Determine the flow rate (kg/min) of slurry entering the scrubber.
   (c) Estimate the solid-to-liquid mass ratio in the slurry leaving the scrubber.
   (d) Estimate the feed rate (kg/min) of fresh ground limestone to the blending tank.
   (e) What are the flow rates (kg/min) of inerts, $CaSO_3$, $CaCO_3$, fly ash, and water in the wet solids removed from the filter? What fractions of the $CaSO_3$ and $CaCO_3$ are dissolved in the liquid portion of the wet solids?
   (f) Estimate the rate (kg/min, L/min) at which the filtrate is recycled to the blending tank. At what rate (kg/min, L/min) is makeup water added to the blending tank?

14.7. At what rate is heat removed from the furnace? Assuming that all of the heat removed from the furnace is used to generate steam (i.e., none is lost to the surroundings), estimate the rate of steam generation in the power cycle.

14.8. Determine the effect of the percent excess air fed to the boiler furnace by calculating the rate of steam generation (kg/min) for air flow rates that are 5% and 25% in excess of that theoretically required. Speculate on the reason for choosing 15% excess air in the prescribed process by giving one possible reason for not using less air and one for not using more.

14.9. Determine the temperature of the flue gas as it leaves the heat exchanger (air preheater) following the boiler. Estimate this value for the two alternative air flow rates corresponding to 5% and 25% excess oxygen.

14.10. Compare the release of SO_2 in the scrubbed flue gas (determined in Problem 14.6) with the EPA limit of no more than 520 nanograms SO_2 per joule of heat input to the boiler. (*Note:* If you determined that meeting this requirement requires greater than 90% removal of SO_2, you would, in practice, have to revise your previous calculations.)

14.11. Power plants of the type described here operate with an efficiency of about 39%; that is, for each unit of heat released with the combustion of coal, 0.39 unit is converted to electrical energy. From this efficiency and the specified power output of 500 MWe, determine the following:

(a) The coal feed rate (kg/h).
(b) The air feed rate (kmol/min, standard cubic meters/min, cubic meters/min).
(c) The flow rate of each component in the gas leaving the furnace (kmol/min, kg/min).
(d) The rate of steam generation (kg/h).

14.12. For the required coal feed rate, scale the results from Problem 14.6 to determine for each scrubber train the actual flow rates (kg/h) of wet solids and filtrate from the filter, the slurry flow rates entering and leaving the scrubber, the flow rates of gas (kg/h and m³/min) entering and leaving the scrubber, and the flow rates (kg/h) of fresh water and limestone fed to each blending tank.

14.13. Why is the scrubbed flue gas reheated before it is sent to the stack?

14.14. The following are alternatives by which the scrubbed flue gas can be reheated: (1) bypassing the scrubber with a fraction of the flue gas leaving the air preheater and mixing this stream with cleaned flue gas; (2) burning natural gas and blending the combustion products with the scrubbed flue gas; and (3) using steam from the power cycle either to heat air that is blended with the scrubbed flue gas or to heat the flue gas in a heat exchanger.

(a) Give a reason for rejecting Alternative 1.
(b) In evaluating Alternative 2, assume that the natural gas consists entirely of methane at 25°C and that it is burned adiabatically with 10% excess air that has the same conditions as the air fed to the furnace. If the combustion products are blended with the cleaned flue gas, at what rate would methane would have to be burned to raise the stack-gas temperature to the desired value?
(c) How much more coal (kg dry coal/h) would have to be burned for Alternative 3 if the amount of heat released by burning coal (kJ/kg dry coal) is the same as determined in Problem 14.7? Suggest two process alternatives by which the heat can be transferred to the flue gas.

14.15. Assuming that option 2 in Problem 14.14 is used, construct a graph of the velocity of the stack gas (1 atm, 80°C) versus the diameter of the stack. Cover a range of velocities from 5 to 15 m/s.

14.16. What is the flow rate of cooling water (kg/h and L/h) required to condense the low-pressure steam in the power cycle?

14.17. The pump used to transport the steam condensate in the power cycle from the condenser to the boiler has an efficiency of 80% (i.e., 80% of the work done by the pump is on the condensate). What size pump (kW and horsepower) is required for the estimated water flow rate? What assumptions have you made in these calculations?

14.18. We assumed a 39% efficiency for this power plant, meaning that for each unit of heat released with the combustion of coal, 0.39 unit is converted to electrical energy. Using an energy balance around the power-generating system of turbines, reheaters, and condensate separators, estimate an energy efficiency defined as 100 × the electrical energy generated (500 MWe) divided by the change in enthalpy of the steam flowing through the power-generating system.

14.19. Why does the EPA standard for new power plants have two criteria? Using your own words, explain them in terms that can be understood by an informed public. Include in your discussion the impact of the regulation on the use of clean coal.

14.20. If the 520-ng/J criterion had been followed, what would be the emission rate of SO_2 in kg/day?

Computational Techniques

This appendix introduces several mathematical concepts and methods that have widespread applicability in the analysis of chemical processes. The presentation presumes a knowledge of elementary calculus, but not of linear algebra or numerical analysis. The student who wishes a broader or deeper treatment of the subjects discussed is advised to refer to a numerical analysis reference.

Appendix A.1   THE METHOD OF LEAST SQUARES

In this section we outline a statistical technique for fitting a straight line to y versus x data. You should be aware, however, that in doing so we are barely scratching the surface of the field of statistical analysis; we will not discuss techniques for fitting functions of several variables, for determining quantitatively the uncertainties associated with a fit, and for comparing alternative functions used to fit a given data set.

Suppose you measure y at four values of x, plot the data on a graph of y versus x, and draw a line through the data points.

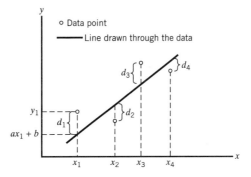

If the line you draw is $y = ax + b$, then at an abscissa point x_i ($i = 1, 2, 3$ or 4) the measured value of y is y_i, and the corresponding value of y on the line is $ax_i + b$. The vertical distance d_i from the i th data point to the line (called the i th **residual**) is therefore

$$d_i = y_i - (ax_i + b), \qquad i = 1, 2, 3, 4 \tag{A.1-1}$$

If d_i is positive then the i th data point must be above the line (why?), if d_i is negative the point is below the line, and if d_i equals zero the line passes through the point. A line is said to fit the data well if the values of most of the residuals are close to zero.

There are several ways to determine the line that best fits a set of data, which differ primarily in their definitions of "best." The most common method is the **method of least squares**.

Suppose there are n plotted points $(x_1, y_1), (x_2, y_2), \ldots, (x_n, y_n)$, so that a line $y = ax + b$ drawn through the points yields a set of n residuals $d_1, d_2, \ldots, d_n$. According to the method of least squares, *the best line through the data is the one that minimizes the sum of the squares of the residuals.*[1] The task is therefore to find the values of a and b that minimize

$$\phi(a, b) = \sum_{i=1}^{n} d_i^2 = \sum_{i=1}^{n} (y_i - ax - b)^2 \tag{A.1-2}$$

You can obtain expressions for the best values of a and b in terms of known quantities by differentiating the equation for ϕ (Equation A.1-2) with respect to both a and b, setting the derivatives equal to zero, and solving the resulting algebraic equations for a and b. The results of these calculations are as follows. If we define

$$s_x = \frac{1}{n} \sum_{i=1}^{n} x_i \qquad s_{xx} = \frac{1}{n} \sum_{i=1}^{n} x_i^2$$
$$s_y = \frac{1}{n} \sum_{i=1}^{n} y_i \qquad s_{xy} = \frac{1}{n} \sum_{i=1}^{n} x_i y_i \tag{A.1-3}$$

then

1. Best line: $y = ax + b$

Slope:
$$a = \frac{s_{xy} - s_x s_y}{s_{xx} - (s_x)^2} \tag{A.1-4}$$

Intercept:
$$b = \frac{s_{xx} s_y - s_{xy} s_x}{s_{xx} - (s_x)^2} \tag{A.1-5}$$

2. Best line through the origin: $y = ax$

Slope:
$$a = \frac{s_{xy}}{s_{xx}} = \frac{\sum x_i y_i}{\sum x_i^2} \qquad \text{(The intercept equals 0.0.)} \tag{A.1-6}$$

Once you determine a and b, you should plot the line $y = ax + b$ on the same graph as the data to get an idea of how good the fit is.

EXAMPLE A.1-1 **The Method of Least Squares**

Two variables, P and t, are related by the equation

$$P = \frac{1}{mt^{1/2} + r}$$

The following data are taken:

P	0.279	0.194	0.168	0.120	0.083
t	1.0	2.0	3.0	5.0	10.0

Calculate m and r using the method of least squares.

SOLUTION The equation may be rewritten in the form

$$\frac{1}{P} = mt^{1/2} + r$$

---

[1]We could also choose the best line as the one that minimizes the sum of the absolute values of the residuals, or the sum of the fourth powers of the residuals. Using the squares simply provides computational convenience.

so that a plot of $1/P$ versus $t^{1/2}$ should be a line with slope m and intercept r. From the tabulated data,

$y = 1/P$	3.584	5.155	5.952	8.333	12.048
$x = t^{1/2}$	1.0	1.414	1.732	2.236	3.162

$$\frac{1}{P} = mt^{1/2} + r$$

$$\Downarrow y = 1/P, \; x = t^{1/2}$$

$$y = mx + r$$

Evaluate the quantities of Equation A.1-3:

$$s_x = \tfrac{1}{5}(1.000 + 1.414 + 1.732 + 2.236 + 3.162) = 1.909$$
$$s_y = 7.014$$
$$s_{xx} = 4.200$$
$$s_{xy} = 15.582$$

Hence, from Equation A.1-4,

Slope: $\qquad\qquad\qquad\qquad m = \dfrac{s_{xy} - s_x s_y}{s_{xx} - (s_x)^2} = 3.94$

and from Equation A.1-5,

Intercept: $\qquad\qquad\qquad\quad r = \dfrac{s_{xx} s_y - s_{xy} s_x}{s_{xx} - (s_x)^2} = -0.517$

so that the final result is

$$P = \frac{1}{3.94 t^{1/2} - 0.517}$$

A check on the results is to plot $1/P$ versus $t^{1/2}$, showing both the data points and the line

$$\frac{1}{P} = 3.94 t^{1/2} - 0.517$$

If the choice of this function to fit the given data was reasonable and if no mistakes were made in the calculations, the data points should be scattered about the line. This is in fact the case, as shown in the following diagram.

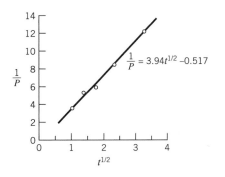

TEST YOURSELF

1. A line $y = 3x + 2$ has been fit to a set of data including the points $(x = 1, y = 4)$ and $(x = 3, y = 13)$. What are the residuals at these two points?
2. What is the definition of the best line through a set of data points that forms the basis of the method of least squares?

3. Could the method of least squares be used to fit a straight line to data points that fall on a distinct curve? (*Corollary:* Does the line that "best" fits the data necessarily fit the data well?)

4. An alternative to the method of least squares might be to minimize the sum of the residuals rather than the sum of squares of the residuals. What would be wrong with this method?

Appendix A.2   ITERATIVE SOLUTION OF NONLINEAR ALGEBRAIC EQUATIONS

A.2a   Linear and Nonlinear Equations

Given below is the van der Waal's equation of state:

$$(P + a/\hat{V})(\hat{V} - b) = RT$$

Solving this equation to calculate P for a given $\hat{V}$ and T is easy, while solving for $\hat{V}$ for specified values of P and T is relatively difficult.

What makes an equation easy or difficult to solve is its **linearity** or **nonlinearity** in the unknown variable. Equations that contain unknown variables raised to the first power only (x, but not x^2 or $x^{1/2}$), and that do not contain products (xy) or transcendental functions ($\sin x$, e^x) of unknown variables, are called **linear equations**. Equations that do not satisfy these conditions are called **nonlinear equations**.

For example, if a, b, and c are constants, and x, y, and z are variables,

$$ax + by = c \text{ is linear}$$

$$ax^2 = by + c \text{ is nonlinear (contains } x^2)$$

$$x - \ln x + b = 0 \text{ is nonlinear (contains } \ln x)$$

$$ax + by = cx \text{ is linear}$$

Linear equations that contain a single unknown variable have one and only one solution (one **root**).

$$7x - 3 = 2x + 4 \implies x = 1.2$$

$$\left.\begin{array}{l} P\hat{V} = RT \\ P = 3, \ R = 2, \ T = 300 \end{array}\right\} \implies \hat{V} = RT/P = (2)(300)/(3) = 200$$

In contrast, nonlinear equations that contain a single unknown variable may have any number of real roots (as well as imaginary and complex roots). For example,

$$x^2 + 1 = 0 \quad \text{has no real roots}$$

$$x^2 - 1 = 0 \quad \text{has two real roots } (x = +1 \text{ and } x = -1)$$

$$x - e^{-x} = 0 \quad \text{has one real root } (x = 0.56714\ldots)$$

$$\sin x = 0 \quad \text{has an infinite number of real roots } (x = 0, \pi, 2\pi, \ldots)$$

The roots of some nonlinear equations, such as the second of the equations given above, can be obtained directly using simple algebra, but most nonlinear equations must be solved using an iterative or trial-and-error technique.

A single equation containing several variables may be linear with respect to some variables and nonlinear with respect to others. For example,

$$xy - e^{-x} = 3$$

is linear in y and nonlinear in x. If x is known, the equation may easily be solved for y, while the solution for x from a known variable of y is much harder to obtain. Another example is the three-term virial equation of state:

$$P\hat{V} = RT\left(1 + \frac{B(T)}{\hat{V}} + \frac{C(T)}{\hat{V}^2}\right)$$

where B and C are known functions of temperature. This equation is linear in P and nonlinear in $\hat{V}$ and T. It is consequently easy to solve for P from given values of T and $\hat{V}$ and difficult to solve for either $\hat{V}$ or T from given values of the other two variables.

Most of the problems you are called on to solve in this text reduce to one or two linear equations in as many unknowns. The hard part of the problems, if there is a hard part, is deriving the equations; solving them is a matter of simple algebra. However, many process problems involve nonlinear equations. Techniques for solving such problems are the subject of this section.

TEST
YOURSELF

Classify the following single-variable equations as linear or nonlinear, considering a, b, c as constants.

1. $3x + 17 = 23x - 12$
2. $3x = a(\ln x) + b$
3. $x \exp(x) = 14$
4. $axy - b^2 = cy/x$
   (a) x is known
   (b) y is known
5. $14x \cos(y) - 8/z = 23$
   (a) x and y are known
   (b) x and z are known
   (c) y and z are known

A.2b   Graphical Solution

In this and the next several sections, we will discuss methods for solving one nonlinear equation in one unknown. Extensions to multivariable problems will be presented in Section A.2i.

Suppose you have to solve an equation of the form $f(x) = 0$ — that is, find the root or roots of the function $f(x)$. [Any equation can be written in this form by bringing every term to the left side. For example, $x = e^{-x}$ becomes $f(x) = x - e^{-x} = 0$.] An obvious solution technique is to plot $f(x)$ versus x and to locate by graphical interpolation the point at which the curve crosses the axis.

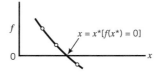

There are several problems with this technique. It is a manual method; it is relatively slow; and it is not very precise. Its principal advantage is that it enables you to see how f varies with x, which is particularly useful when you are dealing with functions that have several roots.

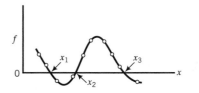

The points x_1, x_2, and x_3 are all *roots* (solutions) of the equation $f(x) = 0$. The computational techniques we will describe shortly would locate one or another of these roots, depending on the initial guess; however, for all but simple polynomial functions a plot is the only convenient method to detect the existence and approximate location of multiple roots. Therefore, unless you know that only one root exists, or you know the approximate location of each root you need to determine, a good procedure is to plot f versus x and use the plot to determine initial estimates for more precise root-finding methods.

TEST YOURSELF

1. Use a graphical argument to justify the statement that a linear function has only one root.
2. How many roots does the function $f(x) = x - \exp(-x)$ have? [*Suggestion:* Sketch plots of $f_1(x) = x$ and $f_2(x) = \exp(-x)$ versus x, and use these plots to obtain your answer.]

A.2c Spreadsheet Solution

If you have access to a spreadsheet program, finding solutions of nonlinear single-variable equations is relatively easy. If the equation has the form $f(x) = 0$, you need only enter a guessed value of x in one cell of the spreadsheet, insert the formula for $f(x)$ in an adjacent cell, and then vary the value in the first cell until the value in the second cell is close enough to zero to meet a specified convergence criterion. The next example illustrates this approach.

EXAMPLE A.2-1 **Spreadsheet Solution of a Nonlinear Equation**

Estimate the solution of the equation $x = e^{-x}$ using a spreadsheet.

SOLUTION The first step is to express the equation in a form $f(x) = 0$ by bringing all terms to one side of the equation. The result is

$$f(x) = x - e^{-x} = 0$$

We might set up the spreadsheet as follows, taking an initial guess of 1.0 for the solution of the equation.

	A	B
1	x	f(x)
2	1	0.632121

The formula entered in Cell B2 would be $= A2 - \exp(-A2)$. As the value of x in Cell A2 is changed, the value of $f(x)$ in Cell B2 changes accordingly. The strategy is to find the value in Cell A2 that drives the value in Cell B2 satisfactorily close to zero. If this is done, the following result is obtained:

	A	B
1	x	f(x)
2	0.56714	-5.2E-06

The desired solution is $\boxed{x = 0.56714}$, for which $f(x) = -0.0000052$. If we wanted a more precise solution we could add a sixth significant figure to the given value of x, but it is rare to need even five significant figures.

The solution is even easier to obtain if the spreadsheet program is equipped with a *goalseek* tool. Once the first of the spreadsheets shown above has been constructed, select *Goal Seek* (it can normally be found under the pull-down "Tools" menu), and enter **B2** as the target cell, 0.0 as the target, and **A2** as the variable cell. The spreadsheet will then search for and (usually) converge on the solution within a fraction of a second.

One limitation to this method (and to all other numerical methods for solving nonlinear equations) is that once you have found one solution, you cannot be sure that there are no additional solutions. The way to determine the existence of multiple roots is to evaluate $f(x)$ over a wide range of x values and find the intervals in which $f(x)$ changes sign (see the second figure in the previous section). Initial guesses can then be made within each of these intervals and the spreadsheet used to determine the roots precisely.

A.2d   Regula-falsi Method

In this and the next subsection, we outline algorithms for finding roots of single-variable equations of the form $f(x) = 0$. The first procedure, termed the **regula-falsi method**, is appropriately used when an analytical expression for the derivative of f with respect to x is not available—as, for example, when $f(x)$ is obtained as the output of a computer program for an input value of x. The algorithm is as follows:

1. Find a pair of values of x—x_n and x_p—such that $f_n = f[x_n] < 0$ and $f_p = f[x_p] > 0$.
2. Estimate the value of the root of $f(x)$ from the following formula:

$$x_{\text{new}} = \frac{x_n f_p - x_p f_n}{f_p - f_n} \qquad \text{(A.2-1)}$$

and evaluate $f_{\text{new}} = f[x_{\text{new}}]$.

3. Use the new point to replace one of the original points, keeping the two points on opposite sides of the x axis. If $f_{\text{new}} < 0$, replace the old x_n and f_n with x_{new} and f_{new}. If $f_{\text{new}} > 0$, replace x_p and f_p with x_{new} and f_{new}. (If $f_{\text{new}} = 0$, you have found the root and need go no further.)
4. See if the new x_n and x_p are close enough for convergence to be declared (see Section A.2h). If they are not, go back to step 2.

What you are doing with this procedure is the algebraic equivalent of drawing a straight line between the two points $[x_n, f_n]$ and $[x_p, f_p]$ on a plot of f versus x and using the intersection of this line with the x axis as the next estimate of the root.

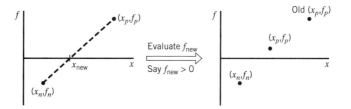

Successive points determined in this manner clearly approach the x axis (where $f = 0$). The procedure terminates when f_{new} is close enough to zero to satisfy a specified convergence criterion.

The regula-falsi method is the procedure used by many spreadsheet programs in their goalseek algorithms.

TEST YOURSELF

1. Suppose a "black box" computer program gives values of a function $f(x)$ for specified values of x. Unknown to the programmer, the function is

$$f = 4 - (x - 2)^2$$

(a) What are the roots of this function? (You should be able to do it by inspection.)
(b) Suppose the programmer tries values $x_p = 3$ and $x_n = 5$. If she uses the regula-falsi method, what will the next pair be? To which root will the method ultimately converge?

2. Derive Equation A.2-1.

A.2e   Newton's Rule

The next algorithm for finding the root of a function $f(x)$ is **Newton's rule**. It is considerably more efficient than regula-falsi, but only for functions for which the derivative $f'(x) = df/dx$

can be evaluated analytically. The formula for proceeding from one estimate of the root to the next is

$$x_{k+1} = x_k - \frac{f_k}{f_k'} \qquad \textbf{(A.2-2)}$$

where x_k is the kth estimate of the root, $f_k = f(x_k)$, and $f_k' = df/dx$ evaluated at $x = x_k$. As always, you begin by estimating a value of the root, x_1. Successive estimates are then generated from Equation A.2-2, with a test for convergence (Section A.2h) being applied after each estimate is obtained.

The easiest way to understand how Newton's rule works is graphically. Suppose the plot of f versus x appears as follows:

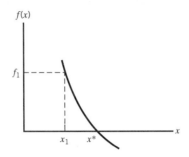

Although it may not be obvious at first glance, Newton's rule is equivalent to choosing a value of x_1 and calculating $f_1 = f(x_1)$, drawing a line tangent to the curve at (x_1, f_1) and using the intersection of this line with the x axis as the next estimate (x_2). As the following diagram shows, the successive values of x generated in this manner $(x_2, x_3, x_4, \ldots)$ may converge on the root x^*, although convergence is not guaranteed.

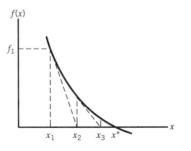

The formula for each estimate (x_{k+1}) in terms of the previous estimate (x_k) may easily be derived. The graphical representation of one step of the procedure is shown on the next page. The slope of the tangent line is $(df/dx)_{x_k} = f_k'$; however, two known points on this line are $(x_{k+1}, 0)$ and (x_k, f_k) so that the slope is also equal to $(0 - f_k)/(x_{k+1} - x_k)$. Equating these two expressions for the slope yields

$$f_k' = \frac{-f_k}{x_{k+1} - x_k}$$

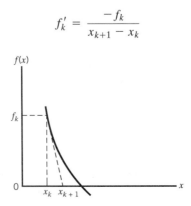

The solution of this equation for x_{k+1} is Newton's rule, Equation A.2-2:

$$x_{k+1} = x_k - \frac{f_k}{f_k'}$$

EXAMPLE A.2-2   *Newton's Rule*

Determine the root of the equation $x = e^{-x}$ using Newton's rule.

SOLUTION

$$f(x) = x - e^{-x}$$
$$f'(x) = df/dx = 1 + e^{-x}$$

When $x = 0$, $f(x)$ is negative, while when $x = 1$, $f(x)$ is positive (*verify*). The root x^* must therefore be between 0 and 1. Try $x_1 = 0.2$ as a first guess.

First Iteration: $x_1 = 0.2$

$$\Downarrow$$

$$f(x_1) = 0.2 - e^{-0.2} = -0.6187$$

$$\Downarrow$$

$$f'(x_1) = 1 + e^{-0.2} = 1.8187$$

$$\Downarrow$$

$$x_2 = x_1 - f(x_1)/f'(x_1) = 0.5402$$

Second Iteration: $x_2 = 0.5402$

$$\Downarrow$$

$$f(x_2) = 0.5402 - e^{-0.5402} = -0.0424$$

$$\Downarrow$$

$$f'(x_2) = 1 + e^{-0.5402} = 1.5826$$

$$\Downarrow$$

$$x_3 = x_2 - f(x_2)/f'(x_2) = 0.5670$$

Third Iteration: $x_3 = 0.5670$

$$\Downarrow$$

$$f(x_3) = 0.5670 - e^{-0.5670} = 2.246 \times 10^{-4}$$

$$\Downarrow$$

$$f'(x_3) = 1 + e^{-0.5670} = 1.5672$$

$$\Downarrow$$

$$x_4 = x_3 - f(x_3)/f'(x_3) = 0.56714$$

The successive estimates of x^* are therefore

$$0.2 \Longrightarrow 0.5402 \Longrightarrow 0.5670 \Longrightarrow 0.56714$$

This is clearly a converging sequence. Depending on how much precision you require, you might stop here or carry out one or two additional iterations. Let us stop here, and say that $x^* \approx 0.56714$.

TEST YOURSELF

1. Could you use Newton's rule to find a root of the equation $x^2 - 3x - 3 = 0$? Would you use it? Why not?

2. Suppose you wish to find a root of $f(x) = 0$, where $f(x)$ versus x appears as shown here.

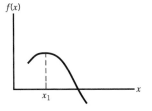

What would happen if you took as a first guess the value of x_1 shown? (What would happen shows that Newton's rule does not necessarily work if a poor choice of x_1 is made.)

A.2f Successive Substitution and Modified Successive Substitution

Problems involving the solution of nonlinear equations can often be expressed in the form

$$x = f(x)$$

where $f(x)$ is a nonlinear function. (*Example:* $x = e^{-x}$.) As shown in Chapter 10, balance equations for multiple unit processes with recycle often fall into this category: x would be the assumed value of a tear stream variable, and $f(x)$ would be the value generated by calculating around the cycle.

The simplest solution method is **successive substitution**. An initial estimate, $x^{(1)}$, is selected; $f(x^{(1)})$ is calculated; and the calculated value is used as the next estimate of the root. The formula is

$$x^{(i+1)} = f(x^{(i)}) \tag{A.2-3}$$

The procedure is repeated until the specified convergence criterion is satisfied.

Sometimes successive substitution works very well, converging in a few steps. Three unsatisfactory convergence patterns are also observed from time to time, however. In the first, the successive estimates oscillate about a central value:

$$37.6, 2.3, 36.8, 2.6, 34.4, 2.9, \ldots$$

The root lies somewhere between 3 and 30, but the procedure will clearly take a large number of iterations to get there.

The problem here is that successive substitution generates overly large steps. Instead of jumping all the way from 37.6 to 2.3, as successive substitution dictates, we should only go part of the way from the first to the second value to obtain our estimate of $x^{(2)}$. To do this, we may use **modified successive substitution** (also called damped successive substitution). The formula is

$$x^{(i+1)} = x^{(i)} + p[f(x^{(i)}) - x^{(i)}] \tag{A.2-4}$$

where p, the **damping parameter**, is a number between 0 and 1. If $p = 1$, the procedure reduces to pure successive substitution, and as p approaches zero, the size of the step becomes smaller and smaller. A few trial-and-error iterations should yield a good value of p for a specific problem.

The second case of slow convergence in successive substitution involves a creeping progression, such as

$$151.7, 149.5, 147.4, 145.6, 143.8, \ldots$$

Again, it appears that the procedure could be converging to a solution, but it is equally clear that it is in no hurry to get there.

The remedy for this problem is to **accelerate** the convergence procedure—to jump over many of the intermediate solutions to which continued successive substitution would lead. The next section outlines Wegstein's method, one of the most commonly used acceleration algorithms.

The third unsatisfactory convergence pattern is instability. For example, if successive substitution yields a sequence like

$$1.0, 2.5, -6.8, 23.5, 97.0,\ldots$$

then successive substitution will clearly not work, no matter how many iterations are attempted. A better first estimate might yield a convergent sequence, or the problem might be intrinsically unstable and must be restructured or solved by a different technique. Texts on numerical analysis outline stability conditions for nonlinear equation solution algorithms; their consideration is beyond the scope of this appendix.

TEST YOURSELF

1. For each of the following sequences of successive estimates of a root, indicate whether successive substitution seems to be adequate or whether you would resort to modified successive substitution or an acceleration method (state which).
   (a) 165, 132, 163, 133, 162, 133,...
   (b) 43, 28, 26, 26.7, 26.71,...
   (c) 21.0, 21.2, 21.4, 21.59, 21.79,...
2. Suppose $x^{(i)} = 14.0$, $f(x^{(i)}) = 13.0$, and you are using modified successive substitution with $p = 0.4$. What is your next estimate of the root?

A.2g Wegstein Algorithm

The procedure outlined in this section encompasses successive substitution and modified successive substitution as special cases, and in addition provides acceleration capability.

1. Begin by choosing $x^{(1)}$. Calculate $f(x^{(1)})$, and let $x^{(2)} = f(x^{(1)})$. (That is, carry out a successive substitution step.) Let $k = 2$.
2. Calculate $f(x^{(k)})$.
3. Check for convergence. If $x^{(k)}$ and $f(x^{(k)})$ are close enough together to meet the convergence criterion, terminate the procedure. If convergence is not achieved, calculate

$$w = \frac{f(x^{(k)}) - f(x^{(k-1)})}{x^{(k)} - x^{(k-1)}} \qquad \text{(A.2-5a)}$$

$$q = w/(w-1) \qquad \text{(A.2-5b)}$$

4. Calculate

$$x^{(k+1)} = qx^{(k)} + (1-q)f(x^{(k)}) \qquad \text{(A.2-6)}$$

5. Increase k by 1 and go back to step 2.

It is not difficult to show that Wegstein's method is tantamount to generating two points on the curve of $f(x)$ versus x and determining as the next estimate the intersection of the line between these two points and the 45° line [at which $x = f(x)$].

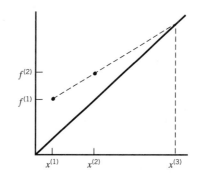

If you examine Equation A.2-6, you will see that if the parameter q equals zero, the procedure reduces to successive substitution; if q is between zero and 1 the procedure is modified successive substitution; and if q is negative the procedure involves acceleration.

TEST YOURSELF

An effort to solve the equation $x = f(x)$ begins with two successive substitution steps:

$$x = 2.00 \rightarrow f(x) = 2.30$$
$$x = 2.30 \rightarrow f(x) = 2.45$$

1. Calculate the next value of x using the Wegstein algorithm.
2. Sketch a plot of $f(x)$ versus x, showing the two given points, and show graphically that the value of x you calculated is correct.

A.2h Convergence Criteria

A problem common to all iterative computational methods is knowing when to quit. An iterative method rarely yields a precise root, but rather gives successive approximations that (if the method converges) approach the root more and more closely. Whether you are doing the calculation by hand or writing a program to do it, you must specify how close is close enough.

Suppose we wish to find a solution of the equation $f(x) = 0$ using a method that gives successive estimates $x^{(1)}$, $x^{(2)}$, and so on. Several different criteria may be used to determine when to terminate the procedure. The most straightforward one calls for termination when the absolute value of $f(x^{(i)})$ falls within a specified interval about $f = 0$:

$$|f(x^{(i)})| < \varepsilon \tag{A.2-7}$$

where ε, the *convergence tolerance,* is chosen to be several orders of magnitude smaller than typical values of $f(x)$ in the range being searched. Decreasing the value of ε leads to a more accurate estimate of the solution but increases the number of steps (and hence computation time) required to get there. There are formal ways to choose the value of ε, but it can be just as easy to choose a value (e.g., 0.0001 times the value of f at the first guessed value of x), find the root, then decrease the value of ε by a factor of 10, search again starting with the previous converged value, and see if the solution changes enough to care about.

The convergence criterion (A.2-7) may yield a false solution if the function $f(x)$ is almost horizontal over a wide range around the root, so that $|f(x^{(i)})|$ may be less than ε (meeting the convergence criterion) when $x^{(i)}$ is still a long way from the root. In this case, one of the following convergence criteria might be more appropriate to use:

$$|x^{(i)} - x^{(i-1)}| < \varepsilon \tag{A.2-8}$$

$$\frac{|x^{(i)} - x^{(i-1)}|}{x^{(i)}} < \varepsilon \tag{A.2-9}$$

Equation A.2-8 is an *absolute convergence criterion.* If the value of x changes by less than ε from one iteration to the next, the procedure is terminated and the last value of x is taken to be the desired root. The strictness of this criterion for a given ε depends on the magnitude of the estimated values. If $\varepsilon = 0.01$, for example, and successive estimates of x are 358,234.5 and 358,234.6, the procedure would not be terminated, even though the estimates are undoubtedly close enough for any realistic purpose. On the other hand, successive estimates of 0.0003 and 0.0006 would lead to termination for the same value of ε, despite the fact that the two estimates differ by a factor of two.

Equation A.2-9, a *relative convergence criterion,* avoids this difficulty. If this criterion is used, a value of $\varepsilon = 0.01$ specifies that the procedure will be terminated when the value of x changes from one iteration to the next by less than 1%, regardless of the magnitude of that value. This criterion does not work if successive estimates of x converge to a value of zero.

A.2i Multivariable Root-Finding Algorithms

Solving n nonlinear equations in n unknowns is usually a difficult problem, and a general treatment is well beyond the scope of this text. In this section we present several approaches without proof or much explanation. For additional details, standard texts on numerical analysis should be consulted.

Three methods that can be used to find the values of $x_1, \ldots, x_n$ that satisfy n simultaneous equations are extensions of methods given previously for single-variable problems. They are (a) successive substitution, (b) the Wegstein algorithm, and (c) the Newton–Raphson method (a multivariable extension of Newton's rule). The example that concludes this section illustrates all three algorithms.

Successive Substitution. Suppose the equations can be cast in the form

$$
\begin{aligned}
x_1 &= f_1(x_1, x_2, \ldots, x_n) \\
x_2 &= f_2(x_1, x_2, \ldots, x_n) \\
&\vdots \\
x_n &= f_n(x_1, x_2, \ldots, x_n)
\end{aligned}
\tag{A.2-10}
$$

(A cycle with n tear stream variables falls into this category.) The successive substitution method consists of assuming values for each of the n unknown variables, evaluating the functions $f_1, \ldots, f_n$, and using the calculated values as the next estimates of the variables. The procedure is terminated when all variable values meet a specified convergence criterion. For example, if $x_i^{(k)}$ is the value of the ith variable at the kth iteration, the procedure might be terminated when

$$
\frac{\left| x_i^{(k)} - x_i^{(k-i)} \right|}{x_i^{(k)}} < \varepsilon, \qquad i = 1, 2, \ldots, n
\tag{A.2-11}
$$

This approach is simple but generally inefficient. The greater the number of variables, the longer it takes for the procedure to converge, if it converges at all. It is generally preferable to use Wegstein's method or the Newton–Raphson method, depending on whether or not the partial derivatives of the functions $f_1, \ldots, f_n$ can be evaluated analytically. (Use Newton–Raphson if they can be, otherwise try Wegstein, but don't be too surprised if it doesn't converge.)

Wegstein Algorithm. If the equations to be solved have the form of Equation A.2-10 [i.e., $x_i = f_i(x_1, x_2, \ldots, x_n)$], guess values for all n variables and apply the procedure of Section A.2g separately to each variable. Terminate when the convergence criteria are satisfied for all variables.

This procedure will work reasonably well if the generating function f_1 depends almost entirely on x_1, f_2 depends only on x_2, and so on (i.e., if there is little interaction among the variables). If this is not the case, convergence will generally be very difficult to achieve.

Newton–Raphson Method. Suppose now that the equations to be solved take the form

$$
\begin{aligned}
g_1(x_1, x_2, \ldots, x_n) &= 0 \\
g_2(x_1, x_2, \ldots, x_n) &= 0 \\
&\vdots \\
g_n(x_1, x_2, \ldots, x_n) &= 0
\end{aligned}
\tag{A.2-12}
$$

The *Newton–Raphson method* is as follows:

1. Estimate (or just guess) values of the n variables $(x_1, x_2, \ldots, x_n)$, calling the estimates $x_1^{(1)}, x_2^{(1)}, \ldots, x_n^{(1)}$. Let $k = 1$ (the number of the iteration).

2. Evaluate the function values $(g_1, \ldots, g_n)$ corresponding to the most recent estimate of the x_i values:

$$g_i^{(k)} = g_i[x_1^{(k)}, \ldots, x_n^{(k)}], \qquad i = 1, 2, \ldots, n \tag{A.2-13}$$

3. If the g_i values are to be used as the basis of a convergence test, declare the procedure to have converged if

$$|g_i^{(k)}| < \varepsilon_i, \qquad i = 1, 2, \ldots, n$$

The convergence tolerance for the ith equation, ε_i, should be a very small fraction of typical values of g_i (e.g., $0.0001 g_i^{(1)}$). If the procedure has not converged, go on to step 4.

4. Evaluate the partial derivatives with respect to each variable

$$a_{ij} = \left(\frac{\partial g_i}{\partial x_j} \right) \quad \text{at} \quad [x_1^{(k)}, x_2^{(k)}, \ldots, x_n^{(k)}] \tag{A.2-14}$$

5. Solve the following set of linear equations for the variables $d_1, d_2, \ldots, d_n$.

$$a_{11}d_1 + a_{12}d_2 + \cdots + a_{1n}d_n = -g_1^{(k)}$$
$$a_{21}d_1 + a_{22}d_2 + \cdots + a_{2n}d_n = -g_2^{(k)} \tag{A.2-15}$$
$$\vdots$$
$$a_{n1}d_1 + a_{n2}d_2 + \cdots + a_{nn}d_n = -g_n^{(k)}$$

If there are only two or three equations, you can solve them by simple algebraic techniques. For larger systems of equations, an equation-solving computer program should be used.

6. Calculate the next set of x_i values as

$$x_i^{(k+1)} = x_i^{(k)} + d_i \tag{A.2-16}$$

7. If the changes in the x values are to be used as the basis of a convergence test, declare the procedure to have converged to $[x_1^{(k+1)}, x_2^{(k+1)}, \ldots, x_n^{(k+1)}]$ if either an absolute or relative convergence criterion is satisfied,

$$|d_i| < \varepsilon_i, \quad i = 1, 2, \ldots, n \quad \text{or} \quad |d_i/x_i^{(k)}| < \varepsilon_i, \quad i = 1, 2, \ldots, n$$

Otherwise, increase the value of k by 1 (so that what was calculated in step 6 as $x_i^{(k+1)}$ is now $x_i^{(k)}$) and return to step 2.

The Newton–Raphson method is based on a linearization of the functions $g_1, \ldots, g_n$ about each estimated set of roots and a solution of the resulting linear equations to get the next estimate. (If you have no idea what all that means, don't worry about it.) It is an efficient procedure to use when analytical partial derivatives of the functions $g_1, \ldots, g_n$ are convenient to evaluate. When there is only one equation ($n = 1$), the algorithm reduces to Newton's rule (Section A.2e).

The next example illustrates the three multivariable nonlinear equation-solving methods described in this section.

EXAMPLE A.2-3 **Solving Multivariable Nonlinear Equations**

Find the solutions of the following simultaneous equations:

$$g_1(x, y) = 2x + y - (x + y)^{1/2} - 3 = 0$$
$$g_2(x, y) = 4 - y - 5/(x + y) = 0$$

1. By successive substitution.
2. Using the Wegstein algorithm.
3. Using the Newton–Raphson method.

In each case, use a starting value ($x = 2$, $y = 2$) and stop when the relative changes in x and y from one iteration to the next are each less than 0.001. (See Equation A.2-11.)

SOLUTION

1. *Successive substitution.* The equations $g_1 = 0$ and $g_2 = 0$ must be rewritten to provide explicit expressions for x and y. One way of doing this is the following:

$$x_c = 0.5[3 - y_a + (x_a + y_a)^{1/2}]$$
$$y_c = 4 - 5/(x_a + y_a)$$

where the subscript "a" stands for assumed and "c" stands for calculated. We assume values of x and y, recalculate x and y using these expressions, and iterate until convergence is achieved. The calculation proceeds as follows:

	Assumed		Calculated	
Iteration	x	y	x	y
1	2.000	2.000	1.500	2.750
2	1.500	2.750	1.156	2.824
3	1.156	2.824	1.086	2.744
4	1.086	2.744	1.107	2.694
5	1.107	2.694	1.128	2.684
6	1.128	2.684	1.134	2.688
7	1.134	2.688	1.133	2.692
8	1.133	2.692	1.132	2.693
9	1.1320	2.6929	1.1314	2.6928

Since the relative changes in x and y in the last iteration are each less than 0.001, the calculation is terminated at this point, and the final values are accepted as the roots of the two given equations.

2. *Wegstein algorithm.* The same functions are used to generate calculated values of x and y from assumed values, only now the equations of Section A.2g are used to generate new assumed values after the first iteration. The results are as follows. (Check the first series of numbers with a hand calculator to make sure you know how to apply the formulas.)

	Assumed		Calculated	
Iteration	x	y	x	y
1	2.000	2.000	1.500	2.750
2	1.500	2.750	1.156	2.824
3	0.395	2.832	0.982	2.450
4	1.092	2.641	1.146	2.660
5	1.162	2.651	1.151	2.689
6	1.150	2.670	1.142	2.691
7	1.123	2.694	1.130	2.690
8	1.136	2.690	1.133	2.693
9	1.1320	2.6919	1.1318	2.6924

In this case, Wegstein's method did not accelerate the convergence. In fact, the large jump in the value of x in iteration 3 (when Wegstein procedure was first used) could have been the first symptom of an instability, but the algorithm recovered well.

3. *Newton–Raphson method.* The necessary formulas are as follows:

$$g_1(x, y) = 2x + y - (x + y)^{1/2} - 3$$
$$g_2(x, y) = 4 - y - 5/(x + y)$$
$$a_{11}(x, y) = \partial g_1/\partial x = 2 - 0.5(x + y)^{-1/2}$$
$$a_{12}(x, y) = \partial g_1/\partial y = 1 - 0.5(x + y)^{-1/2}$$
$$a_{21}(x, y) = \partial g_2/\partial x = 5/(x + y)^2$$
$$a_{22}(x, y) = \partial g_2/\partial y = -1 + 5/(x + y)^2$$

Equations A.2-15 reduce for this two-dimensional problem to

$$a_{11}d_1 + a_{12}d_2 = -g_1$$
$$a_{21}d_1 + a_{22}d_2 = -g_2$$

The procedure is to assume values of x and y; calculate $g_1, g_2, a_{11}, a_{12}, a_{21}$, and a_{22} from the given formulas; solve the two preceding equations for d_1 and d_2; and calculate the new estimates of the roots as

$$x_c = x_a + d_1$$
$$y_c = y_a + d_2$$

The convergence test is then applied, and if the assumed and calculated values are not close enough together, the latter values are used to replace the former ones and the calculation is repeated. The results are shown here.

Iteration	Assumed		Calculated	
	x	y	x	y
1	2.000	2.000	1.130	2.696 (*verify!*)
2	1.130	2.696	1.1315	2.6925
3	1.1315	2.6925	1.1315	2.6925

The superiority of the Newton–Raphson method to others tested is clear in this example and is even more dramatic when more than two equations are to be solved simultaneously. Generally, when analytical derivatives are available, the Newton–Raphson method should be used for solving multiple nonlinear algebraic equations.

Appendix A.3   NUMERICAL INTEGRATION

In Chapter 8, we showed that the enthalpy change associated with the heating or cooling of a substance is evaluated by integrating the substance heat capacity $C_p(T)$ from the initial temperature to the final temperature. This is one of many instances you will encounter in process analysis where an integration is required as part of a problem solution.

It often happens that required values of definite integrals cannot be obtained using the methods of elementary calculus. If, for example, you are called on to evaluate something like

$$\int_0^{10} e^{-x^3} dx$$

you will not find help in a calculus book or a table of integrals—an analytical expression for the integral of $\exp(-x^3)$ simply does not exist.

It is possible, however, to substitute for any mathematical operation such as differentiation or integration a series of arithmetic operations that yield approximately the same result. The arithmetic operations are usually simple but numerous and repetitious and so are ideally suited to computers.

A.3a   Quadrature

The general problem we will discuss is the evaluation of a definite integral:

$$I = \int_a^b y(x)\, dx \tag{A.3-1}$$

There are several possible reasons why you might not be able to evaluate I analytically: $y(x)$ may be a nonintegrable analytical function, such as $\exp(-x^3)$, or it may be a series of tabulated (x, y) data points or a plot of y versus x.

A method of **numerical integration** (or **quadrature**, as it is also called) is required to evaluate I in any of these cases. The specific techniques we will present are algebraic, but the general approach to the problem is best visualized graphically. For the moment, we will suppose that all we have relating x and y is a table of data points, which we may graph on a plot of y versus x.

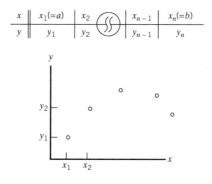

The integral we are trying to evaluate (I of Equation A.3-1) equals the area under the continuous curve of y versus x, but this curve is not available—we only know the function values at the discrete data points. The procedure generally followed is to fit approximating functions to the data points, and then to integrate these function analytically.

The many existing quadrature formulas differ only in the choice of functions to fit to the data points. Two of the simplest approximations are to fit straight lines between successive points and sum the area under the lines, and to fit parabolas to successive triplets of points and sum the areas under the parabolas. These approximations lead to the quadrature formulas known respectively as the **trapezoidal rule** and **Simpson's rule**. We will discuss each in turn.

A.3b   The Trapezoidal Rule

The area under a line through (x_1, y_1) and (x_2, y_2) on a plot of y versus x is easily calculated.

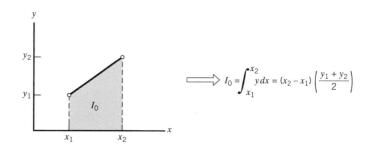

The area under a series of points from x_1 to x_n is obtained by a summation of such terms:

$$I = \tfrac{1}{2}[(x_2 - x_1)(y_1 + y_2) + (x_3 - x_2)(y_2 + y_3) + \cdots + (x_n - x_{n-1})(y_{n-1} + y_n)]$$

$$\Downarrow$$

Trapezoidal Rule: $\quad \boxed{\displaystyle\int_{x_1}^{x_n} y(x)\, dx \approx \tfrac{1}{2}\sum_{j=1}^{n-1}(x_{j+1} - x_j)(y_j + y_{j+1})}$ $\qquad$ **(A.3-2)**

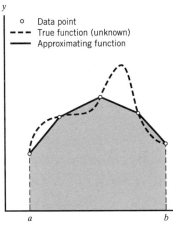

Figure A.3-1 Graphical illustration of the trapezoidal rule.

If the abscissa values of the data points are spaced at equal intervals, then the trapezoidal rule simplifies to

$$I = \frac{h}{2}[(y_1 + y_2) + (y_2 + y_3) + \cdots + (y_{n-1} + y_n)]$$

$$\Downarrow$$

Trapezoidal Rule—Equal Intervals: $\displaystyle \int_{x_1}^{x_n} y(x)\, dx \approx \frac{h}{2}\left(y_1 + y_n + 2\sum_{j=2}^{n-1} y_j \right)$ **(A.3-3)**

where h is the distance between the x values of adjacent data points. Observe that to use the trapezoidal rule, you need not plot anything—simply substitute the tabulated data into Equation A.3-2 or (for equal spacing) Equation A.3-3.

The trapezoidal rule is an approximation, as are all quadrature formulas. Figure A.3-1 illustrates the nature of the error introduced by its use. The integral to be evaluated

$$I = \int_a^b y(x)\, dx$$

is the area under the dashed curve of Figure A.3-1, while the trapezoidal rule, Equation A.3-2, would yield the area under the straight-line segments, which could differ significantly from the correct value of I. Note also, however, that if there were many more data points in the interval between a and b, the approximating series of lines would follow the dashed curve much more closely, and the estimate of the integral would accordingly be more accurate.

A.3c   Simpson's Rule

A second and more accurate quadrature formula is the one most often used. It is applicable only to an odd number of equally spaced data points and is based on fitting parabolic functions to successive groups of three points.

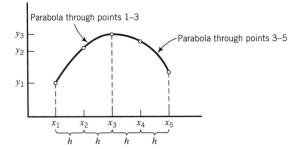

It can be shown after a fair amount of algebra that the area under a parabola through equally spaced points (x_1, y_1), (x_2, y_2), and (x_3, y_3) is

$$I_0 = \frac{h}{3}(y_1 + 4y_2 + y_3)$$

where h is the interval between successive x values. Consequently, the area under a series of such parabolas fitted to n equally spaced points is

$$I = \frac{h}{3}[(y_1 + 4y_2 + y_3) + (y_3 + 4y_4 + y_5) + \cdots + (y_{n-2} + 4y_{n-1} + y_n)]$$

$$\Downarrow$$

Simpson's Rule: $$\int_{x_1}^{x_n} y(x)\,dx \approx \frac{h}{3}\left(y_1 + y_n + 4\sum_{\substack{j=2, \\ 4,\ldots, \\ n-1}} y_j + 2\sum_{\substack{j=3 \\ 5,\ldots, \\ n-2}} y_j \right) \qquad \textbf{(A.3-4)}$$

If you happen to have an even number of data points, you may integrate over all but the first or last subinterval (omit whichever subinterval contributes least to the integral) using Simpson's rule, and over the remaining subinterval using the trapezoidal rule.

EXAMPLE A.3-1   **Simpson's Rule**

The heat capacity of a gas is tabulated at a series of temperatures:

T (°C)	20	50	80	110	140	170	200	230
C_p[J/(mol·°C)]	28.95	29.13	29.30	29.48	29.65	29.82	29.99	30.16

Calculate the change in enthalpy for 3.00 g-moles of this gas going from 20°C to 230°C.

SOLUTION

$$\Delta H(J) = n\int_{20°C}^{230°C} C_p\,dT$$

The data points are evenly spaced in the independent variable (T), so that Simpson's rule can be applied to the integration, but since there are an even number of points the trapezoidal rule must be applied over the first or last temperature interval. Since C_p increases with temperature, we will apply the less accurate trapezoidal rule to the interval from 20°C to 50°C and use Simpson's rule between 50°C and 230°C. If $\Delta T(= 30°C)$ is the temperature interval between data points, Equations A.3-3 and A.3-4 yield

$$\int_{T_1}^{T_a} C_p\,dT \approx \frac{\Delta T}{2}(C_{p1} + C_{p2}) + \frac{\Delta T}{3}[C_{p2} + C_{p8} + 4(C_{p3} + C_{p5} + C_{p7}) + 2(C_{p4} + C_{p6})]$$

$$= 6208 \text{ J/mol}$$

$$\Downarrow$$

$$\Delta H = (3.00 \text{ mol})(6208 \text{ J/mol}) = \boxed{1.86 \times 10^4 \text{ J}}$$

The trapezoidal rule is exact if the function to be integrated is in fact linear in each interval between data points, while Simpson's rule is exact if the real function is parabolic *or cubic* in each interval. (The latter point is far from obvious; for proof, see any reference on numerical analysis.)

A.3d   Numerical Integration of Analytical Functions

When you have an analytical expression for an integrand $y(x)$ but you cannot perform the integration from $x = a$ to $x = b$ analytically, the procedure is to evaluate y at a series of values of x from a to b—that is, to generate a data table—and then to use a quadrature formula such as Simpson's rule to estimate the integral. Now, however, you have the choice of the number of evaluations of $y(x)$ to make.

As a rule, the accuracy of a quadrature formula increases with the number of points in the interval of integration, but so does the required computation time. Choosing the number of points to provide a suitable combination of accuracy and low computation time can be done using sophisticated numerical analysis techniques, but simple trial and error often suffices very well. A common procedure is to evaluate the integral using (say) 9 points, then 17, then 33, and so on ($n_{new} = 2n_{old} - 1$), until successively calculated values agree within a specified tolerance. The last value should be a good approximation to the exact value of the integral.

TEST YOURSELF

Suppose $f(x) = x^3 + 4$. Evaluate $\int_0^4 f(x)\,dx$:

1. Analytically.
2. Using the trapezoidal rule, with points at $x = 0, 1, 2, 3, 4$.
3. Using Simpson's rule, with points at $x = 0, 1, 2, 3, 4$. Explain the relationship between the answers to 1 and 3.

Appendix B

Physical Property Tables

Table B.1 Selected Physical Property Data[a]

Compound	Formula	Mol. Wt.	SG (20°/4°)	T_m (°C)[b]	$\Delta \hat{H}_m(T_m)$[c,j] kJ/mol	T_b (°C)[d]	$\Delta \hat{H}_v(T_b)$[e,j] kJ/mol	T_c (K)[f]	P_c (atm)[g]	$(\Delta \hat{H}_f°)$[h,j] kJ/mol	$(\Delta \hat{H}_c)$[i,j] kJ/mol
Acetaldehyde	CH_3CHO	44.05	$0.783^{18°}$	−123.7	—	20.2	25.1	461.0	—	−166.2(g)	−1192.4(g)
Acetic acid	CH_3COOH	60.05	1.049	16.6	12.09	118.2	24.39	594.8	57.1	−486.18(l) / −438.15(g)	−871.69(l) / −919.73(g)
Acetone	C_3H_6O	58.08	0.791	−95.0	5.69	56.0	30.2	508.0	47.0	−248.2(l) / −216.7(g)	−1785.7(l) / −1821.4(g)
Acetylene	C_2H_2	26.04	—	—	—	−81.5	17.6	309.5	61.6	+226.75(g)	−1299.6(g)
Ammonia	NH_3	17.03	—	−77.8	5.653	−33.43	23.351	405.5	111.3	−67.20(l) / −46.19(g)	— / −382.58(g)
Ammonium hydroxide	NH_4OH	35.03	—	—	—	—	—	—	—	−366.48(aq)	—
Ammonium nitrate	NH_4NO_3	80.05	$1.725^{25°}$	169.6	5.4	Decomposes at 210°C				−365.14(c) / −399.36(aq)	—
Ammonium sulfate	$(NH_4)_2SO_4$	132.14	1.769	513	—	Decomposes at 513°C after melting		699	52.4	−1179.3(c) / −1173.1(aq)	—
Aniline	C_6H_7N	93.12	1.022	−6.3	—	184.2	—	699	52.4	−88.83(l)	−3520.0(l)
Benzaldehyde	C_6H_5CHO	106.12	1.046	−26.0	—	179.0	38.40	—	—	−40.04(g)	—
Benzene	C_6H_6	78.11	0.879	5.53	9.837	80.10	30.765	562.6	48.6	+48.66(l) / +82.93(g)	−3267.6(l) / −3301.5(g)
Benzoic acid	$C_7H_6O_2$	122.12	$1.266^{15°}$	122.2	—	249.8	—	—	—	—	−3226.7(g)
Benzyl alcohol	C_7H_8O	108.13	1.045	−15.4	—	205.2	—	—	—	—	−3741.8(l)
Bromine	Br_2	159.83	3.119	−7.4	10.8	58.6	31.0	584	102	0(l)	—
1,2-Butadiene	C_4H_6	54.09	—	−136.5	—	10.1	—	446	42.7	—	—
1,3-Butadiene	C_4H_6	54.09	—	−109.1	—	−4.6	—	425	42.7	—	—
n-Butane	C_4H_{10}	58.12	—	−138.3	4.661	−0.6	22.305	425.17	37.47	−147.0(l) / −124.7(g)	−2855.6(l) / −2878.5(g)
Isobutane	C_4H_{10}	58.12	—	−159.6	4.540	−11.73	21.292	408.1	36.0	−158.4(l) / −134.5(g)	−2849.0(l) / −2868.8(g)
1-Butene	C_4H_8	56.10	—	−185.3	3.8480	−6.25	21.916	419.6	39.7	+1.17(g)	−2718.6(g)
Calcium carbide	CaC_2	64.10	$2.22^{18°}$	2300	—	—	—	—	—	−62.76(c)	—
Calcium carbonate	$CaCO_3$	100.09	2.93	Decomposes at 825°C						−1206.9(c)	—
Calcium chloride	$CaCl_2$	110.99	$2.152^{15°}$	782	28.37	>1600	—	—	—	−794.96(c)	—

Compound	Formula	Mol. wt.	Sp. gr.	Melting point[b]	Heat of fusion[c]	Boiling point[d]	Heat of vaporization[e]	Critical temperature[f]	Critical pressure[g]	Heat of formation[h]	Heat of combustion[i]
Calcium hydroxide	$Ca(OH)_2$	74.10	2.24	(−H_2O at 580°C)	—	—	—	—	—	−986.59(c)	—
Calcium oxide	CaO	56.08	3.32	2570	50	2850	—	—	—	−635.6(c)	—
Calcium phosphate	$Ca_3(PO_4)_2$	310.19	3.14	1670	—	—	—	—	—	−4138(c)	—
Calcium silicate	$CaSiO_3$	116.17	2.915	1530	48.62	—	—	—	—	−1584(c)	—
Calcium sulfate	$CaSO_4$	136.15	2.96	—	—	—	—	—	—	−1432.7(c), −1450.4(aq)	—
Calcium sulfate (gypsum)	$CaSO_4 \cdot 2H_2O$	172.18	2.32	(−1.5 H_2O at 128°C)	—	—	—	—	—	−2021(c)	—
Carbon (graphite)	C	12.010	2.26	3600	46.0	4200	—	—	—	0(c)	−393.51(c)
Carbon dioxide	CO_2	44.01	—	−56.6 at 5.2 atm	8.33	(Sublimes at −78°C)	—	304.2	72.9	−412.9(l), −393.5(g)	—
Carbon disulfide	CS_2	76.14	$1.261^{22°/20°}$	−112.1	4.39	46.25	26.8	552.0	78.0	+87.9(l), +115.3(g)	−1075.2(l), 1102.6(g)
Carbon monoxide	CO	28.01	—	−205.1	0.837	−191.5	6.042	133.0	34.5	−110.52(g)	−282.99(g)
Carbon tetrachloride	CCl_4	153.84	1.595	−22.9	2.51	76.7	30.0	556.4	45.0	−139.5(l), −106.7(g)	−352.2(l), −385.0(g)
Chlorine	Cl_2	70.91	—	−101.00	6.406	−34.06	20.4	417.0	76.1	0(g)	—
Chlorobenzene	C_6H_5Cl	112.56	1.107	−45	—	132.10	36.5	632.4	44.6	—	—
Chloroethane	C_2H_5Cl	See ethyl chloride									

[a] Adapted in part from D. M. Himmelblau, *Basic Principles and Calculations in Chemical Engineering*, 3rd Edition, ©1974, Tables D.1 and F.1. Adapted by permission of Prentice-Hall, Inc., Englewood Cliffs, NJ.

[b] Melting point at 1 atm.

[c] Heat of fusion at T_m and 1 atm.

[d] Boiling point at 1 atm.

[e] Heat of vaporization at T_b and 1 atm.

[f] Critical temperature.

[g] Critical pressure.

[h] Heat of formation at 25°C and 1 atm.

[i] Heat of combustion at 25°C and 1 atm. Standard states of products are $CO_2(g)$, $H_2O(l)$, $SO_2(g)$, $HCl(aq)$, and $N_2(g)$. To calculate ΔH_c° with $H_2O(g)$ as a product, add $44.01 n_w$ to the tabulated value, where n_w = moles H_2O formed/mole fuel burned.

[j] To convert $\Delta \hat{H}$ to kcal/mol, divide given value by 4.184; to convert to Btu/lb-mole, multiply by 430.28.

(continued)

Table B.1 (Continued)

Compound	Formula	Mol. Wt.	SG (20°/4°)	T_m(°C)[b]	$\Delta\hat{H}_m(T_m)$[c,j] kJ/mol	T_b(°C)[d]	$\Delta\hat{H}_v(T_b)$[e,j] kJ/mol	T_c(K)[f]	P_c(atm)[g]	$(\Delta\hat{H}_f°)$[h,j] kJ/mol	$(\Delta\hat{H}_c°)$[i,j] kJ/mol
Chloroform	$CHCl_3$	119.39	1.489	-63.7	—	61.0		536.0	54.0	-131.8(l)	-3373(l)
Copper	Cu	63.54	8.92	1083	13.01	2595	304.6	—		0(c)	—
Cupric sulfate	$CuSO_4$	159.61	3.606[15°]			Decomposes > 600°C				-769.9(c)	—
										-843.1(aq)	
Cyclohexane	C_6H_{12}	84.16	0.779	6.7	2.677	80.7	30.1	553.7	40.4	-156.2(l)	-3919.9(l)
										-123.1(g)	-3953.0(g)
Cyclopentane	C_5H_{10}	70.13	0.745	-93.4	0.609	49.3	27.30	511.8	44.55	-105.9(l)	-3290.9(l)
										-77.2(g)	-3319.5(g)
n-Decane	$C_{10}H_{22}$	142.28	0.730	-29.9	—	173.8	—	619.0	20.8	-249.7(l)	-6778.3(l)
										—	-6829.7(g)
Diethyl ether	$(C_2H_5)_2O$	74.12	0.708[25°]	-116.3	7.30	34.6	26.05	467	35.6	-272.8(l)	-2726.7(l)
Ethane	C_2H_6	30.07	—	-183.3	2.859	-88.6	14.72	305.4	48.2	-84.67(g)	-1559.9(g)
Ethyl acetate	$C_4H_8O_2$	88.10	0.901	-83.8	—	77.0	—	523.1	37.8	-463.2(l)	-2246.4(l)
										-426.8(g)	
Ethyl alcohol (Ethanol)	C_2H_5OH	46.07	0.789	-114.6	5.021	78.5	38.58	516.3	63.0	-277.63(l)	-1366.91(l)
										-235.31(g)	-1409.25(g)
Ethyl benzene	C_8H_{10}	106.16	0.867	-94.67	9.163	136.2	35.98	619.7	37.0	-12.46(l)	-4564.9(l)
										+29.79(g)	-4607.1(g)
Ethyl bromide	C_2H_5Br	108.98	1.460	-119.1	—	38.2	—	504	61.5	-54.4(g)	—
Ethyl chloride	C_2H_5Cl	64.52	0.903[15°]	-138.3	4.452	13.1	24.7	460.4	52.0	-105.0(g)	—
3-Ethyl hexane	C_8H_{18}	114.22	0.717	—	—	118.5	34.27	567.0	26.4	-250.5(l)	-5407.1(l)
										-210.9(g)	-5509.8(g)
Ethylene	C_2H_4	28.05	—	-169.2	3.350	-103.7	13.54	283.1	50.5	+52.28(g)	-1410.99(g)
Ethylene glycol	$C_2H_6O_2$	62.07	1.113[19°]	-13	11.23	197.2	56.9	—	—	-451.5(l)	-1179.5(l)
										-387.1(g)	
Ferric oxide	Fe_2O_3	159.70	5.12			Decomposes at 1560°C				-822.2(c)	—
Ferrous oxide	FeO	71.85	5.7	—	—	—	—	—	—	-266.5(c)	—
Ferrous sulfide	FeS	87.92	4.84	1193	—	—	—	—	—	-95.1(c)	—
Formaldehyde	H_2CO	30.03	0.815[-20°]	-92	—	-19.3	24.48	—	—	-115.90(g)	-563.46(g)
Formic acid	CH_2O_2	46.03	1.220	8.30	12.68	100.5	22.25	—	—	-409.2(l)	-262.8(l)
										-362.6(g)	
Glycerol	$C_3H_8O_3$	92.09	1.260[50°]	18.20	18.30	290.0	—	—	—	-665.9(l)	-1661.1(l)
Helium	He	4.00	—	-269.7	0.02	-268.9	0.084	5.26	2.26	0(g)	—

n-Heptane	C_7H_{16}	100.20	0.684	-90.59	14.03	98.43	31.69	540.2	27.0	-224.4(l), -187.8(g)	-4816.9(l), -4853.5(g)
n-Hexane	C_6H_{14}	86.17	0.659	-95.32	13.03	68.74	28.85	507.9	29.9	-198.8(l), -167.2(g)	-4163.1(l), -4194.8(g)
Hydrogen	H_2	2.016	—	-259.19	0.12	-252.76	0.904	33.3	12.8	0(g)	-285.84(g)
Hydrogen bromide	HBr	80.92	—	-86	—	-67	—	—	—	-36.23(g)	—
Hydrogen chloride	HCl	36.47	—	-114.2	1.99	-85.0	16.1	324.6	81.5	-92.31(g)	—
Hydrogen cyanide	HCN	27.03	—	-14	—	26	—	—	—	+130.54(g)	—
Hydrogen fluoride	HF	20.0	—	-83	—	20	—	503.2	—	-268.6(g), -316.9(aq, 200)	—
Hydrogen sulfide	H_2S	34.08	—	-85.5	2.38	-60.3	18.67	373.6	88.9	-19.96(g)	-562.59(g)
Iodine	I_2	253.8	4.93	113.3	—	184.2	—	826.0	—	0(c)	—
Iron	Fe	55.85	7.7	1535	15.1	2800	354.0	—	—	0(c)	—
Lead	Pb	207.21	$11.337^{20°/20°}$	327.4	5.10	1750	179.9	—	—	0(c)	—
Lead oxide	PbO	223.21	9.5	886	11.7	1472	213	—	—	-219.2(c)	—
Magnesium	Mg	24.32	1.74	650	9.2	1120	131.8	—	—	0(c)	—
Magnesium chloride	$MgCl_2$	95.23	$2.325^{25°}$	714	43.1	1418	136.8	—	—	-641.8(c)	—
Magnesium hydroxide	$Mg(OH)_2$	58.34	2.4	Decomposes at 350°C	77.4		—	—	—		—
Magnesium oxide	MgO	40.32	3.65	2900	—	3600	—	—	—	-601.8(c)	—
Mercury	Hg	200.61	13.546	-38.87	—	356.9	—	—	—	0(c)	—
Methane	CH_4	16.04	—	-182.5	0.94	-161.5	8.179	190.70	45.8	-74.85(g)	-890.36(g)
Methyl acetate	$C_3H_6O_2$	74.08	0.933	-98.9	—	57.1	—	506.7	46.30	-409.4(l)	-1595(l)
Methyl alcohol (Methanol)	CH_3OH	32.04	0.792	-97.9	3.167	64.7	35.27	513.20	78.50	-238.6(l), -201.2(g)	726.6(l), -764.0(g)
Methyl amine	CH_5N	31.06	$0.699^{-11°}$	-92.7	—	-6.9	—	429.9	73.60	-28.0(g)	-1071.5(l)
Methyl chloride	CH_3Cl	50.49	—	-97.9	—	-24	—	416.1	65.80	-81.92(g)	—

(continued)

Table B.1 (Continued)

Compound	Formula	Mol. Wt.	SG (20°/4°)	T_m (°C)[b]	$\Delta\hat{H}_m(T_m)$[c,j] kJ/mol	T_b (°C)[d]	$\Delta\hat{H}_v(T_b)$[e,j] kJ/mol	T_c (K)[f]	P_c (atm)[g]	$(\Delta\hat{H}_f°)$[h,j] kJ/mol	$(\Delta\hat{H}_c°)$[i,j] kJ/mol
Methyl ethyl ketone	C_4H_8O	72.10	0.805	−87.1	—	78.2	32.0	—	—	—	−2436(l)
Naphthalene	$C_{10}H_8$	128.16	1.145	80.0	—	217.8	—	—	—	—	−5157(g)
Nickel	Ni	58.69	8.90	1452	—	2900	—	—	—	0(c)	—
Nitric acid	HNO_3	63.02	1.502	−41.6	10.47	86	30.30	—	—	−173.23(l) −206.57(aq)	—
Nitrobenzene	$C_6H_5O_2N$	123.11	1.203	5.5	—	210.7	—	—	—	—	−3092.8(l)
Nitrogen	N_2	28.02	—	−210.0	0.720	−195.8	5.577	126.20	33.5	0(g)	—
Nitrogen dioxide	NO_2	46.01	—	−9.3	7.335	21.3	14.73	431.0	100.0	+33.8(g)	—
Nitric oxide	NO	30.01	1.63[18°]	−163.6	2.301	−151.8	13.78	179.20	65.0	+90.37(g)	—
Nitrogen pentoxide	N_2O_5	108.02	—	30	—	47	—	—	—	—	—
Nitrogen tetraoxide	N_2O_4	92.0	1.448	−9.5	—	21.1	—	431.0	99.0	+9.3(g)	—
Nitrous oxide	N_2O	44.02	1.226[−89°]	−91.1	—	−88.8	—	309.5	71.70	+81.5(g)	—
n-Nonane	C_9H_{20}	128.25	0.718	−53.8	—	150.6	—	595	23.0	−229.0(l)	−6124.5(l) −6171.0(g)
n-Octane	C_8H_{18}	114.22	0.703	−57.0	—	125.5	—	568.8	24.5	−249.9(l) −208.4(g)	−5470.7(l) −5512.2(g)
Oxalic acid	$C_2H_2O_4$	90.04	1.90	—	Decomposes at 186°C		—	—	—	−826.8(c)	−251.9(s)
Oxygen	O_2	32.00	—	−218.75	0.444	−182.97	6.82	154.4	49.7	0(g)	—
n-Pentane	C_5H_{12}	72.15	0.63[18°]	−129.6	8.393	36.07	25.77	469.80	33.3	−173.0(l) −146.4(g)	−3509.5(l) −3536.1(g)
Isopentane	C_5H_{12}	72.15	0.62[19°]	−160.1	—	27.7	—	461.00	32.9	−179.3(l) −152.0(g)	−3507.5(l) −3529.2(g)
1-Pentene	C_5H_{10}	70.13	0.641	−165.2	4.94	29.97	—	474	39.9	−20.9(g)	−3375.8(g)
Phenol	C_6H_5OH	94.11	1.071[25°]	42.5	11.43	181.4	—	692.1	60.5	−158.1(l) −90.8(g)	−3063.5(s)
Phosphoric acid	H_3PO_4	98.00	1.834[18°]	42.3	10.54	$(-\tfrac{1}{2}H_2O$ at 213°C)	—	—	—	−1281.1(c) −1278.6(aq, 1H₂O)	—
Phosphorus (red)	P_4	123.90	2.20	590[43 atm]	81.17	Ignites in air, 725°C	—	—	—	−17.6(c) 0(c)	—

Phosphorus (white)	P_4	123.90	1.82	44.2	280	2.51	49.71	—	—	—	—
Phosphorus pentoxide	P_2O_5	141.95	2.387		Sublimes at 250°C			—	—	-1506.2(c)	—
Propane	C_3H_8	44.09	—	-187.69	-42.07	3.52	18.77	369.9	42.0	-119.8(l) / -103.8(g)	-2204.0(l) / -2220.0(g)
Propylene	C_3H_6	42.08	—	-185.2	-47.70	3.00	18.42	365.1	45.4	+20.41(g)	-2058.4(g)
n-Propyl alcohol	C_3H_7OH	60.09	0.804	-127	97.04	—	—	536.7	49.95	-300.70(l)	-2010.4(g)
Isopropyl alcohol	C_3H_7OH	60.09	0.785	-89.7	82.24	—	—	508.8	53.0	-255.2(g) / -310.9(l)	-2068.6(g) / -1986.6(l)
n-Propyl benzene	C_9H_{12}	120.19	0.862	-99.50	159.2	8.54	38.24	638.7	31.3	-38.40(l) / +7.82(g)	-5218.2(l) / -5264.48(g)
Silicon dioxide	SiO_2	60.09	2.25	1710	2230	14.2	—	—	—	-851.0(c)	—
Sodium bicarbonate	$NaHCO_3$	84.01	2.20		Decomposes at 270°C			—	—	-945.6(c)	—
Sodium bisulfate	$NaHSO_4$	120.07	2.742	—				—	—	-1126.3(c)	—
Sodium carbonate	Na_2CO_3	105.99	2.533		Decomposes at 854°C			—	—	-1130.9(c)	—
Sodium chloride	NaCl	58.45	2.163	808	1465	28.5	170.7	—	—	-411.0(c)	—
Sodium cyanide	NaCN	49.01	—	562	1497	16.7	155	—	—	-89.79(c)	—
Sodium hydroxide	NaOH	40.00	2.130	319	1390	8.34	—	—	—	-426.6(c) / -469.4(aq)	—
Sodium nitrate	$NaNO_3$	85.00	2.257	310	Decomposes at 380°C	15.9	—	—	—	-466.7(c)	—
Sodium nitrite	$NaNO_2$	69.00	2.168^{0}	271	Decomposes at 320°C	—	—	—	—	-359.4(c)	—
Sodium sulfate	Na_2SO_4	142.05	2.698	890	—	24.3	—	—	—	-1384.5(c)	—
Sodium sulfide	Na_2S	78.05	1.856	950	—	6.7	—	—	—	-373.2(c)	—
Sodium sulfite	Na_2SO_3	126.05	2.633^{15}		Decomposes			—	—	-1090.3(c)	—

(continued)

Table B.1 (Continued)

Compound	Formula	Mol. Wt.	SG (20°/4°)	T_m(°C)[b]	$\Delta \hat{H}_m(T_m)$[c,j] kJ/mol	T_b(°C)[d]	$\Delta \hat{H}_v(T_b)$[e,j] kJ/mol	T_c(K)[f]	P_c(atm)[g]	$(\Delta \hat{H}_f°)$[h,j] kJ/mol	$(\Delta \hat{H}_c°)$[i,j] kJ/mol
Sodium thiosulfate	$Na_2S_2O_3$	158.11	1.667	—	—	—	—	—	—	−1117.1(c)	—
Sulfur (rhombic)	S_8	256.53	2.07	113	10.04	444.6	83.7	—	—	0(c)	—
Sulfur (monoclinic)	S_8	256.53	1.96	119	14.17	444.6	83.7	—	—	+0.30(c)	—
Sulfur dioxide	SO_2	64.07	—	−75.48	7.402	−10.02	24.91	430.7	77.8	−296.90(g)	—
Sulfur trioxide	SO_3	80.07	—	16.84	25.48	43.3	41.80	491.4	83.8	−395.18(g)	—
Sulfuric acid	H_2SO_4	98.08	1.834[18°]	10.35	9.87	Decomposes at 340°C		—	—	−811.32(l) −907.51(aq)	— —
Toluene	C_7H_8	92.13	0.866	−94.99	6.619	110.62	33.47	593.9	40.3	+12.00(l) +50.00(g)	−3909.9(l) −3947.9(g)
Water	H_2O	18.016	1.00[4°]	0.00	6.0095	100.00	40.656	647.4	218.3	−285.84(l) −241.83(g)	—
m-Xylene	C_8H_{10}	106.16	0.864	−47.87	11.569	139.10	36.40	619	34.6	−25.42(l) +17.24(g)	−4551.9(l) −4594.5(g)
o-Xylene	C_8H_{10}	106.16	0.880	−25.18	13.598	144.42	36.82	631.5	35.7	−24.44(l) +18.99(g)	−4552.9(l) −4596.3(g)
p-Xylene	C_8H_{10}	106.16	0.861	13.26	17.11	138.35	36.07	618	33.9	−24.43(l) 17.95(g)	−4552.91(l) −4595.2(g)
Zinc	Zn	65.38	7.140	419.5	6.674	907	114.77	—	—	0(c)	—

Table B.2 Heat Capacities[a]

Form 1: $C_p[\text{kJ/(mol}\cdot{}^\circ\text{C})]$ or $[\text{kJ/(mol}\cdot\text{K})] = a + bT + cT^2 + dT^3$

Form 2: $C_p[\text{kJ/(mol}\cdot{}^\circ\text{C})]$ or $[\text{kJ/(mol}\cdot\text{K})] = a + bT + cT^{-2}$

Example: $(C_p)_{\text{acetone(g)}} = 0.07196 + (20.10 \times 10^{-5})T - (12.78 \times 10^{-8})T^2 + (34.76 \times 10^{-12})T^3$, where T is in °C.

Note: The formulas for gases are strictly applicable at pressures low enough for the ideal gas equation of state to apply.

Compound	Formula	Mol. Wt.	State	Form	Temp. Unit	$a \times 10^3$	$b \times 10^5$	$c \times 10^8$	$d \times 10^{12}$	Range (Units of T)
Acetone	CH₃COCH₃	58.08	l	1	°C	123.0	18.6			−30–60
Acetylene	C₂H₂	26.04	g	1	°C	71.96	20.10	−12.78	34.76	0–1200
Air		29.0	g	1	°C	42.43	6.053	−5.033	18.20	0–1200
			g	1	°C	28.94	0.4147	0.3191	−1.965	0–1500
			g	1	K	28.09	0.1965	0.4799	−1.965	273–1800
Ammonia	NH₃	17.03	g	1	°C	35.15	2.954	0.4421	−6.686	0–1200
Ammonium sulfate	(NH₄)₂SO₄	132.15	c	1	K	215.9				275–328
Benzene	C₆H₆	78.11	l	1	°C	126.5	23.4			6–67
			g	1	°C	74.06	32.95	−25.20	77.57	0–1200
Isobutane	C₄H₁₀	58.12	g	1	°C	89.46	30.13	−18.91	49.87	0–1200
n-Butane	C₄H₁₀	58.12	g	1	°C	92.30	27.88	−15.47	34.98	0–1200
Isobutene	C₄H₈	56.10	g	1	°C	82.88	25.64	−17.27	50.50	0–1200
Calcium carbide	CaC₂	64.10	c	2	K	68.62	1.19	-8.66×10^{10}	—	298–720
Calcium carbonate	CaCO₃	100.09	c	2	K	82.34	4.975	-12.87×10^{10}	—	273–1033
Calcium hydroxide	Ca(OH)₂	74.10	c	1	K	89.5				276–373
Calcium oxide	CaO	56.08	c	2	K	41.84	2.03	-4.52×10^{10}		273–1173
Carbon	C	12.01	c	2	K	11.18	1.095	-4.891×10^{10}		273–1373
Carbon dioxide	CO₂	44.01	g	1	°C	36.11	4.233	−2.887	7.464	0–1500
Carbon monoxide	CO	28.01	g	1	°C	28.95	0.4110	0.3548	−2.220	0–1500
Carbon tetrachloride	CCl₄	153.84	l	1	K	93.39	12.98			273–343
Chlorine	Cl₂	70.91	g	1	°C	33.60	1.367	−1.607	6.473	0–1200
Copper	Cu	63.54	c	1	K	22.76	0.6117			273–1357

[a] Adapted in part from D. M. Himmelblau, *Basic Principles and Calculations in Chemical Engineering*, 3rd Edition, © 1974, Table E.1. Adapted by permission of Prentice-Hall, Inc., Englewood Cliffs, NJ.

(continued)

Table B.2 (Continued)

Compound	Formula	Mol. Wt.	State	Form	Temp. Unit	$a \times 10^3$	$b \times 10^5$	$c \times 10^8$	$d \times 10^{12}$	Range (Units of T)
Cumene (Isopropyl benzene)	C_9H_{12}	120.19	g	1	°C	139.2	53.76	−39.79	120.5	0–1200
Cyclohexane	C_6H_{12}	84.16	g	1	°C	94.140	49.62	−31.90	80.63	0–1200
Cyclopentane	C_5H_{10}	70.13	g	1	°C	73.39	39.28	−25.54	68.66	0–1200
Ethane	C_2H_6	30.07	g	1	°C	49.37	13.92	−5.816	7.280	0–1200
Ethyl alcohol (Ethanol)	C_2H_5OH	46.07	l	1	°C	103.1				0
			l	1	°C	158.8				100
			g	1	°C	61.34	15.72	−8.749	19.83	0–1200
Ethylene	C_2H_4	28.05	g	1	°C	+40.75	11.47	−6.891	17.66	0–1200
Ferric oxide	Fe_2O_3	159.70	c	2	K	103.4	6.711	-17.72×10^{10}	—	273–1097
Formaldehyde	CH_2O	30.03	g	1	°C	34.28	4.268	0.0000	−8.694	0–1200
Helium	He	4.00	g	1	°C	20.8				0–1200
n-Hexane	C_6H_{14}	86.17	l	1	°C	216.3				20–100
			g	1	°C	137.44	40.85	−23.92	57.66	0–1200
Hydrogen	H_2	2.016	g	1	°C	28.84	0.00765	0.3288	−0.8698	0–1500
Hydrogen bromide	HBr	80.92	g	1	°C	29.10	−0.0227	0.9887	−4.858	0–1200
Hydrogen chloride	HCl	36.47	g	1	°C	29.13	−0.1341	0.9715	−4.335	0–1200
Hydrogen cyanide	HCN	27.03	g	1	°C	35.3	2.908	1.092		0–1200
Hydrogen sulfide	H_2S	34.08	g	1	°C	33.51	1.547	0.3012	−3.292	0–1500
Magnesium chloride	$MgCl_2$	95.23	c	1	K	72.4	1.58			273–991
Magnesium oxide	MgO	40.32	c	2	K	45.44	0.5008	-8.732×10^{10}		273–2073
Methane	CH_4	16.04	g	1	°C	34.31	5.469	0.3661	−11.00	0–1200
			g	1	K	19.87	5.021	1.268	−11.00	273–1500
Methyl alcohol (Methanol)	CH_3OH	32.04	l	1	°C	75.86	16.83			0–65
			g	1	°C	42.93	8.301	−1.87	−8.03	0–700
Methyl cyclohexane	C_7H_{14}	98.18	g	1	°C	121.3	56.53	−37.72	100.8	0–1200
Methyl cyclopentane	C_6H_{12}	84.16	g	1	°C	98.83	45.857	−30.44	83.81	0–1200
Nitric acid	HNO_3	63.02	l	1	°C	110.0				25
Nitric oxide	NO	30.01	g	1	°C	29.50	0.8188	−0.2925	0.3652	0–3500

Nitrogen	N_2	g	1	°C	29.00	0.2199	0.5723	−2.871	0–1500
Nitrogen dioxide	NO_2	g	1	°C	36.07	3.97	−2.88	7.87	0–1200
Nitrogen tetraoxide	N_2O_4	g	1	°C	75.7	12.5	−11.3		0–300
Nitrous oxide	N_2O	g	1	°C	37.66	4.151	−2.694	10.57	0–1200
Oxygen	O_2	g	1	°C	29.10	1.158	−0.6076	1.311	0–1500
n-Pentane	C_5H_{12}	l	1	°C	155.4	43.68			0–36
		g	1	°C	114.8	34.09	−18.99	42.26	0–1200
Propane	C_3H_8	g	1	°C	68.032	22.59	−13.11	31.71	0–1200
Propylene	C_3H_6	g	1	°C	59.580	17.71	−10.17	24.60	0–1200
Sodium carbonate	Na_2CO_3	c	1	K	121				288–371
Sodium carbonate decahydrate	$Na_2CO_3 \cdot 10H_2O$	c	1	K	535.6				298
Sulfur (Rhombic)	S	c	1	K	15.2	2.68			273–368
(Monoclinic)		c	1	K	18.3	1.84			368–392
Sulfuric acid	H_2SO_4	l	1	°C	139.1	15.59			10–45
Sulfur dioxide	SO_2	g	1	°C	38.91	3.904	−3.104	8.606	0–1500
Sulfur trioxide	SO_3	g	1	°C	48.50	9.188	−8.540	32.40	0–1000
Toluene	C_7H_8	l	1	°C	148.8	32.4			0–110
		g	1	°C	94.18	38.00	−27.86	80.33	0–1200
Water	H_2O	l	1	°C	75.4				0–100
		g	1	°C	33.46	0.6880	0.7604	−3.593	0–1500

Table B.3 Vapor Pressure of Water[a]

p_v(mm Hg) versus T (°C)

Example: The vapor pressure of liquid water at 4.3°C is 6.230 mm Hg

	T(°C)	0.0	0.1	0.2	0.3	0.4	0.5	0.6	0.7	0.8	0.9
Ice	−14	1.361	1.348	1.336	1.324	1.312	1.300	1.288	1.276	1.264	1.253
	−13	1.490	1.477	1.464	1.450	1.437	1.424	1.411	1.399	1.386	1.373
	−12	1.632	1.617	1.602	1.588	1.574	1.559	1.546	1.532	1.518	1.504
	−11	1.785	1.769	1.753	1.737	1.722	1.707	1.691	1.676	1.661	1.646
	−10	1.950	1.934	1.916	1.899	1.883	1.866	1.849	1.833	1.817	1.800
	−9	2.131	2.122	2.093	2.075	2.057	2.039	2.021	2.003	1.985	1.968
	−8	2.326	2.306	2.285	2.266	2.246	2.226	2.207	2.187	2.168	2.149
	−7	2.537	2.515	2.493	2.472	2.450	2.429	2.408	2.387	2.367	2.346
	−6	2.765	2.742	2.718	2.695	2.672	2.649	2.626	2.603	2.581	2.559
	−5	3.013	2.987	2.962	2.937	2.912	2.887	2.862	2.838	2.813	2.790
	−4	3.280	3.252	3.225	3.198	3.171	3.144	3.117	3.091	3.065	3.039
	−3	3.568	3.539	3.509	3.480	3.451	3.422	3.393	3.364	3.336	3.308
	−2	3.880	3.848	3.816	3.785	3.753	3.722	3.691	3.660	3.630	3.599
	−1	4.217	4.182	4.147	4.113	4.079	4.045	4.012	3.979	3.946	3.913
	−0	4.579	4.542	4.504	4.467	4.431	4.395	4.359	4.323	4.287	4.252
Liquid water	0	4.579	4.613	4.647	4.681	4.715	4.750	4.785	4.820	4.855	4.890
	1	4.926	4.962	4.998	5.034	5.070	5.107	5.144	5.181	5.219	5.256
	2	5.294	5.332	5.370	5.408	5.447	5.486	5.525	5.565	5.605	5.645
	3	5.685	5.725	5.766	5.807	5.848	5.889	5.931	5.973	6.015	6.058
	4	6.101	6.144	6.187	6.230	6.274	6.318	6.363	6.408	6.453	6.498
	5	6.543	6.589	6.635	6.681	6.728	6.775	6.822	6.869	6.917	6.965
	6	7.013	7.062	7.111	7.160	7.209	7.259	7.309	7.360	7.411	7.462
	7	7.513	7.565	7.617	7.669	7.722	7.775	7.828	7.882	7.936	7.990
	8	8.045	8.100	8.155	8.211	8.267	8.323	8.380	8.437	8.494	8.551
	9	8.609	8.668	8.727	8.786	8.845	8.905	8.965	9.025	9.086	9.147
	10	9.209	9.271	9.333	9.395	9.458	9.521	9.585	9.649	9.714	9.779
	11	9.844	9.910	9.976	10.042	10.109	10.176	10.244	10.312	10.380	10.449
	12	10.518	10.588	10.658	10.728	10.799	10.870	10.941	11.013	11.085	11.158
	13	11.231	11.305	11.379	11.453	11.528	11.604	11.680	11.756	11.833	11.910
	14	11.987	12.065	12.144	12.223	12.302	12.382	12.462	12.543	12.624	12.706
	15	12.788	12.870	12.953	13.037	13.121	13.205	13.290	13.375	13.461	13.547
	16	13.634	13.721	13.809	13.898	13.987	14.076	14.166	14.256	14.347	14.438
	17	14.530	14.622	14.715	14.809	14.903	14.997	15.092	15.188	15.284	15.380
	18	15.477	15.575	15.673	15.772	15.871	15.971	16.771	16.171	16.272	16.374
	19	16.477	16.581	16.685	16.789	16.894	16.999	17.105	17.212	17.319	17.427
	20	17.535	17.644	17.753	17.863	17.974	18.085	18.197	18.309	18.422	18.536
	21	18.650	18.765	18.880	18.996	19.113	19.231	19.349	19.468	19.587	19.707
	22	19.827	19.948	20.070	20.193	20.316	20.440	20.565	20.690	20.815	20.941
	23	21.068	21.196	21.324	21.453	21.583	21.714	21.845	21.977	22.110	22.243
	24	22.377	22.512	22.648	22.785	22.922	23.060	23.198	23.337	23.476	23.616

[a]From R. H. Perry and C. H. Chilton, Eds., *Chemical Engineers' Handbook,* 5th Edition, McGraw-Hill, New York, 1973, Tables 3-3 and 3-5. Reprinted by permission of McGraw-Hill Book Co.

(*continued*)

Table B.3 (Continued)

T(°C)	0.0	0.1	0.2	0.3	0.4	0.5	0.6	0.7	0.8	0.9
25	23.756	23.897	24.039	24.182	24.326	24.471	24.617	24.764	24.912	25.060
26	25.209	25.359	25.509	25.660	25.812	25.964	26.117	26.271	26.426	26.582
27	26.739	26.897	27.055	27.214	27.374	27.535	27.696	27.858	28.021	28.185
28	28.349	28.514	28.680	28.847	29.015	29.184	29.354	29.525	29.697	29.870
29	30.043	30.217	30.392	30.568	30.745	30.923	31.102	31.281	31.461	31.642
30	31.824	32.007	32.191	32.376	32.561	32.747	32.934	33.122	33.312	33.503
31	33.695	33.888	34.082	34.276	34.471	34.667	34.864	35.062	35.261	35.462
32	35.663	35.865	36.068	36.272	36.477	36.683	36.891	37.099	37.308	37.518
33	37.729	37.942	38.155	38.369	33.584	38.801	38.018	39.237	39.457	39.677
34	39.898	40.121	40.344	40.569	40.796	41.023	41.251	41.480	41.710	41.942
35	42.175	42.409	42.644	42.880	43.117	43.355	43.595	43.836	44.078	44.320
36	44.563	44.808	45.054	45.301	45.549	45.799	46.050	46.302	46.556	46.811
37	47.067	47.324	47.582	47.841	48.102	48.364	48.627	48.891	49.157	49.424
38	49.692	49.961	50.231	50.502	50.774	51.048	51.323	51.600	51.879	52.160
39	52.442	52.725	53.009	53.294	53.580	53.867	54.156	54.446	54.737	55.030
40	55.324	55.61	55.91	56.21	56.51	56.81	57.11	57.41	57.72	58.03
41	58.34	58.65	58.96	59.27	59.58	59.90	60.22	60.54	60.86	61.18
42	61.50	61.82	62.14	62.47	62.80	63.13	63.46	63.79	64.12	64.46
43	64.80	65.14	65.48	65.82	66.16	66.51	66.86	67.21	67.56	67.91
44	68.26	68.61	68.97	69.33	69.69	70.05	70.41	70.77	71.14	71.51
45	71.88	72.25	72.62	72.99	73.36	73.74	74.12	74.50	74.88	75.26
46	75.65	76.04	76.43	76.82	77.21	77.60	78.00	78.40	78.80	79.20
47	79.60	80.00	80.41	80.82	81.23	81.64	82.05	82.46	82.87	83.29
48	83.71	84.13	84.56	84.99	85.42	85.85	86.28	86.71	87.14	87.58
49	88.02	88.46	88.90	89.34	89.79	90.24	90.69	91.14	91.59	92.05

T(°C)	0	1	2	3	4	5	6	7	8	9
50	92.51	97.20	102.09	107.20	112.51	118.04	123.80	129.82	136.08	142.60
60	149.38	156.43	163.77	171.38	179.31	187.54	196.09	204.96	214.17	223.73
70	233.7	243.9	254.6	265.7	277.2	289.1	301.4	314.1	327.3	341.0
80	355.1	369.7	384.9	400.6	416.8	433.6	450.9	468.7	487.1	506.1

T(°C)	0.0	0.1	0.2	0.3	0.4	0.5	0.6	0.7	0.8	0.9
90	525.76	527.76	529.77	531.78	533.80	535.82	537.86	539.90	541.95	544.00
91	546.05	548.11	550.18	552.26	554.35	556.44	558.53	560.64	562.75	564.87
92	566.99	569.12	571.26	573.40	575.55	577.71	579.87	582.04	584.22	586.41
93	588.60	590.80	593.00	595.21	597.43	599.66	601.89	604.13	606.38	608.64
94	610.90	613.17	615.44	617.72	620.01	622.31	624.61	626.92	629.24	631.57
95	633.90	636.24	938.59	640.94	643.30	645.67	648.05	650.43	652.82	655.22
96	657.62	660.03	662.45	664.88	667.31	669.75	672.20	674.66	677.12	679.69
97	682.07	684.55	687.04	689.54	692.05	694.57	697.10	699.63	702.17	704.71
98	707.27	709.83	712.40	714.98	717.56	720.15	722.75	725.36	727.98	730.61
99	733.24	735.88	738.53	741.18	743.85	746.52	749.20	751.89	754.58	757.29
100	760.00	762.72	765.45	768.19	770.93	773.68	776.44	779.22	782.00	784.78
101	787.57	790.37	793.18	796.00	798.82	801.66	804.50	807.35	810.21	813.08

Table B.4 Antoine Equation Constants[a]

$$\log_{10} p^* = A - \frac{B}{T + C} \qquad p^* \text{ in mm Hg}, \quad T \text{ in } °C$$

Example: The vapor pressure of acetaldehyde at 25°C is determined as follows:

$$\log_{10} p^*_{C_2H_4O}(25°C) = 8.00552 - \frac{1600.017}{25 + 291.809} = 2.9551$$

$$\implies p^*_{C_2H_4O}(25°C) = 10^{2.9551} = 902 \text{ mm Hg}$$

Compound	Formula	Range (°C)	A	B	C
Acetaldehyde	C_2H_4O	−0.2 to 34.4	8.00552	1600.017	291.809
Acetic acid	$C_2H_4O_2$	29.8 to 126.5	7.38782	1533.313	222.309
Acetic acid*	$C_2H_4O_2$	0 to 36	7.18807	1416.7	225
Acetic anhydride	$C_4H_6O_3$	62.8 to 139.4	7.14948	1444.718	199.817
Acetone	C_3H_6O	−12.9 to 55.3	7.11714	1210.595	229.664
Acrylic acid	$C_3H_4O_2$	20.0 to 70.0	5.65204	648.629	154.683
Ammonia*	NH_3	−83 to 60	7.55466	1002.711	247.885
Aniline	C_6H_7N	102.6 to 185.2	7.32010	1731.515	206.049
Benzene	C_6H_6	14.5 to 80.9	6.89272	1203.531	219.888
n-Butane	$n\text{-}C_4H_{10}$	−78.0 to −0.3	6.82485	943.453	239.711
i-Butane	$i\text{-}C_4H_{10}$	−85.1 to −11.6	6.78866	899.617	241.942
1-Butanol	$C_4H_{10}O$	89.2 to 125.7	7.36366	1305.198	173.427
2-Butanol	$C_4H_{10}O$	72.4 to 107.1	7.20131	1157.000	168.279
1-Butene	C_4H_8	−77.5 to −3.7	6.53101	810.261	228.066
Butyric acid	$C_4H_8O_2$	20.0 to 150.0	8.71019	2433.014	255.189
Carbon disulfide	CS_2	3.6 to 79.9	6.94279	1169.110	241.593
Carbon tetrachloride	CCl_4	14.1 to 76.0	6.87926	1212.021	226.409
Chlorobenzene	C_6H_5Cl	62.0 to 131.7	6.97808	1431.053	217.550
Chlorobenzene*	C_6H_5Cl	0 to 42	7.10690	1500.0	224.0
Chlorobenzene*	C_6H_5Cl	42 to 230	6.94504	1413.12	216.0
Chloroform	$CHCl_3$	−10.4 to 60.3	6.95465	1170.966	226.232
Chloroform*	$CHCl_3$	−30 to 150	6.90328	1163.03	227.4
Cyclohexane	C_6H_{12}	19.9 to 81.6	6.84941	1206.001	223.148
Cyclohexanol	$C_6H_{12}O$	93.7 to 160.7	6.25530	912.866	109.126
n-Decane	$n\text{-}C_{10}H_{22}$	94.5 to 175.1	6.95707	1503.568	194.738
1-Decene	$C_{10}H_{20}$	86.8 to 171.6	6.95433	1497.527	197.056
1,1-Dichloroethane	$C_2H_4Cl_2$	−38.8 to 17.6	6.97702	1174.022	229.060
1,2-Dichloroethane	C_2H_4Cl	−30.8 to 99.4	7.02530	1271.254	222.927
Dichloromethane	CH_2Cl_2	−40.0 to 40	7.40916	1325.938	252.616
Diethyl ether	$C_4H_{10}O$	−60.8 to 19.9	6.92032	1064.066	228.799
Diethyl ketone	$C_5H_{10}O$	56.5 to 111.3	7.02529	1310.281	214.192
Diethylene glycol	$C_4H_{10}O_2$	130.0 to 243.0	7.63666	1939.359	162.714
Dimethyl ether	C_2H_6O	−78.2 to −24.9	6.97603	889.264	241.957
Dimethylamine	C_2H_7N	−71.8 to 6.9	7.08212	960.242	221.667
N,N-Dimethylformamide	C_3H_7NO	30.0 to 90.0	6.92796	1400.869	196.434
1,4-Dioxane	$C_4H_8O_2$	20.0 to 105.0	7.43155	1554.679	240.337
Ethanol	C_2H_6O	19.6 to 93.4	8.11220	1592.864	226.184
Ethanolamine	C_2H_7NO	65.4 to 170.9	7.45680	1577.670	173.368
Ethyl acetate	$C_4H_8O_2$	15.6 to 75.8	7.10179	1244.951	217.881
Ethyl acetate*	$C_4H_8O_2$	−20 to 150	7.09808	1238.710	217.0
Ethyl chloride	C_2H_5Cl	−55.9 to 12.5	6.98647	1030.007	238.612
Ethylbenzene	C_8H_{10}	56.5 to 137.1	6.95650	1423.543	213.091

[a]Adapted from T. Boublik, V. Fried, and E. Hala, *The Vapour Pressures of Pure Substances*, Elsevier, Amsterdam, 1973. If marked with an asterisk (*), constants are from *Lange's Handbook of Chemistry*, 9th Edition, Handbook Publishers, Inc., Sandusky, OH, 1956.

(*continued*)

Table B.4 (Continued)

Compound	Formula	Range (°C)	A	B	C
Ethylene glycol	$C_2H_6O_2$	50.0 to 200.0	8.09083	2088.936	203.454
Ethylene oxide	C_2H_4O	0.3 to 31.8	8.69016	2005.779	334.765
1,2-Ethylenediamine	$C_2H_8N_2$	26.5 to 117.4	7.16871	1336.235	194.366
Formaldehyde	HCHO	−109.4 to −22.3	7.19578	970.595	244.124
Formic acid	CH_2O_2	37.4 to 100.7	7.58178	1699.173	260.714
Glycerol	$C_3H_8O_3$	183.3 to 260.4	6.16501	1036.056	28.097
n-Heptane	$n\text{-}C_7H_{16}$	25.9 to 99.3	6.90253	1267.828	216.823
i-Heptane	$i\text{-}C_7H_{16}$	18.5 to 90.9	6.87689	1238.122	219.783
1-Heptene	C_7H_{14}	21.6 to 94.5	6.91381	1265.120	220.051
n-Hexane	$n\text{-}C_6H_{14}$	13.0 to 69.5	6.88555	1175.817	224.867
i-Hexane	$i\text{-}C_6H_{14}$	12.8 to 61.1	6.86839	1151.401	228.477
1-Hexene	C_6H_{12}	15.9 to 64.3	6.86880	1154.646	226.046
Hydrogen Cyanide	HCN	−16.4 to 46.2	7.52823	1329.49	260.418
Methanol	CH_3OH	14.9 to 83.7	8.08097	1582.271	239.726
Methanol*	CH_3OH	−20 to 140	7.87863	1473.11	230.0
Methyl acetate	$C_3H_6O_2$	1.8 to 55.8	7.06524	1157.630	219.726
Methyl bromide	CH_3Br	−70.0 to 3.6	7.09084	1046.066	244.914
Methyl chloride	CH_3Cl	−75.0 to 5.0	7.09349	948.582	249.336
Methyl ethyl ketone	C_4H_8O	42.8 to 88.4	7.06356	1261.339	221.969
Methyl isobutyl ketone	$C_6H_{12}O$	21.7 to 116.2	6.67272	1168.408	191.944
Methyl methacrylate	$C_5H_8O_2$	39.2 to 89.2	8.40919	2050.467	274.369
Methylamine	CH_5N	−83.1 to −6.2	7.33690	1011.532	233.286
Methylcyclohexane	C_7H_{14}	25.6 to 101.8	6.82827	1273.673	221.723
Naphthalene	$C_{10}H_8$	80.3 to 179.5	7.03358	1756.328	204.842
Nitrobenzene	$C_6H_5NO_2$	134.1 to 210.6	7.11562	1746.586	201.783
Nitromethane	CH_3NO_2	55.7 to 136.4	7.28166	1446.937	227.600
n-Nonane	$n\text{-}C_9H_{20}$	70.3 to 151.8	6.93764	1430.459	201.808
1-Nonane	C_9H_{18}	66.6 to 147.9	6.95777	1437.862	205.814
n-Octane	$n\text{-}C_8H_{18}$	52.9 to 126.6	6.91874	1351.756	209.100
i-Octane	$i\text{-}C_8H_{18}$	41.7 to 118.5	6.88814	1319.529	211.625
1-Octene	C_8H_{16}	44.9 to 122.2	6.93637	1355.779	213.022
n-Pentane	$n\text{-}C_5H_{12}$	13.3 to 36.8	6.84471	1060.793	231.541
i-Pentane	$i\text{-}C_5H_{12}$	16.3 to 28.6	6.73457	992.019	229.564
1-Pentanol	$C_5H_{12}O$	74.7 to 156.0	7.18246	1287.625	161.330
1-Pentene	C_5H_{10}	12.8 to 30.7	6.84268	1043.206	233.344
Phenol	C_6H_6O	107.2 to 181.8	7.13301	1516.790	174.954
1-Propanol	C_3H_8O	60.2 to 104.6	7.74416	1437.686	198.463
2-Propanol	C_3H_8O	52.3 to 89.3	7.74021	1359.517	197.527
Propionic acid	$C_3H_6O_2$	72.4 to 128.3	7.71423	1733.418	217.724
Propylene oxide	C_3H_6O	−24.2 to 34.8	7.01443	1086.369	228.594
Pyridine	C_5H_5N	67.3 to 152.9	7.04115	1373.799	214.979
Styrene	C_8H_8	29.9 to 144.8	7.06623	1507.434	214.985
Toluene	C_7H_8	35.3 to 111.5	6.95805	1346.773	219.693
1,1,1-Trichloroethane	$C_2H_3Cl_3$	−5.4 to 16.9	8.64344	2136.621	302.769
1,1,2-Trichloroethane	$C_2H_3Cl_3$	50.0 to 113.7	6.95185	1314.410	209.197
Trichloroethylene	C_2HCl_3	17.8 to 86.5	6.51827	1018.603	192.731
Vinyl acetate	$C_4H_6O_2$	21.8 to 72.0	7.21010	1296.130	226.655
Water*	H_2O	0 to 60	8.10765	1750.286	235.000
Water*	H_2O	60 to 150	7.96681	1668.210	228.000
m-Xylene	$m\text{-}C_8H_{10}$	59.2 to 140.0	7.00646	1460.183	214.827
o-Xylene	$o\text{-}C_8H_{10}$	63.5 to 145.4	7.00154	1476.393	213.872
p-Xylene	$p\text{-}C_8H_{10}$	58.3 to 139.3	6.98820	1451.792	215.111

Table B.5 Properties of Saturated Steam: Temperature Table[a]

$T(°C)$	P(bar)	$\hat{V}$(m³/kg)		$\hat{U}$(kJ/kg)		$\hat{H}$(kJ/kg)		
		Water	Steam	Water	Steam	Water	Evaporation	Steam
0.01	0.00611	0.001000	206.2	zero	2375.6	+0.0	2501.6	2501.6
2	0.00705	0.001000	179.9	8.4	2378.3	8.4	2496.8	2505.2
4	0.00813	0.001000	157.3	16.8	2381.1	16.8	2492.1	2508.9
6	0.00935	0.001000	137.8	25.2	2383.8	25.2	2487.4	2512.6
8	0.01072	0.001000	121.0	33.6	2386.6	33.6	2482.6	2516.2
10	0.01227	0.001000	106.4	42.0	2389.3	42.0	2477.9	2519.9
12	0.01401	0.001000	93.8	50.4	2392.1	50.4	2473.2	2523.6
14	0.01597	0.001001	82.9	58.8	2394.8	58.8	2468.5	2527.2
16	0.01817	0.001001	73.4	67.1	2397.6	67.1	2463.8	2530.9
18	0.02062	0.001001	65.1	75.5	2400.3	75.5	2459.0	2534.5
20	0.0234	0.001002	57.8	83.9	2403.0	83.9	2454.3	2538.2
22	0.0264	0.001002	51.5	92.2	2405.8	92.2	2449.6	2541.8
24	0.0298	0.001003	45.9	100.6	2408.5	100.6	2444.9	2545.5
25	0.0317	0.001003	43.4	104.8	2409.9	104.8	2442.5	2547.3
26	0.0336	0.001003	41.0	108.9	2411.2	108.9	2440.2	2549.1
28	0.0378	0.001004	36.7	117.3	2414.0	117.3	2435.4	2552.7
30	0.0424	0.001004	32.9	125.7	2416.7	125.7	2430.7	2556.4
32	0.0475	0.001005	29.6	134.0	2419.4	134.0	2425.9	2560.0
34	0.0532	0.001006	26.6	142.4	2422.1	142.4	2421.2	2563.6
36	0.0594	0.001006	24.0	150.7	2424.8	150.7	2416.4	2567.2
38	0.0662	0.001007	21.6	159.1	2427.5	159.1	2411.7	2570.8
40	0.0738	0.001008	19.55	167.4	2430.2	167.5	2406.9	2574.4
42	0.0820	0.001009	17.69	175.8	2432.9	175.8	2402.1	2577.9
44	0.0910	0.001009	16.04	184.2	2435.6	184.2	2397.3	2581.5
46	0.1009	0.001010	14.56	192.5	2438.3	192.5	2392.5	2585.1
48	0.1116	0.001011	13.23	200.9	2440.9	200.9	2387.7	2588.6
50	0.1234	0.001012	12.05	209.2	2443.6	209.3	2382.9	2592.2
52	0.1361	0.001013	10.98	217.7	2446	217.7	2377	2595
54	0.1500	0.001014	10.02	226.0	2449	226.0	2373	2599
56	0.1651	0.001015	9.158	234.4	2451	234.4	2368	2602
58	0.1815	0.001016	8.380	242.8	2454	242.8	2363	2606
60	0.1992	0.001017	7.678	251.1	2456	251.1	2358	2609
62	0.2184	0.001018	7.043	259.5	2459	259.5	2353	2613
64	0.2391	0.001019	6.468	267.9	2461	267.9	2348	2616
66	0.2615	0.001020	5.947	276.2	2464	276.2	2343	2619
68	0.2856	0.001022	5.475	284.6	2467	284.6	2338	2623

[a]From R. W. Haywood, *Thermodynamic Tables in SI (Metric) Units,* Cambridge University Press, London, 1968. $\hat{V}$ = specific volume, $\hat{U}$ = specific internal energy, and $\hat{H}$ = specific enthalpy. *Note:* kJ/kg × 0.4303 = Btu/lb$_m$.

(*continued*)

Table B.5 (Continued)

T(°C)	P(bar)	$\hat{V}$(m³/kg)		$\hat{U}$(kJ/kg)		$\hat{H}$(kJ/kg)		
		Water	Steam	Water	Steam	Water	Evaporation	Steam
70	0.3117	0.001023	5.045	293.0	2469	293.0	2333	2626
72	0.3396	0.001024	4.655	301.4	2472	301.4	2329	2630
74	0.3696	0.001025	4.299	309.8	2474	309.8	2323	2633
76	0.4019	0.001026	3.975	318.2	2476	318.2	2318	2636
78	0.4365	0.001028	3.679	326.4	2479	326.4	2313	2639
80	0.4736	0.001029	3.408	334.8	2482	334.9	2308	2643
82	0.5133	0.001030	3.161	343.2	2484	343.3	2303	2646
84	0.5558	0.001032	2.934	351.6	2487	351.7	2298	2650
86	0.6011	0.001033	2.727	360.0	2489	360.1	2293	2653
88	0.6495	0.001034	2.536	368.4	2491	368.5	2288	2656
90	0.7011	0.001036	2.361	376.9	2493	377.0	2282	2659
92	0.7560	0.001037	2.200	385.3	2496	385.4	2277	2662
94	0.8145	0.001039	2.052	393.7	2499	393.8	2272	2666
96	0.8767	0.001040	1.915	402.1	2501	402.2	2267	2669
98	0.9429	0.001042	1.789	410.6	2504	410.7	2262	2673
100	1.0131	0.001044	1.673	419.0	2507	419.1	2257	2676
102	1.0876	0.001045	1.566	427.1	2509	427.5	2251	2679

Table B.6 Properties of Saturated Steam: Pressure Table[a]

P(bar)	T(°C)	$\hat{V}$(m³/kg)		$\hat{U}$(kJ/kg)		$\hat{H}$(kJ/kg)		
		Water	Steam	Water	Steam	Water	Evaporation	Steam
0.00611	0.01	0.001000	206.2	zero	2375.6	+0.0	2501.6	2501.6
0.008	3.8	0.001000	159.7	15.8	2380.7	15.8	2492.6	2508.5
0.010	7.0	0.001000	129.2	29.3	2385.2	29.3	2485.0	2514.4
0.012	9.7	0.001000	108.7	40.6	2388.9	40.6	2478.7	2519.3
0.014	12.0	0.001000	93.9	50.3	2392.0	50.3	2473.2	2523.5
0.016	14.0	0.001001	82.8	58.9	2394.8	58.9	2468.4	2527.3
0.018	15.9	0.001001	74.0	66.5	2397.4	66.5	2464.1	2530.6
0.020	17.5	0.001001	67.0	73.5	2399.6	73.5	2460.2	2533.6
0.022	19.0	0.001002	61.2	79.8	2401.7	79.8	2456.6	2536.4
0.024	20.4	0.001002	56.4	85.7	2403.6	85.7	2453.3	2539.0
0.026	21.7	0.001002	52.3	91.1	2405.4	91.1	2450.2	2541.3
0.028	23.0	0.001002	48.7	96.2	2407.1	96.2	2447.3	2543.6
0.030	24.1	0.001003	45.7	101.0	2408.6	101.0	2444.6	2545.6
0.035	26.7	0.001003	39.5	111.8	2412.2	111.8	2438.5	2550.4
0.040	29.0	0.001004	34.8	121.4	2415.3	121.4	2433.1	2554.5
0.045	31.0	0.001005	31.1	130.0	2418.1	130.0	2428.2	2558.2
0.050	32.9	0.001005	28.2	137.8	2420.6	137.8	2423.8	2561.6
0.060	36.2	0.001006	23.74	151.5	2425.1	151.5	2416.0	2567.5
0.070	39.0	0.001007	20.53	163.4	2428.9	163.4	2409.2	2572.6
0.080	41.5	0.001008	18.10	173.9	2432.3	173.9	2403.2	2577.1
0.090	43.8	0.001009	16.20	183.3	2435.3	183.3	2397.9	2581.1
0.10	45.8	0.001010	14.67	191.8	2438.0	191.8	2392.9	2584.8
0.11	47.7	0.001011	13.42	199.7	2440.5	199.7	2388.4	2588.1
0.12	49.4	0.001012	12.36	206.9	2442.8	206.9	2384.3	2591.2
0.13	51.1	0.001013	11.47	213.7	2445.0	213.7	2380.4	2594.0
0.14	52.6	0.001013	10.69	220.0	2447.0	220.0	2376.7	2596.7

P	T	$\hat{V}$	$\hat{V}$	$\hat{U}$	$\hat{U}$	$\hat{H}$	$\hat{H}$	$\hat{H}$
0.15	54.0	0.001014	10.02	226.0	2448.9	226.0	2373.2	2599.2
0.16	55.3	0.001015	9.43	231.6	2450.6	231.6	2370.0	2601.6
0.17	56.6	0.001015	8.91	236.9	2452.3	236.9	2366.9	2603.8
0.18	57.8	0.001016	8.45	242.0	2453.9	242.0	2363.9	2605.9
0.19	59.0	0.001017	8.03	246.8	2455.4	246.8	2361.1	2607.9
0.20	60.1	0.001017	7.65	251.5	2456.9	251.5	2358.4	2609.9
0.22	62.2	0.001018	7.00	260.1	2459.6	260.1	2353.3	2613.5
0.24	64.1	0.001019	6.45	268.2	2462.1	268.2	2348.6	2616.8
0.26	65.9	0.001020	5.98	275.7	2464.4	275.6	2344.2	2619.9
0.28	67.5	0.001021	5.58	282.7	2466.5	282.7	2340.0	2622.7
0.30	69.1	0.001022	5.23	289.3	2468.6	289.3	2336.1	2625.4
0.35	72.7	0.001025	4.53	304.3	2473.1	304.3	2327.2	2631.5
0.40	75.9	0.001027	3.99	317.7	2477.1	317.6	2319.2	2636.9
0.45	78.7	0.001028	3.58	329.6	2480.7	329.6	2312.0	2641.7
0.50	81.3	0.001030	3.24	340.6	2484.0	340.5	2305.4	2646.0
0.55	83.7	0.001032	2.96	350.6	2486.9	350.6	2299.3	2649.9
0.60	86.0	0.001033	2.73	359.9	2489.7	359.9	2293.6	2653.6
0.65	88.0	0.001035	2.53	368.6	2492.2	368.5	2288.3	2656.9
0.70	90.0	0.001036	2.36	376.8	2494.5	376.7	2283.3	2660.1
0.75	91.8	0.001037	2.22	384.5	2496.7	384.4	2278.6	2663.0
0.80	93.5	0.001039	2.087	391.7	2498.8	391.6	2274.1	2665.8
0.85	95.2	0.001040	1.972	398.6	2500.8	398.5	2269.8	2668.4
0.90	96.7	0.001041	1.869	405.2	2502.6	405.1	2265.6	2670.9
0.95	98.2	0.001042	1.777	411.5	2504.4	411.4	2261.7	2673.2
1.00	99.6	0.001043	1.694	417.5	2506.1	417.4	2257.9	2675.4
1.01325 (1 atm)	100.0	0.001044	1.673	419.1	2506.5	419.0	2256.9	2676.0

*a*From R. W. Haywood, *Thermodynamic Tables in SI (Metric) Units,* Cambridge University Press, London, 1968. $\hat{V}$ = specific volume, $\hat{U}$ = specific internal energy, and $\hat{H}$ = specific enthalpy. *Note:* kJ/kg × 0.4303 = Btu/lb$_{m}$.

(continued)

Table B.6 (Continued)

P(bar)	T(°C)	$\hat{V}$(m³/kg) Water	$\hat{V}$(m³/kg) Steam	$\hat{U}$(kJ/kg) Water	$\hat{U}$(kJ/kg) Steam	$\hat{H}$(kJ/kg) Water	$\hat{H}$(kJ/kg) Evaporation	$\hat{H}$(kJ/kg) Steam
1.1	102.3	0.001046	1.549	428.7	2509.2	428.8	2250.8	2679.6
1.2	104.8	0.001048	1.428	439.2	2512.1	439.4	2244.1	2683.4
1.3	107.1	0.001049	1.325	449.1	2514.7	449.2	2237.8	2687.0
1.4	109.3	0.001051	1.236	458.3	2517.2	458.4	2231.9	2690.3
1.5	111.4	0.001053	1.159	467.0	2519.5	467.1	2226.2	2693.4
1.6	113.3	0.001055	1.091	475.2	2521.7	475.4	2220.9	2696.2
1.7	115.2	0.001056	1.031	483.0	2523.7	483.2	2215.7	2699.0
1.8	116.9	0.001058	0.977	490.5	2525.6	490.7	2210.8	2701.5
1.9	118.6	0.001059	0.929	497.6	2527.5	497.8	2206.1	2704.0
2.0	120.2	0.001061	0.885	504.5	2529.2	504.7	2201.6	2706.3
2.2	123.3	0.001064	0.810	517.4	2532.4	517.6	2193.0	2710.6
2.4	126.1	0.001066	0.746	529.4	2535.4	529.6	2184.9	2714.5
2.6	128.7	0.001069	0.693	540.6	2538.1	540.9	2177.3	2718.2
2.8	131.2	0.001071	0.646	551.1	2540.6	551.4	2170.1	2721.5
3.0	133.5	0.001074	0.606	561.1	2543.0	561.4	2163.2	2724.7
3.2	135.8	0.001076	0.570	570.6	2545.2	570.9	2156.7	2727.6
3.4	137.9	0.001078	0.538	579.6	2547.2	579.9	2150.4	2730.3
3.6	139.9	0.001080	0.510	588.1	2549.2	588.5	2144.4	2732.9
3.8	141.8	0.001082	0.485	596.4	2551.0	596.8	2138.6	2735.3
4.0	143.6	0.001084	0.462	604.2	2552.7	604.7	2133.0	2737.6
4.2	145.4	0.001086	0.442	611.8	2554.4	612.3	2127.5	2739.8
4.4	147.1	0.001088	0.423	619.1	2555.9	619.6	2122.3	2741.9
4.6	148.7	0.001089	0.405	626.2	2557.4	626.7	2117.2	2743.9
4.8	150.3	0.001091	0.389	633.0	2558.8	633.5	2112.2	2745.7
5.0	151.8	0.001093	0.375	639.6	2560.2	640.1	2107.4	2747.5
5.5	155.5	0.001097	0.342	655.2	2563.3	655.8	2095.9	2751.7
6.0	158.8	0.001101	0.315	669.8	2566.2	670.4	2085.0	2755.5
6.5	162.0	0.001105	0.292	683.4	2568.7	684.1	2074.7	2758.9
7.0	165.0	0.001108	0.273	696.3	2571.1	697.1	2064.9	2762.0

7.5	167.8	0.001112	0.2554	708.5	2573.3	709.3	2055.5	2764.8
8.0	170.4	0.001115	0.2403	720.0	2575.5	720.9	2046.5	2767.5
8.5	172.9	0.001118	0.2268	731.1	2577.1	732.0	2037.9	2769.9
9.0	175.4	0.001121	0.2148	741.6	2578.8	742.6	2029.5	2772.1
9.5	177.7	0.001124	0.2040	751.8	2580.4	752.8	2021.4	2774.2
10.0	179.9	0.001127	0.1943	761.5	2581.9	762.6	2013.6	2776.2
10.5	182.0	0.001130	0.1855	770.8	2583.3	772.0	2005.9	2778.0
11.0	184.1	0.001133	0.1774	779.9	2584.5	781.1	1998.5	2779.7
11.5	186.0	0.001136	0.1700	788.6	2585.8	789.9	1991.3	2781.3
12.0	188.0	0.001139	0.1632	797.1	2586.9	798.4	1984.3	2782.7
12.5	189.8	0.001141	0.1569	805.3	2588.0	806.7	1977.4	2784.1
13.0	191.6	0.001144	0.1511	813.2	2589.0	814.7	1970.7	2785.4
14	195.0	0.001149	0.1407	828.5	2590.8	830.1	1957.7	2787.8
15	198.3	0.001154	0.1317	842.9	2592.4	844.7	1945.2	2789.9
16	201.4	0.001159	0.1237	856.7	2593.8	858.6	1933.2	2791.7
17	204.3	0.001163	0.1166	869.9	2595.1	871.8	1921.5	2793.4
18	207.1	0.001168	0.1103	882.5	2596.3	884.6	1910.3	2794.8
19	209.8	0.001172	0.1047	894.6	2597.3	896.8	1899.3	2796.1
20	212.4	0.001177	0.0995	906.2	2598.2	908.6	1888.6	2797.2
21	214.9	0.001181	0.0949	917.5	2598.9	920.0	1878.2	2798.2
22	217.2	0.001185	0.0907	928.3	2599.6	931.0	1868.1	2799.1
23	219.6	0.001189	0.0868	938.9	2600.2	941.6	1858.2	2799.8
24	221.8	0.001193	0.0832	949.1	2600.7	951.9	1848.5	2800.4
25	223.9	0.001197	0.0799	959.0	2601.2	962.0	1839.0	2800.9
26	226.0	0.001201	0.0769	968.6	2601.5	971.7	1829.6	2801.4
27	228.1	0.001205	0.0740	978.0	2601.8	981.2	1820.5	2801.7
28	230.0	0.001209	0.0714	987.1	2602.1	990.5	1811.5	2802.0
29	232.0	0.001213	0.0689	996.0	2602.3	999.5	1802.6	2802.2
30	233.8	0.001216	0.0666	1004.7	2602.4	1008.4	1793.9	2802.3
32	237.4	0.001224	0.0624	1021.5	2602.5	1025.4	1776.9	2802.3
34	240.9	0.001231	0.0587	1037.6	2602.5	1041.8	1760.3	2802.1
36	244.2	0.001238	0.0554	1053.1	2602.2	1057.6	1744.2	2801.7
38	247.3	0.001245	0.0524	1068.0	2601.9	1072.7	1728.4	2801.1

(*continued*)

Table B.6 (Continued)

P(bar)	T(°C)	$\hat{V}$(m³/kg) Water	$\hat{V}$(m³/kg) Steam	$\hat{U}$(kJ/kg) Water	$\hat{U}$(kJ/kg) Steam	$\hat{H}$(kJ/kg) Water	$\hat{H}$(kJ/kg) Evaporation	$\hat{H}$(kJ/kg) Steam
40	250.3	0.001252	0.0497	1082.4	2601.3	1087.4	1712.9	2800.3
42	253.2	0.001259	0.0473	1096.3	2600.7	1101.6	1697.8	2799.4
44	256.0	0.001266	0.0451	1109.8	2599.9	1115.4	1682.9	2798.3
46	258.8	0.001272	0.0430	1122.9	2599.1	1128.8	1668.3	2797.1
48	261.4	0.001279	0.0412	1135.6	2598.1	1141.8	1653.9	2795.7
50	263.9	0.001286	0.0394	1148.0	2597.0	1154.5	1639.7	2794.2
52	266.4	0.001292	0.0378	1160.1	2595.9	1166.8	1625.7	2792.6
54	268.8	0.001299	0.0363	1171.9	2594.6	1178.9	1611.9	2790.8
56	271.1	0.001306	0.0349	1183.5	2593.3	1190.8	1598.2	2789.0
58	273.3	0.001312	0.0337	1194.7	2591.9	1202.3	1584.7	2787.0
60	275.6	0.001319	0.0324	1205.8	2590.4	1213.7	1571.3	2785.0
62	277.7	0.001325	0.0313	1216.6	2588.8	1224.8	1558.0	2782.9
64	279.8	0.001332	0.0302	1227.2	2587.2	1235.7	1544.9	2780.6
66	281.8	0.001338	0.0292	1237.6	2585.5	1246.5	1531.9	2778.3
68	283.8	0.001345	0.0283	1247.9	2583.7	1257.0	1518.9	2775.9
70	285.8	0.001351	0.0274	1258.0	2581.8	1267.4	1506.0	2773.5
72	287.7	0.001358	0.0265	1267.9	2579.9	1277.6	1493.3	2770.9
74	289.6	0.001364	0.0257	1277.6	2578.0	1287.7	1480.5	2768.3
76	291.4	0.001371	0.0249	1287.2	2575.9	1297.6	1467.9	2765.5
78	293.2	0.001378	0.0242	1296.7	2573.8	1307.4	1455.3	2762.8
80	295.0	0.001384	0.0235	1306.0	2571.7	1317.1	1442.8	2759.9
82	296.7	0.001391	0.0229	1315.2	2569.5	1326.6	1430.3	2757.0
84	298.4	0.001398	0.0222	1324.3	2567.2	1336.1	1417.9	2754.0
86	300.1	0.001404	0.0216	1333.3	2564.9	1345.4	1405.5	2750.9
88	301.7	0.001411	0.0210	1342.2	2562.6	1354.6	1393.2	2747.8
90	303.3	0.001418	0.02050	1351.0	2560.1	1363.7	1380.9	2744.6
92	304.9	0.001425	0.01996	1359.7	2557.7	1372.8	1368.6	2741.4
94	306.4	0.001432	0.01945	1368.2	2555.2	1381.7	1356.3	2738.0

96	308.0	0.001439	0.01897	1376.7	2552.6	1390.6	1344.1	2734.7
98	309.5	0.001446	0.01849	1385.2	2550.0	1399.3	1331.9	2731.2
100	311.0	0.001453	0.01804	1393.5	2547.3	1408.0	1319.7	2727.7
105	314.6	0.001470	0.01698	1414.1	2540.4	1429.5	1289.2	2718.7
110	318.0	0.001489	0.01601	1434.2	2533.2	1450.6	1258.7	2709.3
115	321.4	0.001507	0.01511	1454.0	2525.7	1471.3	1228.2	2699.5
120	324.6	0.001527	0.01428	1473.4	2517.8	1491.8	1197.4	2689.2
125	327.8	0.001547	0.01351	1492.7	2509.4	1512.0	1166.4	2678.4
130	330.8	0.001567	0.01280	1511.6	2500.6	1532.0	1135.0	2667.0
135	333.8	0.001588	0.01213	1530.4	2491.3	1551.9	1103.1	2655.0
140	336.6	0.001611	0.01150	1549.1	2481.4	1571.6	1070.7	2642.4
145	339.4	0.001634	0.01090	1567.5	2471.0	1591.3	1037.7	2629.1
150	342.1	0.001658	0.01034	1586.1	2459.9	1611.0	1004.0	2615.0
155	344.8	0.001683	0.00981	1604.6	2448.2	1630.7	969.6	2600.3
160	347.3	0.001710	0.00931	1623.2	2436.0	1650.5	934.3	2584.9
165	349.8	0.001739	0.00883	1641.8	2423.1	1670.5	898.3	2568.8
170	352.3	0.001770	0.00837	1661.6	2409.3	1691.7	859.9	2551.6
175	354.6	0.001803	0.00793	1681.8	2394.6	1713.3	820.0	2533.3
180	357.0	0.001840	0.00750	1701.7	2378.9	1734.8	779.1	2513.9
185	359.2	0.001881	0.00708	1721.7	2362.1	1756.5	736.6	2493.1
190	361.4	0.001926	0.00668	1742.1	2343.8	1778.7	692.0	2470.6
195	363.6	0.001977	0.00628	1763.2	2323.6	1801.8	644.2	2446.0
200	365.7	0.00204	0.00588	1785.7	2300.8	1826.5	591.9	2418.4
205	367.8	0.00211	0.00546	1810.7	2274.4	1853.9	532.5	2386.4
210	369.8	0.00220	0.00502	1840.0	2242.1	1886.3	461.3	2347.6
215	371.8	0.00234	0.00451	1878.6	2198.1	1928.9	366.2	2295.2
220	373.7	0.00267	0.00373	1952	2114	2011	185	2196
221.2 (Critical point)	374.15	0.00317	0.00317	2038	2038	2108	0	2108

Table B.7 Properties of Superheated Steam[a]

P(bar) (T_sat.°C)		Sat'd Water	Sat'd Steam	Temperature (°C)→ 50	75	100	150	200	250	300	350
0.0	$\hat{H}$	—	—	2595	2642	2689	2784	2880	2978	3077	3177
(—)	$\hat{U}$	—	—	2446	2481	2517	2589	2662	2736	2812	2890
	$\hat{V}$	—	—	—	—	—	—	—	—	—	—
0.1	$\hat{H}$	191.8	2584.8	2593	2640	2688	2783	2880	2977	3077	3177
(45.8)	$\hat{U}$	191.8	2438.0	2444	2480	2516	2588	2661	2736	2812	2890
	$\hat{V}$	0.00101	14.7	14.8	16.0	17.2	19.5	21.8	24.2	26.5	28.7
0.5	$\hat{H}$	340.6	2646.0	209.3	313.9	2683	2780	2878	2979	3076	3177
(81.3)	$\hat{U}$	340.6	2484.0	209.2	313.9	2512	2586	2660	2735	2811	2889
	$\hat{V}$	0.00103	3.24	0.00101	0.00103	3.41	3.89	4.35	4.83	5.29	5.75
1.0	$\hat{H}$	417.5	2675.4	209.3	314.0	2676	2776	2875	2975	3074	3176
(99.6)	$\hat{U}$	417.5	2506.1	209.2	313.9	2507	2583	2658	2734	2811	2889
	$\hat{V}$	0.00104	1.69	0.00101	0.00103	1.69	1.94	2.17	2.40	2.64	2.87
5.0	$\hat{H}$	640.1	2747.5	209.7	314.3	419.4	632.2	2855	2961	3065	3168
(151.8)	$\hat{U}$	639.6	2560.2	209.2	313.8	418.8	631.6	2643	2724	2803	2883
	$\hat{V}$	0.00109	0.375	0.00101	0.00103	0.00104	0.00109	0.425	0.474	0.522	0.571
10	$\hat{H}$	762.6	2776.2	210.1	314.7	419.7	632.5	2827	2943	3052	3159
(179.9)	$\hat{U}$	761.5	2582	209.1	313.7	418.7	631.4	2621	2710	2794	2876
	$\hat{V}$	0.00113	0.194	0.00101	0.00103	0.00104	0.00109	0.206	0.233	0.258	0.282
20	$\hat{H}$	908.6	2797.2	211.0	315.5	420.5	633.1	852.6	2902	3025	3139
(212.4)	$\hat{U}$	906.2	2598.2	209.0	313.5	418.4	603.9	850.2	2679	2774	2862
	$\hat{V}$	0.00118	0.09950	0.00101	0.00102	0.00104	0.00109	0.00116	0.111	0.125	0.139
40	$\hat{H}$	1087.4	2800.3	212.7	317.1	422.0	634.3	853.4	1085.8	2962	3095
(250.3)	$\hat{U}$	1082.4	2601.3	208.6	313.0	417.8	630.0	848.8	1080.8	2727	2829
	$\hat{V}$	0.00125	0.04975	0.00101	0.00102	0.00104	0.00109	0.00115	0.00125	0.0588	0.0665
60	$\hat{H}$	1213.7	2785.0	214.4	318.7	423.5	635.6	854.2	1085.8	2885	3046
(275.6)	$\hat{U}$	1205.8	2590.4	208.3	312.6	417.3	629.1	847.3	1078.3	2668	2792
	$\hat{V}$	0.00132	0.0325	0.00101	0.00103	0.00104	0.00109	0.00115	0.00125	0.0361	0.0422
80	$\hat{H}$	1317.1	2759.9	216.1	320.3	425.0	636.8	855.1	1085.8	2787	2990
(295.0)	$\hat{U}$	1306.0	2571.7	208.1	312.3	416.7	628.2	845.9	1075.8	2593	2750
	$\hat{V}$	0.00139	0.0235	0.00101	0.00102	0.00104	0.00109	0.00115	0.00124	0.0243	0.0299
100	$\hat{H}$	1408.0	2727.7	217.8	322.9	426.5	638.1	855.9	1085.8	1343.4	2926
(311.0)	$\hat{U}$	1393.5	2547.3	207.8	311.7	416.1	627.3	844.4	1073.4	1329.4	2702
	$\hat{V}$	0.00145	0.0181	0.00101	0.00102	0.00104	0.00109	0.00115	0.00124	0.00140	0.0224
150	$\hat{H}$	1611.0	2615.0	222.1	326.0	430.3	641.3	858.1	1086.2	1338.2	2695
(342.1)	$\hat{U}$	1586.1	2459.9	207.0	310.7	414.7	625.0	841.0	1067.7	1317.6	2523
	$\hat{V}$	0.00166	0.0103	0.00101	0.00102	0.00104	0.00108	0.00114	0.00123	0.00138	0.0115
200	$\hat{H}$	1826.5	2418.4	226.4	330.0	434.0	644.5	860.4	1086.7	1334.3	1647.1
(365.7)	$\hat{U}$	1785.7	2300.8	206.3	309.7	413.2	622.9	837.7	1062.2	1307.1	1613.7
	$\hat{V}$	0.00204	0.005875	0.00100	0.00102	0.00103	0.00108	0.00114	0.00122	0.00136	0.00167
221.2(P_c)	$\hat{H}$	2108	2108	228.2	331.7	435.7	645.8	861.4	1087.0	1332.8	1635.5
(374.15)(T_c)	$\hat{U}$	2037.8	2037.8	206.0	309.2	412.8	622.0	836.3	1060.0	1302.9	1600.3
	$\hat{V}$	0.00317	0.00317	0.00100	0.00102	0.00103	0.00108	0.00114	0.00122	0.00135	0.00163
250	$\hat{H}$	—	—	230.7	334.0	437.8	647.7	862.8	1087.5	1331.1	1625.0
(—)	$\hat{U}$	—	—	205.7	308.7	412.1	620.8	834.4	1057.0	1297.5	1585.0
	$\hat{V}$	—	—	0.00100	0.00101	0.00103	0.00108	0.00113	0.00122	0.00135	0.00160
300	$\hat{H}$	—	—	235.0	338.1	441.6	650.9	865.2	1088.4	1328.7	1609.9
(—)	$\hat{U}$	—	—	205.0	307.7	410.8	618.7	831.3	1052.1	1288.7	1563.3
	$\hat{V}$	—	—	0.0009990	0.00101	0.00103	0.00107	0.00113	0.00121	0.00133	0.00155
500	$\hat{H}$	—	—	251.9	354.2	456.8	664.1	875.4	1093.6	1323.7	1576.3
(—)	$\hat{U}$	—	—	202.4	304.0	405.8	611.0	819.7	1034.3	1259.3	1504.1
	$\hat{V}$	—	—	0.0009911	0.00100	0.00102	0.00106	0.00111	0.00119	0.00129	0.00144
1000	$\hat{H}$	—	—	293.9	394.3	495.1	698.0	903.5	1113.0	1328.7	1550.5
(—)	$\hat{U}$	—	—	196.5	295.7	395.1	594.4	795.3	999.0	1207.1	1419.0
	$\hat{V}$	—	—	0.0009737	0.0009852	0.001000	0.00104	0.00108	0.00114	0.00122	0.00131

[a]Adapted from R. W. Haywood, *Thermodynamic Tables in SI (Metric) Units,* Cambridge University Press, London, 1968. Water is a liquid in the enclosed region between 50°C and 350°C. $\hat{H}$ = specific enthalpy (kJ/kg), $\hat{U}$ = specific internal energy (kJ/kg), $\hat{V}$ = specific volume (m³/kg). *Note:* kJ/kg × 0.4303 = Btu/lb_m.

(*continued*)

Table B.7 (Continued)

P(bar) (T_{sat}.°C)		Temperature (°C)→ 400	450	500	550	600	650	700	750
0.0 (—)	$\hat{H}$ $\hat{U}$ $\hat{V}$	3280 2969 —	3384 3050 —	3497 3132 —	3597 3217 —	3706 3303 —	3816 3390 —	3929 3480 —	4043 3591 —
0.1 (45.8)	$\hat{H}$ $\hat{U}$ $\hat{V}$	3280 2969 21.1	3384 3050 33.3	3489 3132 35.7	3596 3217 38.0	3706 3303 40.3	3816 3390 42.6	3929 3480 44.8	4043 3571 47.2
0.5 (81.3)	$\hat{H}$ $\hat{U}$ $\hat{V}$	3279 2969 6.21	3383 3049 6.67	3489 3132 7.14	3596 3216 7.58	3705 3302 8.06	3816 3390 8.55	3929 3480 9.01	4043 3571 9.43
1.0 (99.6)	$\hat{H}$ $\hat{U}$ $\hat{V}$	3278 2968 3.11	3382 3049 3.33	3488 3132 3.57	3596 3216 3.80	3705 3302 4.03	3816 3390 4.26	3928 3479 4.48	4042 3570 4.72
5.0 (151.8)	$\hat{H}$ $\hat{U}$ $\hat{V}$	3272 2964 0.617	3379 3045 0.664	3484 3128 0.711	3592 3213 0.758	3702 3300 0.804	3813 3388 0.850	3926 3477 0.897	4040 3569 0.943
10 (179.9)	$\hat{H}$ $\hat{U}$ $\hat{V}$	3264 2958 0.307	3371 3041 0.330	3478 3124 0.353	3587 3210 0.377	3697 3296 0.402	3809 3385 0.424	3923 3475 0.448	4038 3567 0.472
20 (212.4)	$\hat{H}$ $\hat{U}$ $\hat{V}$	3249 2946 0.151	3358 3031 0.163	3467 3115 0.175	3578 3202 0.188	3689 3290 0.200	3802 3379 .0211	3916 3470 0.223	4032 3562 0.235
40 (250.3)	$\hat{H}$ $\hat{U}$ $\hat{V}$	3216 2922 0.0734	3331 3011 0.0799	3445 3100 0.0864	3559 3188 0.0926	3673 3278 0.0987	3788 3368 0.105	3904 3460 0.111	4021 3554 0.117
60 (275.6)	$\hat{H}$ $\hat{U}$ $\hat{V}$	3180 2896 0.0474	3303 2991 0.0521	3422 3083 0.0566	3539 3174 0.0609	3657 3265 0.0652	3774 3357 0.0693	3892 3451 0.0735	4011 3545 0.0776
80 (295.0)	$\hat{H}$ $\hat{U}$ $\hat{V}$	3142 2867 0.0344	3274 2969 0.0382	3399 3065 0.0417	3520 3159 0.0450	3640 3252 0.0483	3759 3346 0.0515	3879 3441 0.0547	4000 3537 0.0578
100 (311.0)	$\hat{H}$ $\hat{U}$ $\hat{V}$	3100 2836 0.0264	3244 2946 0.0298	3375 3047 0.0328	3500 3144 0.0356	3623 3240 0.0383	3745 3335 0.0410	3867 3431 0.0435	3989 3528 0.0461
150 (342.1)	$\hat{H}$ $\hat{U}$ $\hat{V}$	2975 2744 0.0157	3160 2883 0.0185	3311 2999 0.0208	3448 3105 0.0229	3580 3207 0.0249	3708 3307 0.0267	3835 3407 0.0286	3962 3507 0.0304
200 (365.7)	$\hat{H}$ $\hat{U}$ $\hat{V}$	2820 2622 0.009950	3064 2810 0.0127	3241 2946 0.0148	3394 3063 0.0166	3536 3172 0.0182	3671 3278 0.197	3804 3382 0.211	3935 3485 0.0225
221.2(P_c) (374.15)(T_c)	$\hat{H}$ $\hat{U}$ $\hat{V}$	2733 2553 0.008157	3020 2776 0.0110	3210 2922 0.0130	3370 3045 0.0147	3516 3157 0.0162	3655 3265 0.0176	3790 3371 0.0190	3923 3476 0.0202
250 (—)	$\hat{H}$ $\hat{U}$ $\hat{V}$	2582 2432 0.006013	2954 2725 0.009174	3166 2888 0.0111	3337 3019 0.0127	3490 3137 0.0141	3633 3248 0.0143	3772 3356 0.0166	3908 3463 0.0178
300 (—)	$\hat{H}$ $\hat{U}$ $\hat{V}$	2162 2077 0.002830	2826 2623 0.006734	3085 2825 0.008680	3277 2972 0.0102	3443 3100 0.0114	3595 3218 0.0126	3740 3330 0.0136	3880 3441 0.0147
500 (—)	$\hat{H}$ $\hat{U}$ $\hat{V}$	1878 1791 0.001726	2293 2169 0.002491	2723 2529 0.003882	3021 2765 0.005112	3248 2946 0.006112	3439 3091 0.007000	3610 3224 0.007722	3771 3350 0.008418
1000 (—)	$\hat{H}$ $\hat{U}$ $\hat{V}$	1798 1653 0.001446	2051 1888 0.001628	2316 2127 0.001893	2594 2369 0.002246	2857 2591 0.002668	3105 2795 0.003106	3324 2971 0.003536	3526 3131 0.003953

Table B.8 Specific Enthalpies of Selected Gases: SI Units

$\hat{H}$(kJ/mol)
Reference state: Gas, P_{ref} = 1 atm, T_{ref} = 25°C

T	Air	O_2	N_2	H_2	CO	CO_2	H_2O
0	−0.72	−0.73	−0.73	−0.72	−0.73	−0.92	−0.84
25	0.00	0.00	0.00	0.00	0.00	0.00	0.00
100	2.19	2.24	2.19	2.16	2.19	2.90	2.54
200	5.15	5.31	5.13	5.06	5.16	7.08	6.01
300	8.17	8.47	8.12	7.96	8.17	11.58	9.57
400	11.24	11.72	11.15	10.89	11.25	16.35	13.23
500	14.37	15.03	14.24	13.83	14.38	21.34	17.01
600	17.55	18.41	17.39	16.81	17.57	26.53	20.91
700	20.80	21.86	20.59	19.81	20.82	31.88	24.92
800	24.10	25.35	23.86	22.85	24.13	37.36	29.05
900	27.46	28.89	27.19	25.93	27.49	42.94	33.32
1000	30.86	32.47	30.56	29.04	30.91	48.60	37.69
1100	34.31	36.07	33.99	32.19	34.37	54.33	42.18
1200	37.81	39.70	37.46	35.39	37.87	60.14	46.78
1300	41.34	43.38	40.97	38.62	41.40	65.98	51.47
1400	44.89	47.07	44.51	41.90	44.95	71.89	56.25
1500	48.45	50.77	48.06	45.22	48.51	77.84	61.09

Table B.9 Specific Enthalpies of Selected Gases:
American Engineering Units

$\hat{H}$(Btu/lb-mole)
Reference state: Gas, P_{ref} = 1 atm, T_{ref} = 77°F

T	Air	O_2	N_2	H_2	CO	CO_2	H_2O
32	−312	−315	−312	−310	−312	−394	−361
77	0	0	0	0	0	0	0
100	160	162	160	159	160	206	185
200	858	875	857	848	859	1132	996
300	1563	1602	1558	1539	1564	2108	1818
400	2275	2342	2265	2231	2276	3129	2652
500	2993	3094	2976	2925	2994	4192	3499
600	3719	3858	3694	3621	3720	5293	4359
700	4451	4633	4418	4319	4454	6429	5233
800	5192	5418	5150	5021	5195	7599	6122
900	5940	6212	5889	5725	5945	8790	7025
1000	6695	7015	6635	6433	6702	10015	7944
1100	7459	7826	7399	7145	7467	11263	8880
1200	8230	8645	8151	7861	8239	12533	9831
1300	9010	9471	8922	8581	9021	13820	10799
1400	9797	10304	9699	9306	9809	15122	11783
1500	10590	11142	10485	10035	10606	16436	12783
1600	11392	11988	11278	10769	11409	17773	13798
1700	12200	12836	12080	11509	12220	19119	14831
1800	13016	13691	12888	12254	13036	20469	15877
1900	13837	14551	13702	13003	13858	21840	16941
2000	14663	15415	14524	13759	14688	23211	18019

Table B.10 Atomic Heat Capacities
for Kopp's Rule[a]

Element	C_{pa}[J/(g-atom·°C)]	
	Solids	Liquids
C	7.5	12
H	9.6	18
B	11	20
Si	16	24
O	17	25
F	21	29
P	23	31
S	26	31
All Others	26	33

[a]D. M. Himmelblau, *Basic Principles and
Calculations in Chemical Engineering,* 3rd
Edition, Prentice-Hall, Englewood Cliffs,
NJ, 1974, p. 270.

Table B.11 Integral Heats of Solution and Mixing at 25°C

r(mol H_2O/mol solute)	$(\Delta\hat{H}_s)_{HCl(g)}$ kJ/mol HCl	$(\Delta\hat{H}_s)_{NaOH(s)}$ kJ/mol NaOH	$(\Delta\hat{H}_m)_{H_2SO_4}$ kJ/mol H_2SO_4
0.5	—	—	−15.73
1	−26.22	—	−28.07
1.5	—	—	−36.90
2	−48.82	—	−41.92
3	−56.85	−28.87	−48.99
4	−61.20	−34.43	−54.06
5	−64.05	−37.74	−58.03
10	−69.49	−42.51	−67.03
20	−71.78	−42.84	—
25	—	—	−72.30
30	−72.59	−42.72	—
40	−73.00	−42.59	—
50	−73.26	−42.51	−73.34
100	−73.85	−42.34	−73.97
200	−74.20	−42.26	—
500	−74.52	−42.38	−76.73
1 000	−74.68	−42.47	−78.57
2 000	−74.82	−42.55	—
5 000	−74.93	−42.68	−84.43
10 000	−74.99	−42.72	−87.07
50 000	−75.08	−42.80	—
100 000	−75.10	—	−93.64
500 000	—	—	−95.31
∞	−75.14	−42.89	−96.19

[a]From J. C. Whitwell and R. K. Toner, *Conservation of Mass and Energy,* pp. 344–346.
Copyright © 1969 by McGraw-Hill, Inc. Used with permission of McGraw-Hill.

ANSWERS TO TEST YOURSELVES

p. 9

1. A ratio of equivalent values of a quantity expressed in different units.

2. $(60 \, \text{s})/(1 \, \text{min})$

3. $(1 \, \text{min}^2)/(3600 \, \text{s}^2)$

4. $(1 \, \text{m}^3)/(10^6 \, \text{cm}^3)$

p. 11

1. **(a)** $(1000 \, \text{mm})/(1 \, \text{m})$; **(b)** $(10^{-9} \, \text{s})/(1 \, \text{ns})$;
   (c) $(1 \, \text{m}^2)/(10^4 \, \text{cm}^2)$; **(d)** $(1 \, \text{m}^3)/(35.3145 \, \text{ft}^3)$;
   (e) $(9.486 \times 10^{-4} \, \text{Btu/s})/(1.341 \times 10^{-3} \, \text{hp})$

2. m/s; cm/s; ft/s

p. 13

1. 2 N; $(2/32.174) \, \text{lb}_f$

2. No

3. 1 kg; same; less

4. $2 \, \text{lb}_m$; same; less

p. 15

1. **(a)** 1.22×10^4 (3 s.f.); **(b)** 1.22000×10^4 (6 s.f.);
   (c) 3.040×10^{-3} (4 s.f.)

2. **(a)** 134,000 (3 s.f.); **(b)** 0.01340 (4 s.f.); **(c)** 4200 (3 s.f.)

3. **(a)** 3 s.f.; **(b)** 2 s.f.; **(c)** 3 s.f., 11.2; **(d)** 2 s.f., 12

4. **(a)** 1460; **(b)** 13.4; **(c)** 1.76×10^{-7}

5. **(a)** 4.25–4.35; **(b)** 4.295–4.305;
   (c) 2.7775×10^{-3}–2.7785×10^{-3};
   (d) 2450–2550; **(e)** 2499.5–2500.5

p. 18

1. Breakdowns, routine or unplanned shutdowns, or installing new equipment in the second week. (Many other possibilities exist.)

2. 35.5 or 35 batches/week.

3. 40 batches/week. The second week was clearly abnormal and shouldn't influence the prediction.

p. 19

1. $\overline{V} = 237.4 \, \text{cm}^3/\text{s}$, range $= 21 \, \text{cm}^3/\text{s}$, $s_V^2 = 66.3 \, \text{cm}^6/\text{s}^2$,
   $s_V = 8.1 \, \text{cm}^3/\text{s}$

2. $\dot{V} = 237.4 \, \text{cm}^3/\text{s} \pm 16.2 \, \text{cm}^3/\text{s}$

p. 22

1. All additive terms have the same dimensions. No. Yes.

2. $\text{m}^{-2} \cdot \text{s}^{-2}$

p. 23

3. A multiplicative combination of factors with no net units; st^2/r or r/st^2.

4. $a(\text{lb}_f)$; Q is dimensionless.

p. 23

1. Substitute in Equation 2.7-1.

2. Correct; too high; too low; too low.

p. 27

1. $y = a(x^2 - 2)$

2. **(b)** Plot $1/y$ versus $(x - 3)^2$: $a = \text{slope}$, $b = \text{intercept}$.
   (c) Plot y^3 versus x^2; $a = \text{slope}$, $b = -\text{intercept}$.
   (d) Plot $\sqrt{x}/\sin y$ versus x: $a = \text{slope}$, $b = \text{intercept}$.
   (e) Plot $\ln y$ versus x: $b = \ln(y_2/y_1)/(x_2 - x_1)$,
   $\ln a = \ln y_1 - bx_1$, $a = e^{\ln a}$.
   (f) Plot $\ln y$ versus $\ln x$: $b = \ln(y_2/y_1)/\ln(x_2/x_1)$, $\ln a = \ln y_1 - b \ln x_1$, $a = e^{\ln a}$.

p. 29

1. **(a)** $P = at + b$; **(b)** $P = ae^{bt}$;
   (c) $P = at^b$; **(d)** $y^2 = 3 + a \exp(b/x^2)$;
   (e) $1/F = a(t^2 - 4)^b$.

2. **(a)** P versus t on a semilog paper;
   (b) P versus t on log paper;
   (c) P^2 versus t^3 on semilog paper;
   (d) $1/P$ (or P) versus $(t - 4)$ on log paper.

p. 44

1. Dimensionless.

2. $0.50 \, \text{g/cm}^3$; $2.0 \, \text{cm}^3/\text{g}$; $31 \, \text{lb}_m/\text{ft}^3$; 1.5 g; $36 \, \text{cm}^3$

3. Yes.

4. No—possibly different reference densities used for each.

5. $\rho_{H_2O(s)} < \rho_{H_2O(l)}$; $\rho_{NBA(s)} > \rho_{NBA(l)}$

6. When T rises, the mercury in the thermometer expands. At higher temperatures the same mass thus occupies a greater volume, meaning that the density of mercury ($= m/V$) *decreases*.

p. 46

1. $10.0 \, \text{cm}^3/\text{s}$

2. 159.5 g/min

3. Same; same; greater at outlet.

p. 47

1. 100 mL/min; 100 g/min

2. Flowmeters—see Figure 3.2-1.

3. Too low (the gas is much less dense, so it must flow at a much higher rate than the liquid to raise the float to the same position).

p. 49

1. (a) 6.02×10^{23} molecules;
   (b) M grams

2. The molecular weight of the species expressed in tons.

3. (a) 1 lb-mole, 2 lb_m
   (b) 2 lb-moles, 2 lb_m

4. 2000

5. 50 kmol/h

p. 52-a

1. (a) $80/81$; **(b)** 0.5

2. 0.25 lb_m A/lb_m ; 0.75 lb_m B/lb_m ; 0.333 mole A/mole; 0.667 mole B/mole; 100 lb_m A/min; 100 lb-moles B/min; 400 lb_m/min; 150 lb-moles/min

p. 52-b

1. n/V (mol/L)

2. nM/V (g/L)

3. $(20/C_A)$ (L)

4. $(120 \, c_A)$ g/h

p. 54

1. $x = 125 \times 10^{-9}$ kg C_6H_5OH/kg liquid (or g/g or lb_m/lb_m)

2. 0.125 mg C_6H_5OH

3. 125×10^{-6} g C_6H_5OH/L. (Solution density $\approx$ 1 kg/L.)

p. 56-a

1. See Figures 3.4-1 and 3.4-2 and Equation 3.4-2.

2. The fluid pressure is higher at the bottom than at the top (hydrostatic head effect). No. Maybe. Yes. (The answers depend on how large the tank is.)

3. No. Convert 1300 mm Hg to (say) dynes/cm^2, then multiply by 4 cm^2 to calculate F(dynes).

4. 79 mm Hg

p. 56-b

1. No

2. Pressure relative to a vacuum; pressure relative to atmospheric pressure.

3. 735 mm Hg (absolute); 20 mm Hg of vacuum.

4. 4 in. Hg; 33.9 in. Hg

p. 59

1. See Figure 3.4-3; 0–7000 atm; gauge.

2. See Figure 3.4-4.

3. (a) True; **(b)** true; **(c)** false

4. −14 mm Hg

p. 62

1. Immerse in ice–water mixture, mark mercury level as 0°. Immerse in boiling water, mark level as 100°. Divide 0 to 100 interval into 100 equal subintervals, label appropriately.

2. 1°C

3. 1°C

p. 84

1. Semibatch, transient

2. Batch, transient

3. Semibatch, transient

4. (a) Continuous, transient;
   (b) continuous, steady state

p. 89

3. Steady-state, either no reactions or no net mole change on reaction (e.g., $A \rightarrow B$ but not $A \rightarrow 2B$ or $A + B \rightarrow C$).

4. Steady-state, A is nonreactive.

5. Steady-state, no density change from inlet to outlet. (Good approximation for liquids and solids, effectively requires no reactions and constant temperature and pressure for gases.)

p. 93

2. $\dot{m}_T = 250(1 - x)/60$

3. $n = (75)(1.595)/(154)$

4. $\dot{m} = 50 + \dot{m}_{dg}$; $\dot{m}_{co} = 0.25\dot{m}_{dg}$; $y = 0.75\dot{m}_{dg}/(50 + \dot{m}_{dg})$

p. 95

1. $(atoms)_{in} = (atoms)_{out}$ for each atomic species; multiply all stream amounts by a constant factor; an assumed amount of an input or output stream.

2. (a) Flow rates are 1000, 20,000, and 21,000 (all kmol/h), mole fractions are unchanged.
   (b) Flow rates are 200, 100, and 100 (all lb_m/min), mass fractions are unchanged.

p. 98

1. H_2: $(5 \, lb_m)_{in} = (1 \, lb_m + 4 \, lb_m)_{out}$; O_2: $(5 \, lb_m)_{in} = (4 \, lb_m + 1 \, lb_m)_{out}$; total mass: $(10 \, lb_m)_{in} = (10 \, lb_m)_{out}$.

2. Balance on B, solve for $\dot{m}_1$; balance on C, solve for x; total mass balance, solve for $\dot{m}_2$.

p. 117

1. Yes

2. 4

3. (4 moles H_2O produced)$/$(6 moles O_2 consumed)

4. $(400)(6)/(4) = 600$

5. 200 mol/min

p. 119

1. C_2H_4

2. 100%

3. 50 kmol O_2; 100 kmol C_2H_4O; 50 kmol

4. 50 kmol C_2H_4; 75 kmol O_2; 50 kmol C_2H_4O; 25 kmol

5. 0.80; 0.40; 40 kmol

p. 125

1. 0.90

2. 80%

3. 16 mol B/mol C

4. 80 mol, 10 mol

p. 128

1. Three independent molecular species (C_2H_4, C_4H_8, N_2). Two independent atomic species (N and either C or H).

2. **(a)** $CH_4 + 2O_2 \rightarrow CO_2 + 2H_2O$
   (b) $CH_4 + \frac{3}{2}O_2 \rightarrow CO + 2H_2O$
   (c) $C_2H_6 + \frac{7}{2}O_2 \rightarrow 2CO_2 + 3H_2O$
   (d) $C_2H_6 + \frac{5}{2}O_2 \rightarrow 2CO + 3H_2O$

   Since (b) can be obtained as (a) $- \frac{1}{2}[$(c) $-$ (d)$]$ (*verify*), the four equations are not independent.

p. 134

1. 60 mol; 0.60

2. 120 mol; 0.48

3. 40 mol $CH_4 = 100$ mol $CH_4 - \xi \Rightarrow \xi = 60$ mol
   130 mol $O_2 = 250$ mol $O_2 - 2\xi \Rightarrow \xi = 60$ mol
   60 mol $CO_2 = 0$ mol $O_2 + \xi \Rightarrow \xi = 60$ mol

4. Four molecular species balances (CH_4, O_2, CO_2, H_2O). Three atomic species balances (C, H, O).

5. **(b)** I = O. $(250)(2)$ mol O in $= [(130)(2) + (60)(2) + (120)(1)]$ mol O out
   (c) I = O + C. 250 mol O_2 in $= 130$ mol O_2 out $+ 120$ mol O_2 consumed
   (d) G = O. 120 mol H_2O generated $= 120$ mol H_2O out
   (e) I = O. $(100)(4)$ mol H in $= [(40)(4) + (120)(2)]$ mol H out

p. 135

1. Overall $= 100/110 = 0.909$ mol A consumed/mol A fed; single pass $= 100/200 = 0.500$ mol A consumed/mol A fed.

p. 138

1. Overall conversion $= 0.833$ (83.3%), single-pass conversion $= 0.25$ (25%).

2. Customers want B, not a mixture that contains mostly A. Makes no sense to pay for 200 mol of A (fresh feed) and then discard 140 mol of it.

3. C would keep building up in the system. Take off a purge stream from the recycle.

4. The cost of the reactor that would be required to achieve an 83.3% conversion in a single pass could be much greater than the cost of the separation and recycle equipment.

p. 145

1. 21% O_2, 79% N_2; $79/21 = 3.76$ moles N_2/mole O_2

2. 25% H_2, 25% O_2, 50% H_2O; 50% H_2, 50% O_2

3. 20; 0.95; $5/95 = 0.0526$

p. 146

1. 200 mol O_2/h

2. 200 mol O_2/h

3. (4.76×200) mol air/h

4. $(2 \times 4.76 \times 200)$ mol air/h

5. 50%

p. 190

1. 255 cm^3/s. If T is raised, mass flow rate remains constant but volumetric flow rate increases slightly. Look up the density of liquid water at 75°C and divide it into 200 g/s.

2. $P_h = \rho g h$, and ρ_{Hg} varies with temperature. The difference would be extremely slight.

3. $V_{tot} = \dfrac{m_1}{\rho_1} + \dfrac{m_2}{\rho_2} + \cdots + \dfrac{m_n}{\rho_n}$;
   $\dfrac{1}{\rho} = \dfrac{V_{tot}}{m_{tot}} = \dfrac{m_1}{m_{tot}\rho_1} + \dfrac{m_2}{m_{tot}\rho_2} + \cdots = \dfrac{x_1}{\rho_1} + \dfrac{x_2}{\rho_2} + \cdots$

p. 193

1. A relation between absolute pressure, specific volume, and absolute temperature of a substance. $P\hat{V} = RT$. High T, low P.

2. (c), (e). The mass and mass density of CO_2 are each greater by a factor $(MW_{CO_2}/MW_{H_2}) = 22$.

3. (a) and (c).

$$\dot{V} = \frac{\dot{n}RT}{P} = \frac{\dot{m}}{MW}\frac{RT}{P}$$

Let E denote ethylene and B denote butene. $MW_B = 2MW_E \Rightarrow \dot{V}_E = 2\dot{V}_B$.
$\hat{V} = RT/P$, which is the same for B and E.
Mass density: $\rho = \dot{m}/\dot{V} = (MW)P/RT \Rightarrow \rho_B = 2\rho_E$

4. $RT/P = [0.08206 \text{ L·atm}/(\text{mol·K})](200 \text{ K})/(20 \text{ atm}) = 0.8206$ L/mol < 5 L/mol. From Equation 5.2-3, the error is likely to be greater than 1%.

p. 196

1. See Table 5.2-1.

2. $V_{new} = V_{old}/2$; $V_{new} = 2V_{old}$

3. Decreases (n is unchanged, V increases); nothing

4. (a)

p. 197

1. (b), (d)

2. 5 bar; 50 m³; p_{H_2} increases, v_{H_2} is unchanged.

3. Greater than

p. 200

1. Vapor

2. (a) $P_a < P_b$; (b) ρ_{va} ? ρ_{vb}; (c) $\rho_{la} > \rho_{lb}$

3. Gas. Supercritical fluid.

p. 206

1. The equation to determine $\hat{V}$ for given values of T and P is a cubic equation.

2. Critical temperature and pressure (Table B.1), Pitzer acentric factor (Table 5.3-1).

3. (b), (a), (c)

p. 210

1. No; $T_r = (-190)/(T_c + 8)$, $P_r = 300/(P_c + 8)$, look up z on charts, calculate $V = znRT/P$.

2. Need a different chart for each species.

3. See p. 207.

p. 212

See Example 5.3-4. Nonpolar compounds with similar critical properties.

p. 238

1. Distillation. The naphthas come off from the top of the column, the lubricating oils from the bottom, and the heating oils from the middle.

2. Evaporation, filtration, centrifugation

3. Evaporation, reverse osmosis (membrane filtration at high pressure)

4. Condensation, absorption;   **5.** Adsorption.

p. 243

1. −5°C, 3 mm Hg

2. −56.6°C, 5.112 atm

3. All CO_2 solidifies at 1 atm, solid melts at 9.9 atm and −56°C, liquid boils at 9.9 atm and −40°C.

4. 1 atm; 9.9 atm

5. −78.5°C; −56°C, −40°C;   **6.** No

p. 246

1. Cox chart (Figure 6.1-4); Antoine equation (Equation 6.1-4)

2. Plot p^* versus $1/T_{absolute}$ on semilog paper, draw line through points, and extrapolate to $1/T_4$.

3. The plot is more likely to be linear on the Cox chart.

p. 249

1. Eq. 6.2-1

2. (a) 2; (b) 2; (c) 4; (d) 3

p. 253

1. Yes; yes.

2. 200 mm Hg; 600 mm Hg; 200/960; look up or calculate the temperature at which $p^*_{acetone} = 960$ mm Hg.

3. (a) The temperature to which a gas must be cooled before any constituent condenses; superheated, saturated.

(b) $y_{H_2O} = p^*_{H_2O}(T_0)/P_0$ (i) nothing; (ii) condense; (iii) condense; (iv) nothing

(c) $p^*_{H_2O}(T_{dp}) = y_{H_2O}P_0$. Look up the temperature at which the vapor pressure is $p^*_{H_2O}$.

(d) $T_{dp} = T_0 - T_{sh}$; $y_{H_2O} = p^*_{H_2O}(T_{dp})/P_0$.

p. 254

1. 82°C

2. 50%

3. $s_m = 0.111$, $s_p = 44.4\%$

p. 258

1. Raoult's law: $p_A = x_A p^*_A$, where x_A is the mole fraction of A in the liquid phase; most likely to be valid as $x_A \to 1.0$.

2. Henry's law: $p_A = x_A H_A$; most likely to be valid when $x_A \to 0$.

3. A solution for which either Raoult's law or Henry's law is obeyed for all species at all solution compositions.

4. Henry's law for x_{CO_2} and p_{CO_2}, Raoult's law for x_{H_2O} and p_{H_2O}. Look up Henry's law constant for CO_2 (*Perry's Chemical Engineers' Handbook,* 7th Edition, p. 2-125) and vapor pressure of H_2O (Table B.3 of this book).

p. 264

1. The temperature at which the first bubble of vapor forms if the liquid is heated at the given pressure. The temperature at which the first drop of liquid forms if the vapor mixture is cooled at the given pressure.

2. 92°C; 0.70 mol benzene/mol (from Figure 6.4-1a)

3. 99°C; 0.30 mol benzene/mol (from Figure 6.4-1a); decreases

4. Increase (see Equation 6.4-4); increase (see Equation 6.4-6)

5. The hydrostatic head of the liquid must be added to the pressure at the liquid surface. Convert 5 ft of water to atm, add it to 1 atm, and look up the boiling point of water at the corrected pressure.

6. The nonlinear equations cannot be solved explicitly for I.

p. 268

1. 380 g. The added salt will not remain undissolved.

2. 55°C. Increasing quantities of KNO_3 crystals precipitate (come out of solution).

3. See p. 267. Magnesium sulfate tetrahydrate.

4. $120.4/138.4 = 0.870$

5. Above 40°C the crystals that precipitate are hydrated salts.

p. 270

1. A solution property that depends only on concentration, and not what the solute and solvent are. Vapor pressure, boiling point, and freezing point.

2. 850 mm Hg. Solute is nonvolatile, nondissociative, and nonreactive with the solvent; Raoult's law holds.

3. Greater. $p^* = (1000 \text{ mm Hg})/0.85 = 1176 \text{ mm Hg}$.

4. It lowers the freezing point of water, so it keeps ice from forming at temperatures where it would otherwise form.

5. Antifreeze lowers the freezing point and raises the boiling point of water, so the radiator is less likely to freeze in winter and to boil in summer.

p. 273

1. See definition at bottom of p. 271. Extraction is transference of a solute from one liquid solvent to another.

2. Less soluble; $m_{VA} \gg m_W$

p. 274

1. A tie line connects compositions of two phases in equilibrium.

2. H_2O-rich phase—95.0% H_2O, 2.5% acetone, 2.5% MIBK; MIBK-rich phase—92.5% MIBK, 2.5% H_2O, 5.0% acetone. Mass ratio of MIBK phase to H_2O phase $= \dfrac{0.950 - 0.450}{0.450 - 0.025} = 1.18$.

p. 278

1. In absorption a gas species dissolves in a liquid; in adsorption a gas or liquid species adheres to the surface of a solid.

2. An adsorbate is a species adhering to the surface of a solid adsorbent.

3. Concentration equals partial pressure times a constant (and vice versa).

4. Toxic species in the air are adsorbed on the surface of the carbon. Nonactivated carbon has much less surface area so the mask would become saturated in much less time.

p. 316

1. Kinetic, potential, internal; heat, work

2. Heat is only defined in terms of energy being transferred.

3. $E_i + Q - W = E_f$

p. 318

1. The mass flow rates are the same; $\rho_{\text{outlet}} < \rho_{\text{inlet}}$; $V_{\text{outlet}} > V_{\text{inlet}}$.

2. $\Delta E_p > 0, \Delta E_k > 0$

p. 319

1. Closed system: no mass crosses system boundaries. Open system: mass crosses system boundaries. Adiabatic system: no heat transferred to or from system.

2. $Q = 250 \text{ J}$
   $W = -250 \text{ J}$

3. $\Delta U = -50 \text{ kcal}$

4. If the substance is a liquid or a solid, or a gas under nearly ideal conditions, it is reasonable to neglect the dependence of U on pressure.

p. 321

1. $\dot{V}_{\text{in}} = \dot{V}_{\text{out}}$

2. $P_{\text{in}} > P_{\text{out}}$

p. 322

1. 6000 cal

2. 1000 cal/min

3. The specific volume and pressure: $\hat{H} = \hat{U} + P\hat{V}$

p. 324

1. $\dot{W}_s = 0$

2. $\dot{Q} = 0$

3. $\Delta \dot{E}_k = 0$

4. $\Delta \dot{E}_p = 0$

p. 327

1. A property whose change of value in any process depends only on its initial and final states and not on the path between them.

2. **(a)** 0;
   (b) 5000 J/kg;
   (c) $\Delta \hat{H} = \hat{H}_A(\text{v, } 0°\text{C, 1 atm}) - \hat{H}_A(\text{v, } 30°\text{C, 1 atm})$
      $= (5000 \text{ J/kg} - 7500 \text{ J/kg}) = -2500 \text{ J/kg}$
   (e) No—$\hat{H}$ is a state property.

p. 337

1. Incompressible fluid, negligible heat transfer, and no internal energy changes due to anything but friction.

2. Above, plus no friction or shaft work.

p. 359

$\hat{H} = -2751$ kJ/kg. $\hat{U} = -2489$ kJ/kg.

p. 361

2. Lower P isothermally to 1 atm, cool at 1 atm to 80.7°C, condense at 80.7°C and 1 atm, cool liquid at 1 atm to 25°C, raise pressure to 5 atm.

3. Keeping pressure constant at 1 atm, cool water to 25°C, dissolve NaOH in water at 25°C, raise solution to 50°C.

4. Keeping pressure constant at 1 atm, cool O_2 to 25°C, mix O_2 and CH_4 at 25°C, carry out reaction at 25°C, raise product gas to 300°C.

p. 366

1. (a) ii; (b) i; (c) iii

2. Determine z for C_2H_6 at each of the system conditions. If z is close to 1 for both states, it may be reasonable to neglect $\Delta\hat{H}$.

p. 368

1. $C_v = (\partial\hat{U}/\partial T)_v$,
   $C_p = (\partial\hat{H}/\partial T)_P$

2. (a) exact;
   (b) inaccurate;
   (c) good approximation

3. $\Delta H = (5\text{ g})[0.5\text{ cal/(g·°C)}](20°C) = 50$ cal

p. 371

1. 28.5 J/(mol·°C)

2. 5 cal/(mol·°C); 7 cal/(mol·°C)

3. (a) 22.44 kJ/mol;
   (b) −25.78 kJ/mol;
   (c) −2.22 × 10⁵ Btu/h (Neglect pressure effects.)

p. 373

1. $CaCO_3(s)$: $C_p = 26 + 7.5 + 3(17) = 84.5$ J/(mol·°C)

2. $\Delta H = \{(2\text{ kg})[2.5\text{ kJ/(kg·°C)}]$
   $+ (1\text{ kg})(1.8\text{ kJ/(kg·°C)})\}(10°C) = 68$ kJ
   $\Delta\hat{H} = 68$ kJ/3 kg $= 23$ kJ/kg

3. $\Delta\hat{H} = 25$ cal/g

4. $(C_p)_{\text{mixture}} = [(0.50)(1.00) + (0.50)(0.54)]$ cal/g $= 0.77$ cal/g

p. 381

1. Yes; no

2. (a) 0; (b) 900 J/mol
   (c) Cool the vapor to 75°C; condense at 75°C; cool the liquid to 25°C.

3. $\Delta\hat{U}_f \approx 5500$ cal/mol
   $\Delta U_v = \Delta\hat{H}_v - RT = 26{,}710$ cal/mol

p. 382

1. Equation 8.4-3

2. Equation 8.4-4

3. From Equation 8.4-6, slope of line is $-\Delta\hat{H}_v/R$.

4. From Equation 8.4-7, slope of tangent to curve at $p^* = 1$ atm is $-\Delta\hat{H}_v/R$.

5. Equation 8.4-8

6. $\Delta\hat{H}_v(T_2) = \int_{T_2}^{T_1} C_{pl}\, dT + \Delta\hat{H}_v(T_1) + \int_{T_1}^{T_2} C_{pv}\, dT$

p. 392

Air at 25°C, $h_r = 20\% \Longrightarrow h_a = 0.0040$ kg H_2O/kg dry air; $T_{wb} = 12.5°C$; $T_{dp} = 0.5°C$; $\hat{V}_H = 0.85$ m³/kg dry air, $\hat{H} = (35.00 - 0.27)$ kJ/kg dry air $= 34.73$ kJ/kg dry air

p. 395

1. (a) $(C_p)_{\text{gas}}$, $(C_p)_{\text{liquid}}$, and $(\Delta\hat{H}_v)_{\text{liquid}}$ are independent of temperature, and the enthalpy change of unevaporated liquid is small compared to the heat of vaporization and the enthalpy of the gas.
   (b) Yes; (c) no

2. $T_{as} = 11°C$; $h_a = 0.0069$ kg H_2O/kg dry air; $h_r = 70\%$; $\hat{H} = (31.60 - 0.06)$ kJ/kg dry air $= 31.54$ kJ/kg dry air

p. 397

1. (a) $\hat{H} = -40$ kJ/mol A; (b) $\hat{H} = 20$ kJ/mol A

2. $Q = \Delta H = (5)(-40)$ kJ $= -200$ kJ (evolved)

3. $Q = 5(-60 + 40)$ kJ $= -100$ kJ (evolved)

p. 402

1. −97 Btu/lb_m solution

2. 77°F (the reference temperature for H_2SO_4)

3. 190°F. 65 wt% H_2SO_4. (Highest temperature on a line between the feed points.)

4. (a) −60 Btu/lb_m solution. (b) 30 Btu/lb_m solution. The heat of solution at 77°F (or the energy required to heat the solution from 77°F to its adiabatic mixing temperature of about 195°F).

p. 406

1. 130°F, 0.15

2. 600 Btu/lb_m

3. Vapor fraction $\approx (0.50 - 0.18)/(0.88 - 0.18) = 0.46$

p. 441

2. The second sentence is wrong and the first sentence is correct, but only if the reactants and products are at the same temperature. The energy released by the breaking of reactant bonds and forming of product bonds must be transferred from the reactor to keep the products at the

same temperature; otherwise it remains in the reactor and raises the product temperature. If the reactor is adiabatic, the products at the higher temperature would be at the *same* energy level as the reactants at the lower temperature.

p. 445

1. $H_{products} - H_{reactants}$ when stoichiometric quantities of reactants are fed, the reaction proceeds to completion, and reactants and products are at the same temperature and pressure. As above, with reactants and products at 25°C and 1 atm.

2. **(a)** -20 kJ/mol A reacted; **(b)** exothermic; **(c)** withdrawn; **(d)** higher

3. The latent heat of vaporization of C_6H_{14} at 25°C and 1 atm

4. $\Delta \hat{U}_r(T) = \Delta \hat{H}_r(T) + 2RT$

5. $\Delta \hat{H}_r = \sum\limits_{products} |v_i| \hat{H}_i - \sum\limits_{reactants} |v_i| \hat{H}_i$

   In general, $\hat{H}_i = \hat{U}_i + P\hat{V}_i$; for liquids and solids, $\hat{H}_i \approx \hat{U}_i$; and for ideal gases, $P\hat{V}_i = RT$, so that $\hat{H}_i = \hat{U}_i + RT$. To obtain the desired result, substitute for $\hat{H}_i$ in the expression for $\Delta \hat{H}_r$, and recognize that

   $$\Delta \hat{U}_r + \sum\limits_{products} |v_i| \hat{U}_i - \sum\limits_{reactants} |v_i| \hat{U}_i$$

p. 447

1. See last paragraph of Section 9.2.

2. Multiply Equation 2 by -2 and add the result to Equation 1

   $2A + B - 2A - 2D \rightarrow 2C - 2C - 6E$

   $\Downarrow$

   $B + 6E \rightarrow 2D$, $\Delta \hat{H}_r = \Delta \hat{H}_{r1} - 2\Delta \hat{H}_{r2} = 3000$ kJ/mol

p. 448

1. The formation reaction of CO is $C + \frac{1}{2}O_2 \rightarrow CO$, so the heat of formation is $-\frac{1}{2}$ times the heat of the given reaction, or -110.5 kJ/mol. (This checks with the value in Table B.1.)

2. The heat of vaporization of propane at 25°C.

3. $\Delta \hat{H}_r^\circ = (\Delta \hat{H}_f^\circ)_{CO_2} + 2(\Delta \hat{H}_f^\circ)_{H_2O(v)} - (\Delta \hat{H}_f^\circ)_{CH_4}$

p. 464

1. **(a)** $\Delta \hat{H}_f^\circ (\text{soln}, n = 50) = (\Delta H_f)_A - 100 - 10$
   $= -110$ kJ/mol A
   (b) $\Delta \hat{H}_f^\circ (\text{soln}, n = \infty) = -115$ kJ/mol A

2. **(a)** $\hat{H}(25°C) = -110$ kJ/mol A
   (b) $H(25°C) = -550$ kJ

p. 466

1. $LHV = 2658$ kJ/mol
   $HHV = 2658$ kJ/mol $+ 5(44.013$ kJ/mol$)$
   $= 2878$ kJ/mol

2. $HHV = (0.40)(143$ kJ/g$) + (0.60)(55$ kJ/g$)$
   $= 90.2$ kJ/g

3. For coal:

   ($150/ton)(1 lb$_m$/15,000 Btu)(1 ton/2000 lb$_m$)
   $= 5.0 \times 10^{-6}$ $/Btu

   For natural gas:

   x($/ton)(1 ton/2000 lb$_m$)(1 lb$_m$/23,000 Btu)
   $= 5.0 \times 10^{-6}$ $/Btu

   $\Downarrow$

   $x = $230/ton

p. 469

1. The product temperature when the fuel is burned completely in an adiabatic reactor.

2. Heat loss through reactor walls, incomplete reaction.

3. With an air feed, the heat released from combustion of a fixed amount (say, 1 mol) of fuel goes into heating the nitrogen in the air as well as the reaction products and excess oxygen, whereas only the latter must be heated for a pure O_2 feed. If a fixed quantity of heat is added to a greater amount of material, the resulting temperature rise must be less.

p. 471

1. **(a)** Ignition—abrupt increase in the rate of an oxidation reaction.
   (b) Ignition temperature—the temperature at which ignition occurs when a combustible mixture is heated slowly.
   (c) Ignition lag—time between a mixture reaching its ignition temperature and the occurrence of ignition.
   (d) Flammability limits—composition limits outside of which ignition or explosion cannot occur.
   (e) Flash point of a liquid—temperature at which a liquid gives off enough vapor to form an ignitable mixture with the air above the liquid surface.

2. **(a)** An explosion or ignition; no reaction would occur.
   (b) Yes; the reaction would cease.
   (c) There is a region between the emerging jet (pure methane) and the bulk of the room air (almost no methane) within which the methane fraction falls between the flammability limits. The flame persists in this region.

p. 473

1. A flame is a combustion zone in which various species undergo transitions from states of high energy to states of lower energy. The energy lost in the transitions is given off as visible light.

2. In low-temperature flames, which occur when the fuel and oxygen are not well mixed, carbon particles are formed

and remain unburned, but the heat in the reaction zone raises them to a temperature at which they glow incandescently. The result is a yellow flame. At higher combustion temperatures, various intermediate species are formed and excited to high-energy states, and emit blue light as they return to lower states. The result is a blue flame.

3. In a detonation, a pressure front (shock wave) forms and propagates at supersonic speed, rapidly compressing and igniting the flammable mixture and giving the appearance of an instantaneous combustion. In the vicinity of the detonation, the force of the shock wave can demolish a building or propel a bullet. The noise is the vibration of the eardrums caused by the shock wave.

4. The rate of the reaction depends strongly on temperature; at room temperature it is immeasurably low.

p. 507

1. (a) 1; (b) x_1

2. Ten variables (flows of A, B, C, and D at inlet, A, B, and C in vapor product, and B, C, and D in liquid product) − 4 balances (A, B, C, and D) − 2 relations (given split of B, equilibrium relation for C) = 4 degrees of freedom.

p. 510

1. Mixer—3, separator—4

2. 5 degrees of freedom (7 local degrees of freedom − 2 ties)

p. 548

1. (a) Accumulation, input, output (although the mass balance will yield the result that accumulation = 0);
   (b) all but generation;
   (c) accumulation, output, consumption

2. $dm/dt = 10 \text{ kg/h} - 10 \text{ kg/h} = 0 \text{ kg/h}$

p. 551

1. $dy = (2 - t)\,dt \implies \int_1^y dy = \int_0^t (2 - t)\,dt$

$$\implies y = 1 + 2t - \frac{t^2}{2}$$

2. $\dfrac{dy}{2 - y} = dt \implies \displaystyle\int_1^y \frac{dy}{2 - y} = \int_0^t dt$

$$\implies -\ln(2 - y)]_1^y = t$$

$$\implies \ln\left(\frac{1}{2 - y}\right) = t$$

$$\implies \left(\frac{1}{2 - y}\right) = e^t$$

$$\implies y = 2 - e^{-t}$$

3. $\dfrac{dy}{2 - y} = (2 - t)\,dt \implies \displaystyle\int_1^y \frac{dy}{2 - y} = \int_0^t (2 - t)\,dt$

$$\implies -\ln(2 - y)]_1^y = 2t - \frac{t^2}{2}$$

$$\implies \ln\left(\frac{1}{2 - y}\right) = 2t - \frac{t^2}{2}$$

$$\implies \left(\frac{1}{2 - y}\right) = \exp\left(2t - \frac{t^2}{2}\right)$$

$$\implies y = 2 - \exp\left(-2t + \frac{t^2}{2}\right)$$

p. 560

1. Neglect changes in $\dot{E}_k$ and $\dot{E}_p$; no mass accumulation; $\hat{U}$ and $\hat{H}$ independent of P; no phase changes or chemical reactions; no spatial variations in T; constant C_v and C_p.

2. When the expression for U_{sys} is differentiated to yield Equation 11.3-9, if C_v varies with temperature (and hence with time) another term of the form $M(T_{\text{sys}} - T_{\text{r}})(dC_v/dt)$ would have to be added.

p. 609

1. $-1, +2$

2. The line for which the sum of the squares of the residuals is a minimum.

3. Yes. (Corollary—no.)

4. Positive and negative deviations from the line would cancel out, possibly making a terrible fit look good.

p. 611

1. Linear

2. Nonlinear

3. Nonlinear

4. (a) Linear; (b) nonlinear

5. (a) Linear (multiply equation by z); (b) nonlinear; (c) linear

p. 612

1. A straight line on a plot of f versus x can only intersect the x axis at one point.

2. One root. (The intersection of a 45° line through the origin with a curve that begins at 1 when $x = 0$ and decreases, approaching 0 as x approaches infinity.)

p. 613

1. (a) $x = 4$ and $x = 0$; (b) $x_p = 15/4$; $x_n = 5$. Converge to $x = 4$.

2. The equation for a line on an f versus x plot through (x_n, f_n) and (x_p, f_p) is $f = f_n + [(f_p - f_n)/(x_p - x_n)](x - x_n)$. The intercept of this line on the x axis, which we will call x_{new}, is obtained by setting $f = 0$

and solving the resulting equation for x. The result is Equation A.2-1.

p. 615

1. Yes. No. You can solve for the roots directly.
2. The value of $f'(x_0)$ would be 0, and the correction term f/f' would blow up.

p. 617

1. **(a)** Modified successive substitution; **(b)** successive substitution; **(c)** acceleration
2. 13.6

p. 618

1. 2.60
2. A line on an f versus x plot through $(2, 2.3)$ and $(2, 3.245)$ intersects the 45° line at $x = 2.60$.

p. 626

1. 80
2. 84
3. 80. The function is cubic, so Simpson's rule is exact.

ANSWERS TO SELECTED PROBLEMS

CHAPTER TWO

2. **(c)** 120 hp

6. 4.3×10^5 miles

8. **(b)** 2.6 kg

10. 25 m^3

14. **(b)** 0.135 m/s^2

18. **(a)** $s_T = 2.3°C$

20. **(a)** $s_X = 2.2$

24. **(a)** 88.8 cm/s (or 0.888 m/s)

26. **(b)** 1.13 g/cm^3

28. **(b)** $t(C = 0.1 \text{ mol}/L) = 1.12$ min (extrapolated), 1.70 min (exact)

30. **(d)** $xy = 2e^{3y/x}$

32. **(b)** 109 kg/h

34. **(a)** $k = 9.3 \times 10^{-3}$ min^{-1}

36. $k = 1.573$ (dimensionless). Answer may vary slightly.

38. **(b)** Runs 1, 2, 3, 4 $\Rightarrow$ $b = 0.52$. Answer may vary slightly.

40. **(c)** $t = 209.5$ h

42. **(b)** For Lab 1, $k = 0.0062$ s^{-1}

44. $a = 6.536$

CHAPTER THREE

2. **(b)** 360 lb$_m$/ft^3

4. $68.42 in France

6. **(a)** 445 L

8. (SG)$_{solution}$ = 1.5

10. **(b)** 1.00×10^3 N

12. **(b)** 4.6 kg Ile/h

14. **(f)** 1.08×10^6 g C

16. **(b)** 8715 lb$_m$/h

18. 0.300 g CaCO$_3$/g suspension

20. **(c)** 3.84×10^{-5} kg CaSO$_4$

22. **(a)** $SG = 0.903$

24. 0.917 g/cm^3

26. **(a)** Sample 1: 15.6 mole% CH$_4$, 6.2 wt% CH$_4$

28. **(c)** 87.7 s

30. 0.155 g

32. **(e)** 0.737 atm

36. $P = 32.9$ psi

38. **(b)** $F_{min} = 2.0 \times 10^5$ N

40. $P_{atm} = 775$ mm Hg

42. **(a)** $x = 10$ cm $\Rightarrow R_{water} = 424$ cm

44. **(b)** 6.54 psig

46. **(a)** 393 mm Hg

48. **(c)** 153°F

50. **(b)** 3.26°C/s

52. **(b)** 2.6 lb$_m$ CO

54. At 94°C, $k = 0.707$ L/(mol · min)

CHAPTER FOUR

4. **(d)** $x_{O_2} = 0.21 \, n_2/(n_1 + n_2)$

6. **(b)** Two unknowns must be specified

8. **(c)** .22 (22%) of the large eggs are broken

10. 207 gal.

12. **(b)** 5.80×10^3 mol CH$_3$OH/h

14. **(a)** 6480 lb-moles/h

16. **(b)** 2.96 L 20% solution/L 60% solution

18. **(b)** 8.8×10^7/year

20. **(a)** 1.2×10^{-3} mol H$_2$O(v)/mol

22. **(c)** $x_P = 0.20, m_P = 100.0 \Rightarrow n_A = 3.29, n_B = 1.10$

24. **(b)** velocity = 8.33 m/s

26. **(c)** 423 kg SO$_2$/min removed from feed gas

28. 3 balances on overall system, 2 on Unit 1

30. **(c)** 39.8% salt in the effluent from the 4th evaporator

32. **(b)** 28.6 kg of 42% concentrate

34. **(b)** Recycle ratio = 1.60

36. **(b)** 9.71 kg C$_6$H$_{14}$ recycled/kg C$_6$H$_{14}$ fed

38. **(c)** 3796 kg steam fed to stripper

40. **(c)** 17.6% excess NH$_3$

42. $\dot{\xi} = 56.2$ mol/s

44. 7380 kg 80% H$_2$SO$_4$

46. **(b)** $X_A = X_B = 0.496$

48. **(c)** $X_{CO} = 0.156$

50. **(c)** Fractional yield of C$_2$H$_5$Cl = 0.875

52. 1533 kg ore

54. For the 1st feed mixture, $y_A = 0.2027, y_E = 0.0393$

56. **(b)** CH$_3$OH enters the reactor at the rate of 50.0 kmol/h

58. **(c)** 85.9 kmol/h recycle

60. **(a)** S.P. Conversion = 14%

62. **(b)** 2.535×10^4 kmol/h n-C$_4$H$_{10}$ in emulsion recycle, 2220 kmol H$_2$SO$_4$/h and 1200 kmol H$_2$O/h in acid recycle, 1040 kmol i-C$_4$H$_{10}$/h recycled from distillation column

64. (b) 66.5 kmol/h

66. 250 kmol/h

68. (b-iii) 9.1 mole% CO_2

70. (c) 18.6% excess air

72. (a) Desired percent excess air = 11%

74. 10.7% CO_2, 9.8% H_2O

76. (b) 3.72×10^5 mol air/tonne coal

78. 30% bypass

80. (b) Minimum percent excess air = 117%.

CHAPTER FIVE

2. Bulk density = 3.26 g/cm^3

4. (c) 35.4 kg $CuSO_4$ (anhydrous)

6. (b) 9.3% error

8. (b) 555

10. 165 m/s

12. Unknown MW = 3.9

14. (a) 881 cm^3/s

16. 71.2%

18. 1.52

20. (c) 58.5 g/L

22. (a) y_B (9 am) = 0.62×10^{-6}

24. (b) \$1,048,000/month

26. (b) 3595 mol diluting air/s

28. (b) 3.3 cm H_2O

30. 4.0×10^4 m^3(STP) air/h

32. (a) 346 kPa

34. (c) 1.2×10^5 Å

36. (b) $x = 0.5 \Rightarrow V = 2.16 \times 10^4$ L

38. (b) 71% conversion of propylene

40. (b) 196 kg acetone vaporized/h

42. 0.150 mol Cl_2/mol

44. 1.8% Fe_2O_3

46. (a) 4.08×10^5 L gas/min leaves the condenser

48. (b) f_{SO_2} = 0.602 at 600°C

50. 846 SCMH recycled

52. 25.7 mole% CO_2, 6.5% CO, ...

54. (a) Reactor effluent (C): 49.4 kmol/h,
26% H_2, 53% CH_3OH, 0.03% H_2O

56. 135 m^3/h

58. P_{SRK} = 7.3 atm

60. 1.57×10^4 L/h

62. (b) 37.4 lb$_m$

64. (b) $z = 1.6$

66. 320°F

68. 74.3 mol

70. (b) $P > 2.1 \times 10^4$ atm   **(c)** 5 cycles

72. (b) 34,900 gallons

74. 0.041 m^3

76. 126 ft^3/min

78. (b) 29.9 m^3/h

80. 1.63 kmol/min

82. (b) $C_{11}H_{20}O_{10}$

CHAPTER SIX

2. (b) 0.235 g vapor/g total

4. 58.8 kJ/mol

6. T_b = 116°C

8. 8.0×10^{-4} mol

10. (b) 6.85°C

12. (b) 99.7% condensation

14. At 70°F and 50% r.h., ρ_{air} = 1.196 g/L

16. 5.07×10^4 ft^3/min

18. 0.310 m^3 outlet air/m^3 feed air

20. (c) 0.479 lb-mole octane lost

22. (c) 2.52°C

24. (b) 29°C

26. (b) 96% recovery if the refrigerant is used

28. 1.81 gal. condensed/min

30. (b) Cool to 7.8°C

32. (a) 0.0148 kmol CH_3OH/min

34. $\dot{V}_{in}$ = 2.09×10^5 ft^3/h

36. (b) $\dot{V}_{inlet}$ = 2590 ft^3/h

38. T_{dp} = 58.8°C

40. (b) x = 0.00536 mol C_3H_8/mol

42. (a) 1097 mol leave converter, 15.5% N_2O

44. (b) 8.33×10^{-2} m^3 gas feed/kg liquid feed

46. 0.0711 mol benzene/mol

48. x_{CH_4} = 1.46×10^{-4} mol CH_4/mol

50. (c) 127 cm^3

52. 85.85°C

54. (a) P = 151 psig

56. (a) 0.096 mol C_6H_6/min

58. (b) 22.6% benzene $\Rightarrow T_{bp}$ = 108.1°C

60. (d) 0.39 m

62. (d) x_B = 0.400 $\Rightarrow y_B$ = 0.795,
P = 444 mm Hg

66. (c) 0.46 mol vapor/mol liquid

68. (d) 64.5°C

70. 79.9°C

72. (c) 0.19 lb-mole H_2O/lb-mole

74. 32.3%

76. (c) 23.0 kg A

78. 0.428 lb_m crystals/lb_m liquid

80. 2044 kg/h recycled

82. (c) 64 lb_m recycle/lb_m fresh feed

84. $T_m = 1.9°C$

86. 83.5 g/mol, 6.38 kJ/mol

88. 4.4 g styrene in ethylene glycol phase

90. (a) The extraction requires 13,000 kg of hexane

92. (b) 0.2965 kg alkaline solution/kg acidified broth

94. Water-rich phase: 21% acetone, 3% MIBK, 76% water (answers may vary)

96. (c) For System 2, $\beta_{aw} = 41.8$ (answers may vary)

98. (a) 33.1 grams of silica gel

100. (b) 1.13 hours (68 minutes)

CHAPTER SEVEN

2. (b) 3000 MW

4. (a) $\dot{E}_k = 1.40 \times 10^{-5}$ hp

6. (b) $u = 25.4$ ft/s

8. $\dot{E}_{max} = 3.43 \times 10^6$ kW · h/wk

10. (b) $Q = -765$ J

12. 2338 J/mol

14. 9.0×10^3 cal

16. (a) (i) 941 cm³, (ii) 19.5 J, (iii) 64.3 J

18. (c) $\Delta\dot{E}_p = -\dot{W}_s, \dot{W}_s > 0$

20. (a) 246 L/s

22. 477 m/s

24. (a) 6.02×10^7 J/s

26. 5.80×10^4 kW

28. (b) 2.09 kg steam/s

30. (a) 300 kJ/h

32. (b) 12 kW

34. (a) Every 13 seconds

36. (a) 0.296 evaporates

38. (b) 2920 kJ/s

40. 587 kJ/h

42. (a) 2.63×10^4 cal

44. (c) 5.70×10^4 kJ

46. 38 g

48. (a) 0.53 g evaporated

50. 2.6 g evaporated

52. (a) 0.894 m/s

54. (a) 122 L/min

56. 76.2 m³/min

58. 0.62 kW

CHAPTER EIGHT

2. (c) $\Delta\hat{U} = 2160$ J/mol

4. (d) 31.71 kJ/mol

6. (c) -110.7 kJ/mol

8. 7193 kW

10. (b) $C_p \approx 0.223$ kJ/(mol · K)

12. (c) Total daily cost = \$4.72

14. (a) -2156 kW

16. 1833 Btu/h

18. -0.07643 kJ/g

20. 1810 m²

22. (c) 3.41 m³/s

24. (c) 0.400 m³ steam/m³ propane

26. (b) Relative humidity = 51.2%

28. 1.39×10^6 J/day

30. (b) 290 kW

32. (a) 792°C

34. $\Delta\hat{H}_{analytical} = -1730$ cal/g

36. (c) 60.12 kJ/mol

38. -1.90×10^4 kW

40. (a) 1.95×10^5 kJ/h

42. $\Delta\hat{H}_v(100°C) \approx 38.2$ kJ/mol

44. (c) 33.4 kJ/mol

46. (d) -181 kJ/m³ air fed

48. 733 kW

50. (b) 257 kW

52. (a) 2.06×10^4 kW

54. (d) $T_{f2} = 34°C \implies T_{a1} = 504.6°C$

56. (b) 163 kg/h additional steam

58. (c) 3733 kg/h fresh water

60. (c) 0.51 g evaporate

62. (a) 1760 kW

64. (d) 1.16×10^5 kJ/h

66. (b) $x_F = 0.5, T_f = 110°C$, $P = 760$ mm Hg $\implies T_{evap} = 52.2°C$

68. (a) 0.840 kg steam generated, 973 kJ transferred from condenser

70. $T_{dp} = 77.3°F, \hat{H} = 43.9$ Btu/lb_mDA

72. (d) $Q = -31$ J

74. (b) 1.35 tons of cooling

76. (b) 0.0119 kg H_2O added/min

78. (b) 67 m³

80. (b) -15.5 kW

82. (b) 11.09 kJ/mol HCl

84. 6.09×10^4 J/mol H_2SO_4

86. (b) -471 kJ/L product

88. (b) 291.4 g $H_2O(l)$, 547.3 g $H_2O(s)$

90. (c) 60 kJ/h transferred from the crystallizer

92. (b) $T_{max} = 125°C$

96. (b) $140°F$

98. Liquid: 10.5 g NH_3, 64.5 g H_2O

CHAPTER NINE

2. (c) -1.53×10^5 kW

4. (c) $+122$ kJ

6. (a) 245.4 kJ/mol

8. (c) -1.26×10^5 kJ/h (evolved)

10. (b) $\Delta\hat{H}_c^\circ = -5090$ kJ/mol

12. (c) $\dot{Q} = -0.0339$ kW

14. (c) $Q = 237.3$ kJ

16. (c) 8.11×10^4 kJ/min.

18. (c) $Q = 11,863$ kJ

20. (b) -81.5 kW

22. (b) -6.44×10^4 Btu/h

26. (c) -279 kW (process), -265 kW (reactor)

28. (b) $\mathbf{A}-3.67 \times 10^5$ kJ/h, $\mathbf{F}-1.34 \times 10^6$ kJ/h, $\mathbf{C}-5.65 \times 10^4$ kJ/h

30. (c) 55.9 kmol recycled/h

32. (a) 80.1% conversion

34. (b) 59.2 kJ/mol feed

36. (a) 0.417 mol C_2H_4/mol CH_4 consumed

38. (c) $n_0 = 9.0$ mol O_2

40. -55.0 kJ/mol, -177.9 kJ/mol

42. (a) $\Delta\hat{H}_r^\circ = 222.4$ kJ/mol

44. (a) 177 kJ/mol $(NH_4)_2SO_4$

46. (a) $94.5°C$

48. (a) LHV $= 843$ kJ/mol

50. 72.5% CH_4, 18.8% C_2H_6, 8.7% CO

52. (a) 2.5 gal/h

54. (b) 40.3 kJ, 534 kJ

56. (b) 2.13×10^4 kJ/m^3 fuel

58. (c) -655 kW

60. (b) 1.16×10^5 kg steam/h

62. (a) 99% excess air

64. (b) $2871°C$

66. (a) $1672°C$

68. (d) For Run 2, $T_{ad} = 1738°C$

70. (a) $\dot{m}_B = 30.3$ tonnes/d, $\dot{m}_F = 11.2$ tonnes/d, $\dot{Q}_1 = 8.04 \times 10^7$ kJ/d, $\dot{Q}_4 = 4.7 \times 10^7$ kJ/d

CHAPTER TEN

2. (b) $n_1, n_2, n_3, x_1, x_2, x_3$

4. (a) 7 degrees of freedom

6. (b) 1.80 mol recycle/mol fresh feed

10. (c) For the given conditions, $\dot{Q} = -23.7$ kW

12. (b) T $= 1560°C$

CHAPTER ELEVEN

2. (a) $n_p' = 0.3743$, $n_p(0) = 7.5$

4. (d) $t = 120$ s, $n_{O_2} = 0.30$ lb-mole

6. (d) 26.5 min.

8. (d) C_{SO_2} (2 min.) $= 3.82 \times 10^{-7}$

10. (e) 90% conversion at $t = 4.6$ min.

12. (c) 20.6 seconds

14. (b) 3.5×10^{-3} min^{-1}

16. (b) $k_2 = 0.115$ L/mol

18. (b) 2.7 hours

20. (b) 4.29 kW

22. (c) $T_b = 30°C \Rightarrow t = 100$ min

24. (b) 200 g

26. (b) $54.8°C$

28. (a) $dC_{S2}/dt = 0.080(C_{S1} - C_{S2})$

30. (c) $dx/dt = -(\dot{n}_V/N_L)[-x + ax/(x + b)]$

Index

LIMITED USE LICENSE AGREEMENT

This is the John Wiley and Sons, Inc. (Wiley) limited use License Agreement, which governs your use of any Wiley proprietary software products (Licensed Program) and User Manual(s) delivered with it.

Your use of the Licensed Program indicates your acceptance of the terms and conditions of this Agreement. If you do not accept or agree with them, you must return the Licensed Program unused within 30 days of receipt or, if purchased, within 30 days, as evidenced by a copy of your receipt, in which case, the purchase price will be fully refunded.

License: Wiley hereby grants you, and you accept, a non-exclusive and non-transferable license, to use the Licensed Program and User Manual (s) on the following terms and conditions only:

a. The Licensed Program and User Manual(s) are for your personal use only.

b. You may use the Licensed Program on a single computer, or on its temporary replacement, or on a subsequent computer only.

c. The Licensed Program may be copied to a single computer hard drive for playing.

d. A backup copy or copies may be made only as provided by the User Manual(s), except as expressly permitted by this Agreement.

e. You may not use the Licensed Program on more than one computer system, make or distribute unauthorized copies of the Licensed Program or User Manual(s), create by decompililation or otherwise the source code of the Licensed Program or use,copy, modify, or transfer the Licensed Program, in whole or in part, or User Manual(s), except as expressly permitted by this Agreement.

If you transfer possession of any copy or modification of the Licensed Program to any third party, your license is automatically terminated. Such termination shall be in addition to and not in lieu of any equitable, civil, or other remedies available to Wiley.

Term: This License Agreement is effective until terminated. You may terminate it at any time by destroying the Licensed Program and User Manual together with all copies made (with or without authorization).

This Agreement will also terminate upon the conditions discussed elsewhere in this Agreement, or if you fail to comply with any term or condition of this Agreement. Upon such termination, you agree to destroy the Licensed Program, User Manual (s), and any copies made (with or without authorization) of either.

Wiley's Rights: You acknowledge that all rights (including without limitation, copyrights, patents and trade secrets) in the Licensed Program (including without limitation, the structure, sequence, organization, flow, logic, source code, object code and all means and forms of operation of the Licensed Program) are the sole and exclusive property of Wiley. By accepting this Agreement, you do not become the owner of the Licensed Program, but you do have the right to use it in accordance with the provisions of this Agreement. You agree to protect the Licensed Program from unauthorized use, reproduction, or distribution. You further acknowledge that the Licensed Program contains valuable trade secrets and confidential information belonging to Wiley. You may not disclose any component of the Licensed Program, whether or not in machine readable form, except as expressly provided in this Agreement.

WARRANTY: TO THE ORIGINAL LICENSEE ONLY, WILEY WARRANTS THAT THE MEDIA ON WHICH THE LICENSED PROGRAM IS FURNISHED ARE FREE FROM DEFECTS IN THE MATERIAL AND WORKMANSHIP UNDER NORMAL USE FOR A PERIOD OF NINETY (90) DAYS FROM THE DATE OF PURCHASE OR RECEIPT AS EVIDENCED BY A COPY OF YOUR RECEIPT. IF DURING THE 90 DAY PERIOD, A DEFECT IN ANY MEDIA OCCURS, YOU MAY RETURN IT. WILEY WILL REPLACE THE DEFECTIVE MEDIA WITHOUT CHARGE TO YOU. YOUR SOLE AND EXCLUSIVE REMEDY IN THE EVENT OF A DEFECT IS EXPRESSLY LIMITED TO REPLACEMENT OF THE DEFECTIVE MEDIA AT NO ADDITIONAL CHARGE. THIS WARRANTY DOES NOT APPLY TO DAMAGE OR DEFECTS DUE TO IMPROPER USE OR NEGLIGENCE.

THIS LIMITED WARRANTY IS IN LIEU OF ALL OTHER WARRANTIES, EXPRESSED OR IMPLIED, INCLUDING, WITHOUT LIMITATION, ANY WARRANTIES OF MERCHANTABILITY OR FITNESS FOR A PARTICULAR PURPOSE.

EXCEPT AS SPECIFIED ABOVE, THE LICENSED PROGRAM AND USER MANUAL(S) ARE FURNISHED BY WILEY ON AN "AS IS" BASIS AND WITHOUT WARRANTY AS TO THE PERFORMANCE OR RESULTS YOU MAY OBTAIN BY USING THE LICENSED PROGRAM AND USER MANUAL(S). THE ENTIRE RISK AS TO THE RESULTS OR PERFORMANCE, AND THE COST OF ALL NECESSARY SERVICING, REPAIR, OR CORRECTION OF THE LICENSED PROGRAM AND USER MANUAL(S) IS ASSUMED BY YOU.

IN NO EVENT WILL WILEY OR THE AUTHOR, BE LIABLE TO YOU FOR ANY DAMAGES, INCLUDING LOST PROFITS, LOST SAVINGS, OR OTHER INCIDENTAL OR CONSEQUENTIAL DAMAGES ARISING OUT OF THE USE OR INABILITY TO USE THE LICENSED PROGRAM OR USER MANUAL(S), EVEN IF WILEY OR AN AUTHORIZED WILEY DEALER HAS BEEN ADVISED OF THE POSSIBILITY OF SUCH DAMAGES.

General:

This Limited Warranty gives you specific legal rights. You may have others by operation of law which varies from state to state. If any of the provisions of this Agreement are invalid under any applicable statute or rule of law, they are to that extent deemed omitted.

This Agreement represents the entire agreement between us and supersedes any proposals or prior Agreements, oral or written, and any other communication between us relating to the subject matter of this Agreement.

This Agreement will be governed and construed as if wholly entered into and performed within the State of New York. You acknowledge that you have read this Agreement, and agree to be bound by its terms and conditions.

The accompanying CD
—INTERACTIVE CHEMICAL PROCESS PRINCIPLES (ICPP)—
is a guide and toolkit for students using this text
in an introductory chemical engineering course. It also contains reference materials
that should be useful throughout the chemical engineering curriculum.

ICPP HAS THE FOLLOWING COMPONENTS:

Index of Learning Styles A student's learning style is the way the student tends to take in and process information. The Index of Learning Styles (ILS) is a 44-item questionnaire used to assess preferences on four learning style dimensions (active/reflective, sensing/intuitive, visual/verbal, and sequential/global). By completing this questionnaire, you can get information about your learning style preferences, which in turn suggests the types of teaching and studying that work well for you and the types that don't. We also offer suggestions for the most effective way to use this CD based on your learning preferences. We suggest you complete this questionnaire and make use of the results early in your course studies.

Instructional Tutorials This section of ICPP contains six instructional tutorials, each designed to be worked through at a different point of the introductory course. The tutorials describe systems, pose problems, and guide you through the solutions. There are several ways to approach the tutorials. The easy way, and the least valuable, is simply to look for the answers rather than attempt to solve the problems. The only way to get the full benefit of the tutorials is to try to solve the problems yourself first, and then to check yourself by looking at the correct solutions.

Visual Encyclopedia of Chemical Engineering Equipment Many of the problems in the text make reference to common chemical process equipment, such as heat exchangers, pumps, separation process units (absorption towers and distillation columns), and various types of chemical reactors. In most cases, you won't have to know how these various pieces of equipment work to solve the problems, but you may avoid confusion by getting a good picture of how they work before you begin the analysis. By clicking on the equipment type in the Visual Encyclopedia (developed by Professor Susan Montgomery and her students at the University of Michigan and distributed by permission from The Regents of the University of Michigan), you will obtain a photograph and/or a cutaway view of the equipment item, along with a short description of how it works and where in the chemical industry it tends to be used. Note: The CD must be in the CD drive in order to access this encyclopedia.

E-Z Solve This is a powerful and easy-to-use equation-solving and graphing program. You can use it to solve a set of linear or nonlinear algebraic equations of the type you will encounter in almost every problem in Chapters 2-10 of the text. You can also use it to solve linear or nonlinear ordinary first-order differential equations of the type found in Chapter 11. The set of already solved examples is provided in the subfolder Examples, which you can access from the File|Open menu. Before you first attempt to use the program to solve a problem, you should click on Help on the menu bar and then on Tutorial on the pull-down menu that appears. Working through the tutorial is the easiest way to learn to use the application.

Physical Properties Database This section contains an on-line version of the physical property tables in Appendix B of the text. On the Physical Properties screen, you can look up any of the properties listed in Table B.1 of the text (molecular weight, specific gravity, etc.) for a specified species. On the Heat Capacities screen, you can find the heat capacity of a specified species at a specified temperature and 1 atm. You can also access a polynomial function for the heat capacity, like those listed in Table B.2 of the text, by copying the function name, pasting it into E-Z Solve, and entering a desired temperature as an argument. In the Steam Tables window, you can look up the specific volume, specific internal energy, and specific enthalpy of liquid water, saturated steam, or superheated steam at a specified temperature and pressure. You can also copy and paste the corresponding functions into E-Z Solve.

NAVIGATING ICPP

The installation creates icons in the Windows Start Menu and you can access any component from the Start Menu by clicking on the icons. The Main Menu that provides links to each component; this provides the easiest way to access each component of ICPP.

SYSTEM REQUIREMENTS

The software will run on Windows-compatible PCs, under Windows 95, 98, or NT 4.0. Recommended configuration is 486 DX 66, 16 MB RAM (64 MB RAM for NT), SVGA monitor (16-bit colors with 640x480 minimum resolution), 4X CD-ROM, video card, sound card, and speakers. A Web browser (Netscape Navigator 4.0 or Microsoft Internet Explorer 4.0) must be available to access the Index of Learning Styles component.

About 10 MB of free hard disk space will be required.

Installation instructions are printed on the CD.